Microbiology

in Patient Care

third edition

Marion E. Wilson M.A., Ph.D.

Assistant Director for Microbiology, Public Health Laboratory Services, New York City Department of Health

Formerly, Assistant Director of Laboratories (Microbiology and Serology), St. Luke's Hospital, New York City; Instructor, Department of Microbiology and Immunology, State University of New York, Downstate Medical Center, Brooklyn, N.Y.; Science Instructor, School of Nursing, Massachusetts General Hospital, Boston, Mass.

Helen Eckel Mizer R.N., A.B., M.S., M.Ed.

Associate Professor, Department of Nursing, Western Connecticut State College, Danbury, Conn.

Formerly, Science Instructor (Microbiology), St. Luke's Hospital School of Nursing, New York City; Senior Medical Research Assistant, Harvard School of Public Health, Boston, Mass.

Josephine A. Morello A.M., Ph.D.

Director of Clinical Microbiology Laboratories, Professor of Pathology and Medicine, The University of Chicago, Chicago, Ill.

Formerly, Director of Microbiology, Harlem Hospital Center, New York City; Assistant Professor of Microbiology, Columbia University College of Physicians and Surgeons, New York City; Research Associate, The Rockefeller University, New York City; Instructor of Biology, Simmons College, Boston, Mass.

MACMILLAN PUBLISHING CO., INC.
New York
Collier Macmillan Publishers
London

Macmillan Publishing Co., Inc.
866 Third Avenue, New York, New York 10022

Collier Macmillan Canada, Ltd.

Library of Congress Cataloging in Publication Data

Wilson, Marion E
  Microbiology in patient care.

  Includes bibliographies and index.
  I. Medical microbiology. I. Mizer, Helen Eckel, joint author. II. Morello, Josephine A., joint author.
  III. Title. DNLM: 1. Microbiology—Nursing texts.
QW4.3b/W751m
QR46.W749 1979   616.01   78-6020
ISBN 0-02-428310-X

Printing: 1 2 3 4 5 6 7 8   Year: 9 0 1 2 3 4 5

Acknowledgments: The quotations that appear on the dedication page and on page 4 are from *The Unseen World,* by René Dubos, The Rockefeller University Press, New York, 1962, page 3 and the Foreword, respectively.

The photographs appearing on the cover, and on the title pages for each part and section, were supplied by and are used by permission of the following:

Cover (electron micrograph showing numerous gonococci attached to the surface of epithelial cells from the urethra of a man with symptoms of gonorrhea for 24 hr. Gonococci are between adjoining cells. [26,500 ×]): Michael E. Ward and Peter J. Watt: Adherence of *Neisseria gonorrhoeae* to Urethral Mucosal Cells: An Electron-Microscopic Study of Human Gonorrhea, *J. Inf. Dis.,* **126**:601–604, 1972.

Part One (scanning electron micrograph of staphylococci showing their typical grapelike clustering: 16,000 ×): William G. Barnes, Ph.D., John D. Arnold, M.D., and Arthur Berger, Veterans Administration Hospital, Kansas City, Mo.

Section I (Antoni van Leeuwenhoek [about 54 years of age]; steel engraving from a portrait by J. Verkolje in the Amsterdam Rijksmuseum): Rijksmuseum, Amsterdam, The Netherlands.

Section II (a diagnostic microbiologist): Dr. Josephine A. Morello, The University of Chicago, Chicago, Ill. Photo by Gordon Bowie.

Section III (epidemiologists examining a bat that will be tested for rabies): Dr. Alan Beck, Bureau of Animal Affairs, Department of Health, The City of New York. Photo by *N.Y. Daily News.*

Part Two (dark-field photomicrograph of leptospires; 1600 ×); Drs. J. D. Fulton and D. F. Spooner, London School of Hygiene and Tropical Medicine, London, England.

Section IV (effect of a mask on the aerosol from a sneeze): Marshall W. Jennison: The Dynamics of Sneezing—Studies by High Speed Photography. *Sci. Monthly,* **52**:24–33, 1941.

Section V (home canning: frequently incriminated in botulism): Ayerst Laboratories, New York, N.Y.

Section VI (teen-agers are among the prime targets of the sexually transmitted diseases): Western Connecticut State College, Danbury, Conn. Photo by *The WestConn Echo,* Lubas.

Section VII (scene in an operating room): Danbury Hospital, Danbury, Conn. Photo by Elizabeth Wilcox©.

Appendixes (a surgical scrub): Danbury Hospital, Danbury, Conn. Photo by Elizabeth Wilcox©.

TO

Antony van Leeuwenhoek:

amateur

In passing, it is worth noting that the word *amateur* etymologically means "one who loves." Almost any intelligent and industrious person can become a true scientist, if he is an amateur in the original sense of the word and really loves what he does.

RENÉ DUBOS

Preface

Many important developments in clinical microbiology and immunology have evolved since the second edition of this textbook was published. The rate and significance of the progress made seem even more dramatic in a retrospective view of the ten years that have elapsed since the text made its first appearance. In approaching the current revision, and in reviewing a decade of experience reported by many who have used this book in classroom and clinical settings, the authors (now increased to three in number) have renewed their conviction that it should remain oriented primarily to the interests of those who are or will be directly involved in patient care.

Accordingly, this third edition again directs itself to the goal of interpreting the significance of clinical microbiology for those students in the health field who will be actively engaged in following and using its principles, without overemphasizing details that a student pursuing a career in microbiology must master. For the latter, the course is clearly mapped in classroom and laboratory, and a wealth of technical material is available. For the nursing student, however, and for the many others entering the allied health professions (e.g., dental hygienists, dietitians, hospital sanitarians, inhalation therapists, operating room or cardiopulmonary technicians, optometric technicians, physical therapists, and physicians' assistants), the science of microbiology is often of less interest than its clinical and epidemiologic applications. We have addressed ourselves to these special interests, designing the text to provide the nursing or other paramedical *student* with a basic knowledge of the principles of microbiology and epidemiology, and the *practitioner* with clinical and epidemiologic information essential to the management and control

of infectious diseases. It is our belief that the fundamental purpose of such instruction should be to prepare the health professional to play an informed, intelligent role in infection control — i.e., in those aspects of patient care that are based on an understanding of microbiology and epidemiology.

The order of presentation of material is essentially the same as in previous editions. Part One is intended for classroom use, its three sections covering the basic principles of medical microbiology: Section I, the nature and behavior of microorganisms; Section II, the interrelationships of microbes and the human host, in health and disease; Section III, principles of prevention and control of infectious diseases.

Part Two describes important human infectious diseases. This part is designed to be of value, to both student and graduate, as a source of rapidly obtainable, specific information concerning the clinical nature, epidemiology, and management of particular diseases. The four sections of Part Two group diseases according to the most probable (or usual) routes of entry of the microbial agent, and to the nature of the infecting microorganism (bacterial, viral, rickettsial, fungal, and parasitic diseases are discussed in separate chapters of each section). It is hoped that this arrangement will be useful in locating information and, more important, will help the student to learn to associate the infectious agent of a given disease with its portal of entry and communicability.

New developments have necessitated extensive revisions, and many chapters have been reorganized, updated, or completely rewritten. Changes in Part One include the presentation of new concepts concerning the behavior and classification of microorganisms (Chapters 2, 3, and 4), the interactions of microbe and host (Chapters 5, 6, and 7), antibiotic therapy (Chapter 10), and the rationale of immunization programs (Chapter 11). Discussions of the mechanisms of bacterial reproduction and genetic behavior have been revised, but without undue expansion of this material. In this regard, the relationship between the newer knowledge of bacterial genetics and current understanding of the factors involved in microbial resistance to antibiotics is explained. The pathogenicity and virulence of microorganisms *vis-à-vis* the human host are discussed in the light of information accumulating regarding the mechanisms of microbial attachment to host cells, colonization, and the role of toxic microbial products in the pathogenesis of disease. Among the chapters dealing with host-microbe interactions, Chapter 7 has been completely rewritten and expanded to present current concepts in immunology — i.e., the basic processes of humoral and cell-mediated immunity, and their role in immunity, hypersensitivity, autoimmune diseases, and immunodeficiencies.

In Part Two, every chapter has been revised and updated. Some major events of the last few years have been described, such as the sudden appearance of Legionnaires' disease as an infectious disease entity that has probably existed for some time without prior recognition (Chapter 13), the controversial "swine flu" immunization program of 1976 and its final disruption by the apparent correlation of the use of the vaccine with the Guillain-Barré syndrome (Chapter 14), and the improved characterization of hepatitis A and B viruses by new immunologic and electron microscopy techniques (Chapters 19 and 28). Several chapters have been reorganized: Chapter 16 reflects current classifications of the systemic and subcutaneous mycoses; Chapter 18 distinguishes acute bacterial gastrointestinal infections from bacterial toxemias or systemic infections acquired through the alimentary route; Chapter 20 presents the parasitic diseases in a revised format. Discussion of the sexually transmitted diseases has been expanded to present recent concepts in the diagnosis, treatment, and control of syphilis and gonorrhea; and emphasis has been given to the appearance of penicillin-resistant strains of the gonococcus, an im-

portant new challenge to the management of the gonorrhea epidemic. In the interests of both space and improved organization, Chapter 29 of the previous editions ("Infections of Endogenous Origin") has been eliminated. Some of the diseases originally described therein (infectious mononucleosis, toxoplasmosis, pneumocystosis) have been relocated, and the nature and implications of endogenous infections are discussed in several contexts elsewhere.

An important feature of Part Two is the inclusion of the current recommendations of the Center for Disease Control in Atlanta, Georgia, concerning the use of isolation techniques in hospitals. These recommendations are found in tabular form in Chapters 12, 17, 21, and 25.

As in previous editions, each chapter of Part Two concludes with a summarizing table outlining salient features of the disease entities described therein: agent, entry route(s), incubation and communicable periods, laboratory diagnosis, prevention, therapy, and management. In addition, there are new tables on the classification of related groups of organisms and their diseases.

The illustrative material includes new photographs, graphs, and line drawings. Among the electron micrographs, some were chosen to demonstrate developments that have helped to clarify mechanisms of bacterial reproduction, microbial attachments to host cells, or phagocytic processes.

We have again used questions at the end of each chapter throughout the text to assess the student's ability to apply classroom knowledge of microbiologic principles to patient care situations.

Recommendations for additional readings complete each chapter. Highlighted among them are the inexpensive *Scientific American Offprints* (published by W. H. Freeman & Co., 660 Market St., San Francisco, Calif. 94104).

Finally, the Appendixes have also been revised. They now include serologic procedures and skin tests of diagnostic value, aseptic nursing precautions, and the bacteriologic control of sterilizing equipment.

M. E. W.
H. E. M.
J. A. M.

Acknowledgments

The authors take particular pleasure in gratefully acknowledging their many debts to the colleagues, friends, and families who have so freely shared their expertise or given special support in encouragement of our efforts. Their contributions have been invaluable, and their interest has sustained us throughout the long, and often arduous, process of writing and producing a book.

Because of our physical separation and the diversity of our interests and contacts, the people to whom we owe our thanks fall naturally into separate geographic groups. We have therefore chosen to cite them according to our individual bases of operation:

In New York City, a group of microbiologists affiliated with the Public Health Laboratories of the Department of Health, The City of New York, have rallied to the support of this book from the time of its first edition. Among these, the late Mr. Martin Weisburd, Research Scientist, gave immeasurable quantities of enthusiasm, expertise, and time to the effort of polishing the contents of the third edition and directing its emphasis appropriately. Dr. Yvonne C. Faur, Dr. Sam Schaefler, and Dr. Irene Weitzman, Senior Research Scientists, also kept watch over parts of the manuscript, steering it constructively in the accurate updating of several important sections. Our thanks also go to Dr. Balkrishena Kaul, Research Scientist, whom we consulted about the intricacies of gas chromatography; Dr. John Marr, Director of the Bureau of Preventable Diseases, and Dr. Alan Beck, Director of the Bureau of Animal Affairs, both of the New York City Department of Health, who helped in the search for appropriate illustrative material. Also standing behind us with overall encouragement were Dr. Bernard Davidow, Assistant Commissioner for Laboratories, and Dr. Paul S. May, Deputy Assistant Commissioner, who smoothed the path when the going was rough. Special mention must be made of the generosity of Mr. Hugo Terner, parasitologist at Montefiore Hospital (Division of Gastroenterology), New York City, who contributed several of his remarkable photographs of animal parasites. Last but by no means least of this laboratory group to whom we owe so much is Mrs. Belle Barzman, whose incomparable skills in the preparation of manuscripts are matched only by her secretarial competence and her cheerful determination to see the authors through anything. And finally, the New York City author wishes to acknowledge the long-standing patience of her family and friends, who endured silence and closed doors on so many occasions, and who always helped to make the work possible.

In Danbury, Connecticut, several colleagues participated generously in the preparation of this book. We are particularly indebted to Ms. Paula Chipman, Assistant Professor, Department of Biology, Western Connecticut State College, who took the time to read parts of the manuscript and offer constructive criticism; and to Dr. Ramon Kranwinkel, Associate Pathologist, Danbury Hospital, who contributed a number of photomicrographs taken specifically for our use in this text. The special talents of Ms. Elizabeth Wilcox, medical photographer, are evident in the many new photographs illustrating hospital techniques. Two of the fine line drawings of Otto Schmidt of the Westchester Community College appear again in this edition (Figures 2–16 and 6–10).

In Chicago, special thanks are extended to those colleagues who freely provided time and advice in their areas of expertise. These include Ms. Marjorie

Bohnhoff, Dr. William Causey, Dr. William Janda, Dr. Stephen Lerner, and Dr. John W. Rippon. Mr. Gordon Bowie, pathology photographer, University of Chicago, is responsible for many of the elegant photographs of the laboratory and culture material (Figures 4-6, 4-9, 4-11, 5-4, 8-1, 8-2, 8-3, 13-6, and 13-7, as well as the photograph on page 105).

Other colleagues throughout the United States and Canada, as well as in such distant sites as England, France, The Netherlands, Brazil, and Japan, generously provided new illustrations, and we offer them our special thanks. Their names are cited in the legends for numerous photographs throughout the text, and also on the copyright page.

Many others have been continuously generous in their cooperation, forbearance, and understanding, but no one could have given more than Ms. Katherine Eierdanz, Edwin Mizer, or Paul and Elaine, who have grown up with "The Book." We are all indebted to them for their enduring support.

This edition, like the others before it, owes its life, in the end, to Macmillan Publishing Co., Inc., without whose courtesy and cooperation it could not have come to fruition. With the warmest admiration, we acknowledge its medical editor, Ms. Joan C. Zulch, upon whose expertise, patience, and friendship we have depended from the beginning; and its production supervisor, Ms. Patricia Larson, whose skill with manuscripts knows no bounds.

Contents

Part One

Basic Principles of Microbiology

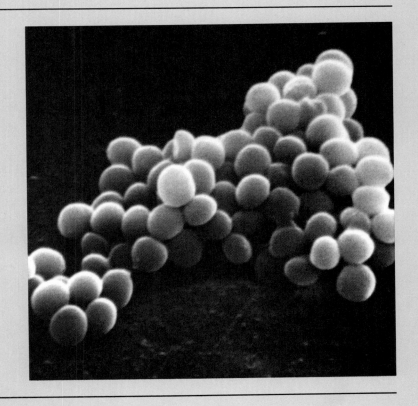

Section I
Character of Microorganisms

1 Introduction: Microbiology, Past and Present

I was thus reminded that we are indebted to our predecessors for the knowledge which enables us to go beyond the limits of their knowledge, and that we find our fulfillment in our successors.

Detlev W. Bronk

The history of developments in medical practice within the last 125 to 150 years has been one of explosively rapid progress. The extent of this progress, within its relatively brief time span, seems all the more remarkable when one compares it with the developments of any comparable time period prior to the first half of the nineteenth century. The pace of medical advance was slow throughout the previous centuries for lack of the technology and the experimentation needed to provide a clear understanding of the basic nature of disease. Although the microscope was invented in the early part of the seventeenth century, its potential value to medicine only began to be realized when it was applied, some 200 years later, to the study of diseased human tissues and cells. The pace of progress began then to quicken and soon to surge in the late 1800s when it was discovered that minute, living organisms, visible only under the microscope, were at the root of such diseases as diphtheria, tuberculosis, typhoid fever, cholera, and others — afflictions that had plagued human populations, sometimes decimating them, throughout recorded history.

Suddenly a new understanding of the mechanisms of many such diseases was at hand, and a new science, later to be called microbiology, began to emerge. It was to become a vital root, itself, of modern medical practice and health care, and an important source of continuing progress.

This new science derived its name from its focus on the study of those living organisms that are too small to be seen except through the enlarging lenses of the microscope (*micro* = small; *bios* = living; *-ology* = study). The organisms newly discovered, being so minute, were called *microbes,* or *microorganisms.* Actually, they had been seen through one of the earliest microscopes invented, long before their importance in disease was recognized. Since that time it has become ever more apparent that the invisible microbial world is enormous and affects our lives enormously.

Microorganisms share all segments of the world we live in, being involved together with plants, animals, and human beings in the great biologic cycles of force and counterforce that characterize the ongoing physical life of our planet. Myriads of microbial species are constantly active in our natural environment. Many also live in continuous association with us on our body surfaces, or on animals and plants. These relationships lead to biologic interplays and interdependencies essential to our welfare, but they may also militate against us if we lose our physiologic balance.

Medical Microbiology

Among the vast numbers of species of microorganisms with which our lives are bound, there are a relative few whose properties and interactions with the human body may lead to development of disease. These are of concern in the field of *medical microbiology* where great progress has been made in defining the nature and properties of harmful microorganisms and the disease processes they induce. Such studies have led to the understanding that *infection* usually represents a balanced relationship between microbes and other living things, whereas *infectious* (or *microbial*) *disease* results from a more damaging interplay between microorganisms with injurious properties and the infected body's responses to them. It has also become clear that the properties of a particular microbe that are responsible for harmful effects in one individual or species do not necessarily induce injury in another, confirming the recognition of microbial disease as a series of interactions between two living organisms, each exerting opposing forces on the other, with resultant injury to the infected (or "host") organism. This concept provides an important basis for the application of microbiologic principles in medicine, nursing, and all patient care.

The management and control of infectious diseases today require the concerted efforts of people trained in several disciplines: chemists, clinical microbiologists, epidemiologists, nurses, pathologists, and physicians. Successful *treatment* of a microbial disease depends on its prompt, accurate *diagnosis*. This is the physician's area of skill and responsibility, supported by evidence provided by laboratory specialists. The *control* and *prevention* of these diseases are also essential, however, because they are so frequently communicable from one person to another and can involve whole populations in epidemics. Here the epidemiologist applies his special knowledge of microorganisms, of the manner in which they are transferred from man to man (or among animals, and from animals to man), of their capacity for survival in the natural environment, and of their reaction to changing environmental conditions.

The professional nurse, as a member of this team of specialists, has a highly important role to play in each aspect of the management of infectious disease—in diagnosis, treatment, and control of active infection, and in its prevention. In a very real sense, the nurse's professional education and experience combine basic elements of each specialty involved, and she applies them continuously in her practice. She must be conversant with diagnostic signs of infectious disease and with methods of treatment, particularly with regard to the patient's response to antimicrobial drugs. She must be able to participate in making the diagnosis, through accurate reporting of the patient's symptoms seen from her particularly close point of view, and by assuring well-timed, careful collection of specimens ordered for laboratory studies to identify the cause of infection. Using her knowledge of the nature of microbial agents of infectious disease, the nurse must apply appropriate precautionary measures in her bedside care of every patient so that communicable diseases cannot be spread from one to another, and problems of new infection are not superimposed upon patients who are ill with other conditions.

Control of active infectious disease requires a working knowledge of the usefulness and applications of physical and chemical agents that suppress or kill microorganisms, as well as familiarity with sources of potentially dangerous microbes, routes by which they spread, and their portals of entry into the human body. *Prevention* of infectious disease hinges on adequate control of these sources and transmission routes. It also frequently depends on specific measures designed to improve individual resistance to infectious microorganisms. Specific resistance is spoken of as *immunity* and plays a vital role in prevention.

While the nurse's responsibilities extend across all areas of infectious disease management, above all she must be a good epidemiologist, competent in her knowledge of the nature of infectious diseases and of available means for controlling or preventing them. In every field of nursing and patient care, whether it be practiced in hospitals, in the public health agency, in physicians' offices, or in industry, a knowledge of basic principles of microbiology is essential. The necessity for such a background of preparation in the nursing curriculum is further highlighted today by the dramatic ease, speed, and frequency of human travel that bring distant diseases closer to home, open up new pathways for communicability of infections, and create new problems in their control.

The Background of Microbiology

As all living organisms are linked together by their interrelated biologic mechanisms, so those who would study any one group of living things usually find the field of their scientific interest interlocked with many others. In beginning the study of microbiology it should be rewarding, therefore, to take a brief look back at the setting within which it evolved and to see how it has been developed by people of widely divergent concerns.

Biologists and Tools

The broad field of biology embraces all sciences that inquire into the nature of life — that is, into the structure, character, and behavior of living forms. These inquiries have been going on insistently since the beginning of recorded history. For centuries, botanists, zoologists, chemists, and physiologists have sought to define life in its many forms, at every observable level. Where the search has been hindered for lack of a method for accurate observation, efforts often have been concentrated on devising better tools or more refined techniques. In general, these efforts have constantly enlarged both our visual and our intellectual range, so that increasingly minute structural features have been visualized, or more subtle details of function have become better understood (Figs. 1–1 to 1–8). So it was that the world of microscopic life, long imagined but never before seen, finally became visible late in the seventeenth century, when a curious eye looked through an enlarging lens at a suspension of material that contained living microbes, swimming and moving about.

It was a simple beginning, of little apparent consequence at the time. The observer was a Dutch merchant, a draper with a hobbyist's interest in lens grinding and optics. His name was Antony van Leeuwenhoek. Although he was an amateur, his skill with lenses was extraordinary. He was not the first to invent a microscope. That achievement had come earlier in his century and no doubt aroused his interest in lenses, which he preferred to make for himself. More important, he developed an untiring interest in the microscopic world they revealed to him as he examined drops of lake water, pepper infusion, scrapings of tartar from his teeth, fecal suspensions, and other fluids. His lenses gave him magnifications up to about 300 diameters, with which he could see "incredibly small" living creatures abounding in these materials. With great objectivity, he recorded his observations of these little "animalcules," as he called them, in a long series of letters written to the Royal Society of London over a period of 50 years. In a letter dated 1683, and another in 1692, he included drawings that clearly depict forms identifiable as bacteria.

Since van Leeuwenhoek was not a scientist, and indeed had received relatively little schooling, he may have been unaware that the existence of microorganisms had been postulated but unconfirmed by medical men concerned with the causation of diseases spread by "contagion." Nonetheless, the objective nature of his reports to the Royal Society left no doubt of the validity of his observations. His contribution was all the more remarkable in view of his personal simplicity. Certainly he could never have dreamed that he would one day come to be known as the founder of a new science of far-reaching influence.

Controversies and Delays

Microbiology thus made a tentative beginning but was not to become a science in the disciplined sense for nearly two hundred years. Although Leeuwenhoek's animalcules were believable and could be confirmed, they were not taken seriously for a time. One of the reasons for this was that the simple observation of microbial life was difficult to interpret without some means of cultivating microorganisms and studying their properties. Certainly their relationship to some of the diseases of man, animals, or plants was not immediately recognizable and received little attention.

Over the ages there had been no dearth either of strong popular beliefs and superstitions or of astute, educated theories concerning the cause of contagious diseases, but these ideas had to be subjected to scientific testing before real progress could be made. It had been recognized from ancient times that diseases can spread from person to person, and many forms of control were devised accordingly, meeting with more or less success according to their basis in pure superstition or clear observation. Efforts to explain the nature of such diseases, and the dreadful epidemic form they so often took, led to some discerning theories in a number of instances. Fracastorius in the sixteenth century, and Kircher in the seventeenth, both wrote with perceptive conviction and foresight of living but invisible agents as the possible cause of contagions. At the time, however, neither these nor conflicting ideas could be proved.

Through the centuries the question of the origin of life itself was interwoven with this problem. From the time of the early Greek philosophers the theory that life could be generated from nonliving matter had enjoyed strong support. By 1700 scientific experiments had dispelled belief in the spontaneous generation of animals of visible proportions, but when microorganisms were discovered, many people continued to apply the theory to this form of life. Thus, until the middle of the nineteenth century, Leeuwenhoek's little animalcules remained in the center of a controversy concerning their origin and were often the subjects of poorly designed experiments that led to claims of further proof of spontaneous generation. For example, flasks of vegetable broth that clouded with the growth of microbes after several days of exposure to air were thought to provide such proof. Nothing was known of the presence of bacteria in the air, or of their adherence to dust particles settling from air into open flasks containing nutrient material. Furthermore, nothing was known of the ability of some bacteria to form heat-resistant spores, so that experiments in which bacterial growth could be demonstrated even in flasks of heated broth were similarly open to misinterpretation.

In spite of the work of several investigators whose well-controlled experiments led to results that refuted the theory of spontaneous generation, the controversy continued until 1860 when Louis Pasteur, a French biochemist and physicist, conducted a series of classic experiments that demonstrated that no growth can occur in flasks of broth properly protected from contamination by airborne microorganisms. In one type of experiment, he introduced beef bouillon into a long-necked flask, then drew the neck of the flask out into a long, sinuous "swan-neck"

shape. When he boiled the liquid in the flask, vaporization forced the air out of the opening at the end of the long neck. Upon cooling of the liquid, air was pulled back into the flask but had to pass over condensed moisture in the neck that washed it free of dust and other particles, so that the bouillon remained clear and free of growth. Control flasks were also heated, but immediately afterward the long necks were broken off, and as a result air pulled into the fluid during cooling was not first washed but delivered dust particles with their burden of viable microorganisms directly into the nutrient fluid, which soon became cloudy with microbial growth (Fig. 1–9). In another series of experiments, Pasteur used an idea of earlier workers and introduced cotton plugs into the necks of his flasks. The cotton acted as a filter, effectively straining out dust particles and keeping the nutrient in the flask free of microorganisms derived from the air. (Cotton plugs are still used in test tubes or flasks, but screw caps and slip-on metal or plastic closures serve the same purpose of preventing air contamination and are easier to prepare and apply.)

When Pasteur presented the results of these experiments, the conclusion that life at every level must be self-reproducing, rather than spontaneous, began to take precedence and the way was cleared for the development and application of scientific methods to study the nature of microscopic organisms and the mechanisms of their self-duplication.

Bacteriology's Golden Age

Pasteur's greatest contributions evolved from his demonstrations of the relationship of certain microorganisms to the spoilage of wines and of others to a disease of silkworms. These were matters of great economic importance to France at the time. The experimental proof that they were of microbial origin led Pasteur to a similar approach in his study of anthrax in sheep and of certain human infections. The successful identification of particular microorganisms as the agents of individual diseases of plants, animals, and human beings then led him to reexamine and reformulate the old "germ theory of disease," which attributed the cause of infection to microbial agents. He brought sound scientific evi-

[Text continued on page 12.]

Microscopes:

Fig. 1-1. Antony van Leeuwenhoek at work with the microscope he invented. His lenses gave him magnifications up to about 300 diameters, revealing microbial forms that amazed him and that he called little "animalcules." (© 1959, Parke, Davis & Company.)

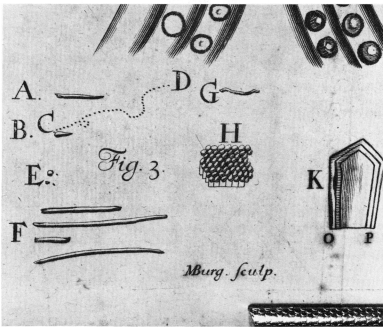

Fig. 1-2. Some of Leeuwenhoek's "animalcules" as he illustrated them in letters written to the Royal Society in London (a 50-year-long correspondence). These organisms were seen in scrapings from his teeth, *A* through *G* being clearly recognizable as bacterial forms. *A* is a rod-shaped bacterium, or bacillus (pl., bacilli). *B* is a shorter, motile rod, the dotted line between *C* and *D* indicating the path of its motion. *E* depicts small, spherical bacteria, or cocci (sing., coccus). *F* represents some long, filamentous bacilli, and *G* appears to be a spirochete with loose, flexible coils, of a type commonly found in the normal mouth. Compare the size and shape of the bacillus in *A* (enlarged by Leeuwenhoek for the drawing) with the photomicrograph of bacilli in Figure 1-4. (Courtesy of Royal Society, London.)

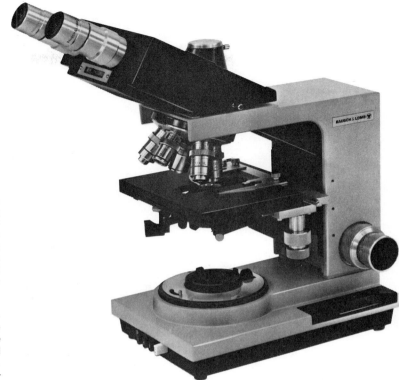

Fig. 1–3. A binocular compound microscope. This is a precision instrument used by today's medical microbiologist. With the oil-immersion lens and a 10× ocular, microorganisms can be magnified approximately 1000×. (Courtesy of Bausch & Lomb.)

Fig. 1–4. Rod-shaped bacteria (bacilli) viewed through the type of compound microscope shown in Figure 1–3, at a total lens magnification of 1000×. Although their shape is distinctive (compare with Leeuwenhoek's drawing, Fig. 1–2, A), their size at this magnification is minute by contrast with that seen by electron microscopy (Figs. 1–6 and 1–8). (Courtesy of Dr. Ramon Kranwinkel, Danbury Hospital, Danbury, Conn.)

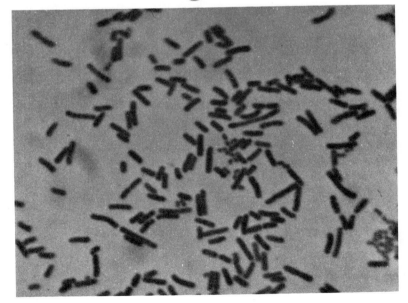

Microscopes:

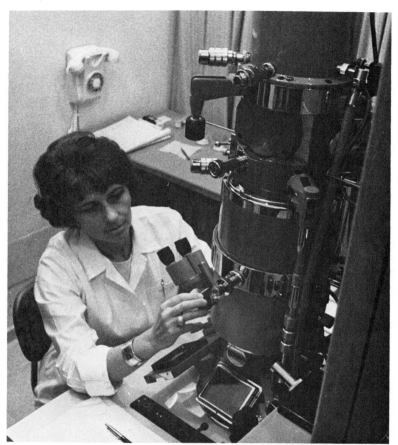

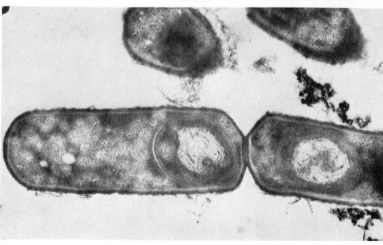

Fig. 1–5. The electron microscope provides magnifications from $10,000\times$ to $200,000\times$ or more, depending on the type of equipment used. In this instrument a stream of electrons, rather than light rays as in an ordinary microscope, is directed at the object to be examined. Many advances in medicine have paralleled the evolution of the microscope. (Reproduced from *World Health,* June, 1976. WHO photo by T. Farkas.)

Fig. 1–6. An electron micrograph of bacilli in which internal structures are clearly visible at this magnification ($32,000\times$). (Reproduced from Ellar, D. J., and Lundgren, D. G.: Fine Structure of Sporulation in *Bacillus cereus* Grown in a Chemically Defined Medium. *J. Bacteriol.,* **92**:1759, 1966.)

Fig. 1–7. A scanning electron microscope. The object being examined is visible on the screen. A magnification range from 15× to 10,000× can be obtained with this instrument, further enlargement being achieved photographically. (Courtesy of International Scientific Instruments, Santa Clara, Calif.)

Fig. 1–8. Rod-shaped bacteria (bacilli) as seen with the scanning electron microscope. Although the final magnification in this photograph (10,000×) is about one-third that of Figure 1–6, note the three-dimensional quality here. The scanning electron microscope does not permit visualization of the internal structures of microorganisms, but it provides a new perspective of surface structures, spatial relationships, and dimensions (see Chap. 4, p. 68). (Reproduced from Klainer, Albert S., and Geis, Irving: *Agents of Bacterial Disease.* Harper & Row, Hagerstown, Md., 1973.)

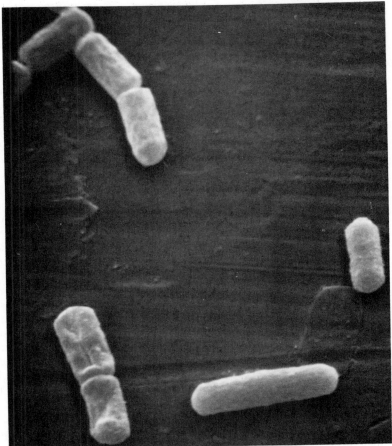

Fig. 1–9. Louis Pasteur conducting a historic experiment, proving that bacteria are not spontaneously generated. (© 1959 Parke, Davis & Company.)

dence to bear on this point for the first time, and this soon came to the attention of the great English surgeon Joseph Lister, who applied the theory in his surgical practice.

Lister speculated that if Pasteur was right about the ubiquitous presence of microorganisms in the air, and if these could fall into a flask of nutrient that supported their growth, such organisms could also enter open surgical wounds and cause the sepsis and wound deterioration that so often led to surgical deaths in those days. With this idea in mind, Lister began to use a phenolic solution in the operating room, using it for his hands and instruments and spraying it into the air around his patients. As a result, his surgery was attended by a very low fatality rate as compared with that of other surgeons who used no precautions to prevent complications they did not believe in as "infections."

At about the same time during the 1860s, but without prior knowledge of Pasteur's work, a German physician named Robert Koch began to make laboratory discoveries of the organisms responsible for various notable human diseases. Together with his students and associates, Koch developed many of the technical methods of cultivation, staining, and animal experimentation vital to the study of micro-

organisms and the interpretation of their behavior as agents of disease. A few years before (in 1840), Jacob Henle, another German scientist, had again outlined the germ theory of disease, stating that its proof would be provided by a series of necessary tests, which included (1) recognition of the infectious microbe in every case of human disease but not in healthy persons, (2) laboratory isolation of the organism from the patient and its pure cultivation, free of other microbes, (3) demonstration of its capacity to produce disease in healthy laboratory animals, and (4) recovery of the organism in pure laboratory culture from infected animals that had displayed typical disease following their inoculation. Henle had no technical means to prove the validity of these "postulates." It was Koch who developed the techniques necessary to the application of these principles (which have since become known as "Koch's postulates"), and armed with such methods he and his associates provided, in rapid succession, proof of the microbial origin and nature of such major human diseases as tuberculosis, diphtheria, typhoid fever, cholera, and gonorrhea.

Thus the field of microbiology achieved scientific status. However, the greatest excitement was generated in the area of medical bacteriology, its advances being so obviously important to human welfare. The contributions of Pasteur, Lister, and Koch, taken together, established the truth of the germ theory and provided the basis for today's understanding of infectious diseases. The laboratories of Pasteur in France and of Koch in Germany became centers of attraction for students from around the world who came to learn methods and ideas and who, in their turn, made further contributions that marked the end of the nineteenth century as the "golden age of bacteriology."

The Microbial World Today

In the two and one-half centuries that have elapsed since Leeuwenhoek's first glimpse of bacteria and protozoa, microbiologic investigations have resulted in the accumulation of a vast body of knowledge that has contributed to man's understanding of the nature of all life and lightened immeasurably the burden of human disease.

The microbial world, as we recognize it today, contains a tremendous variety of organisms, exhibiting an amazing diversity of size, structure, and activity. Microorganisms of one type or another are to be found wherever life is possible — in soil, in water and air, and associated with plant, animal, or human life. Many of their vital functions of growth and reproduction, respiration, synthesis, and excretion, as well as their responses to environmental stimuli, have been found to have the same basic character as those of more structurally complex forms of life. Indeed, the study of microorganisms has served more and more frequently to elucidate many fundamental biologic mechanisms operative in higher organisms as well. Biologists working in the fields of genetics, biochemistry, and biophysics have found microorganisms to be useful and profitable tools in their studies of the nature of life processes. Furthermore, it has become ever more obvious that microbial forms of life make an essential contribution to the maintenance of balanced relationships among all living things.

In the field of medical microbiology today, emphasis is placed not only on the relatively small group of microbes that adversely affect human health as the cause of infectious diseases, but also on the equally important interplay of factors that affect the ability of humans to maintain their balance in a world they share with so many microbial types. In nursing and paramedical professions, more is now required than a familiarity with the microorganisms of medical importance. The fundamental principles of microbiology and of host resistance to infection, as they are currently understood, apply to every phase of patient care. The crisis of infection and the threat of epidemic have been minimized for the modern world by the development of specific measures for their treatment, control, and prevention, but the problem of chronic disease continues.

It should also be understood that new problems of infection have arisen from many of the modern methods of medical and surgical treatment that successfully relieve or cure serious diseases once considered hopeless. The widespread use of antimicrobials has not eliminated infectious microorganisms, but indeed has often made it possible for new types that are resistant to these drugs to emerge.

Prolongation of life for patients with debilitating organic diseases has created a population abnormally susceptible to infection for whom the risk of infection is life-threatening. The development of sophisticated surgical techniques for the treatment of cardiac defects and problems of kidney function, or for removal of extensive tumors, simultaneously demands complete protection of patients from the added burden of infection that may be acquired during surgery or in the difficult period of postoperative recovery.

For the professional nurse, as for every other member of the medical team, an understanding of the potential of infection, existing or threatened, forms an important basis for most procedures and for the total approach to every patient.

Microbiology in Patient Care

In this text a primary effort is made to provide the nurse and others involved in patient care with the most practical and applicable information regarding microbial diseases, including bacterial, rickettsial, viral, and fungal infections and diseases caused by animal parasites.

The emphasis is twofold with regard to the nursing

care of the patient ill with an infectious disease. First, a knowledge of the suspected or diagnosed disease process itself, and of its agent, is essential so that its course from onset to outcome can be anticipated to the patient's best advantage. Second, the communicability of the infection is of vital concern. The nurse must understand the common sources of infection, as well as pertinent routes of entry to and transfer from the body, so that she can help to limit each case of active infection, preventing its spread to other patients in a hospital, to other members of a household, or to those occupied in the care of the patient, whether or not they are in direct contact.

The infectious diseases are presented in Part Two of the text, in an order based primarily on currently available information as to the most probable route of entry of the microbial agent and secondarily on routes of exit and transmission, or on the nature of the organisms involved — that is, fungi, bacteria, viruses, protozoa, or helminths. If students learn to associate the infectious agent of a given disease with its routes of communicability and entry, their techniques as graduates will be oriented toward the probable nursing requirements of the patient and the simultaneous protection of others (including themselves) from the possibility of infection transfer.

Before considering individual diseases, however, the student must learn some basic facts about microbial behavior in general, the fundamental relationships that operate between microbe and host in health or disease, and the general principles of prevention and control as they apply to most infectious diseases. The microbial world is outlined broadly in the following chapter, and the succeeding chapters of Part One describe the basic principles of microbiology.

Questions

1. What role does the nurse play in the diagnosis, control, and prevention of infection?
2. Briefly describe Antony van Leeuwenhoek's contribution to microbiology.
3. Describe Pasteur's experiments that ended the "spontaneous generation" controversies.
4. How did Lister apply Pasteur's work on the "germ theory of disease" to surgery?
5. What contributions to the science of microbiology are attributed to Robert Koch?
6. What knowledge is necessary in the nursing care of a patient ill with an infectious disease?

Additional Readings

*Brock, Thomas D. (ed.): *Milestones in Microbiology.* American Society for Microbiology, Washington, D.C., 1975.

Dobell, Clifford: *Antony van Leeuwenhoek and his "Little Animals."* Dover Publications, Inc., New York, 1960.

DuBos, René J.: *Louis Pasteur, Free Lance of Science.* Charles Scribner's Sons, New York, 1976.

LaRiviere, Jan Wilhelm: Our Oldest Ancestors, the Microbes. UNESCO *Courier,* p. 26, July 1975.

*Nightingale, Florence: *Notes on Nursing: What It Is and What It Is Not.* Dover Publications, Inc., New York, 1969.

Pickering, George: *Creative Malady.* Oxford University Press, New York, 1974.

Vallery-Radot, René: *Louis Pasteur, A Great Life in Brief.* Alfred C. Knopf, New York, 1958.

Wollman, Elie L.: The New World of Microbiology. UNESCO *Courier,* p. 5., July 1975.

Woodham-Smith, Cecil: *Florence Nightingale.* McGraw-Hill Book Co., Inc., New York, 1951.

*Available in paperback.

The Microbial World Defined

2

In medical microbiology all the organisms known to be capable of causing human disease are of interest. As a group they are quite diverse in biologic nature, size, and complexity of structure. In fact, some of them do not meet the definition of a true microorganism as an entity. However, when we are dealing with infectious disease, we are concerned not only with recognition of the organism involved but also with the nature of the illness and the principles of treatment, prevention, and control. These basic principles are similar for all infectious diseases. For this reason any causative organism, whatever its nature, falls within the province of the medical microbiologist and of all others responsible for patient care.

In terms of size and structural complexity, the diverse group of organisms associated with human disease covers a broad range. The *largest* among them — the *parasitic worms* — are invertebrate animals of visible dimensions and *not* microorganisms in any sense, although their developmental forms (eggs and larvae) are microscopic in size. The *smallest* of the group — the *viruses* — are mere particles, comparable in size to molecules and structurally simpler than any microorganism. Between these two extremes are the *true* microorganisms — *bacteria, protozoa,* and *fungi* — distinguished from all other organisms by the simplicity of their structural organization.

To begin the study of this varied group, let us consider briefly the nature of all living things and then see how those we are interested in can be categorized.

The Nature of Living Things

Basic Structures

The higher forms of plants and animals are complex organisms composed of many cells that make up differentiated tissues and organs specialized for particular functions. The *protists,* which will be discussed in detail later in this chapter, are composed of single cells or are simple multicellular organisms. In plants, animals, and protists, the fundamental unit of structure and function is the individual cell. All single cells have basic features and activities in com-

15

mon: they are bounded by a *cell membrane;* they contain either an organized *nucleus* held within a nuclear membrane, or functional nuclear material lying free; and they are composed of a colloidal mixture of proteins, lipids, and nucleic acids called *protoplasm* (from the Latin words *proto* meaning "first" and *plasm* meaning "molded"). Protoplasm may also be referred to as *endo*plasm (*endo* = within) or *cyto*plasm (*cyto* = cell).

Figure 2-1 is a diagram of a single cell showing these basic components as well as some additional features. The presence or absence of other internal structures determines the specialized functions of cells and further characterizes them. Differences in cellular structure, composition, and function distinguish plant and animal cells, determining their divergent means of growth and reproduction, sources and utilization of nutrient, and distinctive tissue formations.

The microorganisms, however, are distinguished from all other living things by their simplicity. As mentioned, they are either unicellular individuals, each being a complete unit capable of performing all vital functions within a single cell; or, if multicellular, differentiation of their tissues is minimal and primitive. Among the organisms of medical interest, three groups, *bacteria, protozoa,* and *fungi,* are recognized as true microorganisms. The *bacteria* are unicellular, with the simplest type of cell organization. The *protozoa* are also single-celled but they have a complex type of cell that permits a very wide range of cellular activities. The *fungi* are either unicellular (yeasts) or multicellular (molds). The cells of molds are complex, but differentiation of their tissues is primitive.

Prior to the discovery of microorganisms almost nothing was known of the evolutionary origins of plants and animals, and forms intermediate between

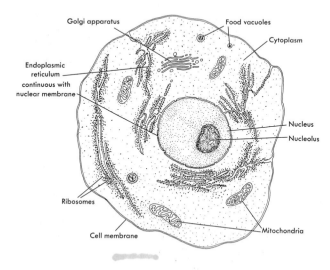

Golgi apparatus   Food vacuoles

Cytoplasm

Endoplasmic
reticulum
continuous with
nuclear membrane

Nucleus

Nucleolus

Ribosomes

Mitochondria

Cell membrane

Fig. 2-1. A single cell. This diagram illustrates some typical structures found in many cells. Additional structures, or variations in these, characterize different types of cells, including microorganisms. *Cell membrane:* mediates exchange of nutrient with the cell's surrounding environment; composed of three layers, but only the outer two are visualized by electron microscopy, giving it the appearance of a double track. *Nucleus:* contains the cell's genetic material and controls its properties and activities; bounded by a *nuclear membrane. Nucleolus:* a mass of ribonucleic acid (RNA) contained within the nucleus. *Cytoplasm* (protoplasm): contains numerous granules called *ribosomes* (composed of RNA and protein) that play a role in the manufacture of certain proteins under the direction of the nuclear material. *Endoplasmic reticulum:* a network of labyrinthine membranes, continuous with the nuclear membrane, that directs the flow of materials within the cell; note that the ribosomes tend to adhere to these and other membrane surfaces. *Mitochondria:* membranous little "organs" (organelles) in which the respiratory enzymes of the cell bring about oxidations that yield energy for cellular functions. *Golgi apparatus:* a stack or clump of membranes where certain proteins and cell wall materials are stored. *Food vacuoles:* membrane-bound bodies involved in the digestion of food particles. Compare this figure with Figures 2-2 and 2-3 showing two types of microbial cells.

them had not been recognized. Accordingly, from the beginning, efforts to categorize microorganisms stressed those characteristics that related them most closely to the lowest known forms of either plant or animal life. The cellular organization and activities of protozoa are very like those of animal cells and they were so classified. Fungi and bacteria have many features common to the simplest plants and were assigned to that group. With continuing study, however, it became apparent that microorganisms as a group possess combined characteristics of both plant and animal cells. Thus, their classification as one or the other is, in many instances, awkward and arbitrary. It has long been suggested, therefore, that microorganisms be placed in a separate group, distinct from plants or animals, until their interrelationships are better understood.

Let us see now how classifications are established and how they serve to clarify the nature of living things.

Principles of Classification

Since Aristotle's time naturalists have been describing and naming plants and animals. Efforts to assemble this vast accumulating volume of information, in a way that reveals meaningful relationships, have always been an important part of the study of living things.

Classification, or *taxonomy,* is a science in its own right. Its purpose is to unify and summarize existing knowledge of the nature of living forms and to provide biologists with a common language through which this knowledge can be communicated precisely. Classification recognizes both the differences and the similarities of organisms, indicating possible origins or developmental pathways. It not only documents structure (anatomy) but relates it to function (metabolic, reproductive, or other processes).

Nomenclature, or the system for naming organisms appropriately, depends on the classifying scheme employed. The system in common use today was developed in the eighteenth century by a Swedish botanist named Carl von Linné (translated as Carolus Linnaeus in the common Latin language of nomenclature). The Linnaean system has been greatly modified and enlarged but remains as the basis for modern classifications. It begins by dividing the living world into two great kingdoms: *Kingdom Planta* and *Kingdom Animalia,* and continues to group related organisms in closer and smaller clusters as their similarities become more detailed (see below).

All plants, or all animals, are grouped in a separate	*Kingdom*
In each kingdom related classes constitute a	*Phylum*
In each phylum related orders constitute a	*Class*
In each class related families constitute an	*Order*
In each order related tribes constitute a	*Family*
In each family related genera constitute a	*Tribe*
In each tribe related species constitute a	*Genus*
Identical organisms constitute a	*Species*

To illustrate this system of naming organisms, we can classify man first as an obvious member of the animal kingdom. Then by describing his structure and characteristics in more and more detail we can place him in the phylum Chordata (subphylum Vertebrata), class Mammalia, order Primates, family Hominidae, genus *Homo,* species *sapiens.*

Note that the scientific name finally assigned to an organism consists of the two words designating genus and species, both printed in italics, e.g., *Homo sapiens* for man. The first letter of the genus name is always capitalized, and the species name is always written in small letters.

We shall take a brief look now at the animal and plant kingdom classifications and see how microorganisms fit into or between them.

The Animal Kingdom

The first phyletic divisions of the animal kingdom separate it into two large groups, containing *vertebrate* and *invertebrate* animals. The earliest distinction that occurs between these two types appears during embryologic development. In the developing *vertebrate* animal, a structure called the notochord materializes, from which the backbone is later derived. Embryologic evidence, together with anatomic and functional distinctions, provides the basis for further subdivisions among higher vertebrates. External anatomic features taken alone may be misleading, as in the case of the fish and the whale, for instance. These two are both anatomically equipped to live in water, but they have little else in common, the whale being a mammal and therefore of a higher order and class than the fish.

Some of the factors that determine animal classifications are cellular differentiation, if any; body symmetry and axis orientation (radial or bilateral, round or flat); number and type of embryonic cell layers; and type of body cavity and body segmentation. The highest member of the animal kingdom is man, in terms of both structure and function. Although the human body develops from a single cell, formed by the union of two single cells, the cellular subdivisions that subsequently occur result in a highly complex, multicellular animal. As we pass down through the animal kingdom to the *invertebrates,* the animals become structurally more simple, but in function they are still capable of very complex activities at the level of their differentiated organs.

The largest of the single-celled microorganisms have many properties in common with animal cells. They have long been classed in the animal kingdom in their own phylum, the *Protozoa* (*proto* = first; *-zoa* = animal). They are among the oldest persisting forms of life, probably derived from still more primitive cells, and probably giving rise in turn to the animal kingdom (Fig. 2–4, p. 21). A few of them are associated with important human diseases.

The term *metazoa* (the Greek *meta* means "after"), by contrast, does not refer to a single phylum but to all later or more highly organized animals, with multicellular bodies differentiated into tissues and organs. Among the lower invertebrate metazoans there are a number of animal parasites, the worms or *helminths,* of medical importance. These are described in the last section of this chapter.

The Plant Kingdom

The vast world of plants is divided into several phyla, their classifications being based, as in the animal kingdom, on anatomic structures and developmental and evolutionary relationships. The more primitive plants, which have no vascular system and are without roots, stems, or leaves, are contained in the phylum *Thallophyta.* This group is an enormous one in itself, containing hundreds of thousands of species, ranging from unicellular microorganisms to great ropy seaweeds. The *thallus,* or body, of the multicellular plants in this group may be differentiated into structural parts, but the only true specialization of function is associated with reproductive structures.

The first major division among thallophytes separates them into two subphyla: the *algae,* which contain chlorophyll and are relatively independent of other forms of life because they can utilize simple inorganic substances as nutrient (they are said to be *autotrophic,* or self-nourishing, because of this), and the *fungi,* which do not have chlorophyll and are dependent on complex organic foods derived from other living things, dead or alive (they are called *heterotrophic,* meaning that their food comes from others).

This traditional classification of the fungi is shown in Table 2–1. There are five major subdivisions of true fungi. Most of the fungal organisms associated with human disease fall in one, the *Deuteromycotina,* or imperfect fungi.

Microorganisms

We have seen that, unlike true plants and animals, microorganisms are single-celled individuals or simple multicellular organisms. Current knowledge of the structure and behavior of microbial cells has led to the widely held belief that they evolved from ancestral cells common also to plants and animals. The simplest microorganisms appear not to have changed very much in at least two billion years as evidenced by electron microscope studies of fossil bacteria. The more complex microorganisms such as algae and protozoa appear to be transitional forms from which multicellular organisms in turn evolved.

Thus, microorganisms are thought of as protists ("first life"), or as prototypes of higher organisms. It was in 1866 that a German zoologist, E. H. Haeckel, first proposed that they be assigned to a third kingdom, the *Protista,* subdivided between the *lower*

Table 2–1. Classification of Fungi as Thallophytes

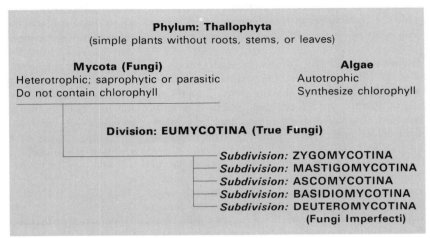

Phylum: Thallophyta
(simple plants without roots, stems, or leaves)

Mycota (Fungi)	**Algae**
Heterotrophic; saprophytic or parasitic	Autotrophic
Do not contain chlorophyll	Synthesize chlorophyll

Division: EUMYCOTINA (True Fungi)

Subdivision: ZYGOMYCOTINA
Subdivision: MASTIGOMYCOTINA
Subdivision: ASCOMYCOTINA
Subdivision: BASIDIOMYCOTINA
Subdivision: DEUTEROMYCOTINA
(Fungi Imperfecti)

protists, having the simplest cell structure, and the *higher protists,* having more complex cellular organization. This distinction rests primarily on the type of cell structures displayed by the two groups.

Lower Protists. The lower, more primitive protists display a type of cell called *procaryotic* because the nuclear material is not contained within a membrane (*pro* = before; *caryo* refers to the nucleus, being derived from the Greek word, *karyon,* for "nut" or "kernel"). Other internal structures, if they are present, also appear rudimentary and are not distinctly compartmented by membrane boundaries (Fig. 2–2).

The *bacteria* are classified in this group together with the *blue-green algae* (*Cyanobacteria*). True algae are the chlorophyll-containing protists, most of

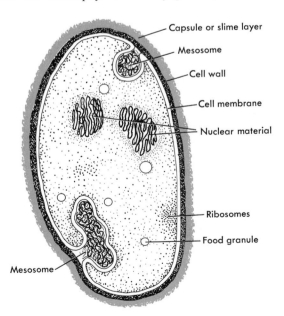

Capsule or slime layer
Mesosome
Cell wall
Cell membrane
Nuclear material
Ribosomes
Food granule
Mesosome

Fig. 2–2. Diagram of a procaryotic cell (a bacterium). Note the lack of a nuclear membrane and the absence of compartmented structures. *Cell membrane:* appears as a double track surrounded by a rigid *cell wall* that may in turn be covered with a slime layer or *capsule. Nuclear material:* consists of fibrils of deoxyribonucleic acid (DNA) that comprise the cell's genetic apparatus, lying free in the cytoplasm. *Mesosome:* a sac formed by and filled with inward folds of the cell membrane (cf. Fig. 2–7); plays a role in cell respiration similar to that performed by the mitochondrion of higher cells (cf. Figs. 2–1, 2–3) and is involved in cell wall formation. *Food granules:* food reserves such as glycogen, lipid, or phosphate may be stored as cytoplasmic granules by various bacteria. *Ribosomes:* granules composed of a large amount of ribonucleic acid (RNA) and protein, involved in the manufacture of certain cellular proteins; they are smaller than the ribosomes in eucaryotic cells.

which are more complex than the blue-green bacteria and are classed in the higher group. Seaweeds are a kind of algae, as are the familiar green furry scums seen so often on the surface of still ponds.

Higher Protists. This group of microorganisms, as well as all true plants and animals, exhibit a *eucaryotic* type of cell (having a true, *"eu-,"* nucleus). In this type of cell the nucleus is a more complex, distinctive structure bounded by a membrane. Other internal structures are also membranous (Fig. 2–3).

There are four major groups of unicellular or simple multicellular organisms classed as higher protists: (1) all the *algae,* (2) *slime molds,* (3) *fungi,* and (4) *protozoa.* Figure 2–4 illustrates the protist classification, the interrelationships of these organisms, and their possible evolutionary position with respect to plants and animals. It is thought that early, unknown procaryotic ancestral cells gave rise to bacteria and blue-green algae and that these have continued to the present day without much change. Eucaryotic cells also emerged from those ancestral lines. The higher algae may then have become the forerunners of the true plants as well as of the earliest types of protozoa (through loss of chlorophyll). From the latter there arose the protozoan forms we know today and also the multicellular animals. This diagram further suggests that slime molds and fungi may be traced to an origin from protozoa.

The microorganisms of greatest importance in medical microbiology are described briefly in the remaining sections of this chapter. Later in the text, the infectious agents of disease are reviewed and classified in more detail (Chap. 4). In Part Two the discussion of each infectious disease includes a description of the causative organism.

The Simplest Microorganisms

We have seen that *bacteria* are procaryotic protists. Classified with them are the *rickettsiae* and *chlamydiae,* two groups of extremely small microorganisms which, unlike other true bacteria, can live only within the cells of the animals they infect. The *viruses* also lead a strictly intracellular existence, but they are classified separately because they are structurally very different from protists, being greatly simplified.

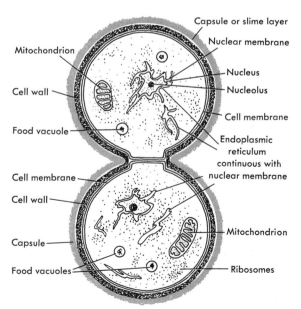

Fig. 2–3. Diagram of a eucaryotic cell (a dividing yeast cell). Note membranous boundaries of nucleus and other internal structures. Compare the form and structures of this cell with those shown and described in Figures 2–1 and 2–2.

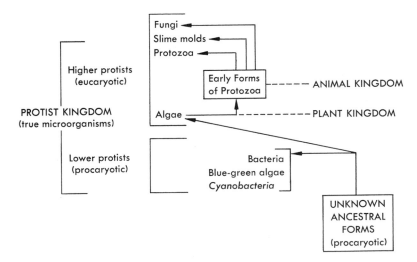

Fig. 2-4. Possible evolutionary relationships of microorganisms, animals, and plants. (Adapted from Jawetz, E.; Melnick, J. L.; and Adelberg, E. A.: *Review of Medical Microbiology,* 10th ed. Lange Medical Publications, Los Altos, Calif., 1972.)

Bacteria

To visualize most bacteria one must use the higher powers of magnification of a good light microscope and enlarge them about 1000 times. Usually it is necessary to stain them to see their surfaces distinctly. Clear visualization of internal structures requires the still greater magnifications of the electron microscope (see Chap. 4).

Morphology. The recognition and classification of the many thousands of bacterial species begin with a study of their structure and form, that is, their *morphology.* Their size and shape, the patterns they form as they divide and multiply, and their cellular structures provide the first obvious clues to their identification.

Size. The measurements of bacteria and other microorganisms are expressed in the smallest units of the metric system. Table 2-2 lists these units and relates them to the meter and to the inch (see also Fig. 2-13, p. 33).

Bacteria vary rather widely in size. The smallest among them, excluding the rickettsiae and chlamydiae, measure approximately 0.5 to 0.75 μm in diameter (500 to 750 nm), but some may be as fat as 1.5 μm (1500 nm). In length they range from 1 to 6 μm (1000 to 6000 nm). The optical tools and stains required to examine them are discussed in Chapter 4, together with laboratory methods for cultivating them.

Shape. Three basic shapes are displayed among bacteria (Fig. 2-5, *A–F*): (1) Individual cells may be *spherical.* These round cells are called *cocci* (sing., *coccus,* a Greek word for "berry") (Fig. 2-5, *A–C*). (2) Individual cells may be shaped like straight little sticks. These are called *bacilli* (sing., *bacillus,* a Latin word for "staff" or "rod") or, simply, rods, to avoid confusion with reference to those that are members of the genus *Bacillus* (Fig. 2-5, *D*). Individual cells may be *spiraled* in shape, and are termed *spirilla* (sing., *spirillum,* from the Latin word for "coil"). These are short little forms, with one to three fixed curves in their rigid bodies. Those with only one curve resemble a comma (Fig. 2-5, *E*).

The *spirochetes* ("animated hairs") are higher forms of spiraled microorganisms. They are much

Table 2–2.   The Smallest Metric Units

Unit	Symbol	Metric Equivalent	Inch Equivalent	Former Term*	Former Symbol
Meter	m		39.37 in.	Same	Same
Centimeter	cm	1/100 m	1/2.5 in.	Same	Same
Millimeter	mm	1/10 cm	1/25.4 in.	Same	Same
Micrometer	μm	1/1000 mm	1/25,400 in.	Micron	μ
Nanometer	nm	1/1000 μm	1/25,400,000 in.	Millimicron	mμ
Angstrom	Å	1/10 nm	1/254,000,000 in.	Same	Same

*Two changes made to unify terminology.

longer and more tightly coiled than spirilla, their bodies are very flexible, and they propel themselves by rapid rotation around their long axes as well as by undulatory motions (Fig. 2–5, *F*). Such complicated movements suggest that these organisms possess more advanced cellular mechanisms than other bacteria; yet they compare with lower protists in nuclear simplicity.

Group Patterns.   Not only is the shape of an individual cell useful in the recognition of bacteria, but the patterns formed by cells grouping together as they multiply are often characteristic also. The most

frequent method of reproduction among bacteria is asexual binary fission; that is, each individual cell splits in half, forming two new cells. As they accumulate, distinctive groupings may be formed:

Coccal Patterns.   Among the cocci, fission may occur at different planes within the round cell. Cocci that split along one plane only tend to arrange themselves in pairs (*diplo*cocci) (Fig. 2–5, *A*), or in chains of varying length (*strepto*cocci) (Fig. 2–5, *B*). When the division occurs alternately in each of two planes, groups of four (tetrads) or cubelike packets of eight are the characteristic result. Haphazard splitting on several planes produces an irregular cluster of cocci,

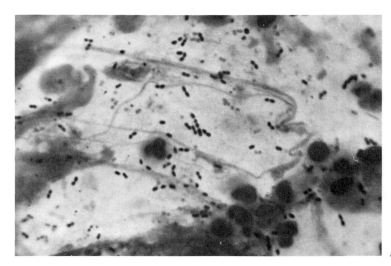

A

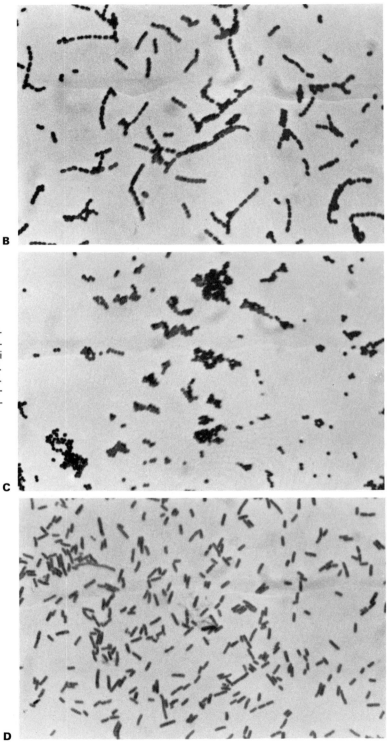

Fig. 2–5. Photomicrographs illustrating morphologic characteristics of some bacteria. *A*. Cocci (spheres) in pairs (diplococci). *B*. Cocci in chains (streptococci). *C*. Cocci in irregular clusters (staphylococci). *D*. Straight rods (bacilli).

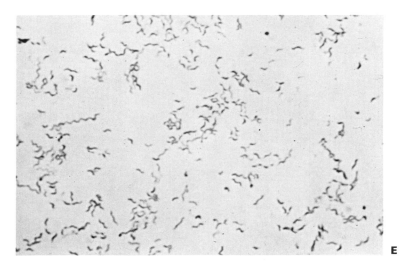

E

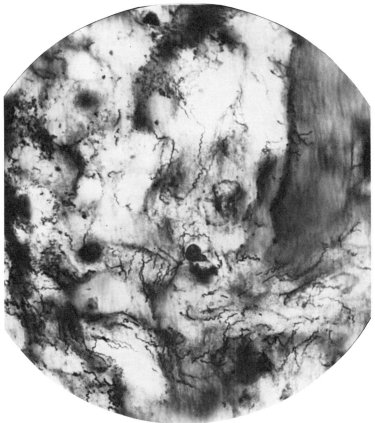

Fig. 2–5 (cont.). *E.* Curved rods (vibrios). *F.* Spirochetes (flexible spirals). (*A, B, C,* and *D* courtesy of Dr. Ramon Kranwinkel, Danbury Hospital, Danbury, Conn. *E* courtesy of Dr. Leon J. LeBeau, University of Illinois Hospital at the Medical Center, Chicago, Ill. *F* reproduced from *Health News,* Jan. 1971; courtesy of New York State Health Department, Albany, N.Y.)

F

bunched together like grapes (*staphylo*cocci) (Fig. 2-5, *C*).

BACILLARY PATTERNS. The patterns formed by bacilli are more limited because they split only across their short axes. They may appear as end-to-end pairs (diplobacilli), or they may line up in chains (streptobacilli). Sometimes, the two new rods formed by the split of a parent cell bend away from the point of fission and lie for a time in V, X, or Y patterns. Very short, small rods can resemble cocci (the term *coccobacillary* is often used for this), but since they divide on one plane only, they do not form packeted or clustered groups (Fig. 2-5, *D*).

SPIRILLA. Like bacilli, the spirilla split on their short axes. When they remain close to each other after division, they may align in short chains (Fig. 2-5, *E*), sometimes in long, sinuous formations. The spirochetes always appear as individual cells, although they may clump and tangle together (Fig. 2-5, *F*).

Cell Wall. The cell wall of true bacteria is rigid (Fig. 2-2, p. 19), but permeable to passage of liquid nutrient material into the cell and to outward passage of substances produced within the cell. The composition of bacterial cell walls varies among species, but they all consist primarily of carbohydrate-protein complexes, and some have a lipoprotein layer as well (*lipo* = fat). Differences among bacteria in the chemical structure of their walls account for their various reactions to the stains used to visualize them under the microscope. Staining techniques are therefore quite valuable in their identification (see Chap. 4, p. 75).

Bacterial cell walls can be removed by laboratory methods that leave the cell membranes intact. Such stripped cells continue to be metabolically active. In nature, the integrity of bacterial cell walls is sometimes lost under the influence of extreme chemical changes in their environment. Substances such as antibiotics and disinfectants can effect such changes, causing bacteria exposed to, but not killed by, them to lose their wall components. Organisms thus denuded have very plastic, variable shapes (they are called "L" forms to commemorate the Lister Institute, where they were first described). One group of bacteria called *Mycoplasma* apparently do not possess the capacity to form cell walls under any circumstances. Their shapes and dimensions are consequently extremely variable. Because they never develop walls, but behave like true bacteria in other respects, the *Mycoplasma* have been placed in their own genus among the bacteria. They are of interest because they have been identified as agents of certain infectious diseases (see Chaps. 4 and 15).

Often the bacterial cell wall is surrounded by a viscous polysaccharide or protein coating, referred to as a slime layer or as a *capsule* (Fig. 2-6). The production of capsular material may be a function of

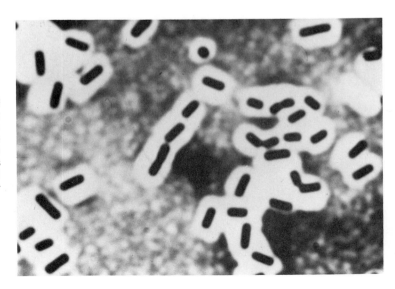

Fig. 2-6. A photomicrograph of encapsulated bacilli (*Clostridium perfringens*). The India ink staining technique for wet mounts (see Chap. 4, p. 75) has been used to demonstrate the clear capsules that surround the dark bacterial cells. (Courtesy of Mr. John Mayhew, Milwaukee, Wis.)

the cell wall, or of some component within the cell, but in any case it is under the direction of the cell's nuclear apparatus. The capsule can play a vital role in protecting the cell against its environment, a matter of great importance in some bacterial infections in which the defense mechanisms of the body must deal with capsular substance.

Cell Membrane. Lying beneath the thick outer wall is the fine membrane that mediates the transport into the cell of materials passing through the wall. Close examination of the membrane with the high magnifications permitted by the electron microscope reveals that at various points it is deviated and folded in upon itself (*invaginated*), forming a rounded sac filled with its own convolutions. This structure is called a *mesosome* (Figs. 2–2, p. 19 and 2–7). Many of the respiratory functions from which the cell derives the energy required for its vital processes take place in the mesosomes or else-

Fig. 2–7. In this electron micrograph of a portion of a bacterial cell, the cell membrane (arrow) is clearly visible beneath the outer cell wall. Two invaginations of the membrane can be seen, one of which has formed a rounded sac filled with membrane convolutions. This sac is called a *mesosome*. (Reproduced from Ellar, D. J.; Lundgren, D. G.; and Slepecky, R. A.: Fine Structure of *B. megatherium* During Synchronous Growth. *J. Bacteriol.,* **94**:1189, 1967.)

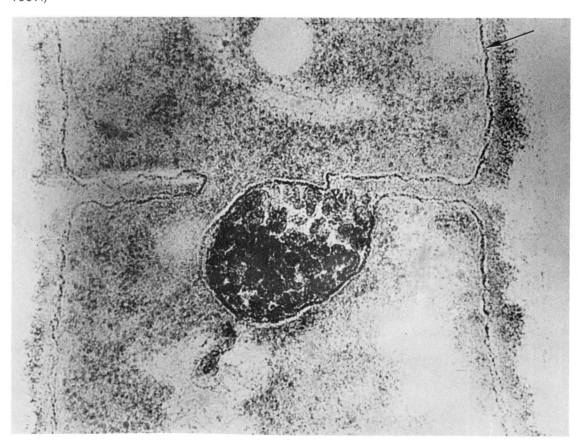

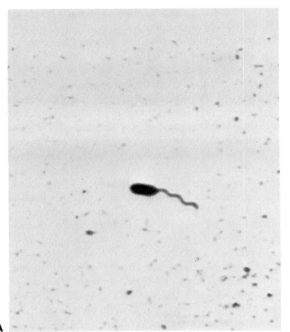

A

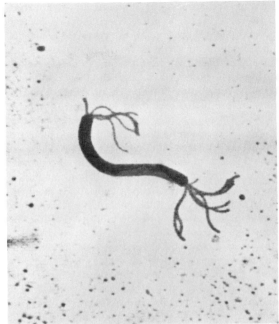

B

Fig. 2–8. Photomicrographs of flagellated bacteria. (2000×.) *A. Pseudomonas aeruginosa* has a single (unipolar) flagellum and is said to be monotrichous. *B. Spirillum serpens* is lophotrichous (tufted), with bipolar flagella. *C.* This *Proteus* species is peritrichous, or flagellated, around its entire surface. (*A* courtesy of Dr. Einar Leifson, Wheaton, Ill.; *B* and *C* reproduced from Leifson, Einar: *Atlas of Bacterial Flagellation.* Academic Press, Inc., New York, 1960.)

where on the cell membrane. (In the eucaryotic cells of higher organisms this function is performed by specialized membranous bodies, within the cytoplasm, called *mitochondria,* Figs. 2–1, p. 16, and 2–3, p. 20).

Flagella. Many bacteria possess *flagella* (sing., *flagellum*), which are fine threadlike appendages that provide the cell with the capacity for active motion, or *motility.* Originating in the cytoplasm, flagella are each composed of a single strand of a protein molecule that repeats itself in a spiral pattern, extending through the cell membrane and wall to the exterior. They are extremely fine in diameter (about 25 nm), but usually as long as, or much longer than, the cell itself (Fig. 2–8). Because they are so fine, special

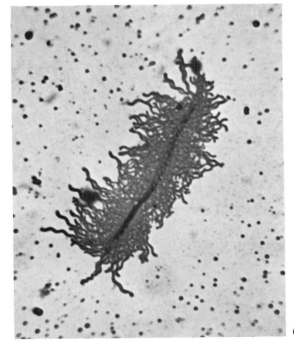

C

staining techniques are required to demonstrate them with the light microscope. Not all bacteria possess flagella. In general, the cocci are not flagellated and are therefore nonmotile, but most spirilla and many bacilli are motile, some possessing many flagella. The number of flagella and their position on the cell (at one or both poles, or circumferential) may be useful features in identification of species.

Pili. Some bacteria display numerous filamentous surface appendages called *pili,* or *fimbriae.* These are very fine structures that can be visualized only with the special staining techniques and high magnifications of electron microscopy. They differ from flagella in that they are finer in diameter, shorter, not undulant, and do not provide motility, but like flagella they are composed primarily of protein. They appear to provide bacterial cells with the ability to adhere to inert surfaces as well as to other cells. Some bacteria have a specialized "sex" pilus through which nuclear material (DNA) can be transferred to another bacterial cell in a process known as bacterial conjugation (Fig. 3–8, p. 61). "Sex" pili are also the sites for attachment of certain viruses (bacteriophages) that infect bacterial cells (see pp. 35 and 60).

Internal Structures. The dense cytoplasm of bacterial cells contains deoxyribonucleic acid (DNA), which is the functional nuclear material, and granules of ribonucleic acid (RNA) in a matrix of protein, carbohydrate, lipid, and other organic substances.

NUCLEUS. Bacteria and other procaryotic cells do not have a true nucleus compartmented within a nuclear membrane. They do possess nuclear material, however, which is visible by electron microscopy as a mass (sometimes two or more) of fibrils (Fig. 2–2, p. 19). These fibrils are composed primarily of DNA, lying free in the cytoplasm.

In all living cells, DNA is the fundamental substance of heredity. Chemically, the DNA molecule is a linear chain of nitrogen bases known as purines and pyrimidines (adenine, cytosine, guanine, and thymine), linked together by alternating sugar (deoxyribose)-phosphate groups. Structurally, it occurs in the form of a double helix produced by the coiling of doubled linear strands (Figs. 2–9 and 2–10). DNA is the chemical substance of *chromosomes,* which in turn are the structural carriers of *genes.* Genes are the individual units of heredity, the determinants that control the properties of organisms. In a eucaryotic cell nucleus, the threadlike chromosomes undergo duplication as the cell prepares to divide. When the cell divides by *mitosis,* each daughter nucleus receives an identical set of chromosomes with identical genes. Sexual reproduction in eucaryotes involves a fusion of two unlike cells with a subsequent exchange of hereditary material. Prior to fusion, the nucleus of each of the two cells must have undergone a reductive division in which the number of chromosomes is halved (*meiosis*). A nucleus containing one set of chromosomes is called a *haploid* nucleus. The fusion of two haploid nuclei then produces a diploid nucleus containing two sets of chromosomes.

In bacterial cells, the "nucleus" consists of a single DNA molecule (a double helix), or chromosome, which is quite long and highly folded. When more than one nuclear mass occurs in a single cell, each probably contains only one DNA molecule. Bacteria divide by *binary fission*; that is, each cell splits into two new daughter cells, each of which contains the same components as the original parent cell. Thus, bacterial division requires the duplication of the DNA chromosome (in a process analogous to mitosis in eucaryotes) and of all other cellular components. It begins with the separation of the two double strands of DNA and the synthesis (manufacture) along each of a new complementary strand. The two new double helices that are formed are each identical with the original (Fig. 2–9). As new DNA is produced and redistributed into the two new halves of the cell, a cross-wall is created by invagination of the cell membrane (see Fig. 2–7; the mesosome appears to play a role in this). When fission occurs, each new cell is a replica of the other and of the parent that produced them both.

The nitrogen bases, or *nucleotides,* as they are called, are paired in a sequential arrangement along the DNA chain. This sequence is distinctive for each type of organism, determining the structure of its proteins and all of its properties.

RIBOSOMES. Ribosomes, so called because they contain a large amount of ribonucleic acid (RNA), are the sites of protein synthesis. They appear as granular structures in the bacterial cytoplasm, often

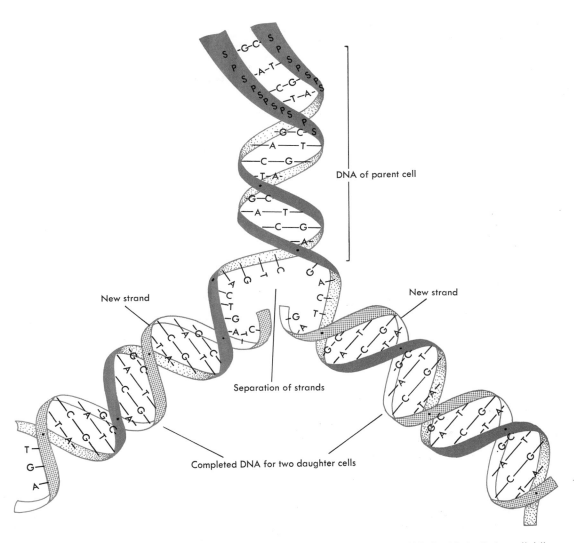

Fig. 2–9. Diagram showing duplication of a DNA double helix in a dividing bacterial (procaryotic) cell. Nitrogen bases (*A* = adenine, *C* = cytosine, *G* = guanine, *T* = thymine) link the deoxyribose (*S* = sugar)–phosphate (*P*) groups in a repeating pattern along the chain. The pattern is distinctive for each type of bacterium and is duplicated precisely for the progeny of each bacterial species. In the drawing above, individual linkages of the parent cell DNA are identical with those of each of the two new daughter cell DNA molecules.

DNA of parent cell

New strand

New strand

Separation of strands

Completed DNA for two daughter cells

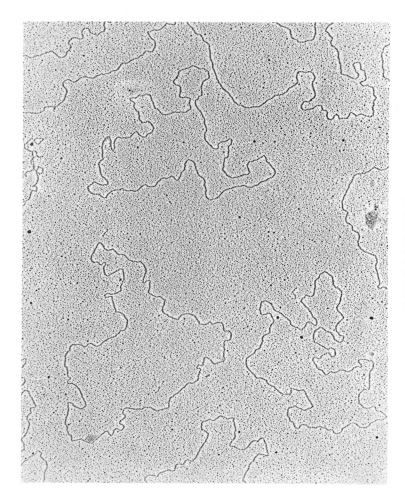

Fig. 2-10. An electron micrograph of several molecules of bacterial DNA, each occurring in a closed loop (circular DNA). Close inspection reveals the double-stranded nature of this molecule. (32,430×.) (Courtesy of Dr. Richard Novick, The Public Health Research Institute of the City of New York, Inc.)

clustered at sites along the cell membrane (Fig. 2-2, p. 19). As stated above, nuclear DNA determines the structure of cellular proteins. It does so by producing a kind of ribonucleic acid called "messenger" RNA, which is a single nucleotide strand having a sequence of nitrogen bases complementary to one of the strands in the double helix. On the ribosome, other "adaptor" molecules of ribonucleic acid called *transfer* RNA are linked with amino acids (the component units of protein). As "messenger" RNA and transfer RNA come together on the ribosome, the former directs the linkage of the amino acids into specific proteins needed for the cell's structures and functions.

SPORES. During their normal growth cycle, certain rod-shaped bacteria are capable of passing into a resistant stage through the production of a *spore* (Fig. 2-11). The vegetative (actively growing) bacterial cell loses moisture and condenses its contents within a thick, double-layered, inner wall, forming a round or ovoid body. The remaining empty vegetative shell then falls away. Bacterial spores represent a resting and protective stage, in which the cell can resist such adverse conditions as excessive temperatures, humidity, or drying. The transforming process is called *sporulation,* and when it is later reversed under favorable growth conditions, each spore germinates into a single vegetative cell, capable of

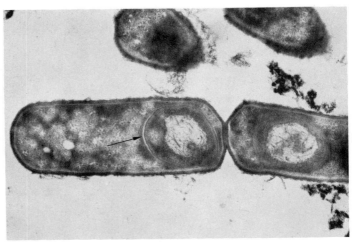

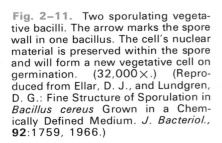

Fig. 2–11. Two sporulating vegetative bacilli. The arrow marks the spore wall in one bacillus. The cell's nuclear material is preserved within the spore and will form a new vegetative cell on germination. (32,000×.) (Reproduced from Ellar, D. J., and Lundgren, D. G.: Fine Structure of Sporulation in *Bacillus cereus* Grown in a Chemically Defined Medium. *J. Bacteriol.*, **92**:1759, 1966.)

function and reproduction in the normal manner once more. Spores are capable of surviving for long periods in soil and dust, and are quite resistant also to effects of boiling or of exposure to live steam. Vegetative bacterial cells, on the other hand, are readily killed in a few minutes in boiling water. These differences are of great importance in the establishment of methods of sterilization (a process that kills all living organisms).

Rickettsiae and Chlamydiae

The rickettsiae (named for H. T. Ricketts who discovered them) were once thought to be transitional forms between bacteria and viruses, being intermediate in size and, like viruses, capable of growth only within higher living cells. The chlamydiae are still smaller than rickettsiae and were once classified as "large viruses." Both groups are now classified with the bacteria, for, although they are much smaller and strictly dependent on an intracellular life, their structures and properties indicate that they are true microorganisms.

Rickettsiae. *Structurally,* rickettsiae display all the features of bacteria described in the preceding section, except that they are not flagellated and they do not sporulate. They have a typical bacterial cell wall, contain DNA and RNA, and divide like bacteria. They are very short little rods ($0.3 \times 1 \ \mu$m), cocco-bacillary or nearly coccal in shape (Figs. 2–12 and 2–13). Their patterns are *pleomorphic* (of variable form); that is, they may occur singly or in pairs, sometimes in short chains, occasionally in long filaments. Because of their small size they require special stains and the highest magnifications of the light microscope for visualization.

In the laboratory, rickettsiae survive and grow only in animal cells grown in the test tube (tissue culture), or in intact animals such as embryos of fertilized chicken eggs, suckling mice or guinea pigs, or other experimental animals. They cannot grow and be studied on artificial nutrient media for they lack the ability to convert nonliving organic substances into a form they can utilize. Intact cells must do this for them. Once localized within living cells, rickettsiae characteristically grow and multiply in the cytoplasm surrounding the nucleus (Fig. 26–1, p. 567).

In nature, rickettsiae occur chiefly as harmless parasites of insects such as lice, ticks, and mites. Sometimes the insect becomes diseased and dies of rickettsial infection, but more often the relationship is not damaging. Furthermore, infected insects transmit rickettsiae to their progeny through their eggs. Rickettsiae can be transmitted to man through the bite of infected blood-sucking insects. When this happens, the human infection is likely to result in serious disease. Rickettsial diseases in humans include typhus fever (not to be confused with typhoid fever), Rocky Mountain spotted fever, rickettsialpox, and Q fever (see Chaps. 4, 15, and 26).

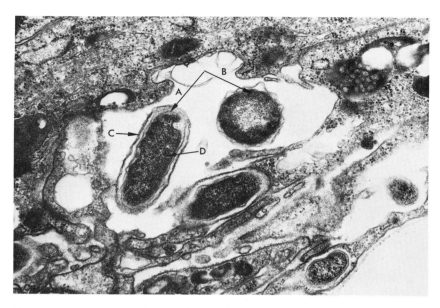

Fig. 2–12. Electron micrograph of rickettsiae lying intracellularly within tissues of their wood tick vector. Note similarity in shape (bacillary and coccal forms) to other true bacteria (arrows *A* and *B*). The cell walls of these organisms, and their cytoplasmic ribosomes, are clearly visible at 28,000 × (arrows *C* and *D*). (Courtesy of Drs. Lyle P. Brinton and Willy Burgdorfer, Rocky Mountain Laboratory, USPHS, Hamilton, Mont.)

Chlamydiae. The word *chlamydia* means "thick walled." The very small size and intracellular life of these organisms originally led to their classification as viruses (formerly "bedsoniae"), but it is now recognized that they differ from viruses in a number of respects. Their thick cell walls are of the bacterial type in structure and composition, and, like bacteria, they possess both DNA and RNA, multiply by binary fission, and contain ribosomes. Chlamydiae are nonmotile, coccoid organisms, ranging in size from about 0.2 µm to 1 µm (200 nm to 1000 nm). They display a unique developmental cycle within the higher living cells they infect. When a single chlamydial particle (called an "elementary body") is first taken into such a cell, a membranous vacuole forms around it. The small chlamydia reorganizes into a larger particle called an "initial body." Within the vacuole the initial body grows and divides by fission. As division repeats, a mass of small particles forms an "inclusion body" within the cell. When freed from the cell, each of these new small particles may infect another cell.

Chlamydiae are associated with several human diseases, including trachoma and psittacosis, as well as a venereal disease called lymphogranuloma venereum (see Chaps. 15 and 22).

Viruses

The viruses are so small and so simplified in structure that they challenge our concepts and definitions of cellular life. They are infectious agents capable of multiplication, but only within complete living cells. In size, they range from 20 to 300 nm in diameter (0.02 to 0.3 µm), the smallest being comparable to a large molecule (Figs. 2–13 and 2–14). Structurally they are not complete cells but merely subunits. Visualized by electron microscopy, some viruses display morphologic helical or polyhedral symmetry,

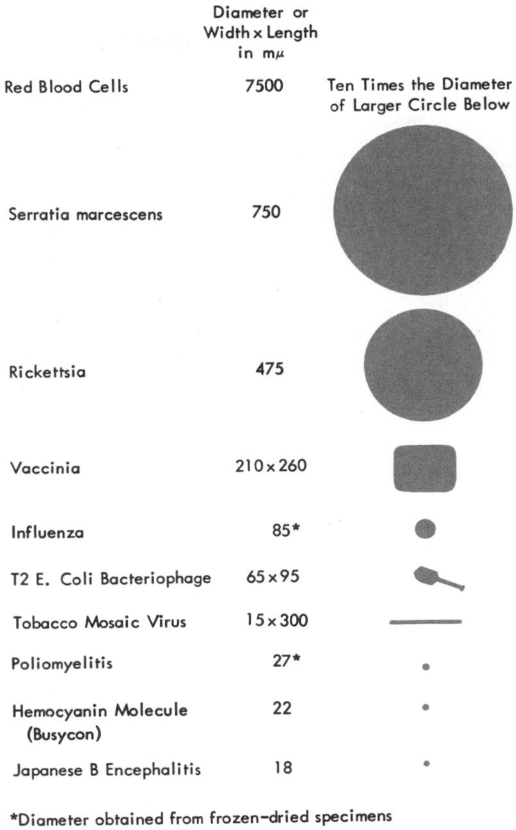

	Diameter or Width x Length in mμ	Ten Times the Diameter of Larger Circle Below
Red Blood Cells	7500	
Serratia marcescens	750	
Rickettsia	475	
Vaccinia	210 x 260	
Influenza	85*	
T2 E. Coli Bacteriophage	65 x 95	
Tobacco Mosaic Virus	15 x 300	
Poliomyelitis	27*	
Hemocyanin Molecule (Busycon)	22	
Japanese B Encephalitis	18	

*Diameter obtained from frozen-dried specimens

Fig. 2–13. In the diagram above compare the size of a red blood cell with that of various microorganisms. Note that the symbol mμ (millimicron) is equivalent to the current symbol nm (nanometer). (Reproduced from Dubos, R.: *The Unseen World.* Rockefeller University Press, New York, 1962.)

whereas others are more complex in shape. Bacterial viruses (bacteriophages) resemble tadpoles, having a globular "head" with a "tailpiece." (See Figs. 2–14, *B,* and 3–7 [p. 60].)

Morphology. A single virus particle appears to represent the final essence of cellular genetic material, wrapped only in a protein coat rather than cellular trappings. The inner substance, or *core,* consists of strands of a nucleic acid (DNA or RNA, but not both in one virus). The protein coat, or *capsid,* is made up of symmetrically arranged subunits, each of which is called a *capsomere.* The capsid may be overlaid by outer additional layers of asymmetrically arranged organic material cohering as an *envelope* (containing proteins, lipids, carbohydrates, and other organic substances in varying proportions). The envelope,

when it is present, is not an organized structure. The entire virus particle, with or without an envelope, is called a *virion* (Fig. 2–15).

Such a particle is stripped of all cellular machinery, all independence, and all function except that involved in reproduction and genetic continuance. When it enters a living cell, the virion's protein coat is discarded at the door (as it penetrates the cell's surface layers) or is removed by cellular enzymes within the cell, and the virus core settles down in the cytoplasm or, later perhaps, in the nucleus. Viral DNA or RNA then begins to direct the metabolic activity of the cell toward the synthesis of viral components (DNA or RNA, and capsid protein), sidetracking the cell from its own functions. In a short time a quantity of new virions is produced and released from the infected cell (which may burst),

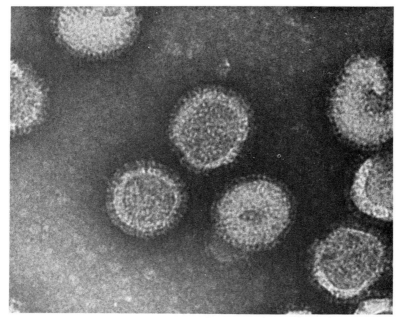

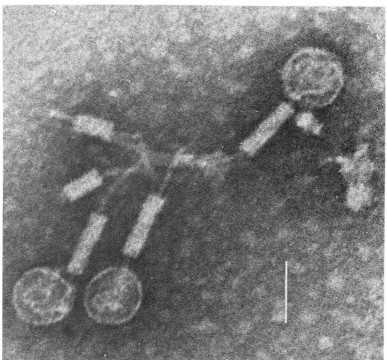

Fig. 2–14. Electron micrographs of two viruses. *A.* Influenza virus (504,000×). (Courtesy of Dr. John Lawlis, Merrell-Laboratories, Swiftwater, Pa.) *B.* Bacteriophages (250,000×) (bar = 60nm). (Reproduced from Eklund, M. W.; Poyski, F. T.; Peterson, M. E.; and Meyers, J. A.: Relationship of Bacteriophages to Alpha Toxin Production in *Clostridium novyi* Types A and B. *Infect. Immun.*, **14**:793, 1976.)

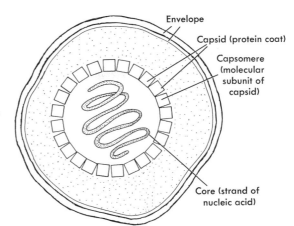

Fig. 2–15. Diagram of a complete virus particle, or virion.

and the process is repeated as adjacent cells are infected.

Virus Infections. The *nature* of viruses has yet to be fully clarified. Their simplicity suggests that they are not individual microorganisms but either (1) components of normal cells that have gotten out of control, (2) descendants of subcellular units that infected the earliest cellular organisms and evolved with them, or (3) products of degenerating bacteria that gave up independence for a specialized parasitic life.

In the laboratory, viruses can be propagated only in living cells, either of intact animals or of fertilized eggs, or in tissue culture. In nature, they are widely distributed as infectious agents of plant, animal, and human diseases, and there is a large group called *bacteriophages* that infect and can destroy bacterial cells. (The Greek word *phagein* means "to eat"; therefore, a bacteriophage is a virus that "eats" bacteria.)

The viruses that infect plants do not also cause animal or human disease. Among those that infect animals, some are restricted to certain species but others are associated with a wide variety of animals. Like the rickettsiae, some viruses are transmitted to man or to animals through the bites of infected insects. Many virus diseases, however, are passed directly from person to person, or from animals to man, without an intermediary.

There are many viruses that are limited not only to an intracellular life but to particular tissue cells (e.g.,

nerve, lung, liver) in certain animal species. Such specific requirements greatly affect the natural patterns of disease. Also, viruses may lie dormant for long periods in their intracellular position, interfering little if at all with the cell's activities or the infected individual's well-being. In this *latent* condition the virus is not multiplied, but remains silent until some change affects either cell or virus, inducing new activity in the latter. The ordinary "cold sore" exemplifies this situation, its viral agent being present but latent in many persons until some stress such as fever or sun exposure activates it. Obvious infection subsides as conditions return to normal for the cell, and the virus returns to latency. Silent virus infections are apparently quite frequent, for many viruses have been isolated from entirely healthy individuals, human or animal.

In recent years a great deal of evidence has accumulated pointing to the involvement of viruses in cancers of animals. Some naturally occurring animal cancers have been demonstrated to be caused by viruses, and cancers have been experimentally induced in animals with viral agents. As yet the evidence is not conclusive that human cancers are associated with viruses. The possibility seems very strong, however, for there are many similarities between human viruses and those that cause animal tumors, and there are ordinary human and animal viruses that can produce experimental animal cancer. Furthermore, benign human tumors, such as warts, are of known viral origin.

Classification. Viruses are now classified by a system of their own, based on their content of DNA or RNA and other chemical composition, size, morphology, and other measurable properties. This classification recognizes the fact that they are not true, complete cellular units, as are other microorganisms, but are nonetheless capable of replicating themselves within living tissue and so must be considered as viable units.

The problem with virus classification has been that available information concerning measurable properties is not complete for all viruses. Unless or until a virus causes a disease in humans, animals, or plants its existence may go unrecognized. Moreover, we know that many viruses can cause more than one type of disease in humans (or other organisms), depending on where in the body they localize. In most instances where a one-virus, one-disease relationship exists, the virus has been named for the disease (e.g., mumps virus, measles virus, or poliomyelitis virus). Viruses have also often been named for the geographic area in which they were discovered (Coxsackie virus, West Nile virus), or for the type of animal with which they are importantly associated (equine encephalitis virus). Table 2–3 briefly summarizes some of the important virus groups and their human diseases, indicating their classification as to some morphologic features and nucleic acid content.

For practical medical or epidemiologic purposes, the human viruses and their diseases can be grouped according to outstanding clinical properties or to their mode of transmission. Thus, viruses that characteristically involve the skin and produce a rash can be classed together (measles, German measles, roseola), as are those that produce pocks on the skin (smallpox, chickenpox, and herpesviruses). Similarly, viruses that require an insect (arthropod) to transmit them from man to man (yellow fever) or from animals to man (equine encephalitis) are grouped together as arboviruses. This classification is detailed in Chapter 4.

Table 2–3 Classification of Some Important Human Viruses

Virus Group	Human Diseases	Envelope	Capsid Symmetry	Nucleic Acid
Adenovirus Papovavirus	Respiratory infections Warts	Absent		
Herpesvirus	Chickenpox (varicella) Shingles (zoster) Cold sores (herpes simplex)	Present	Icosahedral*	DNA
Poxvirus	Smallpox (variola)	Present	Complex	
Reovirus	Respiratory and intestinal infections	Absent	Icosahedral	
Picornavirus	Polio, intestinal infections, colds			
Togavirus	Encephalitis, yellow fever	Present		RNA
Orthomyxovirus Paramyxovirus Rhabdovirus	Influenza Measles, mumps Rabies	Present	Helical†	

*An icosahedron is a polyhedron having 20 faces (the Greek word *eikosi* = 20; *-hedron* is a crystal or geometric figure).
† A helix is a spiral or coiled form.

Complex Microorganisms

The protozoa and fungi are eucaryotic, higher protists, as we have seen. Their individual cells are considerably more complex and larger than bacterial cells. Both are very widely distributed in nature, and both groups include genera that can produce disease in plants, animals, and man.

Protozoa

The animal kingdom probably began with the single-celled protozoa. Today there are many thousands of species of these simple animal cells, with wide variations among them as to biologic properties, size, shape, and structure. They are the largest of the unicellular microorganisms, but they range greatly in size. Some are just barely visible without the help of a microscope; others approach the size of the largest bacteria and must be magnified nearly 1000 times to be seen clearly. They are structurally simple but their physiologic activities — food intake and utilization, excretion, locomotion, reproduction — are quite complex.

The protozoa occupy an entire phylum of the animal kingdom and are subdivided into four subphyla on the basis of the structural means of locomotion possessed by their mature forms (Fig. 2–16):

Sarcodina — the simple *ameboid* forms that move by bulging and retracting their protoplasm in any direction.

Ciliophora — possess *cilia*, or fine, minute hairs that cover the outer cell surface. Rapid, rhythmic beating of the cilia moves these cells along with efficient speed.

Mastigophora — possess long, whiplike *flagella* projecting from their bodies. The lashing of these threadlike extensions provides rapid motility.

Sporozoa — have no external means of locomotion but some immature forms display an ameboid, gliding motility. This group has a complex reproductive cycle, maturing as *spores* or *sporozoites*.

Each of these groups contains organisms associated with human disease. Some protozoan infections are minor in nature, but there are a number that are quite serious and important (see Chap. 4).

Sarcodina (Amebae). The cell bodies of amebae are constantly changing in shape as they glide about and are appropriately described by the Greek word for change, *amoibe*.

Morphology. Active "vegetative" amebae are called *trophozoites* (*tropho* = taking nourishment). They have a thin cell membrane, a granular cytoplasm containing ingested food and food vacuoles, and a well-defined nucleus within a demarcated membrane. (They range in diameter from about 10 to 60 µm.) The granular *endo*plasm is sharply demarcated from the clear, homogeneous *ecto*plasm (*ecto* = without) that separates it from the cell membrane (Fig. 2–16, *A*). The external membrane seems to function as the eyes, ears, nose, and fingers of the cell, responding sensitively to the slightest change in the surrounding environment. The cell moves continuously, the membrane extending in a fingerlike projection in one direction, then another, the endoplasm flowing into each bulge, then retracting and surging elsewhere in an endless quest. The cell protrusions are called *pseudopods,* or false feet, because of the motion they provide. They also provide the means of food ingestion, reaching out to surround and engulf nutrient particles (amebae are not restricted to soluble nutrient as are bacteria). Once in the cell the food is taken into vacuoles where it is digested and metabolized, the waste being excreted through the cell membrane.

Reproduction and Encystment. Amebae reproduce asexually by simple binary fission following mitotic division of the nucleus. Each trophozoite, when it divides, gives rise to two new cells. Under adverse environmental conditions, some amebae are capable of passing into a *cyst* stage. The trophozoite rounds itself up, the membrane thickens and becomes hyaline (glassy), internal activities slow down, and the cell rests until more favorable conditions arise. During *encystment* mitotic nuclear divisions occur, with the production of several nuclei within one cyst. When *excystment* occurs, the trophozoite emerges with its multiple nuclei and undergoes fission and

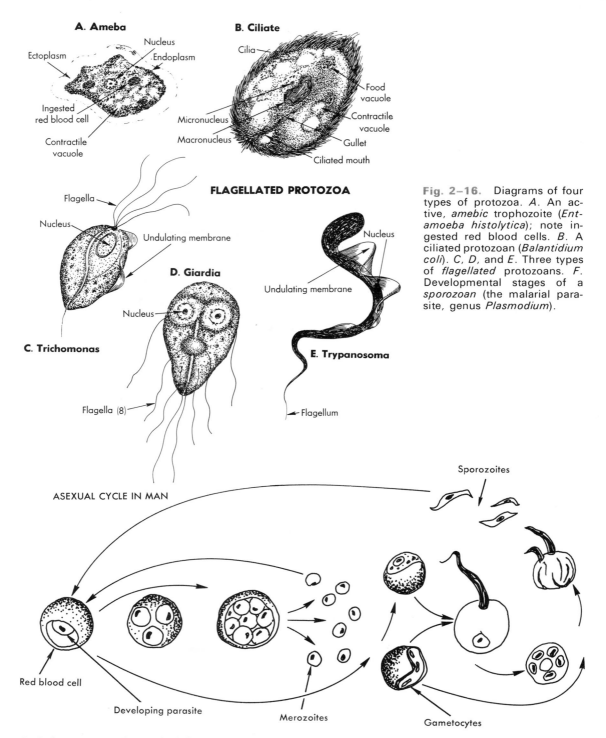

A. Ameba

Ectoplasm

Nucleus

Endoplasm

Ingested red blood cell

Contractile vacuole

B. Ciliate

Cilia

Food vacuole

Micronucleus

Contractile vacuole

Macronucleus

Gullet

Ciliated mouth

FLAGELLATED PROTOZOA

Flagella

Nucleus

Undulating membrane

D. Giardia

Nucleus

C. Trichomonas

Nucleus

Undulating membrane

E. Trypanosoma

Flagella (8)

Flagellum

Fig. 2–16. Diagrams of four types of protozoa. *A.* An active, *amebic* trophozoite (*Entamoeba histolytica*); note ingested red blood cells. *B.* A ciliated protozoan (*Balantidium coli*). *C, D,* and *E.* Three types of *flagellated* protozoans. *F.* Developmental stages of a *sporozoan* (the malarial parasite, genus *Plasmodium*).

ASEXUAL CYCLE IN MAN

Sporozoites

Red blood cell

Developing parasite

Merozoites

Gametocytes

F. A Sporozoan. The malarial parasite's life cycle

SEXUAL CYCLE IN MOSQUITO

nuclear division, eventually forming new uninucleate trophozoites.

Amebic Infection. Many free-living species of amebae abound in soil and water everywhere in the world, but there are a few that are parasitic, that is, dependent on higher organisms. A number of such species may live in the human mouth or intestinal tract, causing little or no harm. However, there is one species of great medical importance that can produce severe intestinal damage, namely, *Entamoeba histolytica,* the cause of amebic dysentery in man. The activities of this organism in the bowel may erode the intestinal lining as it nourishes itself by ingesting red blood cells (Fig. 2–17).

Ciliophora (Ciliates). These organisms have a permanent ovoid shape of fixed dimensions. The largest among them, the paramecia, can be detected with the naked eye, but most are microscopic. Their cili-ated surfaces keep them in constant swimming motion in a watery environment. They have more complex structures than amebae, including a mouth and gullet (also lined with cilia that sweep food in), an excretory pore, and two nuclei, one much larger than the other. The smaller or *micro*nucleus is responsible for cell reproduction and the *macro*nucleus directs the cell's other activities (Fig. 2–16, *B*). Reproduction is usually asexual but may also occur by conjugation. Some but not all ciliates display an encystment stage, the cysts also containing two nuclei.

Most ciliates are free-living. There is one species that inhabits the intestinal tracts of animals and may, infrequently, infect man. This is *Balantidium coli,* a large ciliate (70 × 200 μm) that can cause a chronic, recurrent dysentery if it gains access to the human intestinal tract.

Mastigophora (Flagellates). The members of this group all possess flagella, or long ectoplasmic

Fig. 2–17. *Entamoeba histolytica,* trophozoite. The organism is in the process of ingesting red blood cells in this preparation. (2500×.)

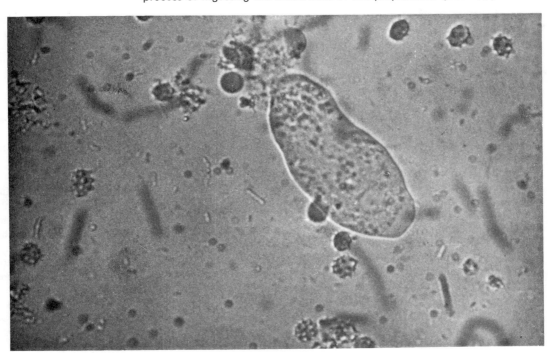

threads that propel them with great rapidity in a fluid milieu. A number of flagellate species are capable of infecting man. These fall into two groups: the *intestinal* flagellates that live in the bowel or on other mucosal surfaces of the body; and the *blood and tissue,* or *hemo*flagellates (*hemo* = blood). The latter are transmitted to man by biting insects and live in the soft tissues. These two groups differ morphologically as well as biologically.

Intestinal Flagellates. Morphologically these organisms have a fixed, ovoid shape (about 10 × 20 μm), with three to eight flagella that propel them. Species of the genus *Trichomonas* (Fig. 2–16, *C*) belong to this group and are characterized by having a short undulating membrane as well as four flagella on the anterior rounded end. They have a single nucleus and occur only as trophozoites, not having a cyst stage. *Giardia,* another intestinal flagellate, has two nuclei (note the bespectacled appearance this gives them in Fig. 2–16, *D*) and eight flagella but no additional locomotive membrane. *Giardia* species frequently display a cyst stage with four nuclei.

These two genera are often found in the human intestinal tract where they may live harmlessly. *Giardia* is becoming more frequently associated with intestinal disorders, when conditions permit it to multiply to very large, irritating numbers. *Trichomonas* is also commonly found on the genital mucosa of both sexes. If it multiplies excessively there, it can cause very troublesome irritations, particularly in women.

Hemoflagellates. Medically these are the most important flagellates. When biting insects deposit them in the body, they enter the bloodstream and are carried to various sites. They multiply asexually, inducing damage in the soft tissues where they localize (liver, brain, or other organs). The two major genera of this group are *Trypanosoma* and *Leishmania.* The former, when mature, are thinly elongated (2 × 20 μm) with a single flagellum, a long undulating membrane, and a central nucleus (Fig. 2–16, *E*). These organisms have a complex cycle of asexual development, some stages occurring in the insect, others in infected man or animals. *Leishmania* also have several developmental stages, including a flagellated form similar to a trypanosome but lacking an undulating membrane, and a small, nonflagellated oval form that lives intracellularly in man. African sleeping sickness (trypanosomiasis) and kala-azar (leishmaniasis) are examples of hemoflagellate diseases.

Sporozoa. The life cycle of sporozoa is complex, involving two alternate means of reproduction. Developmental stages of an *asexual* type of nuclear division occur in a vertebrate animal such as man, but other transitional forms are produced by *sexual* union and nuclear division, usually in insects. The cells resulting from such sexual fusion and division are called *sporozoites,* hence the name of this group. Blood-sucking insects, notably some species of mosquito that prey on vertebrates infected with sporozoans, ingest sexual forms of the parasite during the blood meal. Further reproduction in the mosquito leads to the production of sporozoites, which are then injected into the blood of the next animal upon which the mosquito feeds. A reservoir of infection is thus maintained by cyclic transmission of the organism from one animal to another. *Malaria* is an outstanding example of a sporozoan disease of man, caused by one of several species of the genus *Plasmodium.* It is transmitted by the *Anopheles* mosquito (Fig. 2–16, *F*).

Fungi

The fungi are an extremely large and diverse group, distributed universally. They range in size and complexity from the simple unicellular yeasts to the multicellular molds and mushrooms.

Fungus Morphology. The body of a multicellular fungus is composed of a mat of long, branching, filamentous tubes containing cytoplasm, with nuclei strung out at irregular intervals. In some fungi, these tubes are interrupted at numerous points by septations, or cross-walls. The entire mat is referred to as the *mycelium* (*mykes* is the Greek word for fungus), and the individual filaments that compose it are called *hyphae.* The hyphae form originally from the elongation of a germinating, reproductive cell, microscopic in size, called a *spore.* Some of the yeasts, which are classified with the true fungi, have spores

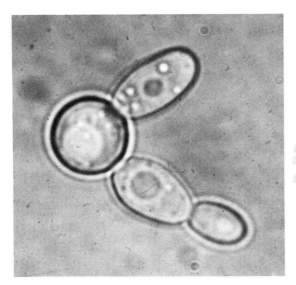

Fig. 2–18. A photomicrograph of a budding yeast cell (*Saccharomyces* species). (Reproduced from Schneierson, S. S.: *Atlas of Diagnostic Microbiology.* Abbott Laboratories, North Chicago, Ill., 1974.)

that fail to produce a mycelium and merely pinch off daughter spores at their germinating tips (Fig. 2–18). These yeasts are, therefore, unicellular.

As more and more mycelium is produced by the growing, branching hyphae, the plant reaches visible proportions. Its eventual size will depend on the continuing availability of nutrient and on other environmental conditions. Most molds do not grow to a height of more than 0.5–1 cm, but they extend later-ally along the surfaces of, and downward into, the material supporting them. That portion of the plant extending downward and into the medium is called the *vegetative mycelium,* and that which extends above the surface is the *aerial mycelium.* Usually the reproductive structures are found in the aerial portion, and their presence contributes to the powdery or fuzzy look of the mold.

There are five subdivisions of true fungi: *Zygomycotina, Mastigomycotina, Ascomycotina, Basidiomycotina,* and *Deuteromycotina* (the *Fungi Imperfecti*). These subdivisions are based on both visible and microscopic features of their mycelia, hyphae, and spores and, most important, on the type of sexual reproduction displayed. The *Fungi Imperfecti* are so called because their sexual stages are rare or have not yet been found. Most of the fungi that cause human disease fall in this group, but each of the others also has some members of medical importance except the *Mastigomycotina.* Some of the latter produce disease in plants but none has infected man.

Zygomycotina. This group displays a type of sexual reproduction in which a *zygote* is formed from the fusion of a pair of cells derived from the same or a different plant (Fig. 2–19, *A*). Germination of this *zygospore,* as it is called, leads to the production of a new generation. Asexual spores are also formed by the fungi of this subdivision (Fig. 2–20). The mycelium is characteristically nonseptate. It gives rise to specialized hyphae that extend aerially, each developing at its tip a round, saclike structure called a *sporangium.* Within this sac many asexual *sporangio-*

Fig. 2–19. Some types of sexual spores of fungi. *A* is a zygospore, or zygote, commonly formed by zygomycetes; *B* is an ascus containing four sexual spores of an ascomycete; *C* and *D* represent more complex ascus formation seen in the higher ascomycetes; *E* is a basidiomycete (a mushroom), and *F* shows its basidium bearing basidiospores. (Adapted from Conant, N. F.; Smith, D. T.; and Callaway, J. L.: *Manual of Clinical Mycology,* 3rd ed. W. B. Saunders Co., Philadelphia, 1971.)

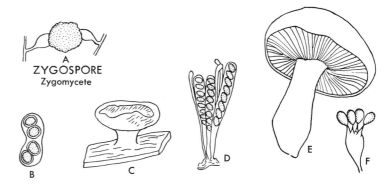

ASEXUAL SPORES
Conidia

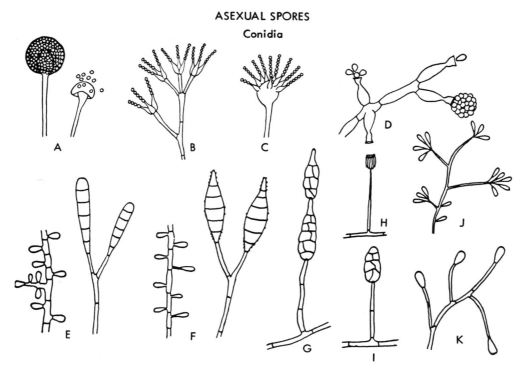

Fig. 2–20. These drawings illustrate several types of asexual spores (conidia) produced by different fungi. The term "conidia" (sing., *conidium*) is broadly used to denote asexual spores borne directly on the fungal hypha or on specialized spore-producing structures called conidiophores. (Reproduced from Conant, N. F.; Smith, D. T.; and Callaway, J. L.: *Manual of Clinical Mycology*, 3rd ed. W. B. Saunders Co., Philadelphia, 1971.)

spores develop, and when these are released by rupture of the sac (Fig. 2–20, *A*), each may germinate to produce new hyphae and a new mycelium.

One of the most familiar members of the *Zygomycotina* is the common black mold of bread, *Rhizopus nigricans*. Some other members of the genus *Rhizopus* are associated occasionally with human disease, usually through opportunistic circumstances. This is true also of species of *Mucor* and *Absidia*, two other genera of this group (see Mucormycosis, Chap. 16).

Ascomycotina. This is the largest of the fungus subdivisions. It includes the true yeasts as well as some mildews and molds. The characteristic sexual spores of this group occur, always in even numbers

(from two to eight), within a sac called an *ascus,* from which they take the name *ascospore* (Fig. 2–19, *B*). The asexual spore of the yeasts is a budding cell called a *blastospore* (Fig. 2–18). Many ascomycetes that are molds have specialized hyphae (*conidiophores*) that produce asexual spores called *conidia*, borne in chains or singly as shown for *B, C,* or *K* in Figure 2–20. "Conidia" is a general term referring to a kind of asexual sporulation seen in many molds.

Bakers' yeast, and other yeasts used in brewing beer or in wine making, are well-known examples of ascomycetes. Some of medical importance are included in a genus of yeastlike organisms known as *Candida*. Several species of *Candida* are often found living harmlessly on human mucosal surfaces, among other microorganisms there, but if circumstances permit overgrowth of these yeasts, or if they have an opportunity to find their way into deeper tissues, severe infection can result. The yeast *Cryptococcus,* which is associated with serious diseases of the lung, brain, and other vital organs, formerly was classified as an ascomycete. Recently, however, its sexual stage has been found and this organism is now known to be a basidiomycete.

Basidiomycotina. The familiar mushrooms and toadstools are included in this group, together with the rusts and smuts that cause disease in plants. A club-shaped structure called a *basidium* produces the typical sexual spore, a *basidiospore.* In the case of the mushrooms, the vegetative mycelium develops underground and pushes upward a compact mass, or button, that develops into the well-known little umbrella-like structure. The basidia form on the gills underneath the umbrella, producing basidiospores that drop to the ground, giving rise to new plants if the right conditions exist (Fig. 2–19, *E, F*). With the exception of *Cryptococcus,* this group contains no members capable of causing *infectious* disease in man, but some mushrooms are medically important because they contain a substance that is poisonous if eaten.

Deuteromycotina (Fungi Imperfecti). This is a large and very mixed group of fungi in which a recognizable sexual stage has not been observed and whose final classification is uncertain. There are many types of asexual spore production among its members, and this forms an important basis for identification of species, together with other microscopic details and visible appearance (Fig. 2–20).

Most of the fungi important in human mycotic diseases are grouped with the *Fungi Imperfecti.* A number of them cause infections involving only superficial tissues, that is, the skin and mucous membranes. "Athlete's foot" and ringworm are examples of such infections. There is also a small group of fungi associated with serious systemic diseases, in which various vital organs, including the brain, may be involved. Among these are diseases such as histoplasmosis, blastomycosis, and coccidioidomycosis for which the mortality rate is high because the tissue involvement may become extensive and treatment is difficult.

The dramatically useful role played by some fungi in many bacterial diseases of man has been continuously extended since the famous ascomycete, *Penicillium* (Fig. 2–21), was first demonstrated to be capable of producing a substance, *penicillin,* that is

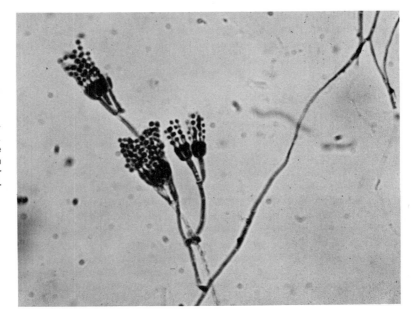

Fig. 2–21. *Penicillium notatum* growing in a slide culture. (236×.) Compare with Figure 2–20, *B*. (Courtesy of Dr. John W. Rippon, The Pritzker School of Medicine, The University of Chicago.)

strongly antagonistic to many bacteria. Since that discovery, the search for such antibacterial agents, commonly called *antibiotics,* among fungi and other organisms has extended around the world and reached stupendous proportions. Untold numbers of fungi, and fungal products, have been studied exhaustively in laboratories everywhere in the effort to produce drugs that can be safely used in the treatment of human infectious diseases. A large majority of such products have had to be discarded either because they are toxic in themselves to human beings or because they are relatively inefficient in their action on bacteria. However, a handful of clinically useful, relatively safe antibiotics has emerged from the search and has become an inseparable part of the medical approach to infectious disease.

Animal Parasites

There are a number of parasitic, invertebrate worms, or *helminths* (the Greek word for worm, derived from one meaning "to roll"), that cause human and animal diseases. In their fully developed adult forms most worms are quite large enough to be seen without help of a magnifying lens, except to visualize internal structures. Usually, the recognition of their immature stages (*ova* or eggs, and *larvae*) in the excretions or tissues of infected patients is required for diagnosis. Microscopic and sometimes other microbiologic techniques are applied to study them. It was pointed out at the beginning of this chapter that the basic principles of prevention and control of helminthic diseases are similar to those applicable in true microbial infections.

Morphology

The helminths are soft-bodied invertebrate animals whose anatomic structures include differentiated muscle tissue; nervous, excretory, and reproductive systems; and often a simple digestive tract. The size range for the entire group is from nearly microscopic dimensions (the adult pinworm is just barely visible without a magnifying lens) to a meter or more in length (some adult tapeworms can be measured with a yardstick).

There are two major groups of helminths: the *nemathelminths,* or roundworms, and the *platyhelminths,* or flatworms. They differ in anatomic respects as well as in their reproductive methods and life cycles (see also Chap. 4 and Chap. 9).

Nemathelminths. The roundworms or nemathelminths (also often called *nematodes*) are cylindric with bilateral symmetry. The outer skin, or integument, of roundworms is a tough cuticle that may be smooth, rough, or spiny. Their bodies are not segmented, and they are pointed at both ends. A few members of this group, notably the hookworms, have hooks or cutting plates around the mouth for attachment to the intestinal wall of the animal or human whose body they have entered. Roundworms have a complete digestive tract and excretory system, as well as a nervous system. The sexes are usually separate. The larger female produces eggs constantly, as many as 200,000 a day in some cases. After fertilization larval development begins, the egg being encased in a shell by that time. When the larvae are hatched, they may go through several subsequent stages before the adult form is reached.

The cyclic development of roundworms from adult to adult, through egg and larval forms, varies greatly among species. In the sections of Part Two where helminthic diseases are discussed, the life cycle of each worm is considered in its relation to human infection, as well as to methods of prevention and control.

Among the important nematodes associated with human disease are *hookworm* and *Ascaris* (Fig. 2-22), which infect the intestinal tract, the *trichina* worm, whose larvae become lodged and encysted in muscle tissue, and the *filarial worms,* whose adult and larval forms both live in the blood and various tissues.

Platyhelminths. The flatworms or platyhelminths also display bilateral symmetry and organized tissues differentiated for special functions, that is, simple muscle and nerve tissue, digestive and excretory organs, and reproductive systems. They are oriented dorsoventrally, hence the term *flat.* They vary greatly in shape and size, some being thinly elongated and segmented, others being nonsegmented and thickly ovoid. On the basis of such anatomic differences, the

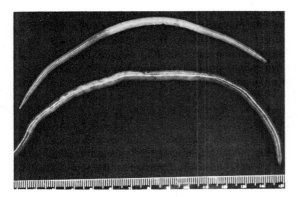

Fig. 2–22. Adult forms of *Ascaris lumbricoides*, a nematode or roundworm. The ruler measures 15 cm.

flatworms are subdivided into two major groups: the *cestodes,* which are tapeworms, and the *trematodes,* also often called flukes.

Cestodes. The bodies of tapeworms are flattened, elongated, and composed of individual segments, each of which contains both male and female sex organs. This combination of sexes in one individual is termed *hermaphroditism.* The series of segments begins with a head, or *scolex,* equipped with muscular suckers, and sometimes with a circlet of hooks as well, which permit attachment to the inner wall of the intestine of an animal, or of man. Just below the scolex is an area of growth, called the neck, from which arises a series of maturing segments, or *proglottids,* each containing fully developed sex organs, nervous and excretory systems, but no digestive tract. At the far end of the series, mature segments become filled with eggs and are said to be *gravid.* The whole length of a tapeworm is called the *strobila,* its full reach varying with the type of worm from three to four proglottids to several hundred (Fig. 2–23). Eggs are extruded from the segments into the intestines, or the gravid proglottids may break off and be passed from the body in the feces, together with free eggs. Sometimes long chains of proglottids are passed, providing very visible evidence of infestation. While the scolex and neck remain attached to the bowel wall, however, regeneration of segments continues and the strobila can grow again, even if all proglottids have broken away.

The eggs of tapeworms must pass through stages of larval development before the adult form is reestablished. For the most part these intermediate stages require one or more additional animals in which they can develop. For this reason, the latter are spoken of as *intermediate hosts,* while the animal that harbors the adult cestode is called the *final,* or *definitive, host.* Most of the tapeworms that infest man live in his intestinal tract, in the adult form. This is true of the so-called beef, pork, and fish tapeworms, and man is therefore one of the definitive hosts for these. In one important case, man may find himself an intermediate host for a small tapeworm that lives harmlessly, in the adult form, in the intestinal tracts of dogs. The larval form of this worm, called *Echinococcus,* derived from eggs passed by the infected dog, penetrates into the tissues of an intermediate host, such as man, and develops structures called *cysts,* or *hydatids* (fluid-filled sacs). The condition that results is spoken of as hydatid disease.

Trematodes. The trematodes, or flukes, fall into two groups on the basis of anatomic and functional differences: those that are flattened, thickly leaf-shaped, and hermaphroditic; and those that are elongated, round, and bisexual. None of the flukes is segmented but, like the tapeworms, they have only a rudimentary digestive system. All of them have a

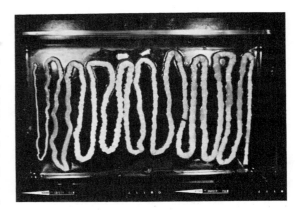

Fig. 2–23. This tapeworm (*Taenia saginata*) is 285 cm long. The entire strobila (fully developed worm) with the scolex (head) was recovered from a case of human infestation. The dish measures 27.5 cm × 40 cm.

tough outer cuticle and muscular suckers for attachment to the host. In their often-complicated life cycles, man and other mammalian animals are the definitive hosts for the adult forms, which inhabit and produce eggs in various sites of the body, while consecutive larval stages require one or more aquatic animals to complete their development.

The common names for the flukes refer to those sites in the body where the adult usually lives and produces its eggs. On this basis they are divided into groups known as blood flukes (the *schistosomes*), intestinal, liver, and lung flukes. Structurally, the blood flukes fall in the group mentioned above that are elongated, round, and bisexual. The adults are quite small, comparable in size to the smallest roundworms. They live within blood vessels, the female producing her eggs in small capillaries of such organs as the large intestine, liver, or urinary bladder. All the other flukes have a thickly ovoid shape and are hermaphroditic. The intestinal fluke lives in the lumen of the upper part of the small bowel, the liver fluke in bile ducts, and the lung fluke in lung tissue itself. The eggs produced by flukes at these various sites may often be trapped in tissues where they cannot develop further but produce defensive reactions in the host. Many eggs, however, may find an exit route from the body, through feces, urine, or sputum. If and when they are deposited in water, the ova hatch into larval forms, which then pass through several generations in intermediate aquatic hosts. Human infestation is incurred by ingestion of larvae in infected fish, crustaceans, or water plants (these hosts differ for the several types of flukes), or by direct penetration of skin by a larval form swimming freely in infected water. Development of the adult then proceeds in the final host, and the production of eggs continues the cycle.

The insects that have great importance in the transmission of many infectious diseases, helminthic or microbial, are described in Chapter 9 in the discussion of sources and routes of infection.

Questions

1. How are microorganisms distinguished from all other living things?
2. How are living things within the plant kingdom classified?
3. What organisms are considered lower and higher protists?
4. What are the basic shapes and group patterns of bacteria?
5. List the function of each of the following structures in bacteria: cell wall, cell membrane, mesosome, flagella, spores, nucleus, ribosome, capsule, pili.
6. How do rickettsiae and chlamydiae differ from other bacteria?
7. How are viruses classified? What is a virion? How does it function?
8. How are protozoa classified? List an example of a disease producer in each group.
9. How are fungi classified? List an example of a disease producer in each group.
10. How are helminths classified? List examples of disease producers in each group.

Additional Readings

Barghoorn, Elso S.: The Oldest Fossils. *Sci. Am.,* **224**:30, May 1971.
Beck, Walter J., and Davies, J. E.: *Medical Parasitology,* 2nd ed. C. V. Mosby Co., St. Louis, 1976.
Dubos, René: *The Professor, the Institute and DNA.* The Rockefeller University Press, New York, 1976.
————: *The Unseen World.* The Rockefeller University Press, New York, 1962.
†Holland, John J.: Slow, Inapparent and Recurrent Viruses. *Sci. Am.,* **230**:32, Feb. 1974, #1289.
Hughes, Sally Smith: *The Virus.* Neale Watson Academic Publications, Inc., New York, 1977.
Jawetz., E.; Melnick, J. L.; and Adelberg, E. A.: *Review of Medical Microbiology,* 12th ed. Lange Medical Publications, Los Altos, Calif., 1976.
*Mikat, Dorothy M., and Mikat, Kurt W.: *A Clinician's Dictionary Guide to Bacteria and Fungi,* 3rd ed. Eli Lilly and Co., Indianapolis, Ind., 1976.

†Morowitz, H. J., and Touretellotte, M. E.: The Smallest Living Cells. *Sci. Am.,* Mar. 1962, #1005.
Poindexter, J. S.: *Microbiology, An Introduction to Protists.* Macmillan Publishing Co., Inc., New York, 1971.
Porter, John Roger: Perfect Match with Animal, Vegetable and Mineral. UNESCO *Courier,* p. 24, July 1975.

*Available in paperback.
†Available in *Scientific American* Offprint Series. (See Preface.)

3

Microbial Life

The Biosphere

Geologists speak of our planet as displaying three zones: the *lithosphere,* or outer solid part composed of rock and soil; the *hydrosphere,* the aqueous envelope of the earth, containing its many bodies of water; and the *atmosphere,* the gaseous envelope or surrounding mass of air. Biologists speak of another zone, the *biosphere,* meaning that portion of the planet where life exists. The great mass of earth's living things spreads around and across its surface in a thin irregular layer. This mass teems with an enormous variety of forms that attest to earth's life-supportive environment. The biosphere itself helps to sustain this environment with a constant cycling of energy and chemical supplies among its diverse living organisms, although man's overwhelming activities increasingly threaten these vital exchanges. Microorganisms have always played a profoundly important role, together with plants and animals, in the cyclic maintenance of earth's environmental qualities.

Plants, Animals, and Microbes

In sharing the planet's life systems, plants and animals maintain these for each other by taking out, transforming, and restoring essential factors. There would, for example, be no atmospheric oxygen available for the respiration of animals and man if it were not for the metabolic activities of green plants. Plants owe their color to the green light-sensitive pigment chlorophyll. This pigment is the key to the process by which plants utilize the energy conveyed by light (solar energy) to drive their manufacture of carbohydrate from water and atmospheric CO_2. Their chemical production of carbohydrate, mediated by light and light-sensitive chlorophyll, is called *photosynthesis* (*photo-* means light, and *synthesis* means manufacture). Oxygen is a by-product of the reaction and is returned to the atmosphere, while the carbohydrate provides energy for other plant processes. Meantime, animals in their respiration take oxygen from the air they inhale, utilize it in their own vital chemical affairs, and return CO_2 (a by-product of their reactions) to the atmosphere on their exhalations. Thus oxygen and CO_2 are cycled to support the differing needs of plants and animals, and the solar energy, briefly fixed in photosynthesis, is returned into space as heat. A constant flow of many such exchanges among living things assures the availability of vital materials.

Microorganisms are a ubiquitous, essential part of the biosphere, living in soil, in water, and in close association with higher forms of life. Whatever their immediate environment, their influence upon it is far reaching. In the soil, some microorganisms live on the roots of plants, providing them with inorganic nitrogen compounds synthesized from atmospheric nitrogen. Plants are unable to utilize nitrogen as an

48

element and could not survive without the chemical transformations of this substance accomplished by their microbial root and soil companions. The nutrient quality of soil is also maintained for living plants by microorganisms that decompose dead plants and animals, chemically converting their components and restoring them to utilizable form. In oceans, rivers, and lakes, microbial activities similarly assure rotations of nutrients essential to higher aquatic forms. In the bodies of animals, microorganisms living on surface tissues often contribute materially to the host's well-being. For example, microorganisms in the gastrointestinal tract of ruminant animals (e.g., cows or sheep) digest the cellulose of grass and hay, making nutrition possible. Also, substances that are manufactured by intestinal organisms can be utilized by animals in their own growth processes. As we know, microbes can also produce harmful and destructive effects. They can not only cause disease in higher organisms, but they are responsible for costly deteriorations of food and many materials.

To understand the interacting role of microorganisms in our world, let us define some of their major nutritional patterns and see how these explain their activities.

Nutritional Patterns

In order to grow and carry on their vital processes, all forms of life must have a source of usable nutrients and energy. The key nutritional element in all living structures is carbon, which must be converted to the complex organic materials of protoplasm. Energy is available directly from sunlight or from the chemical breakdown of organic and inorganic compounds.

Originally, organisms were classified simply as either *autotrophic*, being relatively independent of other forms of life in their nutrition, or *heterotrophic*, deriving their food from organic matter manufactured by other living things (*hetero* = other; *auto* = self; *troph*- implies nourishment). To better characterize nutritional patterns, however, organisms are now classified according to their principal carbon source and the nature of their energy source. Thus, *photosynthetic autotrophs* such as plants, algae, and some bacteria use CO_2 from the atmosphere and light from the sun to manufacture organic building blocks. The group of *chemosynthetic autotrophs,* which consists of only a few genera of bacteria, also uses CO_2 as its principal carbon source, but its energy is obtained from the chemical oxidation of inorganic compounds rather than from light.

The heterotrophs cannot utilize inorganic carbon directly, but must use organic compounds such as carbohydrates, proteins, and lipids (fats) as their major carbon source. A small group of bacteria comprise the *photosynthetic heterotrophs,* which rely on sunlight for their energy source. Man, animals, protozoa, fungi, and most bacteria are *chemosynthetic heterotrophs.* They depend on the chemical energy released from a variety of oxidation-reduction reactions of organic compounds (Table 3-1).

Table 3-1. Comparison of Autotrophic and Heterotrophic Organisms

Nutritional Type	Principal Carbon Source	Energy Source	"Life Style"
Photosynthetic autotroph	CO_2	Sun	Independent
Chemosynthetic autotroph	CO_2	Oxidation of inorganic compounds	Independent
Photosynthetic heterotroph	Organic compounds	Sun	Dependent, saprophytic
Chemosynthetic heterotroph	Organic compounds	Oxidation-reduction of organic compounds	Dependent, saprophytic, or parasitic

Heterotrophs that are capable of living on the *dead* organic material produced by or derived from other organisms are said to be *saprophytic* (*sapro* = dead or rotten). This term also implies that the organism absorbs organic nutrient in soluble form, whereas a heterotroph capable of ingesting solid food is termed *holozoic* (*hol-* = whole). Among the true microorganisms, amebae and many other protozoa are holozoic. Their membranous walls and specialized structures provide them with the capability of ingesting particulate food and excreting wastes. Many bacteria and fungi are saprophytic, their rigid cell walls being capable of exchanging only soluble materials with their environment.

Saprophytic microorganisms vary quite widely in their nutritional needs. Many can survive in a wide range of environments so long as essential nonliving organic nutrient is available. Those that maintain a residence in the bodies of animals or plants, deriving support at the expense of their hosts, are referred to as *parasites*. Some saprophytes may take up a parasitic life if the opportunity presents or live less dependently, in the environment, according to circumstances or their specific nutritional needs.

When parasitic organisms damage their hosts sufficiently to disrupt normal functions and cause illness, they are said to be *pathogenic* ("disease producing"). Many of them have rather strict requirements for the inanimate organic nutrients associated with the hosts they parasitize. Such organisms are said to be *fastidious* in that they cannot survive in an environment that does not provide the precise nutritional factors they require. They can be cultivated artificially on complex inanimate nutrients, but in nature they are usually restricted to parasitism. There are some microbial pathogens that are even more demanding in their parasitism in that they cannot utilize their hosts' prefabricated nutrients but can survive only within *living* host cells. Such organisms (e.g., rickettsiae, chlamydiae, and viruses) can perform so few functions for themselves that they are bound to a life of intracellular parasitism, requiring specific substances produced within the actively metabolizing cells of their hosts. For this reason they are called *obligate intracellular parasites*. In the laboratory, this type of organism can be cultivated only in living cell or tissue cultures, or in intact experimental animals.

Microbial Activities

At the cellular level all forms of life possess similar fundamental mechanisms. The respiratory and metabolic processes of plant and animal cells, and of microorganisms, have certain common vital components. Among these, the most important chemical mediators of every cell's activities are its *enzymes*. These are organic catalysts that promote the uptake and utilization of raw materials, the synthesis of materials needed for cell energy and function, and the degradation of unneeded substances that then become available again in the environment. As catalysts, enzymes initiate and direct chemical reactions without taking a direct part in them. They not only act within cells, but may be secreted by cells into the surrounding milieu. Here they may act on solid food particles to make them soluble for absorption through cell walls or in other ways make the environment more supportive to the life of the cell. Enzymes are usually quite limited, that is, *specific,* in the kind of reaction they catalyze. Some are involved only in the digestion of protein components, some act only in specific steps in the breakdown of particular carbohydrates, others on fatty substances. With these limitations on their activity, a battery of different enzymes is required for the many metabolic processes of both multicellular and unicellular organisms.

The metabolic requirements, mechanisms, and activities of microorganisms are as closely related to their enzymatic components as are those of larger forms. Furthermore, many different microbial and higher organisms contain the very same enzymes, or enzyme systems — a fact that is not surprising when one considers the uniform distribution of *substrates* (substances on which enzymes act specifically) such as carbohydrates, lipids, and proteins.

The following sections review some of the important activities, and indicate a few enzymic reactions, of true microorganisms.

Fungi

The yeasts and molds are heterotrophic saprophytes, widely distributed in nature. Most of the fungi that cause human disease have a primary habitat in the soil but can become parasitic if given the opportunity. The pathogenic fungi are a very small group as compared with the vast numbers of free-living species that do not display parasitism. There are many that cause plant diseases, however, and some of the molds are well known for their ability to grow on and destroy nonliving organic substrates, such as those in fabrics and foods. Mildew and bread molds are good examples.

The yeasts and their metabolic processes have been known and put to good use for hundreds of years. With the help of their enzymes, yeast cells utilize glucose for energy, converting it, under *aerobic* conditions (in which free molecular oxygen is available), into carbon dioxide and water. If free oxygen is not available (that is, conditions are *anaerobic*), these organisms use a different set of enzymes to break down the sugar, producing ethyl alcohol and other low molecular products.

The *aerobic* utilization of glucose by yeasts is of value in bread making. In the leavening process, the release of carbon dioxide from the breakdown of the sugar contributes to the rising of the bread. In this connection it is of interest to note that the main energy food for many animal cells, including those of man, is also glucose, which is oxidized through the activity of enzymes in muscle cells to produce carbon dioxide and water and, most important, energy, as shown in simple outline below:

$$C_6H_{12}O_6 \xrightarrow[\text{enzymes}]{\substack{\text{aerobic} \\ \text{yeast}}} CO_2 + H_2O + \text{Energy}$$

Glucose   *Carbon*   *Water*
               dioxide

The *anaerobic* process (*fermentation*) by which yeasts convert glucose to alcohol leads to the production of wines and other alcoholic beverages:

$$C_6H_{12}O_6 \xrightarrow[\text{enzymes}]{\substack{\text{anaerobic} \\ \text{yeast}}} CH_3CH_2OH$$

Glucose   *Alcohol*
                   (ethyl)

In wine making, grapes are the natural source of the carbohydrate, glucose. Strains of the winemaker's yeast also are found in nature growing with other microorganisms on the skins of grapes, this combination being called the "bloom." In Europe, the vintner still relies on the bloom to produce a natural fermentation of the crushed grapes, which gives the wine its characteristic flavor, or individual bouquet. In bad years, when the bloom may be overgrown by extraneous molds, the winegrower can resort to fermentations effected by a laboratory-grown pure culture of yeast. The latter method is usually employed in the production of North American wines. True vintage wines are produced by natural fermentation by the bloom, so that knowledge of the locale and year of the grape harvest is important to their purchase.

Protozoa

Many protozoa lead saprophytic or holozoic lives in environments rich in organic substances and teeming with bacteria, one of their favorite foods. They abound in freshwater ponds and bogs as well as in salt marshes. In sewage treatment plants, the activity of abundant protozoa helps to control excessive bacterial growth. Many species live saprophytically in the intestinal tracts of animals or man. Some are obligate parasites but not all of these are pathogenic; that is, their activities do not necessarily damage their hosts.

Among the free-living ameboid protozoa there are some orders that are of interest because of their production of chalky or silica shells (foraminifera and radiolaria). When these organisms die, their shells do not decay but become fossilized. As they are deposited and accumulate in a given area, extensive formations may pile up. The well-known white cliffs of Dover, England, are largely composed of such shells (Fig. 3–1). Rocks and limestone may also contain abundant deposits of fossil ameboid forms.

Bacteria

As we have seen, some bacteria are autotrophic, but many are saprophytic heterotrophs. These are ubiquitously distributed in soil and water and in the intestinal tracts and on the surface tissues of animals and human beings, thriving for the most part on nonliving organic material. Some of the pathogenic bacteria are obligate parasites, there being two im-

Fig. 3–1. The white cliffs of Dover, England, are largely composed of the fossilized shells of certain ameboid protozoa. (Courtesy of British Tourist Authority, New York, N.Y.)

portant genera, the *Rickettsia* and the *Chlamydia,* noted for strictly intracellular parasitism.

Environmental Activities of Bacteria. Bacteria are tremendously versatile, not only in their utilization of organic substances and the conversion of these in synthetic processes, but in their ability to bring alternate metabolic mechanisms into play in adjustment to the demands of their environment. If the supply of a particular substrate dwindles, or if other environmental changes occur, bacteria can meet these problems, probably by calling into play alternate enzyme systems or by changes in the constituents of their cell walls. Another mechanism of bacterial survival is seen in the genetic mutations and selections that result in populations more adaptive to a changed environment (see pp. 59–61). This becomes of great practical importance when human infections are treated with antibiotics or when physical and chemical agents, such as heat, disinfectant solutions, or ultraviolet radiation, are used to control the growth of bacteria. Bacteria may survive such measures by changing their method of utilization of a particular substrate, by turning to another kind of available nutrient, or by entering into a resting stage (i.e., by sporulation).

In the natural world, the metabolic activities of bacteria maintaining and nourishing themselves on organic components of animal wastes or dead animals and plants play an indispensable role in the maintenance and nourishment of all life on our planet. The decay and putrefaction of such materials are due in large part to the action of bacterial enzymes that break them down to forms they and

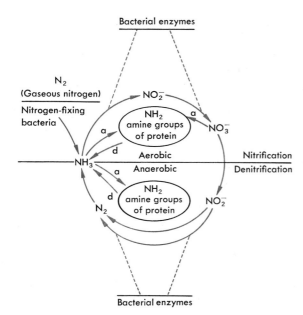

Fig. 3–2. The nitrogen cycle, showing pathways of nitrification (aerobic) and denitrification (anaerobic). *a* = assimilation of ammonia or nitrate into protein amines; *d* = deamination of protein.

other organisms can absorb and transform into new compounds for their own cellular structures and functions.

A brief review of the nitrogen cycle may serve to illustrate this point. Nitrogen gas is a major component of the earth's atmosphere (76 percent by weight) and the ultimate source of the nitrogen compounds of living organisms. It has been previously mentioned (p. 48) that some microorganisms (called "nitrogen fixing") can synthesize elementary nitrogen first into the inorganic nitrogen compound ammonia, and then into nitrites and nitrates needed by plants. These inorganic substances can then be incorporated, by plants, into organic compounds containing carbon, hydrogen, oxygen, nitrogen, and other elements by a process known as *assimilation.* Proteins, which are vital constituents of all living cells, are made up of complex combinations of *amino acids,* the latter being organic acids that contain an *amine* group, NH_2, derived from ammonia, NH_3. Degradation of protein, or some portion of the molecule, occurs through removal of amine groups

by the process of *deamination.* In nature, oxidative bacterial enzymes play a role in the aerobic process of *nitrification,* or the transformations required to produce nitrites and nitrates from ammonia. These inorganic nitrogen compounds can then be assimilated by plants in the formation of the amino acid "building blocks" of proteins. Conversely, bacterial enzymes play a role in the degradation of proteins and inorganic nitrogen compounds. In the anaerobic process of *denitrification,* nitrates and nitrites are transformed to gaseous nitrogen, N_2, as the cycle continues. This concept is illustrated diagramatically in Figure 3–2.

The presence and activities of bacteria in the natural environment often go unsuspected because they are invisible. Sometimes, however, the growth of a pigmented species may become noticeable because of its color (Fig. 3–3). Bacterial pigments are products of intracellular metabolism. They may remain within the cell, coloring the accumulating growth, or

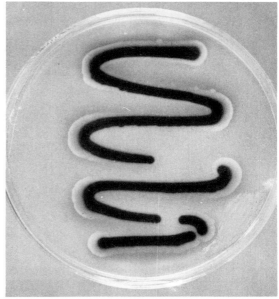

Fig. 3–3. Some strains of *Serratia marcescens* are noted for their ability to produce a red pigment. In this black-and-white photograph the pigmented streak of growth contrasts sharply with the colorless agar background. (Courtesy of Dr. Ramon Kranwinkel, Danbury Hospital, Danbury, Conn.)

be secreted into the environment. In the latter case the surrounding area often becomes vividly stained. A common free-living saprophytic bacterium noted for production of a bright red intracellular pigment is called *Serratia marcescens*. It was given this name because it was thought to be miraculous in origin when it was seen growing on the eucharistic bread in medieval cathedrals (which were often dank and dark and offered excellent conditions for microbial growths of all kinds). In recent years this organism has been increasingly associated with human infections, but usually it is nonpigmented in such cases. It is an excellent example of an opportunistic saprophyte that may become dangerously parasitic under the right circumstances, causing serious infections particularly in hospitalized patients whose resistance is lowered because of other illness.

Economic Uses of Bacterial Activities. The activities of bacteria have been put to a variety of productive uses. Bacteria, like yeasts, utilize carbohydrates for food, converting them into acids, alcohols, and gases, in aerobic and anaerobic processes directed and controlled by enzymes. The ability to use a particular carbohydrate, or other organic compound, differs among species of bacteria, depending on whether or not they possess an enzyme specifically capable of acting on that substrate and changing it chemically. This in turn directs the choice of a particular strain of organism for the job one may wish to have done. In the making of wines, or other alcoholic beverages such as hard cider, fermentation of the sugar to alcohol is a job usually assigned to yeasts, as we have seen. If desired, the alcohol may be further degraded into acetic acid, or vinegar, by introducing a culture of bacteria of the *Acetobacter* genus whose oxidative enzymes conduct the transformation, as illustrated by the simplified equations below:

$$C_6H_{12}O_6 \xrightarrow[\text{enzymes}]{\substack{\text{anaerobic} \\ \text{yeast}}} CH_3CH_2OH$$
$$Glucose \qquad\qquad Alcohol$$
$$\text{(ethyl)}$$

(alcoholic fermentation)

$$CH_3CH_2OH \xrightarrow[\substack{Acetobacter \\ \text{enzymes}}]{\text{aerobic}} CH_3COOH + H_2O$$
$$Alcohol \qquad\qquad Acetic \qquad Water$$
$$\text{(ethyl)} \qquad\qquad acid$$

Another group of organisms called lactobacilli characteristically produce lactic acid in their fermentation of glucose. For this reason the lactobacilli are useful to the dairy industry in the preparation of sour cream and yogurt.

$$C_6H_{12}O_6 \xrightarrow[\substack{Lactobacillus \\ \text{enzymes}}]{\text{anaerobic}} CH_3CH_2COOH$$
$$Glucose \qquad\qquad\qquad Lactic\ acid$$

(lactic acid fermentation)

The production of many popular foods is made possible by such bacterial conversions (Fig. 3–4). Other practical uses of bacterial activities include the preparation of animal hides for leather manufacture, of plant fibers used to make fabrics or rope, of tobacco and tea leaves to improve their flavor, and of industrial alcohols and acids. Many industries of basic importance to our economy have developed by use of the chemical talents of bacteria and other microorganisms for production of materials that are essential in our progress and useful in our pleasures.

Conversely, the bacterial profits of the economic world are balanced against the expenditures of ingenuity, effort, and time required to circumvent or control the less desirable features of bacterial activity. Spoilage of useful foods and other valuable materials is costly and has stimulated major research and development in the field of food preservation. Methods designed to prevent multiplication of putrefactive (decay-producing) bacteria in food must not in themselves be deleterious to flavor, or to nutritional content, and must avoid the use of substances that might be toxic on ingestion by human beings. Natural physical methods of preserving foods include dehydration, cold storage, freezing, and radiation; chemical methods employ the antibacterial effects of salting, pickling in vinegar or in high concentrations of sugar, smoke curing, and so forth. Antibiotics also are used to suppress bacterial growth in some foods, including fish and poultry.

Bacterial Activities and Human Health. In medicine and public health microbial activities are turned to our greatest advantage in (1) the production of antibiotics and (2) the disposal of sewage and wastes.

Antibiotics are products of microbial metabolism that are excreted through their walls into the environment. In the soil, as in the human body where

Fig. 3–4. Cheddaring. A mother culture of *Streptococcus lactis* that has been warmed and allowed to develop the proper acidity in a bulk starter tank is being added to a cheese vat. (Courtesy of U.S. Department of Agriculture, Dairy Inspection and Grading Branch.)

many microorganisms live together, such compounds formed by one species may be quite toxic in their effects on another, or in other cases they may be supportive to the growth of neighboring organisms. The term *antibiosis* (against life) applies to the antagonistic effect (Fig. 3–5). The great names in the discovery and development of antibiotics for medical purposes are Fleming, Florey, and Chain, in England, and Selman Waksman, in the United States, all recipients of Nobel Prizes acknowledging the significance of their work in this field. It was Fleming who first observed the antibacterial effects of the mold *Penicillium*; Florey and Chain who identified the substance, penicillin, responsible for the effect and opened the door to its medical application; and Waksman who began extensive investigations in the United States, discovered streptomycin, and stimulated many major research efforts that followed. Antibiotic production is a major activity of many pharmaceutical companies today (Fig. 3–6). Research in the laboratory modification of antibiotics has led to the successful development of a number of useful drugs, such as ampicillin, oxacillin, methicillin, and other derivatives of penicillin, produced by making chemical changes in the original molecule.

In sewage and waste disposal, advantage is taken of the digestive activities of putrefactive bacteria. One of the common methods for handling sewage is to direct it slowly over beds of gravel and sand so that solid matter has time to settle out and be acted upon by the various bacteria in these beds. As water seeps through, it is cleaned and purified, pathogenic bacteria are removed and die, and the water is made ready again for consumption. Digestion of the settled material by bacteria converts it into a form that can be collected, dried, and used as fertilizer. The purification of water is discussed in more detail in Chapter 11, together with other measures of concern to public health in control of infectious diseases.

The intestinal bacteria of man and animals also break down food residues by digestive processes. Some of these organisms are capable of synthesizing specific growth factors called vitamins that they require in their own metabolism. This provides a sort of built-in supply of vitamins for the human or

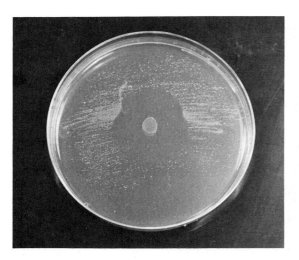

Fig. 3–5. An example of antibiosis. The circular inoculum of alpha-hemolytic streptococci in the center of this plate is inhibiting the growth of a strain of *Corynebacterium vaginale* that was streaked across the entire surface. The latter organism fails to grow in proximity to the *Streptococcus* but shows luxuriant growth on the outer sections of the plate. (Reproduced from Choong, H. P.; Fauber, M.; and Cook, C. B.: Identification of *Haemophilus vaginalis*. *Am. J. Clin. Pathol.*, **49**:590, 1968.)

animal host, who also requires it for certain growth processes. The production of vitamin K, as well as some elements of the vitamin B complex, by intestinal bacteria contributes to the host's well-being.

Pathogenic Effects of Bacteria. It was pointed out earlier that the biochemical flexibility of bacteria presents a challenge to the use of antibacterial agents in treating infectious disease or in controlling bacterial growth in the environment. This versatility also operates in the complex interrelationships of bacteria and the tissues of the hosts they may parasitize. The host's ability to resist the effects of their presence in its tissues and the organism's ability to survive in spite of this constitute a competition at the cellular level. The enzymes of host and parasite may compete for the same substrate, and each may produce substances that are antagonistic for the other. In finding alternate metabolic pathways for utilization of its food, the bacterial cell is often much more

versatile than the cells or tissues of the human or animal body, and this of course provides it with a strong capacity for survival.

Toxins are harmful products of bacterial cell growth and metabolism. When they diffuse out of the bacterial cell into the surrounding medium, they are spoken of as *exotoxins* (diphtheria toxin is an example of this type), but when they are associated with the organism itself, as a part of its chemical structure, they are called *endotoxins.* The important exotoxins, such as those produced by the bacterial agents of tetanus, botulism, and diphtheria, have been isolated and studied extensively. In purified form they are protein in nature and highly lethal to experimental animals. The endotoxins are associated with the outer layer of bacterial cell walls. They are complex large molecules of phospholipid, polysaccharide, and protein. Isolated endotoxins also have toxic and lethal effects on animals, but quantitatively they are much less potent than exotoxins.

The role of toxins in the pathogenicity of microorganisms is described in Chapter 6.

Environmental Requirements for Bacterial Culture. The laboratory identification of bacteria isolated from patients suspected of having infectious diseases is based in large part on demonstration of their biochemical activities, as well as on their microscopic features and growth patterns.

In nature, bacteria obtain the carbohydrate and protein materials, vitamins, and minerals required for all their vital processes from dead or decaying matter or from living, parasitized hosts. To cultivate bacteria in the laboratory, and study these requirements more closely, or identify them through their metabolic behavior, it is necessary to provide substances essential to their survival and growth, in appropriate form. A basic medium for bacterial culture contains meat that has been cooked to break down its proteins partially, to soluble, usable forms; some carbohydrate (usually glucose); and a few simple salts. Some pathogenic bacteria may require additional enrichments, such as the serum proteins or the whole blood of various animal species, special carbohydrates, particular amino acids (constituents of proteins), vitamins, and specific minerals. These special growth factors usually are required only in

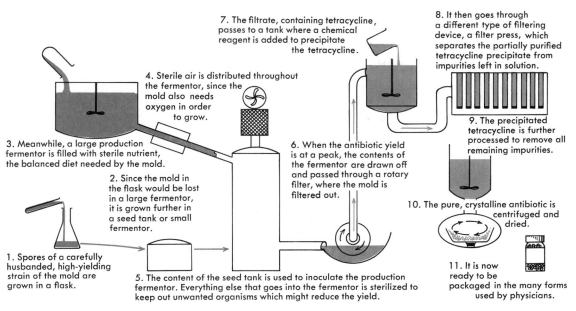

7. The filtrate, containing tetracycline, passes to a tank where a chemical reagent is added to precipitate the tetracycline.

8. It then goes through a different type of filtering device, a filter press, which separates the partially purified tetracycline precipitate from impurities left in solution.

4. Sterile air is distributed throughout the fermentor, since the mold also needs oxygen in order to grow.

3. Meanwhile, a large production fermentor is filled with sterile nutrient, the balanced diet needed by the mold.

9. The precipitated tetracycline is further processed to remove all remaining impurities.

6. When the antibiotic yield is at a peak, the contents of the fermentor are drawn off and passed through a rotary filter, where the mold is filtered out.

2. Since the mold in the flask would be lost in a large fermentor, it is grown further in a seed tank or small fermentor.

10. The pure, crystalline antibiotic is centrifuged and dried.

1. Spores of a carefully husbanded, high-yielding strain of the mold are grown in a flask.

5. The content of the seed tank is used to inoculate the production fermentor. Everything else that goes into the fermentor is sterilized to keep out unwanted organisms which might reduce the yield.

11. It is now ready to be packaged in the many forms used by physicians.

Fig. 3–6. This drawing illustrates, in numbered sequence, the steps involved in the large-scale commercial preparation of a tetracycline, a kind of antibiotic produced by a soil microorganism. Through modern technology, microbes can be induced to yield substances of a quality and quantity not attainable under natural conditions. (Redrawn from *Our Smallest Servants*, Chas. Pfizer & Co., Inc., New York, 1967.)

very small amounts. The final medium may be used in liquid form, or solidified by addition of agar, a nonnutritive substance resembling gelatin in its physical properties.

Environmental requirements of bacteria that must be met in the laboratory include such factors as pH and water content of the medium, atmospheric conditions, temperature, and light.

The pH of the medium, that is, its degree of acidity or alkalinity, must be carefully adjusted because bacteria can be quite sensitive to minor changes in this balance. Most have a narrow pH range (around neutrality, pH 7.0) beyond which they cannot survive. Some can grow at extremes of pH. For this reason, it is sometimes useful to adjust the pH of a medium to a selective point (high or low) that will permit the isolated growth of pH-resistant organisms from a specimen containing a mixture of species many of which may be pH sensitive. Similarly,

growth of *isolated* organisms on media adjusted to pH extremes may be an important clue to their identity.

Water is essential for growth of bacteria, as is soluble food. It is water that keeps food in solution so that organisms can absorb it through their cell walls. Removing water from the environment does not necessarily kill microorganisms, for many are quite resistant to the effects of drying, but it will suppress active metabolism and growth. For this reason, bacteria can be preserved for long periods by quickly freezing them in suspension and then drying them under vacuum, a process known as *lyophilization*.

Oxygen is required in some form by all living organisms, but not necessarily as the atmospheric gas. Bacteria vary in their need for free oxygen. The majority, called *aerobes*, require it for aerobic respiration, through which they obtain energy for growth processes. Some bacteria are greatly inhibited or killed by free oxygen, however. These will grow only in anaerobic environments and are therefore called *obligate* or *strict anaerobes*. They obtain energy through fermentative mechanisms in which oxidations and reductions are accomplished without the mediation of oxygen. Between the two extremes there is a group of bacteria that can adapt to either an aerobic or an anaerobic environment and are said to be *facultative* in this regard. Still others, called *microaerophiles,* require some oxygen but are inhibited by the full amount present in the atmosphere.

Temperature is another critical factor in the growth of bacteria. Some prefer temperatures as low as 10° C (50° F) (these are called *psychrophilic,* or "cold loving"), while others will grow only at temperatures of 50° to 60° C (122° to 140° F) (*thermophilic,* or "heat loving"). Most pathogenic bacteria, however, find normal human body temperature (35° to 37° C) most suitable for their own physiologic activities (they are *mesophilic,* or moderate). Usually there is a 15-to-20-degree temperature range, with maximum and minimum limits and an optimum area between for the growth of particular types of bacteria.

Light is not a requirement for most bacteria, since the majority are not photosynthetic and cannot use light for energy. On the contrary, many species are sensitive to the effects of the ultraviolet component of sunlight. Although laboratory cultures are usually kept in the dark, it is sometimes necessary to adjust light conditions in order to observe a particular property or product of growth in a given species. For example, some bacteria display changes in pigmentation as light conditions are altered.

These requirements are summarized in Table 3-2. In the laboratory, the manipulation of some or all of these factors constitutes a part of the approach to the identification of bacteria and to the study of their metabolic processes. More detailed information is obtained from observations of their morphology, enzyme components, and biochemical activities.

Metabolic processes may be recognized in several ways but the most common approach is to identify one or more enzyme activities. This is done by providing the organisms with some particular compound that is a substrate for the enzyme to be identified and observing the nature of any changes that may occur in that compound. Such observations can be made quite simple by incorporating into the test a chemical indicator of a kind that will provide visible evidence that a change has occurred. Thus, if it is important to know whether or not a particular organism possesses an enzyme for the utilization of lactose, a culture medium is prepared that will support the organism's growth but is free of all sugars other than lactose. If the enzyme for lactose is present, the sugar molecule will be broken down to simpler compounds that are acidic. The pH of the medium is adjusted so that no acid is present to begin with, and the other ingredients of the medium are balanced so

Table 3-2. Environmental Requirements for Bacterial Culture

1. pH	Neutrality (pH 7) optimum for many; acidity or alkalinity can be adjusted to meet needs of individual organisms
2. Water	Essential for all microbial growth
3. Oxygen	*Aerobes* require atmospheric oxygen; *anaerobes* cannot grow in presence of free oxygen; *facultative* organisms grow under aerobic or anaerobic conditions; *microaerophiles* require that atmospheric oxygen be reduced in total amount
4. Temperature	Optimum ranges: *psychrophiles,* 5°–10° C; *mesophiles,* 35°–37° C; *thermophiles,* 50°–60° C
5. Light	Not required for bacterial growth; pigment production may be affected by light; ultraviolet of sunlight may kill bacteria

that little or no acid will be formed during growth if the lactose is not fermented. Finally, an indicator is put in that will provide a visible color change if the pH is lowered by the accumulation of acid products. After the organism has grown in the medium for a few hours, a mere glance at the color of the indicator suffices to determine whether or not lactose fermentation has occurred.

It may be necessary to conduct a number of such tests in narrowing the identification of an organism, first as a member of a large family with widely shared characteristics, then to a smaller tribe or genus with more particular properties, and finally to a species with a unique set of metabolic features.

Genetic Behavior of Bacteria. Genetic variations occur naturally in bacteria as in all living things, and they can also be induced by experimental methods. In either case they occur as a result of changes at sites in the nuclear structure called *genes*. Genes are the functional unit of heredity in all cells. As we have seen in Chapter 2, the fundamental nuclear material is DNA, a double-stranded linear molecule composed of nucleotide bases linked to a deoxyribose-phosphate backbone. For each type of cell the sequences of the bases linked to deoxyribose occur in specific patterns. DNA directs the production of a special type of ribonucleic acid (RNA) that acts as a "messenger" from the nuclear material to determine and order the chemical activities of the cell. It is the sequence of purine and pyrimidine bases in the DNA molecule that determines the structure of the messenger RNA and, in turn, the structure of proteins that regulate the chemical nature of the cell's metabolic processes — that is, the DNA pattern serves as a code for the cell as a whole.

In bacteria, a chromosome is made up of a naked DNA molecule that contains individual coded units, the genes. Each gene determines a specific cellular function. For example, in the species of intestinal bacteria known as *Escherichia coli* the so-called "z" gene is responsible for the biosynthesis of an enzyme that breaks down lactose into glucose and galactose. Other genes determine individual cellular structures or other cellular functions. The totality of genes located in the bacterial chromosome is called the *genome*. Each time a bacterial cell divides by fission, the chromosomal DNA is duplicated so that each new cell contains exactly the same coded genetic molecule (genome) as the parent cell and displays the same characteristics of function and structure.

Genes are subject to change, and when this happens the cellular trait, or traits, for which they are responsible change also. A new characteristic may be added, or an old one lost or altered. If the alteration in the gene does not result in death of the cell or render it incapable of further reproduction, the change will be transmitted through succeeding generations of cells as the gene duplicates at each cellular division.

In addition to the chromosomal DNA, or genome, some bacteria also display *extra*chromosomal bits of genetic material called *plasmids*. Plasmids are independent, self-duplicating genetic elements formed of circular DNA (circular, in that a strand is looped and its two ends are joined together, as shown in Fig. 2–10). *Episomes* are a special class of plasmids that may either exist free in extrachromosomal form or become integrated in the chromosomal DNA. These units play a role in genetic changes in bacteria, as we shall see.

Mutations. Transmissible changes in the characteristics of the cell that are due to gene alterations are called *mutations*. They probably occur among bacteria with the same frequency as in higher organisms, but are manifest more frequently because bacteria are functionally haploid, so that "recessive" mutations are not masked as they may be in diploids. (In diploids, a mutation in a gene of one chromosome set usually has no effect on the functioning of the equivalent gene in the other set of the pair. Thus, the gene whose effect is expressed is called "dominant" to its matched partner gene, which is unexpressed and masked, or "recessive.") Furthermore, the rapid multiplication of bacteria may produce large populations of mutant cells. Mutations are thus seen very commonly among bacteria, affecting one or more characteristics such as cell wall formation, pigment or toxin production, enzyme formation, or their ability to resist the action of antibiotics.

Mutations can be induced experimentally in bacteria, as in other organisms, by radioactive substances, ultraviolet light, or irradiation with x-rays, or by the action of chemical agents such as alkylating agents or hydrogen peroxide.

Transfer of Genetic Material. Three mechanisms have been demonstrated for the transfer of genetic information among bacteria. One of these is mediated by virus infection of the bacterial cells and is called *transduction;* another involves the uptake of nuclear material released from dead cells and is called *transformation;* the third, in which genetic material is transferred directly from one cell to another, is called *conjugation.*

TRANSDUCTION. It will be remembered that viruses contain nucleic acid but they cannot reproduce themselves except within host cells and with the assistance of the nuclear material of the latter. Bacteriophages are viruses that infect bacterial cells (Fig. 3–7). They have structures that appear to be specialized for entry into their host cells. The phage virus

Fig. 3–7. One type of bacteriophage, possessing a globular head and a tailpiece. The head contains the phage genome, enclosed within a capsid shell. The tail serves as an organ of attachment to a bacterial cell and also as a tube through which phage DNA passes into the host cell. (Bar = 60 nm.) (Reproduced from Eklund, M. W.; Poysky, F. T.; Peterson, M. E.; and Meyers, J. A.: Relationship of Bacteriophages to Alpha Toxin Production in *Clostridium novyi* Types A and B. *Infect. Immun.*, **14**:793. 1976.)

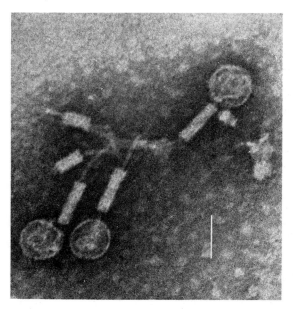

attaches to the cell wall of a bacterium, and its DNA enters the cell, leaving the viral capsid outside. Once inside, the virus nucleic acid takes over, utilizing the bacterium as a source of energy, "building blocks," and propagating mechanisms. When new virus particles are formed, they may contain nucleic acid fragments derived from the host, and these may be transported to another host when the virus is liberated and infects other bacterial cells. If some of the newly infected cells survive the bacteriophage invasion and are able to continue their own reproduction, the genetic material acquired by this transport will be perpetuated in succeeding cell divisions. No change in characteristics of the new cells will be noted if the transfer occurred between genetically identical organisms. But if the transduced genes are different from those originally possessed by the recipient cells, the latter will display new traits accordingly, and so will their progeny.

TRANSFORMATION. A transfer of DNA from one bacterial cell to another is involved in this type of change also, but no intermediary viral agent takes part as in transduction. In this process, fragments of DNA are liberated from a donor cell by lysis (dissolution of the cell) or by an artificial process of extraction and isolation of the material. When a susceptible recipient cell is exposed to this isolated DNA, it absorbs the new molecule and may incorporate it into its own nuclear apparatus. The recipient cell then takes new characteristics as directed by the transforming DNA, and these are inherited by its progeny. Changes that can be induced in certain bacteria in this manner may involve the nature of their capsular material, their fermentative activities, some morphologic features, or their resistance to antibiotics.

CONJUGATION. Bacterial conjugation is a process in which there is a one-way direct transfer of genetic material from one cell to another (Fig. 3–8). The donor or "male" (F⁺) cell carries an F factor (an episome) that can be transferred to a recipient "female" cell that is F deficient. The F⁺ cell has a specialized "sex" pilus. When pairing of an F⁺ and an F⁻ type of cell occurs, there is a transfer of episomal DNA from the former, through the sex pilus, to the recipient. In most instances, the F⁺ characteristic is transferred *in toto* to the recipient cell, which itself then becomes an F⁺, capable of

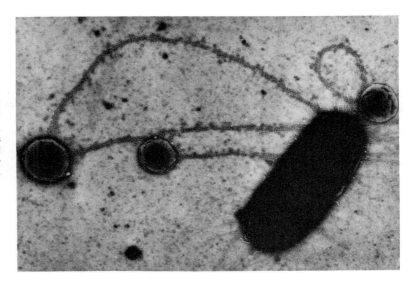

Fig. 3–8. Sex pili connect an F$^+$ donor cell with three small recipient cells. Episomal DNA of the larger cell passes through the pili to the small cells. (Courtesy of Dr. Roy Curtiss III, University of Alabama, Birmingham, Ala.)

transmitting the factor and of converting another F$^-$ to F$^+$. The episomal DNA associated with F$^+$ can thus spread like an infection through a bacterial population, carrying the new genetic trait(s) along.

The conjugal transfer of genetic material is not limited to the transmittal of episomes. In some rare instances the F episome can pick up genetic material from the chromosome (genome) of the donor cell and transfer it to the recipient cell. Also rarely, the F factor may be inserted directly into the chromosome of the recipient cell. When this happens, the recipient cell genome breaks at the site of insertion and a "high-frequency recombination", or *Hfr,* strain is created. Such Hfr strains can then transfer their chromosomal genes in linear sequence into F$^-$ cells by inserting their DNA through the sex pilus. In contrast to the infrequent transfer of chromosomal genes by the F episome, up to 10 percent of the Hfr donor cells can transfer genetic material to recipient cells. Conjugation experiments with Hfr strains have been used to determine the sequence of genes (genetic mapping) in bacteria such as *E. coli.*

Plasmids other than episomes can also be transmitted by conjugation. For some bacterial species, however, plasmid transfer can be accomplished only through transduction by bacteriophage vectors as described above. Transducing viruses may carry plasmid genes, as well as DNA fragments from the chromosome itself, from one bacterial cell to another.

R Factors (Infectious Antibiotic Resistance). R factors are plasmids similar to F factors and can therefore also be transferred from one cell to another by conjugation. However, the term "R factors" refers specifically to those plasmids that carry extrachromosomal genes determining bacterial resistance to antibiotics. Genes that code for resistance to different antibiotics may be carried on the same R factor. For example, an R factor determining resistance to ampicillin may also render the organism resistant to tetracycline and chloramphenicol. Thus, the transfer of R factors from one cell to another may result in the simultaneous transfer of multiple antibiotic resistance. An important characteristic of R factors is that they can be transferred between different bacterial species, as well as between cells of the same species. The spread of R factors among bacteria, termed infectious antibiotic resistance, has been enhanced by excessive use of antibiotics and has become a serious clinical and epidemiologic problem.

Viruses

As obligate intracellular parasites, viruses exemplify the extreme in metabolic simplicity among microorganisms. They possess only the essential mechanisms required for reproduction, and even this function must be mediated through the host cell's genetic mechanisms.

Viruses are known to cause disease in many types of organisms: man and animals, plants, bacteria, fungi. They do so by entering individual host cells and producing changes that lead to cellular malfunctions or death. However, cell infection by a virus is not necessarily lethal, and changes induced in the cell may be beneficial rather than damaging. In any case, the genetic mechanisms of surviving cells infected with virus may be permanently changed, and in this sense viruses may be thought of as agents of heredity as well as of disease.

An example of a plant virus infection that leads to a permanent but nonlethal effect can be found in the variegated tulip. The color variation and streaking that are so admired in tulip blossoms can be induced by a virus (Fig. 3–9). The popularity of the streaked tulip once led to a "tulipomania" in the Netherlands, during the sixteenth and seventeenth centuries. The boom in the tulip market collapsed when growers found that the carefully guarded secret of their variegated flowers lay in the mere rubbing together of a nonstreaked and a streaked bulb. Color variation is now bred in tulips genetically rather than through virus passage, since the latter does produce some stunting of growth as well.

Fig. 3–9. Streaked tulip. The streaking or "break" of these Darwin tulips is caused by a lily-latent mosaic virus. Streaked tulips are now propagated genetically; the virus-infected varieties are regarded as sick and poor-growing individuals and are seldom grown. (Courtesy of Dr. Francis O. Holmes, Rockefeller University, New York, N.Y.)

Questions

1. What is the biosphere?
2. How are plants and animals dependent on microorganisms for survival?
3. Briefly describe the nitrogen cycle.
4. Compare autotrophic and heterotrophic organisms. List several examples of heterotrophs.
5. Why are enzymes essential for all microbial activities?
6. List several beneficial and harmful activities of the following microorganisms: yeasts, molds, protozoa, bacteria, and viruses.
7. What are the environmental requirements for bacterial culture?
8. What is a genome? What is a plasmid?
9. Why is the transfer of genetic material among bacteria important?

Additional Readings

Adler, Julius: The Sensing of Chemicals by Bacteria. *Sci. Am.,* **234**:40, April 1976.

†Amerine, Maynard A.: Wine. *Sci. Am.,* Aug. 1964, #190.

Berg, Howard C.: How Bacteria Swim. *Sci. Am.,* **233**:36, Aug. 1975.

Berger, Terry: 'Tulipomania' Was No Dutch Treat to Gambling Burghers. *Smithsonian,* **8**:70, April 1977.

†Bolin, Bert: The Carbon Cycle. *Sci. Am.,* Sept. 1970, #1193.

Brill, Winston J.: Biological Nitrogen Fixation. *Sci. Am.,* **236**:68, March 1977.

Campbell, Allan M.: How Viruses Insert Their DNA into the DNA of the Host Cell. *Sci. Am.,* **235**:102, Dec. 1976.

†Clowes, Royston C.: The Molecule of Infectious Drug Resistance. *Sci. Am.,* **228**:18, April 1973, #1269.

Cohen, Stanley N.: The Manipulation of Genes. *Sci. Am.,* **233**:24, July 1975.

DaSilva, Edgar J., and Fabian, Fernandez: A UNESCO World Network of Applied Microbiology, UNESCO *Courier,* p. 31, July 1975.

†Delwiche, C. C.: The Nitrogen Cycle. *Sci. Am.,* Sept. 1970, #1194.

†Emerson, Ralph: Molds and Men. *Sci. Am.,* Jan. 1952, #115.

†Kornberg, Arthur: The Synthesis of DNA. *Sci. Am.,* Oct. 1968, #1124.

*Mikat, Dorothy M., and Mikat, Kurt W.: *A Clinician's Dictionary Guide to Bacteria and Fungi,* 3rd ed. Eli Lilly and Co., Indianapolis, Ind., 1976.

Spector, Deborah H., and Baltimore, David: The Molecular Biology of Poliovirus. *Sci. Am.,* **232**:24, May 1975.

†Temin, Howard M.: RNA Directed DNA Synthesis. *Sci. Am.,* Jan. 1972, #1239.

Watson, James D.: *The Double Helix.* Atheneum, New York, 1968.

†Woodwell, George M.: The Energy Cycle of the Biosphere. *Sci. Am.,* Sept. 1970, #1190.

†Available in *Scientific American* Offprint Series. (See Preface.)
*Available in paperback.

4

Tools and Techniques: Identification of Pathogenic Organisms

The primary responsibilities of the diagnostic microbiology laboratory are to examine clinical specimens for the presence of presumptive microbial agents of infectious diseases, to culture and identify the microorganisms in such specimens whenever possible, and, when indicated, to test significant isolated organisms for their susceptibility to antimicrobial drugs. To accomplish reliable results the laboratory must have tools and techniques for visualizing, culturing, and demonstrating the physiologic properties of organisms of microscopic size whose growth requirements are often exacting.

Good microscopes are essential for the proper examination of microorganisms, but microscopic techniques alone cannot distinguish between morphologically similar species that differ in other important properties. Many clinical specimens (such as sputum or feces) contain multiple species of organisms that must be separated and identified by culture techniques. Because the entire environment also abounds with microorganisms, specialized techniques and equipment are required to prevent extraneous organisms from contaminating the work and confusing the results.

Techniques for the initial culture of specimens are designed to support the growth of any pathogenic organisms that might be present. If more than one species grows, each must be isolated from the rest in "pure culture." Final identification of each isolated species is based on studies of staining properties and microscopic morphology, growth requirements, enzymatic activities, and other biochemical distinctions. Antibiotic susceptibility testing can then provide information of value to the physician in choosing appropiate treatment for the patient.

The basic equipment and techniques of the microbiology laboratory, and the general principles of their use, are described in this chapter.

Tools

Microscopes

There are three types of microscopes available for the study of microorganisms. The *compound light microscope* is the most practical tool for the diagnostic laboratory, providing sufficient magnification to observe the size, shape, surface structures, and staining properties of all microorganisms except viruses. The *electron microscope* is a more sophisticated instrument that has made it possible to study virus morphology and to visualize the internal structures of all microbial cells. The *scanning electron microscope* is a research tool that has provided dramatic new observations of cellular surfaces and has shed light on the mechanisms by which microorganisms attach themselves to the surface tissues of animals and humans.

The Compound Light Microscope. This type of microscope has two sets of magnifying lenses, placed at opposite ends of a tube that is mounted on a stand so that it can be raised or lowered. The uppermost or viewing lens is called the *ocular.* The other lens, which is nearest the object being magnified, is called the *objective.* Usually there are three objectives on a revolving nosepiece, capable of magnifying the object 10, 45, and 100 times ($10\times$, $45\times$, $100\times$), respectively (Fig. 4–1). The magnification provided by the objective is further increased by the ocular. If the latter is a $10\times$ lens, which is usual, the combined enlargement provides a total magnification of $100\times$, $450\times$, or $1000\times$.

The platform that holds the slide being viewed is called a *stage* . Just below the centrally placed opening in the stage is the *substage condenser,* which functions to collect and concentrate the light and to direct it upward through the object on the stage. The object absorbs and bends the light, so that when it passes onward through the objective and ocular it is patterned in the image of the object (Fig. 4–2, *A*). The condenser is fitted with an iris *diaphragm* to regulate the amount of light passing into it. Raising or lowering the position of the condenser also changes the amount of light it can collect. Light is reflected from its source into the condenser by a *mirror* attached at the base of the microscope, or from a light bulb built into the base. In general, the higher the magnification desired, the more intense the light must be, but the degree of illumination needed varies also with the density of the object. For this reason, stained material usually requires more light than unstained preparations.

The close study of organisms as small as bacteria requires the use of the $100\times$ objective, immersed in a drop of oil placed on top of the slide holding the object. The oil improves the *resolution* of the objective, that is, its ability to provide sharpness of detail, which is particularly necessary at high magnifications. Resolution and magnification are the two essentials of microscopy, neither being of much value without the other. The resolution of the light microscope is finally limited because as objects diminish in size they can no longer absorb and redirect light — they merely scatter it and are seen with less and less clarity in visible light.

The light microscope can be equipped with a special condenser for *darkfield microscopy.* This condenser directs the light rays in such a manner that only those rays that strike an object on the viewing slide pass through the objective. Thus, bacteria, cells, or other objects present appear brightly lit against a dark background.

In *phase contrast microscopy,* both a special condenser and objectives are attached to the light microscope. This optical system permits discrimination between materials that differ only slightly in thickness and refractive index. Details of cellular structure that are not visible with the ordinary light microscope can be observed with the phase contrast lenses.

The Electron Microscope. When rays of shorter wavelength than those of visible light are directed at microscopic objects, they are bounced off again in patterns determined by the object's structures. Although the human eye cannot see short waves, other sensitive surfaces can record them. A photographic plate coated with a responsive material records the impact of short waves, producing a picture that reveals the design of the object from which they were bounced.

Operating on this principle, the electron microscope sends a stream of short-wave electrons through a vacuum chamber (to avoid their scatter by air) to strike a microscopic object in its path. Instead of absorbing these waves, as a larger object would do with visible light, the object scatters them in a pattern that forms its image. Circular electromagnets attract the scattered electrons still farther outward, thus enlarging the image, which is then focused on a fluorescent screen. Electrons hitting the screen cause visible fluorescence that can be viewed directly and also can be photographed with still further enlargement. The image can be enlarged 200,000 times or

Optical and Mechanical Features of
THE MICROSCOPE

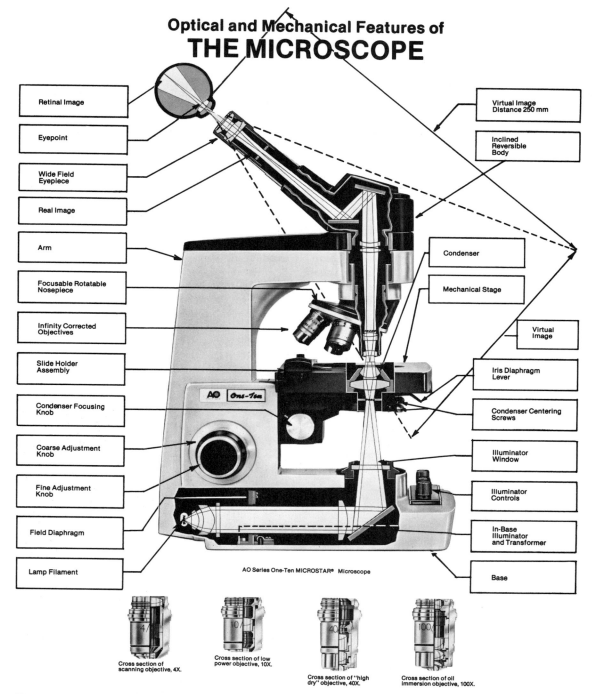

Retinal Image

Eyepoint

Wide Field Eyepiece

Real Image

Arm

Focusable Rotatable Nosepiece

Infinity Corrected Objectives

Slide Holder Assembly

Condenser Focusing Knob

Coarse Adjustment Knob

Fine Adjustment Knob

Field Diaphragm

Lamp Filament

Virtual Image Distance 250 mm

Inclined Reversible Body

Condenser

Mechanical Stage

Virtual Image

Iris Diaphragm Lever

Condenser Centering Screws

Illuminator Window

Illuminator Controls

In-Base Illuminator and Transformer

Base

AO Series One-Ten MICROSTAR® Microscope

Cross section of scanning objective, 4X.

Cross section of low power objective, 10X.

Cross section of "high dry" objective, 40X.

Cross section of oil immersion objective, 100X.

Fig. 4–1. A modern compound microscope with parts labeled for easy identification. The pathway of light is also illustrated. (Courtesy of American Optical Corp., Buffalo, N.Y.)

66

A
Light Microscope

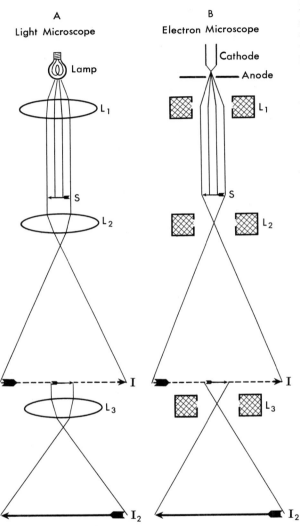

B
Electron Microscope

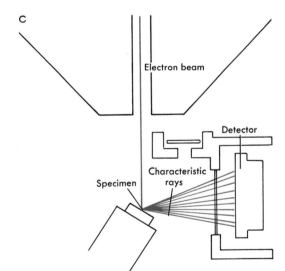

Fig. 4–2. Comparison of the optics of the compound light microscope (A) with the mode of operation of the RCA electron microscope (B) and the ISI scanning electron microscope (C). In the light microscope, a system of glass lenses (L) focuses light rays on the object (S), re-collecting them to form first a primary image (I), then a highly magnified image (I₂). In the electron microscope, a beam of electrons, rather than light rays, is focused by a system of magnetic fields (L) that serves the same function as the lenses of the light microscope. In diagrams A and B, L_1 is the condensing lens, L_2 is the objective lens, and L_3 is the projector lens. The scanning electron microscope (C) also focuses an electron beam on the specimen. Rapid back-and-forth scanning of the beam creates a flying spot of radiation that is collected on and activates a viewing screen to produce an image. (A and B courtesy of Dr. Thomas F. Anderson, The Institute for Cancer Research, Philadelphia, Pa.; C courtesy of International Scientific Instruments, Inc., Santa Clara, Calif.)

more, depending on the equipment used. Thus, if initial resolution is good, the invisible can be brought up into the visual range of the human eye (Fig. 4–2, *B*).

The electron microscope has much greater resolving power than the light microscope. Using the latter, the smallest object that can be seen is about 0.2 μm, while with the electron microscope structures as small as 0.001 μm (1 nm) can be clearly visualized. Objects to be examined must be extremely thin, however. Bacteria are too thick to be examined for a view of their internal structures unless they are first prepared in thin sections. The cells must be fixed and dehydrated in an organic solvent, embedded in plastic, and cut on a special instrument (an ultramicrotome) fitted with a diamond knife. Several slices are prepared and treated with special stains, such as osmic acid, uranium, lead, or permanganate. Such materials scatter electrons well (thus providing good contrast) because they are composed of atoms of high atomic weight. Thin sections of bacteria prepared in this way reveal fine details of internal cellular structures (see Fig. 2–7, p. 26).

If one wishes to see only the outlines of bacteria by electron microscopy, whole cells can be studied using a technique for preparation called shadowing. In this case the specimen is first coated with a thin layer of a heavy metal, such as gold or palladium. The metal is deposited on the object from one side so that treated cells throw a shadow revealing thickness, shape, and dimension (see Fig. 15–3, p. 381, and Fig. 19–2, p. 450).

Negative staining is another common technique used in electron microscopy. Phosphotungstic acid and other stains that do not penetrate cells but scatter electrons are used to pretreat the specimen in this case.

A number of examples of electron micrographs of microorganisms can be found in Chapters 1 and 2 and elsewhere in the text. In these photographs, note how the instrument demonstrates details of both surface and internal morphology of microscopic cells.

The Scanning Electron Microscope. The recent development of the scanning electron microscope has opened up another approach to the visualization of microbial surface structures. The photographic images obtained with this instrument have a remarkable three-dimensional quality (Fig. 4–3), unlike the flat effects seen in electron or light photomicrographs. With this microscope, an electron beam is directed at the specimen to be studied and scans back and forth across it. The material to be examined must first be coated with a thin film of a heavy metal such as platinum. The metallic coating scatters the electrons in an image-forming design onto a viewing screen. The latter is activated by the flying spot of electrons and illuminates the subject, in a picture essentially like that seen on a television screen. Photographs can then be made of the image appearing on the screen (Figs. 1–7 and 4–2, C).

Magnifications ranging from $15\times$ to about $10,000\times$ can be obtained with the scanning electron microscope. However, the instrument can be used to study only the surface structures of an object, for internal details cannot be visualized by this method.

General Equipment

The equipment needed for successful cultivation and identification of microorganisms varies according to the type of organism to be studied, the conditions it requires, and the techniques that must be used to demonstrate its properties. Most diagnostic microbiology laboratories, particularly those in small hospitals, cannot afford the full range of equipment, staff, space, and time required to identify every type of microorganism that might be associated with infectious disease. In that case, the facilities of specialized laboratories, including those of city or state health departments, can be used when difficult organisms are encountered or elaborate techniques are required. However, certain basic items of equipment are essential to the successful operation of every microbiology laboratory, and these will be described here.

Incubators. Most microorganisms have specific temperature requirements for their growth and must be cultivated in a constant-temperature incubator. Modern incubators have an insulated interior chamber, with a thermostatically controlled heating element to provide constant temperature (Fig. 4–4). In addition, large models are often equipped with automatic devices that maintain constant humidity and provide for even circulation of heated air. If the laboratory frequently deals with different types of organisms that vary in their temperature requirements, it must have a separate incubator for each desired temperature range (for *bacteria,* an incubator that holds at 35° to 37° C; for *fungi,* a 28° to 30° C incubator; for *thermophiles,* 50° to 56° C, and for *psychrophiles,* 5° to 10° C incubation).

While microorganisms are under incubation, their atmospheric requirements must be met. *Aerobic* organisms must have adequate access to the ambient air (the closed vessels that contain them must not be airtight). *Microaerophilic* organisms must have lowered oxygen tension and usually also require additional CO_2 in the atmosphere. Modern incubators

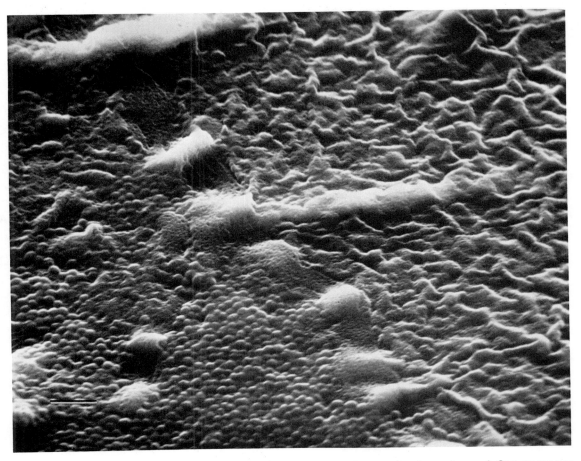

Fig. 4–3. Scanning electron micrograph of a colony of *Streptococcus pneumoniae* growing on a blood agar plate. Note the three-dimensional aspect of individual cocci comprising the colony in the left half of the photo. In the right half, older cocci in the growth have lysed (dissolved), and this area of the colony has collapsed. The bar in the lower left-hand corner represents 10 μm. (Reproduced by permission of the National Research Council of Canada, from Springer, E. L., and Roth, I. L.: Scanning Electron Microscopy of Bacterial Colonies. I. *Diplococcus pneumoniae* and *Streptococcus pyogenes. Can. J. Microbiol.*, **18**:219, 1972. Photo by Dr. Ivan L. Roth, University of Georgia, Athens, Ga.)

can meet this requirement through a device that admits a flow of metered CO_2 and air into the chamber. A simple, inexpensive method for providing increased levels of CO_2 involves the use of a "candle jar." Cultures are placed in a wide-mouthed pickle or mayonnaise jar with a screw-capped top. A candle is also placed in the jar with the cultures. The candle is lighted, and the jar is closed by screwing the lid down tightly. The candle burns briefly within the closed jar, using available oxygen from the air and giving off CO_2. Burning ceases when there is not enough oxygen left to support the oxidation, the

candle goes out, and the jar is then placed in the incubator.

Anaerobic organisms require that all atmospheric oxygen be removed from their environment. Anaerobic incubators have a device for evacuating air from the chamber and replacing it with an inert gas such as nitrogen. A less expensive way of providing anaerobiasis is to place cultures in an "anaerobic jar" fitted with a lid through which air can be evacuated and a hydrogen-containing gas introduced. The lid also contains an electric heating coil and a platinum catalyst fixed to its interior surface. When the lid is heated, the platinum catalyzes the union of hydrogen with any free oxygen remaining in the jar to form water. The final atmosphere in the jar (which is then placed in the incubator) permits growth of obligate anaerobes. A simpler anaerobic jar now in common use contains a "cold" catalyst (palladium-coated alumina pellets) that does not require heating. A commercially available foil envelope containing substances that generate hydrogen and CO_2 is placed in the jar with the cultures. When water is added to the reagents through a small cut in the envelope, gases are generated and released. The jar is immediately clamped tightly; the released hydrogen combines (through the mediation of the catalyst) with the free oxygen in the jar to form water, and the jar is then incubated. The CO_2 given off plays a role in supporting the growth of fastidious anaerobes. The efficiency of anaerobic jars or incubators in providing complete anaerobiasis must be checked each time they are used. A "redox" dye such as methylene blue is a useful indicator because it is colorless when reduced, under anaerobic conditions, and blue when oxidized in the presence of oxygen. An envelope containing a pad soaked in methylene blue is opened and placed with the cultures in the chamber to be incubated. During incubation the color of the pad indicates whether or not anaerobiasis has been achieved and is being maintained.

Media and methods for cultivating anaerobes and other microorganisms are described in a later section of this chapter (p. 76).

Miscellaneous Equipment. Other major items of equipment include refrigerators and freezers for storage of perishable supplies such as culture media, blood, serum, antibiotics, or completed cultures to be saved. Centrifuges are also needed for preparing culture media, or for concentrating liquid specimens prior to culture when this is appropriate. The latter is done when it seems probable that a large volume of specimen may contain only a few of the suspected microorganisms (e.g., urine to be examined for tubercle bacilli). Centrifugation spins the organisms down with the sediment, and the latter can be cultured after the supernatant material is drawn or poured off.

Microbiologic Glassware. For the most part, microbiologic glassware is identical to that used in a chemical laboratory, with its assortment of test tubes, flasks, beakers, pipettes, and the like. One of the items specifically designed for microbiologic purposes is the Petri dish (or plate). It is named for its originator, Richard Petri, a German bacteriologist, and has been a standard item since 1887 when he first designed it. It is simply a shallow, covered dish for the convenient, safe handling of solidified culture media, upon which bacteria and a few other microorganisms can be grown (see Fig. 4–7, p. 80).

Petri dishes and most other vessels are available in disposable plastic. Plasticware is widely used in microbiology laboratories. It must be presterilized by the manufacturer for most microbiologic purposes, for reasons discussed below.

Sterilizing Equipment

All materials and glass or plasticware used in the microbiology laboratory must be scrupulously *clean* and, in most instances, *sterile* prior to their use. Cleanliness is essential because even trace amounts of chemical impurities in Petri dishes or test tubes used for microbial cultures can adversely affect the growth or activities of microorganisms. Sterility is imperative because it eliminates contaminating organisms (from the environment) whose growth in specimen cultures would interfere with the search for significant pathogens. When specimens or cultures are to be discarded, they must be sterilized before disposable items can be thrown away or reusable glassware can be cleaned, to destroy pathogenic organisms they may contain. Sterilizing equipment is therefore essential to safe laboratory operation.

Sterilization. This term implies the destruction of all forms of life, and *sterility* means, in our application, "free of any microorganisms." Sterilization is accomplished by physical or chemical agents, the principles of which are discussed fully in Chapter 10. *Heat* is a principal agent of microbial destruction, and the most reliable. *Dry* heat and *moist* heat have different applications, and laboratory equipment is designed for both.

Dry-Heat Ovens. Dry-oven sterilization of clean glassware and other heat-resistant items is preferable to moist-heat methods because such labware must be dry when used. Residual moisture within vessels used for culture media can affect the critical concentrations of medium ingredients. Laboratory ovens are well insulated, electrically heated, and fitted with thermostatic controls to maintain desired temperature settings. Within the chamber, heated air raises the temperature of all glass or metal surfaces exposed to it, provided the load is arranged to permit even distribution and flow of air. The temperature and time of baking are adjusted to assure the death of even the most heat-resistant organisms, but this adjustment must also take into account the heat-sensitivity of all items in the load. One hour at 175° to 180° C (347° to 356° F), one and one-half hours at 165° to 170° C (329° to 338° F), or two hours at 160° C (320° F) will accomplish sterilization in the oven. (Note that decreases in time must be compensated by increases in temperature.) The choice of temperature is made on the basis of each load's components. When nothing heat sensitive is present, the highest temperature can be used for the shortest time. When cotton plugs or paper wraps are used to protect open vessels, however, they must be heated with care to prevent charring, else their protective value is destroyed. The presence of paper or fabrics in a load dictates the choice of a lower temperature applied for a longer time period in the sterilizing technique.

Steam Sterilizers (Autoclaves). *Moist* heat, applied as *steam under pressure* in an *autoclave,* is used in the laboratory for sterilization of culture media, discarded cultures, dirty glassware, and many other items that would be damaged by dry heat (Fig. 4–5). Modern autoclaves that provide a drying cycle after

Fig. 4–5. Culture media and other supplies in a clinical microbiology laboratory must be sterilized in an autoclave before use. (Courtesy of AMSCO/American Sterilizer Company, Erie, Pa.)

sterilization has been accomplished can also be used for clean glassware.

The autoclave is an indispensable piece of equipment, not only in the laboratory, but in other hospital areas such as operating rooms and central supply areas, where sterilization of equipment and supplies for surgical procedures and other patient uses is of critical importance. In principle, the autoclave is very much like the well-known kitchen pressure cooker, in a larger version with more complicated controls to assure its efficiency and safe operation. Essentially, the autoclave is a chamber that can be sealed and filled with steam that replaces the cooler air within. Steam under pressure is admitted to the chamber so as to displace all the air through a discharge line. When the last of the cool air has been evacuated, the air discharge valve is closed and more steam is admitted until a pressure of 15 to 20 lb per square inch has been reached within the chamber. At this pressure pure steam has a temperature of 121° C. This is a much lower temperature than that used for oven sterilization, but it is much more efficient because moist heat penetrates more quickly than dry heat and more deeply when it is applied under pressure. (In the kitchen, it will be remembered, the advantage of the pressure cooker is that it cooks foods much more quickly and thoroughly than boiling, which provides moist heat without pressure, or oven baking.)

Sterilization can be accomplished in the autoclave at 121° to 125° C (250° to 256° F), with 15 to 20 lb of steam pressure, in 15 to 45 minutes, the time depending on the size, contents, and distribution of the load, as discussed more fully in Chapter 10.

In addition to the uses mentioned above, the autoclave is also satisfactory for sterilization of clean, empty glassware, provided damp cotton plugs, paper wraps, or other fabrics used are permitted to dry undisturbed. Microorganisms deposited on wet porous surfaces can make their way through to the interior along the moist channels between fibers, and contaminate the vessel or material within the wrap.

It is also important to arrange empty vessels in the autoclave for sterilization so that the air within them will not be trapped but can be freely replaced by steam, which is always hotter and lighter than air and does not mix with it easily. This means that empty tubes, flasks, bottles and other such vessels must be laid horizontally, with their plugs or caps very loosely in place, so that the heavier, cooler air within them can run out and down, and so that steam can cover their interior surfaces.

Vessels containing liquids such as culture media or other solutions can be sterilized in an upright position, because these liquids vaporize at autoclave temperatures; the rising vapor displaces the air in the vessel upward and out of the tube or flask, while the pressure of the steam prevents any violent bubbling of the liquid. When the sterilizing period is ended, the steam in the chamber is permitted to escape very slowly, so that the liquid can cool below its boiling point while enough pressure remains to prevent its bubbling over.

Methods for Preparing Glassware

Dirty Glassware. All dirty glass or plasticware from the laboratory that has been used for the collection of patients' specimens, or for any part of the culture procedures, must be sterilized before it can be handled further. The autoclave is always used for this purpose, to assure rapid killing of organisms within the depths of liquid materials and to spare glassware from the damage it would suffer if its contents were baked onto its surfaces with dry heat. To prevent this from happening even in the moist heat of the autoclave, it is best to place empty, dirty glassware in buckets and add enough water to cover its surfaces. After sterilization, plastic items can be discarded but glassware must be scrupulously cleaned and freed of chemical impurities that might, even in trace amounts, adversely affect bacterial cultures. Mechanical dishwashers adapted for laboratory glassware are an integral part of the operations of modern laboratories. These machines provide a final rinsing with distilled or deionized water that removes inorganic residues. The glassware is then allowed to drain dry, or it is dried with heat.

Clean Glassware. The majority of items used in the diagnostic laboratory are made of disposable plastic (presterilized by the manufacturer). After they are used and autoclaved they must be discarded because the plastic is melted by the heat of the sterilizing process (autoclavable plastics are available but costly for routine use). However, reusable glassware items serve a number of valuable purposes and must be processed for recirculation.

All clean glassware, or other items to be resterilized for use in the laboratory, must be carefully closed or covered in a way that will not permit contamination following sterilization. The covers and tops of glass Petri dishes are assembled, and groups of dishes are either wrapped in paper or placed in metal canisters. Test tubes, flasks, and bottles are fitted with either screw-on caps, light metal covers, or cotton plugs tailored snugly so that the air passing through will be efficiently filtered free of dust and bacteria. Pipettes are fitted with a small cotton plug at the mouth end, to protect the user from mischance in drawing up liquid cultures or other contaminated materials. Since mouth pipetting can be extremely hazardous, it is highly recommended that an aspirator bulb be used instead, even with a plugged pipette. The plug also protects the culture from reverse contamination. After plugging, pipettes can be wrapped in paper or placed in suitable covered containers so that they can be picked up for use at the plugged end, without touching the delivery tip. Whether these items are sterilized in the oven or the autoclave, the load must be arranged in the chamber so that dry hot air (in the oven) or pressurized steam (in the autoclave) can reach and sterilize all surfaces.

Filters. One other type of sterilizing equipment frequently used in microbiology laboratories is the *filter.* It is used to remove bacteria or larger microorganisms from fluids or solutions that cannot be heat-sterilized. There are several types of filters in common use, but the principles of their design and use are similar. Filters are made of porous materials (asbestos, diatomaceous earth, sintered glass, cellulose membranes, collodion) of various grades of pore size. When solutions are filtered through them, particulate matter, including microorganisms, is retained according to the relative size of particles and pores. Most viruses and rickettsiae are *filterable;* that is, filters of the finest porosity cannot retain them because of their very small size. In bacteriologic work this is not of practical importance if the filtered solution does not contain living cells that could support the growth and multiplication of any viruses that might be present. Filtration can be used, therefore, to prepare bacteria-free fluids for various purposes. In the preparation of culture media, for example, some of the enriching substances used are heat-sensitive and cannot be added to basal ingredients of the medium prior to autoclaving. Such components as blood serum, vitamin solutions, antibiotics, or carbohydrates can be passed separately through filters of suitable porosity. The filtrate can then be added to the culture medium after its sterilization, using sterile pipettes or syringes. In use, the filter is mounted on a flask (both must be sterilized beforehand) fitted with a side arm. A piece of rubber tubing connects the flask arm to a vacuum. The solution is pulled through the filter into the flask, but care must be taken not to exert so much force that small bacteria are also pulled through.

Techniques

Asepsis

The extensive preparation of glassware and other items, as described above, would be wasted if it were not for the scrupulous care with which the microbiologist subsequently handles his work. He must be meticulous in technique as he handles cultures, to assure his own and his coworkers' safety from their contents and to protect the cultures themselves from contamination by stray organisms from the environment. With media, glass, and plasticware sterile at the start, the microbiologist manipulates every tool and culture so that the microbes he is studying cannot contaminate hands, clothing, or any part of the working area or become mixed with extraneous organisms. His approach is described by the term *aseptic technique,* which means that it is designed to prevent or avoid "sepsis." The latter word refers to any condition, such as putrefaction or infection, resulting from growth of microorganisms, and *asepsis,* accordingly, means the absence of such growth. By learning where, when, and under what circumstances microbes may grow and multiply, one develops an awareness of their presence (a "seeing eye" for the invisible) and an aseptic technique to cir-

cumvent their possible effects. The phrase describes a set of *attitudes,* as well as of methods, that is fundamental and essential for all who are concerned with the diagnosis and care of patients with infectious diseases and the protection of others from the transfer of infection. In the photograph opposite page 104 note the order and cleanliness the microbiologist has established at the laboratory bench. Her aseptic technique assures the integrity of the work as well as her own safety from infectious hazards.

Transfer Technique. The tool most frequently used by the microbiologist for transferring microbes, or the material containing them, from one place of growth to another for examination or further culture is the *inoculating loop or needle.* It consists of a piece of heat-resistant wire, 5 to 7 cm in length, held firmly in a pencil-shaped handle. The wire can be twisted into a small loop at its free end, so that it will hold a drop of liquid, or it can be used as a straight needle. It is heated in the flame of a Bunsen burner, a device that mixes illuminating gas with air to give a narrow cone of very hot flame. Other means of applying a high degree of heat can also be used, such as an electric coil or an artificial gas torch. When the inoculating loop is held to the heat it will quickly glow, and any organisms on its surfaces will be incinerated. After cooling, the sterile wire can then be used to pick up a small quantity of the material to be studied and to transfer it to a glass slide for examination with the microscope or to a Petri dish or other vessel containing culture medium. The loop must be carefully flamed again after use to destroy organisms left on its surfaces. The process of twice flaming is intrinsic to correct use of the loop.

Whenever a culture tube or bottle is opened, the exposed lip of the vessel is passed quickly through a hot flame before the sterile loop is introduced. While the vessel is open, its cotton plug, screw cap, or other type of closure must be held in the hand with the loop in such a way as to prevent contamination of its undersurface by organisms normally present on hands and other surfaces. While the vessel is open, it is held at an angle so that dust particles in the air, with their possible load of bacteria, cannot drop directly into it.

Other tools commonly used in the transfer of organisms from one vessel to another are sterile pipettes, syringes, and swabs. These are useful if larger quantities are to be transferred. Pouring liquids or fluid culture media into other containers is not an acceptable technique if sterility is to be maintained, for it is difficult to guarantee asepsis as the material flows over the lip of one vessel and into another. It is particularly dangerous to attempt this type of transfer of fluid cultures containing living microorganisms.

Microscopy

Bacteria and larger organisms can be examined microscopically in the living state, suspended in a fluid, or in dried stained smears in which the microbes are no longer alive.

Wet Preparations. Microorganisms can be studied in the *living* state for such characteristics as motility, shape, size, and group arrangements. Larger forms, such as yeasts, fungi, protozoans, and helminth ova, when they are in fluid suspensions, or *"wet" preparations,* as they are often called, can be seen with good resolution with the dry lenses. Internal structures can often be identified, even without staining, but to clarify these a dye that is taken up by the cells without distorting them (a "vital" stain) can be used. Weak solutions of methylene blue or of iodine are examples of such stains. Bacteria are too small to be seen with good resolution in wet preparations even with the oil-immersion lens, but their motility as well as their patterns and arrangements (chaining, clustering, packeting, etc.) can be appreciated.

There are two ways of making a wet preparation. A drop of the fluid containing the organisms can simply be transferred to a clean glass slide with the inoculating loop and covered with a thin cover glass. If necessary, the edges of the cover glass may be sealed with a rim of petroleum jelly or colorless nail polish to prevent rapid drying. A longer-lasting mount, with greater depth, can be prepared as a "hanging drop." In this case, the clean, greaseless cover glass is first rimmed lightly with petroleum jelly; then a drop of material for study is placed in the center of the cover. A hollow-ground slide is next inverted so that the concavity is placed directly over the drop on the cover slip. A little gentle pressure

seals the slide and cover together, and when the slide is quickly reinverted, the drop is suspended in the concave well of the slide.

In wet preparations, all particles, including bacteria, suspended in the fluid exhibit a kind of vibratory motion called *brownian movement.* This motion is rapid but nonprogressive and purposeless. It is caused by the continuous movement of molecules of the liquid, which constantly bombard the suspended particles. All bacteria, whether they are truly motile or not, display brownian movement when they are suspended in a wet preparation. In contrast, the activity of motile organisms is progressive and directional and may have a characteristic mode (spinning, undulant, dashing, or sluggish). Bacterial flagella themselves cannot be seen in wet preparations under the compound light microscope, but when motile protozoans, which are much larger than bacteria, are thus examined, locomotive structures (flagella, cilia, pseudopods) can be seen in action. Wet preparations can also be useful for demonstrating the capsules of microorganisms. Some yeasts, for example, possess large capsules that can be visualized by adding a drop of India ink to the preparation. The ink provides a dark background against which the capsule appears as a clear, glassy zone surrounding the organism.

Stained Preparations. Dried, stained smears permit better resolution of bacteria and of internal structures in larger organisms. Stained smears examined with the oil-immersion lens can provide a sharp view of even very small bacteria, when they are thinly spread and well separated. Some part of the cell must retain the color of the stain that has been used; otherwise so much light will pass through the cell that it cannot be seen distinctly. The aniline dyes react intensely with most bacteria. The composition of the cell wall or the cytoplasm determines the staining reaction, which differs with the dye as well as with the type of cell. Staining distinctions are often quite helpful in identification of bacteria.

When material to be studied microscopically has been smeared thinly on the surface of a clean glass slide, using the inoculating loop or a sterile swab, it must be allowed to *dry completely* in air. Drying fixes the material to the slide so that it will not wash off during staining. When dry, the slide is passed quickly through the Bunsen flame to *"heat-fix"* the drop still more firmly to the glass. Excessive heating may distort cells, however, and must be avoided. Air drying, heat fixing, and staining kills the organisms without affecting their morphology unduly.

The *fixed* smear may be stained either with a single dye that reveals individual morphology or with a combination of stains that distinguishes between kinds of organisms. There are several differential staining methods, but the two used most commonly are the *Gram stain* (devised by the Danish physician, Christian Gram, in 1884), and the *acid-fast* or *Ziehl-Neelsen* stain (named for the two men who developed it, in 1892). In both procedures, staining is accomplished in three essential steps: (1) a dye is applied for a time sufficient to stain all bacteria uniformly and is then washed away with water; (2) an alcoholic solvent is applied briefly to remove the dye from all organisms that have not reacted to it permanently; and (3) a dye of a different color than the first is applied, then removed with a final rinse. The second dye counterstains those organisms that were decolorized by the alcohol, but it does not change the color of those that reacted with and retained the first stain. Thus, the result is differential for two kinds of organisms, staining with one color or the other.

Gram Stain. The first dye of this stain is a solution of crystal violet, which stains some bacteria a very dark purple. An iodine solution then follows as a mordant, to "fix" the dye firmly to the cells. Bacteria that retain the purple color through the decolorizing and counterstaining steps are called *Gram positive.* Those that lose the first color and are restained with the pink color of safranin, the counterstain, are called *Gram negative.* These differences are a reflection of dissimilar chemical composition of the bacterial cell walls. The Gram stain is an invaluable ally in identification of unknown bacteria. Very often it is the first step taken in making a bacteriologic diagnosis of an infectious disease, the stain being applied directly to a smear of an appropriate specimen (sputum, urine, spinal fluid) obtained from the patient.

Acid-Fast Stain. The *Ziehl-Neelsen,* or acid-fast, stain requires carbol fuchsin for the first step. This dye, when used with a wetting agent or gently heated

to its steaming point on the slide, can penetrate the fatty substances that characterize the cell walls of some bacteria, such as the tubercle bacillus (the agent of tuberculosis), and stain the cells a bright red. When this dye has been washed away, the second step decolorizes organisms that cannot retain carbol fuchsin when washed with an acid-alcohol solution. The counterstain of the third step is methylene blue. Bacteria such as the tubercle bacillus that retain the red color of carbol fuchsin are said to be *acid fast* because they resist the acid action of the alcoholic decolorizer. Bacteria that wash free of the red dye and take the blue counterstain are non-acid fast. This procedure is very useful for detection of tubercle bacilli in specimens obtained from patients suspected of having tuberculosis. It is also used in identification of these organisms when they have grown in culture.

Other Stains. Large organisms such as protozoa, the ova and larvae of animal parasites, yeasts, and fungi do not usually require any but the simplest stains. They are often studied in wet mounts to which a penetrating stain can be added (iodine for parasites, lactophenol cotton blue for fungi). The fine structures of bacteria (flagella, capsules, spores) require special stains, as do intracellular organisms such as the rickettsiae, or chlamydiae, growing in aggregates within parasitized cells. Stains used to visualize organisms within host cells must differentiate them from the structures of the host cell itself.

Culture Techniques

This section deals primarily with the cultivation of those pathogenic bacteria that are not obligate intracellular parasites. Although rickettsiae and chlamydiae, as well as viruses, require the presence of living cells for their growth in culture, and yeasts, molds, and a few cultivable protozoan species each have their own special requirements, the basic principles for preparation and handling of such cultures are similar to those applied to the study of bacterial populations.

The Nature and Preparation of Culture Media. It was pointed out in Chapter 3 that in nature parasitic bacteria obtain their nourishment from the parasitized host and that their cultivation in the laboratory necessitates the provision of required nutrients and conditions for growth.

Composition of Culture Media. The composition of a good culture medium (plural, *media*) requires careful adjustment of a number of ingredients and factors. In general, nutritional media must contain products of partial protein breakdown, since many bacteria cannot digest intact protein materials. Protein components, such as peptides, peptones, and amino acids, are required for synthesis of bacterial cell protoplasm and for cellular structure. These are commonly provided in the form of partially cooked meat or meat extracts. Bacterial syntheses require energy; therefore, carbohydrates such as glucose, lactose, or starch must be provided. Minerals and vitamins may be added in purified form or as components of some natural food. Minerals are often present as their salts in meat digests, while yeast extract is an excellent source of vitamin B_1. Water is always present in large proportion. Other ingredients can be added for special purposes.

A medium can be made *selective* by incorporating one or more ingredients that will inhibit growh of some organisms while encouaging multiplication of others. A *differential* medium can be prepared by including a substance to which different bacteria respond differently, thus making it possible to distinguish them by their growth characteristics in such a medium. The medium may be used as a liquid broth, or agar may be added to solidify it at temperatures below 45° C (113° F).

Media prepared with meat extracts are usually referred to as *nutrient* agar or broth media, while those prepared with meats that have been slowly cooked or infused with enzymes added to digest them partially are called *infusion* broths or agars. Blood may also be added for further enrichment. For this purpose, whole blood that has been *defibrinated* (the fibrin removed to prevent clotting) is used in a proportion of 5 to 10 percent by volume. For some bacteria, it may be necessary to liberate hemoglobin from the blood cells so that they can utilize it. This can be done by heating the blood gently or adding it to a hot medium. In the process hemoglobin color changes from its customary bright red to a rich brown, so that an agar medium prepared with heated blood is called *chocolate* agar. If an enriched medium without the color or turbidity of blood is desired, other materials can be used, such as human or animal serum, yeast extract, or extra protein components.

pH Adjustment. Before the medium is complete and ready for use, it is important to assure the stability of its acid-alkaline balance, or pH. (pH is a mathematical term used to express the degree of acidity or alkalinity a solution may have.) Most bacteria require that the pH of their environment be neutral, or nearly so, that is, neither actively acid nor alkaline. This means that not only must the finished medium be *adjusted* to neutrality, but it must also be *buffered* against major changes in pH that may occur as a result of bacterial growth in the medium later. As bacterial growth products increase in an artificial medium, from which they cannot diffuse away as they might in a natural environment, large shifts in the pH of the medium toward acidity or alkalinity can occur, and since bacteria are sensitive to such changes the culture may die. For this reason, it is desirable to incorporate substances known as *buffers* into the medium. Buffers are compounds capable of reacting with free acids and alkalis and, in so doing, of binding them so that they are no longer in active solution. Continuing buffer action in the culture throughout the period of bacterial growth therefore prevents large pH changes in the medium.

The microbiologist often wants to be sure that a culture remains stable in pH, or he may be testing for the ability of an organism to produce a pH change by utilizing some ingredient of the medium. For these reasons, indicator dyes, which are sensitive to free acid or alkali and assume a different color at various pH values, are frequently added to culture media. A shift in pH as the culture grows will then produce a visible color change to inform the observer of this event. It is necessary to know the color range for each indicator in response to pH changes. By comparing the color of the test with the standard colors for the indicator at each pH value, an estimate of the degree of acidity or alkalinity developing in the test can be made. Finer measurements can be made with electrical pH meters when it is necessary to be more exact.

The Final Step. The finished product must, of course, be free of any extraneous organisms from the general environment. Such organisms would contaminate the medium with their own growth products, compete for nutrient with the species being studied, and make accurate interpretation of results impossible. The *final step,* therefore, in the preparation of culture media is *sterilization,* usually in the autoclave, as described earlier, or by filtration if heat-sensitive ingredients are involved. It should also be pointed out that in the preparation of media there should be no lapses of time between the various steps taken prior to sterilization that would permit contaminating microorganisms to establish themselves and begin their growth. Subsequent sterilization of the media will kill these organisms, but it will not restore changes they may have produced in nutrient ingredients or remove impurities they may have added.

Uses of Culture Media. Media can be used in either solid or liquid form and solid media may be employed in a number of ways.

Solid Media. At the time of preparation, an agar medium may be dispensed into a number of test tubes, which, after sterilization, are laid at an angle to cool. When the agar has solidified, as it will do at about 40° C, the medium will have a sloping surface, called a *slant.* When slants are inoculated, the surface is streaked with the inoculating wire, and the latter may also be plunged, or stabbed, to the bottom of the tube through the solid butt of medium below. Most frequently, agar media are *plated* by pouring them into Petri plates. This is always done after the medium has been sterilized, in tubes, flasks, or bottles, and has cooled to about 50° C (or has been melted down again after solidification). The material to be cultured can be placed in the Petri dish or added to the agar before it is poured, if desired. The result, in this case, is the growth of organisms throughout the medium, in its depths as well as on the surface, and the preparation is called a *pour plate.* If the medium is poured before it is inoculated, the material for culture is streaked across its hardened surface with the inoculating loop or a sterile swab. This will result in growth of organisms across

the surface along the lines drawn by the tool, and the preparation is called, in this case, a *streak plate.*

Tubed agar media can also be solidified in an upright position, for inoculation by a *stab* of the wire down through the column. Such a *stab culture* may be useful in demonstrating the characteristic way a particular organism will grow along the path of the wire, depending on whether it is motile and can diffuse outward from the stab through the medium and on its ability to grow at varying depths where oxygen availability becomes a factor.

For anaerobic cultures, agar media may be prepared (or purchased) in *roll tubes.* These are made by placing a small amount of an appropriate melted agar medium in tubes while a constant flow of oxygen-free gas (such as a mixture of N_2, H_2, and CO_2) is directed into them to provide an anaerobic atmosphere. The tubes are then stoppered with rubber caps that are impermeable to oxygen and rolled until the agar hardens in a thin film around their interior surfaces. Roll tubes are inoculated by removing the stopper and quickly placing the tube under a cannula through which oxygen-free gas is flowing. The cannula is located directly over a mechanical rotator on which the tube sits. Two drops of specimen are placed in the bottom of the tube and distributed by first inserting an inoculating loop to the bottom of the tube, then pressing it gently against the agar surface and pulling upward as the tube rotates. The stopper is replaced quickly, and the tube is incubated. Since the interior of the tube is filled with oxygen-free gas, an anaerobic incubation chamber is not needed. If anaerobes and/or facultative anaerobes were present in the specimen, they will grow along the agar surfaces streaked by the inoculating loop. Figure 4–6 illustrates this technique.

Liquid Media. A medium used in liquid form is usually called a *broth.* It may be prepared in test tubes in small quantities (5 to 10 ml) or in flasks or bottles. Sometimes a small shallow layer of broth in a large flask is desirable, providing generous oxygenation of the culture. Conversely, to reduce access to the atmosphere the broth may be dispensed in tall, narrow tubes. For some purposes it may be useful to layer a little broth over part of a solid agar slant, so that organisms will have a very wet but solid surface for growth.

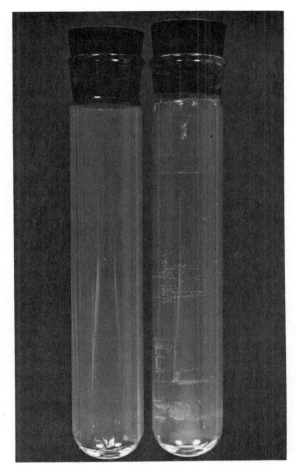

Fig. 4–6. Roll tubes for anaerobic culture. In the roll tube on the right anaerobic organisms have grown along the agar surfaces streaked by an inoculating loop. The tube on the left is uninoculated.

Specimen Culture

The work of the diagnostic microbiologist begins with the arrival in the laboratory of a specimen. It must be accompanied by a request from the patient's physician for a culture. However, the results of that culture will be influenced, even before the specimen reaches the laboratory, by the method and timing of its collection and transport. If these are inadequate

or incorrect, the results may be also, and, even more unfortunately, this source of error may pass unnoticed so that the results are not subjected to proper interpretation. This matter of specimen collection, often performed at the patient's bedside by a nurse or physician rather than by the microbiologist, is of such importance that it will be emphasized many times in this text.

Direct Smears and Preliminary Information. Time and appropriate action are of the essence in medical microbiology. Specific measures for treating the patient seriously ill with an acute infection may be dependent on news from the laboratory regarding the nature of microorganisms, if any, being isolated from the patient's specimens and their response in culture to the action of clinically useful antibiotics. Between the time cultures are started and identifications are complete, the laboratory can provide tentative, preliminary information concerning its results to guide those who are caring for the patient. Even incomplete results can be useful to the physician and the nurse in making decisions as to whether or not the patient should be isolated from others or treated in some specific way. For example, stained smears made immediately from the specimen itself may provide important clues to the nature of organisms present (Gram positive or negative; cocci or bacilli; chained, clumped, or paired; intra- or extracellular), and a good smear report may guide the physician to his first choice of treatment. Direct smears may also guide the microbiologist in choosing the direction cultures should take.

Initiation of Culture. The nature of the specimen, the kind of information desired from it, and the smear reading all determine the type of culture procedures that are initiated. As we shall see, many areas of the human body are parasitized by microorganisms that cause no harm, but there are differences in the kinds of organisms found in different areas because environmental conditions for them vary from one section or tissue of the body to another. Thus a particular kind of specimen, such as a throat swab or a sample of feces has its own "normal flora." These organisms will grow out in the media as a *mixed culture,* in which any pathogenic stranger that may be present must be recognized. The suspected pathogen must then be separated from the others and persuaded to grow alone, in *pure culture,* so that its appearance, properties, and behavior can be identified (Fig. 4–7). Furthermore, not only do the organisms normally found in certain specimens vary from those of others, but the kind of pathogenic organism that can be expected in different specimens varies also. For example, one would not expect to find the diphtheria bacillus in a fecal sample, because it is not likely to survive in the bowel, or the agent of a venereal disease in a surgical wound culture. On the other hand, certain types of specimens, such as blood, urine, and spinal fluid, are normally sterile in the healthy body and contain microorganisms only by accident or as a result of an infectious process. Therefore, any organism found in such a specimen may be suspect.

The initiation of culture procedures is thus carefully guided by the specimen, its source, and the diagnostic possibilities. A battery of different media is inoculated with the specimen, the whole group being chosen to provide a wide range of bacterial requirements — that is, *enrichment* for fastidious organisms, *selective* nutrient to reduce competition and permit the suspected organism to flourish, a *differential* medium to indicate different kinds of organisms growing upon it (Fig. 4–8), a medium free of oxygen for *anaerobes,* and so on. The widest possible screening, consistent with practical limitations on time and space, is provided to cover the widest possible range of microbiologic and clinical expectation.

Day-to-Day Readings. Each day after culture of specimens is started, the microbiologist makes observations of the growth occurring in each of the media inoculated and *plans the next, shortest step* that will narrow identifications to a final decision, as quickly as possible. The work is organized in a logi-

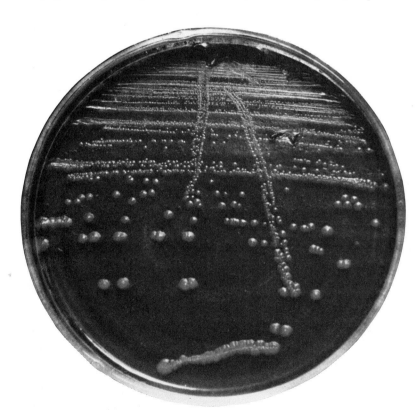

Fig. 4–7. A pure culture of *Corynebacterium diphtheriae* streaked out on an agar plate. (Courtesy of Dr. A. E. Bolyn, The National Drug Company, Swiftwater, Pa.)

cal sequence designed to establish a pattern of facts from a number of clues, both positive and negative, in much the same way as a detective sets out to find evidence and fit it together.

Plate Observations. Plated agar media are examined for growth, and particular note is made of its appearance. Bacteria multiplying on agar surfaces form visible masses called *colonies.* The colonial growth of different types of bacteria is often quite characteristic in its appearance on the various kinds of agar plated. The surfaces of colonies may be rough or smooth; their edges may be even and entire, filamentous, or eroded; and their elevation from the surface may be convex, peaked, or patterned in some way. In addition, bacterial enzymes, or other products of colonial growth, may react with the surrounding medium in some observable manner. On blood agar, for example, some bacteria may cause destruction, or *lysis,* of the red blood cells adjacent to them, leaving clear, colorless zones around colonies growing on an otherwise red and turbid medium. Such a reaction is referred to as *hemolysis* (hemo: blood), and the organism is said to be *hemolytic.* When destruction of red blood cells in hemolytic zones is complete, the reaction is called *beta hemolysis* (Fig. 13–3, p. 320); when it is partial or incomplete, it is termed *alpha hemolysis.* This distinction in hemolytic properties is often important, in differentiating species of streptococci, for example.

Pigment production may be a marked feature of growth, remaining within individual cells to color the whole colony in some cases, in others diffusing away to change the original color of the medium. The utilization of some particular ingredient in a selective or differential medium will also be obvious because of the presence of an indicator reacting visibly to changes produced by bacterial growth (Fig. 4–8).

Broth Observations. In the liquid media included in the original set of cultures for a specimen, bacterial growth may appear diffusely homogeneous, or granular; it may occur in layers; or the bacteria may tend to remain clustered together in colonizing groups, appearing as small puffballs floating in the medium. Some broth media contain substances that reduce or remove free oxygen from solution so that

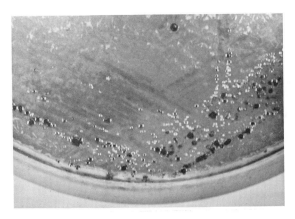

Fig. 4–8. *Salmonella* colonies on bismuth sulfite agar, which is selective because it inhibits many enteric organisms but supports the growth of the pathogenic *Salmonella.* It also differentiates *Salmonella* colonies, which take on a characteristic dark-green or black color, often with a metallic sheen.

anaerobic organisms can grow in their depths but aerobic types also find it possible to grow near the surface where atmospheric oxygen is still available. Facultative bacteria will grow at all levels.

Smear Observations Microscopic examination of stained smears or wet mounts made from the growth in liquid or on solid media provides another guide to further steps toward final identification. Motility, Gram-stain reactions, individual cell morphology, and patterns of arrangement of the bacteria in these cultures lead to decisions as to the next test or medium to use. If original cultures contain more than one species of bacterium, however, these must be separated and grown singly in pure culture before further efforts can be made to identify them, since their reactions to a test medium may vary and lead to confusion of results.

Isolations. Separation of bacterial species for pure cultures is best accomplished on plated media, on which colonies tend to be individual and separated if the streaking technique has not allowed them to crowd together. Isolated colonies are picked from the surface with the loop or straight wire and transferred to another medium where they can grow alone, displaying such properties as the medium supports. Broth cultures are streaked out on agar media of appropriate composition, using the inoculating loop for transfer and spreading the drop out carefully over the surface to permit growth of isolated colonies. The temperature and atmosphere of incubation are chosen for this second set of cultures, or *subcultures,* as they are called, according to indications in the original media as to the nature and preferences of the organisms being studied.

Final Identification. Pure subcultures of isolated organisms are examined after 18 to 24 hours of incubation, and if they provide sufficient information, final identifications are made. If not, further subcultures may have to be made for tests of additional properties. Carbohydrate media are inspected for evidence of fermentation, provided by the color of the pH-sensitive indicator and by the presence or absence of gas.

Other types of enzyme activity, characteristic for a group or a species, can be demonstrated rather quickly and simply with pure subcultures. Some strains of staphylococci, for example, produce an enzyme called coagulase because it is capable of coagulating fresh blood plasma. To test for this enzyme, plasma is mixed in a test tube with a loopful of organisms from an agar plate colony and observed for the formation of a clot after one or two hours of incubation at 35° C. Demonstration of this property, together with colonial morphology, pigment production, and microscopic appearance, might lead to the report of a "hemolytic *Staphylococcus aureus,* coagulase positive," an organism commanding respect for its ability to cause serious infections and for the ease with which it spreads from one person to another. Another example of an enzyme easily identified is the urease possessed by certain organisms commonly encountered in cultures of stools. As its name implies, this enzyme is capable of breaking down urea to ammonia and carbon dioxide. When the organisms are incubated in a urea broth containing a color indicator, the breakdown is recognized by the color response of the dye to the highly alkaline ammonia as it is released.

Many other such tests are available, but usually the systematic application of a few designed to reveal key characteristics is sufficient to make identi-

fications within 24 to 72 hours of the time the culture was originated. Meantime, the microbiologist keeps careful notes and records of tests and results, so that final reports can be made with clear factual support.

Serologic Identification of Microorganisms. Conclusive identification of microorganisms can often be obtained by serologic methods. These are based on the fact that the protein and certain other constituents of microbial cells are "antigenic" — that is, they stimulate the human or animal body to produce "antibodies" during the time of active infection, when the tissues are exposed to bacteria or other organisms. The antibodies are often present in circulating blood as a component of serum. When a sample of blood is taken from an animal or person who has been infected, the serum is separated in the laboratory and tested with the microorganism suspected as the agent of infection. In the test tube, antibodies react specifically in certain visible ways with microbial antigens that stimulated their production in the body. This type of reaction is used to confirm the identity of a microorganism when the nature of the antibody in a given serum sample is known, or conversely to determine the presence and nature of antibodies in an "unknown" serum tested with known organisms. The term *serology* refers to such studies of serum for its antibody content. The application of serologic methods comprises an important part of the work of the microbiology laboratory. The principles involved, and the nature of antigens and antibodies, are discussed in Chapter 7.

Antibiotic Susceptibility Testing

Antibiotic tests can be performed as soon as individual organisms can be isolated in pure culture from the specimen. For several reasons it is not good practice to test mixed cultures for their response to antibiotics. It may be impossible to determine the response of each organism in a mixture to a particular antibiotic; the mixed species may compete with each other for growth on the medium, adding to the antagonism exerted by the drug on one or the other; or the species present may be mutually supportive so that the antibiotic's effect on one or another is lessened. In any of these cases, results are difficult to interpret and may be quite misleading to the physician. It is usual, therefore, for the laboratory to prepare the specimen for culture and to wait for isolated bacterial growth, which can then be tested. An exception to this general rule may be made if the situation is urgent, and if stained smears of the specimen reveal the presence of only one type of organism or of one that is largely predominant. In this case, antibiotic testing can be applied directly to the initial culture. However, it is always best to repeat the test later, using a pure culture of each clinically significant organism isolated from the specimen to confirm the first results.

A number of good methods exist for testing the antibiotic susceptibility of isolated organisms. Some are highly quantitative but expensive to perform; others are semiquantitative but more practical and faster.

Disk Method (Agar Diffusion). The method most frequently used is called the *filter-paper disk agar diffusion method.* Small paper disks of uniform size and shape are impregnated (by the manufacturer) with antibiotic. Each disk contains a known concentration of one such agent. The organism to be tested is first "seeded" on the surface of a suitably enriched nutrient agar plate; that is, it is streaked in a way that will permit it to grow confluently across the entire plate. A number of disks containing different antibiotics are then placed on the plate, spaced at regular intervals from each other. The plate is incubated for 18 to 24 hours, under conditions of temperature and atmosphere required by the organism, and examined the following day. During incubation, the antibiotic in each disk diffuses into the agar, so that the medium circumferential to the disks contains active concentrations of drugs. Meantime, the organism seeded across the plate grows everywhere except in those areas where it encounters a sufficient concentration of an antibiotic capable of interfering with its growth activities. When the plate is examined next day, "zones of inhibition" may be seen around disks that contained antibiotics inhibitory to the organism; that is, such disks are surrounded by clear zones where no growth has occurred. The zones are obvious because they are rimmed by confluent growth beyond the areas of drug activity where the organism has grown uninhibited (Fig. 4–9). Disks containing antibiotics that did not affect the organism at all are surrounded by growth that extends right to their rims.

Several factors have an important influence on the results and must be considered. The medium must be supportive for the organism but have no effect on the antibiotic itself; the latter must be soluble, so that it can diffuse into the medium from the disk; and growth on the plate must be neither so abundant as to obscure any antibiotic effect nor so scant and spotty as to magnify the antagonistic action of the drug on any disk. Standardization of each factor involved, i.e., the amount and density of the inoculum, the medium, the activity and concentration of drug in each disk, and measurement of the diameter of each observed zone of inhibition, provides a degree of reliability of results that is not otherwise possible. A method of testing that specifies the techniques for standardization of each step of the procedure has been developed and is recognized as the model for accuracy and reproducibility of results. This method is known as the *Bauer-Kirby technique.* When alternate methods are used, they should be demonstrated to be at least as accurate and precise as the standard recommended technique.

When the Bauer-Kirby technique is followed, final measurements (these can be made with a ruler, caliper, or template) of the zones of inhibition around antibiotic disks determines whether the tested organism should be reported as susceptible (S), of *intermediate* (I) susceptibility, or resistant (R) to each tested drug. A reference table of zone sizes obtainable with individual antibiotics tested against standard types of bacterial strains is used. If the size of a given zone observed in the test is as large as, or larger than, that shown in the table for the given organism and antibiotic the result is recorded as susceptible. If there is definite inhibition but the zone is not as large as that expected for a completely susceptible strain of the organism, the result is interpreted as intermediate. If the zone is too small, or there is no evidence of any inhibition, the organism is reported as resistant to the drug in question.

When the variables in the disk agar diffusion method are suitably and carefully controlled, and zone sizes properly interpreted, the reported results provide the physician with an almost quantitative estimate of the response of the tested organism to each antibiotic in the battery of drugs used in the

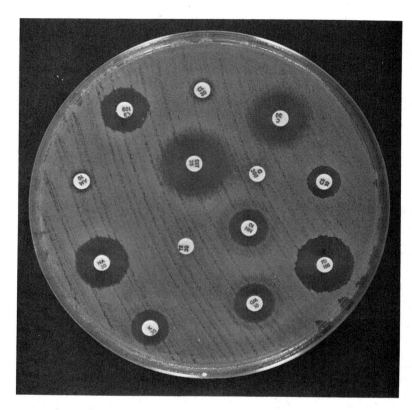

Fig. 4–9. An antibiotic susceptibility test performed by the Bauer-Kirby technique. The microorganism being tested was seeded across the entire plate before the disks impregnated with antibiotics were placed on the surface. Standardized procedures were followed to ensure reliable and reproducible results. After a standard incubation period the plate is now ready to read. Note the even confluence of growth of the test organism across the plate except in areas where it is inhibited by antibiotics diffusing from the disks into the adjacent medium. The clear areas surrounding many of the disks are zones of inhibition of microbial growth. Each zone must now be measured to determine whether the organism is susceptible (S), resistant (R), or intermediate (I) in its susceptibility to each tested drug by referring to a table of zone sizes obtained for standard bacterial strains tested against individual antibiotics. There are two disks on the plate in this photograph that are surrounded by confluent growth and show no zone of inhibition. The organism will be reported as resistant to the antibiotics on these two disks. Two other disks on this plate show very small zones of inhibition. Results for the latter may also be interpreted as resistant when these zone sizes are compared with those listed in the reference table for the same antibiotics tested against a standard strain of this type.

laboratory method. This battery is chosen on the basis of knowledge exchanged between microbiologists and physicians concerning clinically useful antimicrobial agents and the feasibility of using them in routine disk testing. The predictable responses of various microorganisms to certain antibiotics are also taken into account in setting up the battery.

When there is a good background of such collaborative understanding, the laboratory can provide the physician first with reliable information as to the microorganism(s) involved in the patient's disease, and then (or often simultaneously) with an accurate guide to the best choice of antimicrobial therapy.

Broth Dilution Method. Dilution in broth of the antibiotics being tested provides a more fully quantitative measurement of drug effect. In this method, a solution of antibiotic is added in varying concentration to tubes containing a broth nutrient. Each tube is then inoculated with the organism of interest, and incubated. The organism will grow only in those tubes containing a concentration of the drug that is too little to inhibit it. The *lowest* concentration of antibiotic that prevents visible growth in the nutrient is called the *minimum inhibitory concentration* (MIC). Antimicrobial agents can thus be titrated rather exactly for the most effective quantitative dose required to inhibit the organism in the test tube. The clinical value of this information lies in the reasonable prediction of the therapeutic dose needed to control the patient's infection.

This technique can also be applied to the serum, urine, or other body fluid of a patient being treated with a given antibiotic. The patient's specimen is added in varying amounts to tubes of broth nutrient, the organism being tested is added, and the tubes are incubated. If the specimen contains enough of the antibiotic that has been administered, it will inhibit the organism's growth. The concentration of antibiotic present in the specimen can be judged by the highest dilution of the test material that suppresses the organism.

Automated Methods. Numerous efforts have been made to develop automated methods for antibiotic susceptibility testing. One such method provided an electronic probe that "read" each plate for clear, inhibitory zones, measuring and recording individual zone sizes in a digital readout. The reliability of this method was greatly affected by minor fluctuations in the turbidity of the basal agar medium.

Recently, another principle for automation of antibiotic testing has been developed, and the instrument is now commercially available. Justification of its expense has been based on the reduction of labor involved for laboratories performing large numbers of daily susceptibility tests, and on a shortened time for reporting results. This instrument (Autobac I™) was developed by Pfizer Diagnostics (Fig. 4–10). It provides automated measurement of the susceptibility of each microorganism to as many as 12 antibiotic disks within three to five hours. A standardized amount of the organism is inoculated into a liquid nutrient, which is then distributed evenly into each of 13 chambers contained in a plastic cuvette. A different antibiotic disk is placed in each of 12 of the chambers, the thirteenth being a growth control. Many cuvettes can be prepared with different organisms and mounted in an incubator-shaker chamber of the machine. Agitation assures distribution of the antibiotic within each cuvette chamber and prevents bubble formation that would interfere with later readings. After three to five hours of incubation (depending on the organism), an optical device measures the amount of light scattered by the bacteria in each cuvette chamber. Bacteria growing uninhibited in an optically clear nutrient, as in the control chambers, cause increasing cloudiness and, thus, increasing light scattering. In chambers where inhibition of growth has occurred, light scattering is reduced in proportion to the degree of inhibition. The instrument measures the amount of light scattering in the growth control chamber; compares it with the readings obtained for each of the 12 antibiotic chambers; translates the readings into interpretations of resistance, intermediate susceptibility, or susceptibility; and prints out R, I, or S on the reporting ticket (preprinted with the names of the tested antibiotics) inserted for each cuvette. The limitations of this system include its expense, numerous delicate controls, and unavailability of some types of antibiotic disks for usage, the fact that it cannot be applied to slow-growing organisms or to anaerobes, and its space requirements.

Special Techniques

Quantitative Cultures. It is often important to know not only *what* organisms are present in a particular specimen, but also *how many.* This is because the mere presence of an organism in the specimen, even though it may be of a potentially pathogenic variety, does not necessarily mean that it is the cause of the patient's disease. It may simply be an organism normally present in the area of the body from which the specimen was derived, or a contaminant from the environment, but its occurrence in large numbers may indicate its significance.

This question is of particular interest in the case of

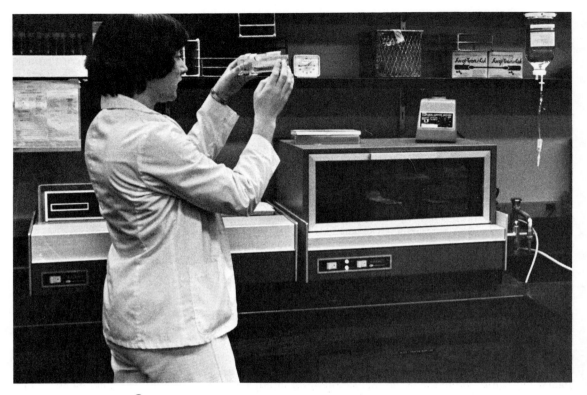

Fig. 4–10. Autobac I(TM) is an instrument that provides automated measurement of microbial susceptibility to antibiotics. The microbiologist is preparing a cuvette in which one microorganism can be tested against twelve different antibiotics. A number of such cuvettes can be used to test different organisms simultaneously. After 3 to 5 hours of incubation of the cuvettes in the shaker chamber on the right, the instrument measures, interprets, and prints out results. (Courtesy of Danbury Hospital, Danbury, Conn. Photo © Elizabeth Wilcox.)

urine samples. Specimens of urine for culture are ordinarily not collected directly from the patient's bladder through a sterile catheter but are voided into a sterile bottle. This means that they may contain, in small numbers, organisms normally found on the perineum. If, however, the sample is collected with a clean technique and reaches the laboratory quickly, before any organisms present in it have an opportunity to multiply directly in the urine, a numerical *count* of the bacteria growing out in culture offers a better evaluation of their significance than a mere report of their presence.

A quantitative culture is prepared by placing a measured quantity of the sample on one or more solid media, counting the number of colonies that appear subsequently, and translating the figure arithmetically to express the number of organisms present in 1 ml of sample. For example, if a loop standardized to deliver one one-hundredth of 1 ml of urine (0.01 ml) is used to inoculate the surface of an agar plate, and if this inoculum yields a total of 150 colonies, the count for the specimen would be 150×100, or 15,000 colonies in 1 full ml. If the specimen is turbid, appearing to contain large numbers of organisms that would be difficult to count by direct sampling, measured dilutions may be prepared and the plating done from these. It is usual to make three tenfold dilutions (1:10, 1:100, and 1:1000) in sterile water or saline and to plate out each in duplicate so that an average can be obtained. Plating can be done by either a pour plate or a streak plate method, and size of inoculum can be adjusted as desired. If larger quantities than the loop can deliver are to be inoculated, a sterile pipette is used to measure the amount sampled.

In the case of quantitative urine cultures, it is usually sufficient for the laboratory to report the approximate range of the count rather than an exact figure, such as "less than 10,000 organisms per milliliter" or "more than 10,000 organisms per ml," the latter being considered the lowest consistent criterion of active urinary infection. It cannot be overemphasized, however, that the reliability of a urine colony count depends on the adequacy of specimen collection and the speed with which the culture can be initiated. If a time lapse is unavoidable before the culture can be started, the specimen should be refrigerated to prevent bacterial multiplication in the interim.

Pathogenicity Tests. Other special methods may be employed in the microbiology laboratory when routine procedures provide inadequate information or inconclusive results. For example, it may be important to study an organism's pathogenicity, that is, its actual ability to produce disease in an experimental animal. Such a test may be indicated particularly if the organism has been isolated from a specimen in which one might normally expect to find nonpathogenic members of the same species. To demonstrate that the isolate in question is indeed pathogenic, the laboratory selects a suitable experimental host. The organism itself, or one of its toxic products, is injected and the consequences to the animal observed. Sometimes it is also necessary to infect a second animal as a control. Treatment of the control animal with an agent specific for the infectious agent being studied then follows. If the control animal survives but the untreated subject dies of the effects of the organism injected, the pathogenicity of the agent has been demonstrated. If neither animal dies or becomes ill, the test organism is reported as nonpathogenic. If the control animal dies, the test is considered invalid and must be repeated.

In some instances only one animal need be used in a pathogenicity test. This is true if the material to be tested can be injected locally into an area of skin or applied to the conjunctiva, the covering membrane of the eye. If a skin site is injected, an adjacent area can be used for the control treatment; in a conjunctival test, the other eye serves as the control.

Pathogenicity tests can also be conducted without using experimental animals. For example, the pathogenicity of the diphtheria bacillus depends on its ability to produce a toxic substance that causes the most serious symptoms of the disease. This toxin is an antigen and will react visibly with its antibody if the two are brought in contact. For this test, a strip of filter paper impregnated with serum containing specific antibody (antitoxin) is embedded in an agar plate, and an isolated strain of a suspected diphtheria bacillus is streaked across the agar plate at cross-angles to the strip. The diphtheria toxin is an exotoxin (see Chap. 3, p. 56). If produced by the test strain, it will diffuse into the agar medium from the area of growth along the streak and meet the antibody diffusing from the paper strip. A line of visible precipitation will be formed at the juncture (see Fig. 7–12, p. 171, and Chap. 13, p. 327), and the strain is reported to be pathogenic (toxigenic). Nonpathogenic strains of the diphtheria bacillus do not produce toxin and therefore show no lines of precipitation in this test. A known culture of a toxigenic strain of the diphtheria organism is always included to assure that the test and reagents are working properly.

Newer Developments. A number of special methods for demonstrating particular properties of interest in the identification of microorganisms have been developed in recent years. Some are of proven value while others are undergoing continuing evaluation, or modification, with respect to their practicality for the diagnostic laboratory. The final determinants for the laboratory in selection of appropriate methods, and their successful performance, include the nature of medical problems with which it is primarily involved (those of a general hospital differ from those of a specialty or research practice), the funds available for its facilities of space and equipment, and, most important of all, the training and skill of its personnel.

Radioimmunoassay. Radioimmunoassay (RIA) is a sensitive method for the detection and measurement of antigens in the serum component of circulating blood. The test antigen is exposed to antibody in a serologic or "immune" test and will combine with, or bind, it if it is specific for that antibody (see Chap. 7, p. 173). A radioactively labeled antigen known to be specific for the antibody is then added. If the antibody has been bound by the test antigen, it cannot combine with the labeled one, but if the test antigen was not specific, the antibody will be free to combine with the second antigen. A scanner measures the amount of radioactive antigen that may have been bound by antibody, thus indicating also how much, if any, unlabeled antigen first combined with and inactivated the antibody. RIA is used for the detection of barbiturate antigens in blood and, in microbiology, for identification of viral antigens (hepatitis, herpes) as well as bacterial toxins and capsular substances. Its value as a diagnostic tool in microbiology has been limited by the fact that some microbial antigens are shared by different organisms so that the test for them may not be specific enough for diagnosis of a particular disease. Conversely, sometimes more than one specific microorganism is involved in an infectious process so that RIA detection of one microbial antigen may not be conclusively diagnostic.

Radiometric Methods. Radiometric techniques in microbiology involve the use of radioactively labeled substrates, such as carbohydrates, that are metabolized by microorganisms. If glucose, for example, is labeled with a radioactive carbon isotope (that is, a carbon atom with an atomic weight of 14 rather than 12, emitting short-wave rays) and used as a bacterial nutrient, the CO_2 that is formed as a by-product of its metabolism will also have the radioactive carbon. A device sensitive to radiation then measures the release of the labeled CO_2, providing early detection of bacterial growth.

Radiometric methods have been used primarily for the detection of bacteria in blood cultures (Fig. 4–11), for which conventional methods can be time

Fig. 4–11. Radioactive glucose can be used as a bacterial nutrient in blood cultures. If bacteria are present, they will metabolize the glucose, releasing radioactive carbon dioxide. The instrument above has a device sensitive to radiation that measures the released labeled carbon dioxide and thus detects bacterial growth.

consuming and uncertain, and for limited bacterial identification based on changes in labeled substrates utilized in specific metabolic activities. They have also been applied to antibiotic susceptibility testing, where the degree of inhibition of glucose metabolism by antimicrobial drugs can be measured if the glucose is labeled with radioactive carbon.

The potential value of radiometric methods must be weighed against their expense and complexity and their relative insensitivity in detecting very small numbers of bacteria or metabolically inert microorganisms.

Gas-Liquid Chromatography. Chromatography is a process of separating two or more chemical compounds by allowing the mixture to pass through a solid (stationary) adsorbent such as silica, diatomaceous earth, paper, or a liquid, so that each compound is adsorbed and then released from the adsorbent at a rate related to its physical properties. In the earliest applications of the technique it was used to separate dyes and, later, colored plant pigments, hence the term *chroma*tography. Today its many uses in the separation of mixtures do not necessarily involve colors, but the term remains.

The principle of gas-liquid chromatographic separation of unknown components of a sample is their distribution between a stationary bed, or *phase,* of large surface area (in a coiled column) coated with a liquid, and a second mobile phase, which is an inert gas flowing through the stationary bed. Specific differences in the physical characteristics of individual components of the sample are reflected in differences in their affinity for the liquid phase. In other words, they are selectively retarded in the solvent, forming separate bands in the gas that then sweeps them beyond the stationary column into a "detector." In one type of detector, the bands are ionized in a hydrogen flame. A recorder indicates the degree of ionization of each band as it passes through, thus measuring the concentration of the ionized compound. The time it takes for each partitioned band to appear after the mixture is first injected is the retention time. The combined measurements of *ionization,* recorded in the form of a peak, and *retention time* characterize each separated component of the sample and permit its identification. The recorded result is called a chromatogram. Figure 4–12 is a diagram of a chromatograph and a chromatogram.

In microbiology, gas-liquid chromatography has been increasingly applied as a convenient tool for the identification of the organic acids produced by microorganisms in fermentations. Since fermentative pathways are often distinctive for different groups of organisms, identification of metabolic products can be very useful in distinguishing microbial species. The method has been particularly applied to the identification of anaerobic bacteria whose organic acid products identified in chromatograms may be as characteristic as fingerprints. The method is relatively inexpensive, the technical maneuvers are simple, and both speed and accuracy in the differentiation of anaerobes can be increased over conventional methods of culture. The chromatograph is not a casual diagnostic tool, however, and its use depends on the degree of sophistication obtainable or desirable within an individual laboratory.

Fig. 4–12. Diagram of the essential parts of a chromatograph (left) and of a chromatogram (right). Note that the chromatogram shows three components to be present in the tested mixture. Compound B is present in the largest concentration (highest peak); the amount of compound A is about half that of B; the concentration of compound C is about half that of A. Compound A has the shortest retention time and therefore probably the lowest molecular weight, whereas compound C has the longest retention time and the highest molecular weight of the three components.

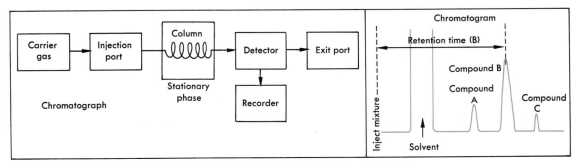

Use of Computers. Computer systems have found widespread use throughout clinical laboratories, particularly in large hospital centers. In the diagnostic microbiology laboratory, as in clinical chemistry or hematology, they provide a means for rapid transmission of laboratory results to patients' charts as well as a highly efficient system for retaining laboratory data (see Fig. 8-2, p. 196). Periodic tabulations of these data can be of great value in maintaining vigilance as to the types of microorganisms being isolated from patients within the hospital at large or in certain units, such as surgical, pediatric, or maternity wards. Similarly, a watch can be kept on the patterns of responses of various organisms to the antibiotics in common use within the hospital, so that any gradual or sudden emergence of resistant strains can be detected promptly. Early warnings of outbreaks of infections can also be obtained from computerized data, together with information as to possible origins of recurring infections from sources in the hospital or in the surrounding community.

These and many other potential applications of computer systems make them a valuable adjunct in the laboratory approach to patient care.

Identification of Pathogenic Organisms

The pathogenic organisms responsible for human diseases fall into five major categories: bacteria, viruses, fungi, protozoa, and helminths. Brief classifications of the medically important members of each group are presented in the following tables.

Bacteria. Tables 4-1 and 4-2 (pp. 91-97, 98) list the important bacterial pathogens, including the rickettsiae and chlamydiae. The classifications are taken from *Bergey's Manual of Determinative Bacteriology,* but the order of presentation and the descriptive terminology are the responsibility of the authors of this text. Major identifying features are shown for each group of pathogens in the tables.

Rickettsiae and Chlamydiae. The rickettsiae and chlamydiae are recognized as true bacteria, although they are much smaller in size and are obligate intracellular parasites. They have cell walls and internal structures like those of the larger bacteria. Their classification is shown in Table 4-2 (p. 98).

Viruses. Table 4-3 (p. 99) reviews the general methods of virus identification and presents a clinical and epidemiologic type of classification of those important in human disease.

Fungi. *Laboratory Identification of Fungi.* Fungi may be isolated from a variety of clinical specimens representing the focus of fungous infection (sputum, spinal fluid, pus aspirated from lymph nodes or other lesions, skin scrapings). They can often be visualized in wet mounts of such specimens. Stains are usually not required for such preparations, but potassium hydroxide solution is used to clear away tissue cells and debris, making the fungi more prominent. These are complex, differentiated organisms whose microscopic structures vary greatly in size and shape. They are identified by the morphology of their reproductive spores and by mycelial features. The fungi grow slowly in culture, producing large colonies whose characteristics are also useful in identification. Some are "dimorphic," or "diphasic," growing as yeasts at 35° C and as molds at room temperature. In addition to mycologic methods, the fungous diseases can sometimes be diagnosed by identification of the patient's antibodies or by skin tests with fungal antigens (see Chap. 7).

Table 4-4 (p. 100) reviews the morphologic features of the important pathogenic yeasts and fungi.

Protozoa. The protozoa associated with human disease are listed in Table 4-5 (pp. 101-102). The classification presented here groups them in the lowest phylum of the animal kingdom. Individual morphology is described in chapters dealing with the protozoan diseases.

Helminths. The parasitic worms are classified in Table 4-6 (pp. 102-103). The morphologic features of these organisms are described in those chapters where the diseases they cause are discussed.

Table 4–1. Medically Important Bacteria: Kingdom *Procaryotae*

Classification	Disease
Group: Gram-negative rods, *strictly aerobic*	
Family I. PSEUDOMONADACEAE	
Genus I: Pseudomonas	
P. aeruginosa	Burn, wound, and systemic infections
P. pseudomallei	Meliodosis
Genera of Uncertain Family Small, delicate rods; require special enriched culture media	
Genus Brucella	Brucellosis
Bordetella	
B. pertussis	Whooping cough
Francisella	
F. tularensis	Tularemia
Group: Gram-negative rods, *facultatively anaerobic*	
Family I. ENTEROBACTERIACEAE Motile or nonmotile; differentiated by carbohydrate fermentations and by antigenic properties; some commensalistic in intestinal tract with pathogenic potential for extraintestinal tissues; some are pathogenic parasites acquired through alimentary route	
Genus I: Escherichia	
E. coli	Epidemic diarrhea of infants; urinary tract and wound infections
Genus IV: Salmonella	
S. typhi	Typhoid fever
S. enteritidis	Gastroenteritis, paratyphoid, and enteric fevers (many varieties of this species may cause these diseases)
S. cholerae-suis	Septicemia, multiple abscesses
Genus V: Shigella	
Sh. dysenteriae	Bacillary dysentery
Sh. boydii, flexneri, and sonnei	Severe diarrheic disease
Genus VI: Klebsiella	
K. pneumoniae (Friedländer's bacillus)	Pneumonia

Table 4–1. *(Cont.)*

Classification	Disease
Genus VII: *Enterobacter*	
E. cloacae } *E. aerogenes* }	Normal intestinal tract; may cause urinary tract infections
E. agglomerans* (see *Erwinia*, Genus XII in family *Enterobacteriaceae*)	Has caused systemic infection when introduced in contaminated solutions injected intravenously
Genus IX: *Serratia*	
S. marcescens	Urinary and respiratory tract infections
Genus X: *Proteus*	Urinary tract and wound infections; infant diarrhea
Genus XI: *Yersinia*	
Y. pestis (formerly *Pasteurella pestis*)	Plague
Y. enterocolitica	Intestinal tract of animals and man; may cause diarrhea, mesenteric lymphadenitis, abscesses, meningitis
Genus XII: *Erwinia*	
E. herbicola (a plant pathogen) also recently classified as *Enterobacter agglomerans**	Has caused systemic infection when introduced in contaminated solutions injected intravenously

Family II. VIBRIONACEAE

Classification	Disease
Genus I: *Vibrio* (curved rods)	
V. cholerae	Cholera
V. parahaemolyticus	Marine environment, seafoods; may cause acute enteritis
Genus II: *Aeromonas*	Associated with systemic and intestinal infections

Genera of Uncertain Family

Classification	Disease
Genus *Haemophilus*	
H. influenzae	Respiratory tract infections; infant meningitis; conjunctivitis
H. ducreyi	Chancroid (a venereal disease)
Pasteurella	Systemic infections (rare)
P. multocida	Animal bite wound infections
Streptobacillus	
S. moniliformis	Rat bite fever, "Haverhill" fever
Calymmatobacterium	
C. granulomatis	Granuloma inguinale (a venereal disease)

*Ewing W. H., and Fife, M. A.: Enterobacter agglomerans, *the Herbicola-Lathyri Bacteria*. U.S. Department of Health, Education, and Welfare, Center for Disease Control, Atlanta, Ga., 1971.

Table 4–1. (*Cont.*)

Classification	Disease

Group: Gram-negative rods, *strictly anaerobic*

Family I. BACTEROIDACEAE

Genus I:	*Bacteroides*	Wound infections, rectal and brain abscess, endo-carditis
Genus II:	*Fusobacterium*	
	F. nucleatum	Vincent's angina (trench mouth), together with *Treponema vincentii*, a spirochete; anaerobic infections of head and respiratory tract
Genus III:	*Leptotrichia*	Normal oral mucosa
	L. buccalis	

Group: Gram-negative cocci and coccobacilli, *strictly aerobic*

Family I. NEISSERIACEAE

Genus I:	*Neisseria* (diplococci)	
	N. gonorrhoeae	Gonorrhea (a venereal disease)
	N. meningitidis	Meningitis
	N. lactamica	Occasional cause of meningitis, bacteremia
	N. sicca }*	Normal nasopharynx
	N. flavescens }	
Genus II:	*Branhamella* (diplococci)	
	B. catarrhalis (formerly placed in the genus *Neisseria*)	Normal nasopharynx; sometimes cause chronic upper respiratory infections, or meningitis
Genus III:	*Moraxella* (coccobacilli)	Conjunctivitis

Group: Gram-positive cocci; *aerobic and / or facultatively anaerobic; some strictly anaerobic*

Family I. MICROCOCCACEAE
Arranged in clusters or packets; saprophytic or parasitic; identified by coagulase, hemolysin, and pigment production

Genus II:	*Staphylococcus*	
	S. aureus	Skin abscesses; impetigo, wound infections, pneumonia, and other systemic infections; enterotoxin-producing strains cause food poisoning; exfoliative toxin causes "scalded skin" syndrome in infants

*These and other *Neisseria* species of the normal nasopharynx are often reported as *N. pharyngis*.

Table 4–1. (*Cont.*)

Classification	Disease
S. epidermidis	Normal commensal of skin but may cause local or systemic infections if introduced by contaminated sutures or instruments
Family II. STREPTOCOCCACEAE Chaining cocci; commensalistic and/or pathogenic	
Genus I: Streptococcus	
S. pneumoniae (formerly Diplococcus pneumoniae)	Pneumonia, other systemic infections
S. pyogenes (usually beta-hemolytic)	Scarlet fever, sore throat, erysipelas, impetigo, puerperal fever, wound and burn infections, and other serious diseases
S. faecalis ("enterococci")	Normal intestinal tract; sometimes cause respiratory and urinary tract infections, subacute endocarditis
S. bovis	Intestinal tract of cows and other animals; causes human endocarditis
S. agalactiae	Mastitis in cows; in humans, neonatal septicemia and meningitis (infected mothers may also have serious systemic disease)
Viridans group (alpha hemolytic species that produce greening of blood agar); no species classifications	Normal respiratory tract; chronic mucosal infections; subacute endocarditis
S. mitis	Normal respiratory tract; may cause endocarditis, dental caries
S. salivarius	Normal oral mucosa
S. mutans	Major cause of dental caries; may cause endocarditis
Family III. PEPTOCOCCACEAE Cocci in chains or clusters, commensalistic and/or pathogenic; strictly anaerobic	
Genus I: Peptococcus	Abscess of skin, respiratory and intestinal tracts
Genus II: Peptostreptococcus	Abscess of respiratory and intestinal tracts
Group: Gram-positive, *endospore-forming* rods	
Family I. BACILLACEAE	
Genus I: Bacillus (aerobic)	
B. anthracis	Anthrax
B. subtilis	Common saprophyte

Table 4–1. *(Cont.)*

Classification	Disease
Genus III: *Clostridium* (anaerobic)	
Cl. botulinum	Botulism
Cl. tetani	Tetanus
Cl. perfringens ⎱	
Cl. novyi	
Cl. septicum	Gas gangrene
Cl. histolyticum ⎰	

Group: Gram-positive, *nonsporing* rods

Family I. LACTOBACILLACEAE

Genus I: *Lactobacillus*	Normal oral, intestinal, and genitourinary mucosa
Genera of Uncertain Family	
Genus *Listeria*	
L. monocytogenes	Congenital infection of infants; meningitis; bacteremia in debilitated patients
Genus *Erysipelothrix*	
E. rhusiopathiae	Swine erysipelas; in humans, lesions of skin of hands (erysipeloid), septicemia, endocarditis

Group: Gram-positive rods; some acid-fast, some filamentous; some form reproductive spores (not endospores); some show true branching; saprophytes and parasites

Coryneform Group

Genus I: *Corynebacterium* (non-acid-fast, nonbranching, beaded)	
C. diphtheriae	Diphtheria
C. pseudodiphtheriticum ("diphtheroids")	Normal nasopharynx, skin commensal
C. vaginale	Vaginitis and urethritis

Order I. ACTINOMYCETALES

Family I. ACTINOMYCETACEAE

Genus I: *Actinomyces* (filamentous branching rods, non-acid-fast)	
A. israelii	Actinomycosis

Table 4-1. *(Cont.)*

Classification	Disease
Family II. MYCOBACTERIACEAE	
Genus I: *Mycobacterium* (mostly non-branching, acid-fast rods)	
M. tuberculosis	Tuberculosis
M. bovis	Tuberculosis of cattle, humans
Mycobacteria of Runyon Groups I-IV: (*M. kansasii, M. intracellulare M. avium* and others)	Disease resembling tuberculosis
M. marinum	Skin nodules or ulcers ("swimming pool granulomas")
M. leprae (Hansen's bacillus)	Leprosy
Family VI. NOCARDIACEAE	
Genus I: *Nocardia* (filamentous, branching, some weakly acid-fast)	
N. asteroides	
N. brasiliensis	Nocardiosis
Family VII. STREPTOMYCETACEAE	
Genus I: *Streptomyces*	Soil organisms: many produce antibiotic substances such as tetracyclines, chloramphenicol, streptomycin
Group: *Spiral* bacteria	
Family I. SPIRILLACEAE	
Genus I: *Spirillum*	
Sp. minor	Rat bite fever
Genus II: *Campylobacter*	
C. fetus	Cause of abortion in cattle; in humans, bacteremia, endocarditis, enteritis
C. sputorum	Normal human mouth
Group: *Coiled* bacteria; long, slender, flexible; rotate on long axis with corkscrew motion; also undulate; not easily cultivated on cell-free media; saprophytic and parasitic	

Table 4–1. *(Cont.)*

Classification	Disease
Order I. SPIROCHAETALES	
Family I. SPIROCHAETACEAE	
Genus III: Treponema (even coils)	
T. pallidum	Syphilis (a venereal disease)
T. pertenue	Yaws (nonvenereal)
T. vincentii (synonym of Borrelia vincentii; see Borrelia, Genus IV in family Spirochaetaceae)	Normal oral mucosa; associated with fusobacteria in Vincent's angina (trench mouth)
Genus IV: Borrelia (loose coils)	
B. recurrentis	Relapsing fever
B. vincentii (synonym of Treponema vincentii)	Normal oral mucosa; associated with fusobacteria in Vincent's angina (trench mouth)
Genus V: Leptospira (tight coils)	
L. interrogans (formerly L. icterohemorrhagiae)	Infectious jaundice (icterohemorrhagia or Weil's disease); ''aseptic'' meningitis
Group: Bacteria that *lack cell walls;* extremely pleomorphic; filterable; require complex cell-free media or grow in tissue culture or embryonated eggs	
Order I. MYCOPLASMATALES	
Family I. MYCOPLASMATACEAE	
Genus I: Mycoplasma	
M. pneumoniae	Primary atypical pneumonia
M. hominis and others	Normal mucosa; may cause urethritis, cervicitis, arthritis; pleuropneumonia in cattle

Classification	Disease

Group: *Obligate, intracellular* bacteria (larger than true viruses); many require arthropod hosts; grow in tissue culture, embryonated eggs, or experimental animals; require special stains for light microscopy; also identified by serologic methods

Order I. RICKETTSIALES

Family I. RICKETTSIACEAE

Genus I: Rickettsia
Arthropod Vector:

Louse	*R. prowazeki*	Epidemic typhus fever
Flea	*R. typhi*	Endemic murine typhus
Tick	*R. rickettsi*	Rocky Mountain spotted fever
Tick	*R. conori*	Boutonneuse fever
Mite	*R. tsutsugamushi*	Scrub typhus (tsutsugamushi)
Mite	*R. akari*	Rickettsialpox

Genus III: Coxiella
Tick (but this agent not dependent on arthropod) *C. burnetii* Q fever

Family II. BARTONELLACEAE

Genus I: Bartonella
Sandfly *B. bacilliformis* (pleomorphic coccobacilli; parasitize erythrocytes as well as tissue cells) Oroya fever, verruga peruana

Order II. CHLAMYDIALES

Family I. CHLAMYDIACEAE

No arthropod hosts; small coccoid microorganisms; form inclusions within a membrane in cytoplasm of parasitized host cells; can be stained with aniline dyes (Gram-negative); formerly called ''large viruses'' or *Bedsonia*

Genus I: Chlamydia

Ch. trachomatis Trachoma; lymphogranuloma venereum (LGV, a venereal disease); inclusion conjunctivitis (''swimming pool'' conjunctivitis); urethritis; pneumonia in infants

Ch. psittaci Psittacosis, ornithosis

Table 4–3.   Identification and Classification of Viruses

Identification

Viruses are small, filterable, transmissible agents of infectious disease. They are obligate intracellular parasites, cultivable only in living cells (tissue culture, embryonated eggs, or animals). Electron microscopy is required for visualization of individual viral particles. Some viruses form aggregates within parasitized cells, and these inclusion bodies may be seen with the light microscope when they are stained by special techniques. The classification of the viruses is based on their chemical composition, size, morphology, and other physical and chemical properties (see Chap. 2).

For the convenience of those whose primary interest lies in the clinical picture of viral infections, the following grouping relates the viruses of medical importance to either their route of entry, arthropod transmission, or type of disease produced:

Clinical and Epidemiologic Classification

Respiratory Viruses
Influenza virus
Parainfluenza viruses
Adenoviruses
Rhinoviruses
Respiratory syncytial
  (RS virus)
Mumps virus

Enteroviruses
Poliomyelitis virus
Coxsackie viruses
ECHO viruses
Infectious hepatitis virus

Poxviruses
(Dermatropic)
Smallpox (variola virus)
Cowpox virus
Vaccinia virus

Herpesviruses
(Dermatropic)
Chickenpox (varicella virus)
Herpes zoster virus
Herpes simplex virus

Exanthem Viruses
(Dermatropic and viscerotropic)
Measles (rubeola virus)
German measles (rubella virus)

CNS Virus
(Neurotropic)
Rabies virus

Arboviruses
(Arthropod-borne)
(Viscerotropic)
Yellow fever virus
Dengue fever virus
Colorado tick fever virus
Sandfly fever virus
(Neurotropic)
Eastern equine encephalitis
  virus
Western equine encephalitis
  virus
St. Louis encephalitis virus
Japanese B encephalitis virus

Table 4–4. Important Pathogenic Fungi

Organisms	Morphologic Features	Diseases
Yeasts or Yeastlike	Yeasty soft colonies	
Cryptococcus neoformans	Encapsulated budding cells	Pneumonia, meningitis, other tissue infections
Candida albicans	Budding cells, pseudomycelium, and chlamydospores	Skin and mucosal infections, sometimes systemic
Systemic Fungi		
Histoplasma capsulatum	In tissues, intracellular and yeastlike In culture at 37° C, a yeast In culture at room temperature, a mold with characteristic chlamydospores	Histoplasmosis is primarily a disease of the lungs; may progress through the reticuloendothelial system to other organs
Coccidioides immitis	In tissues, produces spherules filled with endospores In culture, a cottony mold with fragmenting mycelium	Coccidioidomycosis is usually a respiratory disease; may become disseminated and progressive
Blastomyces dermatitidis	In tissues, a large thick-walled budding yeast In culture at 37° C, a yeast In culture at room temperature, a mold	North American blastomycosis is an infection that may involve lungs, skin, or bones
Paracoccidioides brasiliensis	In tissues, a large yeast showing multiple budding In culture at 37° C, a multiple budding yeast In culture at room temperature, a mold	Paracoccidioidomycosis (South American blastomycosis) is a pulmonary disease that may become disseminated to mucocutaneous membranes, lymph nodes, or skin
Subcutaneous Fungi		
Sporothrix schenckii	In tissues, a small Gram-positive, spindle-shaped yeast In culture at 37° C, a yeast In culture at room temperature, a mold with characteristic spores	Sporotrichosis is a local infection of injured subcutaneous tissues, and regional lymph nodes
Cladosporium Fonsecaea Phialophora	In tissues, dark, thick-walled septate bodies In culture, darkly pigmented molds	Chromoblastomycosis is an infection of skin and lymphatics of the extremities caused by any one of several species
Madurella Allescheria Phialophora	Tissue and culture forms vary with causative fungus	Mycetoma (maduromycosis, madura foot) is an infection of subcutaneous tissues, usually of the foot, caused by any one of several species
Superficial Fungi		
Microsporum species Trichophyton species Epidermophyton floccosum	These fungi grow in cultures incubated at room temperature as molds, distinguished by the morphology of their reproductive spores	Ringworm of the scalp, body, feet, or nails

Table 4–5. The Pathogenic Protozoa*

Kingdom: ANIMALIA

Phylum: **PROTOZOA**

Unicellular organisms; cell membrane; well-organized nucleus enclosed in a nuclear membrane

Subphylum: **MASTIGOPHORA**

Class: **ZOOMASTIGOPHOREA**

Flagellated organisms; reproduce by longitudinal binary fission; two groups of medical importance:
Flagellates of the intestinal and genital mucosa.
—*Giardia lamblia* (sometimes causes a mild diarrheal disease)
—*Trichomonas hominis* (nonpathogenic inhabitant of the intestinal tract)
—*Trichomonas vaginalis* (an agent of vaginal pruritus)
Flagellates of the blood and tissues (arthropod-borne)
—*Leishmania donovani* (the agent of *kala-azar,* a visceral leishmaniasis)
—*Leishmania tropica* (the agent of *Oriental sore,* a cutaneous infection)
—*Leishmania braziliensis* (the agent of *espundia,* a mucocutaneous disease)
—*Trypanosoma gambiense* (the agent of *''African sleeping sickness,''* an encephalitis)
—*Trypanosoma rhodesiense* (another agent of *African sleeping sickness*)
—*Trypanosoma cruzi* (the agent of *Chagas' disease,* a systemic infection)

Subphylum: **SARCODINA**

Class: **RHIZOPODEA**

Move by means of *pseudopodia* (fingerlike projections of cytoplasm); reproduce asexually, frequently pass into cyst stages
—*Entamoeba histolytica* (the agent of amebic dysentery)
—*Entamoeba coli* (nonpathogenic inhabitant of the intestinal tract)
—*Iodamoeba butschlii* (nonpathogenic inhabitant of the intestinal tract)
—*Dientamoeba fragilis* (nonpathogenic inhabitant of the intestinal tract)

Subphylum: **SPOROZOA**

Possess no means of locomotion; reproduce by alternating sexual multiplication (*sporogony*) and asexual multiplication (*schizogony*); parasitic forms only

Class: **TELOSPOREA**. *Subclass:* **HAEMOSPORINA**

Parasitize erythrocytes and tissue cells of man and animals
—*Plasmodium vivax*
—*Plasmodium malariae*
—*Plasmodium falciparum*   The agents of *malaria*
—*Plasmodium ovale*
—*Plasmodium knowlesi*

Class: **TOXOPLASMEA**: Reproduce by binary fission or sporogony
—*Toxoplasma gondii*
—*Pneumocystis carinii*   Intra- and extracellular parasites that sometimes
—*Sarcocystis lindemanni*   cause human disease

Table 4-5. *(Cont.)*

Kingdom: ANIMALIA *(Cont.)*

    Subphylum: **CILIOPHORA**

        Locomotion is by means of cilia; possess a macronucleus and a micronucleus; reproduce by fission or by sexual conjugation; may be free-living or parasitic

      Class: CILIATEA

        Balantidium coli (produces ulcers in colon)

*Selected from Classification of Animals Which Parasitize Man, Produce Venenation, or Serve as Vectors of Human Pathogens, in *Craig and Faust's Clinical Parasitology,* by Ernest C. Faust, Paul F. Russell, and Rodney C. Jung, 8th ed. Lea & Febiger, Philadelphia, 1970, pp. 20–28.

Table 4-6. The Pathogenic Helminths*

Kingdom: ANIMALIA

    Phylum: **PLATYHELMINTHS** *(Flatworms)*

        Multicellular invertebrate animals; bilaterally symmetrical but lack a body cavity; usually flattened dorsoventrally

      Class: **TREMATODA** *(Flukes)*

        Parasitic worms; body unsegmented; hermaphroditic reproduction (usually)

        Family: SCHISTOSOMATIDAE

          The only bisexual trematodes; life cycle requires vertebrate and aquatic hosts in sequence

          — *Schistosoma mansoni*
          — *Schistosoma haematobium* }   Agents of schistosomiasis (visceral infestation)
          — *Schistosoma japonicum*

        Species from Other Trematode Families

          Hermaphroditic flukes; life cycles require vertebrate and aquatic hosts in sequence

          — *Fasciola hepatica* (liver fluke)
          — *Fasciolopsis buski* (intestinal fluke)
          — *Paragonimus westermani* (lung fluke)
          — *Clonorchis sinensis* (liver fluke)

      Class: **CESTOIDEA** *(Tapeworms)*

        Parasitic, hermaphroditic, segmented worms; head or *scolex* equipped with suckers, sometimes with hooklets as well; body of adult, or *strobila,* is a chain of sexually complete units, called *proglottids*

        Order: CYCLOPHYLLIDEA (4 suckers on scolex; embryo mature when egg is laid)

          Family: TAENIIDAE (Definitive and intermediate vertebrate hosts, in sequence)

          — *Taenia saginata* (beef tapeworm)
          — *Taenia solium* (pork tapeworm)
          — *Echinococcus granulosus* (agent of *hydatid* disease)

Table 4–6. *(Cont.)*

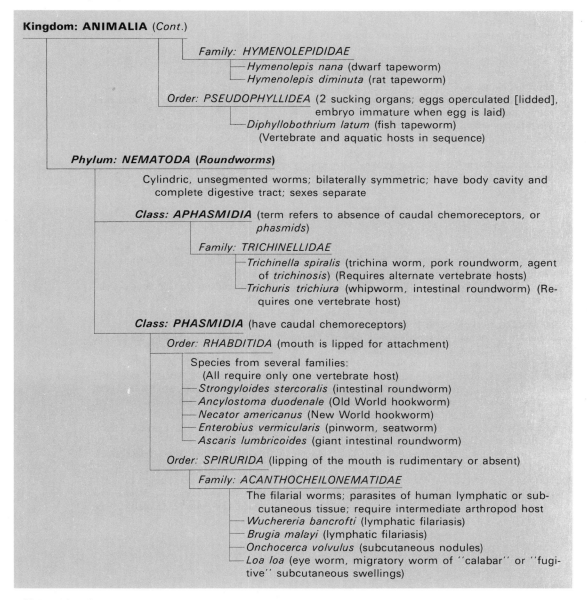

Kingdom: **ANIMALIA** *(Cont.)*

Family: HYMENOLEPIDIDAE
— *Hymenolepis nana* (dwarf tapeworm)
— *Hymenolepis diminuta* (rat tapeworm)

Order: PSEUDOPHYLLIDEA (2 sucking organs; eggs operculated [lidded], embryo immature when egg is laid)
— *Diphyllobothrium latum* (fish tapeworm)
(Vertebrate and aquatic hosts in sequence)

Phylum: **NEMATODA (Roundworms)**

Cylindric, unsegmented worms; bilaterally symmetric; have body cavity and complete digestive tract; sexes separate

Class: **APHASMIDIA** (term refers to absence of caudal chemoreceptors, or *phasmids*)

Family: TRICHINELLIDAE
— *Trichinella spiralis* (trichina worm, pork roundworm, agent of *trichinosis*) (Requires alternate vertebrate hosts)
— *Trichuris trichiura* (whipworm, intestinal roundworm) (Requires one vertebrate host)

Class: **PHASMIDIA** (have caudal chemoreceptors)

Order: RHABDITIDA (mouth is lipped for attachment)

Species from several families:
(All require only one vertebrate host)
— *Strongyloides stercoralis* (intestinal roundworm)
— *Ancylostoma duodenale* (Old World hookworm)
— *Necator americanus* (New World hookworm)
— *Enterobius vermicularis* (pinworm, seatworm)
— *Ascaris lumbricoides* (giant intestinal roundworm)

Order: SPIRURIDA (lipping of the mouth is rudimentary or absent)

Family: ACANTHOCHEILONEMATIDAE

The filarial worms; parasites of human lymphatic or subcutaneous tissue; require intermediate arthropod host
— *Wuchereria bancrofti* (lymphatic filariasis)
— *Brugia malayi* (lymphatic filariasis)
— *Onchocerca volvulus* (subcutaneous nodules)
— *Loa loa* (eye worm, migratory worm of ''calabar'' or ''fugitive'' subcutaneous swellings)

*Selected from Classification of Animals Which Parasitize Man, Produce Venenation, or Serve as Vectors of Human Pathogens, in *Craig and Faust's Clinical Parasitology,* by Ernest C. Faust, Paul F. Russell, and Rodney C. Jung, 8th ed. Lea & Febiger, Philadelphia, 1970, pp. 20–28.

Questions

1. Name the parts of a light microscope and give the function of each.
2. How is the total magnification determined? What is resolution?
3. Describe briefly how an object viewed with an electron microscope is enlarged.
4. Why is a wire inoculating loop used to transfer microorganisms in the laboratory?
5. Briefly describe the Gram stain. Why is it a differential stain?
6. Why is careful preparation of laboratory glassware and media important? Describe some of the uses of culture media.
7. What factors involved in specimen collection can influence the results of culture in the laboratory?
8. Outline the flow of work in the microbiology laboratory in handling a specimen submitted for "smear, culture, and susceptibility."
9. Describe the Bauer-Kirby technique. Why is it necessary to perform antibiotic susceptibility tests on pure cultures of isolated organisms?
10. What major reference do microbiologists rely on for the classification of microorganisms?

Additional Readings

Alpert, Nelson L.: Pfizer Autobac 1™. Instrument Series: Report #42. *Lab World,* April 1976.

Anaerobic Bacteria and Disease. The Upjohn Co., Kalamazoo, Mich., 1975.

Approved Standard. *Performance Standards for Antimicrobial Disc Susceptibility Tests.* National Committee for Clinical Laboratory Standards, Los Angeles, Calif., 1975.

Bauer, A. W.; Kirby, W. M. M.; Sherris, J. C.; and Turck, M.: Antibiotic Susceptibility Testing by a Standardized Single Disc Method. *Am. J. Clin. Pathol.,* **45**:493, 1966.

Buchanan, R. E., and Gibbons, N. E. (eds.): *Bergey's Manual of Determinative Bacteriology,* 8th ed. The Williams & Wilkins Co., Baltimore, Md., 1974.

Crewe, Albert: A High-Resolution Scanning Electron Microscope. *Sci. Am.,* **224**:26, April 1971.

Davis, Larry E.; Caldwell, G. C.; and Hess, H.: Test Tube Epidemic. *Am. J. Nurs.,* **70**:2139, Oct. 1970.

Everhart, Thomas F., and Hayes, T. L.: The Scanning Electron Microscope. *Sci. Am.,* **226**:54, Jan. 1972.

*Lennette, Edwin H.; Spaulding, E. H.; and Truant, J. P. (eds.): *Manual of Clinical Microbiology,* 2nd ed. American Society for Microbiology, Washington, D.C., 1974.

Lorian, Victor (ed.): *Significance of Medical Microbiology in the Care of Patients.* The Williams & Wilkins Co., Baltimore, Md., 1977.

Morello, Josephine A., and Graves, M. H.: Clinical Anaerobic Bacteriology. *Lab. Management,* **15**:20, April 1977.

*Schneierson, S. Stanley: *Atlas of Diagnostic Microbiology.* Abbott Laboratories, North Chicago, Ill., 1974.

*Wilson, Marion E.; Weisburd, M. H.; Mizer, H. E.; and Morello, J. A.: *Laboratory Manual and Workbook in Microbiology: Applications to Patient Care,* 2nd ed. Macmillan Publishing Co., Inc., New York, 1979.

*Available in paperback.

Section II
Character of Microbial Diseases

Man and Microbes in Coexistence

Ecology

An outstanding feature of life in earth's biosphere is interdependency. As John Donne said of the human condition "No man is an island unto himself. . . ," so we can say that no living organism can survive without support systems provided by other forms of life. The science of *ecology* is concerned with the biologic interrelationships that exist in the natural world, linking organisms with each other and with their environments.

An understanding of ecologic relationships requires study of the environmental pressures that may influence the coexistence of biologic species or affect the direction of their evolution. As the human species has developed its enormous technologic and cultural capabilities, it has created many new pressures on the natural world. Steadily increasing human populations and their needs have introduced many complexities into the maintenance of fragile ecologic balances.

The kind of life a given geographic area can support depends on what it can offer in the way of food and water, as well as on the range of its atmospheric conditions. The influence of temperature, humidity, and altitude on terrestrial life is most obvious when one considers the extremes of conditions prevailing in deserts or mountainous regions, polar or tropical areas, coastal land or inland plains. The flora and fauna of such regions differ to an extreme that reflects their ability to utilize available nutrient under conditions that prevail. Aquatic life is subject, in the same way, to regional variations in temperature, salt concentration, oxygen supply, and many other factors. The life of a freshwater lake or mountain stream is vastly different from that of an ocean, with its high concentration of sodium chloride and other salts, just as the life of a mountain slope differs from that of the valley below. Between extremes, from the Himalayas to the jungles of Peru, from the Arctic Ocean to the Sargasso Sea, many subtler variations occur in plant and animal populations sensitive to minor differences within regions of the same general type, to factors introduced by man, and to pressure created by shifts in the balance of such populations as they respond differently to environmental changes.

Microbial Ecology

The ability of any given area to support an animal and plant population is dependent upon the activities of microorganisms in soil. The decaying, putrefactive, and fermentative processes of these microorganisms release vital chemical elements, such as nitrogen, carbon, potassium, and phosphorus, from the organic components of dead material in the area. The released elements are essential to living plants for synthesis and growth, and the plants, in turn, are a source of food for animals. When animals and plants die, the soil microorganisms again break down their complex organic substances, thus perpetuating the cyclic turnover of matter and energy.

The populations of microorganisms residing together in the soil are themselves dependent on the available food material, the temperature, oxygen concentration, and pH of the soil, and the metabolic products of microbial neighbors. As we have seen, these metabolic wastes are sometimes antagonistic (*antibiotic*), sometimes beneficial to adjacent species. When the food or water content of soil fluctuates, microbes must adjust to the new conditions or be eliminated. As acids, carbon dioxide, and other products of microbial activity accumulate, further adjustments must be made. Some organisms may flourish under the changed conditions, while the activity of others may be diminished. Marked and lasting shifts in microbial ecology may produce resultant changes for the higher life of the area. Because their growth periods are much longer than those of microorganisms, higher plants and animals may suffer severe and lasting damage if their microscopic chemists, cooks, and purveyors dwindle off, or vanish from the scene. Without the saprophytic bacteria, yeasts, and molds of the soil — the "chefs of the underworld" — to transform the substance of dead organisms into elements essential for the living, our world would become cluttered with death, and life would cease altogether.

Ecologic Patterns

The biologic interrelationships of living things are recognized in four general patterns, scaled in terms of the dependencies inherent in their associations. At one end of the scale, there may be complete *independence* between two different kinds of organisms living side by side but indifferent to each other. The other three patterns each involve a close relationship in which some type of interdependency is displayed between two diverse types of organisms. These associations are described collectively by the Greek word *symbiosis,* meaning literally "a living together." The dependency may be mutual, in that each organism derives some benefit, trivial or vital, from the other, the pattern being known as *mutualism.* (In common practice, the terms *symbiosis* and *mutualism* are often used interchangeably.) If only one of the two associates is benefited by the relationship, while the other is unaffected by it, the term *commensalism* is used, and the partners are referred to as *commensal* organisms. In the third type of association, one organism derives some degree of support from another at the latter's expense, this being the pattern of *parasitism.* These concepts are illustrated in Figure 5-1, *A–D.*

Mutualism

Mutualistic relationships exist among microorganisms cohabiting the human body and between these and their host. Bacteria living in the intestinal tract receive their primary support from food materials present there. Some species may benefit each other by digesting or synthesizing substances important to both, according to their alternate capacities. In turn, some of the products synthesized by certain bacterial species are of use to the human host, who can absorb them from the intestine and put them to use. The bacterial production of vitamin K is an example of this kind, the vitamin being essential to man for synthesis of a component involved in blood clotting. Humans derive vitamin K from dietary sources, however, and are not ordinarily dependent upon intestinal bacteria for its synthesis.

The mutualism that exists between nitrogen-fixing bacteria and certain plants has been alluded to pre-

Fig. 5-1. Biologic relationships. Using fruit pickers for an analogy, the diagrams above illustrate the following concepts: *1. Independence:* each of two fruit pickers easily meet their daily requirement working independently of each other on different trees (or the same tree). *2. Mutualism:* two pickers working together on the same tree can increase their yields by combining efforts to harvest the best areas of the tree. *3. Commensalism:* the commensalist enjoys the fruit of his working partner's labor without making a contribution of his own, but also without significantly affecting the partner's output. *4. Parasitism:* the parasite reaps all the benefit at the expense of his working partner, who is exhausted and cannot maintain even his normal quota.

viously (pp. 48 and 53). Nitrogen is an essential element for protein synthesis. Although it is abundant in the earth's atmosphere, elementary nitrogen is nutritionally useless to most living organisms until it is "fixed" or combined with other elements such as oxygen or hydrogen. An inadequate supply of usable nitrogen is a serious limitation to the growth of plants and animals, and its scarcity for agriculture is an important factor in today's global problem of hunger. Industrial nitrogen fixation, for the production of chemical fertilizers, is accomplished through the manufacture of ammonia from nitrogen and hydrogen. This process, however, demands large sources of energy, notably from natural gas, and is thus too expensive in every way to satisfy the world's growing needs. Since the largest supply of fixed nitrogen is of biologic origin, attention has been focused recently on the nitrogen-fixing bacteria in an effort to clarify and improve their natural role in the chain of nutritional resources.

Only a few genera of bacteria (including some of the blue-green algae) are capable of fixing nitrogen. Some can do this only through their mutualistic associations with plants (probably the best-known relationship is between leguminous plants and bacteria of the genus *Rhizobium*) whereby they manufacture ammonia at the starting point of the natural nitrogen cycle (see Fig. 3–2, p. 53). The bacterial enzyme nitrogenase, which mediates the fixation of nitrogen with hydrogen, is now being studied intensively. Knowledge of its structure, genetic regulation, and possible transfer to other microorganisms, its properties and activities, may lead to major improvements in agriculture. If additional varieties of plants and bacteria could be genetically manipulated to support each other in the synthesis of nitrogenase (controlled in some bacteria by a gene that can be experimentally introduced into another species via plasmid transfer), natural fixation could be greatly increased. Similar approaches to other biochemical mechanisms that regulate bacterial enzymes and activities involved in the nitrogen cycle are being studied. The potential benefits of such research lie in increasing the world's food supply.

In converse situations, the antagonisms exerted by microorganisms upon each other in the natural world are also of importance to plant, animal, and human life because they serve to control microbial populations and prevent overgrowths of species that might be harmful in large numbers (Fig. 5–2). The demonstration of particular mutualisms and antagonisms among organisms holds many implications for man in his efforts to control his environment to his own advantage. The use of antibiotic or chemical substances in the control of infectious diseases; deliberate introduction into the body of one species of organism that may supplant or control the growth of others; or a specific attack on one member of an interdependent pair to destroy the other indirectly—all these are approaches that may have value. There is an ongoing interest in the relative merits of these methods in the control of insect pests as well as of microorganisms. For example, it has been shown that termites have a mutualistic relationship with certain flagellated protozoans that live in their intestinal tracts (Fig. 5–3). When the termite has foraged successfully for a woody meal, the protozoa in its intestine produce enzymes that digest cellulose fibers, thus benefiting the termite. In return, the protozoa are housed and fed; neither can get along without the other. If the flagellates are eliminated from this association, the termites die within 10 to 20 days; when the flagellates are reinstated, the insects survive. These facts suggest that it may be possible to control the insect by measures designed to eliminate its mutualist companion.

Commensalism

Commensalism is a very common type of association between man and his microbial populations. Organisms commensal with man are supported by him but do not affect his welfare one way or the other. Commensalism differs from parasitism in that the microorganism derives support from nonliving organic substances on the host's surfaces without offering harmful competition to living cells and tissues of the body. On the other hand, the relationship offers no benefit to the host, who remains indifferent to it.

Such relationships are subject to change with changing opportunities for the microorganism. Many commensals living harmlessly on the body's surfaces have a capacity for parasitism, if circumstances permit a competitive association with the body's tissues,

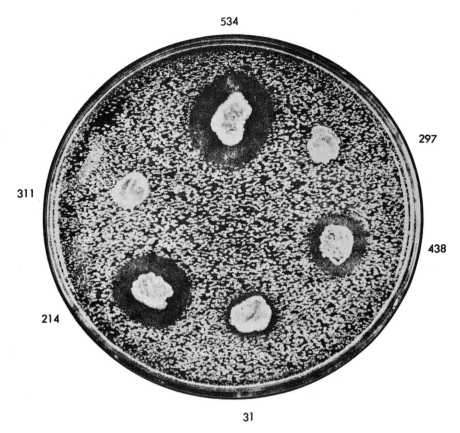

534

297

311

438

214

31

Fig. 5–2. Antagonism of coliform bacteria against shigellae. The entire surface of an agar plate was inoculated with *Shigella paradysenteriae.* After drying, several points on the surface of the agar were inoculated with different cultures of coliform organisms isolated from human stools. Coliform strains 297 and 311 did not affect the growth of *Sh. paradysenteriae.* Coliform strain 31 exerted slight antagonism, but strains, 534, 438, and 214 produced large zones of inhibition of the growth of this species of *Shigella.* (Reproduced from Halpert, S. P.: The Antagonism of Coliform Bacteria Against Shigella. *J. Immunol.,* **58**:153, 1948.)

just as parasitic organisms may adjust to a more independent life. Microorganisms may also display commensal relationships, as shown in Figure 5–4.

Parasitism

Parasitism implies some degree of harmful effect induced in the host by an organism that itself bene-fits from the association. When a parasite actually lives in the tissues or cells of the host, it is called an *endoparasite,* as in the case of the hookworm (Fig. 5–5) and other helminths or microbial parasites. Large organisms (ticks or lice) that attach themselves to the skin or outer mucosal surfaces of a human or animal host, and take their nourishment from living host tissue or blood, are called *ectoparasites.* The period of attachment may be quite temporary, as in

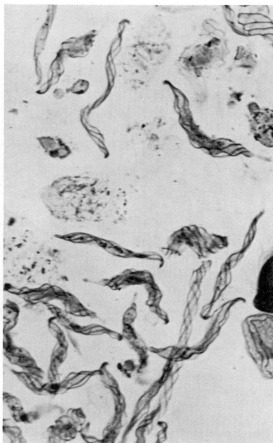

A
B

Fig. 5–3. The termite and its intestinal protozoan partner provide an excellent example of *mutualism.* The termite (*A*) lives on a diet of wood. The protozoa (*B*) in the insect's intestinal tract secrete enzymes for the digestion of cellulose (in the photomicrograph, the protozoa are seen as slender, undulant flagellates in a smear of the termite's intestinal contents). Either organism would die of starvation without the other. (*A* courtesy of U.S. Department of Agriculture, Washington, D.C.; *B* courtesy of Carolina Biological Supply Co.)

the case of a biting mosquito, or more durable, as in the latching on of ticks or the deeper burrowing of mites. The term *infestation* refers to the presence of ectoparasites on a host, as opposed to *infection* by endoparasites.

The nature and degree of harm suffered by the parasitized host may be slight and unrecognized, or serious enough to threaten life. In other words, parasitism is distinct from disease, the latter depending on characteristics displayed by both parasite and host in their responses to each other. From the parasite's point of view, parasitism is successful when a state of balance exists in its relationship with the host, the latter furnishing required support while

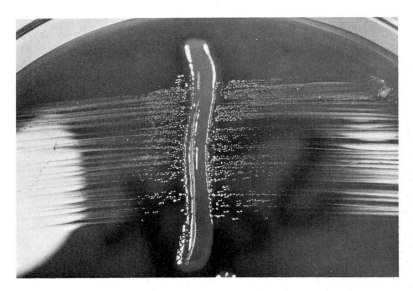

Fig. 5–4. Bacterial satellitism (an example of commensalism). A strain of *Haemophilus influenzae* was first streaked many times across the surface of this blood agar plate; then a strain of *Staphylococcus aureus* was inoculated in a single vertical streak. It can be seen that the *Haemophilus* strain (the tiny, glistening, translucent colonies) has grown most luxuriantly where it is in close proximity to the *Staphylococcus,* its growth dwindling and disappearing as it is drawn outward along the horizontal streaks. This is because one of the growth factors required by the fastidious *Haemophilus* strain is not available in the agar medium but is produced by the *Staphylococcus* in the area of its growth down the central streak. *Haemophilus* grows only where this growth factor is accessible to it and thus appears to be a "satellite" of the *Staphylococcus.* The latter is unaffected by this commensal partnership, whereas the *Haemophilus* reaps a vital benefit.

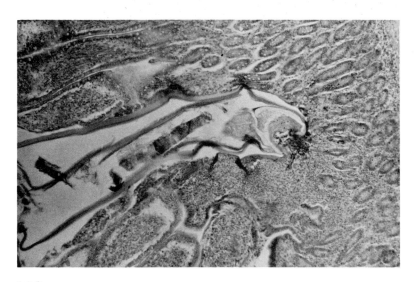

Fig. 5–5. Photomicrograph of a hookworm attached to the intestinal wall of its host.

continuing to function within reasonably normal limits.

The host requirements of parasites are sometimes quite specific, sometimes broad, with regard to the species of host that can be parasitized or to the kind of host tissue within which they can localize. Thus, the parasites of plants are generally not capable of living in animal tissues, while animals may share their parasites to a wide extent. Man acquires many of his parasitic worms from the animals with which he associates or whose flesh he eats. As we have seen in Chapter 2, some helminths have complex cyclic requirements for more than one host, requiring particular tissues in each for their support and development. Cell and tissue specificity is often quite marked among viruses and rickettsiae also, while bacteria, yeasts, and molds are frequently more versatile in their ability to localize in various host tissues.

It should be appreciated that a parasite not obligated to a living partner can support itself, if need be, on nonliving organic matter. As we have seen, many parasitic species of bacteria, yeasts, and molds, when taken from the human body on which they have been dependent, can be cultivated in the laboratory on nonviable nutrient media. The microbiologist's purpose is to grow the organisms of interest in pure culture, so that each may be studied in isolation, without interference from others. There are no pure cultures in nature, however, and relationships that exist between living organisms are not static. Between the two ecologic extremes of a free-living, saprophytic existence and one of obligate parasitism, gradient conditions of symbiosis among living things reflect their diverse abilities to respond to each other and to changing environmental pressures.

Man and His Microbes: The Body's "Normal Flora"

From the moment of his birth the human being lives in the midst of the microbial world. Associations between man and microorganisms are continuous and fluctuating, their contacts never-ending. There is a constant traffic of microorganisms entering the body, in air inhaled, in food eaten, on surfaces of things touched or placed in the mouth; and leaving it in excretions from the intestinal tract, mucosal surfaces, and skin. In normal circumstances, this traffic flows only over the surfaces of the body; that is, the skin and mucosal linings of respiratory, alimentary, and genitourinary tracts. Any daring excursions into deeper tissues beneath these linings are met, often with lethal effect, by the healthy body's defensive mechanisms. The skin and mucosal surfaces, however, not only tolerate the presence of microorganisms, but offer many conditions suitable for their *colonization* by saprophytes, that is, for local growth and multiplication by commensal microorganisms. Abundant nonviable organic nutrient — dead or dying cells in the outer layers of skin, the secretions of skin and mucosal glands, food that passes into and along the intestinal tract — is available for microorganisms that take up permanent or temporary residence on the body. These colonizations are influenced, just as they are in the world around us, by the local environmental conditions of different body areas (type of nutrient, pH, moisture, darkness, oxygen supply, temperature). Other factors may also be involved in this ecology, such as individual or racial differences among humans and the geography and sociology of human populations.

Resident and Transient Flora

Commensal organisms living on skin and mucosal surfaces are referred to, collectively, as *resident,* or *normal,* flora. The nature and variety of these microbial populations are often distinctive for different regions of the body, being influenced by the structure and physiology of surface lining cells. They fluctuate also in response to such host factors as age, activities, or skin type, as well as to external environmental influences. However, within the limitations imposed by such factors, there appears to be a rather fixed and well-defined normal microbial flora distributed over body surfaces, which tends to reestablish itself whenever it is disturbed. Members of the resident flora are capable of parasitism and may become pathogenic if the opportunity arises, but their potential for inducing active or severe disease is small under ordinary circumstances. They are harmless on surface tissues and may be "opportunists" causing disease only if introduced into the bloodstream or tissues in some numbers. Usually predisposing factors that lower the normal host defenses (accidental or surgical injury, malnutrition, other infection) set the stage for such opportunism on the part of resident flora.

Since microorganisms are ubiquitous, and their contacts with the body constant, it is not difficult for new organisms to deposit on body surfaces. However, most new acquisitions do not linger permanently. They may be washed away easily from external areas, they may die in competition with resident organisms, fail to survive conditions presented by the body itself, or simply be ushered out in the body's excretions. For obvious reasons, organisms that are present on the body temporarily, for a few hours or days, but do not become firmly entrenched, are referred to as *transient* flora. Occasionally, under suitable conditions, some transients may hold their place long enough to become resident commensals or establish themselves in a parasitic relationship with the host. In that case, depending on their properties and on the resistance with which they are met, active disease of the host may ensue.

Skin

The anatomic structures and physiologic functions of skin vary from one part of the body to another, and resident flora of different areas reflect these modifications. In addition, constant exposure of skin to contact with the environment assures continuous exchange of transient microorganisms.

Resident Flora. The most common resident commensals of skin are bacteria. As seen with the scanning electron microscope, they occur in small scattered colonies. There may be a few hundred individuals in these colonies, but they tend to get spread or smeared over the skin surface. They adhere to squamous epithelial cells of the outer cornified skin layer, extending downward between these cells into the mouths of hair follicles and gland openings. Scrubbing and washing can reduce their numbers but never eliminate them completely. Skin bacteria metabolize the secretions of sweat glands, often producing odorous substances. Bathing and the use of skin deodorants can prevent this effect.

Among the common skin bacteria, various strains of staphylococci usually predominate. Most of these are classified together as *Staphylococcus epidermidis.* Other bacteria abundant on skin include aerobic and anaerobic diphtheroid bacilli; aerobic spore-forming bacilli; aerobic and anaerobic streptococci of several types; and Gram-negative rods of intestinal origin.

The resident flora is modified by such anatomic features as the presence of hair; folded skin surfaces in close contact, characterized by abundant secretions of sweat and sebaceous glands; proximity to mucosal surfaces such as those of the mouth, nose, and throat, and anal or genital regions; and the nature and condition of clothing worn habitually. Yeasts and fungi, as well as many bacterial species, are often present in hairy parts of the body, under nails, and in skin folds (axilla and groin). Intestinal bacteria such as enterococci (fecal streptococci) and Gram-negative enteric bacilli of the coliform type (*Escherichia coli*) may abound on skin of groin and buttocks. Harmless acid-fast bacilli of the *Mycobacterium* genus may be found in areas rich with sebaceous secretions, such as the external ear canal or inguinal areas. Organisms from the mouth, nose, and throat easily find their way, at least transiently, to skin of the hands and face. These surfaces, therefore, often predominantly display strains of staphylococci and streptococci. The soles of the feet, being highly cornified, support a rather limited bacterial flora, but soft folds of skin between toes, with their sweat and sebaceous secretions, provide good conditions for many species of bacteria, yeasts, and molds. The habitual wearing of shoes contributes to these conditions by eliminating drying effects of air and by maintaining a constantly dark, moist atmosphere in which secretions accumulate. Clothing for other parts of the body may produce similar effects, particularly if it is close fitting, nonabsorbent, or infrequently washed or cleaned.

Transient Flora. Habits and occupational activities greatly influence *transient* flora of the individual host. Transient microorganisms on skin, being superficially located and held in place by sweat, oil, and grime, can be washed away easily with soap and water, aided by scrubbing. Therefore, people who wash their hands frequently and keep their bodies and clothing scrupulously clean display fewer transient organisms than others less assiduous in these efforts. Nurses, physicians, and all persons caring for patients with infectious diseases must take particular care of their own skin and clothing, to reduce the risk of personal infection with transient microorganisms and to avoid transferring them through patient contacts. Physiologic factors (pH; the fatty acids of sebaceous secretions; antibacterial substances present in tears, sweat, and mucus) also assist in elimination of transient organisms from skin and mucosa.

The harmless resident bacteria of the skin can also help to prevent colonization by pathogenic transients. For example, certain pathogenic strains of *Staphylococcus aureus* may cause outbreaks of infection in hospital nurseries. One method of handling such episodes has been the deliberate colonization of infants with a nonpathogenic strain of *S. aureus* that interferes with the growth and establishment of the pathogen, preventing further spread of dangerous infection.

Scrubbing Effects. The reduction of *resident* skin organisms is not simply or permanently accomplished by mechanical or physiologic means. Vigorous hand scrubbing by physicians and nurses preparing for surgery may eliminate transients and some superficial resident commensals, but a proportion of residents located in deeper layers of the skin remains, no matter how prolonged the period of scrubbing may be. Antibacterial agents applied during the scrub assist in removal of superficial organisms but are incapable of eliminating all deeper residents. For this reason, surgeons, nurses, and other members of the surgical team wear presterilized rubber gloves to prevent possible transferral of lingering organisms to the tissues of the patient undergoing surgery. The gloves in turn protect the hands from recontamination. From the moment the scrubbed hand is gloved, however, its diminished resident flora is rapidly replenished from the deeper skin layers, so that by the time a prolonged operative procedure is complete the flora of the hands may be restored to its usual numbers. It is particularly important, therefore, that meticulous care be taken to avoid contamination of the sterile surfaces of gloved hands by touching unclean surfaces, and to prevent punctures and nicks of gloves through which viable microorganisms may pass.

Other precautions taken by operating room personnel to prevent the transfer of resident skin commensals include the wearing of a sterile cap to cover the hair, a sterile mask for the mouth and nose, and a sterile gown to cover clothing. The patient's skin, at the operative site, is also carefully prepared beforehand, shaved clean of hair, and scrubbed with a suitable antibacterial agent to reduce the patient's own resident skin flora to the minimum possible numbers.

Genital Tract

The external genitalia of both sexes display a large variety of resident microorganisms. The normal flora includes saprophytic species of spirochetes, yeasts, and bacteria such as staphylococci, Gram-negative enteric bacilli, fecal streptococci, and lactobacilli. Flagellated protozoans of the genus *Trichomonas* may also reside normally in this region. Along the mucous membranes as they extend inward to the genital organs themselves, the numbers and variety of organisms normally present diminish and disappear. The terminal area of the urethra in both males and females is commonly colonized with bacteria as is the area into which it opens. Urine or other secretions collected for culture at the urethral opening may, therefore, be contaminated with organisms normally present on the urethra and external genital mucosa. Under certain circumstances, these organisms may extend inward along the mucosal surfaces into the bladder and, in females, the vagina. However, microbial extension into the vagina is usually sharply limited by the acidity of vaginal secretions. Soon after birth the vaginal canal becomes colonized with lactobacilli. These organisms enjoy and help to maintain an acid pH, which persists in the vagina for a few weeks after birth. The pH then shifts to neutrality and remains relatively neutral until puberty, the lactobacilli being replaced by a mixed group of bacteria. From puberty to menopause, lactobacilli again predominate and contribute to a normally acid vaginal pH through their metabolic production of acid from carbohydrate. They also serve to discourage the entrance of yeasts, protozoans, and bacteria from the external mucosa into the vaginal canal in any large numbers, although the normal vaginal flora may include a few other organisms, such as anaerobic spore-forming bacilli, aerobic and anaerobic streptococci, and staphylococci. After menopause the acidity of the vagina diminishes, lactobacilli dwindle again in numbers, and a mixed flora takes over once more.

If the normal ecology of the vagina is disturbed, as in antibiotic therapy of an infectious process elsewhere in the body that coincidentally suppresses vaginal lactobacilli, other organisms such as *Trichomonas,* yeasts, or various bacteria may increase greatly, producing a troublesome irritation, or vaginitis.

As the mucosal lining of the female genital tract extends upward into the cervix, it becomes sterile. Normally the cervical mucosa is free of a resident flora and discourages its establishment through its secretions, which contain lysozyme, a substance having antibacterial activity.

Respiratory Tract

Upper. In the upper respiratory tract, the mouth, throat, and nose offer fertile conditions for survival and growth of many kinds of microorganisms. These areas are colonized during, or soon after, birth by several bacterial species, among which alpha-hemolytic (or *viridans* type) streptococci, staphylococci, Gram-negative cocci of the genera *Neisseria* and *Branhamella,* and diphtheroid bacilli predominate. Such organisms are probably passed along from the respiratory mucosa of the mother and all others who may handle and care for the infant. Yeasts also may occur in the mouth, the baby acquiring them from the skin of the mother's breast or from bottle nipples. In the newborn infant's mouth, with its scarcity of bacterial species to maintain control over growth of yeasts, the latter may multiply rapidly and extensively, damaging the oral mucosa and producing a condition known as "thrush" (see Fig. 16–9, p. 401).

At the time the infant's first teeth begin to erupt, other organisms establish themselves in the gingival tissues. These include anaerobic spirochetes, fusiform bacilli, and lactobacilli. Most of these organisms remain in the mouth for life.

Resident Mouth Flora and Dental Caries. The role of the normal flora of the mouth in the development of dental caries has been rather well established. Caries is a process of tooth decay in which demineralization of the surface enamel first occurs and is followed by decomposition of the underlying dentin and cement. Bacterial activities are responsible for both processes. Damage begins with the formation of "plaque" on the enamel surface. "Plaque" consists of gelatinous deposits of high-molecular-weight polysaccharides called glucans. These are produced by streptococci (notably *S. mutans* and anaerobic species) from sucrose and other sugars. Bacteria adhering to the sticky "plaque" then ferment carbohydrates, producing high concentrations of acid that demineralizes the adjoining enamel. This bacterial fermentation is largely due to streptococci and lactobacilli.

Damage to the enamel is followed by bacterial digestion of the protein matrix of the dentin and cement. Proteolytic enzymes of actinomycetes and other bacilli play a role in this. The glucans produced by streptococci and some diphtheroids can also cause soft tissue damage and resorption of bone, as seen in periodontal disease.

Genetic, nutritional, hormonal, and other factors also play a part in the development of caries. Its control involves good mouth hygiene, attention to nutrition with adequate protein and limited carbohydrates, and physical removal of "plaque." The use of fluoride, ingested in water or applied directly to teeth, helps to increase the resistance of enamel to the effects of bacterial acid production. Current studies of *S. mutans* are directed toward prevention of its adherence to smooth surfaces, either by interference with its production of the gelatinous glucan or by antibody reaction with a site on the surface of the bacterial cell through which it binds to glucan on the enamel surface.

Resident Flora of the Pharynx and Nose. The normal flora of the throat, tonsils, and trachea includes aerobic and anaerobic streptococci, staphylococci, *Branhamella,* and *Neisseria* species, members of the genus *Haemophilus,* diphtheroids, anaerobic Gram-negative rods (*Bacteroides* and *Fusobacterium* species), and spirochetes. Other organisms may come and go transiently, some of them associated with a high potential for producing local or generalized disease. Bacterial species such as *Streptococcus pneumoniae* (associated with bacterial pneumonia), the diphtheria bacillus, and the meningococcus (a species of *Neisseria* capable of producing severe disease) often enter the body by way of the respiratory tract, localizing first in the throat. The same is true for many viruses associated with infectious disease. The mere presence of such organisms in the throat does not necessarily imply that disease will ensue for they may be quickly suppressed by the resident flora, or by the body's own physiologic mechanisms.

In health, the nose harbors only a few microorganisms, usually staphylococci and diphtheroids, with occasional strains of streptococci. Frequent colonization of the nose by strains of *Staphylococcus aureus* is a matter of concern with respect to the possible spread of staphylococcal infection from healthy individuals to others whose resistance to the organism is low for some reason. The possible transfer of such strains from hospital personnel to patients who may offer the organism unusual opportunities for entry and localization is made more troublesome by the fact that many nasal strains colonizing hospital personnel are resistant to antibiotics.

Lower. Very few microorganisms are capable of surviving the passage downward from the trachea to the lower respiratory tract, or of establishing themselves permanently there. The bacterial species normally recoverable from sputum, when it is raised from the bronchioles and bronchi, usually reflect the resident flora of the tracheal, tonsillar, and pharyngeal surfaces, since sputum must pass over these to be collected for culture. When infectious disease of the lower respiratory tract exists, however, its agent may be present in sputum and must be separated, in laboratory cultures, from normal throat flora also present.

Intestinal Tract

The intestine of the newborn infant is sterile, but this situation is only momentary. Many species of bacteria and yeasts take up residence in the bowel from the time the infant begins to feed, in a progression that varies with age, eating habits, and different segments of the intestinal tract.

The esophagus contains only transient organisms introduced in food and saliva. The normal acidity of the stomach prevents the establishment of microorganisms there, but as the pH rises and becomes alkaline in the duodenum and jejunum, a more and more abundant resident flora implants, reaching its peak in the large intestine.

In breast-fed babies, the predominant organisms in the intestinal tract are Gram-positive anaerobic rods of the genus *Bifidobacterium*. These organisms seem to protect the infant against infection with harmful intestinal pathogens. The flora is more mixed in bottle-fed infants, and as children begin to eat a variety of foods, the character of the intestinal flora approaches that of adults. Throughout life diet has a continuing influence on the variety and number of microorganisms in the bowel and feces. The upper regions of the intestinal tract contain about 10,000 organisms per milliliter of contents, predominantly streptococci, lactobacilli, and diphtheroids swallowed in food or saliva. In the lower intestinal tract (colon), the numbers of bacteria increase to more than ten billion per gram of intestinal contents. Here, anaerobic bacteria are most prevalent, outnumbering aerobes 1000:1. The major anaerobic bacilli present are the Gram-negative *Bacteroides* and *Fusobacterium* species, and Gram-positive *Bifidobacterium, Eubacterium,* and *Proprionibacterium* species (anaerobic diphtheroids). Anaerobic streptococci are also common.

Among the aerobic residents of the adult large intestine, the most abundant are Gram-negative coliform bacilli (related to *Escherichia coli*). Other aerobic organisms present include fecal streptococci (enterococci), bacilli of the *Proteus* and *Pseudomonas* groups, some lactobacilli, and yeasts.

Under normal conditions, the ecology of the resident intestinal flora is self-balancing. Competitions and mutualisms between microorganisms, and between these and their host, serve to maintain the status quo. If the intestinal tract becomes diseased in some way, however, as by tumor, obstruction, or infectious disease, the normal flora may change greatly. Antibiotic drugs, particularly if taken orally, can also induce marked shifts in the normal intestinal flora, by suppressing drug-susceptible organisms of the intestine. Other commensals normally held in check by those that have been suppressed may then thrive to an unusual degree, and this, in turn, may induce an uncomfortable gastrointestinal irritation, or even more severe disorders.

Conjunctiva

The conjunctiva, the mucous membrane that lines the eyelids and is reflected onto the eyeball, is also a site for bacterial commensals. The resident flora consists of a few kinds of bacteria such as diphtheroid bacilli, *Branhamella,* and *Neisseria* species, staphylococci, streptococci, and some small, microaerophilic, Gram-negative bacilli. The numbers of transient organisms introduced by hands or handkerchiefs are kept to a minimum by the washing action of tears, as well as by their antibacterial component, lysozyme. Injury to the normal membrane barrier, caused by frequent or hard rubbing or by foreign bodies, may permit entry of transient or resident organisms into the tissue beneath, and active disease of the area may then result.

Table 5-1 summarizes the indigenous microbial flora to be found in various areas of the human body.

Infection and Infectious Disease

From the point of view of the host, *infection* can be defined as a state in which the host harbors microorganisms that survive and multiply in or on his tissues. Commensalism is included in this definition, and therefore it can be said that the human host lives all of his life in a state of infection. The definition also includes, but does not necessarily imply, *infectious disease.*

The *potential* for disease exists when microorganisms live in a parasitic relationship with their host or shift to this type of association from a commensalistic or saprophytic one. The extent of harm caused by a parasite may be clinically unrecognizable, or perfectly obvious through a range of symptoms from mild to severe, but in any case it depends on *the properties and functions of both the parasite and the host.* When the damage is sufficiently marked to become clinically manifest, the term *infectious disease* then applies, meaning a noticeable state of abnormality induced by a living organism.

Microbial Factors

The manner in which a disease originates and develops is referred to as its *pathogenesis,* from the Greek words for disease (*pathos*) and origin (*genesis*). Accordingly, the word *pathogenicity* denotes the ability of a microorganism to produce infectious disease. The degree of pathogenicity possessed by a parasitic organism is expressed by the term *virulence,* which is a function of two principal attributes: *invasiveness,* or the ability to get into host tissues, survive, multiply, and spread; and *toxigenicity,* or the ability to produce substances that are toxic to host cells and tissues. Pathogenic microorganisms differ in virulence, according to their relative display of such properties.

Host Factors

The severity of an infectious disease, and the total degree of damage done, reflect the virulence of its pathogenic causative (etiologic) agent, and also the host's capacity for defense against invasion and toxicity. The host's defense mechanisms function to protect the body from entry by microorganisms and to eliminate or control those that do succeed in entering. Unbroken skin and mucous membranes offer natural barriers to intrusion, and within the body many specific defenses are operative, such as *phagocytosis* (the ingestion of foreign particles, including microorganisms, by white blood cells and certain other tissue cells), the *inflammatory response* (the process by which tissue reacts to injury in order to limit and control it locally), and the *specific immune response* (the ability to resist a particular infectious agent, or its toxic products, either by the development of immune substances [*antibodies*] that react specifically with the agent or its products or by increased activity of specialized cells of the immune system [*cell-mediated immunity*]). The kind of tissue

Table 5-1.   Microbial Flora Indigenous to the Human Body*

Anatomic Location	Microorganisms	Comments
Skin	Staphylococci (*S. epidermidis, S. aureus*) Streptococci (viridans, nonhemolytic strep, and enterococci) Corynebacteria (diphtheroid bacilli) Gram-negative enteric bacilli (*Escherichia* and *Enterobacter* species) Mycobacteria (acid-fast bacilli) Yeasts and fungi	In addition, exposed skin may harbor many transients, which may remain for hours, days, or weeks; transients are readily removed but reduction in numbers of residents is temporary at best
Eye Conjunctiva	*Staphylococcus epidermidis* Streptococci (viridans and nonhemolytic strep) *Branhamella catarrhalis* *Neisseria* species Corynebacteria (diphtheroid bacilli) Microaerophilic Gram-negative bacilli	Washing action of tears and their content of lysozyme help to control microorganisms
Nose and throat	Staphylococci (*S. epidermidis, S. aureus*) Streptococci (viridans, nonhemolytic strep, and enterococci) *Streptococcus pneumoniae* *Branhamella catarrhalis* *Neisseria* species Corynebacteria (diphtheroid bacilli) *Haemophilus* species Anaerobes	Action of nasal ciliary cells, swallowing, mucous secretion, and lysozyme of saliva help to control microorganisms
Alimentary tract Mouth	Staphylococci (*S. epidermidis, S. aureus*) Streptococci (viridans, nonhemolytic strep, enterococci, *S. mutans*) *Lactobacillus* species Fusiform bacilli Spirochetes *Actinomyces* species Yeasts	Swallowing action, lysozyme of saliva, good dental hygiene help to control microorganisms; nutrition important in preventing dental caries
Esophagus	No indigenous flora	Organisms swallowed with food or saliva remain only temporarily
Stomach	No indigenous flora	Gastric acidity is too high for microbial growth

Table 5–1. *(Cont.)*

Anatomic Location	Microorganisms	Comments
Intestines (adults)	Anaerobic bacilli (species of *Bacteroides, Fusobacterium, Bifidobacterium*) Anaerobic streptococci Gram-negative enteric bacilli (species of *Escherichia, Enterobacter, Proteus, Pseudomonas*) Streptococci (enterococci) Yeasts Protozoa	The intestines contain much larger numbers of microorganisms than any other area of the body
Genital tract	Staphylococci (*S. epidermidis, S. aureus*) Streptococci (enterococci and non-hemolytic strep, sometimes viridans) Lactobacilli Gram-negative enteric bacilli (species of *Escherichia, Enterobacter, Proteus, Pseudomonas*) Anaerobic bacilli (*Clostridium* species) Mycobacteria (acid-fast bacilli) Spirochetes Yeasts Protozoa (especially *Trichomonas* species)	In the female, lysozyme content of cervical secretions and the acidity of the normal adult vagina control the numbers and variety of microorganisms establishing in this area; lactobacilli predominate between puberty and menopause

*In the bacteriologic culture of specimens obtained from these areas of the body the search for pathogenic organisms, or for commensals that have assumed a pathogenic role, is complicated by the variety of indigenous bacterial species encountered in such specimens.

involved in the infection and the nature of its injury are of particular importance to the host. The vital functions of certain areas of the brain or the action of the heart, for example, may be seriously handicapped by a small injury, while other organs, such as lungs, liver, or kidneys, may continue to function with some efficiency even when extensively damaged.

Microbe and Host Together

Most human infectious diseases result from the introduction of pathogenic *transient* microorganisms into the tissues. On normal skin and mucosal surfaces, the transient pathogen must survive the activities and antagonisms of host cells, as well as the activities and antagonisms of commensal organisms already present, competing with the latter for nutrient. At the same time, the normal resistance of the host is tested: the virulent qualities of an infecting pathogen may be exceptional, or it may arrive in unusual numbers; an unnatural route of entry may be opened up, as in the case of surgical or accidental trauma to skin or mucosa; or the host's defensive mechanisms may have been diminished by preexisting circumstances, such as organic disease, fatigue or other stress, or the extremes of age. Under such circumstances, the host's own commensals, as well as transient pathogens, may be afforded a better opportunity to establish themselves within tissues, and the possibility of the development of active disease is enhanced. In the ensuing struggle for survival, the total interplay of forces between an infecting microbe and the host may lead to a complete victory of one over the other in a relatively short time (*acute* infection), or to a prolonged relationship (*chronic* infection), precariously balanced between them. The symptoms displayed by the host vary accordingly from none at all (*subclinical* infection) to those of a sudden, acute process or a slow, insidious one. Between the extremes, there are many variations in the severity, as well as the duration, of symptoms.

Frequency of Infection Versus Infectious Disease. When the distribution of a particular pathogenic species is studied, one may find that large numbers of normal healthy people harbor the organism in question (such as poliomyelitis virus, respiratory viruses, strains of staphylococci and streptococci), greatly outnumbering those who are made ill by it. This wide prevalence of infection in the human population, together with the relative infrequency of clinically active infectious disease, presents a picture whose contours resemble those of an iceberg. The insidious feature of an iceberg is the large mass that floats unseen beneath the surface of the sea, the visible ice giving little indication of the shape or extent of the mass below (Fig. 5-6). For a given type of infection, the number of people with recognizable

Fig. 5–6. For a given type of infection, the number of people with recognizable *disease* represents the visible part of the iceberg, above the surface, while the much greater number who are transiently or permanently infected by the particular agent involved, but who display *no symptoms* of illness, can be compared to the great invisible mass of ice in the sea below. (Courtesy of Department of the Navy, Washington, D.C.)

disease represents the visible part of the iceberg, above the surface, while the much greater number who are transiently or permanently infected by the particular agent involved, but who display *no symptoms* of illness, can be compared to the great invisible mass of ice in the sea below.

Questions

1. Define ecology.
2. Of what value is bacterial nitrogen fixation to our society?
3. Define parasitism, commensalism, mutualism, and symbiosis. List an example of each and show how it fits the definition.
4. What is the difference between resident and transient flora? Of what importance are they in surgery?
5. List examples of microbial flora indigenous to the skin, genital tract, upper and lower respiratory tract, upper and lower alimentary tract, and conjunctiva. What is the significance of the normal flora?
6. Define pathogenicity and virulence. How do they influence the severity of an infectious disease? What host factors are involved in resistance to infectious disease?

Additional Readings

Andrews, Michael L. A.: *The Life that Lives on Man.* Taplinger, New York, 1977.
* *The Biosphere:* A Scientific American Book. Scientific American Editors, W. H. Freeman and Co., San Francisco, 1970.
* Carson, Rachel: *Silent Spring.* Fawcett World, New York, 1973.
* Klainer, Albert S., and Geis, I.: *Agents of Bacterial Disease.* Harper & Row. Hagerstown, Md., 1973.
† Marples, Mary J.: Life on the Human Skin. *Sci. Am.,* **220**:106, Jan. 1969, #1132.
* Rosebury, Theodor: *Life on Man.* Berkley Medallion Editions, New York, 1970.

* Available in paperback.
† Available in *Scientific American* Offprint Series. (See Preface.)

Microbe Versus Host

6

The microbial factors of importance in establishment of infection and production of infectious disease are considered in this chapter. The process of infection begins at the site of entry of the host by an infectious microbe. Therefore, the *means of entry* to the human host that are available to microorganisms are of particular importance to those responsible for the prevention and control of infectious disease. These entry routes are discussed in the first part of the chapter. *The properties and activities* of microorganisms associated with their pathogenicity and virulence (i.e., with their ability to entrench themselves in human tissues and cause damage) are described in the second half of the chapter.

Infection Entry Routes

General Considerations

Since the point of entry marks the beginning of infection and also affords one of the last opportunities for its prevention, it is important to understand its possibilities. Once infection is established, the ultimate site(s) of localization of the parasite determine whether or not it can make a later exit from the body, through what portal(s) it may leave, and by what route(s) it may be transmitted to another individual. Nursing and epidemiologic techniques designed to prevent infectious disease, or limit its spread from one person to another, are based on a knowledge of the entry, exit, and transmission routes taken by pathogenic microorganisms, in general or in particular.

Reducing the human (or mammalian animal) body to a simple anatomic concept, we can think of it as a cylinder traversed by a central tube, the alimentary canal (Fig. 6–1). Two major blind sacs, the respiratory and urogenital tracts, diverge inward, one at the upper end of the canal, the other from the region near the lower opening. The external body surface is covered by skin, which, with its relatively impermeable cornified, nonviable, outer layer, provides some protection and insulation from the environment as well as sensory communication with it. This protective covering is interrupted only at the

123

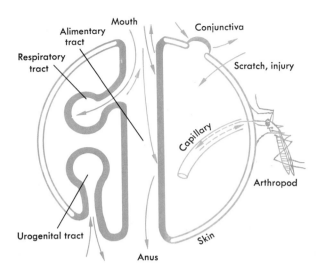

Fig. 6–1. Bodily surfaces as sites of microbial infection and shedding. (Reproduced from Mims, Cedric A.: *The Pathogenesis of Infectious Diseases.* Academic Press, London, 1976.)

site where the eyes are placed and at the openings of the alimentary, respiratory, and urogenital tracts. The external covering of the eyes themselves is a thin transparent layer of viable cells called the conjunctiva. The inner tracts are lined with living cells that are structurally modified for their respective functions. The conjunctiva and inner surface linings are covered by mucous secretions that protect them and keep them moist. For this reason, they are commonly called *mucous membranes,* or *mucosae* (sing. = *mucosa*).

The skin and inner surface linings of the body offer pathways for infection by microorganisms, although the skin is less easily penetrated than the thinner membrane surfaces. All together there are four major avenues through which microorganisms may gain entry into the body: (1) they may enter with inhaled air into the *respiratory tract;* (2) they may be ingested with material taken into the mouth and *gastrointestinal tract;* (3) they may themselves infect or penetrate the *skin or superficial mucous membranes* and be limited to these tissues or spread from there to deeper sites; and (4) they may be deposited *directly into tissues* beneath the skin and mucosae when these barriers are penetrated or traumatized (injured) by some other agent. The latter avenue is referred to as the *parenteral route,* implying direct access to the body tissues by such means as puncture and injection (by a biting or blood-sucking

insect vector or an artificial vector such as a hypodermic needle), by more extensive trauma (caused by accident or surgery or the bites of animals), or by extension from mucosal surfaces (respiratory, alimentary, or urogenital).

Establishment of Infection. The entry route does not necessarily determine the site in which the infective organism will ultimately establish itself and multiply. The areas in which the parasite may gain a foothold are dictated by its own properties, by the biochemical environment afforded by host tissues, and by the host's many defensive mechanisms. In order to colonize or penetrate the body's surface tissues, microorganisms must first become attached. A number of specific microbial attachment mechanisms have been clarified in recent years, although some are yet unknown. We have seen, for example (Chap. 5, p. 116), that *Streptococcus mutans* attaches to tooth enamel through its synthesis of glucans and by the action of a binding site on its cell surface. In the case of poliomyelitis and some other virus infections, viral capsid protein reacts with a specific receptor on the host tissue cell. *Mycoplasma pneumoniae,* the bacterial agent of "atypical" pneumonia, attaches by a "foot" to a specific acid component on the surface of a respiratory epithelial cell. Electron microscopy indicates an apparent role of pili in the attachment of the gonococcus (the agent of gonor-

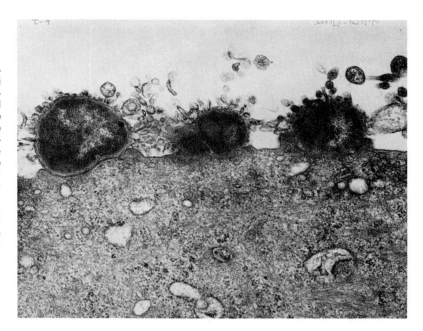

Fig. 6–2. Electron micrograph showing gonococci closely attached to the surface of a urethral epithelial cell. The membrane of the host cell appears to be pushed up around the organisms to form cushionlike structures, as an initial stage of engulfment. (46,000×.) (Reproduced from Ward, Michael, and Watt, Peter J.: Adherence of *Neisseria gonorrhoeae* to Urethral Mucosal Cells: An Electron-Microscopic Study of Human Gonorrhea. *J. Infect. Dis.,* **126**:601, Dec. 1972.)

rhea) to the surface of urethral epithelial cells (Fig. 6–2). Experimental infections of rabbits with cholera indicate that the bacteria (*Vibrio cholerae*) adhere via their surface coats directly to the tips of the brush border of the villi in the intestinal tract (Fig. 6–3).

Once having entered, attached by some means, and colonized at some site in the body, a pathogenic microorganism may induce damage in one or both of two ways: (1) it may *remain localized at or near the site of entry* and damage the surrounding tissues mechanically or physiologically, or (2) either the organism itself or its toxic products, or both, may *disseminate through the body,* producing injury to tissues far removed from the entry point. The parasite itself may have *invasive* qualities enabling it to spread through the tissues, or it may be carried passively in the bloodstream or lymph and thus be distributed widely. Microbial toxins are also disseminated by way of blood and lymph flow from a local site of colonization where they are produced. The general possibilities for localization along the entry route, or for dissemination therefrom, are discussed for each of the four avenues of entry.

Respiratory Entry

The respiratory tract offers the easiest, and most frequently traveled access route for infectious microorganisms. It is also the most difficult one to control. Infections acquired by this route have a world distribution because of face-to-face contacts among people (see Section IV).

Microorganisms are taken into the nose or mouth by inhalation of droplets of moisture containing them. Such droplets are sprayed into the air by the talking, coughing, or sneezing of other persons whose oral, nasal, and pharyngeal secretions contain many of the commensal or parasitic organisms that they happen to harbor. The quantity of droplets reaching the air, and the distance to which they are scattered, depend on the force and frequency with which people talk, cough, or sneeze (see the illustration on page 295).

The larger droplets of mucous spray settle rapidly from the air and are not likely to be inhaled. Fine droplets, however, lose their moisture quickly by evaporation and linger in the air with their burden of

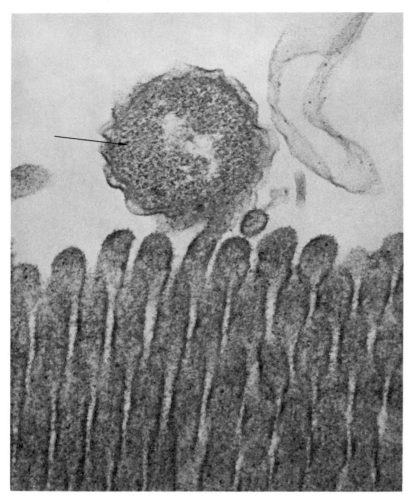

Fig. 6–3. Electron micrograph showing a cross-section of *Vibrio cholerae* (arrow) adhering directly to the edge of the intestinal brush border. The outer membrane of the vibrio is in contact with the surface of a microvillus. (130,000×.) (Reproduced from Nelson, E. T.; Clements, J. D.; and Finkelstein, R. A.: *Vibrio cholerae* Adherence and Colonization in Experimental Cholera: Electron Microscope Studies. *Infect. Immun.,* **14:**543, 1976.)

microbes. When microorganisms enter the nose or mouth, they may lose the competition for nutrient with other organisms already present in the upper respiratory tract, or they may be destroyed by the activities and secretions of host cells. Many survivors of these hazards are caught in and moved along with mucus secreted by cells lining the upper respiratory tract. The ciliated epithelial cells keep the mucus moving and draining into the throat, where it is either swallowed or expectorated, or into the nose, from which it can be blown out. Microorganisms that are swept into the gastrointestinal tract may be killed by the acidity of gastric secretions or by com-

petition encountered farther along from the intestinal commensals.

The organisms present in inhaled air are thus subjected to *filtration, mechanical removal,* or *rerouting,* before the air is carried down into the deeper reaches of the respiratory tract. Only a relative few reach the lungs. Those that survive and entrench locally may induce damage in the respiratory tract or elsewhere, if they or their toxic products, or both, are disseminated from that site.

Localization in the Respiratory Tract. The most familiar example of an inhalation infection localized in the *upper* respiratory tract is the common cold. Several viral agents have been associated with this mild disease. The localization of virus within mucosal cells of the pharynx and nasopharynx and its interference with their normal functions result in the characteristic nasal irritation, sore throat, and profuse watery discharge that replaces normal mucus secretion. *Bacterial* pathogens that commonly localize in the upper respiratory tract include the coagulase-positive staphylococci, the beta-hemolytic streptococci, pneumococci, and *Haemophilus influenzae*. Among these, beta-hemolytic streptococci are probably the only significant cause of acute bacterial pharyngitis, although *H. influenzae* can produce a severe epiglottitis in children.

Members of the normal intestinal flora, such as coliform bacilli or fecal streptococci, may also find their way to the throat, but are usually of little consequence unless aspirated into the lower respiratory tract.

Many of the bacterial species introduced transiently on inhaled air may establish as commensals in the normal nose or throat. However, if the pharyngeal membranes are injured by a virus infection, excessive dryness, or some other factor that interferes with their normal functions and secretion of mucus, commensals may have the opportunity to multiply excessively. They may then add to the preexisting damage as their metabolic products increase in concentration. In such a situation the offending commensals are often called "secondary invaders," or "opportunists."

In the lower respiratory tract, infectious diseases acquired by inhalation include such notable examples as influenza and other pneumonias of *viral* origin; tuberculosis, psittacosis, and the pneumonias caused by *bacterial* pathogens, such as pneumococci, staphylococci, or the plague bacillus (pneumonic plague); and fungous diseases such as histoplasmosis or coccidioidomycosis. Pulmonary disease may also be caused by members of other major groups of organisms (rickettsiae, protozoa, and helminths) but the entry route in these cases is not usually through droplet inhalation.

As air carrying infectious droplets passes from upper to lower respiratory passages, microorganisms are filtered out so that few remain in the air that finally reaches the alveolar spaces of the lungs. Depending on the organism and its environmental requirements, localization may occur at different points along the way (see Fig. 6–4). A cold virus producing symptoms in the upper respiratory passages may pass downward from its original focal point and infect cells of the bronchial linings, but the influenza virus probably does not find a suitable locale until it reaches the smaller branches of the bronchial tree. The tubercle bacillus, when inhaled, usually reaches the alveolar bed and localizes there, but it may also penetrate lung tissue, reach adjacent lymphoid tissue, and be carried from there in lymph and blood. The tissues of the upper respiratory tract are generally resistant to tuberculous infection, but if the organism is ingested, or if it reaches the mouth and throat in sputum coughed up from lesions in the lower respiratory tract, it may sometimes localize there, inducing local damage to the larynx, pharynx, or oral mucosa.

Pathogenic cocci may reach the lungs directly, localizing en route in bronchial passages or in the lung beds, and then spread peripherally through adjacent lung tissues. Spores of pathogenic fungi, such as *Histoplasma* and *Coccidioides* species, must reach the depths of the lungs in fair numbers, and in finely disseminated particles, to establish primary pulmonary disease. Fine dust from the soil, in areas where these organisms are prevalent, carries the infective spores.

Dissemination of Microorganisms or Their Toxins from the Respiratory Tract. Following entry by inhalation, infective organisms can have serious effects in tissues far removed from the respiratory portal. This may happen if (a) the pathogen penetrates or is carried to other parts of the body, or (b) toxigenic substances secreted by the organism multiplying within the respiratory tract are distributed through the body by the blood or lymph.

a. *Penetration and spread* of an infectious organism are well exemplified by the meningococcus (*Neisseria meningitidis*), the bacterial agent of a severe systemic infection. This organism commonly finds a respiratory entry and localizes in the nasopharynx. It may reside there as a harmless transient or it may produce a local infection that is usually inapparent. However, it can reach the bloodstream directly from this site, be circulated through the

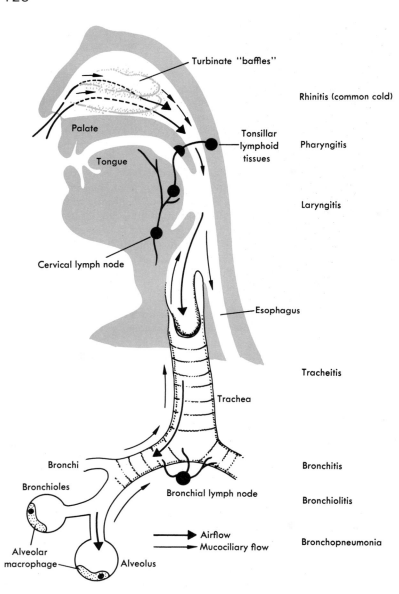

Turbinate "baffles"

Rhinitis (common cold)

Palate

Tonsillar
lymphoid
tissues

Pharyngitis

Tongue

Laryngitis

Cervical lymph node

Esophagus

Tracheitis

Trachea

Bronchi

Bronchitis

Bronchioles

Bronchial lymph node

Bronchiolitis

Airflow
Mucociliary flow

Bronchopneumonia

Alveolar
macrophage

Alveolus

Fig. 6–4. Principal sites of infection in the respiratory tract. (Reproduced from Mims, Cedric A.: *The Pathogenesis of Infectious Disease.* Academic Press, London, 1976.)

body, and relocalize in vital areas such as the brain or adrenal glands. Multiple lesions may be produced also in skin, joints, eyes, ears, and lungs, the symptoms of disease depending on the site of localization. In the brain, the characteristic injury is to the *meninges,* the inner membranous coverings of the brain and spinal cord, which become greatly inflamed. Inflammation of this tissue, whatever the cause, is called *meningitis,* hence the name *meningococcus,* although this organism is not the only infectious agent capable of producing meningeal injury.

There are many other systemic infections whose

agents find a primary point of entry and localization somewhere in the respiratory tract and are later disseminated to other tissues. This is the situation with the viral agents of poliomyelitis and infectious hepatitis, which may localize initially in the throat, then in the bowel, and be carried to other vital tissues. It may occur also in tuberculosis, in pulmonary anthrax, and in fungous diseases of the lungs. As in meningococcal infection, disease effects are produced by the disseminated organism itself and by tissue reaction to it, with symptoms referable to the location of various foci of infection.

b. *The dissemination of toxigenic substances* produced locally by microorganisms in the respiratory tract can also result in systemic disease. When microbial toxins circulate in the blood and damage tissues, the condition is spoken of as a *toxemia,* the word meaning "toxins in the blood." (When organisms themselves are present in the blood, words such as *bacteremia* or *viremia* are used [see Chap. 8. p. 191]).

Diphtheria and scarlet fever are well-known examples of diseases originating as local infections in the nose or throat, but characterized also by serious effects of toxemia. In both instances, the infective organism entrenches near its entry point in the pharynx. An acute sore throat results from the particular damage each organism causes to the membranes or underlying tissues, but, even more important, each secretes a characteristic *exotoxin* that is distributed widely by the bloodstream. These and other bacterial exotoxins have distinctively individual properties, which affect different cells and tissues in different ways. The diphtheria toxin damages the heart muscle and nerve tissue, as well as other organs, by interfering with protein synthesis in these tissues. The streptococcal toxin associated with scarlet fever affects the blood vessels and epithelial cells of the skin, producing the typical scarlatinal rash. (See Tables 6–1 and 6–2.)

Gastrointestinal Entry

The gastrointestinal tract may offer admission to members of any of the microbial groups. Entry may be effected via the many items and materials that are taken into the mouth, but the most common sources of infection by this route are food, water, milk, and fingers.

It has been pointed out that many microorganisms that are swallowed do not survive their journey through the intestinal tract. Some are eliminated in the mouth by the lysozyme activity of saliva; others are killed by the stomach's acidity. In the small intestine, the activity of digestive enzymes, capable of breaking down protein material, may destroy the surface proteins of bacterial cell walls. The possibilities for infectious disease arise when pathogenic microorganisms manage either to *localize in the bowel,* producing damage at this site; penetrate the intestinal wall and *disseminate;* or *produce substances toxic* for human cells from their site in the bowel.

Localization in the Gastrointestinal Tract. The most common infections of the bowel itself are caused by protozoa, helminths, some species of bacteria, and, in infants, viruses. Localization of such organisms occurs in either the small or large bowel, the site usually being characteristic for the parasite in question. Mucosal cells may be damaged and eroded, with consequent invasion of deeper tissues in some cases. Symptoms often include nausea and vomiting, diarrhea, intestinal spasm with accompanying pain, and, in some cases, sloughing of the mucosal lining with bleeding into the bowel.

Among the *protozoa* that may infect the bowel, the one that is associated with serious human disease is *Entamoeba histolytica,* the agent of amebic dysentery. *Helminth* infections of the bowel are caused by the common roundworms (pinworm, hookworm, *Ascaris*) and the tapeworms.

Important intestinal diseases of *bacterial* origin include gastroenteritis induced by strains of the genus *Salmonella;* bacillary dysentery caused by bacilli of the *Shigella* genus (often called shigellosis); and cholera, a severe disease of the bowel caused by a small curved rod, *Vibrio cholerae.*

Yeast and fungous infections of the bowel are usually transient, their agents being relatively incapable of invading intestinal or other tissue from this entry site. One group worth mentioning here, however, are the yeastlike members of the genus *Candida,* which may infect and damage the mucosal surfaces of the mouth and throat as well as the lower bowel.

Dissemination of Microorganisms from the Gastrointestinal Tract. Typhoid and paratyphoid fevers afford excellent examples of *bacterial* diseases whose agents enter the body through the alimentary canal, usually in contaminated water or food, but which penetrate the intestinal lining directly. They localize first in the lymphoid tissue of the bowel wall and are disseminated from there by lymph and blood to other organs, notably the liver, gallbladder, and kidneys. The typical symptoms of typhoid and associated fevers are produced by this systemic invasion, not by localization in the bowel.

Entamoeba histolytica may also erode its way into the bowel wall, enter the portal circulation, and reach the liver, where it induces abscess formation. From this site, it can extend directly through the diaphragm into the lungs, or be carried from such sites to establish itself in other areas including the brain.

In some *helminthic infections,* adult worms living in the intestinal tract produce eggs or larvae capable of penetrating the intestinal wall. They are picked up in the bloodstream and distributed widely. When deposited in human tissues, these larval forms cannot complete their development into mature adults. Some of them die as a result of host reactivity and later become calcified, but others remain alive in the tissues for long periods. *Trichinosis* is a roundworm infestation of this type. The serious nature of this disease is related to larval invasion of muscle tissue (Fig. 6–5).

The agents of two very important *virus* diseases also commonly localize in the bowel, as well as in the upper respiratory tract, but produce disease by invasion of other tissues. These are the poliomyelitis virus, which invokes its most serious effects when distributed to cells of the central nervous system, and the virus of infectious hepatitis, which invades and damages liver cells. A number of other viruses have been implicated in acute gastroenteritis. The most common include the *enteroviruses,* so called because of their enteric, or intestinal, localization. This group, which includes the poliomyelitis virus, falls in a category now known as *picornaviruses*. Others involved are *adenoviruses, reoviruses,* and *rotaviruses*. The latter have recently been identified as important agents of gastroenteritis in infants. The viruses associated with intestinal infection are discussed in Chapter 19.

The Toxins of Intestinal Pathogens. Most of the bacterial pathogens that damage the intestinal tract

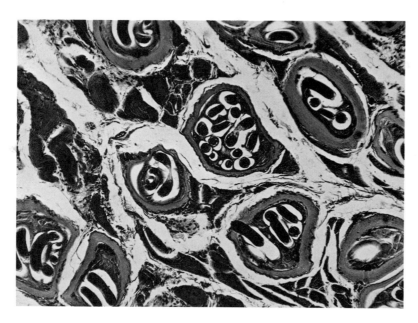

Fig. 6–5. Calcified cysts of *Trichinella spiralis* in human muscle. (300×.) (Courtesy of Dr. Kenneth Phifer, Rockville, Md.)

(they are often called "enteric pathogens") have been assumed to do so through the action of their *endotoxins*. These substances are components of bacterial cell walls that usually are not liberated into the environment until the cell disintegrates (see Table 6–1). Endotoxins are known to produce marked physiologic effects on the body including fever, increased numbers of white blood cells, decreased blood pressure, hemorrhage, and even shock and death. Enteritis caused by *Salmonella* and most species of *Shigella* is considered to be endotoxic disease, but the exact relationship of endotoxin to the damaging effects produced is still not known.

In certain diseases of the intestinal tract, the primary damage is caused by *exotoxins*. In these cases the toxin remains localized in the bowel with the organism and is not absorbed from the intestine. The vibrios that cause *cholera*, for example, multiply in the small intestine, producing a type of exotoxin called *enterotoxin* because it is toxic for the bowel. This toxin binds to cells of the villi, causing cellular enzyme imbalances that result in an enormous hypersecretion of chloride, bicarbonate, and water. The effect is massive diarrhea with acidosis. In *bacillary dysentery*, the bacterial agent (*Shigella dysenteriae*) not only has an endotoxin but also secretes an exotoxin whose effects on the mucosal cells of the large bowel result in severe diarrhea. This species of *Shigella* is distinct from other members of the genus whose effects on the bowel are primarily endotoxic. Some strains of *Escherichia coli*, once regarded as harmless intestinal commensals, also have been found to produce enteric disease. In some instances their production of an enterotoxin (exotoxin) leads to a cholera-like intoxication. They are also the cause of a *Shigella*-like dysentery and are now thought to be one of the primary agents of "traveler's diarrhea."

Bacterial Food Poisoning. The term *bacterial food poisoning*, used so frequently, refers to one of three types of disease induced by the ingestion of food contaminated with bacteria or their toxic products.

The first type is an intestinal *infection* caused by members of the genus *Salmonella*. When these organisms are ingested with contaminated food or water, they may survive and multiply in the bowel. The enteritis that results is produced by this active infec-tion, but the organisms are often eliminated quickly, and their effects are therefore acute, but short-lived. In some cases, however, they may continue to reside in the bowel without producing symptoms, thus constituting a continuing threat of infection transfer to contacts of the infected individual. In some instances, following an initial enteritis, these *Salmonella* species may invade the enteric lymphoid tissue and disseminate to other organs such as the liver and gallbladder, producing fever and systemic symptoms like those of typhoid fever (i.e., paratyphoid fever).

The second type of food poisoning is *not* an infection, but results from the ingestion of food in which an exotoxin-producing microorganism has been multiplying. *Staphylococcus aureus*, for example, is notorious for its ability to produce an enterotoxin. When this organism is present as a contaminant in food and has an opportunity to multiply therein, the toxin thus preformed in the food is responsible for the acute gastroenteritis that follows its ingestion. The toxin is a severe irritant to the gastric and intestinal mucosae, but symptoms subside as soon as the last of it has been eliminated and do not recur. *Clostridium perfringens*, an anaerobic spore-forming bacterium, is another example of an organism that can produce an exotoxin in contaminated food, under the right circumstances, and can be responsible for this same type of food poisoning. *Bacillus cereus*, an aerobic spore-former, is also associated with food poisoning, but the role of its exotoxin has not been clearly defined.

The third type of bacterial food poisoning is known as *botulism*. In this case, also, a preformed bacterial exotoxin is ingested with food. This toxin is produced by the anaerobic soil organism *Clostridium botulinum*. If the organism is present in canned or vacuum-packed foods, having survived the cooking and canning process by passing into a heat-resistant spore stage, the spore may germinate and multiply under the conditions of anaerobiasis provided in the sealed container. In doing so, it secretes a substance that is a powerful neurotoxin. When it is ingested with contaminated food, the botulinal toxin is absorbed from the bowel (unlike staphylococcal enterotoxin) and exerts far-reaching, often fatal, neurotoxic effects. The organism itself, also ingested, has no effect on the bowel, nor does it penetrate the intestinal wall. The absorbed exotoxin affects motor nerves in such a way that nerve impulses cannot be transmitted to muscle, and the affected muscles cannot contract. The resulting paralysis of intestinal and respiratory musculature is severely incapacitating and may be quickly fatal (see Chap. 18).

Skin and Surface Membrane Routes

The skin and *external* mucosa offer another possible portal of entry for microorganisms, but the variety of pathogenic microbes that find their way by these paths is more limited than in the two routes discussed above. We have seen that skin surfaces have a large commensal flora, varying with the anatomic area, sex, and age. These surfaces are normally impervious to the many microorganisms with which they come in contact, and most of these are easily washed away or mechanically removed. Breaks in the skin or mucosa or some unusual capability on the part of an infecting organism apparently is required before microorganisms can enter and establish themselves (see Section VI).

Localization in the Surface Tissues. Local infectious disease of the skin or mucous membranes may be caused by *yeast and fungi* (*Candida* species, ringworm, "athlete's foot"), by *bacteria* (notably staphylococci or streptococci), by a few *viruses* (herpes infection, warts) or by the larval forms of *helminths* (the "ground itch" induced by hookworm larvae, "swimmer's itch" caused by larvae of *Schistosoma*).

Differences in the type of skin lesion produced by various kinds of microorganisms are related in part to the capacities and opportunities of the organism to localize in particular anatomic areas or in certain layers of the epidermis or dermis, and in part to the kind of reaction they elicit from the host. Thus, the so-called "superficial" fungi infect only the outermost layers of skin or its appendages, some characteristically involving the hairy skin (ringworm of the scalp); others the smooth skin surfaces, skin folds (between fingers or toes, axilla or groin), or nails. Similarly the herpes virus of the "cold sore" shows a preference for cells of the oral mucocutaneous border. Often the host reaction is minimal to very superficial infections, but if the organism involves deeper layers of the dermis, more vigorous defenses are brought to bear. Helminthic larvae entering the skin from contaminated soil or water and wriggling about under the epidermis, unable to penetrate further, elicit a good deal of response from host cells whose job it is to deal with foreign intruders. (These are the *phagocytes*, cells of the blood and tissues that ingest foreign particulate matter.) Large organisms are not easily inactivated, however, so that the reaction around them tends to be diffuse until they die and can be walled off.

Bacterial infections of the skin generally involve structures in the deeper layers of the dermis and elicit greater and more effective host reaction. Bacteria are generally not capable of penetrating intact skin, but they enter hair follicles or ducts of sweat and sebaceous glands. The secretions of these glands discourage the establishment of microorganisms, for they have an acid pH and contain antimicrobial substances, such as fatty acids and lysozyme. When bacteria do establish in these sites, by virtue of their own invasive properties or some local abnormality in the skin or its functions, the cells of the dermis make every effort to localize the infection and prevent its spread. They are assisted in this by phagocytes marshaled to the area by the bloodstream. Thus, the typical lesion of a staphylococcal infection of the skin is a "pimple" or "boil," a small abscess circumscribed by a defensive wall of material laid down around it by the host. The pus that forms in an abscess contains many phagocytes, which ingest the bacteria, while the tissue formed around the area keeps them localized. This type of host response is described more fully in Chapter 7.

The conjunctiva is normally kept moist and healthy by the continuous secretion of tears and the frequent action of the eyelids which mechanically keep the eyes clear of foreign particles. Infections of the conjunctiva may result if defects in the lacrimal gland or lid disease interferes with these cleansing mechanisms. More frequently, however, minor injuries to the conjunctival surface from foreign bodies rubbed in by contaminated fingers are responsible for conjunctivitis. The most common bacterial conjunctivitis, called "pink-eye," is spread by fingers and may travel rapidly through a household or schoolroom. Pink-eye is caused by *Haemophilus aegyptius* (a species closely related to *Haemophilus influenzae*), a fastidious Gram-negative rod once called the Koch-Weeks bacillus.

Chlamydiae are responsible for the severe conjunctival disease known as trachoma, also easily transmitted by fingers or contaminated objects. Chlamydial agents grouped in the species *C. trachomatis* also cause inclusion conjunctivitis and a venereal disease called lymphogranuloma venereum. The term *TRIC agents* is often used for this group, in a shorthand taking the first two letters of *tra*choma and the first letters of the words *i*nclusion *c*onjunctivitis. TRIC agents causing conjunctivitis can also be spread by inadequately chlorinated, contaminated swimming pool water. Pool water may not only deposit microorganisms as it flows over the conjunctiva but also may irritate the eyes somewhat so that the tendency is to rub them. The combined effect of minor mechanical and chemical irritation probably assures entry and attachment of a TRIC agent, if present, with a resulting "swimming pool conjunctivitis." The genital tract is probably involved not only in the venereal transmission of TRIC agents but also in environmental contamination as well as transferal by hands or to the newborn (see Chap. 22, pp. 510–11, and 515–16).

Conjunctivitis due to an *adenovirus* can also be spread by swimming pool water. This virus is best known, however, for its association with a conjunctivitis commonly seen in shipyard workers and called "shipyard eye" (see p. 516). Shipbuilders often get foreign bodies in their eyes and go to the medical office for treatment. The physician himself may be responsible for transmission of the viral infection if the instrument he uses for extraction of the foreign body is not sterilized between cases. This is an example of an *iatrogenic disease*, or one caused by a person giving medical treatment (*iatros* is a Greek word meaning medical treatment or healing, as in ped*iatric*; the suffix *-genic* means to cause or produce). The viral agent of "shipyard eye" is only one of many adenoviruses, most of which are associated with infections of the upper respiratory tract or other mucous membrane surfaces (see Table 14–1, p. 357).

Dissemination of Infection Through or from a Superficial Entry Point. Some of the organisms mentioned above, particularly streptococci and staphylococci, may produce substances that enable them to spread through the skin, often involving wide areas. This is the case in *erysipelas,* a streptococcal infection of the skin characterized by a diffuse and rapidly spreading involvement of dermal tissues, with an accumulation of fluid (edema) but little pus formation. In *impetigo,* an infection caused by staphylococci or streptococci, separately or together, superficial blisters form on the involved skin, break easily, and liberate infected fluid which spreads to involve adjacent areas. The invasive quality of such infections affords further risk of extension to deeper tissues. The organisms may enter lymphatic or blood vessels within the lesion and be distributed through the body, inducing severe damage to more vital tissues.

The venereal diseases constitute the most important and common group of systemic infections acquired by direct contact with infectious agents which enter through superficial mucous membranes. These are diseases associated with sexual intercourse, the causative organism being transmitted directly from an infected to an uninfected individual. The genital mucosa offers the usual route of entry to transmitted microorganisms but is not always involved in the consequences.

There are several microbial types of venereal disease. Syphilis, perhaps the most serious in its long-range effects, is caused by a spirochete, *Treponema pallidum.* Gonorrhea (caused by a Gram-negative diplococcus, *Neisseria gonorrhoeae,* also called the gonococcus) is a venereal disease that has spread out of control and today is rampant among all sexually active age groups. Another genital infection on the increase is caused by a *herpesvirus* (herpes progenitalis, also called herpesvirus type 2). There are at least three other somewhat less common, but nonetheless troublesome, venereal diseases (all bacterial in nature). These three are chancroid, or "soft chancre," caused by *Haemophilus ducreyi,* which is also known as Ducrey's bacillus; granuloma inguinale, caused by a bacillary agent called *Calymmatobacterium granulomatis;* and lymphogranuloma venereum, caused by a chlamydial organism of the TRIC group discussed above for their association with conjunctival diseases. Organisms of the bacterial genus *Mycoplasma* are also known to be associated with venereal transmission and have been isolated from "nonspecific" infections of the cervix, urethra, and prostate. Their role as agents of such infections and of certain types of reproductive failures has been suspected but not established beyond doubt.

The nature and symptoms of the systemic diseases induced by different venereal agents are widely varied and depend on the tissues involved following dissemination of the organisms from their local entry site. These diseases are discussed in Chapter 22.

Parenteral Entry

Finally, we must consider the possibility of entry of infection directly into the *deeper* tissues of the body or into the bloodstream, by (1) penetration of, or (2) deep injury to, the skin and mucosal barriers.

1. Penetration to Deep Tissues. Without significant injury to or infection of the skin or mucosa, penetration may be accomplished in one of three general ways: (a) direct entry of the organism itself into the blood capillaries or lymphatic vessels of the skin, without localization in the dermis; (b) injection of the organism by a biting or bloodsucking insect; and (c) injection of organisms by a contaminated needle, syringe, or solution.

a. ***Direct entry of an organism*** to the parenteral tissues, as an active rather than a passive accomplishment, is relatively rare. Intact skin and external mucous membranes present a strong barrier to microorganisms, but injured or infected skin may permit access. The health of the skin and surface membranes is of great importance, particularly when the risk of infection is high, as in the handling of infectious material from sick patients.

The best examples of organisms with an active capability for penetration of human skin are found among the helminth larvae, notably the hookworms and *Schistosoma* (the blood flukes). Infective hookworm larvae on the surface of moist soil or *Schistosoma* larvae swimming in water following their emergence from their intermediate snail hosts attach

to the surface of the skin, wriggle through, and are picked up in the bloodstream from the capillary beds in the lower layers of the skin. From there they find their way to tissues in which development into the adult form may occur.

b. ***Injection of an organism by a biting or bloodsucking insect*** is the access route for a number of human diseases. The entire group of parasites that may gain entry in this fashion includes members of each of the major classes of microorganisms, with the exception of the yeasts and fungi.

As an accidental host, man is occasionally infected by certain species of rickettsiae that infest ticks or mites and ordinarily are passed around only among wild rodents and other small animals that these insects parasitize. Man, if he happens to be around, is also attractive to ticks and mites, and through their bites he may become infected with their rickettsiae. Two diseases acquired in this way are Rocky Mountain spotted fever and the Oriental infection known as tsutsugamushi. On the other hand, diseases such as malaria and yellow fever are *primarily human infections,* kept going in the human population by mosquito vectors, which transmit them from one person to another (Fig. 6-6). Between these two extremes there are a number of diseases vectored by a variety of insects (mosquitoes, biting flies, lice, ticks, and mites), whose microbial agents find suitable reservoirs in several animal species as well as man.

The microbial agents of diseases transmitted to man by biting insects include *protozoa* (such as the organisms responsible for malaria, the trypanosomes of "African sleeping sickness," and species of *Leishmania*); *roundworms* of the group known as filaria; a few *bacteria* (such as the plague bacillus and *Bartonella*); the *rickettsiae* of typhus fever, rickettsialpox, and others mentioned above; and a number of *viruses* (yellow fever, dengue fever, and Eastern and Western equine encephalitis viruses). In addition, the relapsing fevers caused by *spirochetes* of the genus *Borrelia* are vectored from man to man by body lice or from infected animals to man by ticks.

Many of these diseases occur naturally only in tropical or semitropical areas of the world, primarily because of climatic restrictions on the insect vector.

Fig. 6-6. A mosquito of the genus *Aedes* transmits yellow fever. (Courtesy of Center for Disease Control, D.H.E.W., Atlanta, Ga.)

However, geographic boundaries no longer prevent the mixing of human populations as they once did, so that the recognition and treatment of these diseases may be important in any area to which people have traveled after becoming infected. Also, the prevention and control of such diseases have become an ever more pressing concern in the interest of assuring the welfare of human populations struggling to reach civilized standards of health.

c. *The injection of organisms by a contaminated needle, syringe, or solution* is an ever-present possibility wherever these items are used. Pathogenic, or merely "opportunistic," organisms may be introduced in this way into parenteral tissues or the bloodstream, sometimes with disastrous results. The virus of serum hepatitis, or type B hepatitis, is a good example. (This is another disease that is frequently iatrogenic in origin.) Because the route of entry is unnatural, however, some organisms may not be able to survive. Also, some microorganisms appear to be totally incapable of pathogenicity and are easily eliminated by the host's defenses. In some instances, the host may react to protein antigens of the organism and develop symptoms of *hypersensitivity* (see Chap. 7). Constant maintenance of strict techniques of sterilization and asepsis offers the only assurance that infection cannot be acquired by this route.

2. Deep Injury to the Skin and Mucosal Barriers. Trauma to the skin or mucous membranes, from surgery, accident, or the bite of an animal, opens up many possibilities of entry by microorganisms of all types.

During surgery, these possibilities are strictly limited or eliminated by the sterilization of instruments, linens, and clothing, and by the aseptic techniques of the operating room. Commensals of normally low pathogenicity from the skin or mucosal surfaces may take advantage of the surgical situation, however, and enter to infect the wound or more distant tissues. Surgical entry of the bowel, or its perforation in areas ulcerated by infection or other causes, creates the risk of exposing the peritoneal cavity to microorganisms usually confined to the intestinal tract, and peritonitis may result. Similarly, major or minor wounds in the mouth resulting from tooth extraction or other oral surgery may admit normal mouth flora to deeper tissues or to the bloodstream, with consequent risk of systemic infection.

The bite of an animal introduces microorganisms that may be normal residents of the animal's mouth but are pathogenic for human tissues. They may also have been pathogenic for the animal, which bites because it is ill and thus transmits infection. Among microorganisms that can be transmitted to human beings in this way the one with the most serious potential is the rabies virus. Rabies in man is almost always fatal, although its development may be prevented. Following the bite of a rabid animal, injections of a related but nonvirulent virus that stimulates the body to produce antibodies against the rabies virus must be given. If antibody is formed before the virus enters the cells of the brain, where it causes irreversible effects, the rabies virus is inactivated and the threat of disease avoided. (The development of this method for prevention of rabies was another of the contributions of Pasteur.) Rabies is often associated with dogs, which acquire it from the bites of other animals. Wild animals of the forest, and bats, appear to constitute the reservoir of this disease.

Other accidental wounds, such as lacerations, gunshot wounds, or crushing injuries, may also introduce a variety of microorganisms into deep tissues from the body's surfaces or from those of the penetrating or crushing objects. The risk of infectious disease is always increased in such instances, not only because the barriers are broken, but because the local tissue defenses are handicapped by the injury. Organisms may thus have a much better opportunity to multiply and penetrate deeper before the tissue has a chance to heal and defend itself. Normal blood supply to the injured tissue may be cut off or seriously reduced if blood vessels are damaged or compressed by tissue reaction. This means that substances normally carried in the blood to an area of injury to take part in its defense and healing are prevented from reaching it in adequate concentration. This would include antibiotics administered to assist in combating infection. For these reasons, the prompt surgical cleaning of such wounds, with removal of foreign objects, dead tissue, or blood clots, is of foremost importance in the prevention of infection and the promotion of healing.

The Toxins of Wound Pathogens. Organisms of particular importance with respect to their production of powerful exotoxins should be noted here as possible contaminants of deep traumatic wounds. These are anaerobic, spore-forming bacilli, members of the genus *Clostridium.* These organisms are frequent members of the normal intestinal flora. They are also present in the intestinal tracts of many animals, including dogs, horses, and cattle, and therefore they are widely distributed in soil. Because they can form resistant spores, they may survive for very long periods in this stage and be found in dust as well as soil indoors or outdoors. Consequently, they are often present on objects with which wounds are inflicted, such as rusty nails, sharp sticks, animal claws, knives, and bullets. *Clostridium* spores are also frequently found on the skin or bed linen of sick patients, a fact of some importance to the patient with a surgical or accidental wound, if the wound permits the conditions necessary for their growth (see Chap. 27). The damage they may induce is a consequence of their germination as actively reproducing bacilli that produce toxins. The latter exert their effects locally in the wound, or they may be disseminated.

Local clostridial toxicity is exemplified by the clinical condition known as gas gangrene. The species of most importance in this very serious infection is *Clostridium perfringens* (once named the "Welch" bacillus, or *Cl. welchii*). Several exotoxins are produced by this organism as it grows anaerobically within injured tissue. These substances have a disintegrating effect on adjacent living host cells, so that additional areas of dead tissue are provided for the growth of the organism, more toxins are produced, and the infection spreads, often with great rapidity, threatening large areas and even life itself. This is an example of infection spreading locally by virtue of the effects of exotoxin within the area.

Disseminated clostridial toxicity is exemplified by the disease known as *tetanus,* or lockjaw, caused by the species *Clostridium tetani.* In this case also the germination of tetanus spores within a contaminated anaerobic wound leads to multiplication of the organisms and exotoxin production. However, neither the toxin nor the organism exerts much local effect, this species being incapable of tissue invasion. The serious consequences of tetanus arise from the absorption of exotoxin from the local area and its extension through the bloodstream or along peripheral motor nerve trunks to the spinal cord. It is a *neurotoxin,* which interferes with normal transmission of nerve impulse to muscle, so that severe spasm and convulsive contraction of voluntary muscles result from its action. When muscles of the face, jaw, and neck become "locked" in contraction, the mouth cannot be opened, and swallowing is difficult or impossible, hence the term *lockjaw.* Involvement of the muscles of respiration may have fatal consequences.

The possibilities of growth of clostridial contaminants in traumatic wounds with elaboration of exotoxins offer another urgent reason for prompt surgical removal of dead tissue, the restoration of an adequate blood supply, and aeration of the wound.

Damaging Microbial Properties

Obviously the production of infectious disease in man does not depend solely on the opportunity for entry of the body by microorganisms. If this were true, illness would be the rule, rather than the exception that it is. The biologic resistance of the human host with its many defenses provides a partial explanation for the fact that infection is more common than infectious disease. The other half of the explanation lies in the fact that the great majority of the microorganisms with which man is surrounded has few or none of the properties associated with pathogenicity and virulence.

Pathogenicity

In the preceding chapter it was stated that the *pathogenicity* of an organism is related to its ability to produce disease, as reflected in its *virulence.* Following their entry into the body, pathogenic microorganisms display the degree of their virulence in terms of two general qualities: *invasiveness,* or ability to attach, multiply, and spread in the tissues; and *toxigenicity,* or production of substances injurious to human cells and tissues. Among the microbes pathogenic for man, the properties that add up to virulence are not always fully understood, but a number of factors appear to make a definite contribution to the pathogenesis of some infectious diseases.

Virulence

The measurement of virulence is commonly expressed in terms of the numbers of a particular organism that are required to kill a given species of experimental animal (or its cells in tissue culture), under well-defined experimental conditions. Factors that must be defined in this measurement are related to variations in both the host and the pathogen. The host, or its isolated cells in the culture tube, must be susceptible to the organism. The intact animal must be infected via an entry route that permits the organism to establish itself and induce disease (the route of administration may be intravenous, intramuscular, intraperitoneal, intracerebral, etc.). The *numbers* of organisms introduced, and the *period of time* they require to induce disease, or death, or both, must also be defined. Virulence is then expressed in quantitative terms for the organism: as the *minimum lethal dose* (MLD), or the smallest number of organisms inducing disease and death of a specified host, or its cells, in a stated time; or the *50 percent lethal dose* (LD_{50}), which allows for variations in host resistance and states the numbers of organisms required to kill 50 percent of all the infected animals or cell cultures within a certain time.

These terms reflect both the invasiveness and the toxigenicity of a pathogenic organism under controlled conditions and help to explain differences in virulence as they are observed under natural conditions. Thus, the natural occurrence of infectious disease in an individual host may be the result of entry of large numbers of relatively *avirulent* organisms, or small numbers of a *virulent* one.

Invasiveness

Specific explanations of the invasive character of virulent organisms have been found in many instances. In general, they relate to the ability of an organism, once it has gained access to the body of a suitable host, to (1) grow and multiply, (2) protect itself against the defenses of the host, and (3) penetrate and spread through the tissues.

1. Factors Involved in Growth and Multiplication Within the Host. All the properties of an organism that contribute to its production of disease depend first on its ability to utilize available nutrient within the host's tissues and to multiply there. If vital nutrients are not present, or not in available form or sufficient quantity, the invading organism cannot grow and therefore appears to be avirulent. Sometimes the failure to grow reflects the presence or absence of oxygen at the entry site (depending on the organism's requirements); the prevailing pH of the area; the degree of competition offered for a particular growth factor (such as a vitamin or a certain mineral) by the host's own cells or by commensal organisms resident in the area of entry; or the host's deficiency with respect to particular nutrient requirements of the microorganism. Conversely, the successful utilization of available nutrient by the invading organism may deprive host cells of growth factors vital to their own health, so that tissue injury may be the result of such starvation, whether or not other microbial properties are involved.

2. Factors That Protect Microorganisms from Host Defenses. Phagocytosis is one of the important processes by which the body deals with foreign particles or microbes that find their way into tissues. The invasiveness of some bacterial pathogens appears to be related to their ability to resist phagocytosis or to destroy the phagocytes themselves.

The ability to resist the action of phagocytes is due in some cases to the nature of surface components present on the bacterial cells. Many species of bacteria are encapsulated by either viscous polysaccharide material (e.g., strains of pneumococci) or polypeptides (anthrax bacilli), or possess surface proteins (streptococci). Such components of the cell either prevent their ingestion by phagocytes or, as is the case with the waxy cell wall of the tubercle bacillus, enable them to survive it. The products of bacterial growth may be protective in other instances. For example, the familiar *coagulase* associated with some strains of staphylococci is an extracellular enzyme secreted by the organism under optimal conditions for its growth. It induces clotting of the plasma that exudes from capillaries into an area of tissue injury. When staphylococci are themselves producing this injury, and the infected area in which they are growing contains coagulase they have produced, they become surrounded by a fibrinous clot as plasma seeps in and is coagulated. The organisms are thus walled away from the action of phagocytes, as well as from other host defenses.

The ability to kill phagocytes is associated with the production of a substance that is toxic for leukocytes, which is called *leukocidin*. Leukocytes are white blood cells, some of which are highly phagocytic. These cells enter an injured area from adjacent capillaries, but may be inactivated and killed by leukocidin secreted into the tissue by invading organisms, such as actively growing streptococci and staphylococci.

Phagocytic cells contain many small cytoplasmic granules called *lysosomes* (see Chap. 7). These are little membranous sacs containing various enzymes that are important in intracellular digestion. One of these enzymes, *lysozyme*, is capable of lysing some bacteria by breaking down a component of their cell walls, after they have been engulfed by the phagocyte. In some instances, the tables may be turned by virulent microorganisms taken up by phagocytes. For example, a hemolytic toxin (*hemolysin*) of some streptococci is not only capable of destroying host red blood cells but may also destroy phagocytes that have engulfed the organisms. Hemolysin release within the phagocyte has an immediate effect on the lysosome granules, causing them to explode and discharge their content of digestive enzymes into the cell's own cytoplasm. The "suicidal" result is that the phagocyte's own enzymes digest and liquefy its cytoplasm and the cell dies. Staphylococcal hemolysins too can kill phagocytes, but the virulence of staphylococci is also related to their production of a non-hemolytic leukocidin that acts on phagocytic cell membranes, breaking down the lysosomal sacs with the same lethal effect.

3. Factors That Contribute to the Penetration and Spread of Microorganisms.

A number of the extracellular products of bacterial growth have been associated with specific kinds of damage to host tissue that permit spread of the offending organism. Like the staphylococcal coagulase, mentioned above, several of these products are enzymes for which the substrates are components of host tissue. Some of these enzymes are listed below.

Collagenase, an enzyme that disintegrates the proteins of collagen tissue.

Mucinase, an enzyme that digests the mucin produced by many mucosal lining cells.

Hyaluronidase, an enzyme capable of breaking down hyaluronic acid, the gel-like matrix of connective tissue.

Lecithinase, an enzyme destructive for lecithin, a component of many human cells including red blood cells.

Kinase, an activator of an enzyme present in plasma. When activated, the enzyme, called fibrinolysin, is capable of dissolving fibrin clots.

Hemolysins, substances capable of destroying red blood cells. Some hemolysins, such as lecithinase, are enzymes; others are not (Fig. 6–7).

The production of one or more of such substances by a microorganism multiplying within host tissues creates additional injury to the latter, disrupting their normal architecture, interfering with their defensive activities, and promoting the spread of the organism to adjacent tissue. Blood and lymph vessels within the injured area may be eroded in the process, so that organisms can enter the blood and lymph streams to be distributed to other parts of the body. The flow of blood into the infected region is often reduced or cut off during the course of such events, so that the concentration of both nutrient and defen-

Fig. 6–7. A photomicrograph of rabbit erythrocytes prior to (*A*) and five minutes after (*B*) injection of the animal with purified pneumolysin (a hemolysin of the pneumococcus). Note crenation and shrinking of the red blood cells caused by the hemolysin. (Reproduced from Shumway, C. N., and Klebanoff, S. J.: Purification of Pneumolysin. *Infect. Immun.,* **4:**388, 1971.)

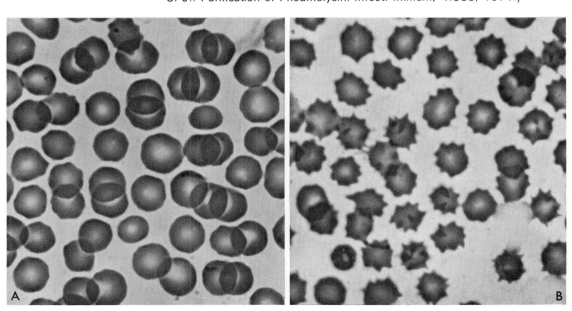

sive substances brought by the blood is further diminished.

One excellent example of a cycle of destruction to tissue set up by the enzymes of an invading organism is seen in the case of gas gangrene, caused by *Clostridium perfringens*. In this case, the anaerobic bacilli get their start in traumatized tissue to which the supply of blood and oxygen is low. As they multiply, they produce many exotoxins among which, in this case, are enzymes with *collagenase* and *lecithinase* activity. These substances break down healthy tissue on the margins of the injured area, enlarging the damage, releasing more nutrient for the organisms from host cells, and extending the zone of reduced blood supply. Continuing growth of the organisms means continuing production of enzymes and rapid extension of tissue injury.

The spread of streptococci through the tissues is often associated with their production of *hyaluronidase*. Many strains of streptococci and staphylococci secrete *hemolysins* (called, appropriately, streptolysins or staphylolysins) as well as *kinase* (streptokinase, staphylokinase), the latter preventing their restriction within fibrin clots formed in the course of the body's defensive inflammatory response (see Chap. 7). *Mucinase* is produced by influenza virus attached to mucosal cells of the respiratory tract. By altering the surface characteristics of mucin, changing it from a viscous to a thin watery substance, mucinase helps to assure the penetration and attachment of influenza virus to ciliated epithelial cells (Fig. 6–8), thus contributing to its invasiveness.

Toxigenicity

Endotoxins are components of the cell walls of Gram-negative bacteria that are released into the surrounding area, usually after death of the cell. Chemically, endotoxins are composed of lipid and carbohydrate. The lipid portion is responsible for the physiologic effects of endotoxin (e.g., fever, blood vessel changes that may lead to hemorrhage, shock, and death). The carbohydrate portion is the antigen that stimulates the production of circulating antibody against endotoxin (see Chap. 7). Infections caused by the endotoxin-containing Gram-negative bacteria are often associated with lots of tissue and blood vessel damage. Although many of the symptoms of these diseases are referable to the release of endotoxin, conclusive evidence as to its role is lacking. Table 6–1 shows some distinctions between endotoxins and exotoxins.

Exotoxins, as we have seen, are substances, usually protein in nature, produced and secreted by bacteria during their growth. Only a few bacterial pathogens

Table 6–1.   Distinctions Between Bacterial Endotoxins and Exotoxins

Endotoxins	Exotoxins
1. Components of bacterial cell walls	1. Secreted by or released from bacterial cells into surrounding medium
2. Stimulate production of antibodies of questionable protective value (Chap. 7)	2. Stimulate production of antibodies (antitoxin) of high protective value (Chap. 7)
3. Toxic effects nonspecific with regard to microbial agent; produce similar symptoms (fever, weakness, generalized aches)	3. Toxin effects specific to each exotoxin; clinical symptoms distinctive
4. Endotoxic diseases: typhoid and paratyphoid fever, gonococcal and meningococcal disease	4. Exotoxic diseases: botulism, cholera, tetanus, gas gangrene, scarlet fever, diphtheria (see Table 6–2)
5. Antitoxins not available	5. Antitoxins available for treatment (botulism, tetanus, diphtheria)

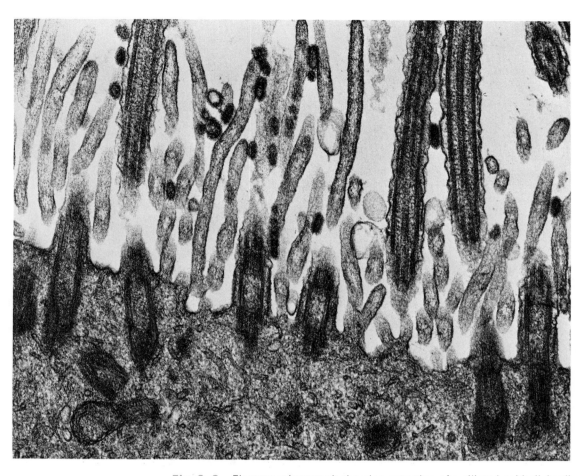

Fig. 6–8. Electron micrograph showing a portion of a ciliated epithelial cell from an organ culture of guinea pig trachea after incubation with influenza virus for one hour at 4° C. Virus particles are attached to cilia and to microvilli. (Reproduced from Mims, Cedric A.: *The Pathogenesis of Infectious Disease.* Academic Press, London, 1976. Photograph courtesy of Dr. R. R. Dourmashkin, Clinical Research Center, Middlesex, England.)

secrete exotoxins. These toxins differ from each other chemically as well as in their biologic effects, each producing a specific type of injury to particular cells and tissues (Table 6–1). The important toxin-producing bacteria have all been mentioned in connection with their usual entry routes into the human body and the manner in which their toxins induce disease. These diseases are summarized briefly in Table 6–2.

Other Factors Contributing to Infectious Disease

In addition to chemical injuries that may be caused by products and secretions of microorganisms, a good deal of *mechanical* damage may be done. As foreign bodies their presence in the tissues is irritating and may be mechanically disruptive. A mass of

Table 6–2. Diseases Caused by Bacterial Exotoxins

Disease	Type of Toxin	Nature and Site of Injury
Bacterial Food Poisoning		
Staphylococcal enteritis	Enterotoxin	Severe irritation of gastric and intestinal mucosa
Botulism	Neurotoxin	Interrupts motor nerve impulses; muscle paralysis (e.g., respiratory muscles)
Cholera	Enterotoxin	Stimulates enzyme that causes hypersecretion of water, chloride, bicarbonate from small intestine
Scarlet fever	Scarlatinal exotoxin (erythrogenic)	Affects epithelial cells of skin and mucosa of upper respiratory tract
Diphtheria	Exotoxin, blocks protein synthesis	Interferes with protein synthesis in many tissues, especially cardiac muscle, nervous tissue, kidney
Tetanus	Neurotoxin	Interrupts motor nerve impulses; muscle spasm (e.g., face, neck, and respiratory muscles)
Gas gangrene	Lecithinase	Destroys lecithin in muscle tissue locally, or in red blood cells
	Collagenase	Destroys collagen protein in muscle tissue locally

worms in the intestinal tract, for example, may cause intestinal obstruction. The presence of helminth larvae in organs such as liver, bone, or muscle may disrupt normal architecture and interfere with tissue function. They may also block the flow of blood through a vessel and thus diminish the nutrition of adjacent tissues. These physical effects are often aggravated further by the nature of the host's reaction to a foreign body. The supply of blood to an injured area usually is increased, and both cells and fluid from the blood exude into the tissue, causing swelling. This increases the pressure and further disrupts the functions of the tissue. As the reaction proceeds, fibrinous tissue is laid down around the area (Fig. 6–9). When the infection subsides, the damaged region may be filled in with fibrinous scar tissue, then further solidified with deposits of calcium, or it may be completely cleared and returned to normal.

Another form of injury that may result from the host's own reactions to an invading organism is known as *hypersensitivity*. If the host becomes overreactive to a microorganism or its products, the tissue at the site of reaction is often damaged. Hypersensitivity is developed through the same mechanism that leads to immunity. It is discussed in Chapter 7, together with other host responses to infection. Table 6–3 summarizes the several mechanisms of microbial pathogenesis.

Unanswered Questions Concerning Microbial Disease

Many explanations have yet to be found for the pathogenesis of some infectious diseases. Virulence has been defined as a combination of invasiveness and toxigenicity; yet neither of these is synonymous

Fig. 6-9. An adult trematode, or fluke (*Clonorchis sinensis*), established within a bile duct. (10×.) Its presence there is injurious to the architecture and function of the duct. The host's inflammatory reaction and the fibrosis that occurs later also narrow or obstruct the duct. (Courtesy of Dr. Kenneth Phifer, Rockville, Md.)

Table 6-3. Mechanisms of Microbial Pathogenesis

Routes of Entry	Microbial Activities (Pathogenicity and Virulence)

A. Respiratory tract (droplets and droplet nuclei)
B. Gastrointestinal tract (milk, water, food, fingers, fomites)
C. Intact skin and mucosa
D. Parenteral entry
  1. Trauma
    a. Accidental
    b. Surgical
  2. Injection
    a. Accidental (syringe and needle)
    b. Insect vectors
	A. Attachment and multiplication at entry site
	  1. Mechanical damage
	  2. Metabolic competition
	  3. Chemical damage (toxins)
	  4. Allergenic effects (host hypersensitivity)
	B. Invasion and multiplication at local or distant sites; factors 1, 2, 3, and 4 as above
	C. Toxigenic factors
	  1. Exotoxins
	  2. Endotoxins
	  3. Enzymes (e.g., lecithinase, coagulase, hyaluronidase)

Fig. 6–10. The balance between the microbe and the host is an ever-changing one. The virulence of the organism is pitted against the resistance of the host. What factors tip the seesaw in favor of the host? In favor of the organisms?

with production of disease. For example, some viruses may widely invade the human body without producing apparent ill effects; some endotoxic bacteria are harmless commensals of the intestinal tract (e.g., strains of *Escherichia coli*) yet their close relatives are correlated with severe diarrheal disease, especially in infants; and some organisms display strict host specificity with regard to virulence (as in the case of the strain of tubercle bacillus that is virulent for man and the guinea pig but does not produce disease in the rabbit).

The evaluation of an organism as the virulent agent of a given disease once rested on the conditions or "postulates" originally outlined by Robert Koch in the late 1800s. These require the repeated isolation of the organism from cases of the clinical disease having similar or identical symptoms; its repeated recognition in pure laboratory culture; its ability to produce the characteristic disease in a suitable experimental animal; and its reisolation from the infected animal. These criteria satisfactorily established the microbial agents of many infectious diseases, particularly those caused by bacteria and higher organisms, but they are difficult to apply to viruses, and they do not explain virulence as such, or host specificity.

It may be that, like infection, virulence is an expression of the combined properties of the host, or a host species, and the pathogen. Infection is a dynamic state in which two organisms continuously oppose each other. In a simple analogy, it might be said that the relationship between a parasitic microorganism and the host that harbors it is rather like a seesaw, the position of the board depending on the weight each opponent can bring to bear on the other (Fig. 6-10). It depends, also, on the balance point, and whether its placement allows either adversary an advantage over the other. Thus, a normal, healthy individual balanced against an organism of low virulence may easily hold his own, but a person whose resistance has been lowered for some reason might have less success with the same organism. In the same way, virulent microbial species may induce more or less serious disease, depending on the host's ability to control the situation. Their failure to do so at all in a certain type of host may reflect the weight of biochemical properties brought to bear by one type of host as compared with another.

We have considered some of the factors that add weight to the parasite's position on the board; now let us turn our attention to the human host and the power with which he can balance or tip the board in his own favor.

Questions

1. List the routes through which microorganisms gain entry to the human body.
2. Give examples of diseases that may be acquired through each entry route.
3. What determines where or whether an organism will ultimately establish itself and multiply in the human body?
4. How do pathogenic microorganisms display their virulence following entry into the body?
5. List the extracellular products of bacterial growth that contribute to penetration and spread of microorganisms.
6. Compare endotoxins and exotoxins.

Additional Readings

*Dubos, René: *Mirage of Health*. Harper & Row, New York, 1971.
McNeill, William H.: *Plagues and Peoples*. Anchor Press, Doubleday, New York, 1976.
Mims, Cedric A.: *The Pathogenesis of Infectious Disease*. Academic Press, Inc., London; Grune & Stratton, New York, 1976.
*Youmans, Guy P.; Paterson, P. Y.; and Sommers, H. M.: *The Biologic and Clinical Basis of Infectious Diseases*. W. B. Saunders Co., Philadelphia, 1975.
*Zinsser, Hans: *Rats, Lice and History*. Little, Brown and Co., Boston, 1975.

*Available in paperback.

7

Host Versus Microbe

To maintain its position and balance in a world in which many antagonizing biologic forces are constantly brought to bear upon it, the human body has had to develop many defensive mechanisms. It has probably done so continuously from the time of its earliest evolution, as have other living organisms. The problems created by the inroads of parasitic microorganisms have in particular necessitated development of protective processes. Some of these are *nonspecifically* directed against *any* foreign substance or particle that may find its way into the body, whereas others are *specifically* developed in response to particular microorganisms or their products. Most of these responses of the body, if adequate and successful, contribute to health and well-being, but they may also lead to self-damage if the specific (i.e., *immune*) mechanisms become overreactive, causing hypersensitivity or autoimmune diseases. Deficiencies in the immune system may seriously impair the body's capacity to resist even minor infections.

In this chapter we will consider the most important factors, both nonspecific and specific, that enable the human host to resist infection and describe some of the diseased conditions resulting from disorders of the immune system.

Nonspecific Resistance to Infection

The *general* resistance offered by the healthy body to most microorganisms is nonspecific in two connections: it does not operate against any one organism only, but is effective against microbes in general; and in many particulars operates as well against nonliving foreign bodies, functioning also to inactivate chemical substances that do not necessarily originate in living organisms.

The most important and best-defined host factors involved in nonspecific resistance are associated with functions of the skin and mucous membranes, cellular and fluid elements of the blood, cells of the reticuloendothelial system (RES), and the lymphatic system. The phagocytic cells of the blood, RES, and lymphatic system play a vital role in the body's defense. The mechanisms of phagocytosis, the combination of factors involved in the defensive inflammatory response, the role of interferon in resistance to viral infection, and other less well-defined factors will be discussed here.

Entrance Barriers

The skin and mucous membranes offer the first barrier to microorganisms coming into contact with the body. If these structures are intact and functional, entry to deeper tissues is denied to most microbes. This is due to both the architectural and the physiologic properties of these tissues, as well as to the role of phagocytes and plasma components constantly supplied to them by the bloodstream.

Architecture of Skin and Mucous Membranes. The structure of the surface tissues is such that a continuous sheath of closely bridged cells covers all exposed surfaces, supported beneath by dense connective tissue. In addition, the outermost surface of the skin is layered over with *keratin,* a protein substance that is not only tough and dense, but waterproof.

At the opening of the respiratory tract, the nasal cavity has a mucociliary lining, similar to that which covers much of the lower respiratory surfaces. Ciliated epithelial cells compose this lining. They are bathed by mucus secreted from individual cells (goblet cells) and from subepithelial glands. Microorganisms or other foreign particles carried into the lower airways are trapped in mucus and swept up to the back of the throat by the beating action of this "mucociliary escalator." In the nose, the nasal hairs as well as the ciliary lining also sweep particles to the back of the throat, from whence they are swallowed. In the upper part of the nasal cavity, the "baffle plates" formed by the turbinate bones are covered by a ciliated mucosa that acts as a trap for small particles. At the lower end of the respiratory tract the alveoli do not have ciliated cells or mucus, but they are lined by phagocytic cells that engulf foreign particles that may manage to progress that far. In the mouth, the salivary gland secretions provide a mechanical flushing action, and the movements of the tongue and the reflexes of swallowing ensure that most ingested or inhaled particles are carried to the stomach.

Physiology of Superficial Tissues. The physiologic defenses of the skin include its secretions, notably those of the sebaceous and sweat glands, which have an antimicrobial effect. Fatty acids of the sebaceous gland secretions discourage some fungi, as well as bacteria, from colonizing on the skin or scalp. The acid and salt concentrations of sweat provide a similar protection. *Lysozyme* is an antimicrobial substance present in mucus, tears, saliva, and other body secretions, as well as in phagocytes (Chap. 6, p. 132). This is an enzyme that can break down some bacterial cell walls by splitting an important protein-carbohydrate constituent therein.

In the stomach, the acidity of gastric secretions destroys many microorganisms that are swallowed. Farther along the intestinal tract, the action of the body's digestive enzymes, together with microbial competition from commensal flora, keeps the surviving newcomers in check. Large numbers of commensal bacteria are closely associated with the intestinal wall, maintaining permanent residence either in the mucous layer or attached to the epithelial cells. Pathogenic bacteria probably must find a foothold for attachment if they are to survive in this milieu of mucus, acid, enzymes, bile, and the metabolic products of competing organisms. The constant motion of the intestinal tract, with contraction and expansion of the villi, keeps the material in the lumen moving along, making attachment more difficult. However, some pathogenic enteric bacteria may multiply in the lumen, their increasing numbers enhancing the opportunities for attachment. *Salmonella* species and pathogenic strains of *E. coli* can attach to the microvilli that form the brush border of intestinal epithelial cells, as we have seen that cholera vibrios do (Fig. 6–3, p. 126). The enteroviruses (e.g. poliovirus) must also be bound to receptor sites on the surface of intestinal lining cells if they are to penetrate and survive, since they can multiply only within living cells. Whether or not disease follows attachment and penetration depends on microbial multiplication and spread, damage from exo- and endotoxins, and the body's defensive reactions to such factors. Normally, most ingested transient microorganisms are either destroyed in the intestinal tract or finally removed with passage of fecal material from the body.

In the genitourinary tract, the surface tissues are

similarly covered by a mucous film. The direction of mucus flow, its pH, salt concentration, and lysozyme content, as well as the activity of resident commensals, tend to discourage the localization of transient organisms. The pathogenic organisms of the venereal diseases, however, have a special opportunity for entry during sexual intercourse with an infected partner and are then usually able to attach to and penetrate the genital mucosa, establishing new infection.

When the skin or mucous membranes are damaged or diseased, their ability to withstand infection, or prevent the penetration of microorganisms to other tissues, is compromised. Even minor wounds and abrasions may open up routes of entry for microbes. Moisture is important to the integrity of surface structures, healthy skin being relatively dry, mucous membranes normally moist. Skin becomes macerated in the continuing presence of excessive moisture, such as may accumulate under tight, wet dressings. The risk of infection by organisms trapped in such moisture is increased, while at the same time damage to the skin's structures decreases its effective resistance to infection. Conversely, excessive dryness of mucosal surfaces, particularly those of the respiratory tract, also predisposes to infection. Malfunction of important structures, such as may result from the occlusion of glandular ducts to the surface (those of lacrimal, salivary, sebaceous, or sweat glands), offers still another opportunity to microorganisms to cross the surface barriers.

Blood Barriers

Blood is a vital and dynamic factor in the body's resistance to infection. Not only does it constantly supply oxygen and nutrient to all tissues, but it carries away waste products of cellular metabolism that would become toxic if they were to accumulate. If circulation is diminished or cut off from any area, tissue injury or death (necrosis) may result, and the involved area becomes more susceptible to the activities of any microorganisms that may enter it. Blood is comprised of a fluid portion called *plasma,* and of *cellular* constituents. Both have important functions in responding to foreign substances, including microbes and their products.

Plasma. The fluid portion of blood, called *plasma,* has many components. It is a colloidal solution of proteins in water, containing also carbohydrates, lipids, vitamins, hormones, electrolytes, and dissolved gases — all of which are carried to and exchanged in the tissues nourished by the bloodstream. Antibiotics and other drugs are also carried in the plasma. The proteins present in largest quantity are *albumin, fibrinogen,* and *globulins. Albumin* is responsible for much of the osmotic force of the blood and is thus important in regulating the exchange of water between plasma and the intercellular compartment, or the fluid space between cells. *Fibrinogen* is a vital factor in blood clotting, being converted by thrombin into fibrin. This conversion also occurs as a part of the inflammatory response. *Globulins* occur in three major groups known as alpha, beta, and gamma globulins. Alpha and beta globulin fractions are involved in the transport of other proteins, combining with them or with other substances, so that they can be circulated for metabolic needs. *Gamma globulins* are of greatest importance in immunity. During the development of immunity (i.e., specific resistance), gamma globulins known as *immunoglobulins,* or *antibodies,* are synthesized. These circulate in the blood or other fluids to provide "humoral" immunity (*humor* is a Latin word meaning moisture and was used in medieval medicine to denote a body fluid or juice).

Antibodies are produced in specific response to particular foreign substances and become widely distributed in body fluids, including plasma. Infectious microorganisms usually stimulate antibody production, so that the concentration of antibodies in plasma characteristically rises during the course of an infection. Other plasma components that play a nonspecific role in infection occur normally. Their concentration is not increased in response to infection, and their activities are not directed against any one organism but against infectious agents in general. These nonspecific substances may participate, however, in the reactions that occur between antibodies and microorganisms, supplementing the antibody effect or bringing it to completion.

One such nonspecific constituent of plasma is *complement,* so called because it makes an important contribution to (that is, it complements) the effect of antibodies in some of their reactions. Complement is a complex of proteins (some of which are globulins) normally present in plasma. Acting alone it has no destructive effect on foreign cells. If these cells, microbial or other, have stimulated the body to produce antibodies, the latter combine specifically with them. Complement then also attaches to the combination of cell and antibody and brings about the destruction, or lysis, of the cell. Complement activity requires the presence also of *calcium* and *magnesium,* both of which are normally present in the blood in ionized form.

Since complement is intimately involved in the interactions of antigens and antibodies, it is discussed later in this chapter under the development and functions of the immune system (p. 163).

Cellular Constituents of Blood. The cellular components of blood of importance in nonspecific resistance are the white blood cells, or *leukocytes*. Of these the principal types are granulocytes, monocytes, and lymphocytes. The *granulocytes* are further classified as neutrophils, eosinophils, and basophils depending on their staining characteristics. All of them have multilobed nuclei (they are said to be polymorphonuclear, although this term, shortened to *polymorph,* is generally reserved for the neutrophils) and contain numerous granules called *lysosomes,* which are "bags" of hydrolytic enzymes and other degradative substances. The common differential stains used to study blood cells distinguish neutrophils as having neutral-staining or azure-colored granules; eosinophils, with acidophilic, red granules; and basophils, with basophilic or dark blue granules. Neutrophils are the most numerous of the white blood cells and are most important for their phagocytic activity, as will be discussed. Eosinophils play a role in allergies. Their numbers increase in allergic reactions, including those resulting from certain parasitic infections. It is thought that they contain antagonists to substances causing blood vessel dilatation in the allergic response, and that they also are able to degrade antigen-antibody complexes that cause vessel and tissue damage. Basophils, like the mast cells of the tissues, produce histamine, a vasodilator released in allergic reactions. They also contain heparin, an anticoagulant that prevents blood clotting.

The *polymorphonuclear neutrophils* (PMN's) are the most abundant of the body's phagocytic cells.

They are produced continuously in the bone marrow and discharged in large numbers into the circulating blood. There they survive for only a few days, but their loss is balanced by new arrivals from the bone marrow, which reserves a large supply against emergency or sudden demand. PMN's are constantly available at any point of entry of foreign material into any tissue to which there is an adequate blood supply. The presence of a foreign substance, especially if it is a living microorganism, provides a chemical stimulus that attracts phagocytes to the scene. This attraction is spoken of as *chemotaxis.* The polymorphonuclear cells have an ameboid type of motility. When attracted by a chemotactic stimulus, they migrate out of the capillaries, make their way to its source, and begin the attempt to eliminate the foreign object by, quite literally, devouring it. (The Greek word for eating is *phagein.*) Moving as amebae do, these phagocytes (or *microphages* as they are sometimes called) surround and engulf particulate matter, digesting it if possible (Figs. 7–1 and 7–2). The process of phagocytosis is discussed under a separate heading.

The *monocytes* also arise from stem cells in the bone marrow. These are mononuclear cells, larger than PMN's, that are precursors of *macrophages.* Monocytes are also chemotactically attracted out of the circulating blood to sites of infection in the tissues where they become enlarged, mature, phagocytic macrophages. They are not so numerous as polymorphs, nor is there a large reserve of them in the bone marrow. Macrophages are widely distributed throughout the body; however, many are in fixed positions within the reticuloendothelial system, as we shall see.

Lymphocytes play a profoundly important role in immunity and hypersensitivity. They are involved in the inflammatory response of nonspecific resistance, but their specific function is the production of the antibodies responsible for humoral immunity and of substances (*lymphokines*) that play a part in cellular (cell-mediated) immune responses.

Reticuloendothelial System (RES)

The term *reticuloendothelial system* refers to the phagocytic cells that are widely distributed in various

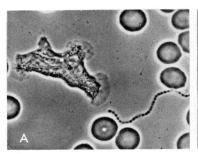
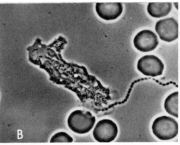
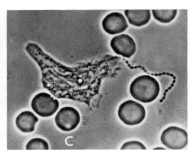

Fig. 7–1. Phagocytosis. *A*. A polymorphonuclear cell approaches a long chain of cocci. *B*. Thirty seconds later phagocytosis is in progress. *C*. After another 65 seconds the phagocytic cell is still in good condition and phagocytosis is proceeding. (Reproduced from Wilson, A. T.: The Leukotoxic Action of Streptococci. *J. Exp. Med.,* **105**:463, 1957. Courtesy of Rockefeller University Press and Dr. Grove D. Wiley, Alfred I. DuPont Institute, Wilmington, Del.)

tissues of the body. Some are fixed in the *reticulum,* a network of loose connective, interstitial tissue; some in the *endothelium,* the lining of vessels and sinusoids (capillary expansions). Fixed macrophages are also abundant in lymphoid tissue, spleen, bone marrow, lungs, liver, and central nervous system. In some of these anatomic sites they have particular designa-

tions: in the central nervous system they are called *microglia;* the phagocytic cells lining the blood vessels of the liver are called *Kupffer's cells;* in the lungs, the phagocytic cells fixed in the alveolar linings are called *dust cells* because they take up particles of dust from the inhaled air, as well as microorganisms.

Some of the phagocytes of the RES are called

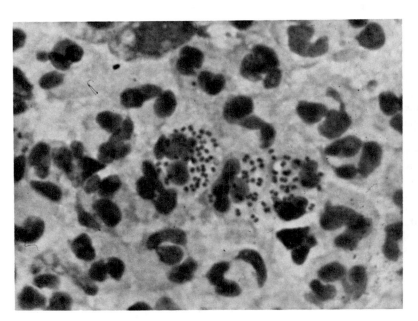

Fig. 7–2. Photomicrograph of a stained smear of genital exudate from a patient with gonorrhea. Many polymorphonuclear leukocytes can be seen, some of which have phagocytosed the infecting gonococci (*Neisseria gonorrhoeae*). Gonococci are characteristically seen as intracellular diplococci in such smears. (1000×.) (Courtesy of Center for Disease Control, D.H.E.W., Atlanta, Ga.)

wandering cells because they move about with ameboid activity, responding to chemotactic stimuli. The wandering cells include the histiocytes of connective tissue and the monocytes attracted into the tissues from the bloodstream.

Whatever their location, the principal function of the RES cells is phagocytosis and removal of foreign or useless particles, living or dead. They have a great capacity for clearing a local scene, not only of microorganisms and other unwanted intruders, but also of the body's own cellular debris, including dead or dying leukocytes, red blood cells (Fig. 7–3), and the fragments of other disintegrating cells. When the battle of infection is over, it is the macrophages and the fixed RES cells that clean up the rubble and help to restore order, so that tissues can function normally once more.

Lymphatic System

Lymphoid tissue is scattered throughout the body. Nodes of the upper respiratory tract (Fig. 7–4), cervical region, gastrointestinal system, genitourinary tract, and the inguinal area are strategically located to pick up microorganisms that get past the main entry barriers. Here most of them are filtered out by phagocytes that line the numerous branching channels within the node. Some may escape to the efferent lymph vessel, but are filtered out in the next or succeeding nodes. The lymphatic vessels themselves can be penetrated readily by bacteria, but such intruders are quickly drained to a node in the area and filtered out there.

Altogether then, the phagocytes of the blood, the RES, and the lymphatic system offer a highly efficient nonspecific defense against microorganisms, functioning continuously to maintain the safety and sterility of body tissues (Fig. 7–4). When organisms are involved whose virulent properties include the capacity to multiply inside phagocytes and destroy them, then other defense mechanisms are needed more urgently, and sometimes disease becomes persistent, even progressive. Also, lymphoid tissues constantly engaged with microbial encounters at a major entry point such as the upper respiratory tract may become chronically infected. In this way, the effective function of tonsils and adenoids may be re-

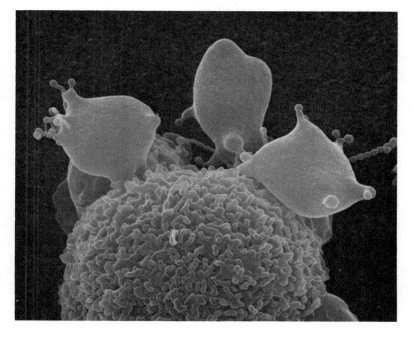

Fig. 7–3. A scanning electron micrograph of three opsonized erythrocytes in contact with the folds of a macrophage. The cells of the RES (reticuloendothelial system) clear a local scene not only of microorganisms and other intruders but also of the body's own cellular debris, such as erythrocytes. (8600×.) (Reproduced from Orenstein, Jan M., and Shelton, Emma: Membrane Phenomena Accompanying Erythrophagocytosis. *Lab. Invest.*, **36**:363, 1977. © 1977, U.S.-Canadian Division of the International Academy of Pathology.)

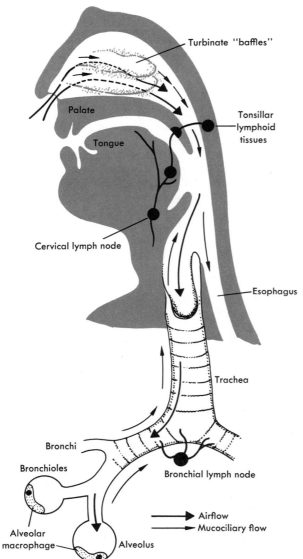

Turbinate "baffles"

Palate

Tongue

Tonsillar lymphoid tissues

Cervical lymph node

Esophagus

Trachea

Bronchi

Bronchioles

Bronchial lymph node

Airflow
Mucociliary flow

Alveolar macrophage

Alveolus

Fig. 7–4. Mechanisms of nonspecific resistance to respiratory tract infection. (Reproduced from Mims, Cedric A.: *The Pathogenesis of Infectious Disease*. Academic Press, London, 1976.)

duced, and these tissues may then constitute a possible source of infection to other areas via their efferent lymphatics. Under these circumstances, surgical removal of tonsillar and adenoidal tissue is often recommended by physicians, after duly con-

sidering the consequences of losing any protective value they may yet afford.

The lymphatic tissues of the body are also highly important in immune reactions, as discussed in the next section of this chapter.

Phagocytosis

Probably most microbes that manage to penetrate the body's outer barriers and reach the tissues find themselves imprisoned within phagocytes (Fig. 7–2). Contact with polymorphs may be a chance encounter, but PMN's respond swiftly by chemotaxis when soluble microbial products are released in the tissues. Attachment of the foreign particle to the phagocytic cell surface, and then its ingestion, is greatly enhanced by complement, a nonspecific component of plasma, acting together with specific antibodies that may also be circulating in plasma. Antibodies are large immunoglobulin molecules that combine specifically with surface constituents of microorganisms, forming a sticky coat on their surfaces. Motile organisms then become immobilized, microbial activities conducted in the cell wall are slowed down, and the microbes can no longer take in or put out soluble materials in a normal manner — they are, in a word, stuck. Thus inactivated, microorganisms are more appetizingly prepared for the phagocytic feast and more easily ingested, and digested, by the phagocytes. The antibodies that have performed this helpful catering service are called *opsonins*, from the Greek word meaning "to prepare food for." The enhancement of phagocytosis via antibody coating is termed the *opsonic effect.* This process of opsonization may also be helpful in the destruction by phagocytes of the body's own worn-out cells (Fig. 7–3).

When antibody is not available, phagocytosis is facilitated by the presence of fibrin, which serves as a sort of trap to pin down microorganisms. The same effect may operate on rough surfaces, formed by the polymorphonuclear leukocytes themselves when they adhere to the walls of small blood vessels or the alveoli of the lungs, as they sometimes do in infection. Bacteria passing by in the bloodstream, or in inhaled air, are caught on irregular edges, and "surface phagocytosis" then occurs (Fig. 7–5).

Once engulfed, ingested microbes or other particles are at once encircled within a membranous

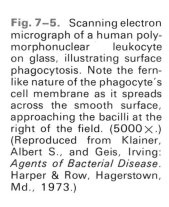

Fig. 7–5. Scanning electron micrograph of a human polymorphonuclear leukocyte on glass, illustrating surface phagocytosis. Note the fernlike nature of the phagocyte's cell membrane as it spreads across the smooth surface, approaching the bacilli at the right of the field. (5000×.) (Reproduced from Klainer, Albert S., and Geis, Irving: *Agents of Bacterial Disease.* Harper & Row, Hagerstown, Md., 1973.)

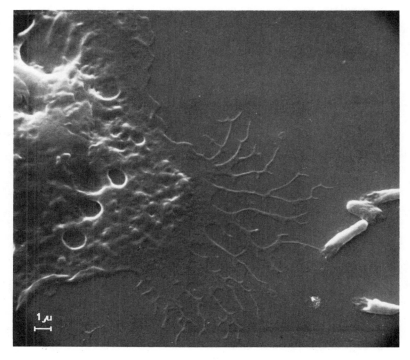

vacuole called a *phagosome*. The lysosomal granules (see Chap. 6, p. 138) of the polymorph migrate to this vacuole, fuse their own membranes with its wall, and form a *phagolysosome*. Various enzymes within the lysosomal sac (acid hydrolases, myeloperoxidases) then proceed to kill the microorganisms and digest the microbial cell contents. Residue from this process is finally extruded from the phagocyte (Fig. 7–6).

Cytoplasmic granules other than lysosomes contain cationic (positively charged) proteins that also have a killing (bactericidal) action on ingested bacteria. Polymorphs probably possess other mechanisms as well for dealing with ingested microorganisms, but these are not yet fully understood.

Phagocytosis in macrophages is essentially the same as that described for PMN's but there are some differences. Although macrophages display chemotaxis, they respond to different chemical stimuli than those that attract PMN's, and they do not migrate through tissues so readily. Their lysosomal enzymes are different; i.e., they do not contain the myeloperoxidase enzyme or other bactericidal proteins found in PMN granules. These factors account for differences in the distribution of the two types of phagocytes in tissues and for the fact that some types of organisms killed by polymorphs may survive and grow within macrophages, and vice versa. Also, macrophages normally contain fewer granules than do polymorphs, but in the immune process they are activated by lymphocytes to produce more. They have a much longer life in the tissues than do PMN's

and are often involved, as shown previously, in digesting the remains of dead or dying leukocytes or other body cells (Fig. 7–3).

Although macrophages and polymorphs are specialized for phagocytosis, this function is a basic type of activity for other body cells as well. Epidermal cells of the skin, intestinal epithelial cells, or vascular endothelial cells, for example, can be shown to take up foreign particles. However, such cells are not so well equipped as the true phagocytes for killing and digesting microorganisms they have ingested.

We have seen that pathogenic microorganisms may resist phagocytosis or survive it once taken up. Capsulated organisms, for example, may be difficult for phagocytes to deal with unless first coated with capsular antibody. Streptococcal hemolysins and staphylococcal leukocidins destroy the lysosomal membrane, releasing the PMN's digestive enzymes with disastrous effects on the cell's own cytoplasm (Chap. 6, p. 138). Indeed, the virulence of some pathogens, such as tubercle bacilli or brucellae, is associated with their ability to multiply within phagocytes. In the process they may kill the leukocytes, but first they may be carried within these cells through the bloodstream to other tissue sites, far from the point at which they were picked up. When such organisms emerge from the cells they have destroyed, they may localize in the new site, if conditions there are conducive to their growth; they may be picked up again by other phagocytes; or they may be met by more specific defenses.

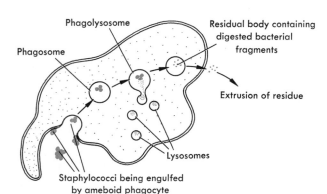

Fig. 7–6. Diagram showing phagocytosis and digestion of bacteria.

Inflammatory Response

The body's normal response to infection or tissue injury is called *inflammation.* It constitutes a marshaling of all specific and nonspecific defense mechanisms.

The inflammatory response is designed to localize foreign substances at the site of their entry, to prevent their penetration into other areas, and to assist in the restoration of injured tissues to normal function.

As the inflammatory response begins at the site of injury, the walls of small blood vessels in the vicinity dilate and become more permeable, so that plasma, rich in immunoglobulins, fibrinogen and other proteins, and phagocytic cells pass through more readily. The phagocytes stick to the lining of blood vessels, then pass between the endothelial cells by a process known as *diapedesis,* and migrate to the affected tissues. Initially, PMN's are the most numerous cells present, but in later stages of inflammation, monocytes and tissue macrophages predominate. In chronic infections, the larger phagocytes of the tissues fuse together to form giant cells, sharing their many nuclei. Even if they cannot destroy ingested organisms, these fused cells assist in segregating them from adjacent tissue. Segregation proceeds most efficiently through fibrosis, i.e., the continuing formation of a fibrous connective tissue wall around the site. Within the active zone, plasma components exert their effects and cellular efforts continue, the battle's outcome depending on the host's vigor weighed against the nature and extent of the injury (Fig. 7–7). Many microorganisms die under these conditions, but some may remain viable though quiescent for long periods of time.

The collection of fluid and cells that forms within the area of injury is called the *inflammatory exudate.* This material may vary a good deal in consistency and character. When it is thick and yellow with large numbers of leukocytes, it is called a *purulent,* or *suppurative,* exudate (more commonly, *pus*). A microbial agent that elicits this type of response is said to be pus-producing, or *pyogenic.* When the exudate contains more plasma than cells and is therefore thin, it is called a *serous* exudate. If it contains sufficient fibrinogen to form a fibrin clot, it is referred to as a *fibrinous* exudate.

If it occurs superficially in skin or mucous membranes, the inflammatory reaction usually becomes quickly manifest with the local development of redness, heat, tenderness, and swelling, as the blood supply to the area increases and as exudate accumulates, pressing on the adjacent nerve supply. Individual tissue cells are compressed as the spaces between them fill up. As the area becomes walled off with fibrinous clots and a fringe of connective tissue, it grows more tense and hard to the touch. An *abscess* is forming, reddened on its periphery by excessive blood supply, yellowing at the center where the cellular exudate is heaviest. Eventually it may rupture spontaneously through the surface (*suppurate*), or it may be opened surgically to drain its contents. Abscess formation in the deeper tissues and organs of the body follows much the same course, provoking symptoms of tenderness or tissue dysfunction. A deep abscess may rupture and disseminate its contents to adjacent tissues unless it can be approached and drained surgically. However, the inflammatory reaction may proceed quite effectively to its own conclusion. Most successfully, inflammation ends with complete solution of the problem before permanent tissue damage is done; the exudate is absorbed, cellular debris is completely phagocytized, fibrinous deposits are removed, and the area returns to a state of normal function. Lesser degrees of success may involve permanent disfigurement of the involved tissue by its partial or total replacement with fibrous scar tissue (sometimes calcified), or an intermediate state of affairs in which a viable but subdued organism is held in check by chronic inflammation. In the latter case, the outcome is measured in terms of the host's ability to react defensively over a long period of time, as well as the capacity of the infectious microorganism to survive the host's strengths and to revive its own when the opportunity presents. This is often the situation in chronic infectious diseases that may become arrested or remain slowly progressive, such as tuberculosis and brucellosis.

The inflammatory reaction is, of course, a most vital defense, providing early response to, and warning of, infection. It is mediated by all the circulating antimicrobial factors, nonspecific and specific, cited above (polymorphs, monocytes, lymphocytes, macrophages, fibrinogen, complement, antibodies); a

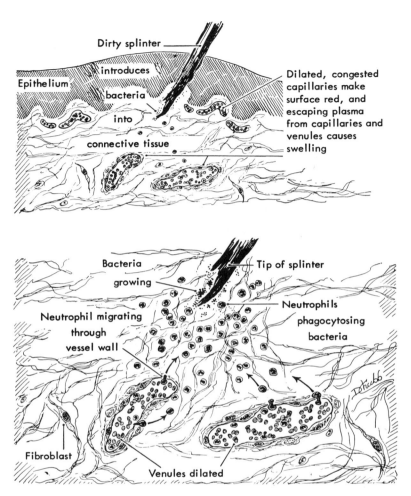

Fig. 7-7. Injury to human tissues is characteristically met by an inflammatory response. Here a splinter has become lodged in the skin. Small blood vessels in the area become dilated and congested with cells. Ameboid, phagocytic neutrophils migrate from these vessels toward the site of injury. They surround and attempt to ingest both the splinter and the bacteria it has introduced into the subcutaneous tissue. Plasma escaping from the swollen blood vessels causes swelling. Redness of the local skin area results from the dilation of these vessels. (Reproduced from Ham, A. W.: *Histology,* 7th ed. J.B. Lippincott Co., Philadelphia, 1974.)

rising concentration of host metabolic products within a walled-off area, with a lowering of pH; and a rise in local temperature — all injurious to microbial cells. The price paid by the host depends on the degree of mechanical injury induced locally by exudate and swelling and of chemical damage caused by toxic products of disintegrating cells and microorganisms. Such toxic substances may be carried off in the bloodstream to exert damaging effects on distant tissues. A secondary fall in the leukocytes of the peripheral blood is sometimes a consequence of this dissemination.

Another change in the blood commonly occurring as a result of inflammation in the tissues is an increased *erythrocyte sedimentation rate* (*ESR*). This abnormality may possibly be due to changed relationships in the plasma proteins. As the gamma globulin fraction of plasma increases with antibody production, the changed ratio of this protein to albumin allows more rapid settling of erythrocytes.

New Factors Appearing in Blood. Some of the new substances arising in blood during the inflammatory and immune responses to infection also appear to be nonspecific in character. Two of these are cited below.

Interferon. Interferon is a protein produced by infected tissue cells, and by certain lymphocytes, in response to double-stranded RNA, which is a characteristic viral product. Viruses containing this nucleic acid are the most frequent inducers of interferon production, but it is also elicited by some rickettsial, bacterial, and protozoan infections. Probably any infected cell (epithelial, nerve, muscle, or parenchymal) can manufacture this nonspecific substance if suitably provoked. Interferon is released from infected cells, enters new uninfected cells, and codes them to synthesize another protein. This new protein remains in the uninfected cell, blocking new infection by inhibiting the action of viral or microbial RNA. Interferon does not act directly on the infecting virus, or interfere with its entry into the new cell, but it does prevent its subsequent replication in the protected cell. This protection lasts for about a day.

Although interferon is nontoxic and nonantigenic (which means that it is also nonallergenic), with a broad range of antiviral activity, efforts to apply it to the treatment of viral diseases have been disappointing. It no doubt plays a role in the body's natural defense against viruses and in recovery from infection, but only limited success has been achieved in clinical use. One problem is that interferons produced by different animal species are not effective in other species. This means that, to be useful in human infection, interferon should be produced in human cells grown in tissue culture, a difficult and expensive procedure. A synthetic RNA preparation has been used to provoke interferon production in cell cultures and in animals, but it must be given in high concentrations, with only minimal results in achieving protection against reinfection. The best results have been obtained with superficial infections of the skin, conjunctiva, or respiratory tract, but to protect against these, interferon must be given before virus infection begins.

C-Reactive Protein. A new globulin called *C-reactive protein* appears in the blood during the inflammatory process. Its name reflects the fact that it is reactive with (is precipitated by) a component of the pneumococcus known as C carbohydrate. Its presence in the blood is not always related to pneumococcus infection, however, for it is increased by any noninfectious inflammation and appears during the course of other microbial invasions as well.

Other Nonspecific Factors

A number of biologic as well as environmental factors appear to influence the human host's ability to resist infection. Age and hormonal and genetic constitution are involved. These in turn are influenced by external factors, such as nutrition, stress, and physical environment.

Age. Age is a determining factor for a number of reasons. The changing incidence of infectious diseases from childhood to adulthood and old age is largely due to the operation of immune mechanisms, as we shall see later, and to the ability of the host to produce effective antibody. The latter function changes progressively in a definite relationship to age. As age progresses, structural, metabolic, and functional changes that occur in the host's tissues and cells undoubtedly affect their ability to maintain a balance with microorganisms as they are encountered. Profound shifts in hormone production that occur at puberty, during pregnancy, or at menopause also influence the health of tissues and, therefore, their resistance.

Hormones. Hormones also appear to influence susceptibility to infection independently of age. This is true particularly of those of the pancreas, thyroid, adrenal, and pituitary glands. Thus, the deficiency of pancreatic hormone (insulin) that occurs in diabetes seems to lower resistance, for diabetic patients are commonly confronted with complicating infections, such as tuberculosis, fungous diseases, or troublesome urinary tract infections. Production of thyroid and adrenal hormones in amounts above or below normal may likewise lead to changes in susceptibility or in the infectious process itself. The regulatory role of pituitary hormone on thyroid and adrenal output and in control of reproductive organ activity implies at least an indirect effect on the response to infection. The most clearly demonstrable of these hormone effects is that of excessive adrenal hormone in suppressing the inflammatory reaction and the production of antibody. This is a serious consideration in the use of adrenal hormone therapy for certain organic diseases, for with suppression of two of the body's essential defensive mechanisms susceptibility to infection is markedly increased. Furthermore, if the efficiency of the inflammatory reaction is decreased, preexisting infection, such as tuberculosis, may no longer be well contained but may become generalized.

Genetic and Racial Factors. Genetic factors are often correlated with incidence of infection in certain types of individuals or among groups. For individuals the explanation sometimes lies in inherited abnormalities, usually physiologic, involving one or more of the defense mechanisms. Among races of people, such differences as hormone production, metabolic factors, and environmental or nutritional factors may affect nonspecific resistance to infection.

Nutrition. The nutritional status of the host contributes to his ability to resist infectious disease in several ways. Nutritional deficiencies lead to an increased risk of infection in many instances, because the malnourished individual cannot maintain his tissues in a state of normal, healthy function. This may mean that his body cannot produce an adequate quantity or quality of some of the substances involved in defense, such as plasma proteins (complement, antibody), hormones, and external secretions. Good nutrition alone may not spare the host from infection with virulent organisms but it may reduce the opportunities for microorganisms of low pathogenicity to initiate infectious problems. On the other hand, obligate intracellular parasites probably *require* host cells in a good nutritional state to serve their own metabolic interests. In general, however, dietary deficiencies lead to a decreased resistance to infection by diminishing the normal functions of the body.

Stress. Whether physical or emotional, stress frequently results in a lowering of resistance to infection. The adrenal gland responds to stress with an increased output of hormone, and this results in suppression both of antibody formation and of the inflammatory reaction. In addition, stress may produce such effects as increased blood pressure, with resultant changes that diminish the blood supply to some tissues. Such a combination of factors may well afford additional opportunities for infection by microorganisms normally held in check by the body's defenses, particularly if the conditions of stress are prolonged.

Climate. Climate and other factors of the physical environment also affect resistance, particularly in individuals subjected to sudden marked changes to which they have not adapted. Very cold climates where the humidity is low, for example, are irritating to the unaccustomed respiratory mucosa and may reduce its efficiency in resisting even microorganisms of low pathogenicity. In warm, moist climates the skin may become macerated by perspiration, especially on the feet and in the axillary or inguinal folds, and be more susceptible to fungous and bacterial infections. Warm areas of low humidity and normal to high atmospheric pressure, on the other hand, offer benefits to patients with respiratory diseases such as tuberculosis, because the lungs are better aerated and have a better blood supply under these conditions than in chilly sections where humidity runs high and barometric pressure low.

Specific Resistance to Infection: The Immune System

Development and Functions

Immunity can be defined as a state of *specific, relative resistance* to infection. The words *relative resistance* describe all that we have observed up to this point of the shifting, uneasy relationship that the host maintains with parasitic microorganisms. This leaves us short of the full concept of immunity until the word *specific* is added, which now implies that whatever degree of resistance may exist, it is directed against a particular type of microorganism. It is this specificity of resistance that is the important characteristic of immunity.

In describing the nonspecific mechanisms of resistance to infection, we have spoken continuously of the body's reaction to "foreign" substances or particles. It is the *immune system* that enables the body to recognize a foreign agent as not being "self," and to take specific action for neutralizing, killing, and eliminating it. This action begins early in the processes of nonspecific resistance and involves some of the same cells and systems, notably lymphocytes and the lymphoid tissues. In some instances, the activity of the immune system leads to tissue damage as seen in allergic disorders and other hypersensitive states, or in the autoimmune diseases in which there is a failure to recognize "self" resulting in an attack on the body's own cells and tissues. These problems are discussed in the last section of this chapter.

Self-recognition by the immune system involves distinguishing the characteristic proteins or protein-linked components of body cells and tissues from those of an alien substance or agent introduced into the tissues. This distinction provokes certain lymphocytes to produce *antibodies* specifically directed against the foreign protein, or *antigen,* that has entered the picture. Other lymphocytes become "sensitized" to the invading antigen and react with it directly as well as by releasing a number of chemical factors called *lymphokines* that are designed to destroy it. This dual nature of lymphocytes is responsible for *humoral immunity,* mediated by antibodies (immunoglobulins) circulating in the blood, and for *cellular immunity,* induced by intact sensitized lymphocytes and their products.

Let us now consider the nature of antigens and antibodies, and of the lymphocytes involved in these immune reactions, before turning our attention to the mechanisms of humoral and cellular (cell-mediated) immunity.

Antigens and Antibodies.

Antigens. An antigen is most simply defined as a foreign substance, usually protein in nature, that when introduced into the tissues of an animal host stimulates the production of antibodies and/or sensitized lymphocytes. To do this, an antigen must reach tissues where it is in contact with antibody-producing cells. Simple ingestion of foreign proteins, for example, does not generally result in antibody formation because they are broken down in the digestive tract before their components are absorbed into the tissues. Antigenic molecules must be sufficiently large (the molecular weight must be at least 10,000) to be retained in colloidal suspension, rather than in true solution, for adequate contact with lymphocytes. Most substances of this size are proteins, but poly-

saccharides and other complex molecules may also function as antigens. Some molecules that are too small to be antigenic may become so if they are linked to protein. Lipids are antigenic only when combined with proteins or polysaccharides, or both. A molecule that is not antigenic of itself, but becomes so after linkage with a protein, is called an *incomplete antigen,* or *hapten.*

The chemical structure of an antigen determines the specificity of its interaction with antibody. Various chemical groups on a protein or other antigen molecule are responsible for its antigenicity; that is, they are the *antigenic determinants* of the molecule, accounting for its effectiveness in stimulating antibody formation and for the specificity of its reactions. Although the basic constituents of all living protoplasm are proteins, carbohydrates, and lipids, each species is chemically and antigenically unique. This is not surprising in view of the complexity of protoplasmic components. Proteins in particular afford unlimited variety because their constituent units, the amino acids, may interlock in many variations of combination, as do letters of the alphabet. The "letters" of different proteins may be the same, but the "words" formed by letter combinations are distinct. Thus the distinctive proteins of one species are quite unique from, and foreign to, those of another. Since foreign proteins behave antigenically upon their introduction into the tissues of animal hosts, this means that the protein components of any dissimilar species can provoke an antibody response in human tissues should they reach them. Obviously, this cannot happen often in nature, except in the case of microbial proteins, which find their way into the appropriate situation with relative ease.

MICROBIAL ANTIGENS.   Many constituents of microorganisms either are of protein nature, are linked to protein, or are large carbohydrate molecules and therefore antigenic. The bacterial cell has been described as a mosaic of antigens in three dimensions. Furthermore, many of the substances produced by microorganisms as metabolic by-products are antigenic. A given organism may contain or produce many antigens that differ one from another and can induce the formation of different antibodies. In a bacterial cell, for example, the constituents of several structures or products may be distinctive as antigens. These would include:

Somatic antigens of the cell body (soma) itself. These are sometimes called "O" antigens. In some bacteria they are the endotoxins of the cell. Some cellular antigens occur only on the surface of the cell and are called *surface* antigens.

Capsular antigens in species that form a capsule.

Flagellar antigens (also called "H" antigens) in motile species.

Exotoxins and other protein substances secreted by some species.

The different antibodies that are produced in response to such characteristic antigens in the infected patient often provide diagnostic proof of an infectious disease. For example, typhoid fever is caused by a flagellated bacillus with a characteristic surface antigen. Identification of antibodies forming in the patient's serum during the course of disease will reveal that they were specifically invoked by the "H" antigen of the typhoid bacillus, its "O" antigens, and a surface antigen designated "Vi" to associate it with *vi*rulence. Antibodies are given the same name as that of the antigen to which they respond. In this example, therefore, the patient's serum would contain typhoid O, H, and Vi antibodies.

Antibodies. The definition of an antibody is the converse of that given for an antigen: it is a specialized protein substance produced by an animal host in response to a foreign antigen in its tissues, and it is capable of reacting specifically with that antigen.

Antibodies are produced primarily by *plasma cells,* which are derived from highly specialized lymphocytes (B cells; see p. 162) responding to the presence of antigen. They are immunoglobulins (Igs) released from these cells and found circulating in the blood, associated with the gamma globulin fraction of plasma. Structurally, immunoglobulins have four polypeptide (multiply linked amino acids) chains arranged in a Y pattern. Each molecule has two chains of heavy molecular weight (H chains) and two of light molecular weight (L chains) linked together by disulfide bonds (Fig. 7–8). The outer end of each arm of the Y has a unique amino acid sequence that is specific for a given antibody molecule. These are the reactive sites that specifically bind the molecule to its antigenic determinant, their composition varying with the nature of the antigen that elicited their

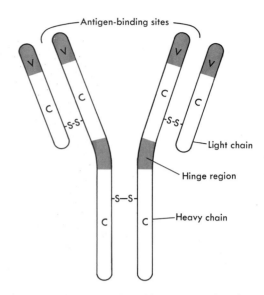

Fig. 7–8. Diagram of the Y structure of an immunoglobulin molecule (IgG). The shaded areas marked *V* indicate sites where amino acid sequences are variable and specifically reactive for antigen. The clear sites marked *C* represent constant amino acid sequences characteristic for an immunoglobulin class. Disulfide bridges are marked *S–S*. The "hinge" region provides the molecule with flexibility, permitting it to swing in or out for matching and binding to antigenic sites.

formation. Most antibody molecules have two such combining sites, that is, they are bivalent. Some may have as many as five, while others are monovalent. The larger the number of combining sites, the more avid and effective is the antibody in its interactions with antigen. Those with only one combining site are called *incomplete* or *blocking antibodies.* The remaining, larger portion of the antibody Y has an amino acid sequence that is constant and characteristic for its class (there are five classes of immunoglobulins, as described below). In the major classes of antibody (IgG and IgM) it is this constant portion of the molecule that activates complement and gives it opsonic properties. In the latter case, when the antigen-binding sites have attached to a microbial surface antigen, coating the microorganism, the free end of the antibody molecule mediates linkage of the coated particle to a polymorph or macrophage, thus promoting phagocytosis.

Antigen-Antibody Specificity. The surfaces of the combining sites of the antibody molecule are reciprocal, or converse, in a physical sense to the structure of the antigen molecule, especially with reference to the latter's determinant groupings. Therefore, when antigen and antibody combine, their surfaces dovetail, like two pieces of a jigsaw puzzle or a key in a lock. It is this complementary arrangement of their surface structures that provides the basis for the specificity of immune reactions.

The reciprocal structure of the combining sites on the antibody molecule with respect to its antigen also accounts for the fact that differences in antigens are "recognized" by their respective antibodies, only specific combinations occurring between them. The term *homologous* is used to designate either member of a specific pair. Related but dissimilar antigens and antibodies are said to be *heterologous* with respect to each other. The specificity of homologous immune reactions is often exquisitely accurate in detecting even subtle molecular differences in the combining pairs. Such reactions manipulated *in vitro* sometimes afford the only means by which protein materials can be demonstrated to differ. The immunity of human or animal hosts to infectious agents is similarly exact, different microbial antigens inducing the formation of specifically different antibodies. Conversely, antigen species similarities are detectable in overlapping antigen-antibody reactions.

Classes of Immunoglobulins. Five different classes of immunoglobulins are found in plasma or serum. (Serum is the clear portion of blood remaining after coagulation and clot formation.) They are named, and abbreviated, according to the types of heavy chains they possess, each designated by a Greek letter: *gamma* = IgG; *mu* = IgM; *alpha* = IgA; *epsilon* = IgE; and *delta* = IgD. Each class possesses different antibody activity. IgG constitutes the major portion (75 to 80 percent) of the Igs found in serum and is the only class capable of crossing the placental barrier. It contains antibacterial, antiviral, and antitoxic antibodies. IgM is the earliest Ig formed in the immune response to infection. It comprises about 7 percent of the antibody content of serum and contains the natural blood group antibodies as well as antimicrobial antibodies. IgA is found in only small amounts in serum but is present in large quantities in the external secretions of epithelial surfaces, where it is called *secretory* IgA (sIgA). It is found in tears, saliva, colostrum, and milk, and the secretions of the bronchial and intestinal tracts, protecting the body's external surfaces against invading microorganisms. IgE includes the antibodies involved in allergic reactions. It is found in very low concentrations in blood but has a marked ability to attach to tissue cells where its reactivity with the provoking antigens causes tissue injury. IgD is also found in slight concentrations in serum, but its function has not been clarified.

The role of the immunoglobulins is described more fully under humoral immunity, and their more important properties are summarized in Table 7-1 (p. 165).

Immune Functions of the Lymphoid System. The body's immune responses to antigenic stimuli are a function of the lymphoid system and its specialized cells. Like other cells of the circulating blood, lymphocytes are derived from *stem cells* located in the bone marrow. Stem cells are the undifferentiated precursors of the mature red and white blood cells found normally in the circulating blood. Their potential for development into red cells or granulocytes comes under several influences. Alternatively, they may develop into lymphoid stem cells. Further differentiation of these into mature lymphocytes produces two distinct types of lymphocyte populations, depending on where in the body the conversion takes place and on the influences operating within that site. One type of population is referred to as *B lymphocytes,* or *B cells,* because in birds (where they were first recognized) the differentiation takes place in a lymphoid organ called a *b*ursa. The avian bursa is associated with the intestinal tract. Its equivalent in man and animals has not been positively identified but is thought to be the lymphoid tissue of the upper and lower intestinal tract, e.g., tonsils, Peyer's patches, and the appendix. The other type of lymphocyte population is comprised of *T lymphocytes,* or *T cells,* so called because their conversion occurs in, and is dependent on, the *t*hymus.

The transformation of stem cells into B and T lymphocytes is probably complete in man in the first few months after birth. These cell populations then migrate to lymphoid tissue. They are found primarily in lymph nodes, but also in the bone marrow, spleen, and gastrointestinal tract. Eventually, it is the B cells that form the antibodies that provide humoral immunity. T cells become sensitized lymphocytes, responsible primarily for cellular immunity, though they may play a role in humoral immunity as well. (See Fig. 7-9.)

B Lymphocytes. B cells contain numerous immunoglobulin molecules but usually only one class of Ig per cell. When they come in contact with a foreign antigen, or antigen-antibody complexes, they are stimulated to proliferate into *clones,* or clusters of new identical cells. They also respond by maturing to *plasma cells,* which are large lymphocytes capable of synthesizing specific antibodies. These immunoglobulins, with their specific antigen-binding sites (Fig. 7-8), are then released to circulate in the blood and lymph streams (often in other body fluids as well). Where they encounter antigen that provoked their formation, they combine with and inactivate it. Since each of the small B cells of the new clones is capable of maturing into a plasma cell, the potential for antibody production is large. When the first contact with a new antigen is made, the *primary* antibody response requires 48 to 72 hours or longer before antibody appears, the immunoglobulin being predominantly of the IgM class. Subsequent contacts

with the same antigen elicit a faster *secondary* antibody response, within 24 to 48 hours, the major immunoglobulin now being of the IgG class.

A small proportion (about 20 percent) of the small lymphocytes found in circulating blood are B cells, but most are restricted within lymphoid tissues. They have a short life-span of a few days or weeks. However, some plasma cells probably revert to small B lymphocytes, to serve as long-lived "memory cells" ready to respond promptly to previously encountered antigens.

B cells alone can produce antibody when there are many copies of the *same* determinant on the antigen molecule. In some cases, however, when the antigen has more than one kind of determinant, as in the case of a hapten antigen linked to a carrier molecule, the cooperation of T cells is needed for antibody production by B cells. B cells react with the hapten itself, but T cells must react with the carrier to induce the antibody response.

T Lymphocytes. The conversion of lymphoid stem cells into T lymphocytes is dependent upon an intact thymus, at least in the early years of life. Later they find their way to the lymph nodes, spleen, bone marrow, and the circulating blood. In the blood, they represent the larger portion (65 to 80 percent) of the small lymphocytes found there. Their life-span is

Fig. 7-9. B cells (*A*) and T cells (*B*) are lymphocytes that play basic roles in man's humoral and cellular immune responses. (Reproduced from Nysather, J. O.; Katz, A.; and Lenth, J.: The Immune System: Its Development and Function. *Am. J. Nurs.,* **76**:1614, Oct. 1976.)

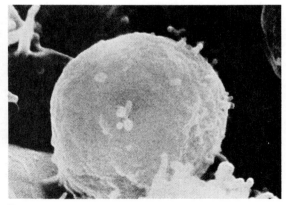

A

B

much longer (months or years) than that of B cells.

Thymus-conditioned T cells do not secrete immunoglobulins. When they react with certain antigens, they become sensitized, or "immunologically committed." They may then interact with those antigens in a way that is protective to the body, providing cellular immunity (often referred to as cell-mediated immunity, or CMI). These interactions are called into play particularly for defenses against intracellular microorganisms or those involved in chronic, persistent infections. CMI reactions are also involved in the recognition by T cells of tumor proteins produced by cancer cells. As cancer cells multiply in the body, their normal protein antigens become altered and are treated as "nonself" by T cells, which then attempt to destroy them. These cells are also responsible for the rejection of foreign grafts and tissue transplants. CMI mechanisms produce harmful effects in the body as well. It is the T cells that induce the changes seen in delayed hypersensitivity, including those of some infectious diseases (e.g., tuberculosis) and of autoimmunity.

The cellular immune response begins when a T lymphocyte becomes sensitized by contact with a specific antigen. On subsequent encounters with the same antigen, T cells may react in a number of ways: like B cells, they may proliferate as new clones; they may exert direct toxic action, especially on antigenic particles such as microorganisms; or they may release a number of chemical factors, called *lymphokines,* that marshal other defensive forces. The various activities of lymphokines are described under cellular immunity.

Role of Complement in the Immune System. Complement (C′) includes a complex series of nine serum proteins, numbered C_1 through C_9. Although it is a nonspecific blood factor, with some activities unrelated to the immune system, C′ plays an important role in mediating and augmenting immune reactions, both in the body (*in vivo*) and in the test tube (*in vitro*). The complement system consists of various precursors of enzymes whose activation is brought about in the presence of an antigen-antibody reaction. The first component, C_1, is activated after combining with a complex of antibody bound to antigen. The C_1 enzyme then acts on the next precursor to form a large amount of C_4. In turn this activates still larger amounts of C_2, and so it continues in a "cascade reaction." Eventually, the initial activation of C_1 generates enormous quantities of the later components, so that the final response is greatly expanded. The cascade is usually initiated by immune complexes involving IgG or IgM antibodies, but it may also be stimulated by nonspecific substances.

The later C′ components exert a number of biologic effects *in vivo*, including inflammation and cell damage. The cascading sequence is controlled, however, by inhibitors and by the instability of the activated components, so that they cannot diffuse to produce widespread damaging effects. Particular effects of C′ components include:

1. Histamine release, a factor in hypersensitivity.
2. Release of chemotactic factors that attract leukocytes to the site of antigen-antibody reactions.
3. Enhanced phagocytosis of microorganisms and their antigens.
4. Activation of the inflammatory response, bringing phagocytes and plasma factors to the scene.
5. Cytolysis, i.e., lysis of microbial cells or of infected tissue cells. Viruses and their antigens and bacteria or bacterial antigens attached to cell surfaces may be coated with specific antibody at such sites and destroyed by C′, together with the infected cell.

It can be seen, in consideration of these factors, that the complement system serves as an efficient bridge between nonspecific and specific resistance factors, as well as between humoral and cellular immunity. Activated by IgG or IgM produced in B cells, the C′ cascade releases, among other things, a chemotactic substance that brings T lymphocytes and macrophages to the site of antigen localization. There the action of T cell lymphokines promotes further macrophage activity and other defenses, as we shall see.

Humoral Immunity

Humoral immunity is conferred by specific antibodies in the circulating blood or other body fluids, acting in conjunction with the complement system in interaction with foreign antigens. These antibodies are produced by plasma cells, derivatives of B lymphocytes, in response to the presence of antigen. They are immunoglobulins designed to react exclusively with the antigen that induced their formation, or with any other having a nearly identical molecular configuration. In binding to particulate antigens, antibodies have an opsonic effect, promoting phagocytosis by polymorphs or macrophages. The bound immune complex also activates, or "fixes," complement. The ensuing series of enzymatic reactions involved in complement fixation leads to antigen destruction and the death of microorganisms.

Humoral immunity can be transferred passively from one individual to another, since antibodies and complement are both present in cell-free plasma. This *passive immunity* can be acquired artificially, by injection of plasma or serum taken from an immune person or animal, or naturally in the case of the fetus who receives maternal antibody transferred across the placental barrier. However, not all immunoglobulins can pass the placenta, those of the IgG class being the only ones capable of doing so.

Among the immunoglobulins, the classes IgM, IgG, IgA, and IgE each serve specialized functions. The function of the IgD class is as yet unknown. *IgM* is the major component of the primary antibody reaction, appearing first in a given immune response. It is a very large molecule with a high molecular weight and therefore restricted to the vascular system, being unable to diffuse through fine membranes. It has more antigen-binding sites than the IgG molecule shown in Figure 7–8 and is therefore more avid in combining with foreign invaders. IgM also has a greater capacity for activating complement than do smaller molecules, and it is a better opsonin. Since this powerful antibody is formed early in response to a new infection, it is particularly valuable in giving the host a head start in the race to prevent an invading microorganism from gaining a permanent foothold or spreading. When the secondary immune response gets underway, IgM is gradually replaced by newly formed IgG antibodies. For this reason, the detection of IgM antibodies for a particular microbial antigen is indicative of either recent or continuing infection. For example, the presence of IgM antibodies for rubella in a pregnant woman with a history of recent rubella-like illness indicates that she did (or does) indeed have rubella, whereas the presence of rubella IgG alone would mean that she is probably not currently infected or has passed the infectious stage of the disease.

IgM antibodies are also the first to be formed during uterine development, if the fetus becomes infected. Antibody formation is possible after the fifth or sixth month and is of the IgM type. Accordingly, the presence of antimicrobial IgM in the cord blood of a newborn is indicative of intrauterine infection, since these antibodies cannot be transferred across the placenta from the mother's blood.

As pointed out previously, the natural blood group antibodies (ABO) are of the IgM class. This class also serves, together with IgG, in providing specific antitoxins directed against the exotoxins secreted by the organisms of diphtheria, tetanus, and botulism.

IgG is the principal antibody of the secondary immune response. Most of this antibody is circulated in the blood, but it is not too large to find its way into other body fluids such as lymph, or into spinal, synovial, and peritoneal fluids. IgG is active against all the major microbial antigens. Since it can be transferred across the placenta, it serves to protect the fetus and newborn against those infections for which the mother has developed this type of antibody (naturally or through artificial immunization). The passive immunity of the newborn is only temporary, but by the time it has waned he is usually capable of producing his own IgG antibodies.

IgA is found to only a slight extent in serum, but secretory IgA (sIgA) is the principal immunoglobulin on mucosal surfaces and in colostrum. These antibodies are important in resistance to infection of the surface tissues of the body, especially those of the respiratory, alimentary, and urogenital tracts. It is found primarily in the secretions of these mucous tissues, increasing in amount during mucosal infections. It enters the blood by way of the lymphatic system, so that serum IgA levels are also raised during such episodes. Maternal sIgA in colostrum and milk helps to boost the passive immunity of the newborn, who has already acquired IgG antibodies via placental transfer *in utero.*

IgE antibodies occur in only slight concentrations in serum but play an important role in hypersensitivity of the immediate (anaphylactic) type. This is a damaging role, inducing tissue injury. The IgE molecule has an affinity for attaching itself, via the leg of its Y, to certain cells of the body, notably mast cells and basophils. When its exposed antigen-specific combining sites, on the open arms of the Y, bind with an antigen, the combination triggers the release of biologically active substances, including histamine, from the mast cell or basophil. The results are damaging to surrounding tissues and responsible (depending on the degree and site of the reaction) for the symptoms of such conditions as anaphylactic shock, allergic asthma or rhinitis (hay fever), urticaria (hives), and allergic (atopic) dermatitis. Attachment of IgE to mast cells and basophils in the skin and other organs sensitizes them to the antigens that induce these allergic responses. Such antigens are referred to as *allergens*. IgE has been called, variously, *reaginic* antibody, *atopic* antibody (atopy refers to allergy, or, literally, "out of place" reactions), *skin-sensitizing* antibody, or *anaphylactic* antibody. *Allergies* are discussed in the last section of this chapter, under hypersensitivity.

Some of the important properties of the four major classes of immunoglobulins are summarized in Table 7–1.

Cellular Immunity

Cellular immunity, or the cell-mediated immune response (CMI), is a function of the T lymphocytes (thymus-conditioned, or T, cells). These cells are present in lymphoid tissue and are also continuously recirculated through the lymphatics to the blood, then via capillaries to the tissues and back to lymphatic vessels. This constant movement assures eventual contact between foreign antigens entering the body and T cells sensitized for specific reactions with them. Each T cell carries on its surface antibody-like receptors that are specific for a given antigen, and there are a few specifically reactive T cells for every antigen to which the body has been exposed.

In response to antigen contact, a sensitized T cell reacts by differentiating to form a "lymphoblast" equipped for new protein synthesis, dividing into clones of identical cells, destroying microbial cells by interacting with their surface antigens, and releasing a number of chemical mediators called lymphokines.

Table 7–1. Important Properties of Immunoglobulins

Property	IgM	IgG	IgA	IgE
Heavy chain	Mu	Gamma	Alpha	Epsilon
Effective binding of antigen*	+ + + +	+ +	+ +	+ +
Complement fixation*	+ + + +	+ +	−	−
Distribution:				
Blood	90%	40%	Serum IgA low	Low proportion
Extracellular fluids	10%	60%	0	0
Secretions	0†	0†	100%	High proportion
Transfer to offspring	0	Via placenta	Via colostrum and milk	0
Major function	Primary immune response; antimicrobial and natural blood group (ABO) antibodies	Secondary immune response; major systemic immunoglobulin; antimicrobial antibodies	Protection of mucosal surfaces	Immediate hypersensitivities (allergies); reaginic antibodies

*+ + + + = strong; + + = moderate; − indicates insignificant or no fixation.
†Increased during inflammatory response.

Lymphokines have various actions, the end result being attraction of phagocytes and promotion of a local inflammatory response. Different lymphokines are characterized by their known activities, e.g., chemotactic factor, migratory inhibition factor (MIF), macrophage activation factor (MAF), lymphocyte transforming (blastogenic) factor (LT), transfer factor (TF), and so on. It appears probable that interferon, with its antiviral activity (see p. 157), is a lymphokine, but this substance is also produced by many other types of cells.

When chemotactic factors have brought blood monocytes and polymorphs, as well as macrophages, to the scene, migration inhibition factor (MIF) prevents them from moving away again. MAF activates macrophages, stimulating them to produce many new lysosomal granules and transforming them to a highly active phagocytic state. Meanwhile, the blastogenic factor (LT) initiates rapid mitosis of sensitized T cells, and transfer factor (TF) converts additional nonsensitized T cells into sensitized T lymphocytes. Some of the newly sensitized T lymphocytes become "killer" cells, producing a toxin (called lymphotoxin or cytotoxin) that reacts specifically with the antigens of microorganisms attached to infected cell surfaces, killing the affected tissue cells in the process. This cytotoxic or cytolytic activity, as it is called, adds to the work of macrophages, which must remove the cellular debris.

CMI may involve the body in some very intense reactions, once the various T lymphocyte populations are sensitized and activated. The ability to generate this response appears somewhat later than humoral immunity. Although it is protective in defending against invading microorganisms, particularly intracellular organisms as they attach to or emerge from tissue cells, or tumor proteins, the delayed hypersensitivity induced also causes tissue injury. CMI responses are generated in all infections, but their scope and intensity vary with different infectious agents (and with different animal hosts). Delayed hypersensitivity is more pronounced in some chronic, persistent infections (tuberculosis and fungal diseases) but may be demonstrated in others as well. The terms *delayed* and *immediate* hypersensitivity refer to the time required to elicit a reaction by the immune system to a given antigen once immunity has been established. Reactions of the immediate type, mediated by humoral IgE antibodies, occur within minutes or a few hours; those of the delayed type, mediated by the T lymphocytes involved in cellular immunity, require from one to three days, sometimes more (see p. 179).

Natural passive transfer of cellular immunity cannot occur because intact sensitized lymphocytes are not able to cross the placental barrier. CMI can be artificially induced, however, by injection of lymphoid cells or their extracts. Such passive transfer is not practical for routine immunization but has been valuable for studying the mechanisms of immunity and hypersensitivity in experimental animals. It is being investigated for its value in treating patients who lack CMI (see immunodeficient diseases, p. 186).

Table 7–2 summarizes some of the factors involved in CMI.

Review of the Immune System. Let us now review briefly the natural course of events in a new microbial infection and follow the dual role of the immune system in dealing with it. Usually the initial number of invading microorganisms is small and the antigenic stimulus slight, but as they multiply this stimulus increases. Chemotactic signals go out from the site of microbial attachment, and phagocytes and antigen-sensitive lymphoid cells (both B and T cells) reach the scene. As the phagocytes begin their attack on the microbial particles, B and T cells begin to divide repeatedly, forming individual clones. B cells differentiate into plasma cells, which then synthesize molecules of IgM with combining sites specifically designed to interact with the new microbial antigens. The inflammatory response is underway, and the capillaries of the area exude plasma with its content of complement. As IgM is formed (this requires a few days), its multiple combining sites attach to antigen on microbial surfaces, forming a sticky coat that immobilizes and agglutinates the living organisms. This immune complex activates the enzyme precursor of complement forming C_1 with a resulting cascade of C′ components. This effect enhances phagocytosis of the immune complexes now being formed more steadily, releases more chemotactic factors, and increases the inflammatory reaction. Complement fixation may also bring about direct cytolysis of the microbial cells.

Table 7–2. Some Important Mechanisms of Cellular Immunity (CMI)

Agent	Activity
Sensitized T lymphocyte	Releases lymphokines
Lymphokines:	
Chemotactic factors	Attract PMNs, monocytes, macrophages
MIF	*Migration inhibition factor:* prevents further migration of phagocytes
MAF	*Macrophage activation factor:* increases phagocytic activity of macrophages
LT	*Blastogenic factor:* induces mitosis (cloning) of T cells (lymphocyte transformation)
TF	*Transfer factor:* converts nonsensitized T cells to sensitized T lymphocytes
Interferon	Blocks virus infection of tissue cells (also produced by many kinds of cells other than lymphocytes)
Cytotoxic T lymphocyte ("killer" cell)	Releases lymphotoxin (cytotoxin), which destroys tissue cells with microbial antigens on their surfaces

If the site of infection is on a mucosal surface, of the upper respiratory tract for example, sIgA is formed and combines with the antigen. In this case, however, there is little or no seepage of plasma into the area and hence no complement is present. Opsoninization and phagocytosis must therefore proceed without complement fixation.

In the meantime, the T cells have cloned, the sensitized T lymphocytes have combined with microbial cells through their surface receptors, lymphokines are being released, and a crescendo of tissue reactions is proceeding. If all these defenses are successful, the invaders cease to multiply, and they succumb to the host's forces. Without further antigenic stimulus, B and T lymphocytes withdraw, leaving the field to the macrophages to clean up. These lymphocytes will "remember" the occasion, however, and if called upon again to meet the same organisms will do so with built-in capacity for more rapid specific responses. Most important, the B lymphocytes begin within a few days of the initial infection to produce IgG antibodies, which replace IgM and may persist in the blood for long periods.

Humoral antibodies and T cell surface receptors are formed against a great variety of microbial components and products, so that in time the normal person's immunity to infection becomes broader (see Mechanisms of Immunization, p. 176). Its strength depends on the nature and frequency of antigenic stimuli as well as on the vigor of the immune response. Deficiencies in either the B or T cell systems, or both, lead to serious problems.

Figure 7–10 presents a diagrammatic summary of the mechanisms of immunity, indicating the role of B and T cells, phagocytes, and complement.

Interactions of Antigens and Antibodies

Serology. The interactions of antigens and antibodies are often observable, or can be made so, whether they occur *in vivo* or *in vitro*. *Serology* is the *in vitro* study of serum for its antibody content, and *serologic reactions* are the interactions of antigens and antibodies occurring in the test tube. These reactions are used routinely to identify an antibody or an antigen when one of these reactants is known.

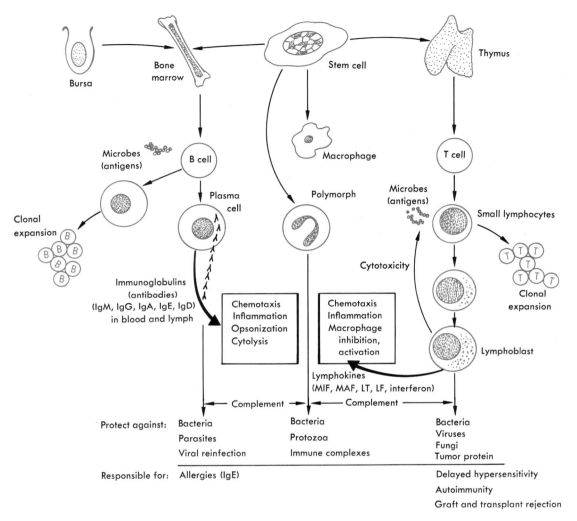

Fig. 7–10. Mechanisms of immunity.

The serology laboratory commonly works with the serums of patients suspected of having infectious disease, testing them with known microbial antigens. The microbiology laboratory, conversely, tests unknown microbial isolates with serums of known antibody content to confirm their identity. Serologic techniques can be performed in a quantitative manner to provide information as to the relative amounts of antigen or antibody present in a given system.

This is done by using one of the two reagents in a constant quantity and diluting the other until it fails to produce further reaction. When a patient's serum is being tested with a particular microbial antigen, for example, the serum is diluted in a serial fashion. A row of tubes containing the saline diluent in even amounts is set up; a measured amount of serum is placed in the first tube and mixed well. Then a measured aliquot of this dilution is placed in the second tube and mixed, this process being repeated down the line so that a graded series of dilutions is obtained. The antigen is then added to each tube in a constant volume. When the tubes are examined for evidence of antigen-antibody combination, the last dilution of serum that shows reactivity is reported, using the reciprocal of the dilution figure. This figure is spoken of as the *titer of the serum* and provides an estimate of the level of antibody present in a unit volume. For example, if the last reactive dilution of the serum was 1:100, the titer is 100.

When serologic diagnosis of infectious disease is attempted by demonstrating the presence of specific antibodies in the patient's serum, it is very frequently necessary to test two samples of serum, one obtained from the patient soon after the onset of his symptoms, the other taken ten days to two weeks later. The reason for this is that, as we have seen, antibody production increases during the course of active infection, being first stimulated as the microbial agent begins to grow and multiply in the body, then rising steadily in response to continuing antigenic stimulus. A patient in the early acute stage of illness may have no detectable serum antibodies against the causative microorganism, or the serum titer may be low, in the range of 20 or 40. Such a level of antibody has little significance, because it might merely reflect a previous vaccination with the microbial antigen or a titer lingering from some previous attack of the infection and does not necessarily indicate the nature of the current disease. If the test is repeated in a week or two, however, and the titer is found to have risen by at least a fourfold increase (80 or 160 as compared with the examples above), this provides serologic evidence of current active response to the microorganism in question. This type of serologic information is often very valuable, both in diagnosing the present disease and in estimating the past experience of the patient with a given antigen.

Distinctions between IgM and IgG antibodies, when appropriate tests are available, help to determine whether a patient is in an early or still active stage of infection, or whether active infection has subsided. Since IgM is produced in the primary immune response, its presence indicates new or very recent infection. The presence of IgG indicates a secondary immune response, whereas the height of the IgG titer observed at two-week intervals provides clues as to whether or not infection is subsiding. IgG titers usually reach a peak during clinical convalescence, waning gradually thereafter, but often persisting at a low level for long periods. Chronic, persistent infection may induce elevated levels of both IgM and IgG antibodies.

When a patient's serum is to be studied diagnostically for its antibody content, particular laboratory tests are ordered to confirm the clinical or bacteriologic evidence of the nature of the patient's infectious disease. These tests are often called by names that reflect the nature of the antibody sought or of the reaction that will be seen according to the techniques of the test. Thus, such terms as *febrile agglutinins, complement fixation,* and *FA (fluorescent antibody)* are often used. The reasons for this terminology will be apparent in the following description of *in vitro* reactions. (The common laboratory tests used in serologic diagnosis are listed in Appendix I.)

In Vitro **Reactions.** Laboratory studies utilize blood plasma or serum of human and animal origin as a source of antibody, and antigens from a variety of sources, including microorganisms. When an antibody combines with its specific antigen, the reaction may have one of several observable results, depending on the nature and condition of the antigen, as well as on the presence of other nonspecific factors. Consequently, antibodies are often described according to the type of reaction that occurs.

Agglutinins. Antibodies that cause a visible agglutination of the antigen are called *agglutinins.* This type of reaction occurs if the antigen is on the surface of or adherent to a large particle (such as a microbial cell, a red blood cell, a latex particle). Since antibodies have multiple combining sites, each antibody links up with at least two antigen molecules. The latter in turn may be linked through antibody to other antigen molecules. This lattice-like structure grows until large clumps form and settle out of suspension. Saline in physiologic concentration must also be present to encourage adherence of particles. The suspending solution becomes clearer as more particles clump and settle out, and the agglutinated masses become visible.

Precipitins. Antibodies that cause a visible precipitation of the antigen are called *precipitins*. The reaction occurs in the same way as that just described, but in this case the antigen is in solution, unassociated with any cell or particle. Soluble free antigen combines with antibody, and in the presence of saline the complexes formed become visible as a fine precipitate settling out of solution. Fine glass capillary tubes are often used to demonstrate precipitin effects, the dense areas of precipitation of combined antigen and antibody being easily seen in a thin column of clear fluid (Fig. 7–11).

Another form of precipitation test called *gel diffusion,* or *immunodiffusion,* involves the use of a supporting agar gel in which the reaction takes place. This involves making wells in the agar gel, placing an antigen solution in one well and serum antibody in an adjacent one. Each diffuses through the agar. Where antigen and antibody meet, if they combine specifically, visible precipitation lines are formed. If they do not combine no precipitate can be seen. Another application of this technique is made in the *Elek plate method* for testing *C. diphtheriae* isolates for toxigenicity (Fig. 7–12). This method, previously described in Chapter 4 (p. 87), detects toxin, produced by a growing strain of the diphtheria bacillus, by the lines of precipitate it forms with antitoxin diffusing from a paper strip laid on the agar.

Complement-Fixing Antibodies. Many antigen-antibody complexes attract, or "fix," complement to their outer surfaces. With few exceptions, the result of this fixation cannot be directly observed. In the case of some bacteria and red blood cells, however, fixation of complement in the presence of specific antibody results in lysis of the organisms or the red cells. The lysis is visible as a clearing of a suspension that had previously been turbid with intact cells. This phenomenon is the basis for a serologic method, the complement-fixation test, used commonly when the

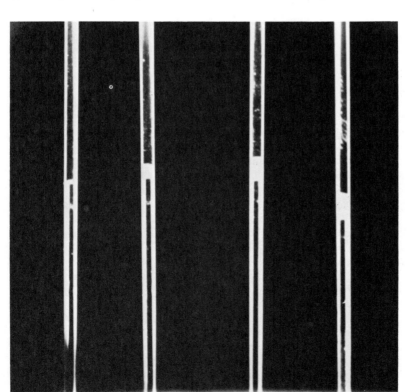

Fig. 7–11. Capillary precipitation. Dense areas of precipitate are visible where antigen and antibody layers have met and combined near the center of the capillary tubes.

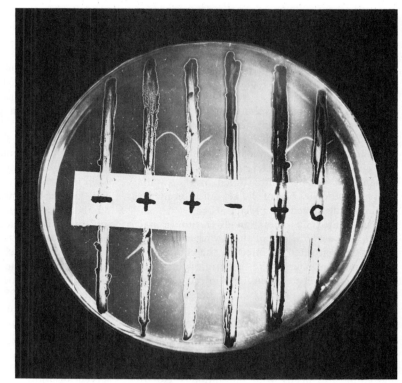

Fig. 7–12. The Elek test (*in vitro* gel diffusion) for diphtheria toxigenicity. A filter paper strip, wet with diphtheria antitoxin, is laid across an agar plate. Strains of *Corynebacterium diphtheriae* are streaked individually at right angles to the strip. Within 24 to 48 hours, lines of precipitate become visible on either side of toxin-producing strains. These lines form where diffusing toxin and antitoxin have met and reacted. In this photograph, three test strains (marked +) and the control (*C*) are toxin producers, and two of the tested strains (marked −) are nontoxigenic. (Courtesy of Dr. George Hermann, Decatur, Ga.)

antigen-antibody-complement combination does not result in visible lysis.

Complement is present in the serum or plasma normally and is therefore readily obtainable from laboratory animals, usually guinea pigs. Because its enzyme precursors are sensitive to heat, whereas antibodies are not, the C′ component of serum can be easily inactivated by gentle heating (in a water bath), leaving immunoglobulins intact. Human serum to be tested for antibody is first treated in this way to remove C′. An unheated animal serum containing active C′ can then be used as a complement source. When complement is added to a test tube in which human serum antibody and a test antigen have been mixed and have specifically combined, C′ is activated as previously described (p. 163) and is fixed, or adherent, to the surface of the antigen-antibody complex. When this happens, fixation of complement can be tested for by adding to the test tube a second antigen-antibody complex that will provide a

visible result if complement is available. The system most often used is sheep red blood cells (S-RBC) coated with anti-S-RBC antibody. If the complement has been fixed by the first combination, none will remain for the second complex and no red cell lysis will be seen in the latter. If the S-RBC do lyse visibly, it means that complement was not fixed to begin with, which in turn signifies that no antigen-antibody combination occurred in the tube before complement was added. Thus, a *positive* complement-fixation test is indicated by a *negative* reaction in the second (indicator) system.

Opsonins. These are antibodies that combine with the surface antigens of microorganisms or other cells and thus enhance their susceptibility to phagocytosis. This process also is facilitated by the presence of complement. It can be demonstrated in the test tube using phagocytes obtained from the blood or from a purulent exudate. When bacteria are coated with specific antibody, exposed to phagocytes, and viewed microscopically, they are seen to be taken up much more rapidly and completely than those in a control preparation containing no antibody. When antibody and complement are both present, opsonization is still more rapid and efficient.

Antitoxins. Antibodies that are induced by toxins and combine with them are called *antitoxins.* Toxins, when separated from the microbial cells that produce them, are soluble antigens. When they combine with specific antitoxin, therefore, they form a precipitate that can be demonstrated by the capillary tube method or by one of the gel diffusion methods previously described for precipitins.

Toxin-antitoxin combination also results in inactivation, or *neutralization,* of the toxin, an effect that can be demonstrated by injecting the complex into an animal. The bound toxin is harmless to the animal, whereas free toxin similarly injected produces injurious or lethal effects (see *In Vivo* Reactions).

Fluorescent Antibodies. The fluorescent technique provides another means of visualizing antigen-antibody interactions. This method involves the preliminary treatment of antiserum with a fluorescent dye that conjugates firmly with the antibody globulin. The antibody is thus "labeled," and when it combines with its specific antigen, the site of combination fluoresces visibly when examined with a microscope fitted with an ultraviolet illuminator. As the ultraviolet rays pass through the specimen, the dye-conjugated antibody fluoresces brightly on the surfaces of cells it has coated. The use of fluorescein-labeled antibody makes it possible to locate microorganisms quickly and identify them in smears from clinical specimens or in tissue sections. Intracellular viruses can also be visualized by this method.

Conjugation of antiserums with fluorescent dyes is expensive and time consuming; therefore, it is impractical to label a battery of antiserums, each directed against a different microbe. To solve this problem, an *indirect* method of fluorescent staining can be used not only to identify unknown microorganisms, but also to determine the presence of specific antibodies in patients' serums. Human serum is first injected into an animal, which responds by producing antibodies to human globulin (itself an antigen). Serum from the animal is then conjugated with fluorescent dye, providing a labeled antihuman antibody. To detect microbial antigens, the preparation is exposed to unlabeled human serum of known antibody content. If a specific antigen-antibody combination occurs, the microbial cells become coated with human globulin, but the reaction is not visible. Now the fluorescent-labeled antihuman-globulin antibody from the immunized animal is added to the preparation. It combines with the human globulin coating the cells, providing a new outer layer that now will fluoresce in ultraviolet light. This indirect method makes it possible to use a number of ordinary, unlabeled human antiserums in screening smears for the localization of antigens. They can then be treated with one labeled antihuman globulin, which points out those cells that are coated with specific human antibody. These techniques are illustrated diagrammatically in Figure 7–13. The effects of "immunofluorescent" staining of microorganisms can be seen in Figure 7–14.

To determine whether antibodies against a specific antigen are present in a patient, his serum is allowed to react with that antigen. The preparation is then treated with labeled antihuman globulin and observed for fluorescence. The principle is the same as just described, but in this case, the identity of the

Fig. 7–13. Diagrammatic illustration of the indirect fluorescent antibody reaction. (Courtesy of Bio-Quest, Division of Becton, Dickinson & Co., Cockeysville, Md.)

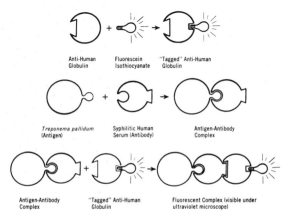

Anti-Human Globulin   Fluorescein Isothiocyanate   "Tagged" Anti-Human Globulin

Treponema pallidum (Antigen)   Syphilitic Human Serum (Antibody)   Antigen-Antibody Complex

Antigen-Antibody Complex   "Tagged" Anti-Human Globulin   Fluorescent Complex (visible under ultraviolet microscope)

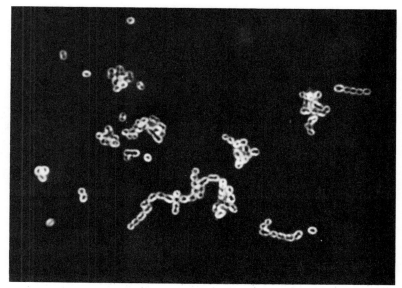

Fig. 7–14. Immunofluorescent staining of group A streptococci makes them appear bright apple green against a black background. (1500×.) (Reproduced from Maloy, L. B.: Lab Helps Prevent Strep Epidemic. *Health News,* December 1970. Courtesy of New York State Department of Health, Albany, N.Y.)

antigen, rather than the antibody, is known. This test is most widely used for the detection of antibodies against *Treponema pallidum,* the causative agent of syphilis (p. 499). In practice, when detecting microbial antigens by fluorescent antibody techniques, specific antiserums from animals rather than humans are used because it is more practical to immunize them against a variety of agents. The fluorescent-labeled antiglobulin serum must then be prepared in a second animal species.

OTHER LABELING TECHNIQUES. In recent years a number of techniques have been developed for labeling either antigens or antibodies with markers that can be detected following antigen-antibody combination. Three of these are described briefly below.

Radioimmunoassay (RIA). The RIA method involves the use of antigen labeled with radioactive iodine. It is generally applied to the identification of soluble (e.g., barbiturate drugs) or virus (e.g., hepatitis) antigens that may be present in a patient's circulating blood. The patient's serum is exposed to antibody that is specific for the known antigen being sought. A preparation of the known antigen is labeled and added to the same tube. If the patient's serum contains the antigen, it will bind the antibody,

leaving the labeled known antigen free. If there is no antigen in the patient's serum, the labeled antigen will be bound to antibody. An anti-gamma globulin is now added to the tube (as in the FA test above) and combines with the antibody. This complex forms a precipitate that is placed in a scintillation counter and analyzed for radioactivity. A count of the radioactive particles in the precipitate indicates whether or not the labeled antigen was bound to antibody. A high count indicates that it was bound because there was no antigen in the patient's serum. A low or zero count is a positive test, indicating that an unlabeled antigen in the patient's specimen combined specifically with the antibody before the labeled one could do so.

Ferritin Labeling. Ferritin is an iron-containing protein that can be conjugated with immunoglobulins. Ferritin-labeled antibodies are used to detect the presence of antigens in tissues studied by electron microscopy. Antigen-antibody complexes appear as dense, dark spheres at their sites of localization.

Immunoperoxidase Technique. In this method the enzyme peroxidase is conjugated with antibody, and the conjugate is added to a specimen suspected of containing an antigen specific for the antibody. If antigen is present and binds with antibody, the peroxidase is also bound. A simple test for peroxidase activity is then applied, using an enzyme substrate that gives a color reaction when it is degraded by the peroxidase tied into the immune complex.

Other Antibody Effects. Other effects of antibody reactions that can be demonstrated *in vitro* include the inhibition of growth of homologous microorganisms in culture or visible changes in the appearance of organisms. Motile bacteria or spirochetes, for example, become immobilized in the presence of homologous antibody. The *Treponema pallidum immobilization* test, or TPI, is an example of this kind of antibody effect. Another is the *quellung reaction,* in which anticapsular antibodies cause an apparent swelling (*quellung*) of the capsules of intact bacteria. This is a form of precipitation, in which the soluble antigen of the capsule precipitates in the combination with antibody. The reaction increases the surface tension and density of the capsule so that it becomes more visible and appears swollen (Fig. 13–2).

Immunoelectrophoresis combines immunodiffusion with electrophoresis, taking advantage of the migratory properties of antibody and antigen molecules in an electric field. Soluble proteins can be separated by an electric current, migrating at different rates according to their charged ions, size, and other properties. A thin supporting base of starch, cellulose acetate, or an agar gel is used for this. When the protein antibodies and antigens present in human serum, for example, have been separated in this way, one can then place a solution of known antibody or antigen parallel to the migration lines formed by electrophoresis. Diffusion of the new material brings it in contact with the separated proteins. Any of the latter that specifically "recognize" the new material and form an immune complex with it can be identified by the lines of precipitation that form at the junction of antigen with antibody.

Counterimmunoelectrophoresis is a method commonly used to detect hepatitis B virus antigen in patients' serums. Here electrophoresis is used to drive the negatively charged virus antigen toward the anode and the antibody toward the cathode. This countermigration in an electric field brings the two in conjunction, again in a thin layer of gel support, with the formation of visible precipitin lines if antigen-antibody combination occurs.

***In Vivo* Reactions.** Many of the reactions demonstrable in the test tube may occur in the body when antibodies combine with antigens in the tissues. Microorganisms agglutinated or lysed by antibody with the assistance of complement *in vivo* are thereby inactivated and prevented from exerting further damage. They are also more readily phagocytosed when coated with antibody. Exotoxins and other harmful microbial products are similarly neutralized by antitoxins produced in response to their antigenic effect. Viruses may be neutralized by antibody if enough is present in the circulating blood to combine with such agents before they have localized within cells. These *in vivo* effects of antibodies have given rise to still another term for them; i.e., they are said to be due to *neutralizing,* or *protective, antibodies.*

Neutralization. Antibody neutralization of an infectious or toxic agent can be demonstrated in laboratory animals in one of three ways:

1. ACTIVE IMMUNIZATION. A virulent microorganism (or its toxin) can be inactivated in the test tube by *physical* or *chemical* means, without destroying its antigenic properties. The inactivated agent is then injected into the tissues of an animal to stimulate the production of specific antibodies. When this *actively immunized* animal is later injected with the living, virulent organism in question, or with unmodified toxin, it will survive this "challenge," while nonimmune control animals succumb to the challenge dose. A specific neutralizing effect of antibody protects the immune animal.

2. PASSIVE IMMUNIZATION. The serum of animals or humans who have a preexisting immunity (acquired naturally or artificially) to a given microorganism or its toxin can be injected into another animal. This *passively immunized* animal is then injected with a dose of the living, virulent organism in question, or its toxin. Homologous antibodies present in the injected serum protect the animal from disease, but nonimmune control animals are damaged by the challenge with active antigen.

3. *In Vitro* NEUTRALIZATION. A toxic or infec-

tious agent can be exposed to specific antiserum in the test tube, where antigen-antibody combination will occur. Neutralization of the antigen by the antibody can then be demonstrated by injecting the complex into a nonimmune animal. This animal will show no evidence of disease or toxicity, while a nonimmune control animal receiving unneutralized antigen alone will succumb to its effects.

Tissue cultures of viruses can also be used for neutralization tests. Living animal or human cells grown in the test tube are infected with known or unknown viruses that will multiply, producing visible effects on the cells, unless neutralized by specific antibody. Known viruses can be used to identify unknown antibody, or vice versa.

Skin Tests. Useful diagnostic demonstration of an antigen-antibody reaction can often be obtained with an *in vivo* skin test. The skin of human beings or of animals is used as a testing site for one of two purposes: (1) to determine the state of immunity of the individual, and (2) to determine whether or not a specified antigen is present in the body.

1. DETECTION OF HUMORAL OR CELLULAR IMMUNITY. Most skin tests are used either for the detection of a specific circulating antibody or to determine the status of cell-mediated immunity (CMI) to a given disease. In either case the homologous antigen is injected intracutaneously, most conveniently in the skin of the ventral surface of the forearm. The best examples of tests for humoral antibodies (IgG) are the *Schick* and *Dick* tests. The former involves injection of diphtheria toxin as the antigen in testing for diphtheria antitoxin immunity. For the Dick test, streptococcal erythrogenic (scarlet fever) toxin is used to detect its specific antitoxin. Both these toxins are injurious for skin if unneutralized by antibody and will produce an area of reddening, swelling, and tension at the local site of injection. Thus, a *positive* skin reaction indicates the absence of humoral immunity, whereas a *negative* reaction results if there is sufficient circulating antitoxin to neutralize the toxin at the injection site.

Skin tests with *allergens,* that is, the antigens responsible for allergies, are also used to detect humoral antibodies—of the IgE class in this case. The interpretation of positive and negative skin reactions is the reverse of that stated above, however. It will be remembered that IgE antibody-antigen complexes trigger the release of histamine and other biologically active substances that cause tissue damage. Therefore, a *positive* skin test with an allergenic antigen signifies that it has reacted with a specific IgE antibody present in the tissue fluids. The reaction is one of immediate hypersensitivity, occurring within minutes or a few hours. A *negative* test, however, indicates that the patient has no IgE antibody for the particular allergen tested.

This latter interpretation of skin test reactions applies also in cellular immunity, but for a different reason. Most of the antigens used in this instance are cell-free extracts of microbial somatic antigens, more or less purified (tuberculin for tuberculosis, brucellin for brucellosis, histoplasmin for histoplasmosis, and so on). If the individual being tested has been previously exposed to the antigen, or is currently infected, sensitized T lymphocytes that are specifically reactive to it will be present throughout the lymphatic system and recirculating in the blood, as we have seen. At the site of a skin test, therefore, these sensitized cells will react with the injected antigen, inducing an inflammatory response that evolves slowly over the ensuing 48 to 72 hours. The intensity of the reaction and the time required for its development reflect the degree of sensitivity. A *positive* test now indicates delayed hypersensitivity and active cellular immunity. A *negative* skin test with a microbial extract antigen provides good evidence that the patient has not been previously infected with the organism in question, or that any exposure he may have had was minimal (provided the test is not given too early in a current infection).

2. ANTIGEN DETECTION. Injection of antibody into the skin to detect antigen present there is generally not practical because antigen localization at any point selected for testing cannot be predicted. One notable exception exists, however, in the case of scarlet fever. In this disease, the *erythrogenic toxin* produced by the causative organism (a streptococcus) localized in the throat is disseminated through the body. In the skin the injury it inflicts on the cells becomes evident as the typical scarlatinal rash develops. The rash and the disease can be specifically diagnosed with a skin test that employs the homolo-

gous antitoxin injected intracutaneously at the center of an area where the rash is most evident. The resultant combination of antibody with toxin neutralizes the latter, and the rash blanches at the injection site within a few hours. This test is sometimes referred to as the *Schultz-Charlton reaction,* being named for the men who developed it. It should be noted that the Dick test, mentioned above, employs the erythrogenic toxin for the opposite purpose of detecting antibody and determining immunity.

A list of the skin tests of diagnostic value, the diseases for which they are useful, and their interpretations is given in Appendix II.

Mechanisms of Immunization

Immunity may occur as a *natural* phenomenon, existing even though there has been no previous contact between the host and the infectious agent. Much more is known, however, about immunity that is *acquired* through biologic or artificial contacts with microorganisms.

Natural Immunity.

Species Immunity. The human host has a natural resistance to many microorganisms that cause disease in other species. There are almost no infectious diseases of plants to which man is susceptible, and vice versa. Similarly, many of the infections that occur in lower animals are not seen in human beings, the reverse being true as well. An example would be the natural immunity of many laboratory animals to poliomyelitis virus. Although this virus can be adapted to grow in the tissues of some other animals, it produces symptoms like those of the human disease only in monkeys or other primates closely related to man.

The basis of species immunity is obscure. Physiologic differences, the biochemical constitution and function of individual tissues, and nutritional and environmental limitations may play a part, individually or collectively, in creating barriers for microorganisms in one species as opposed to another. Many of these factors have their roots in the genetic, inheritable constitution of the species. It may be that these immunities are built in to the inheritable characteristics that distinguish and maintain each species.

Racial Immunity. Within the human species certain races and families are more resistant to some infectious diseases than are other persons. For example, both the incidence and severity of smallpox and tuberculosis have varied markedly among peoples of the world. From the time of their first contact with such diseases, and with repeated exposures for long years of history, Negroes, American Indians, and some other dark-skinned races have found them devastating in impact and outcome. Smallpox and tuberculosis have made deep inroads on white races also, but a racial capacity to resist and survive is far more evident among the latter.

The explanation may have several facets: these groups are subject to different stresses of environmental and nutritional origin; their biochemical heredity may be as disparate as their ethnic histories; the history of exposure is longer by centuries for white "Caucasians" than for the other groups. The latter fact, particularly, suggests that racial immunity, if it has a genetic basis, is a manifestation of biologic adaptation, occurring over a long period of time. When susceptible populations are decimated by disease, and the resistant survivors are capable of reproducing, the biologic characters of resistance are transmitted to succeeding generations and become more and more firmly established in later populations. Ethnic considerations guide and determine the nature of successive matings, so that as inheritable constitutional factors of resistance are passed along, they become numerically more prevalent within the racial framework of the population. On the other side of the coin, the parasite involved must make adjustments of its own, if it is to survive within a resistant host population, and must also possess permanent, inheritable characteristics reflecting selection from previous microbial generations more susceptible to host defenses.

Biochemical definition of inheritable factors that account for racial and family immunity may be as subtle as the hereditary supply of a particular red cell enzyme. A deficiency of this enzyme brings about a change in the hemoglobin molecule, which in turn leads to an increased resistance to one of the malarial parasites.

Individual Immunity. Populations highly susceptible to a particular microorganism may contain individuals who are strongly resistant to it, even though no previous contact with it can be demonstrated. Such *individual immunity* is a kind of biologic variation common among living things. It may be based in part on nonspecifically protective mechanisms, such as good nutrition and hormonal and metabolic health, but these do not account for the specificity of resistance displayed by an individual toward a specified organism, particularly if susceptibility to that organism is the rule in the population. The strongest explanations appear to link this type of immunity with biochemical characteristics, of both parasite and host, that are genetic in origin, but not yet precisely defined.

Acquired Immunity. The immunity that results from the production and activity of antibodies and sensitized cells in the host defending itself specifically against intruding foreign substances is spoken of as *acquired.* The term *active* acquired immunity implies that the host has produced his own antibodies in response to an antigenic stimulus. The antigen may be encountered in a natural way, or introduced in an artificial manner designed to induce an immune response in the host. *Passive* acquired immunity means protection of the host by antibodies he did not produce himself but received from another host, either naturally, by placental transfer, or artificially, by injection.

Active Immunization. Active immunity, acquired *naturally* through infection, may develop whether or not the infection results in clinically recognizable disease or remains subclinical. In either case, the immune response continues for as long as antigenic stimulation persists, but some destruction and removal of humoral antibodies occur at a slow rate, so that there is a maximum peak for antibody levels in the blood. This peak is maintained while the antigen remains in the body, then slowly declines when it disappears. The decline may be quite rapid in some cases, occurring in a matter of days, but in many instances (e.g., diphtheria and smallpox) low levels of antibody are detectable for many weeks, months, or years after the original infection. Cellular immunity may be even more durable.

Reentry of the antigen into the body at a later time is met by a more rapid and abundant production of antibody, as well as a more efficient cellular immune response. This phenomenon is referred to as the *anamnestic* (remembering) reaction. Thus, repeated subclinical infections with a specific organism commonly encountered, such as the poliomyelitis virus, may result in continuous antibody formation at a level sufficient to prevent the disease from ever recurring.

On the other hand, the first attack by a virulent organism may find the body unprepared and unable to avoid disease. Initial production of antibody is not immediate, and protective levels are not reached for at least 10 to 14 days, during which time acute symptoms of illness become manifest. In the days before the discovery of antibiotics, this period was often critical for the patient (as, indeed, it still may be), until antibodies joined the defense to help turn the tide, sometimes very dramatically, and convalescence began. Antibodies are usually at their peak at the time recovery becomes complete, then fall away slowly.

In certain chronic, progressive infectious diseases no durable, protective immunity develops, although the body becomes relatively resistant to superimposed, second infections with the same organism. This is true in the case of tuberculosis and in syphilis, for example. Recovery from, or arrest of, diseases of this type rests with the body's nonspecific defenses, some of the effects of hypersensitivity, and chemotherapy.

Active Artificial Immunization. Many infectious diseases can be prevented by immunizing techniques employing an infectious agent or its products treated or administered in such a way that they will not cause clinical illness. Some of the most commonly used immunizing agents and methods are discussed briefly below.

1. KILLED VACCINES. This term is used for injectable preparations containing microorganisms grown in laboratory culture (or in experimental animals, chick embryos, tissue culture, etc.), separated from the growth medium, and killed by heat or a chemical agent. Formalin and phenol are useful for this purpose. Vaccines are usually injected subcutaneously, less often by intracutaneous or intramuscular routes. As a rule multiple injections are required to produce durable protection, two or three being

spaced over a few weeks' time and one or more "booster" shots (for the anamnestic effect) being given at yearly intervals. The Salk poliomyelitis vaccine and influenza and swine flu vaccines are examples of killed preparations of organisms given in this manner.

2. LIVING, ATTENUATED VACCINES. Virulent microorganisms can be treated by a method that eliminates their pathogenic properties without killing them or inactivating their antigens. An originally virulent organism in this condition is said to be *attenuated*. The most successful methods for attenuation included drying (Pasteur's original rabies vaccine contained dried, attenuated virus [see Fig. 27–2, p. 594]); continued, prolonged passages of the organism through artificial culture media to induce mutations; and cultivation and passage of the organism through hosts other than man. Occasionally advantage can be taken of the close relationship of strains of organisms causing disease in man or an animal host, but not both. The agent of bovine tuberculosis, for example, causes a serious infection in cattle. It is antigenically very similar to the human tubercle bacillus and is also capable of inducing human disease. However, a strain of the bovine bacillus has been attenuated by many years of culture on artificial media in the laboratory and is used as a vaccine against human tuberculosis (BCG). A similar antigenic relationship exists between the viruses of cowpox and smallpox, so that the former, which is relatively avirulent for man, has been used as an effective attenuated immunizing agent for human beings. (This was originally Edward Jenner's discovery and great contribution, early in the nineteenth century. It was he who coined the term *vaccine,* from the Latin word for cow, *vacca.* The strain now used for vaccination is so different from the cowpox virus, it is thought that at some point, in some laboratory, an attenuated smallpox virus was inadvertently substituted.)

Attenuated vaccines usually provide a lasting immunity after a single dose. The rabies vaccine (now attenuated in duck eggs) is an exception, for it must be given in multiple doses over a period of days.

3. TOXOIDS. Toxoids are nontoxic but antigenic substances derived from toxins. Microbial exotoxins, such as those of the diphtheria and the tetanus bacilli, can be prepared in this way, having first been separated from the bacterial cells that produce them. The toxin present in the filtrate of a liquid culture is first treated with heat or formalin to destroy its toxicity, but not its antigenicity. It may then be precipitated with alum to provide an injectable preparation that is slowly absorbed from the tissues, thus assuring a prolonged antigenic effect. Toxoids are usually given intramuscularly in two or three initial doses separated by four to six weeks, followed by an annual booster.

Passive Immunization. Passive immunity is acquired *naturally* through the placental transfer of maternal antibodies (serum IgG) to the blood of the developing fetus *in utero.* The congenital immunity acquired by infants depends on the infections their mothers have had or have been immunized against as well as on the level and protective qualities of the antibodies involved. The mother's antibodies (sIgA) are also contained in her milk, particularly in the colostrum, and these reinforce the breast-fed child's immunity. As maternal antibody declines the child becomes more susceptible to infection, but by this time (at the age of four to six months), his own IgG antibody mechanism becomes functional. His exposure to common microorganisms at this stage is likely to lead to solid immunity against them, and he can be given further protection against dangerous pathogens by active immunization with vaccines and toxoids.

Passive Artificial Immunization. Passive immunity artificially acquired can be useful in providing temporary protection to a susceptible individual who has been exposed to a very hazardous infection. The source of antibodies is immune serum from a person or animal that has been actively immunized by infection or by an artificial method. (Cellular immunity can be passively transferred only by injecting sensitized lymphocytes.) Most commonly, a "pool" of serum from a number of individuals of the same species is obtained, the gamma globulin content is separated out by "fractionation" of the serum, and this "purified" antibody is administered. The concentration of serum by this method assures a high level of antibody in a small volume per dose. Pooled gamma globulins of human origin may contain antibodies to a number of common human diseases, such as poliomyelitis, infectious hepatitis, measles, and others.

The use of passive antibodies in the prevention of disease constitutes an emergency measure only, since the immunity provided does not persist beyond two to three weeks. Susceptible persons who have been exposed or are in risk of exposure to a disease that might have serious consequences for them are suitable candidates for passive immunization. Such persons might include the very old, or the very young, whose immune and nonspecific resistance mechanisms are inadequate. In some instances passive immunization of all contacts of a serious infection, such as infectious hepatitis, is advisable if natural active immunity is not widespread because an effective vaccine is not available or the disease is not common. Passive immunization may prevent the development of individual disease altogether or permit only a mild, attenuated infection to ensue. In the latter case, an active, effective immunity usually follows, as a result of direct antigen stimulation of antibody production. Whenever effective vaccines or other immunizing agents are available, however, active immunization should be provided because it assures better and more durable protection. Another reason for the advisability of active immunization is that the use of animal serums in human beings exposes the latter to foreign proteins whose antigenic activity may lead to a hypersensitive state of the body tissues as a result of an immune response to the antigens of the foreign serum. The "serum sickness" that ensues can be quite devastating to the tissues. Even if no untoward reaction follows the initial serum dose, another injection of the same serum species in a later emergency may produce a violent immediate reaction, sometimes with fatal consequences (see p. 181).

With respect to the treatment of infectious disease, immune serum is of limited usefulness in this era of antibiotic therapy. It is helpful in the treatment of certain infections characterized by microbial exotoxin production, however. In diseases such as diphtheria or tetanus, antibiotic attack on the organism is essential in eliminating it from its site of growth and multiplication, but of no effect in counteracting the activity of exotoxin already produced. In this situation, or in botulism, the immediate administration of antitoxin is indicated, to neutralize toxin before it can attach to tissue cells. Speed is of the essence in such cases, because even antitoxin cannot prevent damage to cells that toxin has reached.

Disorders of the Immune System

Diseases directly attributable to the functions of the immune system may be related either to *hypersensitivity*, mediated through either humoral or cellular factors, or to *immunodeficiencies*, i.e., functional failures of either humoral or cellular mechanisms, or of both. The dual nature of the immune system is again in evidence as we consider these problems.

Hypersensitivity

Characteristics. We have seen that the immune system plays a highly protective role in defending the body against the intrusion of foreign antigens, including those of microbial invaders. This same system may become injurious, however, if its responses are excessive, overreactive, or *hypersensitive*. The consequent damage to the body may be quite serious in some instances, even lethal.

Most of the injury results from interactions of antigens with antibodies, on the one hand, or sensitized cells, on the other, with resulting excessive triggering of biologic responses we have observed operating more normally in immunity: capillary permeability with edema and inflammation, release of histamine and other biologically active substances, lymphokine effects. The general nature and symptomatology of these reactions depend on whether they are mediated by antibodies or by sensitized T lymphocytes. The former are responsible for reactions of the *immediate* type of hypersensitivity, the latter for the *delayed* type. Immediate hypersensitivity includes the systemic reactions of anaphylaxis and serum sickness, and the more localized tissue responses of allergy (atopy). Delayed hypersensitivity provides the mechanism for the "allergy of infection," allergic contact dermatitis, and the autoimmune diseases.

Immediate Hypersensitivity. The manifestations of immediate hypersensitivity depend on several factors: the nature of the antigen, the frequency and route of antigen contact, and the type of immunoglobulin reacting with it. The initial dose of antigen (there may be several doses, or exposures) to which an individual later becomes sensitive is known as the *sensitizing dose.* A later dose of the same antigen creating tissue damage is called the *eliciting dose,* or the *shocking dose* if the injury is profound. Immediate reactions begin very quickly, within minutes of contact with the eliciting dose. They may disappear within an hour or so if there is no widespread, residual damage. If the antigen is introduced directly into the tissues, by injection or an insect sting, the clinical result will be systemic, as in anaphylaxis or serum sickness. When the route of contact is superficial, involving the epithelial tissues of skin or the mucosal linings (respiratory, conjunctival, intestinal), the symptomatology is more localized at those surface sites, as in asthma, hay fever, or allergic rhinitis and conjunctivitis. These local reactions are referred to as symptoms of *allergy* (meaning "altered reactivity"), or *atopy* (out of the way, out of place). They continue while contact with the antigen(s) persists, the symptoms reflecting the frequency of exposure; i.e., they may be constant, fluctuating, seasonal, sporadic, and so on.

The antigens of immediate hypersensitivity are usually called *allergens,* especially with reference to the localized allergies. They may also be labeled more specifically as injectants, inhalants, ingestants, or contactants. They may be complete antigens, in the usual sense, or hapten antigens bound to a large carrier molecule (see p. 159). The best-known allergens of atopy include pollens, dust, animal danders, feathers, and foods. The injectants inducing systemic reactions may be any foreign protein to which the body has been sensitized, but the most common offenders are animal serums (used for passive immunization), drugs (penicillin and some other antibiotics), and bee venom (including that of related stinging insects, e.g., yellow jackets, wasps, and hornets).

The antibodies involved are usually of the *IgE class,* but *IgG* and *IgM* (both acting together with complement) are also implicated in serum sickness. IgE is responsible for anaphylaxis and the common allergies, and for this reason it has many names, the most frequently used being *reaginic, atopic, skin-sensitizing,* or *anaphylactic* antibody. When it is formed in response to the sensitizing dose of antigen, very little IgE circulates in the blood, for it has a tendency to attach to tissue cells, particularly tissue mast cells, and to basophils.

Understanding that the degree and site of injury in hypersensitivity will vary with circumstances, we can say that the eliciting or shocking dose of antigen sets off the following train of physiologic events:

1. Antigen combines with antibody at the site of the latter's attachment to tissue cells.

2. Mast cells and basophils release a number of physiologically active substances from their intracellular granules. Histamine is prominent among these substances.

3. As a result of histamine release (together with other "mediators," e.g., serotonin from blood platelets and kinins formed from plasma proteins), there is a sudden dilatation of blood vessels, contraction and spasm of smooth muscle (particularly in the bronchioles and small arteries), and increased capillary permeability with edema.

If these effects are massive, there may be a sudden drop in blood pressure, with circulatory collapse and shock. Extensive smooth muscle contraction brings about severe bronchospasm, and edema results in laryngeal constriction. These clinical symptoms are characteristic of anaphylactic shock. Lesser or more localized effects occur in serum sickness and in atopic allergies.

Anaphylaxis. Anaphylaxis is a term coined to imply an effect opposite to that of *prophylaxis,* which signifies the prevention of disease. The extreme of injury that can be produced by hypersensitive immune mechanisms is *anaphylactic shock,* which stands in marked contrast to the prophylactic, or protective, effect usually associated with immunity. It occurs as a result of immediate and simultaneous physiologic changes induced by the shocking dose of antigen at many sites in the body. To achieve such an effect, the antigen must be introduced by a parenteral route (that is, directly into the tissues) in a fairly concentrated dose. Intramuscular, intravenous, or intracardiac injections may be required for experimental animals, depending on the purity and concentration of the antigen and the degree to which the tissues were sensitized after the sensitizing dose of antigen was given. The allergens involved may be complete antigens, such as the foreign proteins of the serum of another animal species, or incomplete antigens such as penicillin. The symptoms induced by the shocking dose of such materials in sensitive human beings include sudden difficulty of respiration due to bronchospasm and to edema of the larynx; facial edema; a rapidly increased heart rate due to arterial constrictions, then a fall in blood pressure with vasodilation and shock. A generalized urticarial reaction may occur as well, with giant "hives" developing at the site of injection of the antigen and also over large areas of skin. Intense reactions can result in death within minutes or a few hours.

The patient may recover from an anaphylactic episode if the shocking dose, the degree of his sensitization, the strength of his constitution, or all of these factors permit. Before the administration of potentially shocking agents, the patient should be tested for possible sensitization.

Serum Sickness. This is a systemic type of hypersensitivity that can sometimes be induced by an *initial* dose of sensitizing antigen, but more usually it results from a *second eliciting* dose given some time later. The type of antigen most frequently involved is animal serum given prophylactically to provide immediate passive immunity. Horse serum is commonly implicated simply because many antiserums can be obtained in large quantities by artificially immunizing horses. Diphtheria antitoxin, for example, is prepared in this way. It is used to treat patients with diphtheria infection, to prevent the effects of intoxication, especially in those not actively immune to diphtheria toxin. However, the horse serum used for this treatment may itself be dangerous to the patient. The antitoxin and other protein antigens in horse serum are foreign to human tissues and therefore induce antibody formation. After the initial dose of serum, antihorse antibodies begin to reach a significant level in the blood and tissues in about 8 to 12 days. If the original antigen (horse serum) has been eliminated from the body by this time, no untoward reactions will take place. If sufficient antigen remains, the antibody reacting with it and complement will produce symptoms beginning at the time antibody level is reaching its height. More

frequently, serum sickness is provoked by a second dose of serum, from the same animal species, on some later occasion in life. The symptoms of serum sickness are similar in many ways to those of anaphylaxis, but less severe. They include sudden fever; hives; edema of the face, hands and feet; lymphoid reactions with swelling of many nodes; and often an intense reaction at the site of serum injection. Edema of the laryngeal and other upper respiratory tissues can be a severe threat to normal respiration. The illness continues until all the antigen has been inactivated or eliminated.

Serum sickness may result from administration of antigens other than those associated with animal serums if the hypersensitivity mechanism is invoked. The incomplete antigens of penicillin and other drugs may induce serum sickness in hypersensitive individuals. For this reason, use of a particular antibiotic in treatment of infectious disease may be contraindicated. The patient's history of possible reactions to drugs should be carefully considered before they are administered. Evidence of hypersensitivity can also be obtained in some cases by skin testing.

Atopic Allergy. Unlike anaphylaxis and serum sickness, which require injection of the allergen, atopic allergies occur naturally through the inhalation or ingestion of, or contact with, the antigen. The well-known problems of asthma, hay fever, hives, and other skin rashes exemplify this type of hypersensitivity. In these instances the cellular injury induced by antigen contact with antibody-sensitized tissue is limited in scope. While these diseases are incapacitating, they are not systemic in their effects. The antigen-antibody mechanism constitutes the basis of allergic symptoms but other factors such as emotional stress, hormonal changes, and physical reaction to environmental stimuli may influence the response markedly. The capacity for hypersensitivity probably has a genetic basis, as evidenced by the distribution of allergic disorders in families. Actual sensitization of the individual, however, depends on his contacts with particular antigens.

DIAGNOSIS OF ALLERGY. *Skin tests* are commonly used to determine the nature of the antigen to which an individual is allergic. A solution or extract of the allergen is applied to the skin, either along the line of a short scratch made with a blunt instrument or by intradermal injection. A *positive* test becomes evident within minutes, with reddening (erythema) at the site and wheal formation. The wheal is a raised, blanched area in the center of the injected site, caused by the local collection of edema fluid. Erythema is due to local capillary dilation (it may flare out widely), the wheal to increased permeability of the capillary walls, both resulting from histamine release triggered by the antigen-antibody complex. These reactions reflect the same mechanism that operates in anaphylaxis; they disappear usually within an hour.

Because of the risk of fatal anaphylaxis in sensitive individuals being treated with injectable foreign proteins, such as animal serums and drugs with an allergenic potential, *these agents should never be used without first attempting to demonstrate hypersensitivity by means of a skin test.*

Skin testing for atopic allergies is sometimes inadvisable for exquisitely sensitive persons. In this case a *passive transfer test* can be done by drawing blood from the sensitive patient and separating the serum portion containing reaginic (IgE) antibody. Small quantities of this serum are injected intradermally in a series of sites in the skin of a volunteer. After a few hours, when IgE has had time to attach to local tissue cells, a series of suspected allergens are injected, each in a separate site where serum was placed. A positive reaction displays the typical wheal with erythema. This passive method is called the *Prausnitz-Küstner (P-K)* test.

DESENSITIZATION. Repeated doses of very small quantities of specific allergen elicit the formation of IgG antibodies. These serve as *blocking* antibodies, as they circulate in blood, by combining with free antigen in the blood or tissues before it can form a complex with reaginic (IgE) antibody attached to tissue mast cells. Desensitization can thus relieve or eliminate allergic symptoms for patients suffering from atopic allergies.

It should be remembered that injectable preparations of antigens, or antibodies, are almost never "pure" in the sense that they are free of other substances that may have antigenic activity. Foreign proteins derived from a culture medium in a vaccine preparation, or from animal tissues, can also induce antibody production and activity of the hypersensitive type. In addition to the immediate systemic reactivity described for anaphylaxis or serum sickness, these materials, when injected repeatedly into a particular skin site, can induce a localized tissue sensitivity (Arthus reaction). In such case each injection is followed by a more and more intense local reaction, which may terminate in a hemorrhagic necrosis of tissue in this area. It is advisable, therefore, when a series of desensitizing injections are being given, to vary the site of injection of each new dose.

TREATMENT. Nonspecific relief of atopic allergies can be achieved with antihistaminic drugs. Other types of medications, such as decongestants, vasoconstrictors, bronchodilators, or expectorants, may be useful as indicated. Adrenal hormones (corticosteroids) are administered in severe cases, either to suppress antibody formation or to reduce the intensity of inflammatory responses.

Immune Complex Diseases. There are a number of diseases that appear to be due to the formation of antigen-antibody combinations (*immune complexes*) that fix complement. Antibodies of the IgG or IgM class are involved, since these are the only immunoglobulins that bind complement to any extent (see Table 7-1). The antigens are soluble proteins or nucleic acids, some derived from viruses or bacteria, but not all of them have been identified. The biologic activity of the complexes results from triggering the complement cascade, which induces inflammatory responses damaging to blood vessels and basement membranes. The kidney glomeruli, for example, are involved in the glomerulonephritis that may follow streptococcal infections or that occurs in systemic lupus erythematosus. Rheumatoid arthritis is a chronic inflammatory joint disease caused by deposition of immune complexes on synovial membranes, but the antigen is unknown. Such diseases are classified as being of the immediate type of hypersensitivity because they are mediated by serum antibodies, rather than by CMI.

Delayed Hypersensitivity. The sensitized T lymphocytes of CMI are responsible for the immunologic injury of the delayed type of hypersensitivity. Their proliferation in response to antigenic stimulus, production of lymphokines, and cytotoxicity (see Fig. 7-10) provide the beneficial effects of cellular immunity but can be damaging in some circumstances.

(Immunity is like a fire: it can warm one's house when restricted to the fireplace or burn it down if it goes out of control.)

The term *delayed* originally arose from the fact that positive skin tests with a specific antigen require about 24 hours to become visible, and about 48 to 72 hours to reach their full development, in contrast to the immediate wheal seen in skin reactions involving antibody. Delayed hypersensitivity can be passively transferred by injection of intact sensitized lymphocytes. Although this is not a practical means of routine passive immunization, experimental work is being conducted to determine its value in the treatment of patients with cancer or deficient CMI mechanisms.

Delayed hypersensitivity is operative in microbial allergies ("allergy of infection") and allergic contact dermatitis. These are distinct from the atopic allergies of immediate hypersensitivity, with their constitutional or hereditary basis. CMI is also responsible for the autoimmune diseases and provides the mechanism for foreign graft rejection as well as tumor immunity.

Allergy of Infection. Hypersensitivity is a frequent complication of infections and may sometimes be a major feature of an infectious disease. In many instances, the organism involved may be relatively inert, immobilized within a lesion created by the body's inflammatory response, or have few virulent properties, yet the disease it induces is of long duration, persistent, and chronic because of hypersensitization. Infestation of the tissues by parasitic worms; fungous infections; viral and bacterial diseases—all may have an allergic component. Any type of tissue may be involved, depending on the localization of CMI reactions.

SKIN TESTS FOR MICROBIAL ALLERGIES. Delayed skin reactions are those that develop in not less than 24 to 48 hours following intracutaneous injection of the antigen. After some hours redness and edema appear, and increasing induration (hardening) of the area develops. Vasodilation and infiltration of the site with plasma and leukocytes account for these events. The cellular infiltrate often becomes quite dense, containing both polymorphonuclear and mononuclear cells. Tissue necrosis may occur in the indurated area when the reaction is severe. As the reaction subsides, redness and swelling disappear within a day or so, but the induration may persist for weeks.

This type of skin test result is typical when microbial antigens are used to demonstrate the hypersensitivity of infection. It provides a diagnostic clue as to the nature of an infectious disease or its stage of development. A positive skin test alone, however, is not necessarily indicative of *currently active* infection, but only of the fact that infection (clinical or subclinical) with the agent has occurred *at some time prior to the test.* It may be of assistance in the diagnosis of a large number of diseases, whether of bacterial origin (tuberculosis, brucellosis, lymphogranuloma venereum), fungal (histoplasmosis, coccidioidomycosis), or helminthic (trichinosis). The antigen employed in most instances is a cell-free extract of the organism, containing one or more of its allergenic constituents.

Because of the delay in development of a skin reaction in this type of test, the test site should be examined daily for at least three days to determine the extent of the sensitivity if the result is positive. The size of the reaction, the amount of visible damage to the skin, and the time of its development all provide evidence of the degree of the patient's sensitivity, and a clue as to when his original infection may have occurred. A slight reaction that takes some time to develop in a patient giving no evidence of current disease indicates either a very old infection that has since subsided or a very new one not yet clinically manifest. Conversely, a more rapid and damaging reaction may indicate either a current, active infection or a residual strong sensitivity from a previous encounter with the antigen. Clinical symptomatology and other diagnostic test results must be weighed together in the interpretation of a skin test. The reaction may be interpreted as negative if there is no visible injury (other than that produced by the injecting needle) when the site is examined at 24, 48, and 72 hours. A minor reaction arising and disappearing within the first 24 hours sometimes occurs if the material in which the antigen was prepared is itself irritating.

Allergic Contact Dermatitis (ACD). Contact dermatitis may be induced by many simple chemical agents. The symptoms of ACD may range from those of a minor skin rash to the appearance of large blister-like lesions. These vesicles, or *bullae,* as they are called, are surrounded by an area of erythema; they may ooze or become crusted, and they are usually intensely itchy. Scratching not only may lead to infection of the lesions but may distribute the antigen to other areas of the body.

The most common antigens of ACD include such widely different agents as metals (nickel, mercury), soaps, cosmetics, disinfectants, ointments, and drugs. Plant substances, particularly the oleoresins of poison ivy, poison oak, and poison sumac are notoriously allergenic (Fig. 7–15). Since many of these substances are not composed of large molecules characteristic of complete antigens, it is thought that they function as haptens in combining with proteins of the skin. The skin protein then acts as a carrier molecule, forming a complete antigen but now foreign, with the hapten providing the new antigenic determinant groups. T cells then react to this new alien intruder.

Repeated direct contact with the allergen is generally the cause of ACD, but indirect exposure may also induce it. The oily resin of poison ivy, for example, may be carried on clothing or in the smoke from burning plants. Skin testing to determine the causal agent is usually done by the *patch* test method, rather than by intradermal injection of suspected allergens. For the patch test, the antigen is taped to the skin, either directly, or on a piece of gauze or other inert fabric, and kept in contact for 48 to 72 hours.

Corticosteroids are used to treat severe ACD, but usually treatment is merely topically supportive. The condition can be completely cleared only by avoidance of contact with the allergen.

Autoimmune Diseases. Failure of the immune system to recognize "self" and its activity against the body's own tissue antigens may lead to the development of *autoallergies, or autoimmune diseases.* Not all the factors involved in the pathogenesis of such diseases are fully understood, nor is it always clear whether cellular or humoral immunity is responsible, or both. Possibly an imbalance between them or a defect in one of these mechanisms may contribute to autoimmunity, but certain other mechanisms are also thought to be operative, as follows:

1. Some tissues of the body are relatively isolated from the antibody-producing cells of the blood and lymph. Thus, under normal circumstances proteins

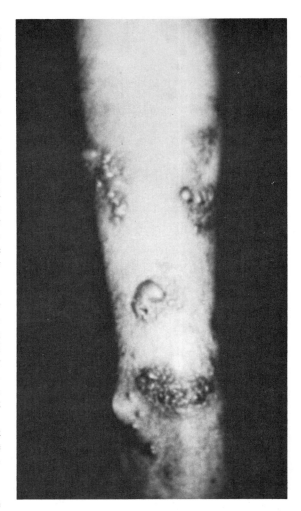

Fig. 7–15. The clinical lesion of poison oak dermatitis is acute and bullous and displays a linear pattern. (Courtesy of Dr. William L. Epstein, University of California Medical Center, San Francisco, Calif.)

of the crystalline lens of the eye, spermatozoa, thyroid, and brain tissue are not encountered by the antibody mechanism. The introduction of such proteins into the circulating blood through some disease process, or surgical procedure, may result in the production of specific antiorgan antibodies by cells that react to them as foreign antigens. Lymphoid cells probably also become sensitized to them. The clinical result may be an allergic endophthalmitis, thyroiditis, or encephalitis, or aspermatogenesis (failure to produce viable sperm), depending on the tissue antigen involved.

2. Antibody formed in response to a specific antigen may cross-react with a different one if both large antigen molecules have one or more constituents in common, "recognizable" by the antibody. Thus, an antibody produced in response to a foreign protein, such as a streptococcal antigen, might also react with a tissue protein of similar molecular structure. This may be the basis of some hypersensitive characteristics of diseases such as *rheumatic fever* and *glomerulonephritis* occurring as sequelae to streptococcal infection.

3. Incomplete antigens may attach to the surface of host cells or combine with host protein to alter its antigenic specificity. An immune response may thus be stimulated by the altered protein, and subsequently cause injury to the cell. Certain blood disorders, such as *"autoimmune hemolytic anemia,"* may be caused by the action of antibodies on antigens of the red blood cells or other elements of the blood, including platelets. In some cases the sensitization has been attributed to the administration of certain drugs (quinidine, sulfonamides) and to their subsequent behavior as haptens.

4. A genetic defect may operate in such a way as to bring about antibody production against some normal human protein. This type of effect appears to have influence in the patient with *rheumatoid arthritis* whose serum contains an IgM or IgG directed against normal human IgG. In other "collagen" diseases (lupus erythematosus, polyarteritis nodosa, scleroderma, and others) there is an autoimmune destruction of the nuclei of certain types of normal human cells.

Graft Rejection. Attempts to replace defective or diseased organs with tissue transplantation are sometimes defeated by the body's own immune mechanisms. This happens if the donor's tissues are sufficiently foreign to the recipient to cause them to be attacked and destroyed by T lymphocytes and, to some extent, antibodies. Donors for transplantation must be chosen on the basis of close matching of their tissue antigens with those of the recipient. Im-

munosuppressive techniques, i.e., procedures that inhibit the immune system, are also used to prepare the recipient for a transplant. X-radiation was once used for this purpose because it destroys lymphoid tissue, but it is permanently destructive and nonselective and has therefore been supplanted by more selective agents. Synergistic combinations of drugs that are anti-inflammatory (corticosteroids) with others that inhibit mitosis, and thus prevent the cloning of B or T lymphocytes, may be given. An antilymphocyte antiserum, containing antibodies to human lymphocytes, can be produced in horses and is an effective immunosuppressant. Before use, the horse serum is treated to concentrate the appropriate immunoglobulin fraction and to eliminate many of the horse antigens that can give rise to serum sickness or anaphylaxis.

The nursing care of immunosuppressed patients is extremely challenging, for such patients have little or no resistance to even minor infections. Their recovery, and the success of transplantation, depend heavily not only on the immunocompatibility of the grafted tissue, but also on their strict isolation in an aseptic environment (see Chap. 10, p. 256). Even so, pneumonia and other infections caused by the patient's own commensalistic organisms may constitute life-threatening complications until such time as immunosuppressant therapy can be discontinued and the immune system's functions can recover.

Tumor Immunity. Cancer cells differ antigenically from normal cells and are therefore subject to attack by the immune system, particularly CMI. Although T cells can and do destroy cancer cells, tumors or malignancies of the lymphatic system may progress nonetheless if their growth outstrips the ability of the immune system to keep pace, just as the battle of infection may be lost in the establishment of infectious disease. The propensity of the cellular immune system to destroy "nonself" tumor cells offers hope that immunotherapy, particularly in patients with immune defects, may be a valuable new approach to the treatment of cancer.

Summary of Hypersensitivity. Some of the important differences between immediate and delayed types of hypersensitivity are summarized in Table 7–3.

Table 7–3. **Differences Between Immediate and Delayed Types of Hypersensitivity**

	Immediate Type	Delayed Type
Mediating system	Humoral immunity	Cellular immunity
Factors involved	Primarily IgE antibodies, without complement IgM and IgG involved with complement in immune complex diseases and serum sickness	T lymphocytes and their lymphokines
Clinical examples	Anaphylaxis, serum sickness, atopic allergies (asthma, hay fever, hives), immune complex diseases (rheumatoid arthritis, systemic lupus erythematosus)	Microbial allergies, allergic contact dermatitis (chemicals, poison ivy, and others), autoimmune diseases (allergic encephalitis, glomerulonephritis, and others), graft rejection, tumor immunity
Tissue events	Capillary dilatation and increased permeability, edema, smooth muscle contraction (histamine, serotonin, and kinin effects)	Inflammatory response, macrophage activation, cytotoxicity (T cell and lymphokine effects)
Passive transfer	Circulating antibodies transferred passively by serum injection	Transferred passively by injection of lymphoid cells or their extracts

Summary of Resistance Mechanisms. We have seen that many of the mechanisms of nonspecific resistance are closely meshed with those of the immune system, and that the dual aspects of specific immunity may often overlap. The most outstanding factors that protect the human host from the inroads of infectious disease are summarized in Table 7–4.

Immunodeficient Diseases

Disorders of the immune system create serious problems in resistance to infection. Immunodeficient persons are wide open to infection, giving almost any microorganism they meet the chance to be a pathogen, or a real pathogen the chance to display more virulence. The incidence of cancer is also much greater among such individuals than in the general population.

Immunodeficiencies may be congenital (primary), iatrogenic (induced by immunosuppressive drugs), or the result of malignancies of the lymphatic system. Those that are not congenital are referred to as secondary immunodeficient disorders. We will classify them here on the basis of their origin in B cell abnormalities of function, T cell defects, or a combination of these.

B Cell Abnormalities. An aberrancy in the production of immunoglobulins (gamma globulins) is called a *gammopathy*. There may be an overabundance, as in *hypergammaglobulinemia,* or a decreased production or increased loss as in *hypogammaglobulinemia.*

Hypergammaglobulinemias. Each class of immunoglobulins is believed to be produced by a different specific clone of B cells. If one type of clone produces an overabundance of its immunoglobulin, the condition is called *monoclonal gammopathy* (MG). Biclonal or triclonal gammopathies are rare, but *polyclonal* gammopathy (PG), i.e., a general increase in all classes of immunoglobulins by many clones, is more common among patients with hypergammaglobulinemias.

PG can develop from a variety of diseases, including infections, diseases of the liver, immune complex

Table 7–4. Host Versus Microbe

Nonspecific Resistance Factors	Specific Resistance Factors

I. Skin and mucosa
  A. Architecture
    Mucociliary elevator
    Keratin
  B. Physiology
    Lysozyme
    Intestinal enzymes
    pH
    Flushing action
	I. The immune system
	  A. Humoral immunity
	    B lymphocytes (produce immunoglobulins [antibodies] of five classes: IgM, IgG, IgE, IgA, IgD)
	  B. Cellular immunity
	    T lymphocytes (produce lymphokines that inhibit phagocyte migration, activate macrophages, kill microorganisms)

II. Blood
  A. Circulates nutrients, removes wastes
  B. Complement
  C. Leukocytes
	II. Natural immunity
	  A. Species
	  B. Racial
	  C. Individual

III. Reticuloendothelial system
  A. Fixed macrophages
  B. Wandering macrophages
	III. Acquired immunity
	  A. Active
	    1. Natural (via infection or disease)
	    2. Artificial (via immunization with vaccines, toxoids)

IV. Lymphatic system
  A. Filtration
  B. Lymphocytes (specific)
	B. Passive
	    1. Natural (via placenta or colostrum)
	    2. Artificial
	      a. Serum antibodies (immunoglobulins)
	      b. Sensitized lymphocytes

V. Phagocytosis
  A. Polymorphs
  B. Macrophages	

VI. Inflammatory response
  A. Localization of agent
  B. Biochemical effects	

VII. New products or changes
  A. Interferon
  B. C-reactive protein
  C. Elevated ESR	

VIII. Normal activity of hormones	

IX. Environmental factors
  A. Host nutrition
  B. Stress
  C. Climate	

disorders such as rheumatoid arthritis or systemic lupus erythematosus, and malignancies. MG disorders are called plasma cell dyscrasias because they result in an overproduction of one type of immunoglobulin from a particular clone of proliferating plasma cells (mature B cells). Multiple myeloma is a malignant disease of plasma cells associated with MG. Large quantities of the immunoglobulin are filtered through the kidneys in this disease, appearing in the urine where it is called Bence Jones protein. Renal disease is a prominent feature of the later stages of multiple myeloma. Because high levels of abnormal immunoglobulins are accompanied by decreased levels of normal gamma globulins, the patient's susceptibility to infection often becomes the final cause of death.

Hypogammaglobulinemias. A decreased synthesis of immunoglobulins, increased catabolism (destruction), or loss through the urinary or intestinal tracts may all lead to hypogammaglobulinemias. Complete failure of immunoglobulin production is a congenital defect called *agammaglobulinemia.* Normally, the fetus and newborn infant can synthesize some IgM, acquiring most of their IgG passively from the mother. Normal babies produce their own IgG by the end of the first year of life. Infants with agammaglobulinemia may appear normal until this time, or until their maternal IgG has disappeared, but then they become abnormally susceptible to infection. Because they have a normal CMI response to infection, hypersensitivities may develop.

Secondary hypogammaglobulinemias develop in adults and children as a result of bone marrow and lymphoid disorders. Increased catabolism causes myotonic dystrophy, and disorders of the gastrointestinal tract or kidneys may lead to dangerous losses of gamma globulins by these routes.

Low levels of immunoglobulins can be treated by injections of gamma globulin, the frequency depending on the severity and/or permanence of the condition.

T Cell Abnormalities. Congenital absence of the thymus gland leaves little hope for an affected infant to survive long or normally. This condition, called *DiGeorge's syndrome,* leaves the infant open to severe bacterial, viral, or fungous infections, as well as more liable to develop cancer. Such children must be raised in complete protective isolation, with no direct physical contact with others. Although their immunoglobulins are normal, the complete absence of CMI provides ample opportunity for microorganisms to invade disastrously.

Even within their short life-spans, thymus-deficient children show an increased incidence of cancer as compared with normal children. These tumors tend to arise in the lymphatic or reticuloendothelial systems, often being of the lymphosarcoma or reticulosarcoma type. Studies of malignancies of the lymphoid system have contributed enormously to an understanding of the monitoring role of T lymphocytes in natural cancer prevention. It is when T cell function is absent or overwhelmed that the incidence of cancer rises.

Combined Immunodeficiency. Congenital absence of bone marrow stem cells results in a dual deficiency of B and T cells that is usually fatal within the first year of life. This total immunologic incompetence is called the *Swiss type* of immunodeficiency. Recently, techniques for the transplantation of bone marrow, with competent stem cells, have had some success in restoring such children to normalcy.

The primary challenge in the nursing care of immunodeficient patients is to prevent infection. Not only must the external environment be controlled (the patient's room, equipment, attending personnel), but the patient's internal environment — his own body — must be kept germ free.

Questions

1. What are the most important host factors involved in nonspecific resistance?
2. How does the blood act as a barrier to microorganisms?
3. What is the role of the reticuloendothelial system (RES) in nonspecific resistance to infection?
4. How does phagocytosis differ for macrophages and polymorphs?
5. What is the inflammatory response? Name two substances that arise in the blood during inflammatory and immune responses and indicate their significance.
6. Define antigen; antibodies; antigen-antibody specificity.
7. List the five classes of immunoglobulins and their different antibody activity.
8. What are B lymphocytes? T lymphocytes?
9. What is humoral immunity?

10. What is cellular immunity?
11. How can cellular or humoral immunity be detected?
12. What is acquired immunity? Passive immunity?
13. What is the chief distinction between hypersensitivity and immunity?
14. What is anaphylaxis?
15. Why are skin tests important?
16. What are autoimmune diseases?
17. What are immunodeficient diseases?

Additional Readings

Beletz, Elaine E., and Couo, G.: Infection and the Elderly Patient. *Nursing '76,* **6**:14, Aug. 1976.

Blount, Mary, and Kinney, A. B.: Chronic Steroid Therapy. *Am. J. Nurs.,* **74**:1626, Sept. 1974.

Burke, Derek C.: The Status of Interferon. *Sci. Am.,* **236**:42, Apr. 1977.

*Burnet, F. M.: *Immunology: Readings from Scientific American. Introductions and Additional Material.* W. H. Freeman and Co., San Francisco, 1976.

*Burnet, F. M.: *Immunology, Aging and Cancer.* W. H. Freeman and Co., San Francisco, 1977.

Capra, J. Donald, and Edmundson, A. B.: The Antibody Combining Site. *Sci. Am.,* **236**:50, Jan. 1977.

†Cooper, Max D., and Lawton, A. R.: The Development of the Immune System. *Sci. Am.,* **231**:58, Nov. 1974, #1306.

Craven, Ruth F.: Anaphylactic Shock. *Am. J. Nurs.,* **72**:718, Apr. 1972.

Donley, Diana L.: The Immune System: Nursing the Patient Who Is Immunosuppressed. *Am. J. Nurs.,* **76**:1619, Oct. 1976.

†Edelman, Gerald M.: The Structure and Function of Antibodies. *Sci. Am.,* **223**:34, Aug. 1970, #1185.

Faulk, W. P.; Demaeyer, E. M.; and Davies, A. J. S.: Some Effects of Malnutrition on Immune Response in Man. *Am. J. Clin. Nutr.,* **27**:638, June 1974.

Good, Robert A., and Fisher, D. W. (eds.): *Immunobiology.* Sinauer Assoc., Inc., Stamford, Conn., 1971.

Ham, Arthur W.: *Histology,* 7th ed. J. B. Lippincott Co., Philadelphia, 1974.

†Hilleman, Maurice R., and Tytell, A. A.: The Induction of Interferon. *Sci. Am.,* **225**:26, July 1971, #1226.

Lister, Joann: Nursing Intervention in Anaphylactic Shock. *Am. J. Nurs.,* **72**:720, Apr. 1972.

†Mayer, Manfred M.: The Complement System. *Sci. Am.,* **229**:54, Nov. 1973, #1283.

Mims, Cedric A.: *The Pathogenesis of Infectious Disease.* Academic Press, London; Grune & Stratton, New York, 1976.

†Niels, Kaj Jerne: The Immune System. *Sci Am,* **229**:52, July 1973, #1276.

Nysather, John O.; Katz, A. E.; and Lenth, J. L.: The Immune System: Its Development and Functions. *Am. J. Nurs.,* **76**:1614, Oct. 1976.

Old, Lloyd J.: Cancer Immunology. *Sci. Am.,* **236**:62, May 1977.

Raff, M. C.: Cell Surface Immunology. *Sci. Am.,* **234**:3, May 1976.

*Rose, Noel R.; Milgrom, F.; and van Oss, C. J. (eds.): *Fundamentals of Immunology,* 2nd ed. Macmillan Publishing Co., Inc., New York, 1979.

Sbarra, A. J.; Silvaraj, R. J.; Paul, B. B.; Strauss, R. R.; Jacobs, A.; and Mitchell, G.: Bactericidal Activities of Phagocytes in Health and Disease. *Am. J. Clin. Nutr.,* **27**:629–33, June 1974.

†Weiss, Jay M.: Psychological Factors in Stress and Disease. *Sci. Am.,* **226**:104, June 1972, #554.

*Available in paperback.

†Available in *Scientific American* Offprint series. (See Preface.)

Diagnosis of Infectious Diseases

The prompt, accurate diagnosis of infectious disease is the primary responsibility of the physician. The early recognition and treatment of disease are of urgent importance to the patient and to all who have been exposed to infection. The family and community contacts of the patient with a communicable disease must be considered and protected from risk insofar as possible. If the patient is admitted to a hospital, appropriate safeguards must be provided to prevent cross-infection of other patients and of members of the hospital staff. Since the route and ease of transmission vary with the nature and stage of infectious disease, diagnostic accuracy may be doubly important. The hospital's nursing service and its laboratory facilities can offer immediate assistance to the physician, so that diagnosis and treatment of the patient may proceed without delay, and suitable precautions may be established for the protection of others.

To play their essential roles well, the nurse and all others involved in patient care must be alert to the symptomatology and communicability of infections. They must also understand the circumstances in which the laboratory's diagnostic role can best be fulfilled. A background of theoretic and practical instruction in the nature of infectious diseases should provide a concept of the rhythms of host-parasite interactions and of the corresponding steps required for diagnosis, therapy, prevention, and control of infection.

In this chapter we shall review some of the broad patterns of infection, their clinical recognition, and the part the laboratory can play in diagnosis.

Clinical Stages of Infectious Disease

Clinical infections are often described as being either *acute* or *chronic,* depending on the rise and fall of symptoms displayed by the host in response to an *active* parasite. In general, acute infections are characterized by a rising pitch of symptoms that continue for as long as the microorganism or its products are actively capable of inflicting damage, then subside and disappear as host defenses become effective. Chronic situations arise when a pathogenic organism is able to survive the host's defenses and to maintain some level of damaging activity over a longer period of time. In either case, infection proceeds in a series of stages reflected by host responses.

Acute Disease

Human responses to acute infection fall into four periods with respect to the activities of the infectious agent and the symptoms produced thereby. These are the *incubation period,* the *prodromal period,* the *acute stage* of active disease, and the *convalescent stage* of recovery.

Incubation Period. At their first encounter, host and parasite appraise each other in the light of their respective capabilities and requirements. The parasite requires adequate living quarters and a little time to establish itself therein. The host meets the intrusion according to his capacity for both nonspecific and specific resistance to the parasite's effect on cellular functions. There are no outward signs of infection during this period when the "incubating" parasite is attempting to establish itself. If it fails in this, the only permanent record of its attempt may be kept in the immune accounts of the host whose lymphocytes have stored away the memory of the encounter. If the organism survives and multiplies, the symptomless incubation period comes to an end when the parasite's activities begin to embarrass host cells. The length of this period varies. It may be very short (two to three days) for some virulent bacteria or quite long for certain viruses and fungi (weeks or months), but the average incubation time for most agents of acute infections lies in the range of 10 to 21 days.

Prodromal Period. In its Greek origin the word *prodromal* means "running before." Its medical use refers to a brief period, usually one or two days, in which the earliest symptoms of trouble appear preceding the development of acute illness. The patient feels mildly uneasy or indisposed. He may have a "scratchy" sore throat, some headache, or gastrointestinal discomfort, but often he is unable to point out the source of this *malaise*. This *gradual onset* of symptoms provides the earliest evidence of the activities of a pathogenic microorganism, or its products, in host tissues. It may reflect the general dissemination of the organism through the body from its portal of entry, or it may coincide with injury developing at the site of localization, for example in the throat or in the gastrointestinal tract. On the other hand, some acute infections have an abrupt onset, illness developing with little advance warning.

Acute Period. This is the period when host-parasite interactions reach full intensity, producing symptoms of illness corresponding to the degree and site of injury. The inflammatory response of the host is in full swing, and immune mechanisms come into play. If this is the host's first encounter with the invading organism, his antibody-producing cells will require time for their response to its antigens, and cannot be expected to be protective in the early days of acute illness. An anamnestic antibody response to a pathogen previously encountered may shorten the period of acute symptoms or alleviate them. Conversely, a preexisting hypersensitivity is likely to contribute to tissue injury and heighten the symptoms of this period. More usually, hypersensitive reactions mark the later stages of chronic infections.

Fever. Fever is perhaps the most characteristic and least variable symptom of infection. It is a nonspecific response to tissue injury, possibly a defensive mechanism, associated also with many noninfectious diseases, such as malignancies, disorders of the endocrine or central nervous system, or autoimmune hypersensitivities. Fever is apparently induced by mechanical or chemical injury to the centers of temperature regulation in the brain. Substances that induce fever (that is, a *febrile* reaction) are called *pyrogens*. The endotoxins of certain bacteria are pyrogenic, as are some constituents of leukocytes themselves. It is possible that injury to leukocytes by bacterial pyrogens results in the liberation of a cellular pyrogenic chemical that stimulates the thermoregulators in the brain.

The pattern of fever seen from day to day in the acute stage of infection may reflect the nature of the disease, its progress, or the manner in which it spreads in the body. In some acute bacterial infections, fever may rise to a peak, continue with little fluctuation while infection persists, then subside as the organism is subdued or eliminated. In other instances, the patient's temperature may fluctuate markedly every few hours, rising and falling between two points. This type of pattern is often seen when "showers" of organisms are released periodically into the bloodstream from some localized site in which they are multiplying, inducing a febrile reaction. The periodicity and height of fever peaks may reflect the frequency and intensity of such releases. When organisms are present in the bloodstream, this fact is denoted by attaching the suffix "-emia" (i.e., "in the blood") to a word descriptive of the microbe involved. Thus, we speak of *bacteremia, viremia,* or, more specifically, streptococcemia, staphylococcemia, and so on. Normally, microorganisms are easily removed from the blood but in fulminating infections they may disseminate rapidly by this route from an original focus to localize in many other sites. This situation is referred to as *septicemia,* meaning sepsis or infection of the blood.

Febrile reactions are also influenced by the age and general condition of the patient. Children, for example, are notorious for their frightening ability to develop sudden high fevers for reasons that may be quite insignificant or extremely serious. Very old people may also display this kind of unstable temperature control. Patients in shock may have no fever or even a depressed body temperature. It should be pointed out, also, that fever is accompanied by a rise in pulse and respiration rates, as well as general metabolic activity. It often induces restlessness, irritability, loss of appetite, headache, or diffuse pains. Some pathogenic microorganisms may find the host's increased body temperature and metabolic activities antagonistic to their own growth requirements, but fever is not necessarily an effective defense mechanism.

Rash. Rash is another frequent characteristic of the acute phase of infectious disease. In some infections, the appearance of a rash and its distribution on the body may be highly typical and diagnostic, as in the case of scarlet fever, measles, chickenpox, or smallpox. The lesions that appear in the skin, upper respiratory tract, and oral mucosa are produced differently in various diseases. The cells of the skin may be one of several sites of localization of the organism itself, as in the "pox" diseases; they may be injured by the effects of a circulating toxin, as in scarlet fever; or they may be involved in the host's hypersensitive response to microbial antigens. *Petechial* lesions (from an Italian word meaning "flea bite") appear in the skin when the walls of its small blood vessels rupture, spilling blood into surrounding tissue. Such rupture may occur if the vessel lining cells are injured by microbial constituents or actually infected by microorganisms. The latter occurs in systemic meningococcal disease or in rickettsial infections, such as typhus fever, which are characterized by the appearance of petechial hemorrhages over large areas of the body. The rash of an infection may be very fleeting in its appearance, or it may be pronounced and persistent. When cells of the skin within a particular area die as a result of infection, the local inflammatory reaction may clear the site of cellular debris, leaving no trace. If the damage has been more severe, inflammation may continue to fibrosis and scar formation. This is the case in smallpox, the healed skin lesions remaining as permanent scars, or pocks.

Localized Symptoms. The symptoms of many acute infections are referable to the original or extending sites of localization of the infectious agent. *Nausea, vomiting,* and *diarrhea* are frequent hallmarks of gastrointestinal infection, together with the pain of intestinal spasm. The symptoms of different kinds of intestinal infections may be quite similar; but the time and nature of their onset, their duration and fluctuations, may furnish important clues as to their possible bacterial, viral, or helminthic etiology. Respiratory infections may also be difficult to distinguish, because a cough, sore throat, or catarrhal discharge is common to many. In a few instances the morphology of a lesion may be characteristic or suggestive; a streptococcal sore throat is typically edematous and red, or "beefy"; in diphtheria a tough pseudomembrane may be formed over the pharynx or the larynx; some virus infections produce small, clear blisters, or vesicles, on the reddened wall of the pharynx. Similarly, the visualization by x-ray of lesions in the lower respiratory tract may be diagnostically useful by revealing their size, shape, density, and distribution. Abscess formation and other manifestations of the inflammatory process, wherever they occur, produce symptoms referable to the local tissue injury. Inflammatory infections of the brain or other central nervous system tissue may be associated with such symptoms as drowsiness, mental confusion, stiffness of the neck, a spastic or flaccid paralysis of involved muscle groups, or other neurologic signs of motor nerve damage.

Convalescent Period. The patient's survival and recovery from the acute phase of his disease depend on the extent and nature of the damage done, on the strength of his own resistance mechanisms, and on the adequacy of his medical support. When it is possible to offer specific therapy aimed directly at elimination of the infectious microorganism, the acute stage may be shortened, the degree of damage lessened, and the strain on constitutional reserves lightened.

The convalescent phase begins with the decline of fever, often accompanied by a feeling of weakness but infinite improvement. It is at this point that antibody production has usually reached a peak and, if the level is sufficiently protective, contributes significantly to the final inactivation of the infecting organism. The persistence of antibody may then provide the individual with a long-lasting immunity. It should be noted that prompt, successful antibiotic therapy may not only abort the disease but also lower the level of antibody production by eliminating the antigenic stimulus at an early stage.

The speed and completeness of recovery in any case are determined by the nature of any injury induced, its extent and permanence or reversibility.

Chronic Disease

A persistent, chronic relationship between a host and a given parasite may remain unnoticeable if the latter exerts no pressures the host cannot readily encompass. More usually, chronicity of infection is characterized by a rise and fall of symptoms referable to the activities of the parasite, or its products, and the host's responses to them. These responses often involve the reactions of hypersensitivity, as described previously.

The symptoms may be similar and acute each time the parasite enters an ascendant phase; they may be of diminished character (subacute) if the host's defenses are more effective; or they may change in quality altogether, and increase in intensity if new tissues become involved. Syphilis affords an excellent example of the latter situation. It is a long and slowly progressive disease whose direction is marked off into three stages, each with a different symptomatology. Tuberculosis is also a chronic disease that may be slowly or rapidly progressive. In both instances, hypersensitive changes contribute to injury and are reflected in the permutation of symptoms as time passes. Thus the pattern of chronic infections reflects the seesaw equilibrium between the two

antagonists, just as it does in acute infections, although the level of reaction is different.

Laboratory Diagnosis

Specimens

The laboratory diagnosis of infectious disease begins with the *proper collection of the right specimen at the optimum time.* For this reason an adequate liaison is essential between the laboratory, the physician, and the nursing service (when the patient is hospitalized). The laboratory depends on the physician for information concerning the possible nature of the infection so that it can apply appropriate methods to isolation of the pathogen. The physician, in turn, may find the laboratory's advice useful with respect to the selection and timing of the specimen. The hospital nurse's role is frequently the essential one of assuring collection of the desired material at the right time, in a specified manner, and its prompt transportation to the laboratory. She must also be alert to the laboratory's dependence on full identification of the specimen, as to patient, type of material, time of collection, and the nature of studies desired. The *individual* nursing unit in a hospital may have many such specimens to prepare in a day; the laboratory receives these and as many more, perhaps, from *each* nursing unit or outpatient clinic in the same day (Fig. 8–1). The patient ill with an infectious disease must depend on physician, nurse, and laboratory to collaborate effectively in the selection, timing, collection, and subsequent handling of his specimen so that prompt, accurate diagnosis and treatment may be assured. Neighboring hospitalized patients must depend on this same group to provide security from cross-infection, particularly during the interval that may elapse between correct diagnosis and appropriate treatment of the infectious case.

Specimen Selection. The many *kinds* of specimens that may be sent to the laboratory for the isolation and identification of significant pathogens include blood, urine, feces, sputum, and pus from draining wounds most commonly, because they are most

Fig. 8–1. If the microbiologist is to be successful in the search for microorganisms of pathogenic significance in patients' specimens, the appropriate clinical material must be collected in suitable containers. Specimens and their accompanying request slips must be labeled correctly and completely.

readily obtained. Infectious microorganisms may be found in a variety of tissues and body fluids from which specimens may have to be taken by surgery, needle aspiration (bone marrow, cerebrospinal fluid, pleural or peritoneal exudates), or intubation (bile drainage). The search for the suspected agent of a given infection may take several directions before it is found, but often clinical signs of localization direct selection of the most appropriate type of specimen. Blood samples are frequently rewarding, particularly when the patient is febrile but does not immediately present symptoms of localizing infection. Specimens of blood repeatedly drawn over the course of a "spike" of fever may contain the organism being "showered" into the bloodstream from a local area of multiplication.

Specimen Timing. Knowledge of the usual clinical course of a given infection, together with careful observation of the presenting signs of the individual case, guides the selection of appropriate specimens for the laboratory. The *timing of their collection* must be geared to the *rise* of symptoms, rather than their fall, if the responsible pathogen is to be recovered in a viable, active state, not significantly affected by host defenses.

It is essential that specimens for laboratory culture be taken *before antibiotic therapy is initiated*. Antibiotics may not only affect the viability of microorganisms in the tissues; they may also remain active in the specimen to interfere with growth or change the metabolic characteristics of organisms sought in culture.

In acute infectious diseases, blood samples drawn during the prodromal period may yield microorganisms being disseminated silently from an original point of entry, but the stage of acute illness is most likely to reveal organisms not only in the blood but in specimens representing an area of tissue localization (a swab from the throat or an excised lesion, sputum, feces, aspirated spinal fluid, pus from an abscess, and so on). Similarly, throughout the course of a chronic infection specimens collected from involved tissues at the height of recurring symptoms offer the laboratory the best opportunity for isolation and identification of the responsible agent.

Correct timing of the collection of blood samples *for serologic studies* is also critical because immunologic diagnosis often requires the demonstration of a *rising* level of antibody as infectious disease proceeds from the acute to the convalescent phase. The presence of antibody in a single serum sample may reflect *either* current infection *or* some previous exposure to the specific antigen, but an increasing quantity of the antibody in serum specimens drawn as the disease progresses provides strong evidence that the antigen is now involved in the infectious process. Since antibody formation can be slow, particularly on first exposure, it is important that blood samples be spaced to reflect the change in the patient's symptoms.

In view of these considerations, the *right* specimen appears to be one dictated by the nature, localization, and duration of the patient's symptoms, while its *optimum timing* corresponds with their development.

Specimen Collection. The *proper collection* of specimens involves not merely their sterile or aseptic transfer into sterile containers, but such equally important factors as a *quantity* adequate for the necessary studies; *material representative of infection* rather than of the commensal, contaminating flora of the region; and *prompt examination* by the laboratory. The volume of the specimen obtained for testing is often critical. For example, small numbers of significant organisms may not be detected unless a large specimen sample is available for concentration. A generous volume is also necessary when several culture methods or serologic tests must be performed on the same specimen. When cultures are to be made of material expected to contain commensalistic organisms, it is essential to collect the specimen so that the *source* of infectious organisms is represented. Thus, a draining abscess should not be sampled at its surface, where normal flora of the skin may be predominant, but from its deep interior where the pathogenic agent can be found. In respiratory infections a sputum sample or a throat swab should be chosen according to localization of lesions, and the inclusion of saliva in the sample should be avoided. The interval between collection and examination of the specimen must be brief to assure the survival of fastidious pathogens, despite the presence of hardier organisms or other components of the specimen deleterious to them (phagocytic action may continue in a sample of pus, urine may be too acid or alkaline, antibiotic residues may be active).

Specimen Handling. It should be noted that clinical specimens are selected and sent to the laboratory for culture because it is expected that they may contain viable pathogenic agents of infectious disease. For this reason, *every such specimen must be considered as itself a potential source of the disease for all who handle it.* Careless manipulation of the specimen when it is placed in its container may contaminate the hands of the physician or the nurse, the outer surfaces of the vessel, or the request slip accompanying it. The messenger who transports the container is then also subjected to the risk of contaminated hands or clothing, as are laboratory personnel who subsequently work with the specimen. Each of these people, in turn, may have many continuing contacts with others—patients, other personnel, friends and family at home—all of whom may form a communicating chain for transfer of infection. It is essential, therefore, that every specimen be handled with thoughtful care, placed in a container of adequate size to retain it without spilling, and fitted with a tight closure. If adjacent surfaces are contaminated by the specimen during its collection, they should be promptly and carefully disinfected with an effective germicidal agent, kept conveniently at hand for this purpose, before those surfaces are touched again. In the microbiology laboratory, all techniques and equipment used in handling specimens and cultures are designed to reduce risk of cross-infection to a minimum, both within the laboratory and beyond it. The

Fig. 8–2. *A.* A computer clerk records the reading for a Gram-stained smear of sputum that is to be cultured. *B.* The preliminary smear report on this specimen will become a permanent part of the patient's record. The information is also stored in the laboratory for future reference if needed.

same principles of aseptic care must be applied at the patient's bedside and throughout the hospital.

Functions of the Laboratory

The laboratory has three essential functions with regard to clinically significant infections: it can (1) establish or confirm the etiology by isolating and identifying the agent; (2) perform antibiotic susceptibility testing on the isolated organism; and (3) conduct serologic tests to determine the patient's specific antibody response. Thus, the laboratory can provide either a microbiologic or a serologic diagnosis, or both, together with valuable information as to the appropriate course of treatment for the patient.

Microbiologic Diagnosis.

Isolation and Identification. Laboratory recognition of a pathogenic organism often begins with the microscopic examination of stained or unstained preparations of an appropriate specimen. An immediate report of findings in stained smears or wet mounts of the material may be very useful to the physician at this point, even if no organisms can be seen or if those that are visualized cannot be completely identified (Fig. 8–2). Preliminary microscopic work can be valuable also in guiding the microbiologist in selecting the optimum cultural method, or other procedure, for a conclusive identification of the organism sought (Fig. 8–3). When smears reveal nothing of significance, clinical information as to the possible nature of the infection becomes particularly important in planning cultures, since there is no one method that can assure isolation of every pathogenic microorganism (see Chap. 4).

Antibiotic Susceptibility Testing. Antibiotic testing may provide information of great clinical value when the causative organism of an infectious disease has been isolated in pure culture. If the physician has instituted chemotherapy before the laboratory's report is received, the result may confirm the original choice of a drug(s) or suggest more appropriate treatment to achieve maximum effects more quickly. In certain types of infection, the laboratory can also perform tests with the patient's serum during the course of treatment to determine whether the serum level of the antibiotic being given is adequate for suppressing or killing the causative organism (see Chap. 4, p. 82).

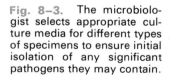

Fig. 8–3. The microbiologist selects appropriate culture media for different types of specimens to ensure initial isolation of any significant pathogens they may contain.

Serologic Diagnosis. Demonstration of specific antibody response in the patient may be obtained by the study of the patient's serum in the serology laboratory or by the application of appropriate skin-testing antigens. As discussed in the preceding chapter, a number of serologic methods can be applied to the recognition of antibody *in vitro* (see pp. 167–76).

Questions

1. List the clinical stages of acute infectious disease.
2. Define pyrogen; bacteremia; viremia; septicemia.
3. What three essential functions does the laboratory have in regard to clinically significant infections?
4. What are the nurse's responsibilities in the collection of microbial specimens?

Additional Readings

*Anaerobic Bacteria and Disease. The Upjohn Company, Kalamazoo, Mich., 1975.

*Lennette, Edwin H.; Spaulding, E. H.; and Truant, J. P. (eds.): *Manual of Clinical Microbiology,* 2nd ed. American Society for Microbiology, Washington, D.C., 1974, Chap. 6.

Mizer, Helen E.: The Tapeworm and the Noodle. *Am. J. Nurs.,* **63**:102, July 1963.

*Schneierson, S. Stanley: *Atlas of Diagnostic Microbiology,* 2nd ed., Abbott Laboratories, North Chicago, Ill., 1974.

*Wilson, Marion E.; Weisburd, M. H.; Mizer, H. E.; and Morello, J. A.: *Laboratory Manual and Workbook in Microbiology: Applications to Patient Care,* 2nd ed. Macmillan Publishing Co., Inc., New York, 1979.

*Available in paper back.

Section III

Epidemiology and Infection Control

9

Sources, Routes, and Patterns of Infectious Diseases

Epidemiology

Epidemiology is the science that deals with the incidence, distribution, and control of disease in a population. It constitutes the summation of all known factors that determine the presence or absence of a disease in a given group. For infectious diseases, these contributing factors include the *pathogenic agent,* the *reactive host,* and *environmental conditions* that *together* determine the occurrence of disease in a specified population over a stated course of time. In its original sense the word *epidemiology* applied to the study of *epidemics,* that is, of infectious communicable diseases that fell upon large numbers of people simultaneously or in rapid succession. The Greek words *epi* (upon) and *demos* (people) combine to provide this earlier concept of the study of mass outbreaks occurring as distinct episodes. Later, when the microbial agents of many infectious diseases were recognized, it became apparent that they do not always produce disease in epidemic patterns but may be implicated also in occasional, *sporadic* cases. Furthermore, an epidemic outbreak may be limited to very small numbers of

people, its size depending in part on the state of resistance to the parasite in question existing at the time in that particular population. It is now apparent, also, that a specific parasite may have a widespread, persistent distribution in a given host population, as indicated by numerous subclinical infections, while the cases of clinical illness attributable to the organism are few, but relatively uniform in number from one time to another. When this is the case, the disease is said to be *endemic,* or "in the people."

The modern concept of epidemiology embraces the entire biography of a disease with respect to its patterns of incidence as these are influenced by the parasite (or other agent), the host populations affected, and environmental factors. The term is often extended to the study of all interlocking influences that determine the occurrence, distribution, and effect of any disease on man's welfare and health. Thus, one may speak of the epidemiology of a particular infection (such as diphtheria or smallpox), or of a noninfectious disease (diabetes or arthritis), or of defined social problems (alcoholism, narcotic addiction, poverty).

In preceding chapters, emphasis has been placed

on the major contributions of parasite and host to the causation and course of individual infectious disease. In this section, we shall discuss influences exerted on host and microorganism by their natural, shared environment, examining the intersecting role of the latter in the epidemiology of infection. With knowledge of the epidemiologic basis of infectious disease, we can then consider how it can best be prevented or controlled by measures designed either to destroy the parasite, control its environmental distribution, improve host resistance, or by any feasible combination of these approaches. Epidemiologic principles form the basis for all methods of controlling infection. They are as applicable to the single case of communicable disease as to large-scale problems of public health.

Communicability of Infection

Many microorganisms are capable of survival under a wide range of environmental and host conditions because of their versatile metabolic mechanisms. However, the ubiquity of microorganisms cannot alone account for the incidence of infectious diseases, else the latter would be overwhelmingly high among human beings and animal hosts. We have seen that the incidence of *disease* versus *infection* can be explained partly by the fact that only a relatively few microorganisms possess the qualities of pathogenicity required to induce host tissue injury and partly by the opposing fact of host resistance mechanisms.

The question then arises: how can we account for the *distribution* of infection and infectious disease, attributable to a given organism of well-defined pathogenic quality, among human or other host populations possessing similar defensive properties? In other words, why are some diseases more *communicable* than others among human or animal populations, or both; and why do some have little impact on a particular population species? The answer is again twofold: it depends in part on the available *sources* of infection from which disease may arise and in part on the ease and direction of the transferral, or *transmission,* of pathogenic microorganisms from such sources to susceptible hosts.

Sources of Infection

Microorganisms reproduce wherever and whenever their capabilities permit them to do so. If the focus of their activities happens to lie within the body of another living organism that succumbs to their virulent effects, the parasites reach the end of the line also, unless they can be transferred by some means to another milieu where they can continue to grow. The site of colonization within the host determines whether or not transmission is possible and, if so, the ease of transferral to other hosts by direct or indirect routes. An organism localized in deep tissues, without access to the body's excretions, may not have a portal of exit under normal circumstances. In this case it will not be an epidemiologic threat to others unless some intermediary agent is able to effect a transfer to another host. A blood-sucking insect, for example, may carry certain microorganisms from the bloodstream of one host to those of another and maintain them in transit. Exudates from the mouth and upper respiratory tract, or discharges from the intestinal and genitourinary tracts, offer the simplest means of escape for many microorganisms that infect the human or other animal host. The total concept of communicability of infection depends, therefore, on the continuing survival of microorganisms from a source of multiplication through their transit to another host.

A local environment or host that supports the survival and multiplication of pathogenic microorganisms is spoken of as a *reservoir of infection.* An infected, though not necessarily diseased, human, animal, or insect host may constitute a *living* reservoir. There are also many environmental media (soil, air, food and milk, water) that may serve as *inanimate* reservoirs of infection. In the full epidemiologic sense, a reservoir can be said to exist when an animate or inanimate medium provides a pathogenic microorganism with adequate conditions for its maintenance over a prolonged period of time and *also* provides opportunities for its transmission to a new, susceptible host. The latter may in turn become a new reservoir, capable of infecting other hosts or an environmental medium, thus lengthening the chain of infection.

Living Reservoirs

Man. The chief reservoir of many important human infections is man himself. Very many people, whether or not they display symptoms of clinical disease, harbor pathogenic organisms and transmit them, directly or indirectly, to others. The most obvious of the human sources of infection are, of course, those people who are recognizably ill with a disease that is communicable. This source is also the most readily controlled, when promptly and accurately diagnosed, for the sick individual can be confined at home or in the hospital. The potential reservoir he represents is then restricted so that transmission of the pathogen is limited or impossible, while treatment of the patient is aimed at destruction of the parasite.

The largest continuing human reservoir, and the most difficult to recognize or control, is comprised of the many people who are subclinically infected with pathogenic agents of communicable disease. Infected individuals with only mild symptoms of illness, or none at all, circulate freely in the community, often communicating their infections as they go. The infected person who displays few or no symptoms referable to the parasite he harbors is often spoken of as a *carrier,* but this term is epidemiologically applicable only if the infection "carried" is transmissible to other persons.

When an acute infection subsides, the causative pathogen may remain alive, though subdued, during the patient's convalescence, for weeks, months, or even years after his recovery. The convalescent may thus become a carrier and remain so for as long as he harbors the viable, transmissible parasite. The epidemiologic control of carriers of important communicable diseases is directed primarily at preventing their formation of environmental reservoirs through contamination of food and water supplies that could involve large numbers of other people.

Animals. Animal reservoirs of infection are a frequent threat to human health in all parts of the world. The infectious diseases of animals that can be transmitted to man are called *zoonoses* (from the Greek words *zoon* for animal, *nosos* for disease). The pathogenic agents of the zoonoses are included in every major group of microorganisms: fungi, bacteria, viruses and rickettsiae, protozoa, and helminths. Human beings may acquire these infections through direct handling of animals and their products, by eating meat or other products from diseased animals, or through the bite of insects that prey on both animals and man. There are a few zoonoses for which infected human beings also serve as reservoirs, capable of transferring their infection directly to other people (as in the case of many *Salmonella* and streptococcal infections) or indirectly through another host or vector (for instance the systemic protozoan diseases, leishmaniasis and trypanosomiasis, transmitted between human beings and animals by biting insects). For the most part, however, man is an accidental, or tangential, victim of the majority of the zoonoses, and if infected he does not constitute a reservoir for such diseases as rabies, anthrax, brucellosis, or trichinosis, because he has no significant role in their transfer to other hosts.

Insects. The insects commonly involved in the transfer of infectious agents from one reservoir to another belong in the invertebrate phylum *Arthropoda* (literally, jointed feet, from the Greek words *arthron* for joint and *pod* for foot). There are several classes of arthropods, but only two, the *Insecta* and the *Arachnida,* contain members that are important as *vectors* of infection. The class Insecta includes flies, mosquitoes, fleas, lice, and the true bugs; the Arachnida include ticks and mites. Examples of important insect vectors are shown in Figures 9–1 to 9–7. Their classification is given in Table 9–1.

Most of these arthropod vectors are parasitic themselves, preying on vertebrate animals as a source of blood, a rich protein nutrient for the insect. An enormous variety of animals — wild birds and domestic poultry, rodents and small game, wild and domestic animals — offer the parasitic insects a source of support. When a biting or blood-sucking insect preys on an infected host, it may pick up infectious microorganisms in its blood meal and later deposit them on or inoculate them into the skin or mucous membranes of another individual of the same or different species. Thus a number of unrelated animal species may be linked together in the epidemiology of an infectious disease through the activities of an insect vector.

Often the parasitic arthropod becomes a true host

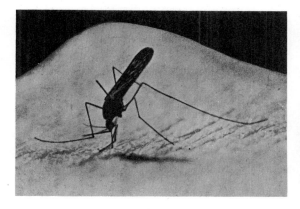

Fig. 9–1. The *mosquito vector* of the malarial parasite. (Reproduced from James, M. T., and Harwood, R. F.: *Herm's Medical Entomology,* 6th ed. Macmillan Publishing Co., Inc., New York, 1969.)

Fig. 9–3. Two adult *fleas,* an egg, and the larval forms of the insect. This type of arthropod can transmit the plague bacillus from rats to man. (Reproduced from James, M. T., and Harwood, R. F.: *Herm's Medical Entomology,* 6th ed. Macmillan Publishing Co., Inc., New York, 1969.)

to the pathogenic microorganism; that is, the insect is itself infected, affording the microorganism another site of multiplication or of development. There are some instances in which a microbial parasite actually requires an arthropod host to support it through developmental changes that are essential to the maintenance of its life cycle. This is the case with

Fig. 9–2. The *tsetse fly,* vector of the agent of African sleeping sickness. (Reproduced from Harwood, R. F., and James, M. T.: *Entomology in Human and Animal Health,* 7th ed. Macmillan Publishing Co., Inc., New York, 1979.)

the malarial parasite, which undergoes a sexual stage of development in its mosquito host. Certain helminthic parasites, such as the fish tapeworm and filarial roundworms, also pass through a portion of their life cycles in an arthropod host. Some pathogenic species of bacteria, viruses, and rickettsiae maintain themselves within their insect hosts throughout the latter's lifetime. The insect itself is not necessarily diseased by this parasitism, and in some instances may even transmit infecting microbes through its eggs to its own succeeding generations. The parasitic type of arthropod vector can thus be considered a *reservoir* of infection when it maintains and transfers the microbial agents of infectious disease to man, animals, or both. Insects that serve as host and reservoir for infectious agents are often referred to as *biologic vectors* to distinguish them from those that function merely as *mechanical carriers* of microorganisms from one reservoir to another.

There are a number of nonparasitic arthropods that have been incriminated as mechanical, or passive, vectors of infection. Chief among these are houseflies and cockroaches. These insects are sapro-

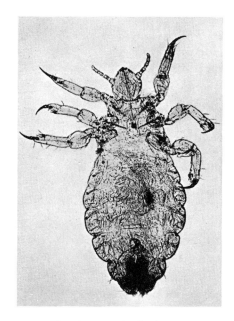

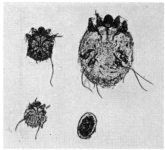

Fig. 9–4. One variety of *tick* (male and female) involved in the transmission of infectious disease. Below, the female is depositing eggs. (Reproduced from James, M. T., and Harwood, R. F.: *Herm's Medical Entomology,* 6th ed. Macmillan Publishing Co., Inc., New York, 1969.)

Fig. 9–6. The human body *louse,* a vector of epidemic typhus fever, a rickettsial disease. (Reproduced from James, M. T., and Harwood, R. F.: *Herm's Medical Entomology,* 6th ed. Macmillan Publishing Co., Inc., New York, 1969.)

Fig. 9–5. An adult *mite* (the largest of the forms shown below) and three developmental stages. This type of insect is involved in the transmission of scrub typhus, a rickettsial disease of the Far East. (Reproduced from James, M. T., and Harwood, R. F.: *Herm's Medical Entomology,* 6th ed. Macmillan Publishing Co., Inc., New York, 1969.)

Fig. 9–7. The *housefly,* a nonparasitic insect that may serve as a mechanical vector of infectious disease. (Reproduced from Harwood, R. F., and James, M. T.: *Entomology in Human and Animal Health,* 7th ed. Macmillan Publishing Co., Inc., New York, 1979.)

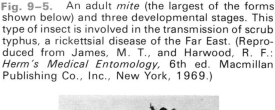

Table 9–1. Medically Important Arthropod Vectors

Kingdom: ANIMALIA

Phylum: ARTHROPODA The insect vectors or hosts associated with important human infectious diseases

Elongated segmented invertebrates, possessing an exoskeleton, bilateral symmetry, paired jointed appendages; two classes contain the important vectors of infectious microorganisms:

Class: ARACHNIDA, Order: ACARI

Vectors or Hosts of Agents Causing:

Head and thorax fused
Abdomen not segmented
No antennae
Adults have 4 pairs of thoracic legs
Mouth has a piercing organ (*hypostome*)

Ticks

Rocky Mountain spotted fever
Q fever
Relapsing fever
Tularemia

Mites

Tsutsugamushi fever (scrub typhus)
Rickettsialpox

(Includes *Sarcoptes* species, the agent of scabies, a mite infestation of human skin)

Class: INSECTA

Distinct head, thorax, abdomen
1 pair of antennae
2 pairs of wings on thorax
3 pairs of legs on thorax
Abdomen segmented

Order: ANOPLURA—LICE

No wings
Flattened dorsoventrally
Jointed antennae
Sucking mouth

Pediculus corporis (body louse)

Epidemic typhus; relapsing fever

Pediculus capitis (head louse)

Phthirus pubis (pubic or "crab" louse)
(Infestation with these ectoparasites is called *pediculosis*)

Order: HETEROPTERA (True bugs)

2 pairs of wings (some wingless)
Piercing and sucking mouths

Reduviid bugs ("assassin" bugs)

Chagas' disease (South American trypanosomiasis)

Bedbugs

Table 9–1. *(Cont.)*

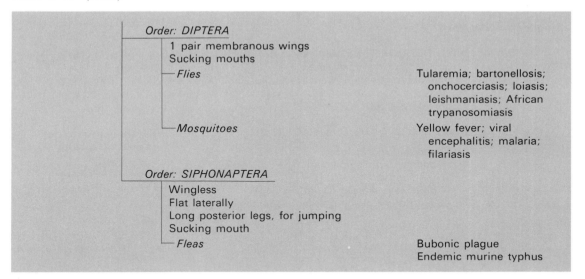

Order: DIPTERA
1 pair membranous wings
Sucking mouths

— *Flies* — Tularemia; bartonellosis; onchocerciasis; loiasis; leishmaniasis; African trypanosomiasis

— *Mosquitoes* — Yellow fever; viral encephalitis; malaria; filariasis

Order: SIPHONAPTERA
Wingless
Flat laterally
Long posterior legs, for jumping
Sucking mouth

— *Fleas* — Bubonic plague
Endemic murine typhus

phytic in their nutrient habits; flies, in particular, are notorious for feeding on, and breeding in, organic wastes, such as human and animal excrement and garbage of all kinds. Unlike their parasitic, biting relatives, houseflies do not feed directly on human beings or animals, but they do visit intimately with them and are capable of transmitting their microorganisms back and forth in a passive transport. Flies may ingest microbial parasites, and although they do not support them for long, they may later deposit them by defecation or regurgitation on any surface they alight upon. They also carry microorganisms around on their hairy legs, depositing them in an invisible trail (Fig. 9–7). The communicability of infections carried in such a mechanical way then depends on a number of other factors, such as the density of flies and of infectious microorganisms in reservoirs available to the passive vector; the speed of transfer of a fastidious pathogen to a new source of support; the point of entry to and susceptibility of a new exposed host. In other words, nonparasitic insects are not true reservoirs of infection, but may contribute passively to the spread of infectious disease.

Inanimate Reservoirs

The principal inanimate, or environmental, reservoirs for pathogenic organisms are *soil, air, food,* and *water.* To the extent of the ability of such elements to support the survival, if not the growth, of pathogens, such reservoirs may play a large part in the epidemiology of many infectious diseases.

Soil. The soil teems with microorganisms, most of which are totally without pathogenicity for man. In general, the only *natural inhabitants* of the soil that may be pathogenic for man are *fungi* and the anaerobic bacterial spore-former *Clostridium botulinum.* The spores of *fungi* are capable of long-term survival on dry surfaces, but the vegetation and reproduction of these simple plants require moist, nutrient soil. The great majority of fungi are saprophytic, but there are a few species of yeasts and molds with a low-grade capacity for parasitism. These may cause infections in man through his direct contacts with the soil, which constitutes their primary reservoir, or with dried earth blown about as dust. The virulent and invasive properties of the pathogenic fungi are not great, and resistance of the human host to them varies, so that contact with the reservoir and the ease of transmission do not alone assure disease. The fungous diseases are frequently referred to as *mycoses*. *Systemic* mycoses, or fungous infections of deep tissues and organs, include such diseases as coccidioidomycosis, blastomycosis, histoplasmosis, cryptococcosis, and a few others. It must be noted that the agents of the *superficial* mycoses (that is, fungous diseases of skin, hair, or nails, commonly referred to as *ringworm* infections) usually have a primary reservoir in infected man or animals and are transmitted from these living sources rather than from the soil.

Clostridium botulinum, a spore-forming anaerobic bacterial species responsible for the acute type of food poisoning called *botulism,* also has a primary reservoir in the soil. The organism can be isolated from the feces of domestic animals that graze on soil containing it in abundance, and from the intestinal tract of fish, but it is very rarely found as a commensal in the human intestine. As described in Chapter 6, botulism results from the ingestion of foods in which the vegetative form of these bacilli has been multiplying anaerobically and producing exotoxin.

The soil may also serve as a *secondary* reservoir for some types of organisms whose *primary source* is the intestinal tract of human beings or animals or both. However, this is true only for those pathogenic agents that are capable of surviving — if not multiplying — in the soil through any period that may elapse before they can enter a new host. The intestinal viruses (such as poliomyelitis virus), for example, cannot remain alive for long outside of viable host cells; consequently, their transfer from contaminated soil by mechanical vectors or by direct means must be very rapid, as there are no durable inanimate reservoirs for them. On the other hand, spore-forming bacteria of the intestinal tract, protozoans, and the eggs or larvae of intestinal helminths do survive in soil, retaining their infectivity long enough to be communicable from a secondary soil reservoir. The aerobic spore-forming bacterium, *Bacillus anthracis,* is highly pathogenic for man and his domestic animals. The anthrax bacillus probably does not multiply continuously in soil, which is not an optimum milieu for its growth, but it forms spores under

adverse conditions. Anthrax spores are very resistant to environmental stresses, remain viable for long periods in soil or in the hides, wool, and hair of infected animals, and germinate into vegetative, reproductive bacillary forms on ingestion or inhalation by a new host. The spores of anaerobic intestinal bacteria of the *Clostridium* genus are similarly capable of a long survival in soil or dust. Some intestinal protozoa can live and reproduce in soil, at least for short periods, in areas that are sufficiently moist, nutrient, and moderately acid. The protozoan agent of amebic dysentery, *Entamoeba histolytica,* does not survive well in soil, although it may find a secondary reservoir in water and food supplies. The eggs of several species of roundworms, and of some tapeworms, find their way to the soil from their primary reservoir in the intestines of infected human beings or animals. The eggs of some species, such as hookworm, may hatch in soil and go through a period of larval development there, the larvae being infective for new hosts.

In the sense that soil generally does not support multiplication of many pathogenic organisms, but merely offers conditions that permit their survival or developmental changes, it is a *limited,* rather than a true, reservoir. The route of transmission of pathogenic bacteria, protozoa, and helminths from soil to a human host may be direct, as in the case of the fungi, or quite indirect by way of other animal hosts.

Air. In its pure state, air does not supply nutrients for any living organisms, but, in practical terms, air is never pure. It is constantly being contaminated by dust from the soil, by various volatile products of human activities (such as those carried by smoke), and by respiratory secretions expelled from the mouth, nose, and throat of human beings (see the illustration on p. 295).

In the course of normal social communications, human beings talk, laugh, clear their throats, and thus gently expel their respiratory secretions into the air. When their respiratory tracts are irritated by infection, or by other agents, they may cough and sneeze, spraying the air over greater distances with large quantities of finely scattered mucus and its content of commensalistic or potentially pathogenic microorganisms. This mucous spray takes the form of *droplets* of various size, containing protein and heavy with moisture. These droplets may initially travel a few centimeters or as far as 1 meter, depending on the force with which they are expelled, and make direct contact with other individuals nearby. When left free in the air, heavier droplets settle out quickly, contaminating objects and surfaces on which they happen to fall. Lighter ones hang in the air for a longer time, gradually losing moisture by evaporation and becoming lighter still, until they are mere *droplet nuclei* of flaked, dry protein, usually with microorganisms still clinging to or held within them. Many pathogenic microorganisms contained within such droplet nuclei are capable of surviving long enough in air, or on the surfaces they settle upon, to remain infectious for other individuals.

Microorganisms discharged from the human or animal body by other routes may also reach the air. Discharges from the intestinal and genitourinary tracts, or from open wounds, contain protein and other microbial nutrient, moisture, and often many potentially infectious organisms. Such material may contaminate the soil or any of the myriad objects with which man surrounds himself, including his clothing. As it dries, its protein particles become less sticky and readily flake off from the surfaces to which they have adhered. They can be floated into the air when disturbed, or picked up on moist hands, carrying any microorganisms that have survived the drying process.

Pathogenic microorganisms cannot multiply in air, but many remain viable while they are protected from total drying within droplet nuclei or soil dust, supported by residual protein and moisture. The air thus becomes an intermediate source of many infectious diseases although, like contaminated soil, it is a *limited, secondary* reservoir. The limiting factors in airborne infection include the capacity of microorganisms to survive drying; the speed and direction with which they are carried by air to new hosts; the availability of vectors, such as insects or contaminated hands; the numbers and virulence of specific organisms; and the susceptibility of exposed hosts.

Food. Food for human and animal consumption is also a source of nutrient for many microbial forms of life. Raw food is probably always contaminated, by microorganisms from the soil or from hands, but is not necessarily a threat to health for two reasons: (1) the average contaminant from such sources lacks pathogenicity for man, and (2) microorganisms are normally not given an opportunity to multiply actively in food that is eaten fresh, or is adequately preserved, before and after cooking. When food is protected from the effects of microbial growth and spoilage, it may be eaten raw or cooked without fear of infection. Such protection is afforded by refrigeration or freezing; adequate canning; smoking, salting, or drying methods; or pasteurization as appropriate.

Unfortunately, contaminated food is all too frequently a primary or secondary reservoir for a variety of serious infections. Foods such as meat, fish, shellfish, and milk that are obtained from *infected animal sources* are a *primary* reservoir for some notable helminthic and bacterial diseases acquired by persons who eat these foods raw or imperfectly cooked or treated. The *contamination* of originally clean food by pathogenic microorganisms can create dangerous *secondary* reservoirs of infectious disease, or production centers for poisonous bacterial toxins, particularly if such contaminated food is improperly preserved, handled, or prepared so that microbial multiplication occurs before it is eaten.

Food As a Primary Reservoir. Foods derived from infected animals or their products afford a direct route of transmission from the animal to the human host's intestinal tract. The human infection that results is not necessarily limited to or even primarily localized in the intestinal tract, but the individual may then also become a reservoir, capable of transmitting the agent to other persons or to animals.

Diseases acquired from the ingestion of infected meat, fish, poultry, eggs, or milk and maintained by an infected human reservoir fall chiefly into two groups: those of (1) *helminthic* or of (2) *bacterial* origin (see Table 9–2, group 4).

1. HELMINTHIC DISEASES. The helminthic diseases include those caused by pork, cattle, and fish *tapeworms*, ingested as larval forms encysted in the muscles of these animals, and the *trematode* (or fluke) infestations acquired by eating fish, crustaceans, or water plants infected with larvae. Although the pathogenic agents and the clinical symptomatol-

ogy of these diseases differ in many respects, their epidemiology is similar with regard to sources and routes of infection. Man and animals both play a part in maintaining these helminths in nature, disseminating eggs into the environment. Their geographic distribution is limited by environmental influences on the cyclic developmental stages of the parasite and its hosts. The social and gastronomic habits of the human host also have their effect. In acquiring these infections, human beings are often the victims of their taste for raw food, or of carelessness in cooking it. Cooks and housewives may find these diseases an occupational hazard if they are given to checking the taste of food before it is completely cooked.

2. BACTERIAL DISEASES. The bacterial diseases acquired from infected food include bovine tuberculosis, salmonellosis, and a number of streptococcal infections. The vast majority of cases of *human tuberculosis* are caused by a human strain of the tubercle bacillus, for which man is the only important reservoir. The bovine strain of the organism is capable, however, of causing tuberculosis in man as well as cattle, human infection being acquired from the meat or milk of infected cows. Lesions of the bovine type usually do not involve the lung but, as a result of spread of the organism from the oropharynx and intestine, may occur in many other tissues of the body, including the skin, lymph nodes, bones, and central nervous system. The disease is not so readily communicable from these localizations as from pulmonary lesions, but man and diseased cattle may both continue the infection. Human cases of bovine tuberculosis have become rare in the United States and other countries where adequate government inspection and control of meats, herds, and milk have been in effect, but the disease remains a problem in some parts of the world.

Salmonella infections are widespread among animals, rodents, poultry, and other birds. They are also entrenched in many human populations. Their transfer among these reservoirs often occurs through the contamination of food supplies, but animal meat and products, notably eggs, can be *directly* infectious. Fresh or dried eggs are a frequent source of *Salmonella* outbreaks, particularly when they are only lightly cooked or eaten raw, as in eggnogs. The use of eggs as a valuable nutrient for patients and con-

valescing invalids may expose individuals who are particularly susceptible to infection. For this reason it is important for hospitals and nursing homes to choose suppliers of eggs from infection-free poultry farms and, when in doubt, to serve cooked eggs only.

Streptococcal strains that can produce human disease may originate in domestic animals such as cattle and horses, although, as in the case of tuberculosis, the majority of streptococcal diseases are of human origin. Streptococcal skin and throat infections may be acquired through handling animals or ingestion of infected milk, or, conversely, the infected human individual may transmit his disease to animals. The dairy farmer who has a streptococcal infection, or is a carrier, may infect his cows, the most common result being udder infections. The subsequent distribution of milk from such sources may lead to an epidemic spread of streptococcal disease, such as septic sore throat or scarlet fever.

FOOD-ACQUIRED DISEASES NOT TRANSMITTED BY MAN. There are a few diseases that man acquires from infected animal food sources but does not transmit to others. For these *man is not a reservoir,* but an accidental victim of infection (see Table 9–2, group 5).

Trichinosis, for example, is a roundworm disease of many wild animals, rodents, and swine. It is a systemic disease, induced by the encystment of larvae in muscles and organs. Man acquires it by eating the flesh of infested animals (domestic pork, wild game animals), ingesting larvae that have survived inadequate cooking. The human disease is also systemic and provides the end of the line for the parasite that has no transferring portal in this case.

A few bacterial diseases, such as *brucellosis* and *tularemia,* also belong in this group of zoonoses that may be transmitted to, but not by, human beings. The reservoirs for brucellosis are domestic animals, such as cattle, swine, sheep, and goats. Human infections may be airborne, from the dust of stables and pastures; they may be acquired through direct contacts with infected animals; or they may be transmitted through milk, cheese, and other dairy products derived from infected animals and distributed to people who would otherwise have no contact with the disease. Tularemia is a natural infection of many wild animals and some domestics. It is particularly frequent in rabbits. Several arthropod reservoirs also exist for this organism, commonly the wood tick, and in some areas a biting fly or a mosquito. The disease is transmitted by these insects, but may also be acquired through handling infected animals or by ingestion of game meat (especially rabbit) that has not been fully cooked.

Food As a Secondary Reservoir. The *contamination* of food by infectious microorganisms entrenched in human and animal populations may occur in a number of simple ways. It may be quite *direct*, from the infected intestine (via feces), skin, throat, or other superficial area to human hands, food, or milk; it may be *indirect*, through fecal pollution of water supplies from which fish and shellfish are harvested or in which foods are washed; or it may be the work of *mechanical vectors*, such as houseflies trafficking from feces to food. As a reservoir, food is most dangerous when it permits active multiplication of a contaminating pathogen. This occurs when it is fully or partially cooked or exposed for long periods to warm temperatures, so that its proteins and other complex nutrients are denatured to forms readily utilized by microorganisms. Since food contamination may occur unpredictably, silently, and invisibly, the handling of food should always provide barriers to the multiplication of pathogenic organisms it may contain. The best protection against growth of contaminants that have already found their way into foods lies in immediate thorough cooking or prompt refrigeration, since microbial pathogens do not thrive at temperature extremes.

When contaminating pathogenic microorganisms are given an opportunity to multiply in foods before they are eaten, the stage is set for bacterial "food poisoning," and one of two possible effects may ensue: (1) an active infection may result from ingestion of organisms, or (2) symptoms of toxicity may follow ingestion of bacterial toxins preformed in food during the period of active microbial growth, prior to the meal.

1. INGESTION OF PATHOGENIC ORGANISMS. A number of important infections, for which man is a reservoir, are acquired from secondary food reservoirs directly or indirectly contaminated by infected persons. Some of these diseases are localized in the intestinal tract and are transmitted from that route,

notably shigellosis, salmonellosis, cholera, and amebiasis. Others, such as typhoid or paratyphoid fevers or viral hepatitis, have systemic localizations and effects, but the organisms may nonetheless be transmitted via liver and gallbladder secretions to the intestinal tract or to the kidneys and be shed in feces or urine. When the pulmonary or upper respiratory tracts are involved in these infections, as they sometimes are, transmission may occur through sputum and mucous contamination of hands or other objects involved in food preparation and service. As pointed out earlier, the symptomatic human case of infectious disease, promptly diagnosed and removed from the active scene, may represent a controlled reservoir. The asymptomatic carrier of such infections, however, can unknowingly continue to create secondary reservoirs in food, milk, and water supplies from which disease may spread ever more widely. This is particularly true in the case of typhoid fever and other *Salmonella* infections, as well as in mild, aborted, or incubating cases of viral hepatitis. These diseases are spread not only by the contamination of prepared foods by food handlers, but even more widely by food exposed to sewage-polluted water. Fish and shellfish harvested from polluted water or vegetables washed in contaminated water are often eaten raw or incompletely cooked. Infected foods, or those contaminated with pathogenic organisms permitted to multiply, may, however, still be safe to eat *if they are cooked sufficiently before the meal* to destroy their microbial content.

2. INGESTION OF BACTERIAL TOXINS. Toxins produced by staphylococci or *Clostridium* species (especially *Cl. botulinum*) actively growing in foods prior to their ingestion produce illness referable to their effects on cells of the intestinal mucosa or, if they are absorbed, on cells of other tissues. Staphylococci are most commonly introduced into food by handlers. *Clostridium* species, including *Cl. botulinum*, are soil inhabitants, whose spores may be present on most fresh vegetables. The spores are not pathogenic, but they are quite heat-resistant, and for this reason may survive light cooking processes that often precede the canning of foods (especially in home canning). Within the anaerobic, sealed can, surviving spores of *Cl. botulinum* can germinate to vegetative bacterial forms, which in turn produce botulinal toxin as they metabolize. The toxin is relatively sensitive to

heat, so that if food containing it is boiled for 15 minutes it may be eaten safely. Unfortunately, canned vegetables, fish, soups, and other foods taken from the can often are cooked very briefly, or not at all, so that the problem of prevention of botulism poisoning lies with the killing of spores during canning. Staphylococcal enterotoxin is more resistant to heat than botulinal toxin and can be destroyed only by prolonged heating (see Chaps. 6 and 18).

Milk. Milk is a special food, rich in protein, carbohydrate, fat, and othe nutrients particularly favorable for growth of contaminating, pathogenic microorganisms. Although milk is sterile when produced by the healthy, lactating animal, it passes through mammary ducts that have a normal commensalistic flora. From that point it is further directed over surfaces and into containers that may be contaminated with bacteria and other microbes of wide variety. For these reasons, milk, whatever its source or subsequent handling, can be expected to contain a mixed microbial flora of environmental origin, most of which have no pathogenic import. The most common bacterial species that contaminate milk, and grow well therein, belong to the genera of *Streptococcus, Lactobacillus,* and *Micrococcus*. Species of these groups abound in milk, but cause no more damage than souring when storage temperature and time permit their overgrowth. The potential threat offered by milk lies in the similar opportunity it may afford pathogenic contaminants to multiply under suitable conditions of temperature and time.

Milk-borne diseases originate either from infected animals or from the hands and methods of milk handlers. The important diseases of *animals* transmitted to human beings through their milk are *tuberculosis* and *brucellosis*. Less frequently encountered are such infections as Q fever, a respiratory disease of rickettsial origin, and foot-and-mouth disease, a viral disease, that can devastate cattle herds but to which man is, fortunately, largely resistant. Human streptococcal infections may originate from animal sources, or from the cross-infection of animal milks by infected handlers, as previously described.

Human diseases most commonly transferred through milk contaminated by handlers or their methods include not only *streptococcal* infections but

also *diphtheria*, both transmitted by respiratory or skin contacts, via contaminated air, equipment, or hands. Infections transmitted and acquired by the gastrointestinal routes are also frequently traced to secondary milk reservoirs. Important diseases of this type include bacillary dysentery, or *shigellosis;* intestinal *Salmonella* infections; and systemic salmonelloses such as typhoid or paratyphoid fever.

Water. Water from natural sources is, like air, seldom pure. Whether it comes to us from ground sources (springs and wells), from surface accumulations (rivers, lakes, reservoirs), or directly from the rain and snow that maintain these supplies, water always contains organic and inorganic material, suspended or dissolved. Rain washes down particulate matter suspended in the air, as well as volatile chemicals from the atmosphere, many of which go into solution in rainwater. Surface and ground pools of water are not only fed by rain, but are also exposed to soil. Consequently, water may normally contain many varieties of microorganisms derived from air or soil, as well as the supporting chemical substrates they need to maintain themselves and to multiply in this milieu. Water may be purified of its microbial or chemical contents by several methods (see Chaps. 10 and 11). Such purification is indicated whenever there is a chance that water has been contaminated by pathogenic organisms or harmful chemicals, or when it has been polluted with materials that give it obviously unpleasant characteristics of odor, taste, or appearance.

Man himself is responsible for the most frequent and hazardous pollution, or contamination, of water supplies. The normal microbial flora of water, derived from soil or air, does not include species pathogenic for man. This flora is composed of many heterotrophic saprophytes and autotrophic varieties of microorganisms. Man contributes a dangerous contamination to water when he permits it to be polluted with his sewage. The water drained from kitchen sinks, laundry tubs, bathtubs, and toilets not only contains high concentrations of soaps and detergents, but teems with microorganisms of pathogenic as well as nonpathogenic varieties, many of which are capable of multiplication in impure water. The possibility of disease transmission arises when sewage contaminates water or food supplies intended for human consumption, especially if subsequent purification is not accomplished. The diseases most commonly transferred in this way, from a water reservoir directly, or via food supplies bathed in contaminated water, are those of enteric origin. Microbial pathogens excreted in feces from an intestinal colonization, or from a lesion draining to the intestinal tract from the liver or gallbladder, include the bacterial agents of *typhoid and paratyphoid fever, shigellosis,* and *cholera;* the viruses responsible for *poliomyelitis* and *infectious hepatitis;* the protozoan agent of *amebic dysentery;* and a number of species of *helminth ova.*

Pathogenic organisms excreted in urine may also survive in water and be transferred through this type of intermediate reservoir. The systemic *Salmonella* infections, such as typhoid and paratyphoid fever, may be transferred through urine. The same is true for hemorrhagic jaundice caused by spirochetes of the *Leptospira* genus and for *Schistosoma* infections of the bladder.

For many pathogenic microorganisms, water is a limited reservoir in that it cannot support their continuing multiplication. Fish and shellfish foods that become infected with waterborne parasitic organisms, bacterial, viral, or helminthic, may of course continue the chain of disease. Water itself, however, does not remain infectious indefinitely unless it is continuously supplied with the organic nutrient required by pathogenic organisms capable of saprophytic existence or is constantly recontaminated.

Endogenous Sources

The term *endogenous* means that something is originating, growing, or proceeding from *within.* In biologic usage, "within" may refer to an entire living organism, such as the human body, or to a tissue segment or an individual cell. An infectious microorganism growing within individual cells or tissues of the body may be tolerated and well contained by a regional set of circumstances, yet raise havoc if it gains access to other types of cells or tissues functioning under other conditions in different parts of the body. For example, a *Salmonella* infection restricted to the intestinal tract may be inapparent, because it is well tolerated by the lining cells of the bowel, or it may produce symptoms reflecting its local toxic effects within the intestine. Should circumstances arise that permit dissemination of *Salmonella* organisms to body tissues beyond the intestinal tract, localization of infection in new sites produces a new and different symptomatology. One might say that a systemic disease, distinct in many ways from the original, has been acquired from an *endogenous* source, rather than from an outside, or *exogenous,* reservoir, although the latter may have been the first locus of infection.

It is equally important to remember that microorganisms living a commensalistic or saprophytic kind of life on the skin or mucosal linings may induce destructive reactions in systemic tissues to which they are foreign. Enteric bacilli, streptococci, and staphylococci derived from endogenous sources within the intestinal tract or the skin, where they may reside unnoticed, can be injurious to other tissues. Accidental trauma, erosions of skin or bowel by ulcerative conditions or tumors, and surgical procedures offer the most common means of transfer of microorganisms from endogenous sources, usually by way of the lymph or bloodstream, to parenteral tissues where their successful localization may result in disease. Since the organisms involved seldom have well-marked properties of virulence, these infections may be subacute and slow to progress, depending also on the nature, function, and reactivity of the infected tissues. (See Chaps. 5 and 6.)

Summary: Diseases and Reservoirs

In Table 9–2, many of the common communicable diseases of man or animals are summarized in groups arranged according to their major reservoirs.

Routes of Transfer of Infection

The communicability of infection depends not only on the nature and accessibility of reservoirs, but also on the routes and mechanisms available to the infectious agent for its transmission to new hosts. We have seen that there are four major avenues through which infecting microorganisms may enter the human body: respiratory tract, gastrointestinal tract, skin and mucosal surfaces, and parenteral route. (See Chap. 6.) These same avenues may also serve as exit

Table 9–2.   The Reservoirs of Infectious Disease

Group 1. Man is the only reservoir.

Diseases:

Viral:	Measles, rubella, mumps, influenza, herpes, poliomyelitis, chickenpox, smallpox.
Bacterial:	Shigellosis, whooping cough, diphtheria, gonorrhea, syphilis, lymphogranuloma venereum, trachoma, pneumonic plague, actinomycosis
Fungal:	Candidiasis (the agent is an endogenous commensal of human oral mucosa).

Transmission: Direct, man to man, no durable environmental reservoirs.

Group 2. Man is the only living reservoir.
Secondary reservoirs exist in the inanimate environment.

Diseases:

Viral:	Infectious hepatitis.
Bacterial:	Typhoid fever, cholera, tuberculosis (human), staphylococcal and streptococcal infections (of human origin).
Protozoan:	Amebiasis.
Helminthic:	Hookworm disease, ascariasis, enterobiasis.

Transmission: Direct from man to man, or indirect from environmental reservoirs:
Food harvested from contaminated water (oysters, clams) or soil (garden vegetables) and eaten raw or insufficiently cooked.
Water for drinking, washing, cleaning, or swimming.
Soil contacts
Surfaces

Group 3. Man and one or more species of arthropods are the reservoirs.

Diseases:

Viral:	Yellow fever (mosquito).
Rickettsial:	Epidemic typhus fever (louse).
Protozoan:	Malaria (mosquito).

Transmission: Arthropod to man, man to arthropod. Direct man-to-man transmission does not occur.

Group 4. Man and one or more species of animals are the reservoirs.
Arthropods involved as biologic or mechanical vectors for some.
Secondary environmental reservoirs may exist for some.

Diseases:

Fungal:	The superficial mycoses: ringworm of hair, skin, nails.
Bacterial:	Salmonellosis, staphylococcal and streptococcal infections (of animal origin), tuberculosis (bovine), clostridial diseases (tetanus, gas gangrene), listeriosis.
Protozoan:	Leishmaniasis, trypanosomiasis.
Helminthic:	Filariasis (tissue nematodes), taeniasis (intestinal tapeworms), schistosomiasis and other trematode diseases.

Transmission: Direct from animal to man or animal.
Direct from man to man.
Indirect from environment, especially food, milk, or water.

Table 9–2. (Cont.)

Except:

Clostridial diseases:	endogenous or indirect from fecally contaminated soil.
Leishmaniasis Trypanosomiasis Filariasis	arthropod to man or animals
	Direct transmission does not occur
Taeniasis:	ingestion of infected beef or pork
Trematode diseases:	ingestion or penetration of larvae from aquatic sources

Group 5. Various animal or arthropod species, or both, are the reservoirs.

(Man may be an accidental victim of these infections, but he is *not* a reservoir for them, since they are rarely, if ever transmitted from infected human beings to others of any species.)

Diseases:

Viral:	Equine encephalitis, rabies
Rickettsial Bacterial	Scrub typhus, murine typhus, Rocky Mountain spotted fever, Q fever, psittacosis, anthrax, brucellosis, tularemia, bubonic plague, leptospirosis, relapsing fever.
Helminthic:	Echinococcosis, trichinosis.

Transmission:

Equine encephalitis:	Horses to mosquito to man. Direct man-to-man transmission does not occur.
Rabies:	Bite of a rabid dog or other animal. Direct man-to-man transmission possible but not documented.
Rickettsial diseases (except Q fever)	Animal to arthropod (ticks, mites, fleas) to man. Direct man-to-man transmission does not occur.
Q fever:	Indirect from animals to man via airborne dust or ingestion of raw milk. Direct transmission from animals to man possible. Direct transmission from man to man does not occur.
Psittacosis:	Direct contact with infected birds, or indirect through airborne dust. Man-to-man transmission direct but rare.
Anthrax Brucellosis Tularemia Leptospirosis	Direct or indirect from animals to man. Dust from infective hides or tissues (anthrax); raw milk (brucellosis), ingestion of infective meat (tularemia), contaminated swimming water (leptospirosis) the usual transmittal agents.
Bubonic plague:	Rat to rat flea to man. Direct man-to-man transmission does not occur except in cases of bubonic plague terminating with plague pneumonia.
Relapsing fever:	Lice and ticks transmit to man. No direct transmission from man to man.
Echinococcosis:	Ingestion of infective ovum from intestinal tract of an animal, usually dogs.
Trichinosis:	Ingestion of pork containing encysted larvae.

Table 9–2. *(Cont.)*

Group 6. Soil is the primary reservoir.
Animals (including birds and fish) may be secondary reservoirs.

Diseases:
  Bacterial: Botulism, nocardiosis
  Fungal: The systemic mycoses: coccidioidomycosis, cryptococcosis, histoplasmosis,
                 sporotrichosis.

Transmission: Usually by contact with contaminated soil; through skin contact, inhalation of
                  airborne dust; or, in the case of botulism, through spore-contaminated foods.

---

routes, depending on the site of localization of the organisms in the infected body and the availability of a transporting mechanism. *Respiratory* and *gastrointestinal* discharges are expelled from the body with ease and frequency. They may spread infection far and wide through droplet contamination of air; through contamination of surfaces, hands, and handled objects; and through establishment of secondary reservoirs of microbial growth in water, milk, or food supplies. Infectious discharges from exposed lesions of the *skin and mucosa* may also be widely disseminated by hands and other mechanical vectors, or by person-to-person contacts. *Parenteral (deep) tissue* infections, not exposed to the surface, cannot be transmitted naturally unless their microbial agents find their way into the body's discharges or into an insect host as it takes a meal.

The transfer of an infectious microorganism from its exit point from one host to its portal of entry into another may be quite direct and immediate, or it may be delayed and sometimes extremely devious (see Table 9–3). Intermediate, inanimate reservoirs may link the travel routes of the organism between living hosts, sometimes at close range, but in some instances over great physical distances (Fig. 9–8). Some of the parasitic protozoa and helminths display complex developmental cycles that require their transfer in sequence from one type of host or reservoir to another.

Transmission Routes from Living Reservoirs

There are two basic modes of transmission of infectious agents from infected human, animal, or insect hosts to susceptible human beings: (1) transfer may be made by *direct contact* with an infected, living host, or (2) it may follow upon *indirect contact* with infectious material derived from a diseased host.

1. Direct Contact. This type of transfer involves close, or frequent, physical encounters between susceptible and infected hosts, the infectious agent being transmitted directly to an appropriate site of entry to a new host.

Infected human hosts may transfer their microbes to other persons through their many physical contacts: hand shaking, kissing, sexual intercourse, or other personal and tactual associations. Droplet transmission may also be quite direct among people in close proximity. The most usual exit and entry sites for infections spread among and by human beings in direct association are the respiratory, oral, and genital mucosa. Venereal diseases are spread almost exclusively by sexual contact. The bacterial agent of gonorrhea may infect the conjunctival mucosa of a newborn infant at the time of its delivery by an infected mother whose vaginal discharge contains the organism. Syphilitic mothers may also transmit their disease to the developing fetus, *in utero*. Children sometimes acquire venereal diseases from infected adults through fondling and kissing. Other infections may also be spread by contaminated hands, either those of the infected individual or those of persons in close contact with him. Surface infections from suppurating wounds, pimples, or boils are readily transferrable by hands as well.

Nurses, physicians, and others who care for infec-

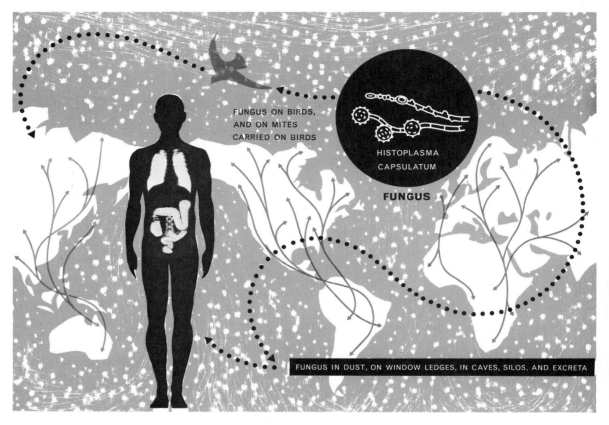

Fig. 9–8. The fungal agent of histoplasmosis may be carried to all parts of the world by birds. (Adapted from *What's New,* Summer, 1962. Courtesy of Abbott Laboratories, North Chicago, Ill.)

tious patients, particularly those whose diseases are directly transmissible, must take every precaution to protect others as well as themselves from contact spread of infection. Medical personnel may carry infectious organisms, in a nearly direct route, on their hands, on their clothing, or even on their faces, from one patient to another. The cardinal principles of good medical and nursing care of all patients include aseptic techniques designed to prevent the spread of infection. These techniques are of particular importance to patients whose susceptibility to infection is high: surgical and obstetric patients; the very old or the very young; and those with underlying, debilitating disorders that undermine the normal resistance mechanisms. (See Chap. 10.)

Infected animals may also transfer their infections to susceptible human hosts who come into direct contact with them. Animal breeders, farmers, dairymen, butchers, hide and leather workers, poultry raisers, pet lovers, and, be it noted, meat eaters — all are subject to exposure to a number of animal infections that are transmissible by a direct route. The most frequent transfers occur by the hand-to-mouth route; by ingestion of infected meat or animal products such as milk or cheese; by direct inhalation of contaminated airborne excretions; or through abraded skin or mucosal surfaces.

Infected insect hosts or vectors may directly introduce infection into human tissues or blood, via the parenteral route. However, the insect-borne diseases are not transmissible between human or animal hosts, their agents being dependent on the contacts, transport, and support of their arthropod hosts for their continuation in nature and for their access to other host species.

2. Indirect Contact. Human infections may be acquired from other infected hosts even without immediate intimate contact. Microorganisms that can survive on surfaces contaminated by sick persons, animals, or infected insects may later find an entry route into susceptible human hosts who touch or handle such items. Clothing, bedding, surgical dressings, instruments, bedpans, paper or linen handkerchiefs, toys, eating utensils, food, drinking glasses, or their contents — all may provide a bridge of indirect contact between infected hosts and new susceptibles. Insect vectors can play an indirect role in the spread of disease by depositing infectious material on food, skin, clothing, or other contact surfaces.

The usual route of entry for infections acquired by indirect contact is oral, following a hand-to-mouth transfer of infective matter from a contaminated surface. Occasionally, entry may occur through the skin or mucosa, particularly if a local injury or lesion permits.

Routine provision for the nursing care of hospitalized patients includes the sterilization of many items that could be an indirect source of infection on contact. Items such as water carafes, glasses, dishes, bedding, bedpans, and thermometers are either disinfected or sterilized after their use to assure that neither acute nor clinically inapparent infections will be transferred by these means. Disposable materials, to be used once and thrown away, greatly relieve the labor of some hospital nursing routines, but care must be taken that such items are disposed of in a manner that will not permit infection to continue or flourish. Incineration is generally the method of choice.

Transmission Routes from Inanimate Reservoirs

We have seen that most inanimate media (soil, air, food, milk, water) serve as *secondary* reservoirs for those infectious microorganisms they can support,

providing the latter with interim means of survival and transport between living, *primary* reservoirs. The saprophytic fungal agents of the systemic mycotic diseases of man and the bacterial species *Clostridium botulinum* alone have a primary reservoir in inanimate sources, such as soil or decaying vegetable matter.

The route of transmission to susceptible human hosts of pathogenic organisms surviving less permanently in inanimate media depends on the nature of man's exposure to, or use of, these secondary reservoirs. Infectious water, food, or milk may reach the gastrointestinal tract by an obviously direct route, or by a less apparent, indirect path, via contaminated objects or skin. Water that is used for washing clothes or utensils, for bathing or swimming, or for irrigating crops may contaminate food, hands, or clothing. Some of the helminthic parasites that undergo a phase of their normal development in water or wet soil are capable of direct penetration of normal human skin (the larvae of hookworms and *Schistosoma* species), while other kinds of pathogenic organisms contaminating these media can enter the body only through injured skin or the alimentary mucosa. Air containing infectious droplets or particles of dust commonly finds a direct route of entry to the respiratory tract, but it may also deposit its burden on surfaces from which a hand-to-mouth transfer can occur later.

Biologic products, including serum, plasma, soluble nutrients (e.g., dextrose or vitamin solutions), drugs, or vaccines may also become contaminated with microorganisms capable of surviving and multiplying in them. When such contaminated materials are injected parenterally into the human body, very serious infections may arise from this direct introduction of microorganisms into the tissues or bloodstream, even though the agents involved may display little or no ability to establish themselves by entry through more readily available portals.

Table 9–3 summarizes the routes of transfer of infection.

Routes of Transfer of Animal Parasites

Animal parasites pass through cyclic, developmental stages of growth, maturation, and reproduc-

Table 9–3. **Transmission Routes**

From Living Reservoirs		From Inanimate Reservoirs	
Direct	**Indirect**	**Direct**	**Indirect**
Kissing	Contact with infectious material derived from diseased host, i.e., sputum, feces, exudate from wound, animal tissues	Ingestion of contaminated food, milk, water	Via contaminated intermediary objects: utensils, clothing, bedding, skin
Hand shaking		Skin penetration by parasite from intermediate reservoir	
Droplet transmission during talking, sneezing		Injection of contaminated biologic products	Usual route is oral following hand-to-mouth transfer of infectious material from contaminated surfaces.
Sexual intercourse	Usual route is oral following hand-to-mouth transfer of infectious material from contaminated surfaces.	Soil contacts	
Insect vectors		Dust inhalation	
Animal contacts			

tion (see Chap. 2). The full circle of changes involved is referred to as a *life cycle*. The continuation of parasitic protozoa and helminths in nature depends on completion of their life cycles, and their transfer to new hosts depends in turn on the transmissibility of a given developmental stage.

Life Cycles of Protozoans and Helminths. Many animal parasites cannot complete their life cycles in one host or reservoir. A human or animal host may be parasitized either by their adult, reproductive forms or by their intermediate, immature forms, sometimes both. The host in which the adult parasite lives and reproduces is called the *definitive*, or *final*, host, and the host in which immature stages of development of the parasite occur is defined as the *intermediate* host. More than one type of intermediate host may be required, in sequence, to complete the life cycles of some parasitic worms. *Alternate hosts* are those that each, individually, can support both the adult and intermediate forms of a parasite but only by alternating, together providing continuity for the organism. Animal parasites vary widely in complexity of development, but basically they display four types of patterns with respect to their host requirements. These are discussed below and illustrated in the accompanying figures.

Type 1. The cycle is maintained by definitive hosts only. Intermediate or alternate hosts are not required, although many of the parasites in this group must have a period of intermediate development in soil. All the *intestinal* roundworms (nematodes) and the *intestinal* protozoa display this simple type of host-host or host-soil-host cycle.

A. HOST-TO-HOST TRANSFER. This type of transmission is possible for the intestinal protozoans and for one of the roundworms: *Enterobius vermicularis,* or *pinworm* (Fig. 9–9).

Entamoeba histolytica, the agent of amebic dysentery and of systemic amebiasis, is the only intestinal protozoan of consistent medical importance. The vegetative trophozoites of this organism invade the mucosal tissue of the large intestine (they may also be carried to other tissues and organs in the body where they induce abscess formation). The ameboid trophozoites may be passed from the body in feces, but they do not survive long outside of the body, and if they are ingested directly they usually are killed by the acidity of the gastric juice or by the action of bile in the upper intestine. The most infective form of this parasite is the *cyst*, which is more resistant to chemical and physical agents than the trophozoite. The cyst is also a reproductive form; with its four nuclei it can give rise, by mitosis, to eight amebae.

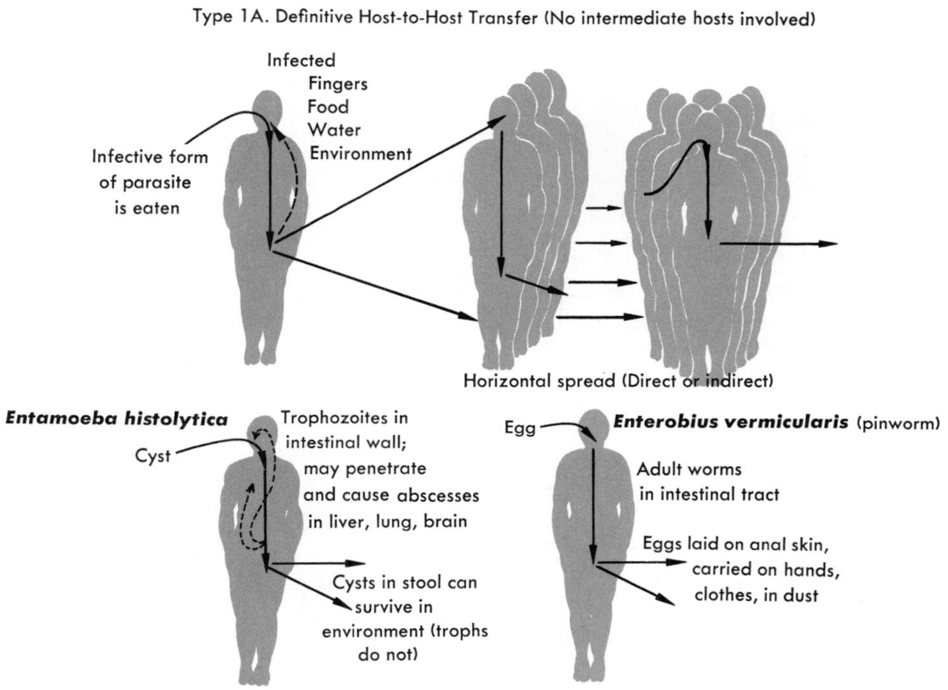

Type 1A. Definitive Host-to-Host Transfer (No intermediate hosts involved)

Infected
Fingers
Food
Water
Environment

Infective form
of parasite
is eaten

Horizontal spread (Direct or indirect)

Entamoeba histolytica

Cyst

Trophozoites in
intestinal wall;
may penetrate
and cause abscesses
in liver, lung, brain

Cysts in stool can
survive in
environment (trophs
do not)

Egg

Enterobius vermicularis (pinworm)

Adult worms
in intestinal tract

Eggs laid on anal skin,
carried on hands,
clothes, in dust

Fig. 9–9. This diagram illustrates the simplest type of life cycle among the animal parasites. Because intermediate hosts are not required, the cycle being complete within each host, direct or indirect horizontal spread to additional hosts may occur.

The patient with an active clinical infection may pass both cysts and trophozoites. Many amebic infections are clinically inapparent, but "carriers" pass cysts and can be dangerous sources of infection until they are recognized. Amebic cysts can survive in water and food, and on cloth or other objects that have been subject to fecal contamination. When ingested, following their direct or indirect transfer, they pass without injury through the stomach and small bowel to the large intestine, where they "excyst," each forming eight trophozoites.

Enterobius vermicularis is the smallest of the roundworms, having the dimensions of a small straight pin — hence its nickname. The adult worms live in the human intestine and copulate there. The gravid female then migrates downward through the large bowel to the anus, where she deposits fertilized ova. Her presence and activities on perianal skin

may cause an intense pruritus. Scratching transfers the ova to hands and fingernails, which pass them in turn to inanimate objects or to mouths. The ova also survive well on contaminated clothing or bedding. When shaken from these, they may remain viable in dust for some time. When ingested, each pinworm ovum develops into an adult in the intestine, and the cycle is renewed.

B. HOST-TO-SOIL-TO-HOST TRANSFER. This is the route for all the other intestinal roundworms. Among them we see three kinds of variation in the basic pattern of the cycle (Fig. 9–10).

Trichuris trichiura, commonly called the *whipworm*, lives as an adult in the intestinal tract. Its ova are shed in feces, but they must reach soil and have several weeks there for further development. They become infective when the larval form developing within each egg is well differentiated. Fingers,

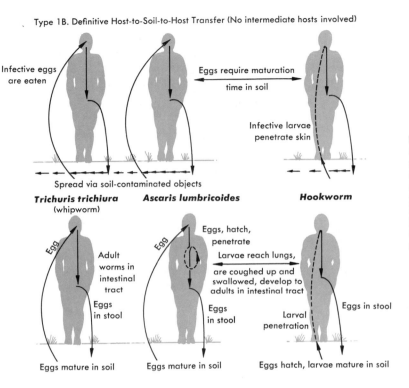

Type 1B. Definitive Host-to-Soil-to-Host Transfer (No intermediate hosts involved)

Infective eggs are eaten

Eggs require maturation time in soil

Infective larvae penetrate skin

Spread via soil-contaminated objects

Trichuris trichiura (whipworm)   **Ascaris lumbricoides**   **Hookworm**

Egg — Adult worms in intestinal tract — Eggs in stool — Eggs mature in soil

Egg — Eggs, hatch, penetrate — Larvae reach lungs, are coughed up and swallowed, develop to adults in intestinal tract — Eggs in stool — Eggs mature in soil

Larval penetration — Eggs in stool — Eggs hatch, larvae mature in soil

Fig. 9–10. The diagram indicates that although intermediate hosts are not required, the immature forms of these parasites must have maturation time in the soil. For this reason they cannot be transmitted directly to other hosts, but indirect horizontal spread does occur.

clothes, toys, or other objects contaminated with infective soil are the usual transferring vehicles. When the eggs are swallowed, they develop into adults in the lower intestine.

Ascaris lumbricoides, the largest of the roundworms (20 to 30 cm in length), resembles *Trichuris* in the cycle of maturation of eggs passed in feces to the soil and subsequently transferred to the mouth. When infective eggs are swallowed, however, the larvae are hatched in the upper intestine, penetrate its wall, and are carried by the bloodstream to the lungs. Here they wriggle through the alveolar walls and up the air passages of the bronchial tree, into the trachea, the larynx, and finally past the epiglottis into the mouth. They are swallowed again, and when they reach the intestine once more they finally develop into adult, egg-producing worms. The migrations of *Ascaris* larvae into the lungs (and sometimes into other organs, via the bloodstream) cause far more serious disturbances than does the intestinal infestation. Insofar as the cycle is concerned, how-

ever, the transfer route is from feces to soil to mouth, the infective form being the ovum containing a differentiated larva.

Hookworm species (*Necator, Ancylostoma,* and *Strongyloides*) display another variation. The hookworm females produce their ova in the intestinal tract. The ova are passed in feces, but must reach warm, moist soil for further development. Under the right conditions, the larvae hatch after a day or two in soil. They develop into active, attenuated, threadlike forms (filariform larvae) that are capable of penetrating human skin directly, usually through the soft areas between the toes and around the ankles of barefooted victims. Once into the dermis, the larvae are carried by the bloodstream through the body to the lungs. As described above for *Ascaris,* the hookworm larvae make their way up the bronchi to the pharynx and are swallowed. Final maturation of larvae occurs in the intestinal tract, and the cycle continues with production of eggs by new adults. *Strongyloides* species follow a closely similar path, but the ova of this worm hatch into larvae in the intestinal tract. Sometimes these larvae develop within the intestine into a stage that can penetrate the wall and migrate in the bloodstream to the lungs or elsewhere, resulting in a new intestinal infection or in systemic lesions, or both. Usually, the larvae pass into the soil, where they develop into infective forms that can penetrate skin. This worm may also live freely in the soil, developing into adult forms that may later produce larvae infective for man.

Type 2. The cycle is maintained by alternate hosts. Adult and larval forms develop in each host supporting this type of parasite, but each new host represents a dead end for the invading organism unless an alternate host comes along. The tissue invading roundworm, *Trichinella spiralis*, the agent of trichinosis, is the only outstanding example of this type of worm that is parasitic for man. The natural reservoirs for the trichina worm are the pig, the rat, and a number of wild carnivores (bears, foxes), man being an accidental victim (Fig. 9–11). The adult worm develops in the intestinal mucosa where it produces larvae. The larvae may also be deposited directly in the mesenteric lymph nodes and lymphatics, from which they reach the thoracic duct

where they enter the bloodstream. They are then carried to all parts of the body, localizing in striated muscle tissue. Here they become encysted and remain permanently encapsulated. The muscle tissue must be ingested by another animal while the encysted larvae remain viable, if the parasite's cycle is to continue.

For man, the infective stage of trichina is encountered in pork flesh. If it is eaten without adequate cooking or curing, viable larvae excyst as the meat is partially digested in the stomach, and adults develop in the upper small intestine. Larvae then reach striated muscle tissue as previously described. Human epidemics of trichina infection occur when numbers of people eat the same infected meat.

In nature, rats acquire trichinosis from ingestion of infected pork flesh (in slaughterhouses or on pig farms). Pigs pick it up again, sometimes from eating rats, but more usually from garbage scraps containing uncooked infected animal flesh.

Type 3. The cycle is maintained sequentially in insect and animal (or human) hosts. Definitive and intermediate hosts in alternate sequence keep the parasites of this group going in nature. Human beings and animals are the intermediate hosts for the *proto-*

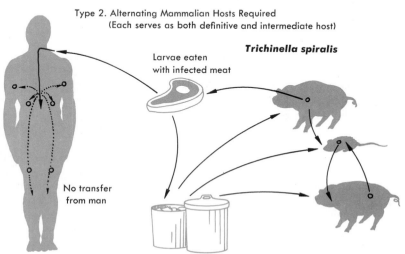

Type 2. Alternating Mammalian Hosts Required
(Each serves as both definitive and intermediate host)

Trichinella spiralis

Larvae eaten with infected meat

No transfer from man

Fig. 9–11. Adult and immature forms of the parasite develop within the same host but have no portal of exit. Infected host tissues must be eaten by an alternate host to perpetuate the parasite's life cycle.

Ingested larvae penetrate intestinal wall, develop into adults and copulate.
Larvae in mucosa and lymphatics reach bloodstream, circulate, encyst in striated muscles.
Muscle tissue must be eaten by alternate host to continue cycle.

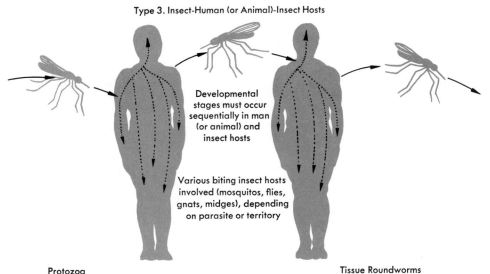

Type 3. Insect-Human (or Animal)-Insect Hosts

Developmental stages must occur sequentially in man (or animal) and insect hosts

Various biting insect hosts involved (mosquitos, flies, gnats, midges), depending on parasite or territory

Protozoa

Plasmodium species (malaria)
Leishmania species (kala-azar)
Trypanosoma species (African sleeping sickness)
(Various developmental forms in human tissues and circulating blood, ingested in blood meal of biting insect in which further development occurs. Insect then infects new host[s].)

Tissue Roundworms

Filarial species
(Adult worms in human tissues produce larvae [microfilariae] that circulate in blood and are picked up by biting insect in which cyclic development continues.)

Fig. 9–12. This diagram illustrates the role of biting arthropod hosts in maintaining the cycles of those protozoans and roundworms that infect the blood and tissues of man and animals.

zoans that parasitize their tissues, the biting or blood-sucking insects that transfer these parasites being their definitive hosts (Fig. 9–12). There are also some *nematodes*, of the group known as *filarial* worms, that are transferred to human tissues by insects. In this case, however, the host roles are reversed.

PROTOZOANS. The blood- and tissue-infecting protozoans of man include species of *Plasmodium* (malaria), *Leishmania* (kala-azar, oriental sore, espundia), and *Trypanosoma* (African sleeping sickness, Chagas' disease). These organisms, and the diseases they cause, can be of overwhelming importance in those areas of the world in which their insect vectors flourish.

Since protozoans are unicellular organisms, the recognition of "adult" and "immature" forms rests on different criteria than those applied to structurally complex worms. Protozoa are capable of reproduction by asexual mechanisms, but are also dependent on "sexual" recombinations for species continuation. Those involved in the blood and tissue infections of man reproduce asexually in the human body, displaying several developmental stages (schizogony). The blood-sucking mosquitoes and biting flies that prey on man for their blood meals pick up these parasites and provide conditions required for their continuing reproduction by sexual fusion (sporogony). For this reason, man is defined as the intermediate host, insects as the definitive hosts, for protozoan parasites of this type. Transferral of these agents of human disease always involves the biologic function, as well as the mechanical intervention, of insects.

NEMATODES. The filarial nematodes live in human tissues as adult worms, producing larval forms that circulate in blood. The larvae, often called *microfilariae*, are ingested by blood-sucking insect hosts that support their development through continuing intermediate stages. When infected insect hosts (mosquitoes and flies) prey again on human beings, they inject infective larval forms into new human hosts. The route of transfer again involves insects, but this time they serve as intermediate hosts, transmitting immature forms that evolve to maturity in the human, or definitive, host.

Type 4. The cycle is maintained by one type of definitive host and one or more intermediate hosts. Definitive and intermediate hosts in alternate sequence also maintain the parasites of this group, but insects are not involved. The flatworms, or *platyhelminths*, display this type of complex life cycle. The group includes the cestodes (intestinal tapeworms and one important tissue-invading tapeworm) and the trematodes, or flukes (Fig. 9–13).

CESTODES. Most of the tapeworms that infest

Fig. 9–13. The fourth type of life cycle involves definitive-intermediate-definitive host sequences as in type 3, but there is no role for biting insects. Man acquires these parasites by ingesting the infective form (except in the case of *Schistosoma* larvae, which can penetrate through the skin).

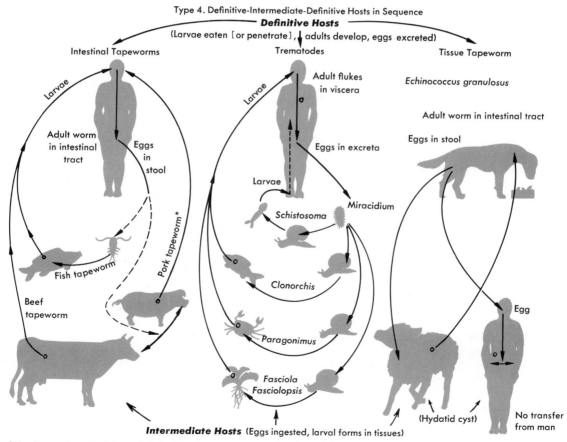

Type 4. Definitive-Intermediate-Definitive Hosts in Sequence
Definitive Hosts
(Larvae eaten [or penetrate], adults develop, eggs excreted)

Intestinal Tapeworms | Trematodes | Tissue Tapeworm

Larvae

Adult worm in intestinal tract | Eggs in stool

Adult flukes in viscera

Echinococcus granulosus

Adult worm in intestinal tract

Eggs in stool

Larvae

Eggs in excreta

Larvae

Pork tapeworm*

Schistosoma

Miracidium

Fish tapeworm

Clonorchis

Beef tapeworm

Egg

Paragonimus

Fasciola Fasciolopsis

(Hydatid cyst) | No transfer from man

Intermediate Hosts (Eggs ingested, larval forms in tissues)

*Man is sometimes the intermediate host for the pork tapeworm (cysticercosis) man live in his intestinal tract, but there is one important variety, with a different capability and cycle, that can invade other tissues.

The intestinal tapeworms of man live as adults in the bowel. Man is the definitive host for these organisms, which include *Taenia saginata* (the beef tapeworm), *Taenia solium* (the pork tapeworm), and *Diphyllobothrium latum* (the fish tapeworm). Eggs produced by these worms are discharged in feces into soil or water, and ingested by another animal. In the intestinal tract of the second animal, the ova hatch into larval forms that penetrate the animal's bowel to invade and encyst in its tissues. In the case of the fish tapeworm, the egg is first eaten by a freshwater arthropod, which is in turn ingested by a fish. The cycle is continued by man and other animals who eat infected flesh of intermediate hosts. Viable larvae eaten in raw or inadequately cooked beef, pork, or fish develop into adult tapeworms in the intestinal tracts of definitive hosts.

In the case of the pork tapeworm, man may become the intermediate host if he ingests the eggs of *Taenia solium*, passed in feces from another person with the intestinal infestation. Larvae hatched in the bowel then penetrate the intestinal wall, reach the bloodstream, and are carried to various tissues where they encyst. This systemic infestation with the larval form of the pork tapeworm is called *cysticercosis*.

The tissue-invading tapeworm, Echinococcus granulosus, has a natural reservoir in dogs, which serve as definitive hosts, and ruminant domestic animals (sheep, cattle), and pigs, the intermediate hosts. The adult worm inhabits the dog's intestinal tract, where it causes no disturbance. The eggs are discharged in the dog's feces and ingested by an intermediate host, ordinarily a domestic animal grazing on contaminated grass or rubble. Ingested eggs hatch in the intestinal tract of intermediate hosts, penetrate the bowel, and develop into immature larval forms within various organs, notably the liver. In the natural course of events, dogs complete the cycle by ingesting larval forms in the flesh of slaughtered animals or those they have killed. Man becomes an accidental victim of infection through his association with infected dogs. The ova discharged in dog feces can be transferred to a man by contaminated hands or food (Fig. 9–13). As an intermediate host, man suffers the same larval invasion of tissues as experienced by other animals, but human infection represents the end of the parasite's cycle.

TREMATODES. The *flukes* also enter man's tissues as larval forms, but they develop there into adults that produce ova, so that man is a definitive host in this case. There are several varieties of parasitic trematodes whose distribution in the world is dependent on the ecology of their intermediate hosts (snails, fish, crabs, and water plants) as well as on the social and gastronomic habits of their human hosts. Adult flukes have different predilections for human tissues. The *Schistosoma* species are called *blood* flukes because the adults live and copulate in blood vessels, but the ova penetrate to surrounding tissues. The liver flukes, *Clonorchis* and *Fasciola*, lodge in the biliary ducts of the liver; intestinal (*Fasciolopsis*) and lung (*Paragonimus*) flukes, similarly, have specific localizations. At these various sites of lodgment, adult flukes produce ova that find their way out of the body via feces, urine, or sputum. They must reach water promptly if further development is to occur. The larval forms (*miracidia*), that hatch from the eggs in water are motile and can swim about but must find a suitable species of snail to continue their intermediate stages of maturation. They reproduce asexually within the snail, producing new forms that emerge to the water again. These motile larvae may be infective for man, capable of penetrating his skin when he swims, bathes, or wades in infected water (e.g., the fork-tailed *cercariae* of *Schistosoma* species). The larval forms of some flukes may encyst on water plants or be ingested by fish, crabs, and other crustaceans that are subsequently eaten by man, either raw or only partially cooked. Whether the larvae enter the body through skin or alimentary tract, they penetrate to small blood vessels and are carried through the body, localizing finally in tissues where they can best develop into adults, to continue the cycle.

Control of Parasitic Diseases. The control of diseases caused by animal parasites is based on a knowledge of their life cycles, so that their routes of transfer can be interrupted. Parasites of type 1 are best controlled by proper disposal of human feces to prevent water or soil contamination. Control of those of type 2 calls for protection of farm animals from infected food, destruction of rodents, and public education on the necessity for proper preparation of meat, especially pork, for the table. The destruction of insects, control of their breeding, and personal protection from insect bites are indicated for parasites having arthropod vectors. Tapeworm and fluke diseases are controlled by the protection of animals, sanitary measures that prevent the fecal contamination of water, and education of the public in personal hygiene and adequate preparation of food. The recognition and treatment of human and animal cases of parasitic diseases may remove some of the sources of infection, but inapparent infections often remain as continuing reservoirs.

Patterns of Infectious Diseases

Epidemiology is the study of the natural occurrence of diseases within a community. In human populations, the occurrence of infectious diseases is characterized by patterns that are either epidemic, endemic, or sporadic. These terms reflect the *numbers* of persons who are clinically ill during a given period of time as well as those who have inapparent infections, the *frequency* with which overt illness occurs, and the *rate of its spread* through a population. The factors that influence these three features of community disease (numbers, frequency, spread) can be considered in three major categories: (1) the degree of effective resistance in the host population, with respect to a particular microbial agent of disease; (2) the sources, transmissibility, and incubation period of the infectious microorganism; and (3) the physical and social conditions of the human environment.

1. Resistance in Host Populations. When the *number of cases* of communicable disease occurring within a stated time period (a week, a month, a year) exceeds expectations based on previous experience in similar periods, the situation is termed an *epidemic*. The extent of the epidemic is determined by the ratio of susceptible versus resistant individuals in the population, and this in turn is influenced by the familiarity of the community with the disease. If the disease is a new one, or if the agent is a new strain of a mutating pathogen, its introduction may result in a large epidemic because the population can be expected to contain a high proportion of susceptible people. Epidemics may also result when a disease is reintroduced to a community that has not experienced it for one or more generations. As the number of susceptibles declines during the course of an epidemic, spread of the disease becomes difficult and finally ceases, unless new susceptibles are continuously introduced. Conversely, as the proportion of immune individuals rises, transmission is steadily reduced, even among the nonresistant. For this reason, artificial active immunization of large numbers of people is often an effective block to the epidemic spread of infection.

On the other hand, even small numbers of cases of infectious disease may constitute an epidemic, when they are seen in communities that are normally free of the disease in question. Small outbreaks may reflect a limited source of infection, or of restricted opportunity for exposure to susceptibles, or they may be the forerunners of larger events to come.

The *frequency* of epidemic disease often reflects the rise, duration, and waning of immunity in a population. Many of the viral diseases of childhood, such as measles and mumps, provide a lifelong immunity. They are seen primarily in young children, rather than in immune adults. They often occur in epidemic patterns, following two- to four-year cycles, as succeeding groups of susceptibles are born into a community or enter it from outside. Cyclic epidemics of influenza, however, usually occur among people of all ages as their immunity wanes and different strains of the virus to which they are not immune become prevalent. By contrast, diseases such as the common cold are seen in fairly high, uniform incidence, because specific resistance to cold viruses is never significantly effective or long-lasting, and there are many agents that cause the cold syndrome.

2. Sources, Transmissibility, and Incubation Period. The *rate* at which an infectious disease spreads is primarily a function of the sources, transmissibility, and incubation period of the agent in question. The respiratory diseases that spread from person to person on an airborne route travel with great speed, particularly when people are exposed to each other in closed areas, such as theaters, schools, barracks, and other communal places. The shorter the incubation period required for manifest disease, the more explosive the epidemic will appear, as in the case of influenza, streptococcal sore throat, or scarlet fever. Similarly, food-, water-, and milk-borne outbreaks may appear explosively, because large numbers of people may be exposed to the same source of infection simultaneously, or within a short period of time. Diseases such as *Salmonella* food poisoning, amebic dysentery, staphylococcal and streptococcal infections, carried in milk, food, or water, all have short incubation periods. They can appear in a population within a matter of hours following ingestion of contaminated material. Other diseases such as infectious hepatitis, the systemic fungal diseases, or parasitic infestations take more time to develop. When these occur in epidemic patterns among people exposed to a common source (those sharing a meal containing virus-infected shellfish or trichina-infected pork, for example, or those working in an area where dust inhalation involves the common risk of fungous infection), the slow development of overt disease may make it difficult to relate individual cases to their point of origin. Arthropod-borne infections may also appear in epidemic distribution when seasonal breeding increases the numbers of insect vectors that have access to an animal reservoir of infection. Outbreaks of equine encephalitis in the United States, for example, are usually associated with a spring or summer increase in the density of mosquito populations.

Diseases become *endemic* in populations with which infectious agents have established some balance. Clinically inapparent infections become the rule in such communities. Sporadic cases of disease occur from time to time, but epidemics do not occur while the equilibrium lasts. The balance can be disturbed by sudden changes in the proportion of susceptible persons, as when armies enter an area of endemic disease, or by environmental stresses that involve large numbers of people, such as those reduced by war to poverty and hunger. The ability of an infectious agent to establish the equilibrium of endemicity in a human or animal population depends on the availability of suitable reservoirs and transmission routes and on the responses of susceptible individuals to infection.

3. Physical and Social Conditions. The immediate conditions of the human environment may influence enormously the patterns and incidence of infectious disease. Low-income groups experience the constant stresses of poor nutrition, crowded and unsanitary living conditions, and the social habits of ignorance. Exposure of such populations to a new and highly virulent infectious agent may have disastrous consequences, of epidemic proportions. More usually, persistent, low-grade infections are the rule in poor communities, these endemic diseases contributing to a decreased resistance to any new diseases that may be introduced. Social customs, particularly in the matter of food and drink, sanitary standards, and even the political stability of a society — all can play a role in the nature and incidence of communicable diseases. We have also seen (Chap. 5) how large an influence the physical environment exerts on man's interrelationships with microorganisms. Climate, weather, and geography determine the ecology of human, animal, plant, and microbial populations. For microorganisms such factors may establish the availability of reservoirs and vectors, of vital importance to their continuation in nature. For human beings, the total environment presents the constant challenge to discover the means of prevention or control of infectious disease, at its source or along the route of transmission.

Questions

1. Define epidemiology. Why should a nurse understand its basic principles?
2. Why are some diseases more communicable than others among human or animal populations?
3. Define reservoir of infection. Give examples of inanimate and living reservoirs.
4. What are endogenous sources of infection? Give examples.
5. List six major reservoirs of infection. Give examples of diseases in each group, indicating those with insect vectors and specifying the insect(s) involved.
6. Name the routes of transfer of infection.

7. Describe briefly the four types of life cycles among animal parasites. List examples in each group, indicating the infective form for, and route of transmission to, man in each case.
8. What factors influence the numbers, frequency, and spread of a community disease?

Additional Readings

Beck, Walter J., and Davies, J. E.: *Medical Parasitology*, 2nd ed. C. V. Mosby, Co., St. Louis, 1976.

Clendening, Logan (compiled with notes by): *Source Book of Medical History*. Dover Publications, Inc., New York, 1960, Chaps. XIX, XXXV.

Fox, John P.; Hall, C.; and Elveback, L.: *Epidemiology, Man and Disease*. Macmillan Publishing Co., Inc., New York, 1970, Chaps. 1–6, 9–11.

Hubbert, William T.; McCulloch, W. F.; and Schnurrenberger, P. R. (eds.): *Diseases Transmitted from Animals to Man*, 6th ed. Charles C Thomas, Publisher, Springfield, Ill., 1975.

*Roueché, Berton: *Annals of Epidemiology*. Little, Brown & Co., Boston, 1967. Introduction, Chaps. 9 and 11.

*Roueché, Berton: *Eleven Blue Men*. Berkley Medallion Editions, New York, 1953.

*Available in paperback.

10 Prevention and Control of Infectious Diseases

General Principles

The prevention or control of infection requires methods that can effectively accomplish one or more of the following results: (1) destruction or control of microbial agents of disease; (2) elimination or control of the sources, routes, or agents of transmission of infection; and (3) partial or full protection of the human host from serious effects of disease through improvement of his resistance, specific treatment of his infections, or both.

The successful application of such approaches, singly or in combination, has led to some dramatic changes in the epidemic patterns of many infectious diseases. Smallpox, for example, has been virtually eliminated within the past decade through the success of a global Smallpox Eradication Program coordinated by the World Health Organization (WHO). Begun in 1967, this program of case detection, quarantine, and immunization is expected to complete the eradication of a once dreadful epidemic disease by the end of this decade (see Chaps. 11 and 14). Diphtheria also has been controlled, although not totally eliminated, in countries where large-scale immunization programs have been maintained. Improved sanitary standards and the development of effective vaccines for poliomyelitis have similarly reduced the incidence and spread of this disease. Where prevention has not been possible, control has been gained over the sources or transmitting agents of infection, as in the case of insect-borne diseases caused by protozoa (malaria), viruses (yellow fever), and rickettsiae (typhus fever). It is also possible to control some infectious diseases by averting their severe consequences. Viral diseases such as hepatitis, measles, or mumps can be aborted or prevented if exposure of nonimmune persons is followed by prompt administration of gamma globulin to provide passive protective antibody. Specific chemotherapy administered immediately upon appearance of symptoms is now usually quite effective for controlling formerly serious diseases caused by pneumococci and streptococci. Venereal diseases (gonorrhea and syphilis) remain uncontrolled at their source, but can be successfully cured by early case detection and adequate chemotherapy. A significant measure of control is obtained when diseases can be aborted by either immunologic or therapeutic means, for not only is the individual case resolved, but the opportunities are removed for further spread of virulent infection from a sick patient.

In this chapter, attention will be given to the principles involved in each of the three important approaches outlined above for prevention and control of infectious diseases. The emphasis here is on problems of individual infection. Application of the principles to protection of the public health are discussed in Chapter 11.

Definitions

Discussion of the measures used to destroy or suppress microorganisms, or to break the chain of their transmission from one host to another, requires the use of technical terms whose meanings should be clear to all who use them. The student should review them before continuing to study the principles of control and should refer to them frequently in subsequent reading of the text to become thoroughly familiar with their proper use. They are arranged in two tables: Table 10–1 lists terms referring to the destruction or suppression of *microorganisms;* Table 10–2 lists those used in the management of *infectious diseases.*

Control of Infectious Microorganisms

To the microbiologist, a dead microbial cell is one that has permanently and irreversibly lost its ability to reproduce, as tested by available techniques for demonstration of its growth, *in vitro* or *in vivo.* Assuming such techniques are always adequate in providing optimal media and other growth conditions in which viable organisms may multiply, the basic information they provide permits the following generalizations:

1. Microorganisms can be irreversibly "killed," or reversibly inhibited, by a number of physical agents as well as by chemical methods.

2. Various microbial groups differ greatly in their response to sterilizing or inhibiting agents. The cells of a particular bacterial species may also vary in this respect, young growing cells being more susceptible than older resting ones, while spores formed by some bacterial species are much more resistant to antimicrobial agents than are their vegetative forms.

3. Sterilization is always a function of time, regardless of the agent employed or microorganism involved, since changes leading to the death of cells occur gradually, and not all the organisms die simultaneously, even in homogeneous cultures.

4. The rate of sterilization is also dependent on the nature, concentration, or intensity of the sterilizing agent, as it is on the nature and numerical density of microorganisms present.

5. The choice of a method of sterilization, disinfection, or inhibition, in a given situation, must take into account a number of other factors that may limit the success or practicality of the procedure. Some of these factors are: the nature and function of the material to be processed (fabrics, rubber, glassware, etc.); conditions that may affect the activity of the antimicrobial agent; and the influence of extraneous factors on the response of microorganisms to the agent (the presence of proteins, lipids, salts, the pH, etc.).

In the following discussion, the more useful of the physical and chemical agents employed for sterilization, disinfection, or microbial inhibition will be described, together with the reasons for, and the limitations on, their use.

Antimicrobial Activity of Physical Agents

Heat. The application of heat is the simplest, most reliable, and least expensive means of assuring sterilization of materials that are themselves not damaged by it. It is a rapid process to which all living protoplasm is susceptible.

Mode of Action. The efficiency of heat as a sterilizing agent is a function of two primary factors operating together: the intensity or *temperature* and the *time* for which a given temperature is maintained. When these two factors are specified, the

Table 10–1. Terms Indicating Destruction or Suppression of Microorganisms

Destruction	Suppression
Sterilization: complete destruction of all forms of microbial life. An *absolute* term; there are no "degrees" of sterility.	**Asepsis:** literally, without infection. Refers to techniques that prevent entry of living microorganisms.
Disinfection: destruction of pathogenic microorganisms on inanimate objects by chemical or physical agents.	*Surgical asepsis:* use of techniques designed to exclude *all* microorganisms.
Concurrent: continuing disinfection of infectious discharges, or of objects soiled thereby.	*Medical asepsis:* use of techniques designed to exclude agents of communicable disease but not necessarily *all* others.
Terminal: final disinfection of all surfaces and objects in an area previously occupied by an infectious patient.	**Sanitization:** removal of pathogenic microorganisms from inanimate objects by mechanical and chemical cleaning. Does *not* imply sterilization or complete disinfection.
	Cleaning: soil removal. Never implies disinfection or sterilization.
	Antibiotic: literally, against life. A microbial product that suppresses or destroys other microorganisms.
Antisepsis: chemical disinfection of skin or other living tissue.	**Antibacterial:** against bacteria. Refers to agents that suppress or destroy bacteria.
	Antimicrobial: against microbes. Refers to agents that suppress or destroy any microorganism.
-cide: noun suffix meaning "killer" or "killing." Adjectival suffix = -*cidal*.	**-stasis:** noun suffix meaning "halted" or "arrested." Adjectival suffix = -*static*.
Germicide: germ-killing agent	*Bacteriostasis:* bacterial growth arrested; multiplication halted, but bacteria not killed and may resume growth when bacteriostatic agent is removed.
Bactericide: kills bacteria	
Sporicide: kills spores	
Fungicide: kills fungi	*Fungistasis:* fungal growth halted.
Virucide: kills viruses	*Virustasis:* viral growth halted.

effect of heat on various species of microorganisms may be compared. The lowest temperature that kills all the organisms in a standardized pure culture within a specified time period is the "thermal death point" of that species, while the time required to sterilize the culture at a stated temperature is its "thermal death time."

The effects and consequently the practical applications of heat depend on whether it is *moist* or *dry.* Many proteins of microbial cells are enzymes that function only within a narrow temperature range. In the intact cell they are held in a fine suspension. When heat is applied in the presence of *moisture,* the structure of cellular proteins is altered, they are coagulated (as the white of an egg is coagulated when it is boiled), and their enzymatic function is destroyed (Fig. 10–1). When the temperature of *dry* heat is raised, cell proteins do not coagulate, but they and other components of the microbial cell are oxidized. At low temperatures dry-heat oxidation is very much slower than protein coagulation in the presence of moisture. Moist-heat sterilization is more

Table 10–2. Terms Used in the Management of Infectious Diseases

Term	Denotes
Contamination	Presence of viable microorganisms on animate or inanimate surfaces; in water, food, or milk; or in human or animal discharges.
Infection	Presence of viable microorganisms in a living host's tissues.
Infectious disease	State of evident disease in a living host resulting from infection.
Communicable disease	Infectious disease caused by an agent transmissible from an animate or inanimate reservoir, directly or indirectly, to a susceptible host. (N.B.: Not every infectious disease is communicable between hosts of the same species.)
*Nosocomial infection	An infection originating and acquired in a medical facility. Includes infections acquired in the hospital but not apparent until after discharge, lingering infection acquired during a previous admission, and infections among staff. Excludes infections present or incubating in a hospitalized patient at the time of admission.
*Isolation	Separation of infected persons, for the period of communicability, in places and under conditions that prevent direct or indirect transmission of the infectious agent to susceptible persons who may spread the agent to others. Applies also to animals.
*Quarantine: *Complete*	Limitation of freedom of movement of healthy persons or domestic animals exposed to a communicable disease, for a time equal to the longest incubation period of the disease, to prevent effective contact with others not so exposed.
Modified	Selective, partial limitation of freedom of movement of persons or domestic animals, on basis of differences in susceptibility or danger of disease transmission, e.g., exclusion of children from school, or restriction of military personnel to post or quarters.
*Communicable period	Period during which an infectious agent may be transferred directly or indirectly from an infected person to another person, animal, or arthropod; or from an infected animal to man.
	In diseases such as diphtheria and scarlet fever, in which the first entry of the pathogen involves *mucous membranes,* the communicable period is from the date of first exposure until the infecting agent is no longer disseminated from those membranes, i.e., from before prodromal symptoms until termination of the carrier state, if such develops.
	In diseases such as tuberculosis, syphilis, and gonorrhea, communicability may occur at any time over a long, intermittent period when unhealed lesions permit discharge of infectious agents through body orifices or from skin surfaces.
	In diseases transmitted by arthropods (e.g., malaria and yellow fever), communicability exists when the infectious agent occurs in the blood or tissues of the infected person in infective form and in numbers sufficient for vector infection. The communicable period for the arthropod vector is that time during which the agent is present in the arthropod tissues in an infective form that can transmit infection.

*Adapted from definitions stated by the Committee on Communicable Disease Control, American Public Health Association, in Abram S. Benenson (ed.): *Control of Communicable Diseases in Man,* 12th ed. A.P.H.A., Washington, D.C., 1975.

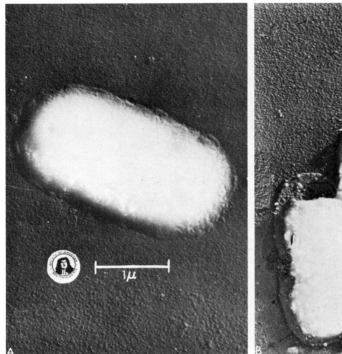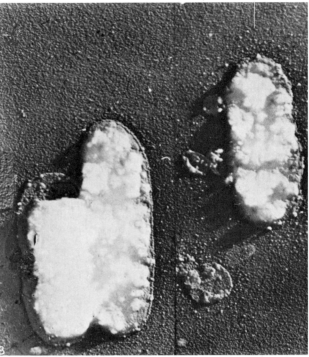

Fig. 10–1. The effect of moist heat on *Escherichia coli*. *A*. An electron micrograph of a typical cell of *E. coli* from a one-hour broth culture. *B*. Three cells of *E. coli* that have been heated at 50° C in saline for ten minutes. The coagulated cytoplasm has become granular and has shrunk away from the cell wall. (Reproduced from Heden, C. G., and Wyckoff, R. W. G.: The Electron Micrography of Heated Bacteria. *J. Bacteriol.*, **58**:153, 1949.)

efficient, therefore, because it can be accomplished more rapidly and at lower temperatures than dry-heat methods. It is also far less injurious to many materials that would themselves be oxidized by the high temperatures required for sterilization by dry-heat methods.

Uses of Moist Heat. The thermal death point for most pathogenic microorganisms lies between 50° and 70° C, with ten minutes' exposure. However, the heat-resistant spores of some species of pathogenic bacteria can survive at 100° C for many minutes, while those of certain saprophytic species can resist up to 24 hours of boiling. The choice of a moist-heat method of sterilization depends, therefore, on the

nature of the job to be done and the temperature that can be produced by the method.

PASTEURIZATION. This process utilizes the minimum degree of moist heat that can be expected to kill nonsporing pathogenic bacteria in milk or other potable liquids within a time that will not spoil the flavor or quality of the beverage. The method was first devised by Pasteur, who found that spoilage of wine could be prevented by heating it to 60° C for a time sufficient to kill fermentative and oxidative organisms that cause souring. Pasteurization of milk has resulted in effective control of a number of important milk-borne diseases, notably, tuberculosis, brucellosis, salmonellosis, and streptococcal diseases of bovine origin. The technique consists of raising the temperature of milk to 62.9° C (145° F) for 30 minutes, or to a somewhat higher temperature (71.6° C or 161° F) for a much shorter period (15 seconds), then cooling it rapidly to avoid flavor loss (Fig. 10–2). This temperature-time relationship is based on the thermal death point of the rickettsial agent of Q fever, one of the most heat resistant of the vegetative microorganisms that may contaminate milk. The nonpathogenic, heat-resistant bacteria that occur naturally in milk are not killed by pasteurization. These include streptococci and lactobacilli that cause milk to sour rapidly at room temperature, or more slowly at refrigerator temperatures.

BOILING. Boiling water provides a very simple means of disinfection under many circumstances. The temperature of boiling water, and of the steam it evolves, is 100° C. Since this is well above the thermal death point of vegetative bacteria, fungi, and most viruses, boiling for 10 to 30 minutes kills all except heat-resistant spores and viruses. Contaminated materials may be *disinfected* in this way, provided there is no reason to suspect the presence of pathogenic spores or viruses. However, it is essential that detergent agents be added to the water when dirty items are boiled to prevent the coagulation of extraneous protein matter (blood, mucus, pus), for bacteria trapped within coagulated particles are protected from the killing effects of temperature and moisture. Clean instruments, syringes, needles, and other metal or glass items may be *disinfected* by complete immersion in boiling water for not less than 30 minutes. Since the method cannot be guaranteed to produce sterility, safer methods must be used in the preparation of equipment when sterilization is essential.

HOT WATER. Water, at temperatures from 60° C to boiling, accomplishes sanitization, and at least partial disinfection, particularly if it is applied with force and agitation in washing and rinsing items to be cleaned. Modern laundry and dishwashing equipment provides far better conditions for disinfection than manual washing of clothing, bedding, or dishes. The action of detergent soaps added to wash water also plays a large role in the inhibition or destruction of microorganisms at elevated temperatures.

Fig. 10–2. A modern pasteurizer. The proper handling of milk *before* and *after* pasteurization reduces the possibility of disease being transmitted by milk. (Courtesy of Crepaco, Inc., Chicago, Ill.)

STEAM UNDER PRESSURE. Pressurized steam is the most reliable form of moist heat for sterilization. The temperature of freely flowing steam is 100° C. It is sometimes used in this form as a sanitizing or disinfecting agent, as in the steam flushing of bedpans in the hospital or of animal cages in the laboratory. Steam becomes a sterilizing medium when its temperature is increased by subjecting it to increasing positive pressure.

A steam pressure sterilizer is essentially a chamber that permits, in sequence, first, the entry of flowing, saturated steam and the exit of displaced air; then closure of the air outlet so that the continued admission of steam produces a rising pressure. As the pressure increases, the temperature of steam rises proportionately to a degree that can be balanced with reasonably short time periods to accomplish sterilization. Temperature and time are the essential sterilizing factors; increasing pressure provides the means for raising the temperature of steam, thus shortening the time required. Table 10–3 illustrates the influence of pressure on the temperature of steam and on the time needed to ensure destruction of heat-resistant organisms (see also Chap. 4, p. 72).

The factor of first importance in successful autoclaving is the complete elimination of air from the sterilizing chamber. Air is cooler, dryer, and heavier than steam, and although an air-steam mixture can be brought to any desired pressure, the mixture will have a lower temperature than that of pure steam at the same pressure. The temperatures stated in Table 10–3 are accurate only for saturated steam unmixed with air.

When steam saturated with moisture enters the autoclave chamber, it quickly condenses on cold surfaces within. This condensation of steam releases a large amount of latent heat (about 540° per cubic foot) and at the same time wets materials exposed to it. The rapid release of heat and moisture has a penetrative effect that quickly raises the temperature of inner, as well as outer, surfaces of items within the sterilizer to that of the surrounding steam. Vessels or containers having hollow air spaces must be covered loosely, however, or placed in horizontal positions (if they do not contain liquid material) to permit air displacement and steam condensation. At the end of the sterilizing period, steam is slowly released from the chamber. When the pressure within the autoclave has returned to normal, a drying period is required for fabrics, clean glassware, and other items that must be free of condensed moisture before use.

Since it takes a few minutes for steam pressure to rise within a closed chamber, the *timing* of an autoclave procedure cannot begin until sterilizing temperature has been reached. The time required to sterilize a particular item or group of materials placed together in a full loading of the autoclave chamber varies with the nature of the load (Fig. 10–3). Steam penetration of thick, bulky, porous articles such as linen packs for the operating room takes longer than steam condensation on the exposed, impenetrable surfaces of metal or glass instruments, which are quickly raised to sterilizing temperatures. Microorganisms buried within the depths of porous materials are not reached so rapidly as those lying on exposed surfaces. For these reasons,

Table 10–3.   Pressure-Temperature-Time Relationships in Steam Pressure Sterilization

Steam Pressure	Temperature		Time
Pounds per Square Inch (Above Atmospheric Pressure)	Centigrade	Fahrenheit	(Minutes Required to Kill *Exposed* Heat-Resistant Spores)
0	100°	212°	—
10	115.5°	240°	15–60
15	121.5°	250°	12–15
20	126.5°	260°	5–12
30	134°	270°	3–5

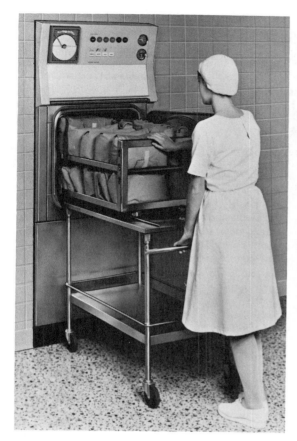

Fig. 10–3. A load of carefully packed and arranged articles is being pushed into the high-vacuum sterilizer. The articles are placed on end so that steam can circulate and permeate the packaged contents. The time necessary for sterilization is dependent on the size and nature of the load. (Courtesy of AMSCO/American Sterilizer Co., Erie, Pa.)

the timing of autoclave procedures must be judged for each load, with critical care.

Materials and items that must be sterilized vary in their own susceptibility to heat. Nonporous materials such as glass or metal withstand it very well, but cloth, rubber, and plastic may be damaged. Time and temperature adjustments must be made with this in mind. Furthermore, the steam penetrability of covering materials must be borne in mind when the autoclave is used for sterilization. Rubber and some

plastic surfaces can be successfully sterilized without deterioration under steam pressure, but they can never be used as covering for other items to be sterilized because they are impenetrable to steam.

The requirements and limitations of steam pressure sterilization may be summarized as follows:
1. The *temperature* of the steam chamber must reach a sterilizing level (121° C). This is possible only if all air originally present has drained out through the discharge outlet.
2. Items to be sterilized must be *packaged* and *arranged* within the autoclave to permit full steam penetration of each package. (Instruments or other objects to be used promptly after sterilization may be autoclaved without covering, provided they are not exposed to undue risk of air contamination upon removal from the sterilizer.)
3. The *timing* of the autoclave procedure must be based on the period during which sterilizing temperature is available and on the nature of the materials in the load. It is dangerous to underestimate time, but it is inefficient and wasteful to prolong it unnecessarily to the point of injury to valuable materials.

Modern autoclaves are equipped with regulating devices for the control of air discharge, pressure, temperature, and timing, which reduce the work of the operator to a minimum. Like all mechanical equipment, however, they require continuing intelligent monitoring. The safe operation of an autoclave is similarly the responsibility of the person who uses it, *on each occasion*.

The final criterion of the safety of steam pressure sterilization is the bacteriologic demonstration of its efficiency in killing heat-resistant organisms (Fig. 10-4). The microbiology laboratory can provide suitable materials for conducting such tests (usually a preparation of living, heat-stable spores), and also perform the necessary culture work, but the actual use of the test material is more critically managed by those who consistently operate the autoclave. The laboratory's collaborative advice should be sought, but the supervision and use of controls remain the concern of those responsible for the end result. When properly carried out, bacteriologic tests can provide information regarding the mechanical efficiency of the autoclave; the adequacy of techniques for packaging individual, typical items; and the pro-

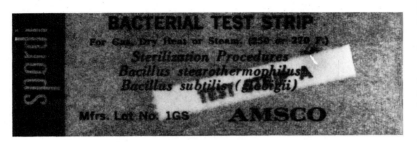

Fig. 10–4. The final criterion for the safety of steam-pressure sterilization is the bacteriologic demonstration of its efficiency in killing heat-resistant organisms. A bacterial test strip containing spores of *Bacillus stearothermophilus* and *Bacillus subtilis* can be used for this purpose. (The test strip is within the glassene paper.) (Courtesy of AMSCO/American Sterilizer Co., Erie, Pa.)

ficiency of the operator in arranging the total load and regulating conditions to permit steam penetration throughout the chamber (see Appendix IV).

Uses of Dry Heat. Sterilization by dry heat has an oxidative effect, not only on microorganisms but also on the materials they contaminate. If the latter are expendable, they may be sterilized by incineration; if they are heat resistant, flaming or hot-air baking may be appropriate.

INCINERATION. This is an effective technique for the disposal of many contaminated objects that cannot be used again, *provided* the temperature within the incinerator is kept sufficiently intense to ensure the prompt ashing of materials at the rate they are fed into it. Overloading of a dry-heat incinerator may result in the "lumping" of partially burned material with condensation of its moisture content, coagulation of protein material, and protection of microorganisms trapped within. Incinerating chambers must be of a size adequate to the maximum load introduced and capable of maintaining temperatures that will burn their largest loads promptly and completely, or sterilization cannot be guaranteed. Various hospital practices in this respect are open to serious question: the design of such equipment is frequently inadequate to the need; its use is turned over to untrained personnel; the safety and efficiency of incineration are seldom subjected to critical review.

FLAMING. Heat-resistant objects may be sterilized by flaming if the heat of the flame is sufficiently intense to bring about immediate vaporization of moisture and prompt oxidation of exposed microorganisms. The microbiologist commonly uses the flame of a Bunsen burner to sterilize his wire inoculating loop or needle. The heat of such a burner will sterilize metal surfaces within seconds, provided they are not covered with moist proteinaceous materials that coagulate before their microbial content can be oxidized. Such materials tend to "clump" in suddenly applied heat and to spatter from the flame, carrying organisms that may still be viable. For this reason, laboratory burners should be equipped with heat-conductive shields to catch spatter and oxidize it completely, so that the air, workbench, or worker will not be sprayed with potentially infectious material.

HOT-AIR BAKING. Baking in an insulated, thermostatically controlled oven is an effective method for sterilizing materials that can withstand high temperatures. Since dry heat does not penetrate well, sterilization requires higher temperatures for longer periods with dry than with moist heat and may be destructive to some materials. Rubber and plastic goods are quickly destroyed by dry heat, but glass and metal withstand it very well. Some materials, such as powders, waxes, oils, glycerin, and petroleum jelly, are not penetrable by moist heat, or would be damaged by moisture, and for these hot-air sterilization is a necessity.

The requirements and limitations of hot-air sterilization may be summarized as follows:

1. Most items that must be baked can withstand *160° to 165° C for 2 hours,* or *170° to 175° C for 1 hour.*

2. Clean, dry glassware, metal items, syringes, and needles can be baked at *170° C for 1 hour,* without injury.

3. Oils, waxes, and jellies are not easily penetrated by moderate heat, but can be damaged by intense heat. If prepared in small, shallow units they can be sterilized by baking for *2 hours at 160° C.*

4. Temperatures *below 160° C* require periods longer than 2 hours and are impractical.

5. Temperatures *above 180° C* cause charring in a very short time.

6. Whatever the temperature and time, each item in the load must be placed so that heated air flows freely among all objects to be sterilized.

Cold. The metabolic activities of microorganisms are slowed at decreasing temperatures to the point of complete inhibition of their growth under freezing conditions. Cold is, therefore, a good bacteriostatic agent. Low-temperature storage can be used to preserve microbial cultures for long periods. Freezing and drying under vacuum (lyophilization) preserves the viability of many bacteria for years.

Refrigeration or freezing effectively delays microbial spoilage of foods, without greatly affecting the quality and flavor of stored items. The growth of spoilage organisms is very slow at normal refrigerator temperatures (0° to 8° C), or inhibited altogether at freezer temperatures (−5° to −25° C). Frozen foods must be kept in the freezer until prepared for the table. It is unwise to refreeze food that has thawed, because microbial growth may produce metabolic products or induce changes in food that could be unpleasant or harmful at the next thawing.

In the natural world, cold inhibits and sometimes even kills some microorganisms of water and soil, but does not sterilize these reservoirs. As a matter of fact, ice from contaminated water has frequently been incriminated as a source of infection when used to chill drinking water, other beverages, or food.

Drying. Many microorganisms are inhibited but not killed by drying. In the absence of moisture, a microbial cell is not capable of growth and reproduction but may remain viable for years, resuming growth when moisture and nutrient are restored. This fact is applied in the laboratory to the preservation of cultures, as mentioned above, by drying them under vacuum from the frozen state. (Freeze drying is also used to preserve serum and other blood products.)

In the hospital situation it is most important to remember the survival capacity of dried microorganisms. Bacterial spores have an extraordinary resistance to drying, but vegetative bacteria, viruses, fungi, protozoa, and helminth ova withstand it also to a lesser degree. Dust, clothing, bedding, and dressings may contain dried remnants of urine, feces, sputum, or pus contamination, with infectious microorganisms still present and viable. Incautious shaking of soiled fabrics, sweeping, dry mopping, or dry dusting all contribute to wide dissemination of microorganisms, many of which may be infectious.

Ultraviolet Radiation. Sunlight contains light rays of varying wavelengths (measured in Angstrom units, Å). In the visible light spectrum, the longest rays (7500 Å) are at the red end of the scale, the shortest (4000 Å) at the violet end. The invisible part of the spectrum contains rays that are longer than those of the visible reds (*infrared*) and others that are shorter than the violets (*ultraviolet*). The ultraviolet spectrum extends from 4000 Å downward in length to about 2000 Å, and the lower end of this scale (2800 to 2500 Å) has a strongly destructive effect on many microorganisms, particularly bacteria. In bacterial cells, ultraviolet rays are absorbed by nucleic acids. The chemical effects on the genetic mechanisms of cells may induce mutations or death.

In the natural world the ultraviolet component of sunlight probably affects many organisms exposed directly to it. However, sunlight as a sterilizing or disinfecting agent is not very reliable, since short rays will not penetrate glass or opaque surfaces but can kill only those organisms most directly exposed.

Ultraviolet radiation by mercury vapor lamps is widely used as a sterilizing agent. Such lamps help in control of airborne infection when placed in confined areas of public use (elevators, schoolrooms, barracks) or in critical hospital areas (operating rooms, nurseries, infectious disease wards). In the laboratory they are used to kill suspensions of microorganisms prepared for vaccines; and for decontamination of working areas. In commercial preparation of sterile biologicals or of foods and containers, ultraviolet radiation can reduce possibilities for microbial contamination. The effectiveness of ultraviolet lamps is limited by the life of the radiation source, whose output should be measured after every thousand hours of use. These lamps continue to glow visibly after their short-ray output has become ineffective, so that reliance on them may be dangerous. Their placement is also a matter of important strategy, if circulating air or surfaces are to be effectively sterilized. People using ultraviolet lamps directly on their work must protect their eyes from exposure to radiation.

Ionizing Radiations. Some types of radiation, of shorter wavelength and greater energy than ultraviolet rays, have a lethal effect on many types of cells (tissue cells, bacteria, viruses, and other microorganisms) because they cause ionization of molecules that lie in their path. X-rays, gamma and beta rays from radioactive materials, and high-speed electrons emitted in a beam from an electron "accelerator" — all are destructive in sufficient dosage. This type of radiation has limited applications in the sterilization of expensive surgical materials that are heat sensitive; in the sterilization of tissues to be used for transplantation; and in preventing food spoilage.

Sonic Disruption. Sound waves that exceed the limit of audibility for human ears, at frequencies of 100,000 per second or higher, have remarkable effects on materials in solution or suspension. In coursing through liquid media they induce very rapid formation and collapse of submicroscopic bubbles, which in turn creates a negative pressure (suction effect) on any particles in suspension in the liquid. Protein materials are coagulated by this action, bacteria are disintegrated, and other types of particles are broken up and dispersed.

Sonic disruption is sometimes used as a method for releasing components of bacterial cells. These cellular constituents can then be extracted for study by simple procedures that do not induce the structural changes imposed by chemical methods. A more practical application of sound wave action is offered for hospital use by manufacturers of ultrasonic cleaning equipment. This type of apparatus is expensive, but provides a safe, effective method for cleaning delicate instruments without damage. Ultrasonic baths can be used only for decontamination of used instruments, which must then be washed free of solvent and resterilized for use.

Filtration. Filters whose average pore size is smaller than that of undesirable microorganisms can be used to free liquid or gaseous media of such particles. The finest filters currently available (thin plastic films, sintered glass) do not necessarily yield *sterile* filtrates, for they cannot withhold the small viruses, but they do remove bacteria and other particles of diameters greater than 1.2 to 1.5 μm. Unglazed porcelain, glass, cellulose acetate, and plastic are particularly effective for filtering liquids because they withhold bacteria and larger microorganisms without absorbing fluid. Asbestos filters and diatomaceous earth are finely porous, but also quite absorptive (see Chap. 4).

Coarser materials such as cotton, gauze, and paper can effectively filter bacteria from air when used correctly. Cotton plugs can prevent passage of bacteria in and out of test tubes, pipettes, needle hubs, or other airways, if they are sufficiently deep and dry, made of nonabsorbent cotton, and not charred or moistened by heat-sterilization techniques. Plugs must be thick enough to occlude the opening of the vessel but not fitted so tightly as to prevent passage of sterilizing hot air or steam through them during sterilization. Paper, gauze, and fabric can be used to wrap items or packs of equipment to be heat-sterilized (Fig. 10–5) because these materials will subsequently act as bacterial filters when packages are exposed to air, hands, or shelf storage. Fabrics and paper are not effective as filters if they are moist or if their physical structure is damaged by overheating. Moisture-filled pores may be considered as filters-in-reverse, for they promote passage of bacteria and other particles along their wet surfaces to the interior, with a kind of capillary attraction. This is why a drying period is an essential part of the process of steam pressure sterilization of wrapped equipment.

Masks used in surgical procedures (or in other situations requiring asepsis) are designed to prevent outward passage of microorganisms into a sterile area from the mouth, nose, and throat, or to protect the wearer from inhalation of infectious organisms. In either case, they are effective only for the period during which they remain unsaturated with exhaled moisture. After that, they are not only useless as filters but dangerously capable of shedding contaminated droplets of moisture in either direction, and should be discarded at once.

Modern forced-air ventilation systems also employ

Fig. 10–5. Instruments are prewashed and carefully wrapped for sterilization. (Courtesy of Danbury Hospital, Danbury, Conn. Photo © Elizabeth Wilcox.)

filtration devices to purify the air of dust. Hospital or laboratory areas where infection is of critical concern can be ventilated with equipment designed to filter out particles of submicroscopic size. However, the efficacy of such equipment is strictly limited by the extent to which it is monitored. Filters must be changed when they become clogged with dust and moisture, or their function is reversed and their action dangerous.

Antimicrobial Activity of Chemical Agents

Principles of Chemical Disinfection. Disinfection has been defined as a process that destroys *pathogenic* microorganisms. It is a chemical process involving interactions between disinfectant agents and various constituents of microbial cells. The result may be inhibitory to further growth of cells while they remain in contact with the chemical agent, but reversible upon its removal. When this is the case, the agent is said to be *static* in its effect, i.e., bacteriostatic, fungistatic, etc. If the reaction produces irreversible changes in vital cellular components, the result is lethal, or *cidal,* and the cells are unable to grow again even when the agent is removed (see Table 10–1).

Like all chemical reactions, disinfection is influenced by the *nature* and *concentration* of the reactants, and by *time.* A number of other factors may also influence the result, notably the physical and chemical properties of contaminating substances that may be present (organic soil, inorganic salt), as well as pH and temperature. The presence of extraneous organic and inorganic materials at the site of reaction between microbial cells and a disinfectant may prevent the desired result either by chemical inactivation of the disinfectant or by physical protection of the microorganisms. Many chemical agents are lethal to pathogenic organisms under controlled conditions arranged in a test tube, but under actual conditions of use in the hospital environment their efficacy may be quite limited by practical circumstances. It is essential, therefore, that those who apply chemical disinfection to the control of infectious disease understand both its principles and its limitations.

Microorganisms Versus Disinfectants. The major classes of microorganisms, as well as certain forms or stages of individual species or groups, are not uniformly susceptible to chemical disinfection. Among the animal parasites, protozoa and the immature, microscopic stages (ova and larvae) of parasitic helminths are susceptible to germicide action. The vegetative forms of fungi are also readily killed by some disinfectants, but their resting spores (chlamydospores) are quite resistant. The vegetative forms of most bacterial species succumb within a short exposure period to active disinfectants. The tubercle bacillus is more resistant, but of all microbial forms, bacterial spores display the greatest degree of resistance to chemical, as well as physical, agents of disinfection. Many viruses can be killed as easily as vegetative bacteria, and by the same germicides, but hepatitis viruses are much more resistant.

The resistance of bacterial and fungal spores to chemical action is related to their thick and relatively impervious walls and to the fact that their condensed protoplasmic contents contain little moisture for chemical reactivity. Tubercle bacilli are rich in lipids, and their waxy cell walls are resistant to water (hydrophobic); hence they are not readily penetrated by germicides in aqueous solutions, but are more easily destroyed by those carried in lipid solvents or by hydrophilic "wetting" agents (see p. 245). Among viruses, those that have some lipid content, such as influenza and herpes viruses, are more susceptible to detergent germicides than are lipid-free viruses, such as the poliomyelitis virus. The latter is destroyed by formalin and by alcohol. The resistance of hepatitis viruses has not been well defined, and for this reason, sterilizing methods, when feasible, are generally preferred to chemical disinfection for their destruction (see Chaps. 19 and 28 for recommendations).

Since microorganisms differ in their response to disinfectants, the choice of the latter for a particular situation must be based in part on the type of microbial contamination to be dealt with. Disinfection of mouth thermometers, for example, must achieve the destruction of all pathogenic vegetative forms of bacteria, including the tubercle bacillus, which might be present in saliva or sputum. When destruction of bacterial spores must be assured, as in the case of instruments to be used for surgery or dressings removed from a *Clostridium*-infected wound, *heat sterilization* is a necessity, for disinfection procedures cannot guarantee killing of all spores. *Chemical sterilization* with strong germicides such as formalin or glutaraldehyde is possible with a long exposure period (several hours). This may be feasible for heat-sensitive items (e.g., lensed instruments) provided they are not themselves damaged by the germicide.

DESTRUCTION VERSUS INHIBITION. The static or cidal effects of disinfectants are an important issue in their choice and use. In many instances, the actual conditions for use of a chemical agent may lead to mere *inhibition* of growth—a dangerous result if destruction is planned or assumed. Microbiologic methods are required for final demonstration of disinfectant action in each situation. With proper techniques, such methods can demonstrate whether or not the use conditions for a disinfectant (its concentration, time of exposure, the presence of extraneous protein) are sufficient not only to prevent the growth of pathogenic microorganisms *while they are in contact with the disinfecting agent,* but to destroy them so that they cannot subsequently grow and multiply *if and when the agent is removed.*

It must be emphasized that the effectiveness claimed for a disinfectant in killing vegetative bacterial species (such as strains of *Staphylococcus* or *Salmonella*) may have no bearing on its ability to destroy fungal or bacterial spores, tubercle bacilli, or some viruses. It should also be recognized that while most classes of microorganisms can be killed within a 10-to-15-minute period by many of the available disinfectants, this may be true only for *optimal,* rather than *average,* conditions.

Disinfectants Versus Microorganisms. The chemical nature of a disinfectant determines its mode of action and its efficacy against different kinds of organisms. Some chemical agents coagulate the microbial cell protein, the effect being rapid and irreversible. The average disinfectant has a less powerful action, however, and denatures cellular proteins by more subtle, slower processes. Oxidation of protein enzymes by halogen compounds and lysis of cells or leakage of their contents induced by phenolics (or their detergent carriers) are examples of disinfectant action.

Unfortunately, the activity of chemical disinfectants is not specific for microbial proteins but is destructive to some degree for all living protoplasm. The stronger and most rapidly effective antimicrobial compounds are limited in their usefulness by their toxicity for human tissues. Strong germicides may be used under prescribed conditions for disinfection or chemical sterilization of some critical items, but less toxic compounds must be chosen for large-scale disinfection, as in hospital housekeeping, or for surgical asepsis.

Other limitations on chemical disinfection include (1) the concentration of microbial cells to be inactivated, (2) the concentration of germicide that will do the job effectively without environmental damage, (3) inactivation of disinfectants by chemical contaminants from the environment, and (4) factors influencing the exposure of microorganisms to contact with disinfectants.

MICROBIAL CONCENTRATION. The number of microorganisms present in material to be chemically treated has a direct influence on the adequacy of any chosen concentration of disinfectant applied. The greater the concentration of microbial cells, the larger must be the germicide concentration, or the longer the time allowed for disinfection.

In practice, it is always necessary to observe this principle by providing thorough physical cleaning of all items and surfaces before germicide treatment. With reduced microbial density, germicides are more efficient within a shorter time.

GERMICIDE CONCENTRATION. In general, the time required for disinfection is shortened as germicide concentration is increased. There are two important limitations to this principle, however. (1) The activity of germicides in aqueous solutions is dependent on the water present. Decreasing the "water of reactivity" below a critical point by raising the germicide concentration may destroy some of the disinfectant activity of the solution. For example, concentrations of ethanol from 10 to 80 percent are increasingly bactericidal, the most efficient action being obtained in the 60 to 80 percent range. Ethanol solutions of 80 to 100 percent volume concentration are little more efficient than 40 percent solutions. (2) Strong disinfectant solutions are corrosive to metals, fabrics, plastics, and skin, yet some may have little antimicrobial activity in weak solution.

The concentration of choice for a given disinfectant is that which displays the most rapid antimicrobial activity, without simultaneous damage to materials exposed to it.

INACTIVATION OF DISINFECTANTS. Germicides may be inactivated by organic or inorganic contaminants on items to be disinfected. The active portions of some disinfectant molecules may combine with human proteins of blood, mucus, or pus. In this bound condition they are no longer free to react with microbial constituents. This factor should be considered in choosing a disinfectant and its concentration for a particular purpose, because some agents are more readily bound than others, by virtue of their chemical structure. The use of strong concentrations of some disinfectants, if compatible with materials to be treated, may assure an excess of free, active molecules despite extensive protein contamination.

Inorganic compounds such as salts, metals, acids, or alkalis can also interfere with disinfectant activity by affecting the pH of reaction, or by combining with the germicide as proteins do. Volatilization may reduce the available concentration of some disinfectants, particularly those with an alcoholic base or those that depend on the activity of a halogen, such as chlorine or iodine. Temperature is not an important factor in hospital disinfection, for most germicides are effective at temperatures ranging from just above to slightly below room temperature.

OTHER FACTORS INFLUENCING DISINFECTION. The presence of protein contamination on items to be disinfected may not only inactivate the germicide but also protect microorganisms from full contact with it. This is particularly true for strong disinfectants that coagulate proteins. Viable microorganisms trapped within coagulated particles remain unexposed to the surrounding chemical. Strong disinfectants that coagulate protein should not be used for the decontamination of body excretions (for example, in home nursing care), unless combined with a compatible detergent that prevents coagulation.

The protective effect of protein offers another major reason for careful scrubbing and cleaning of objects to be chemically disinfected. Situations sometimes arise, however, when it may be unsafe to wash items grossly contaminated with infectious organisms. Surgical instruments covered with pus from infected wounds and heat-sensitive equipment similarly contaminated should not be handled. Instruments can be boiled with detergent disinfectants or be placed in washer-sterilizer equipment. In either case, detergent action prevents protein coagulation during heat sterilization. Instruments so treated must then be carefully cleaned and lubricated before being sterilized again for another use.

The physical properties of objects and surfaces to be chemically disinfected must also be considered, for they influence the speed and ease with which effective results can be achieved. Hard, smooth surfaces are relatively easy to disinfect, but porous materials are not readily penetrated by many germicides. Grooved or hinged instruments and needles, catheters, and irrigation tubing of small bore offer particular difficulties, for their inner as well as outer surfaces must be in contact with the disinfectant. The choice of germicide and its concentration must be balanced against the type of material it is expected to disinfect, and the technique of application must ensure its efficacy under each individual set of circumstances.

Antisepsis or Degermation. The chemical disinfection of skin is usually referred to as *antisepsis*. The term *degermation* is preferable because it recognizes that sterilization of skin is impossible, and that even disinfection can be accomplished only temporarily at the uppermost layers.

Many useful chemical disinfectants are too irritating to skin and mucous membranes to permit their use as antiseptics. At best, the maximum concentrations of safe antiseptics are weak with respect to antimicrobial activity, as compared with disinfectants that can be applied to inanimate surfaces. When applied to skin, antiseptics reduce the numbers of surface organisms markedly and rapidly, but do not penetrate readily to deeper layers, into hair follicles or glandular crypts. The mechanical action of scrubbing, followed by the flushing effect of running water, is an important factor in cleansing the skin surface and in degerming deeper layers of some resident flora.

Soaps, alcohol, alcohol and iodine combinations, and iodophor preparations are among the most widely used antiseptics in surgical practice. For speed of initial degermation, alcohol, alcohol and iodine in combination, and iodophor compounds are among the most effective antiseptics.

Hexachlorophene is an effective skin disinfectant but is not recommended for repeated, routine use because it accumulates in an active residue on skin and has toxic effects if continuously absorbed into the body.

None of these antiseptics can prevent restoration of the resident flora on gloved hands during lengthy surgical procedures. For this reason, gloves that are nicked or punctured in the course of a surgical operation should be changed, particularly if enough time has elapsed to permit some bacterial recolonization on the hands.

A word of caution must be added regarding the storage and handling of antiseptics. Antiseptic solutions ready for use in the operating room for surgical scrub, or for the preoperative preparation of patients' skin, may become contaminated with microorganisms resistant to their chemical action. Sometimes this may happen also with strong disinfectants. If they are actively supporting microbial growth, such solutions constitute a dangerous reservoir of infection, particularly if they are applied in surgical or other hospital situations where asepsis is essential. The assumption that the antimicrobial nature of disinfectants and antiseptics guarantees their sterility is not warranted. Working solutions, as well as stocks from which they are prepared for final use, should be stored and handled under aseptic conditions.

The Evaluation and Selection of Germicides. Adequate laboratory demonstration of the efficacy of disinfectant agents or techniques is not often available in hospitals, but two points should be remembered in evaluating descriptive literature concerning germicides and in choosing one for a specific purpose:

1. Comparison of the activity of a disinfectant with that of phenol may be offered as evidence of value, but this is not always pertinent. Pure phenol (carbolic acid) is an efficient bactericidal agent, limited in use by its toxicity for skin and other tissues. In high concentrations it coagulates protein; in low concentrations it inactivates microbial enzymes by altering surface tension. When used as a comparative standard for judging the activity of other disinfectant compounds, its killing action for a particular organism is determined, under controlled conditions, at a stated concentration and time exposure. The values for concentration and time required to kill the same organism, under the same conditions, are then measured for the disinfectant being tested. If the latter requires a greater concentration or a longer time than phenol, its efficiency is judged to be less than that of phenol. The arithmetic ratio of its activity with respect to phenol, that is, its *phenol coefficient,* will be expressed in a figure *less than one.* If the tested disinfectant kills the organism in question in a lower concentration or a shorter time period than the standard, it will be given a phenol coefficient *greater than one.*

This is interesting laboratory information, but it

provides no clues as to the numerous other factors that may govern the activity of the test disinfectant under the conditions of its actual use. If it is a compound unrelated to phenol in chemical structure and activity, test-tube comparisons with a phenol standard have no practical meaning and leave the following questions unanswered: Does it have a different mode of action at high and low levels of concentration, as does the phenol standard? Is it capable of penetrating porous materials, or should its use be restricted to hard-surfaced items? Will it corrode metal, fabric, or skin? The answers to questions such as these are much more important than consideration of phenol coefficient in evaluating the range of action of a disinfectant.

2. The ideal, all-purpose chemical disinfectant *has yet to be discovered.* Therefore, the selection of any germicide must be based on the use to which it will be put when this is weighed against its properties and expense. As many of the following desirable qualities as are pertinent to the needs of a given situation should be sought when choosing a disinfectant:

a. It should be able to kill all pathogenic microorganisms, including the most resistant types known or suspected to be of concern. If it cannot do this, its spectrum of antimicrobial activity should be clearly defined.

b. It should be capable of killing microorganisms rapidly, at a concentration not destructive to materials that must be treated. Its use dilutions and the time required for their action should be clearly stated for each of its applications and purposes. Its destructive properties, if any, should be described with reference to the kinds of materials affected.

c. Whether or not it is used for skin degermation, it should not be toxic for human tissues. Its physiologic toxicities should be stated with reference to specifications of the Federal Insecticide, Fungicide and Rodenticide Act.

d. It should be impervious to the inactivating effects of soaps, waxes, proteins, or other chemical contaminants; if not, its chemical combining potentialities should be fully stated.

e. It should be penetrating but noncorrosive, odorless, and nonstaining. Any deficiencies in these qualities should be described.

f. It should be stable to pH and temperature changes within reasonable, defined limits.

g. It should be readily soluble in water and stable in solution. If more expensive solvents must be used, their effects and price must be considered.

h. It should be stable for a reasonable period of shelf storage. It is important that the effects of time on its potency be defined, since hospital stocks are frequently purchased in large quantity.

i. It should be simple and uncomplicated to handle and apply effectively, so that even unskilled personnel can use it correctly.

j. It should be easily manufactured at low cost. If this cannot be done, the value of the disinfectant must be judged in the light of the importance and difficulty of the job to be done and the availability of alternative methods.

It must be added that, once the choice is made, it is important to follow the manufacturer's instructions carefully and completely, for otherwise proper analysis of any problems that may arise later is virtually impossible.

Major Classes of Disinfectants and Their Uses. Disinfectants can be classified on the basis of their general chemical nature and activity. Brief descriptions are given on the following pages of the major classes of disinfectants and their uses, i.e., phenols and phenolic derivatives, alcohols, surface-active agents (detergents and soaps), oxidizing agents (halogens), heavy metals, and alkylating agents. Table 10–4 summarizes their applications, advantages, and limitations.

Phenols. The phenols have been popular disinfectants since Lister first used carbolic acid in his operating room. Phenol itself is still commonly used as a standard in measuring the antibacterial potency of other substances, as described earlier. Because it is irritating and corrosive for skin it is seldom used today as an antiseptic or disinfectant, but some of its derivatives have a safer application. Most phenolics currently employed contain a phenol molecule, chemically altered to reduce its toxicity or increase its antibacterial activity, in combination with a soap or detergent. The latter serves to disperse protein, lipid, and other organic matter, reducing the possibility that such material may bind and inactivate the disinfectant.

The best known of the phenolic derivatives combined with soaps or detergents are the *cresols* and the *bis-phenols*. Probably the most familiar members of the *cresol* group are Tricresol, Lysol, Staphene, O-Syl, and Amphyl, these being proprietary names for different formulations of altered phenols useful as *environmental disinfectants*. The *bis-phenols* are so called because the chemical alteration involves the linkage of two phenol molecules. Hexachlorophene is such a compound. The combination of hexachlorophene and some other bis-phenols with soap or detergent provides compounds that are relatively nontoxic for skin but retain good antibacterial activity.

The hospital applications of hexachlorophene have been strictly limited in recent years because of the serious toxic hazards that accompany its continuous use, particularly for cleansing infant skin through which it is readily absorbed. It may be used in some instances to control nursery outbreaks of staphylococcal skin infections, but never in concentrations exceeding 3 percent.

The phenols coagulate protein and are effective in killing the vegetative cells of bacteria. When combined with detergents they are also good tuberculocides, virucides, and fungicides. They are not sporicidal under ordinary conditions of use as surface disinfectants, for they can kill spores only in heated solutions applied for prolonged periods of time. When combined with soaps or detergents, the phenolics are not very soluble in water, but remain active in aqueous concentrations as low as 1 to 5 percent. This combination of qualities makes them inexpensive to use, but, more important, confers residual bactericidal or bacteriostatic effects when their solutions are left to dry on treated surfaces (walls, floors, furniture, skin). As water evaporates from phenolic films, the disinfectant molecule remains active for a time, deterring bacterial growth. These disinfectants can also be used to treat discarded cultures, sputum, feces, and heavily contaminated objects because of their low affinity for organic matter and their antibacterial activity, even in high dilution.

Alcohols. Alcohols that are water soluble and of low molecular weight may be bactericidal under proper conditions of use. Methyl alcohol, the lightweight of this family of compounds, is not a good disinfectant, but ethyl and propyl alcohols have antimicrobial activity in the presence of water. In *absolute* concentrations, 100 percent pure, alcohols have little or no effect on microorganisms, but in the presence of water, *in 50 to 80 percent dilutions,* ethyl, propyl, and isopropyl alcohols are lethal for the tubercle bacillus and for vegetative forms of other bacteria. At these concentrations, alcohols can penetrate cell membranes and precipitate proteins, but they cannot penetrate bacterial spore walls, and they are ineffective against certain viruses.

The moisture present in materials to be treated with alcohol must be considered when decisions are made about the dilution to be used. Dry surfaces may be wiped or washed with 50 to 70 percent alcoholic solutions, but wet materials may require a higher concentration, since they dilute the disinfectant further. The bactericidal effect of alcohol is lost below a 50 percent concentration, although bacteriostasis may be accomplished at dilutions as high as 1 percent. It is important to remember that chemically inhibited organisms may flourish once again when the bacteriostatic agent is removed or evaporates. Alcohols are more volatile than water, and therefore alcoholic solutions become weaker on standing in open containers that permit evaporation. Their bactericidal activity is slowly lost under such conditions, and they become merely bacteriostatic. If contaminated alcoholic disinfectants are used, they may spread viable microorganisms over the wiped surfaces. Then as the alcohol volatilizes, the bacteria freed from its action may grow and multiply.

Alcohols are good lipid solvents and are frequently used to cleanse and disinfect skin surfaces, clearing the latter of oil, debris, and microorganisms. Other disinfectant compounds are sometimes dissolved in alcohol, rather than in water, the resulting *tincture* having added disinfectant properties as well as better grease solvency. (Tincture of iodine is a well-known example.) Alcoholic solutions are drying to the skin since they remove oil and thus are damaging to one of the natural skin barriers. When used on traumatized tissues, they may add further injury by coagulating cellular proteins. The use of alcohols and tinctures as antiseptics is limited for these reasons, but they have valuable applications in degermation of normal skin surfaces (in preparation for venipuncture or minor surgery) and of small, hard-surfaced objects such as thermometers. Because they are expensive, volatile, and difficult to control in standing solution, alcohols are not frequently used for disinfection of large surface areas.

Surface-Active Agents (Detergents and Soaps). These agents have the property of decreasing the tension that exists between molecules that lie on the surfaces of liquid suspensions or solutions. Liquids that have a high surface tension (oils, for example) are not easily penetrated by other materials, nor do they spread into them. They tend, rather, to separate from surrounding substances because of the strong attracting force that holds their surface molecules together. When the surface force of liquids is low, they spread, or "run," more easily. They are said to be "wet" because they flow across and penetrate into adjacent surfaces as pure water does. Liquids with low surface tension are also more "miscible"; that is, they mix with other fluids, and can *be* mixed. Surface-active agents accumulate at interfaces of liquids (that is, at their adjacent surfaces), lowering the surface tension of molecules on both sides of the interface, so that they can run together. The liquids become more miscible, one with the other, and the agent is said to have a "wetting" action. The antibacterial action of soaps and detergents is partly due to their ability to lower the surface tension of liquid interfaces. The membranes of bacterial cells themselves constitute interfaces between the protein-lipid complexes of the cells and their walls and the moist, aqueous surroundings in which they usually lie. When surface-active agents come in contact with microbial interfaces, they remove the molecular barriers that hold cells and surrounding liquid media apart; they form a bridge between the interior contents of cells and the external milieu. Sometimes the bridging compound is itself denaturing for microbial cell contents and induces cellular injury or death. If not, disinfectant chemicals can be mixed with surface-active agents that effect their entry into cells by lowering surface tension.

SOAPS. Soaps are surface-active agents, capable of improving the miscibility of oils with water. For this reason they are very useful cleansing agents for skin and clothing. Soaps also have a lethal action on a few fastidious pathogenic bacteria, notably streptococci, pneumococci, and *Haemophilus* species, but have no chemical effects on most skin microorganisms.

When disinfectant chemicals are *added* to soaps (as in the case of such proprietaries as Lifebuoy, Dial, Praise, and certain liquid soaps, which contain phenolic disinfectants or other compounds), their antibacterial efficacy is increased. The combination of soaps with disinfectants must be carefully gauged, however, to avoid neutralizing the antibacterial portion of combining chemical molecules. For this reason, the *sequential* use of soaps and disinfectants should be carefully considered, for even traces of soap left on skin or other objects may bind disinfectants subsequently applied and thus neutralize, or inactivate, their chemical effects on microorganisms.

DETERGENTS. Natural or synthetic soaps with a high degree of surface activity are called *detergents*. Depending on their chemical structure, they may or may not ionize in water solution. Those that do *not* ionize (neutral detergents) have no usefulness as disinfectants, although they may be good cleansers. Those that *do* form ions in aqueous solution have bactericidal activity, probably because their charged ions concentrate on bacterial cell membranes and disrupt their normal functions. *Anionic* detergents are those that form negatively charged ions (anions). These include the natural soaps (the sodium or potassium salts of fatty acids) and bile salts.

Cationic detergents form positively charged ions (cations). This group includes the *quaternary ammonium compounds,* or "quats." The most familiar proprietary name in this group is Zephiran. Although these compounds have many useful qualities (they are bacteriostatic in low concentrations, non-toxic, soluble in water, inexpensive), they are severely limited by their affinity for many proteins and for soaps. They are also readily bound and inactivated by plant fibers, such as those in gauze and cotton pads. Aqueous solutions of quats bound to such materials can thus become reservoirs for bacterial growth and have been incriminated in numerous serious nosocomial (hospital-acquired) infections. For this reason, and because they lack the broad spectrum of activity (they are neither tuberculocidal nor sporicidal) of other useful disinfectants, their use in hospitals has been generally disapproved.

Oxidizing Agents. These substances inactivate microorganisms and other types of cells by effecting an electron exchange on some of the free radical groups of their protein molecules. This oxidizing effect may be lethal to cells if it inactivates vital intracellular protein enzymes. Chemical compounds that release oxygen readily may have bactericidal effects. Those most commonly used as antiseptics are hydrogen peroxide, sodium perborate, and potassium permanganate. Peroxide and perborate are sometimes used as wound irrigants, to reduce or control local infection, to oxidize and clear away tissue debris, and to prevent anaerobic bacterial growth in traumatized tissue by oxygenation of the area. Potassium permanganate solutions (in high dilutions) are used as urethral antiseptics, and also for treatment of superficial fungous infections of the skin. At safe concentrations, these compounds are not markedly efficient as bactericides and their use in antisepsis is therefore quite limited.

The most powerful oxidizing agents, and the most useful in antisepsis or disinfection, are the *halogens* and their derivatives. Iodine, chlorine, bromine, and fluorine all have bactericidal properties, probably because they combine irreversibly with protein enzymes, inhibiting their activity.

IODINE. This halogen is an effective antiseptic for normal skin and for superficial wounds. It is also very useful in environmental disinfection, for it is active against many vegetative bacterial species, including the tubercle bacillus, as well as fungi and some viruses. It is not an effective sporicide in the use concentrations that must be employed to avoid toxicity. Pure iodine is not readily soluble in water, but it dissolves more easily in solutions of sodium or potassium iodide. The release of free iodine from the iodine-iodide complex of these solutions accomplishes disinfection. A 3.5 percent tincture of iodine in alcohol containing potassium iodide is a very effective skin antiseptic. Tinctures may also be used for disinfecting thermometers and small instruments, provided volatilization of these solutions is pre-

vented. Tinctures used on skin may be damaging if the iodine concentration has increased because of alcohol evaporation. This factor, as well as the staining effects of iodine, has tended to discourage the use of tinctures on skin. Both these limitations are counterbalanced by the rapidly effective action of iodine tinctures as compared with other antiseptics. For minor procedures such as venipuncture the preparation of small areas of skin with iodine alcohol solutions carefully formulated and applied can produce safe results quickly. When large areas of skin must be treated, however, the irritating effects of iodine and alcohol outweigh the advantages. In this case, less toxic iodine preparations or other types of antiseptic agents are used, but more time must be allowed for their action, and mechanical scrubbing becomes more important.

Iodine forms loosely bound complexes with organic molecules. In aqueous solution or suspension, the bound iodine carried by such a complex slowly dissociates and is freed. The combination of iodine and the carrier molecule is called an *iodophor*. Such compounds retain the useful properties of inorganic iodine solutions, but they are less toxic to the skin. The iodophors commonly used for environmental disinfection contain iodine combined with a nonionic detergent (some of the proprietary names include Wescodyne, Ioclide, and Hi-Sine). Used in a concentration of 75 ppm of available iodine, the iodophors may be applied to the general disinfection of floors, walls, and furniture. A concentration of 450 ppm has tuberculocidal activity. Formulations of iodine with polymeric molecules, such as polyvinylpyrrolidone (povidone iodophors), are used for skin degermation. Betadine, Prepodyne, and Isodine are examples of this type of iodophor. The iodophors have a wide range of antimicrobial activity. They should be used in fresh solution, for iodine slowly dissipates from dilutions that are allowed to stand. Fortunately, however, this change is easy to recognize as the original deep-brown iodine color of these solutions fades to yellow with the loss of iodine. Fabrics and skin are not permanently stained by iodophors, but plastic materials treated with them remain discolored. The chief limitation on iodophor detergents is their reactivity with protein and other organic matter. In combining with such material they lose disinfectant activity. It is important, therefore, that surfaces to be treated with iodophors be free of extraneous contamination.

CHLORINE. In gaseous or liquid form, chlorine is a useful and inexpensive bactericidal agent for the disinfection of water and sewage. There are a number of compounds of chlorine that in aqueous solution release it readily as free or available chlorine. Chlorine itself then combines with water to form hypochlorous acid, a potent oxidizing and bleaching agent with a rapid bactericidal action. A few parts per million of available chlorine in drinking water, or swimming pools, are disinfectant. Larger concentrations are required in sewage, because some of the chlorine also combines with organic substances and is then no longer free to react with bactericidal cell proteins.

The most commonly used inorganic compounds of chlorine are hypochlorites of potassium, sodium (e.g., Clorox), or calcium (chlorinated lime). Sodium or potassium hypochlorite in aqueous solution (where it forms hypochlorous acid) can be used to disinfect small objects and surfaces. It is a convenient disinfectant for the sickroom at home, but is seldom used for this purpose in hospitals because, like most other chlorine compounds, it is bleaching, corrosive, and malodorous. Chlorinated lime is prepared by exposing freshly slaked lime to chlorine gas. The resulting compound is unstable and quickly frees the chlorine if exposed to air. It must be kept in tightly sealed containers and be freshly prepared for each use. Solutions of 0.5 to 5 percent chlorinated lime are used for disinfection of dairies, barns, abattoirs, and similar installations. Inorganic chlorine compounds are also used in sanitization of dishes or laundry.

BROMINE AND FLUORINE. These halogens have very limited application in disinfection or antisepsis because they are toxic in useful concentrations. Bromine combinations with certain organic compounds, notably the salicylanilides, have proved to be effective, however, in some environmental uses, as in the treatment of fabrics with a laundry rinse.

Ions of Heavy Metals. Metallic ions such as mercury and silver, are readily taken up by the proteins of living cells, including microorganisms. The effect is highly injurious to the activity of enzyme protein and may lead to rapid death of the cell. This broad-spectrum toxicity means that mercury and silver salts have only limited bactericidal applications. The inorganic salts of these metals, e.g., mercuric oxide, mercuric iodide, and silver nitrate, are used in low concentrations as antiseptics. The mercuric compounds are frequently incorporated in ointments for use in the treatment of minor skin infections. Organic compounds of mercury, such as Merthiolate and Mercurochrome, are also used topically, sometimes for the treatment of superficial infection, sometimes for the degermation of skin prior to surgery. Merthiolate may also be added to microbial vaccines as a preservative. Silver nitrate is used in 1 percent aqueous solution, its best-known application being in the prevention of gonorrheal infection of the eyes of newborn babies.

Alkylating Agents. The antimicrobial activity of this class of disinfectants is due to the substitution of their alkyl groups for the reactive hydrogen on the chemical radicals of microbial proteins, nucleic acids, or enzymes. Alkyl groups are univalent hydrocarbon radicals such as occur in aldehydes ($-CHO$) and similar compounds, including ethylene oxide and beta-propriolactone. Substitution of alkyl groups for the hydrogen in microbial amines ($-NH_2$), hydroxyl radicals ($-OH$), sulfhydryl groups ($-SH$), or carbonyl groups ($-COOH$), for example, disrupts microbial metabolic pathways, resulting in destruction of microorganisms. The chemical structure of some alkylating agents is shown in Table 10–4. Some of these are active as liquid disinfectants, some are used as gases, and some may be used in either the gaseous or liquid phase.

ALDEHYDES. *Formaldehyde* and *glutaraldehyde* are the best-known examples of this group of disinfectants. Formaldehyde has a long history of use both as a gaseous fumigant and as a liquid germicide (formalin). Glutaraldehyde, a related liquid compound, has come into more prominence in recent years.

Formaldehyde and Formalin. Gaseous formaldehyde was once used as a fumigant for closed rooms or instrument cabinets, but this practice has been abandoned because of its irritating and toxic properties. It is damaging to skin and mucous membranes, and its odor, both as a gas and as a liquid, is unpleasant and penetrating, qualities that limit its usefulness as an environmental disinfectant.

Formalin is a 37 percent aqueous solution of formaldehyde gas. It is one of the few chemical disinfectants having sporicidal activity, and it comes close to being a sterilizing agent under the proper conditions. Its avid combining power makes it an excellent tissue fixative for histologic work, but its use as a chemical disinfectant or sterilizer is limited by its noxious properties. It does have practical applications as an instrument soak, however, when used in alcoholic solution with small quantities of added alkali to prevent corrosion. Proprietary solutions such as Bard-Parker Germicide can be employed as a soak for small instruments and transfer forceps, killing spores in 18 hours at room temperature.

Formalin is also commonly used in the preparation of microbial antigens for vaccines. It is valuable for this purpose because it can kill microorganisms or inactivate their toxins without destroying their antigenic properties. Upon injection, such formalinized antigens are capable of stimulating antibody production without risk of infection or toxicity. In antigen preparation, a high dilution of formalin is used for a period of several hours. It is not toxic of itself in vaccines because its combining sites are bound to microbial protein, leaving it incapable of further activity.

Glutaraldehyde. This disinfectant, as it is commonly used in a 2 percent buffered aqueous solution, is as effective as formalin without being noxious. It is cidal for most bacteria within five minutes, kills tubercle bacilli and viruses in ten minutes, and is sporicidal in 10 to 12 hours. It can be considered a chemical sterilizer when used for a minimum of ten hours, or a disinfectant when applied for ten minutes. It is recommended as a soak for lensed instruments, rubber or plastic items, anesthesia or inhalation therapy equipment, sharp instruments and thermometers, and even for soiled instruments from septic cases.

Cidex is the proprietary name for the commercially available glutaraldehyde prepared with a buffer-activator that brings the pH to 8.0. This activated solution remains stable for 14 days if stored at cool room temperature. A newer formulation, *acid* glutaraldehyde (trade name, Sonacide) has properties similar to those of alkaline glutaraldehyde, but is said to destroy spores in one hour if the temperature is raised to 60° C. Although nonirritating when used correctly, glutaraldehyde should not be allowed to come in contact with the skin or eyes, but if this happens, it should be flushed away promptly with a copious flow of water.

ETHYLENE OXIDE (ETO). This gaseous compound is widely used for gas sterilization within a closed chamber similar to an autoclave. ETO is particularly suitable for this purpose because it is lethal for all varieties of microorganisms, is penetrative when under increased pressure, and does not permanently bind to items or materials exposed to it but dissipates slowly when these are removed from the sterilizer. It is a flammable, explosive gas, however, and for this reason it is used in a mixture with inert gases such as carbon dioxide or freon. In this condition it requires three to four hours under pressure to penetrate shallow layers of materials and to sterilize all surfaces within a chamber load. A minimum period of 24 hours at room temperature must be allowed following sterilization for ETO to dissipate from treated materials. The time required, toxicity, flammability, and expense of the equipment needed are limitations to be weighed against its efficacy as a sterilizing agent for sensitive materials. Its use is generally reserved for objects that would be damaged by more efficient methods of heat sterilization or chemical disinfection. Lensed instruments, plastic items, catheters made of synthetic materials, Lucite and plastic components of heart-lung pumps, artificial heart valves and vessel segments, mattresses and other bedding, and many similar articles can be safely sterilized with ethylene oxide, this being the method of choice despite its expense (Fig. 10–6).

BETA-PROPRIOLACTONE (BPL). This compound is an unstable liquid that decomposes within a few hours on exposure to moisture. It has been applied in this form to the sterilization of fluids, such as serums and vaccines, its natural decomposition being an advantage since none of it remains active. Although it is a powerful sterilant, it has not been widely used because it has been shown to be carcinogenic in liquid form.

In its gaseous phase, BPL is used as a fumigant.

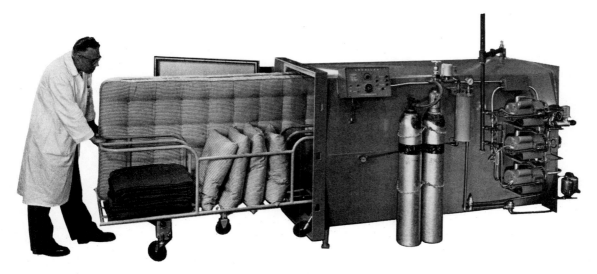

Fig. 10–6. Ethylene oxide can be used to sterilize mattresses, pillows, and other bedding used for patients with known or suspected infectious disease. (Courtesy AMSCO/American Sterilizer Co., Erie, Pa.)

Although less objectionable and more efficient than formaldehyde, BPL gas is irritating to the skin and eyes.

Other Chemical Disinfectants. Other agents with limited bactericidal or bacteriostatic uses include the organic *solvents,* organic *acids,* and mineral *salts.*

ORGANIC SOLVENTS. Solvents such as ether, acetone, chloroform, benzene and toluene, all have some antimicrobial activity, but most of these compounds are not soluble in water and do not have a rapid killing effect on microorganisms. They are most frequently used for their bacteriostatic and fungistatic qualities, as preservatives for protein and other solutions.

ORGANIC ACIDS. These are often used in the preservation of foods, to prevent the growth of bacteria and molds. Vinegar, which is dilute acetic acid, is well known for this use. Benzoic and proprionic acids, or their salts, are also commonly added to foods as preservatives. Boric acid solutions are sometimes used for their mild antiseptic action as irrigants for wounds or for eyes.

MINERAL SALTS. In concentrations rising from 1 to 2 percent, mineral salts become increasingly inhibitory for bacteria and other microorganisms. For this reason salting is commonly used as a method for food preservation.

Control of Environmental Sources of Infection

"It may seem a strange principle to enunciate as the very first requirement in a Hospital that it should do the sick no harm." These words of Florence Nightingale should remain uppermost in the minds of all who are involved in patient care. Safety from infection is of particularly critical importance in hospitals, where it is, unfortunately, often most difficult to guarantee. It requires that all of the principles of asepsis, disinfection, and sterilization be continuously integrated to provide infection control and to prevent the occurrence of nosocomial infection.

The minimum goal that must be met by hospitals is *cleanliness.* This should be the coordinated responsibility of the heads of all major hospital departments, i.e., administration, housekeeping, main-

Table 10-4. Some Major Classes of Chemical Disinfectants*

Class: PHENOLICS (synthetic phenols)

OH

Phenol

Cresols

Proprietary products: Synthetics: Amphyl, Lysol, O-Syl, Staphene, Vesphene, Tergisyl, Armisol

Use dilution: Creosol—2% for 30 minutes; 5% if hard water
Amphyl—0.5–1% (also available as spray)
Lysol—1%
O-Syl—2.5%
Staphene—0.5–2%
Vesphene—1–2%
Tergisyl—2%

Recommended for: Creosol—equipment, linen, excreta
Amphyl—instruments
Lysol—laundry rinse for blankets, linens
O-Syl, Staphene, Vesphene—floors, walls, equipment
Tergisyl, Armisol—environmental uses

Advantages: Not inactivated by organic matter, soap, or hard water (except Creosol); residual effect if allowed to dry on surfaces; high detergency

Limitations: Creosol must be used in soft water and is slow-acting; Lysol and Creosol both have odor

Bis-phenols: Hexachlorophene (G-11)
(Note warning under limitations)

Proprietary products: Liquid detergents—Septisol, pHisoHex, Hex-O-San, Surofene
Hand Creams—Septisol antiseptic skin cream

Use dilution: Septisol—2%; pHisoHex—3%; others as recommended by manufacturer

Recommended for: Skin disinfection, but note limitations

Advantages: Good cleaners and have prolonged antibacterial action

Limitations: Not sporicidal, not tuberculocidal, have slow action, toxic if used continuously and absorbed into the body in increasing quantity (especially through delicate skin of infants); should be rinsed off after use on skin

Class: SOLUBLE ALCOHOLS   ***Ethyl:*** C_2H_5OH ***Isopropyl:*** $CH_3CHOHCH_3$

Use dilution: 70–90%

Recommended for: Thermometers (add 0.2–1% iodine); instruments; skin preparation; hands; spot disinfection

Advantages: Rapidly bactericidal; tuberculocidal

Limitations: Not sporicidal; corrodes metals unless reducing agent added (e.g., 2% sodium nitrite); drying to skin (1:200 cetyl alcohol may be added as emollient); bleaches rubber tile

Table 10–4. (*Cont.*)

Class: HALOGENS   *Chlorines, Iodines*

Iodines: Tincture; iodophors

   Proprietary products: Wescodyne; Betadine; Iobac; Klenzade; Micro-Klene; Virac

   Use dilution: Tincture of iodine: 0.5–2%
                   Wescodyne: 75 ppm (90 ml or 3 oz to 5 gal water), or 450 ppm to kill tubercle bacilli;
                      tincture = 10% Wescodyne in 50% ethyl alcohol
                   Other iodophors: see manufacturer's recommendations

   Recommended for: Tincture—skin preparation; thermometers (see alcohols)
                      Wescodyne—thermometers, utensils, rubber goods, dishes; as a tincture, used
                         for spot disinfection, or for single presurgical scrub
                      Betadine—presurgical skin preparation
                      Others—many designed for specific purposes in eating places or dairies

   Advantages: Iodophors are cleaning and disinfecting; nonstaining; leave residual antibacterial effect;
                 tuberculocidal as tinctures; loss of germicidal activity indicated by fading color

   Limitations: Tincture of iodine stains and is irritating to tissues
                  Iodophors somewhat unstable, inactivated by hard water; may corrode metals, drying to
                  skin as tinctures

Chlorines: Hypochlorites or hydrochlorous acid derivatives (HClO)

   Proprietary products: Hypochlorites: Clorox, Purex, other bleaches
                           HClO: Warexin

   Use dilution: Hypochlorites: strongest concentration recommended by manufacturer; if dry, mix to
                      thin paste
                   HClO: Warexin used in 1.5% aqueous solution

   Recommended for: Floors; plumbing fixtures; spot disinfection; fabrics not harmed by bleaching;
                      Warexin for dishes but not silverware

   Advantages: Tuberculocidal unless highly diluted (Warexin limited)

   Limitations: Bleach fabrics; corrode metals; unstable in hard water; must be freshly prepared; tarnish
                  silver

Class: ALKYLATING AGENTS

Aldehydes: Formaldehyde

$$H-\overset{\displaystyle |}{\underset{\displaystyle O}{C}}-H$$

   Proprietary product: Bard-Parker Formaldehyde Germicide

   Use dilution: Gas, or full-strength solution

   Recommended for: Gas: cabinet disinfection
                      Solution: transfer forceps, instrument soak

   Advantages: Vapor disinfection of delicate instruments; sporicidal, noncorrosive

   Limitations: Requires long period for effective disinfection; odorous; toxic to skin and mucous
                  membranes

Table 10–4. *(Cont.)*

Glutaraldehyde

$$HC{-}CH_2{-}CH_2{-}CH_2{-}CH$$

with O below each terminal CH

Proprietary products: Cidex, Sonacide

Use dilution: Cidex—2% buffered aqueous solution
Sonacide—2% (acid)

Recommended for: Soak for lensed instruments, heat-sensitive equipment

Advantages: Bactericidal, tuberculocidal, virucidal, sporicidal

Limitations: Irritating to skin and eyes

Ethylene Oxide

$$CH_2{-}CH_2$$

with O below forming a ring

Proprietary products: Carboxide, Cryoxide, Steroxide

Use dilution: Gas; exposure time: 4–18 hours

Recommended for: Blankets, pillows, mattresses; lensed instruments; rubber goods; thermolabile plastics; books; papers

Advantages: Harmless to most materials; sterilizes

Limitations: Requires special equipment

*Proprietary products are listed as examples only; many similar products are available under other brand names. Unless otherwise noted, the disinfecting time is ten minutes.

tenance, and nursing. Each area of the hospital, from its garbage disposal facility to the operating room, should be subjected to rigorous, daily cleaning to preclude survival and transmission of pathogenic organisms from possible as well as obvious reservoirs. The cleaning equipment itself should be chosen and applied with a knowledge of the microbiologic needs of hospitals. Dusting cloths, mops, brooms, and dry vacuum cleaners can themselves contribute to the spread of microorganisms when they are used without regard for the principles of disinfection.

With a basic state of cleanliness existing throughout the hospital, *aseptic conditions* can then be established and maintained with relative ease wherever they are indicated. Medical asepsis has been defined as a condition excluding the infectious agents of communicable disease, while surgical asepsis implies the exclusion of all microorganisms (Table 10–1).

More often than not, asepsis is a goal rather than a realized condition, because it is dependent on the coordination of many techniques applied by many people (housekeepers, maintenance staff, nurses, physicians—even patients' visitors on occasion). Not all of these techniques can be fully effective under all circumstances, but when they are conscientiously integrated, they can provide the safe environment patients have a right to expect in modern hospitals.

In learning aseptic techniques it would be helpful to recall Joseph Lister's advice to his students: ". . . you must be able to see with your mental eye the septic ferments as distinctly as we see flies or other insects with the corporeal eye. If you can really see them in this distinct way with your intellectual eye, you can be properly on your guard against them; if you do not see them, you will be constantly liable to relax in your precautions."

Medical Asepsis

Medical asepsis includes all the fundamental concepts of good housekeeping (dust removal, scrubbing, laundering, disinfecting) with some additional techniques applied to special conditions created by infectious illness. Some of these techniques are taught as a part of the general public education concerning communicability of infection: the personal hygiene of careful hand washing and bathing; sanitary handling of food and dishes; proper methods for cooking or refrigerating foods; hygienic attitudes toward body functions and excretions (toilet habits, covering the mouth and nose when coughing or sneezing, adequate disposal of facial tissues and handkerchiefs). More specialized methods are added to these by nurses and physicians directly involved in the care of sick patients. These include all the precautionary measures taken to prevent *direct* transmission of infection from person to person or *indirect* transfer of pathogens by way of instruments, equipment, or any inanimate objects present in the sickroom.

Surgical Asepsis

Surgical asepsis adds the last links to this chain of techniques designed to eliminate the danger of transferring infection. The aseptic techniques that surround surgical procedures must provide the best guarantees that *all* microorganisms have been eliminated from any scene of action intimately involving the body's deep tissues. When injections are given, superficial wounds are sutured, or deep surgery is performed, the body's primary defensive mechanisms are bypassed, and entry is made into tissues whose susceptibility to infection is increased by the injury induced by the procedure. Surgical asepsis is intended to prevent the introduction of any contaminating microorganisms. The area of the patient's skin to be entered is thoroughly cleansed and scrubbed with antiseptic, and if the procedure is to be extensive, surrounding areas are covered with sterile cloth drapes, so that a sterile field is established at the site of surgery. The surgeon and all personnel involved similarly prepare their own skin,

scrubbing hands and forearms, and then with sterile gloves, gowns, caps, and masks they cover all areas from which microorganisms might be shed onto the operative field. Instruments, dressings, and sutures are all sterilized for use in surgery, and any equipment that cannot be presterilized is treated with an appropriate disinfection technique. The operating room itself and its walls, floors, and equipment are cleaned and disinfected before use, and every pattern of activity within the room is designed to maintain the aseptic atmosphere. The ventilation of operating rooms should be such that, with doors closed fresh, filtered air is delivered, circulated, and exhausted continuously.

Once asepsis has been adequately established in a closed, properly ventilated operating room, the number of microorganisms present in the air or on surfaces is reduced to the barest possible minimum. From that point on, as surgery proceeds, microbial contamination of the atmosphere progressively increases, its principal source being the people present in the room. The degree to which such contamination rises, and its potential danger to the patient, depend largely on the technique and skill of attending personnel in controlling their own activities so that the major sources of their microorganisms (nose and throat, hands, uncovered skin and hair) are under continuous, effective restraint. Close guards are kept against breaks in technique: damaged gloves are discarded and replaced; saturated masks are exchanged for fresh ones; backup supplies of sterile instruments, solutions, and other essential items are kept at hand.

It should be emphasized that surgical asepsis is a concept to be applied not only in operating rooms, but in any situation requiring close protection of the patient when the risk of infection is high. It should be maintained with respect to surgical wounds for the first few postoperative days, or until tissues are sufficiently healed to manage their own defenses. It is applied in delivery rooms to prevent sepsis of injured mucous membranes and in nurseries to protect susceptible newborns from any risk of infection. Wherever it is established, surgical asepsis represents a crowning touch to a basic structure of techniques that begin with sanitary cleanliness and end by disinfecting and sterilizing the environment to the fullest extent possible.

Application of Aseptic Techniques in Hospitals

In many hospitals, general policy with respect to infection control is established by a committee, composed of representatives from the medical and nursing staffs, housekeeping and maintenance departments, the laboratory, and administration. The infection control committee can set up uniform guidelines for procedures throughout the hospital, basing these on sound principles of asepsis specifically applied to each local situation. Key members of this committee are the *hospital epidemiologist* and the *infection control nurse*. The former is often a physician with special training in epidemiology, or he may be a clinician specialized in infectious diseases. This individual generally serves as the committee's chairperson, forming a liaison with the medical staff and authorizing final policy. The infection control nurse, or *nurse-epidemiologist* as he or she may be called, monitors the hospital for infection among patients or staff, supervises isolation and precaution techniques, and provides consultation and training for hospital personnel involved in problems associated with infection.

The function of the *microbiology laboratory* in infection control is one of consultation and provision of surveillance to determine the efficacy of aseptic procedures. Laboratory tests, properly designed and applied, can demonstrate whether or not hospital practices accomplish the desired result. In-use testing can be applied to equipment, solutions, floors, and furniture, or even to the concept of a technique. In many situations, it can provide final information concerning the value of preventive measures. The laboratory should also be prepared to combine its diagnostic function with surveillance methods so that outbreaks of particular infections, if they occur, can be quickly recognized and traced to their source. In the latter regard, the assistance of epidemiologic experts from city or state health departments is available and should be utilized. The microbiologist also plays an important role in providing information concerning the microorganisms that are most prevalent in the hospital, or in a given service, and the patterns of their responses to antibiotics. This can form the basis of medical policy for the clinically

judicial use of antibiotics and also provides an index of the efficiency of infection control policies.

To provide some consistency of approach among hospitals with respect to infection control, the Center for Disease Control (CDC) of the U.S. Public Health Service has formulated a national policy. In a manual entitled *Isolation Techniques for Use in Hospitals* (see the references for this chapter), the CDC has described the general principles of infection control, with specific recommendations for precautionary and isolation techniques to be followed for several categories of hospitalized patients with infectious diseases. These recommendations are based on the known pattern of spread of specified communicable diseases. Each category of isolation is printed on a card that can be posted on the door of the patient's room (Fig. 10-7). The card specifies precautions to be taken and, on the reverse side, the disease or condition requiring that type of isolation. Hospitals are encouraged to modify these to meet their own needs.

General Principles of Infection Control. Control measures practiced in hospitals should be based on available information concerning the communicability and mode of spread of infectious diseases. Differences in communicability and possible transmission routes should form the basis for decisions regarding isolation techniques and the types of precautions to be taken in caring for infected patients. If warranted, the first step in control is the isolation of the infectious patient; all remaining precautions are designed to prevent transmission of that patient's infection to others, or of a new communicable disease to that patient.

Isolation of the Infectious Patient. In many instances, patients with communicable diseases should be cared for in private rooms properly equipped for their needs as well as for nursing purposes. (Infants and premature babies can be cared for in individual incubator-isolators, or "isolettes.") A private bathroom should be available, and individual supplies and equipment for the care of the patient should be provided throughout the communicable period. If a basin facility with running hot and cold water is not available, a handwashing basin must be provided

Fig. 10–7. The infection control nurse (right) confers with the nurse who is caring for a patient on enteric precautions. (Courtesy of Danbury Hospital, Danbury, Conn. Photo © Elizabeth Wilcox.)

within the room, together with an effective antiseptic for skin care.

Since isolation presents some problems both for the patient and for the hospital, it should not be imposed if not warranted by the difficulties of controlling transmission. Adults with enteric infections, for example, may be cared for adequately in a shared room, provided appropriate precautions are taken, whereas children should be isolated in a room with a private bathroom. Isolation should be discontinued promptly when the communicable period has passed, or specific treatment has removed the threat of further transmission. The procedures involved in isolation can be time consuming and expensive and may even interfere with some aspects of good patient care. The necessity for putting on and taking off gowns, masks, and gloves may discourage hospital staff from making frequent visits, adding to the patient's burden of solitude.

CONCURRENT DISINFECTION. The techniques of concurrent disinfection are practiced throughout the course of the patient's hospital stay, or until such time as the physician considers him to be no longer infectious. These techniques are designed to assure the immediate destruction of pathogenic microorganisms in the infectious discharges of the patient or on objects soiled by such material. Nondisposable items are first disinfected, with a minimum of handling, then cleaned thoroughly or resterilized. If the patient's infection involves spore-forming bacterial pathogens, contaminated articles to be reused must be sterilized by steam under pressure. Disposable items may be incinerated, or autoclaved if this is more convenient.

TERMINAL DISINFECTION. When the patient has been discharged from the room, terminal disinfection of the room is required. All equipment is either sterilized or carefully disinfected, with particular attention to blankets, mattress, and pillows. All items that can be laundered, including window curtains or drapes (these have no place in an isolation room), are disinfected by good laundry techniques. Plastic covers on pillows and mattresses can be scrubbed with liquid germicide; otherwise ethylene oxide sterilization is recommended for these items. Furniture and floors are washed clean and then treated with an appropriate liquid disinfectant. The walls should also be washed, particularly if the infection was one of serious nature transmitted by air or droplet spread, and the room should be well aired.

Isolation of the Susceptible Patient. Modern medicine has developed many techniques for the treatment of patients with severe organic disease. Open-heart surgery, artificial heart and kidney machines, and organ transplants offer new hope for the management of hitherto incurable conditions. These procedures frequently involve the patient in long and difficult periods of surgery or continuous treatment, during which he is, or becomes, increasingly susceptible to infection. The complications of infection, should it occur in these patients, can be extremely serious and threatening both to the success of treatment and to life itself. Other patients under severe stress offer similar problems of unusual susceptibility to infection: patients with immunodeficiencies or those being treated with immunosuppressive drugs, burn patients, premature babies, and people severely debilitated by age may be candidates for *protective* isolation. Concurrent disinfection is not necessary in such instances, but strict asepsis is maintained throughout the critical period of the patient's care.

The Surgical Patient. Preparation of the patient scheduled for surgery requires adequate disinfection of the skin area to be incised. The hair is first removed, and solvents (soap or alcohol) are used to wash away skin oils and most superficial microorganisms. Final application of an effective antiseptic further reduces the microbial flora of the region. Patients being prepared for abdominal surgery are sometimes treated for a few days beforehand with a combination of antibiotics intended to "sterilize" the bowel contents or to reduce normal intestinal flora to a very low level temporarily. This procedure is reputed to decrease the risk of infection to surrounding tissues of the peritoneum when the bowel must be opened, particularly if prolonged surgery is anticipated or if the patient's defenses are unusually low.

The surgical patient who presents no evidence of overt infection before or after surgery does not require isolation, although the wound itself is isolated by good dressing technique, until normal healing occurs. On the other hand, when an *infective* lesion is incised in the operating room, particular precautions must be taken for concurrent and terminal disinfection of the surgical room and all its equipment. The infective postoperative patient must be managed in such a way that he offers no threat of infection transfer to other patients with unhealed surgical wounds. If his infection is severe, and particularly if it involves spore-forming bacterial pathogens, he should be isolated with full precautions. Under less dangerous circumstances, techniques may be adapted to the needs of the situation. Careful dressing techniques are indicated, with immediate destruction of contaminated dressings. Sterile gloves are used when infected wounds are handled, or dressings may be changed with sterile transer forceps, one for dirty use and one for clean (Fig. 10–8). "Occlusive" dressings are sometimes applied to infected wounds for topical isolation of infectious drainage. This technique employs a waterproof lamina over the dressing, adherent to the patient's skin around the full circumference of the area.

Personnel Techniques. Nurses and physicians who care for patients, and others (including their visitors) who come in contact with them, are sometimes responsible for continuing the chain of transmission of infection. The most obvious and frequent transportation offered to pathogenic microorganisms is by way of *unwashed hands,* but clothing and hair may also serve. It should also be remembered that medical personnel with their own infections (colds, sore throats, infectious skin lesions) can transfer these readily to patients, whose resistance to them may well be lowered by their illness. The basic techniques that can prevent this transmission in large measure are relatively simple, but they should be *conscientiously and consistently practiced.*

HAND WASHING. Proper hand washing is an essential factor in the prevention of disease transmission. Adequate hand-washing facilities should be available in all areas of the hospital for personnel and patients alike (Fig. 10–9). Good antiseptic soaps or detergents should be kept stocked at all hand basins, and these materials themselves must be kept in clean containers, changed frequently. Brush

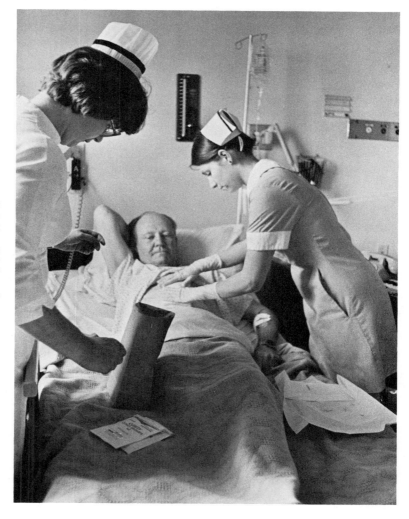

Fig. 10–8. Surgical aseptic techniques are required when changing a dressing to prevent contamination of the incision. The soiled dressings have been placed in a paper bag for proper disposal. (Courtesy of Danbury Hospital, Danbury, Conn. Photo © Elizabeth Wilcox.)

scrubbing of the hands should be avoided, except in surgery, by those who must wash frequently, because it is irritating.

The principles of antisepsis must be built in to hand-washing procedures if they are to be effective. The hands always carry a burden of their own normal flora, together with contamination from other sources, unless they have just been washed. They should be freshly washed when patients or their equipment are to be handled. When infectious pa-

tients are cared for, particularly those in isolation, hand washing is a part of concurrent disinfection and should always be done just before leaving the room. Hand contacts with furniture, telephones, elevator buttons, and doorknobs may transfer infection.

Hand washing should be thorough each time, with attention to the back as well as the palms, wrists, and each side of the fingers. Soap or detergent should be rubbed carefully, the hands being held down from the elbow so that contaminated water will not run up

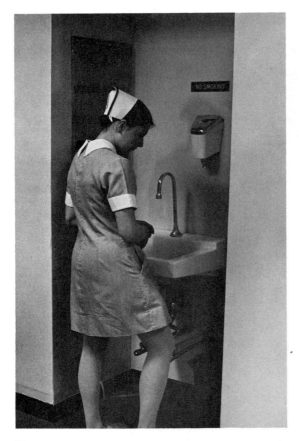

Fig. 10–9. Proper hand washing, a medical aseptic technique, is an essential factor in the prevention of disease transmission. The knee control at this sink eliminates faucets, which are easily contaminated by unwashed hands. (Courtesy of Danbury Hospital, Danbury, Conn. Photo © Elizabeth Wilcox.)

the arm. Rinsing should be complete, preferably under running water. Paper towels for drying are preferable to cloth, particularly in the sickroom, for a wet used towel may retain viable microorganisms.

The fingernails should be kept short and scrupulously clean. Rings and wristwatches should not be worn by those caring for infectious patients.

Gown Technique. The use of gowns is indicated under most isolation conditions, whether the patient is infectious or is being cared for with "reverse pre-

cautions." A clean gown that is ample in size and can be fastened securely protects the clothing and arms from contamination with infectious microorganisms. A fresh one should be available for each person who enters the sickroom on each occasion, and discarded after each use. Gowns are meant to be protective. If they are to fill this purpose, they should be properly fastened and should not be worn if wet or soiled. Moist areas on the gown may involve the clothing beneath and form a capillary bridge for the transfer of microorganisms.

Masks. Masks are necessary only when caring for patients under stringent types of isolation. The patient on respiratory isolation should also be masked if he must be taken from the room for treatment elsewhere in the hospital. Personnel masking is protective for short periods of close contact with the patient. The limitations of the mask should be understood, however. It should be worn correctly and discarded promptly after each use. When a used mask has been touched or handled, the hands should be washed thoroughly.

Personal Hygiene. The nurse, because of her close association with sick people, must be particularly aware of personal hygiene, to protect her own health as well as to avoid transmitting infection to others. Her clothing, hair, and skin should be kept clean and in good condition. One of the major reasons for a nursing uniform and the traditional white duty shoes is that this kind of clothing is easy to clean. When it is soiled this fact is immediately recognized and remedied. The practice of wearing this clothing outside of the hospital is a poor one, because the nurse should come to the patient's bedside in clothes as freshly clean as possible.

Personal care of skin and hair should include use of effective soaps or shampoos, together with nongreasy lubricants that help to avoid excessive drying. Hair (and wigs) should be shampooed frequently, particularly when the nurse is caring for an infectious patient. The coiffure should be simple and not require frequent arrangement with hands that may be contaminated. If hair is long or unruly, it should be controlled by a hairnet.

Above all, the habit of thorough frequent hand washing should be developed early and continued throughout nursing service. It should be a part of personal as well as professional life to guard the hands against contamination with potentially pathogenic microorganisms, derived from one's own body or from others.

Infections Among Personnel. All medical personnel should have physical examinations at regular intervals. This is of particular importance for physicians, nurses, and dietary personnel who should be examined for evidence of active infections of the respiratory and intestinal tracts or the skin. When indicated, they should be isolated and treated as any other patient would be, with due concern for the relative advantages of home or hospital care with respect to the specific infection. When personnel infections are not severe and can be treated on an ambulatory basis, the individuals involved should either be removed from duty altogether or be placed on duty in hospital areas where patient contacts are impossible or unimportant, until the infection has been cured or all danger of its transmission has passed.

Control of the Hospital Environment. The entire physical environment of the hospital must be considered in planning and maintaining effective procedures for infection control. Increasing emphasis is now placed on the architectural design of hospitals with respect to access and transfer routes of infection from physical or human sources. In the meantime, many hospitals have fundamental faults of design that create difficulties in control of traffic patterns, routing of essential hospital services (garbage and waste removal, the flow of clean and dirty laundry, food service), provision of adequate isolation facilities for infectious patients, or protection of areas, such as operating rooms and intensive-care units, in which infection control is of critical importance. In each hospital organization, those responsible for provision of an aseptic atmosphere must consider indigenous problems and devise effective methods to cope with them at every level.

RESPONSIBILITIES. The principles of asepsis should form the basis of procedures applied by *every* hospital department. The problems of ventilation, air cooling, and heating are the province of the engineering and maintenance department, which should be aware of the necessity for providing clean, filtered air, particularly to critical areas. Air filters, and refrigerant solutions used for air cooling, can themselves become reservoirs of multiplying organisms disseminated through hospital ventilators, unless properly maintained.

Sanitary maintenance of the hospital, its floors, walls, furniture, and nontechnical equipment, is generally the responsibility of the housekeeping department, whose manager should be expert on the subject of chemical disinfection, as should the laundry manager. The compatibility of soaps, detergents, and disinfectants and their effective combination in assuring asepsis are a matter of fundamental concern in these areas. Knowledge of the capacity of microorganisms to establish reservoirs in organic wastes (on mops, in spilled garbage, in dirty laundry rooms) is fundamental to hospital safety. The dietary department is also intimately involved in the application of aseptic principles to the preservation, preparation, and service of food to hospital patients. The sanitizing of kitchen equipment and use of appropriate methods for food preparation are incumbent upon the dietitian. The continuing health of food handlers, as well as all other personnel, and their freedom from infection are the responsibility of the hospital's health service physician, with assistance from the microbiology laboratory.

In the medical and surgical care of patients, physicians and nurses working together must consistently apply a procedural knowledge of asepsis. Patients being treated for noncommunicable diseases must not be exposed to risks of complicating infections, while those isolated for infection should be cared for without neglect of the special emotional as well as physical needs created by the circumstances of isolation. Nurses and physicians are the technical practitioners of asepsis, and their special knowledge requires special responsibility. (See Appendix III for a summary of aseptic nursing precautions.)

Control of Active Infectious Disease

Confirmation of the diagnosis of a case of active infectious disease, on the basis of developing clinical symptoms and laboratory findings, often requires some time — a few hours, a day or two, sometimes longer. In the meantime the physician must make decisions, based on probabilities, as to the patient's care and treatment. If the disease could be a communicable one, he may order the patient's isolation, either at home or in the hospital. The isolation may be complete or partial, with appropriate precautions, depending on the clinical and epidemiologic possibilities, as we have seen.

In the meantime, the question of specific treatment of the patient, as well as protection for his susceptible contacts, may be urgent. Therapy for the patient and prophylaxis for his contacts may have to be decided initially on the basis of clinical suspicion, then adjusted when the diagnosis is clear. Two kinds of agents are available for the control of many active infectious diseases: chemical agents, i.e., drugs or antibiotics, for *chemotherapy* or *chemoprophylaxis;* and immune serums for *immunotherapy* or *immunoprophylaxis.*

For the sick patient, chemotherapeutic methods, instituted promptly, are generally most effective. However, in those diseases characterized by the action of potent microbial exotoxins, e.g., tetanus, botulism, or diphtheria, immunotherapy with specific antitoxin may be imperative early in the course of illness.

When it is necessary to protect the contacts of a patient ill with a communicable disease, chemoprophylaxis may be useful in some instances, but in others it may be necessary to consider passive immunization with immune serums (immunoprophylaxis).

Treatment of the Patient

Chemotherapy. The use of chemical agents, applied externally or internally, in the treatment of human disease is an approach that is probably as old as man himself. The scientific approach to chemotherapy of infectious diseases came only after their etiology was discovered in the late nineteenth century. Paul Ehrlich's work in the early years of the twentieth century laid the foundation for the systematic chemotherapeutic approach to microbial diseases. His experiments with arsenical drugs (number 606 in a series of compounds tested was the most successful and famous) in the treatment of syphilis led to his

recognition of the selective activity of drugs on microorganisms as opposed to host cells with which they are associated in disease. He also encountered the development of resistance to drugs by microbial cells and explored means for countering this with combinations of compounds acting in different ways on microbial components. The next important impetus to the chemotherapy of infection came in the 1930s with the discovery that a family of simple organic compounds known as sulfonamides possessed antimicrobial usefulness. In 1940, the demonstration of penicillin, an antibacterial substance produced by the mold *Penicillium,* as a therapeutic agent of great practical value opened a new age in the management of infectious diseases.

The dramatic success of penicillin in the treatment of infection led to the development of many other useful drugs derived from molds and other microorganisms. The term *antibiotic* was applied to those substances, synthesized by microbes, that have an inhibitory or destructive action on other microorganisms. The word was meant to distinguish such agents from natural or synthesized chemical compounds, such as the sulfonamides, used as antimicrobial drugs. In recent years, since the discovery of a successful method for chemical synthesis of penicillin and for altering its molecule to make it more effective against some organisms, the term has lost some of its original distinction, though we still use it.

Although enormous numbers of antibiotics have now been discovered (most of them are produced by fungi and other soil organisms), only a relative few have proved to be safe and effective in chemotherapy. Many are too damaging to host tissues in concentrations that are effective against microorganisms, whereas others are too limited in their antimicrobial range of action. As with other classes of drugs, the criteria for selection of antimicrobial agents to be used in chemotherapy are (1) low toxicity for host tissues; (2) lack of allergenic properties giving rise to host hypersensitivity; (3) a broad spectrum of activity against many different types of microorganisms; (4) cidal rather than static action on microorganisms; and (5) antimicrobial effects to which microorganisms do not display easy adaptation with development of resistance. Not all these properties are displayed by every useful antibiotic, but the best possible combination is sought in the purification and preparation of such agents, which must meet minimum regulatory standards.

Mode and Range of Action of Antimicrobial Drugs. The *mode* of action of antimicrobial agents is through chemical interference with the function or synthesis of vital microbial cell components. This constitutes one basis for the classification of such agents, as shown in Table 10–5. Antibiotics may act by inhibiting one or more of the steps involved in the synthesis of bacterial cell walls (Fig. 10–10), or they may interfere with the permeability of the cell membrane. It will be recalled (Chap. 2) that bacterial ribosomes mediate the synthesis of cellular proteins. Some antibiotics exert their effects because they attach to portions of the ribosome molecule, causing a misreading of the messenger RNA code for protein synthesis. Certain antimicrobial drugs act by interfering with nucleic acid metabolism, and others are inhibitory because they become substitutes, or analogues, for compounds normally used by bacterial cells in synthesis or enzyme function. Antimicrobial agents that act as analogues are commonly called antimetabolites. The sulfonamides, for example, fall in this class because they act as analogues for a vitamin (para-aminobenzoic acid, or PABA) that many microorganisms require for synthesis of an enzyme, folic acid, involved in formation of nuclear bases (e.g., purine and pyrimidine). Substitution of a sulfonamide for PABA interferes with the construction, by the cell, of properly functional folic acid and therefore prevents further growth (Fig. 10–11). Bacterial cells that do not require folic acid for their growth or that can use preformed folic acid are not inhibited by the sulfonamides.

To be useful in chemotherapy, an antimicrobial agent must exert its effects on the pathogenic microorganism causing the disease without seriously damaging host tissues. The toxicity of a particular drug for host cells as well as microbial cells depends in part on whether or not they share the same kinds of components affected by the drug, or whether a given component has the same function in both. For example, some drugs that interfere with nucleic acid metabolism, such as actinomycins, are not selective in this and are too toxic for host cells to be useful in chemotherapy. On the other hand, penicillin, which acts on bacterial cell walls, has little or no toxicity for human or animal cells. Since mammalian cells do not synthesize walls, this may be one reason that penicillin is not toxic for them. The difference in composition of the walls of Gram-positive and Gram-negative bacteria may also account for the greater activity of penicillin against the former than against the latter.

Another classification of chemotherapeutic drugs used for infectious diseases is based on their clinical value with regard to their *range* or *spectrum* of action against various microorganisms. Thus, some agents such as penicillin may have a narrow spectrum, affecting Gram-positive bacteria primarily, yet still be the drugs of choice for most Gram-positive infections. The tetracyclines and chloramphenicol, how-

Table 10–5. Mode of Action of Antimicrobial Drugs

Mechanism	Drugs
Inhibition of synthesis of bacterial cell walls	Penicillins Cephalosporins Cycloserine Vancomycin Ristocetin Bacitracin
Affect permeability of microbial cell membranes	Polymyxins Colistimethate Nystatin Amphotericin
Affect ribosomal protein synthesis	Chloramphenicol Tetracyclines Gentamicin Kanamycin Streptomycin Neomycin Erythromycin Lincomycin Clindamycin
Interfere with nucleic acid metabolism	Rifampin Nalidixic acid
Antimetabolites	Sulfonamides Trimethoprim Aminosalicylic acid Sulfones

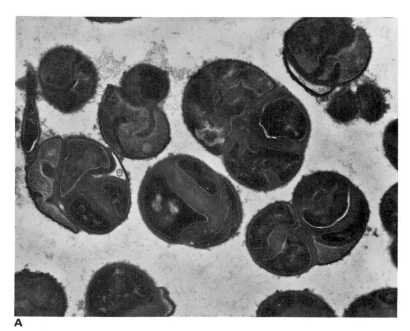

A

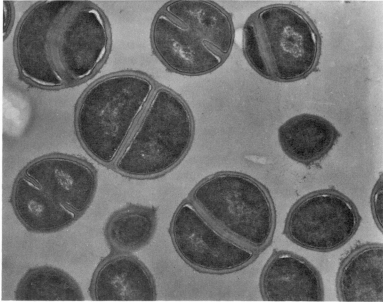

B

Fig. 10–10. The effect of penicillin on susceptible staphylococci. *A.* Staphylococci grown on agar containing one-third of the minimum inhibitory concentration (MIC) of penicillin show irregular shapes and numerous thick cross walls. *B.* The controls are staphylococci grown on nutrient agar without penicillin. Compare the normal cross-walls in the controls with those in experiment A. (Reproduced from Lorian, Victor, and Atkinson, Barbara: Some Effects of Subinhibitory Concentrations of Penicillin on the Structure and Division of Staphylococci. *Antimicrob. Agents Chemother.*, **7**:864, 1975.)

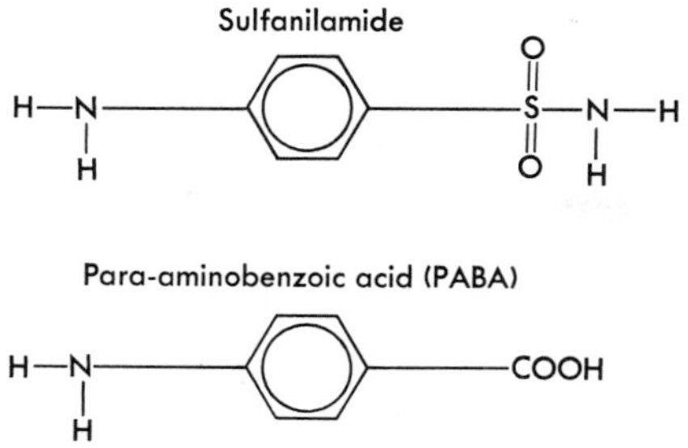

Sulfanilamide

Para-aminobenzoic acid (PABA)

Fig. 10–11. Chemical structures of sulfanilamide and para-aminobenzoic acid (PABA). Note the close similarity that makes it possible for the drug to substitute for the vitamin in bacterial syntheses, thus interfering with cellular metabolism.

ever, are examples of drugs with a broad spectrum of activity against both Gram-positive and Gram-negative organisms as well as rickettsiae. This clinically useful classification of antimicrobial drugs is shown in Table 10–6.

Today, many thousands of antimicrobial agents are known, some from natural sources, some synthesized in the laboratory, and many derived semisynthetically from a natural compound. Often when the active portion of an antibiotic molecule has been isolated and purified chemically, it has been possible to add side chains to the molecular nucleus, thus changing its spectrum or mode of action. In this way, semisynthetic compounds have been produced with greater, or different, antibacterial activity than that of the parent substance. The "synthetic" penicillins have been very important in this regard because they enlarge the clinical usefulness of this antibiotic. Penicillin and its earlier derivative, penicillin G, are themselves destroyed by the action of an enzyme produced by many staphylococci and some Gram-negative bacilli. This enzyme is called *penicillinase*. It acts by splitting the penicillin molecule, thus rendering it inactive. Penicillinase production is responsible for much of the resistance to this drug seen among many bacterial strains. Some synthetic penicillin molecules (oxacillin, methicillin, and cloxacillin) are not destroyed by penicillinase, because they have a slightly different chemical structure. They can be used effectively, therefore, to treat infections caused by penicillinase-producing staphylo-

cocci and other organisms. Ampicillin is a synthetic penicillin that is destroyed by penicillinase activity, but it has a broader spectrum of activity against Gram-negative bacteria than the original antibiotic (Fig. 10–12).

The portion of the penicillin molecule that is split by penicillinase is called a beta-lactam ring, and the enzyme is also often referred to as beta-lactamase, or simply lactamase. This same ring formation occurs in the cephalosporins, a group of drugs chemically similar to the penicillins, with a similar mode of action. Chemical modifications of the parent molecule have produced many modified cephalosporins, some of which are resistant to lactamase and therefore more useful clinically against microorganisms possessing this enzyme (Fig. 10–13).

Problems in Chemotherapy. The principal result of antimicrobial drug action is to slow the rate of growth of young, actively multiplying microbial cells. The use of such agents is not necessarily curative of itself, however. Generally, the desired effect is

Fig. 10–12. Two derivatives of penicillin. Substitution of different side chains on the basic part of the molecule at the point shown produces different penicillins with altered activity.

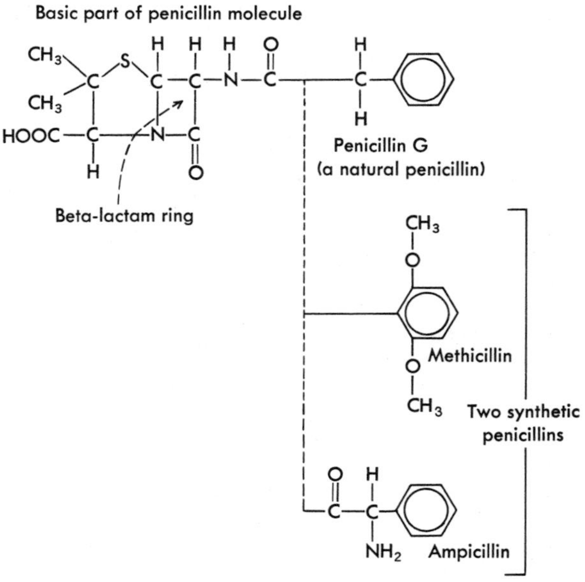

Basic part of penicillin molecule

Penicillin G (a natural penicillin)

Beta-lactam ring

Methicillin

Two synthetic penicillins

Ampicillin

Table 10–6. Choice of Antimicrobial Drugs for Therapy*

Causative Organism	Disease	Drug(s) of First Choice	Alternative Drugs
Gram-negative cocci			
Gonococcus	Gonorrhea	Penicillin	Spectinomycin, tetracycline
Meningococcus	Meningitis	Penicillin	Chloramphenicol
Gram-positive cocci			
Pneumococcus	Pneumonia	Penicillin	
Streptococcus			
Viridans group	Endocarditis	Penicillin	
Hemolytic group A	Scarlet fever,	Penicillin	Erythromycin,
	puerperal fever,	Penicillin	cephalosporin
	erysipelas	Penicillin	
Other Strep.		Penicillin	
Staphylococcus	Furuncles,	Penicillin	
Nonpenicillinase	abscesses,	Penicillin	
producing	pneumonia,	Penicillin	
	meningitis,	Penicillin	
Penicillinase	osteomyelitis,	Oxacillin,	Clindamycin,
producing	bacteremia	methicillin	vancomycin
Enterococcus	Endocarditis	Penicillin or	Vancomycin +
(fecal strep)		ampicillin +	gentamicin
		gentamicin	
Gram-positive bacilli			
Actinomyces	Actinomycosis	Penicillin	Tetracycline
B. anthracis	Anthrax	Penicillin	Erythromycin
Clostridium	Tetanus,	Penicillin	Tetracycline,
	gas gangrene	Penicillin	erythromycin
C. diphtheriae	Diphtheria	Erythromycin	Penicillin
Nocardia	Nocardiosis	Sulfonamide	Tetracycline (minocycline)
Gram-negative bacilli			
Enterobacter	Urinary infection	Gentamicin	Chloramphenicol, tetracycline
Bacteroides fragilis	Wound	Clindamycin	Chloramphenicol
other species	infections	Penicillin	Tetracycline, clindamycin
Brucella	Brucellosis	Tetracycline + streptomycin	Chloramphenicol + streptomycin
E. coli	Sepsis	Ampicillin + gentamicin	Cephalosporin + gentamicin
Haemophilus	Meningitis Pneumonia	Ampicillin (if susceptible)	Chloramphenicol
Klebsiella	Pneumonia	Cephalosporin + gentamicin	
	GU infection	Cephalosporin	Gentamicin

Table 10–6. (*Cont.*)

Causative Organism	Disease	Drug(s) of First Choice	Alternative Drugs
Gram-negative bacilli (*Cont.*)			
Yersinia	Plague	Streptomycin	Tetracycline
Francisella	Tularemia		
Proteus	Urinary tract infections		
mirabilis		Ampicillin	Cephalosporin
vulgaris		Gentamicin	Carbenicillin
Pseudomonas			
aeruginosa	Urinary infection	Carbenicillin	Gentamicin, colistimethate
pseudomallei	Meliodosis	Tetracycline	Chloramphenicol
Salmonella	Typhoid fever	Chloramphenicol, ampicillin	Trimethoprim-sulfamethoxazole
	Paratyphoid		
	Enteritis		
Shigella	Dysentery	Ampicillin	Trimethoprim-sulfamethoxazole
Vibrio cholerae	Cholera	Tetracycline	Trimethoprim-sulfamethoxazole
Acid-fast bacilli			
Mycobacterium			
tuberculosis	Tuberculosis	INH + ethambutol	Rifampin
leprae	Leprosy	Dapsone	Rifampin
Nocardia	Nocardiosis	Sulfonamide	Tetracycline (minocycline)
Spirochetes			
Borrelia	Relapsing fever	Tetracycline	Penicillin
Leptospira	Leptospirosis	Penicillin	Tetracycline
Treponema	Syphilis	Penicillin	Erythromycin
Mycoplasma	Pneumonia	Tetracycline	Erythromycin
Rickettsiae	Typhus fever and others	Tetracycline	Chloramphenicol
Chlamydiae			
Psittacosis	Pneumonia	Tetracycline, sulfonamide	Erythromycin, chloramphenicol
Lymphogranuloma	Venereal infection		
Trachoma	Eye infection		

*Adapted from *Review of Medical Microbiology,* 12th ed., by Ernest Jawetz, Joseph L. Melnick, and Edward A. Adelberg. Lange Publications, Los Altos, Calif., 1976.
See also, *The Medical Letter* (56 Harrison St., New Rochelle, N.Y.), "Handbook of Antimicrobial Therapy," 1976.

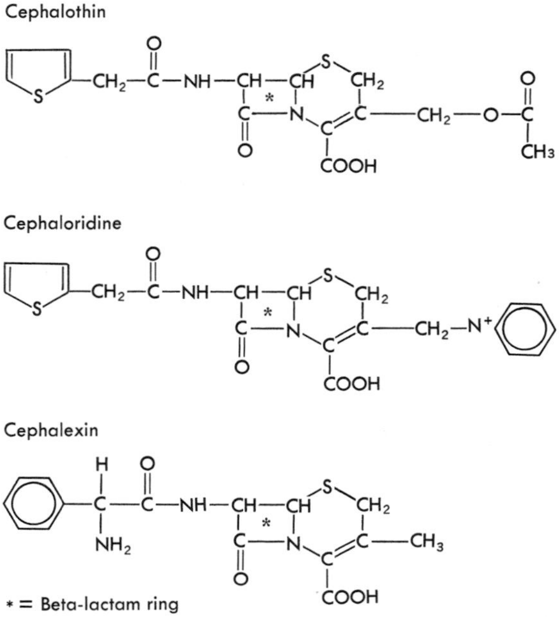

Cephalothin

Cephaloridine

Cephalexin

* = Beta-lactam ring

Fig. 10–13. Chemical structures of some cephalosporins.

to suppress the invading microorganisms sufficiently, or long enough, until the body's nonspecific and specific resistance mechanisms can intervene and ensure complete recovery. Although the dramatic success of chemotherapy has reduced the incidence, morbidity, and mortality of many infectious diseases, certain problems must be emphasized, as discussed below.

Drug toxicity for the host may seriously interfere with the drug's value in therapy. Some drugs, such as penicillin, have few toxic effects and may be given in concentrations and quantities larger than the effective antimicrobial level required without inducing tissue damage. Others must be carefully adjusted in dosage, because their effective concentrations are not much different from their toxic concentrations, allowing little leeway. Different antimicrobial drugs may have characteristic toxicities for certain human tissues: the tetracyclines and other broad-spectrum antibiotics given orally may be irritating to the intes-

tinal mucosa, though this may be partly due to their effect on normal intestinal bacteria; gentamicin can induce renal damage and also injury to the eighth cranial nerve resulting in partial or complete deafness; chloramphenicol in prolonged, high dosage may cause depression of bone marrow with aplastic anemia, or other blood dyscrasias; the sulfonamides may produce renal injury because they are insoluble and tend to precipitate out when excreted through the kidneys. These are some of the reasons that chemotherapeutic drugs should be issued only by a physician's order, prescribed with discriminating care, and taken only under his or her supervision.

Drug allergies, or ***hypersensitivities,*** may be frequent complications with antimicrobial compounds that have some antigenic properties. For example, gentamicin, vancomycin, and penicillin not infrequently produce allergic skin rashes. Allergic effects are the result of previous sensitization of the patient with the drug; the timing and severity of the reaction depend on the route of administration of a repeated dose, the extent of the hypersensitive mechanism, and the tissues involved by it (see Chap. 7). Penicillin, although it is nontoxic, has a frightening allergenic potential for hypersensitized individuals. Reactions among the latter range from skin rashes (manifested as a phase of the serum-sickness type of immediate response) to anaphylaxis, which may be fatal. These hypersensitivities afford another important reason for caution in the use of antimicrobial therapy. Treatment of insignificant infections with antibiotics should be discouraged, particularly if the value of the drug's effect on the specific infectious agent being treated has not been established by laboratory identification of the organism and demonstration of its susceptibility to the drug as a confirming test of relevance.

Development of microbial resistance to drugs presents another problem in chemotherapy. The genetic basis for the appearance of antibiotic-resistant mutants in susceptible bacterial populations was discussed previously. (See Chap. 3.) Continuous use of drugs, particularly in closed environments such as hospitals, tends to suppress susceptible microorganisms and foster the survival of resistant strains, so that the latter become prevalent. This situation is further encouraged by the close personal contacts constantly occurring between hospital patients and personnel who exchange their resistant strains directly or indirectly through the immediate environment. This is the reason for the increase and persistence of drug-resistant strains of staphylococci and many species of Gram-negative enteric bacilli indigenous to the human body. When resistant bacterial strains cause serious infection, the problem of finding adequate means of chemotherapy is proportionately severe.

Combined therapy is sometimes useful in preventing the emergence of resistance in a bacterial strain. A combination of two antibiotics that act by different mechanisms on the causative organisms of the infection to be treated, but do not interfere with each other, may act synergistically to prevent the appearance of resistant mutants.

Changes in the normal flora of the body frequently occur as a result of antibiotic therapy. The interbalance maintained among the microorganisms of the skin and mucous membranes by their own competitive activities may be seriously disturbed if one or more species of drug-susceptible microorganisms of the normal flora are suppressed during the course of antibiotic treatment of an infection. Under these circumstances, drug-resistant organisms may multiply to an extent that becomes injurious, and a new and different infection may begin. These secondary, complicating infections of antibiotic treatment, or *superinfections* as they are also called, occur most frequently on the mucosal surfaces of the mouth and throat, the intestinal tract, and the genitourinary system, but they may extend to systemic tissues as well. Strains of yeast (notably *Candida albicans*), staphylococci, and enteric bacteria are most usually involved and are particularly difficult to treat because of their drug resistance. Antimycotic drugs, such as nystatin, are frequently administered together with the antibiotic of choice when the primary infection is treated, to prevent overgrowth of yeast if the normal balance is upset during therapy.

Surgical Treatment of Infection. When an infectious process is well localized by the inflammatory response of tissues, and encapsulated by fibrotic reaction as in abscess formation, the chemotherapeutic approach may not be helpful without the physical relief of surgery. This is because the cellular barriers thrown up against the causative microorganism may also preclude penetration of the area by an antibiotic in concentrations that are effectively bactericidal, or even bacteriostatic. Under these circumstances surgical excision of the lesion may be indicated, if it lies in an accessible region of the body and does not offer severe anatomic or functional problems to the surrounding tissues. Antibiotics given at the same time serve as a "cover" to prevent new localizations of organisms scattered by surgical manipulation.

Role of the Microbiology Laboratory in Chemotherapy. The diagnostic microbiology laboratory can assist the clinician in selecting effective chemotherapeutic drugs for treatment of infectious disease. The causative organism must first be isolated in pure culture, and then tested for its response to a number of antibiotics, usually by the paper disk method. If the results of treatment with a chosen drug are not satisfactory, the laboratory may test the original organism again by more exacting methods to find a more useful drug or to indicate a more effective dosage of the first one. It may also test fresh specimens from the patient to determine whether the infectious agent has become drug-resistant or whether a superinfection has arisen. Assays of the patient's serum, urine, or other body fluids for antibiotic content may also be useful in guiding the physician in his choice of dosage or route of administration of the drug.

Laboratory methods for antibiotic susceptibility testing are described in Chapter 4.

Immunotherapy of Infectious Disease. Prior to the advent of antibiotics, antisera were frequently used in treatment of certain acute infections, notably pneumococcal pneumonia and bacterial meningitis. Today, however, effective chemotherapy and large-scale active immunization programs have markedly reduced the incidence and serious consequences of such diseases. Passive immunization as a therapeutic measure is now largely limited to those few situations in which bacterial infection is accompanied by the production of highly injurious or lethal toxins. These include diphtheria and the clostridial diseases such as tetanus, gas gangrene, and botulism. Ingested toxin in botulism, or toxin produced by bacteria growing in the tissues in the other instances cited, is not affected by antibiotic drugs. In these situations specific antitoxin given promptly on diagnosis of developing infection or toxemia prevents irreparable toxin damage to tissues. The concurrent administration of antibiotics inhibits further growth — and toxin production — of the invading organisms.

Treatment of the Patient's Contacts: Prophylaxis

Chemoprophylaxis. In general, the use of antimicrobial drugs is not indicated as a means of preventing spread of infection among the contacts of an active case. Such a procedure tends to encourage the emergence of resistance among microorganisms that are responsive to such drugs, and it has no effect at all on viruses and other groups of organisms indifferent to drug action.

There are two notable exceptions to this general rule, however. The first is in the case of *Neisseria* infections, that is, gonococcal or meningococcal diseases. Gonococcal infection (gonorrhea) is venereal in its spread and highly communicable by this route. Meningococci are easily transmissible under crowded conditions by droplet spread from the upper respiratory tract. Severe meningococcal disease (meningitis or bacteremia) can reach epidemic proportions when persons live in close contact (e.g., military recruits). Because the consequences of these diseases are serious when unchecked, and because both these species of *Neisseria* are susceptible to antimicrobial drugs, in certain circumstances chemoprophylaxis of the contacts of known infectious cases is an important means of control. Gonorrheal cases and contacts of patients with proven gonococcal disease are examined for infection and treated with penicillin. Newborn babies are also susceptible

contacts of mothers who may have gonorrheal infection. The prophylactic measure in this case is treatment of the eyes of babies at birth with silver nitrate solution to prevent conjunctival infection with *N. gonorrhoeae*. Persons living in close contact with patients who have meningococcal disease are treated with either sulfonamides or rifampin, depending on the response of the patient's organism to these drugs. Many strains are resistant to sulfonamides because of their previous widespread prophylactic use. Penicillin is very effective in systemic meningococcal diseases but cannot eliminate the carrier state. This is because it is not concentrated in the oropharyngeal secretions sufficiently to inhibit organisms present in the throat. For these reasons, rifampin is the most reliable prophylactic agent for meningococcal contacts.

The *second* exception in chemoprophylaxis is in the use of drugs to suppress or prevent *malarial* infection. In this instance, the infectious source is not the case of human disease but a mosquito reservoir. In areas of the world where infective mosquito vectors abound, human exposure is frequent and sometimes unavoidable. A family of drugs related to or derived from quinine (chloroquine, amodiaquine, primaquine) is effective both in treating malaria and in preventing or aborting its development in the exposed individual.

Immunoprophylaxis. For many infectious diseases the most effective means of control is mass active immunization of the population. Even when this has been accomplished sporadic cases of these diseases may occur from time to time (poliomyelitis, diphtheria, pertussis) and require immediate consideration of the immunity of exposed contacts. Renewed active immunization with vaccines may be indicated to boost lagging immunity or to protect new susceptible members of the community. When exposure to tetanus infection threatens a nonimmune individual, passive immunization with antitoxin may be necessary to prevent disease, but active immunization with toxoid should then follow. (See Chap. 7.)

The use of human gamma globulin as a source of antibodies is indicated following exposure of susceptible individuals for whom the risk of certain diseases is great. This might be true if the disease is frequently accompanied by serious consequences (poliomyelitis, hepatitis, mumps in adults) and active immunity has waned, or if no active artificial immunization is available. The prophylactic use of gamma globulin may also be advisable for individuals to whom any infectious disease might represent unusual risk, such as the very young, the very old, or others whose resistance is seriously compromised by medical or surgical problems.

Questions

1. Define sterilization; asepsis; disinfection; antisepsis; germicide; bactericide; bacteriostasis; antibiotic; sanitization.
2. Define contamination; infection; infectious disease; nosocomial infection; communicable disease; communicable period.
3. Describe the action of heat on microorganisms.
4. When is the sterilizing temperature in an autoclave reached and how is it maintained?
5. What factors are necessary to ensure sterilization of articles in an autoclave?
6. How do chemicals kill vegetative organisms?
7. What qualities should be sought when selecting a disinfectant?
8. List several barriers used to control direct transmission of infectious diseases in hospitals.
9. What are the problems in the use of chemotherapy?

Additional Readings

*Benenson, Abram S. (ed.): *Control of Communicable Diseases in Man,* 12th ed. American Public Health Association, Washington, D.C., 1975.

Boucher, R. M. G.: Advances in Sterilization Techniques — State of the Art and Recent Breakthroughs. *Am. J. Hosp. Pharm.,* **29**:662, Aug. 1972.

*Brock, Thomas D. (ed.): *Milestones in Microbiology.* American Society for Microbiology, Washington, D.C., 1975.

Ethylene Oxide Sterilization, A Guide for Hospital Personnel. Released by the Association for Advancement of Medical Instrumentation Subcommittee on Ethylene Oxide Sterilization, Smith and Underwood, 1023 Troy Ct., Troy, Mich., 1976.

Evaluation of Useful Antimicrobial Chemicals. *Am. Oper. Room Nurs. J.,* **23**:1084, May 1976.

Fisher, E. J.: Surveillance and Management of Hospital Acquired Infections. *Heart Lung,* **5**:788, Sept.–Oct. 1976.

Fox, Marian K.; Langner, S. B.; and Wells, R. W.: How Good Are Hand Washing Practices? *Am. J. Nurs.,* **74**:1676, Sept. 1974.

Garner, Julia S., and Kaiser, A. B.: How Often Is Isolation Needed? *Am. J. Nurs.,* **72**:733, Apr. 1972.

Halleck, F. E.: Hazards of EO Sterilization in Hospitals, *Hosp. Topics,* **53**:45, Nov.–Dec. 1975.

Infection Control in the Hospital, 3rd ed. American Hospital Association, Chicago, 1974.

Isolation Techniques for Use in Hospitals, 2nd ed. U.S. Department of Health, Education, and Welfare, Public Health Service (DHEW Pub. No 76-8314), 1975.

Jenny, Jean: What You Should be Doing About Infection Control. *Nursing 76,* **6**:78, Nov. 1976.

*Kunin, Calvin N. (ed.-in-chief): *Hospital Sepsis.* MEDCOM Inc., Travenol Laboratories, Deerfield, Ill., 1972.

Mizer, Helen E.: The Staphylococcus Problem Versus the Hexachlorophene Dilemma in Hospital Nurseries. *J. Obstet. Gynecol. Neo. Nurs.,* **2**:31, Mar.–Apr. 1973.

Morello, Josephine A.: Assays of Antimicrobial Agents in Serum. *Lab. Med.,* **7**:30, Apr. 1976.

Perkins, John J.: *Principles and Methods of Sterilization,* 2nd ed. Charles C Thomas, Publisher, Springfield, Ill., 1973.

*Available in paperback.

Sanford, P. J.: Disinfectants That Don't. *Ann. Intern. Med.,* **72**:282, 1970.

Speller, D. C. E.; Stephens, M.; Raghunath, D.; Viant, A. C.; Reeves, D. S.; Broughall, J. M.; Wilkinson, P. J.; and Holt, H. A.: Epidemic Infection by a Gentamicin Resistant *Staphylococcus aureus* in Three Hospitals. *Lancet,* **1**:464, Feb. 28, 1976.

Steere, Allen C., and Mallison, George F.: Handwashing Practices for the Prevention of Nosocomial Infections. *Ann. Intern. Med.,* **83**:683, 1975.

Symposium on Infection and the Nurse. *Nurs. Clin. North Am.,* **5**:85, Mar. 1970.

Wenzel, Kathryn: The Role of the Infection Control Nurse. *Nurs. Clin. North Am.,* **5**:89, Mar. 1970.

Wilson, Robert: A Brief Introduction to Sepsis: Its Importance and Some Historical Notes. *Heart Lung,* **5**:393, May–June 1976.

Communicable Disease Control in the Community

11

Public Health Agencies and Services

The organization of federal, state, and city health departments in the United States had a slow and difficult history prior to this century. Until that time, the structure, authority, and functional services of government-directed health agencies were often insecure, because medical authority was itself uncertain about the direction or value of methods to be applied on a public scale, and for lack of political or financial support. A new impetus came at the turn of the century, due to findings in the laboratories of Pasteur in France, and of Koch in Germany, on the nature and transmission of infectious diseases. Young American medical graduates who traveled to Europe for their postgraduate training in the 1890s and 1900s came home with news of some revolutionary answers to old problems of epidemics and their possible control. Since that time, governments around the world have taken increasing responsibility in applying controls on many diseases and in offering other supportive health services as well.

In the United States today, health facilities are provided by government at every level. Since 1953, most federal agencies involved in public health affairs have been coordinated within the Department of Health, Education, and Welfare (DHEW), this organization having cabinet status. The United States Public Health Service, whose origins go back

to the year 1798 when the Marine Hospital Service was founded for American Navy men, now functions under DHEW administration. It conducts research in the field of preventive medicine, provides hospital facilities for servicemen, and gives financial as well as other practical assistance to state and local health departments.

State Health Departments. In the United States, state health departments are autonomous in their authority over health regulations and services within their own geographic areas. Many cities also assume local responsibility for health problems. The New York City Department of Health is one of the oldest of such organizations. It was established in 1866 and became the scene of action of some of the most vigorous and successful battles fought in the United States against such public health problems as diphtheria, smallpox, cholera, and venereal disease. It was the work of local community health departments in the cities and towns that first led to the development of public health services and administration in this country. The Public Health Service now functions as an advisory leader in this field, offering assistance to the states when emergencies or special problems arise or new programs are to be developed.

International Health Organizations. There are several organizations that operate on an international level in the protection of public health, the most

271

important of these being the World Health Organization, with headquarters in Switzerland. The services of the World Health Organization, supported by many nations of the world, include distribution of technical information and assistance, standardization of drugs, and development of international regulations important to control of epidemic diseases. In many countries, WHO experts have helped governments to develop programs for control or eradication of diseases such as smallpox, malaria, yaws, and trachoma. With the achievement of increasing control over the epidemic spread of severe infections, and over the endemic incidence of others, improved standards of public health have contributed greatly to the advancing development of the world's resources.

Voluntary Health Organizations. The development of preventive medicine and public health programs has been furthered in the United States by the outstanding efforts of some privately endowed voluntary organizations, such as the Commonwealth, Ford, and Rockefeller Foundations. Through the provision of research facilities, money, and technical assistance, the great foundations have been influential in promoting medical progress in every part of the world. Other voluntary health agencies, large and small, have been an integral part of this advance. The American Red Cross, the Catholic Sisters of Charity, the American Lung Association, and the American Public Health Association are only a few examples of the many voluntary organizations that now function on a national scale but had their beginnings in local communities struggling with individual health problems. It was at this level also that public health nursing first emerged as a community service.

Public Health Nursing. The first formal nursing training in the United States was given at the New York Hospital in 1798. The first visiting nurse service was provided a few years later, in 1813, in Charleston, South Carolina. An epidemic of yellow fever in that city spurred the Ladies' Benevolent Society to organize a nursing service for the home care of the sick. By 1862, the year Florence Nightingale established a district nursing association in England, the Civil War in the United States had created an urgent need for the medical and nursing care of sick and wounded soldiers. The Women's Central Relief Association, later incorporated into the Sanitary Commission, was organized to provide nursing, and Miss Nightingale herself was an advisor to this group. The Catholic Sisters of Charity also gave invaluable service as volunteer nurses during that war, moving onto the battlefields in the wake of the fighting to find and minister to the wounded. The eventual establishment of public health nursing as a professional practice dates from the late 1800s in New York. The first strong foundations were established by the work of the New York City Mission in providing trained nurses to care for the sick in their own homes, the establishment of the Henry Street District Nursing Service by Lillian Wald and Mary Brewster (Miss Wald is considered the founder of public health nursing in the United States), and the sponsorship of community nursing projects by the New York City Department of Health. Public health nursing as we know it today is an essential function of local, state, and federal health services, with an international role to play as well.

Public Health Controls of Environmental Reservoirs of Infection

The major environmental reservoirs of infectious disease in the public domain are water supplies, sewage, and food, including milk. The sanitary standards and regulations that have been developed during this century for control of the purity of water and food resources have effectively curbed these sources of infection and, as a result, the incidence and epidemic patterns of many communicable diseases, once so prevalent and devastating, have been significantly reduced.

Water and Sewage Sanitation

The principal waterborne diseases are typhoid fever and other salmonelloses, bacillary and amebic dysentery, and cholera. Viruses excreted in feces, notably the agents of infectious hepatitis and poliomyelitis, may also be spread through contaminated water, as may some helminth ova. Fecal contamination of water supplies by man and animals constitutes the major infectious hazard and the main necessity for water purification (see Chap. 9).

In the days before the development of methods to safeguard public water supplies, epidemics of waterborne intestinal disease occurred frequently. Usually they were of short duration, beginning explosively among people sharing a contaminated water supply. Pathogenic organisms do not live long in water, or multiply there. Therefore, epidemics of waterborne infection are usually characterized by the rapid appearance of large numbers of cases (the number depending on the extent to which a water supply is shared) and an almost equally precipitous fall in incidence. In some instances, continuous undetected contamination of water from infectious sewage may cause prolonged epidemics. An epidemic of cholera that occurred in a London parish in the year 1854 provides a dramatic example. During a ten-day period in the autumn of that year, more than 500 people fell ill of cholera and died of the disease. There was a great deal of cholera elsewhere in the city at that time, but these cases were concentrated in one neighborhood. As the epidemic continued, two men named John Snow and John York began a study of the area and were able to prove, by epidemiologic methods only (the bacteriologic nature of the disease was not understood at that time), that the outbreak stemmed from the water of a particular well, known as the Broad Street well. It was also discovered that sewage from the cesspool of a house on Broad Street was the source of pollution of the well, and that there had been a case of undiagnosed intestinal disorder in that house shortly before the cholera outbreak began. Snow traced many of the cases to the use of water from the Broad Street well and also explained the failure of the disease to involve people who lived in the same neighborhood but did not drink from the polluted well (Fig. 11–1).

The incidence of cholera, typhoid fever, and other intestinal infections rarely reaches epidemic proportions today in those countries or communities that have established firm sanitary standards for protection of water and disposal of sewage. These diseases persist in endemic form, however, in many parts of the world. Asymptomatic carriers of intestinal pathogens persist in all populations and are a possible

source of new epidemics at times when there is a break in sanitation methods or systems. Such breaks may be minor, involving one well or one city apartment house, but when major disasters disrupt the functions of an entire city or a large area, the danger of large epidemics becomes immediate. Earthquakes and floods often pose this problem, as does the damage of war.

Sources of Public Water Supplies. Community water supplies are usually obtained today from surface water, dammed in reservoirs that collect it from brooks, streams, and rivers. These are fed by rainfall, or by runoff of excess rain from the ground. Water runs over large areas to its accumulation point in reservoirs and is subject to pollution along much of its course, so that it must be purified before delivery to the community. Local governments find it increasingly difficult to restrict industrial use or population of territory involving their watersheds, and neighboring communities frequently create problems for each other in pollution of shared water supplies. An "upstream" town, for example, may secure clean water from its river as it passes and use the same river, at a lower point, for sewage disposal. Other towns, located downstream from this point, are then subject to hazards of water contamination. Chemical pollution of rivers by industrial wastes similarly affects many communities sharing a common water supply. The need for firmer state and local government controls on the sources of water pollution is now widely recognized.

Ground water accumulations from rain and snow can be tapped by well digging for individual home use. The safety of well water depends on its protection from surface contamination or from the sewage disposal system employed. Wells must be deep enough to assure good filtration of water seeping from the surface. The location of outdoor latrines, septic tanks, or ducts that run sewage off to a distant collection point should be planned with regard to the protection of wells and their water sources.

Water Purification. To prepare it for consumption, water is pumped from reservoirs or rivers into purification plants (Fig. 11–2). It is first screened for large debris, such as leaves, twigs, or dead fish, then

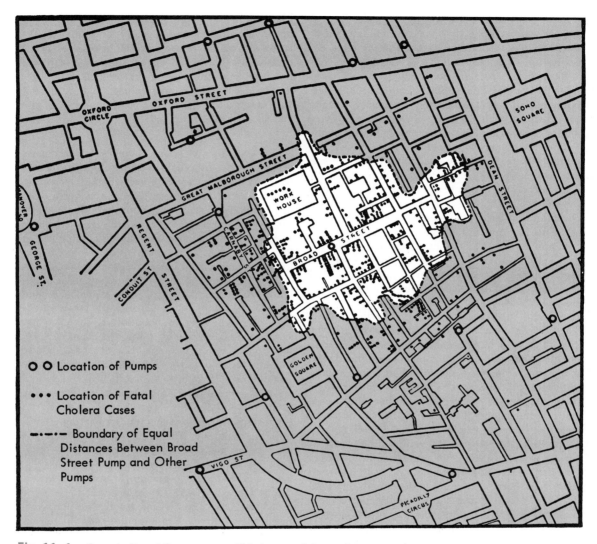

Fig. 11–1. Snow's Broad Street pump. This is one of the earliest examples of disease mapping. By marking each new case of cholera on a map, Dr. John Snow was able to prove that the Broad Street pump was a central point in the London cholera epidemic of 1854. This map-making also helped to establish the fact that cholera is a waterborne disease. (Reproduced from *Spectrum*. Courtesy of Chas. Pfizer, Inc., New York, N.Y.)

pumped into storage reservoirs where it is subjected to the following processes:

1. Aeration increases the oxygen content of the water and speeds oxidative utilization of organic material by bacteria. It also displaces other dissolved gases that contribute unpleasant odors and tastes to polluted water.

2. Coagulation of many soluble chemical contam-

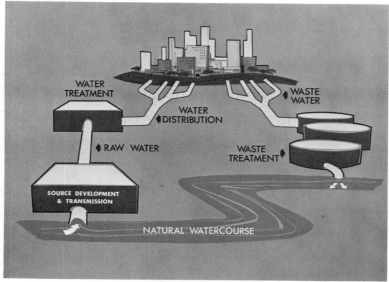

Fig. 11–2. Water drawn from a natural watercourse must be treated in a purification plant to prepare it for consumption, because upstream cities and towns may have polluted it with their wastes. Sewage and waste water should be treated before return to the watercourse to avoid damage to indigenous aquatic life and to protect downstream cities from dangerous pollution. (Courtesy of Center for Disease Control, D.H.E.W., Atlanta, Ga.)

inants is accomplished by adding compounds such as aluminum sulfate. Colloidal complexes are flocculated and eventually settle out of solution.

3. Sedimentation of coagulated or insoluble materials removes remaining coarse contamination.

4. Filtration removes fine particles, including bacteria. Water seeping down through filter beds passes first through slow sand filters, closely packed, then through looser rapid sand filters into coarse gravel beds. Below the gravel, successive layers of pebbles, larger rocks, and tile continue the filtration, delivering relatively pure water into collecting tanks. The upper sand layers of a filtration system accomplish most of the purification. About 98 percent of bacteria and most suspended solids are trapped there. Oxidative processes remove organic matter, and many bacteria are destroyed by the activities of protozoa. Approximately six million gallons of water are filtered per day through an acre of sand bed, a rather slow process. The filters must be cleaned from time to time by scraping off the top inch of sand and replacing it.

5. Chlorination is the final step. The bacterial action of chlorine (see Chap. 10), in a concentration

of 2 ppm, assures disinfection of filtered water and its safety along the piped route to the consumer. Chlorinated lime or sodium hypochlorite may be used as sources of chlorine, but the gaseous form is least expensive and the usual choice.

Sewage Disposal. Raw sewage, with its teeming content of potentially pathogenic intestinal microorganisms, is the most dangerous threat to water supplies. It may have disastrous consequences for fish and wildlife as well as for human beings.

Sewage disposal on a community scale involves its collection in a treatment center. Bulky trash is screened out, and the sewage is then stirred and agitated to permit aeration. Large particles are broken up and bacterial decomposition of organic material is promoted, As solid matter settles out (Fig. 11–3), the "activated" sludge that collects contains products of decomposition, which may be reduced in volume, dried, and sold for fertilizer. The liquid component drawn off from sludge is treated with chlorine and can be pumped off into a river or the ocean as harmless effluent (Fig. 11–2).

In a septic tank, sewage is decomposed largely by anaerobic activities of bacteria, while chemical methods are used for small collecting tanks, as in airplanes and trains.

Fig. 11–3. A primary settling tank for sewage. Before routing into this tank, sewage is first screened to remove large objects, then passed through a chamber where smaller particles are removed. In the tank, organic matter slowly settles out. This sediment is taken off as sludge, treated in holding tanks, and finally dried for use as a fertilizer (but this is not recommended for fertilizing plant foods that will later be eaten without cooking). (Reproduced from *Health News,* Oct. 1960. Courtesy of the New York State Department of Health, Albany, N.Y.)

Public Health Standards. Governmental responsibility for safe standards of community sanitation lies with public health services at local, state, and national levels. Authority is provided by laws or statutes specifying protection of water supplies, adequate disposal of sewage, housing sanitation, inspection of food, and other matters. Federal legislation controls interstate waterways and their use, with primary responsibility left to individual states concerned. In 1948 the Water Pollution Control Act established federal authority, and in 1961 and 1966 amendments to this act designated the Department of Health, Education, and Welfare as the major water resource agency of the government. The amendments extended federal authority to all navigable waters, including coastal waters; authorized federal enforcement procedures in cases of intrastate pollution; and permitted federal action against municipalities in violation of the law.

Bacteriologic Standards for Water. Drinking-water standards were first defined by the U.S. Public Health Service. These recommendations are used by many states as a guide for maintaining water supplies and adequate sewage disposal.*

Standard Methods for the Examination of Water and Waste Water, 14th ed., American Public Health Association, Washington, D.C., 1975.

Health department laboratories maintain a constant surveillance of community water supplies, testing them for chemical as well as bacteriologic purity. The "potable" character of water (that is, its suitability for drinking) is judged by its freedom from odor, taste, color, and harmful chemicals. It should be relatively free of nitrate and other decomposition products of organic matter.

The bacteriologic examination of water includes the enumeration of bacterial colonies growing out on agar plates, in a standard plate count, and tests for the presence of coliform bacteria. The latter provide an index of fecal pollution. Potable water should have a low total count of bacteria (0 to 10 per milliliter of water) and be free of coliform organisms.

A *presumptive test* for coliforms is made by inoculating a test sample of water into tubes of lactose broth containing an inverted tube to trap gas. After 24 hours of incubation, the tubes are examined for the presence of gas as an indication of lactose fermentation. Coliform bacteria ferment lactose rapidly, while other organisms that utilize this sugar require 48 hours or more to do so. The presence of gas is merely presumptive of coliforms, and the test must be confirmed.

The *confirmed test* is done by plating an inoculum from gas-positive lactose tubes onto eosin–methylene blue (EMB) agar, or Endo's medium. These agar media also contain lactose, together with dyes that inhibit nonenteric organisms and change color to indicate colonies that have utilized the lactose. Coliform colonies have a characteristic deep purple color on EMB, and their surfaces take on a metallic luster. The appearance of such colonies confirms the presumptive test.

The *completed test* requires final identification of characteristic colonies. Isolated colonies are inoculated into media that identify the biochemical properties of the coliform group. Stained smears of the organisms complete the proof, if these reveal Gram-negative, nonsporing bacilli.

A more rapid test for the isolation and identification of fecal coliforms has been devised and is gaining widespread usage. This method involves filtration of water through a cellulose acetate membrane filter. The filter is then incubated on a pad soaked in EMB or Endo's medium. Suspect colonies, if present, can be seen growing on the filter within 24 hours. The presence of fecal streptococci (or other intestinal organisms indicating excretory pollution of water) can also be detected by this method.

Standards for water usually refer to coliform counts on the rational assumption that as pollution with excrement increases, the possibility of contamination with pathogenic organisms also rises. When bacteriologic tests reveal unacceptable levels of fecal contamination, appropriate checks are made for possible sources of pollution or for failures in the purification system, with corrective action as indicated.

Chemical analyses are also essential in monitoring the quality of drinking water. Water that is microbiologically safe may not be potable because of chemical pollution with such substances as pesticides, detergents, metallic ions, alkalis, or similar contaminants.

Food Sanitation

Basic Control Measures. Strict cleanliness is important in all stages of food preparation, but must be coupled with adequate refrigeration to prevent spoilage bacteria or pathogens from growing. This is particularly important for foods that have been partially or fully cooked, and for frozen foods. Cooked foods provide better culture media for bacteria than many uncooked foods. Frozen foods are not sterilized by the freezing process, and when they thaw, the rupture of cells and fibers may make them more subject to bacterial growth. Proper cooking of food is necessary to ensure its safety from infectious organisms (or their toxins) that it may contain.

Public health regulations require inspection of food-processing plants, abattoirs, restaurants, dairies, and all other establishments involved in food preparation for adequacy of basic sanitation, including refrigeration methods. In addition, health service agencies at all levels conduct a continuous campaign to educate the general public in the sanitary principles involved.

Meat Inspection. Government controls on the safety of meat include its inspection for bacterial spoilage or other deterioration before it is distributed for retail sale (Fig. 11–4). More important, certain

Fig. 11–4. These government meat inspectors examine sides of beef at a commercial packing company. (Courtesy of U.S. Department of Agriculture, Washington, D.C.)

primary infections of meat can be detected by visual inspection for tissue lesions. This is particularly true for encysted tapeworm larvae in beef or pork muscle, which can be seen and sometimes felt as small, white, hard nodules in the tissue. Unfortunately, fish tapeworms are not so easily detected, nor are larval encystments of the roundworm *Trichinella*, which may infest pigs. Trichinosis remains prevalent in the United States and other countries for three essential reasons: pigs are fed on garbage containing uncooked meat that perpetuates the cycle of the parasite; pork infestation cannot be detected by inspection; and people accidentally or deliberately eat rare

meat or pork products that have not been fully cured or cooked. The only effective methods to control trichinosis lie in the hands of those who raise and feed pigs and those who prepare pork for the table (Fig. 9–11, p. 221).

Control of Milk-Borne Diseases. The important diseases that are transmissible by milk are tuberculosis, brucellosis, and a number of streptococcal infections, as discussed in Chapter 9. In the United States, restraints have been maintained on these infections by three essential measures: (1) *by controlling the source of disease in cattle* through inspection, elimination of infected animals, and vaccination when appropriate, (2) *by controlling the handler* as a secondary source of infection, and (3) *by controlling the milk* by pasteurization and sanitary bottling methods. The first two of these are discussed in the next section of this chapter dealing with various control measures applicable to living reservoirs of infection.

Pasteurization is a very effective precautionary measure for assurance of safe public milk supplies. It is a heat-disinfection procedure (see Chap. 10) that destroys vegetative forms of all pathogenic microorganisms, with the possible exception of hepatitis virus. It does not destroy spores, but these have no apparent importance in milk. Dairy sanitation must be strictly maintained if pasteurization is to be fully effective. This includes grooming of cows, sterilization or disinfection of all equipment, and delivery of milk into a closed presterilized system.

Bottling milk in presterilized glass, plastic, or waxed cardboard containers keeps its content of nonpathogenic organisms at a low level, but milk must be kept under constant refrigeration thereafter. The species of *Lactobacillus* and *Streptococcus* that form the normal flora of milk are reduced in number but not eliminated by pasteurization. Their subsequent multiplication results in the formation of lactic acid from degradation of milk sugars, and this is responsible for milk souring. Pasteurization and refrigeration delay this process.

Bacteriologic standards for milk are determined by laboratory methods that (1) count total numbers of organisms surviving per milliliter of milk, (2) test for the presence of coliform bacteria or for specific pathogens if these are suspected, and (3) determine adequacy of pasteurization by testing for the heat-sensitive enzyme phosphatase, normally present in raw milk (its absence from pasteurized milk indicates that heating conditions were sufficient for enzyme inactivation). Milk is rated as *grade A* if the total bacterial count does not exceed 30,000 per milliliter after pasteurization. *Grade B* milk has counts in the 30,000 to 50,000 per milliliter range, while *grade C* has high counts, in excess of 50,000, and should not be used except in cooking. *Certified milk* is produced under conditions of extremely careful controls over the health of cattle and handlers and on the sanitary conditions of milk collection. It is the purest milk that can be produced, with counts not exceeding 10,000 per milliliter. Public health regulations permit the sale of unpasteurized certified milk in many areas, but in general only pasteurized milk may be marketed for public consumption.

Public Health Controls of Living Reservoirs of Infection

The cardinal principles involved in the prevention and control of infectious disease have been discussed in several contexts throughout previous sections of this text. They are summarized briefly here, together with a description of their broad applications to the protection of the public health.

Control of Human Disease

Artificial Active Immunization. Probably the most important factor in prevention of infectious disease in man and animals is the development of *active immunity*. This may occur naturally following a bout with a pathogenic microorganism, but the protective effect may vary in extent and duration. *Artificial active immunization* through the administration of available vaccines is, therefore, a tool of primary value in the prevention of many human and animal infections, or at least of their more dangerous consequences.

The use of vaccines for the specific protection of human beings dates back to 1796 when Dr. Edward Jenner first experimented, in England, with the idea that people could be protected from serious disease by injection with material derived from others suffering from less severe infection. This was the first milestone in the effective control of smallpox. This disease had been a terrible scourge of mankind for centuries: transmitted readily by person-to-person contact, it had decimated military and civilian populations alike, lost and won wars, and sometimes threatened the survival of afflicted communities. (In Jenner's day, about 30 percent of children born in England died of smallpox before reaching maturity, and on the American continent many Indian populations were virtually wiped out by the inroads of this disease introduced by European settlers.)

Jenner's success with fluid collected from the less dangerous lesions of cowpox as a preventive agent for smallpox eventually led to a study of the basic principle involved and to its controlled application in prevention of many infectious diseases (see Chap. 7). Since then, smallpox itself has been gradually eliminated from its last endemic strongholds in tropical areas (Bangladesh and Ethiopia) and will soon be eradicated completely, thanks to concerted efforts of the World Health Organization. Mass immunization programs have also led to control over such diseases as tetanus, diphtheria, yellow fever, pertussis, and poliomyelitis. In each of these instances, unlike smallpox, *endemicity* of infection has not been eliminated, but the *epidemic* dimensions of these diseases are now controllable.

Successful control of infectious diseases with vaccines depends on a number of factors, such as the global distribution of the pathogenic organism, its entrenchment in animal or arthropod reservoirs, the characteristics of its antigens, and the ease with which these can be provided in safe but potent immunizing preparations. Diseases such as measles, mumps, rubella, influenza, and the venereal infections—distributed worldwide and readily transmissible through direct contacts—are particularly difficult to control without effective vaccines. In latter years the problem has been solved for measles, mumps, and rubella with the availability of vaccines that appear to induce a solid immunity if administered during childhood with suitable dosage schedules. By contrast, influenza vaccines provide only a short-lived protection because of the changing antigenic nature of influenza viruses. The recent efforts to provide mass immunization in the United States against "swine flu," one of the many varieties of influenza virus, illustrate the problems of maintaining the immunity of large numbers of people against microorganisms whose epidemic spread and antigenic character are difficult to predict (see Chap. 14). On the other hand, the agents of gonorrhea and syphilis have thus far defied all efforts to prepare their antigens in a form suitable for mass immunization.

In other instances, diseases such as cholera, plague, typhus, and typhoid fever have been controlled partially through curbs on their sources and partially through selective immunization. Vaccines against these diseases do not induce solid, long-lasting immunity but have been useful for protecting persons at high risk of exposure (military people or travelers entering endemic areas).

Many research efforts continue in the field of vaccine development, both for the improvement of antigenicity in available vaccines and for methods to produce new ones. A number of microbial agents, in addition to those of the venereal diseases, continue to be a challenge, either because they are difficult to cultivate (e.g., hepatitis viruses), or because when grown in substrates suitable for vaccine production they do not possess a useful degree of antigenicity (e.g., trachoma).

As vaccines become available, the remaining question of importance to immunization programs is "Who should be immunized and when?" The most general answer to this is that everyone who is susceptible and likely to be exposed should be protected against diseases that pose a threat to individual or public health.

Children. Immunization in childhood is indicated for those diseases that have a wide distribution, are readily transmissible, and for which effective vaccines are available. Current immunization programs for children include protection against diphtheria, pertussis, tetanus, poliomyelitis, measles, mumps, and rubella. Immunization schedules are based on the capacity of individual vaccines to induce solid immunity after one or more injections. Combined vaccines containing several different organisms or antigens simplify the procedure, but the length of the series of injections required then depends on the antigenicity of each of the individual components. It is important to remember that artificial immunization does not confer lifelong protection, and that booster doses are necessary to maintain the effect, the interval depending on the characteristics of each vaccine preparation currently available (Fig. 11-5 and Table 11-1).

Travelers. People who travel may be exposed to diseases not known or frequently encountered in their own regions, or to which they are not immune for other reasons (e.g., neglect of vaccination). Not only is their own health threatened in this event, but the public health may be compromised by a transmissible infection spread by a traveler along his entire route. In this age of rapid and frequent travel by many people of the world, the necessity for maintaining controls on international transmission routes of infectious disease is greater than ever before. Public health services of each country work to protect their own people, and through international cooperation of governments supporting the World Health Organization this protection extends across national borders. The International Sanitary Regulations of the WHO have standardized requirements for international travel with respect to the infectious problems indigenous to each country, so that travelers can be informed of the immunizations required or recommended for entry into various countries, or for reentry into their own (see Table 11-2).

Since 1976 many countries have dropped their requirement for smallpox vaccination for entering travelers. A few, especially African and Asian nations, still require that evidence of successful vaccination within three years must be presented at time of arrival. A current certificate of vaccination is no longer necessary for persons entering the United States from other countries, including Americans returning from travel.

Yellow fever remains endemic in parts of Central and South America and in Africa. Persons arriving from yellow-fever-infected countries into an area where mosquito vectors may be receptive to the

2 Months
DTP
TOPV

4 Months
DTP
TOPV

6 Months
DTP
TOPV

1 Year
Measles
Tuberculin test
Rubella
Mumps

1 ½ Years
DTP
TOPV

4 to 6 Years
DTP
TOPV

14 to 16 Years
TD
and thereafter
every 10 years

Fig. 11—5. Schedule of immunization for normal infants and children. (Reproduced from Katz, Samuel L.: Childhood Immunizations. *Hosp. Prac.,* **11**:52, Nov. 1976.)

virus, and transmit it, must present evidence of yellow fever vaccination received within ten years.

Cholera is endemic, often epidemic, in many countries of the Mediterranean area and the Near and Far East. Immunization against this disease is not durable, so that travelers coming from infected areas into countries where it is not prevalent may be asked to present evidence of vaccination within six months. The U.S. Public Health Service does not routinely recommend cholera immunization for travelers to countries that do not require it as a condition for entry.

Other immunizations not necessarily required for travel are nonetheless strongly advocated by the Public Health Service. These include diphtheria, tetanus, poliomyelitis, and typhoid fever vaccination

Table 11–1. Recommended Schedule for Immunization of Children

Age	Product (Antigen) Administered
2–3 months	DPT (diphtheria and tetanus toxoids, plus pertussis [whooping cough] bacterial antigen)
	Oral poliovaccine, trivalent
4–5 months	DPT
	Oral poliovaccine, trivalent
6–7 months	DPT
	Oral poliovaccine, trivalent
12 months	Measles vaccine
15–19 months	DPT
	Oral poliovaccine, trivalent
2 years	Mumps vaccine
4–6 years	DPT
	Oral poliovaccine, trivalent
8–10 years	Rubella vaccine
12–14 years	TD (tetanus and diphtheria toxoids)

*Adapted from *Review of Medical Microbiology,* 12th ed., by Ernest Jawetz, Joseph L. Melnick, and Edward A. Adelberg. Lange Medical Publications, Los Altos, Calif., 1976.

under all circumstances of travel; plague immunization if the country to be visited has recently experienced the disease, or if the traveler anticipates exposure in an endemic region; and typhus immunization for persons entering areas where typhus remains endemic.

Military Personnel. Service personnel are immunized routinely against many of the diseases cited above, notably diphtheria, tetanus, and typhoid fever. Poliomyelitis and influenza vaccines are also given when indicated. When personnel are assigned to regions of the world outside of the United States, new immunizations are scheduled according to geographic need. Military and diplomatic personnel stationed in Asian countries must be protected against cholera. Yellow fever immunization may be a necessity even in countries where it is not endemic if the area is considered "receptive" with regard to its mosquito vectors. Malaria, amebiasis, and numerous helminthic diseases are serious problems in tropical areas, but immunization against these infections is not possible, their control unfortunately depending on less specific and often less effective methods.

The great epidemic diseases historically associated with war — smallpox, plague, typhus, diphtheria — have been brought under control by vaccines and immunization programs, so that armies no longer must fight these microbial enemies. Also, wounded soldiers are far less subject to risk of threatening wound infections. Tetanus is under immunologic control, while other complications, such as clostridial gas gangrene, staphylococcal and streptococcal infections can be promptly controlled by modern surgical and chemotherapeutic methods, provided help can be received promptly.

Medical Personnel. In hospitals and clinics, medical personnel are important candidates for controlled immunization programs. Physicians and nurses, bacteriologists and other laboratory personnel, aides, orderlies, and all others who come into close contact with patients should be fully immunized against those diseases for which effective vaccines are available and which they may encounter in their work. This immunity should be carefully maintained by a regular schedule of booster doses. The basic program should include diphtheria, pertussis, tetanus, and poliomyelitis vaccines. Rubella vaccine is not recommended, however, for susceptible (previously

Table 11-2. Immunization for Foreign Travel*

Disease	Required	Recommended	Comments
Cholera	Required by some countries if traveling from infected areas	Not routine	One injection of vaccine; certificate valid for 6 months
Diphtheria		All children and adults	Children: see Table 11-1 Adults: TD (tetanus-diphtheria toxoid) booster every 10 years
Plague		Indicated for travel to interior of Vietnam, Cambodia, Laos, or if special exposure	Two injections of vaccine, 4 weeks apart; third injection 4-12 weeks later; booster in 6-12 months if risk continues
Poliomyelitis		All children and adults, especially if traveling to tropical areas where close contact with local residents will be made	Children: see Table 11-1 Adults: one booster dose of trivalent oral vaccine if previously immunized, or, if not, complete primary series
Tetanus		All children and adults	Children: see Table 11-1 Adults: TD (tetanus-diphtheria toxoid) booster every 10 years
Typhoid fever		Indicated for travel to endemic areas (Africa, Asia, Central and South America)	Two injections of vaccine, 4 weeks apart; 1 booster dose every 3 years
Typhus fever		Special risk groups only	Check current vaccine dosage
Yellow fever	Required by some countries for all entering travelers; by some others if coming from infected areas	Indicated for travel to infected areas (parts of Africa, South America)	One injection of vaccine; 1 booster every 10 years.

*Data from *Health Information for International Travel 1977,* Center for Disease Control, Public Health Service, U.S.D.H.E.W., Atlanta, Ga., HEW Publication No. (CDC) 77-8280. Health conditions and vaccination requirements for travelers are subject to change in all countries. Travelers should consult local health department in advance of travel.

nonimmunized) women of childbearing age in this group, because the live virus vaccine constitutes a risk to the fetus of an unsuspected pregnancy. It may be given immediately after childbirth, when the chances of a new pregnancy are remote, to women who have no rubella antibodies (see Chap. 14).

Control of Human Carriers of Infection. It must be emphasized that man himself can be a dangerous reservoir of infection—never more so than when he goes unrecognized as such. The normal appearance of a healthy carrier of pathogenic organisms, or of a person in the asymptomatic incubation period of an infectious disease, challenges epidemiologic detection.

Detection of the Incubating Disease. The patient incubating a disease declares himself soon enough, but until he becomes symptomatic little can be done to recognize him. Under limited circumstances, as in closed populations (barracks, schools, prisons, even immigration offices), the suspicion of infection may be confirmed by bacteriologic methods. Diphtheria bacilli or meningococci, for example, may be recognized in throat smears and cultures, or enteric pathogens in stool specimens; but in the absence of symptoms of active infection, such findings do not distinguish immune or resistant carriers from susceptible individuals who are incubating a disease. When there is serious risk of individual or epidemic disease, prophylactic measures are taken.

Detection of the Healthy Carrier. When the carrier of a specific infection has been recognized as such, he may be placed under suitable control. If his carrier state represents a threat to others, he may be treated with antimicrobial drugs in an effort to eliminate the infection, or he may be removed from areas or activities in which transmission of the organism is most likely.

Carriers of *Salmonella* or pathogenic strains of *Staphylococcus,* for example, should not be permitted to prepare food for others or to come in close contact with medical or surgical patients. It is sometimes very difficult to eliminate such infections by chemotherapy and the carrier state remains a difficult problem. In the typhoid carrier, the organism may be sequestered in deep tissues, such as the kidney or gallbladder, and it may continue to appear in urine or feces. Drug treatment may not succeed in such cases, and while surgical removal of the gallbladder may effect a cure if that organ is the sole site of colonization, the carrier state may persist in some individuals for long periods.

Diphtheria carriers are treated by the administration of both penicillin and antitoxin, but here too, permanent cure is not always achieved. Mass prophylactic immunization of children against diphtheria has apparently reduced the carrier rate and has also decreased the significance of carriers by creating an immune population, but such programs must be maintained.

Staphylococcus carriers perhaps represent the most difficult problem of all, particularly when they are found among hospital personnel. Staphylococcal cross-infections acquired by hospital patients can be troublesome and serious, particularly in nurseries and other critical patient areas. The strains are difficult, if not impossible, to eliminate from some carriers, although a patient, systematic attack from several directions may succeed. A combination of bactericidal antibiotic therapy, maintained at an active level for a sufficient period of time, and consistent use of disinfectant soaps for skin care may succeed. Beyond this, strategic job placement of hospital personnel carriers, and their careful instruction in personal hygiene techniques, can provide a reasonable solution to the double problem of patient protection and personnel employment.

Food Handlers. Health department sanitary codes generally have very specific requirements regarding preemployment examination of food handlers and their frequent surveillance. Routine preemployment tests include complete physical examination for evidence of communicable disease; chest x-ray or skin test to rule out tuberculosis; serologic tests for syphilis; and stool examination and culture to screen for the typhoid bacillus or any other *Salmonella* species, *Shigella* bacilli, pathogenic amebae, or helminth ova or larvae. These examinations must be repeated at regular intervals after employment. Any symptoms of illness should be reported and treated promptly, with temporary removal from the job if indicated. Sanitary toilet and washbasin facilities must be provided for kitchen personnel, with continuing education and emphasis on the necessity for careful hand washing. Other matters of personal hygiene should also be stressed, such as the cleanliness of hair, skin, and clothing; short, clean fingernails; and handkerchief protection of coughs and sneezes. Cuts and breaks in the skin of the hands should be covered with waterproof dressings when food is being handled, and hands, arms, and face should be examined frequently for the appearance of infected pimples, boils, rashes, or other infectious lesions.

The registration of typhoid carriers is still required by health departments, although this problem has decreased with declining incidence of endemic typhoid fever in the United States. The typhoid carrier, when detected, is specifically prohibited from participation in food handling, and also from care of children or of the sick. Every medical effort is made to clear the carrier of his infection, but until satisfactory results of treatment can be demonstrated, he is kept under surveillance with respect to his possible threat to public health.

Communicable Disease Control. Definitions of the terms *communicable disease, isolation, quarantine,* and *communicable period* are given in Table 10–2 (p. 231). A brief study of these definitions provides a review of current knowledge of the principles operating in infection transfer.

Communicable diseases are not always directly transmissible between hosts of the same species, as we know. Their control depends on knowledge of their sources and routes of transfer. Because we possess much more specific information on these matters today than was available one hundred years ago when the infectious nature of many diseases was first being discovered, the rules of control read very differently nowadays. Isolation and quarantine were once words that involved entire communities in dreaded procedures of social imprisonment. Alarming notices were tacked on the doors of the quarantined; distress of physical segregation was added to the suffering of illness and death; the stench of fumigation and strong chemicals blended with the odors of sick and unsanitary cities — but the results were effective when methods happened to match the requirements of the situation, despite ignorance. Today, isolation and quarantine are methods quietly applied in selective situations, primarily to prevent epidemics rather than to cope with them after the fact.

Isolation of the patient sick with a communicable disease remains an important controlling factor in some instances. The control of diphtheria or pneumonic plague, for example, depends on it, at least until chemotherapy has ended the patient's communicable period and the patient's susceptible contacts have been treated appropriately. Conversely, we have learned that isolation has little apparent effect in controlling infections such as poliomyelitis and the viral pneumonias. The requirement of isolation is left for the most part to the judgment of physicians or of hospital infection control committees, to

be applied or modified along the lines suggested by the Center for Disease Control (see Chap. 10). The advice of local health officials may be sought as well.

Quarantine is still applied to the control of certain communicable diseases, in that the activities of exposed susceptible individuals may be restricted on the grounds that infection may be spread among other nonimmune persons, or to animal and insect hosts, during the incubation period. The necessity for quarantine as a tool in disease control has diminished today, because of the effective mass use of available vaccines and a corresponding reduction in the number of persons remaining susceptible to diseases controlled by this method. More adequate controls on animal diseases also have changed the situation, as have systematic attacks on the arthropod vectors of some infectious diseases. Nonetheless, quarantine remains a useful method of assuring that infectious disease will not be spread during the unrecognizable period of incubation. It is applied selectively: to travelers who are careless of the risk to themselves and others when they fail to obtain appropriate artificial immunization in advance of possible exposure; to schoolchildren and others (such as military personnel) in closed groups with many susceptibles; and, in a reverse manner, to the protection of critically susceptible patients in hospitals.

In the United States the Public Health Service is responsible for control of communicable diseases that might be disseminated through international trade and travel. This agency maintains quarantine stations at all ports, sea and air, where foreign commerce enters the country. All ships and planes, American or foreign, arriving from ports in other countries, are required to obtain health clearance from these quarantine stations. Radio contacts with ships or planes entering debarkation ports usually provide advance warning of overt illness aboard. These precautions are designed to prevent importation of such diseases as cholera, yellow fever, plague, typhus, or anthrax. Vessels arriving from countries where such diseases are prevalent or endemic are fumigated after passengers and crew have disembarked. They are also treated with insecticides to kill any insect vectors surviving the voyage. (Such a procedure might have prevented some of the devastating epidemics of yellow fever experienced by our eastern port cities during the early part of the last century when shipping from the West Indies and South America first became intensely profitable.)

When necessary, quarantine is imposed on individuals without valid certificates of vaccination who have been exposed to the communicable diseases of other countries. Passengers who arrive obviously ill with symptoms suggesting any of the diseases mentioned above are quarantined in hospitals where they can be promptly treated. Quarantine restrictions are strictly enforced, under penalty of fine, and can be lifted only by the responsible health agency.

Case finding and reporting within the community are important aspects of the preventive epidemiologic work of health departments. All local and state health boards require that physicians, hospitals, and similar institutions report certain diseases of a communicable nature. Some endemic diseases, such as measles and chickenpox are characterized by seasonal increases and cyclic epidemics every three to four years. Health agencies want to be forewarned of these rising incidence rates. Other diseases are reported so that the source of infection can be traced, with a view to preventing further transmission. Most such reports can be submitted by mail, but in the case of acute diseases accompanied by threat of epidemic, the local health department should be immediately notified by telephone. Immediate reports should be made when cases of yellow fever, cholera, plague, rabies, anthrax, poliomyelitis, diphtheria, or psittacosis are suspected.

Health departments also require that new cases of tuberculosis or venereal disease be reported. In view of the ease of transmission of such infections, it is important that all contacts of new cases be investigated to ensure early diagnosis and treatment of developing disease. Public health nurses often play an important part in case finding and in encouraging exposed, susceptible people to seek medical advice or treatment.

Methods of prophylaxis for susceptible persons exposed to communicable diseases are discussed in Chapter 10. Both chemoprophylactic and immunologic measures may be useful on a large scale in prevention of epidemics. Mass immunization procedures with vaccines are sometimes appropriate when diseases such as influenza threaten to involve large segments of the population in serious disease. The expense, and the risk of potentially dangerous side effects of the vaccine, must be weighed against the public health benefit. Passive immunization on a large scale, as once applied in epidemics of poliomyelitis, is inappropriate today. This is because much of the population has either natural or artificial active immunity to diseases of epidemic significance. With the wane of certain pathogenic microorganisms from an immunized population, however, this natural stimulus to continuous antibody production declines, so that it becomes correspondingly more important to maintain active immunity by artificial means. Community immunization programs must recognize the necessity for continuing campaigns to hold a basic level of protective immunity with booster schedules for useful vaccines.

Control of Animal Disease

The principles involved in control of infectious disease are the same in their application to animals as to man. The basic premise in either case is that the best way to control disease is to prevent it, using three major approaches: immunization whenever possible, sanitary protection of the environment, and protection of the healthy from exposure to the sick (isolation of communicable infection, quarantine and prophylaxis of those exposed, control of sources of disease).

Control of Diseases of Domestic Animals. In some instances, successful control of domestic animal infection is achieved by methods that cannot be used for humans. "Isolation of infection," for example, sometimes means destruction of infected, sick, or dying animals and vigorous methods for disinfection of carcasses. Quarantine may be applied to an entire herd of animals, without selecting those actually infected. Prophylactic antibiotics are used primarily to prevent the *chance* of infection in normal animals or fowl, as well as to protect exposed individuals (cf., Chap. 10, p. 268). Controlling sources of animal diseases often also involves control of insect vectors that perpetuate many of these infections, as well as protection of animals from infectious human beings. The chain of transmission of many infections (see especially those of group 4 in Table 9–2, p. 213) links man and animals by direct contact, and later through human consumption of animal products, notably milk, eggs, and meat.

Immunologic methods applied to animals include active and passive immunizations and skin testing to detect infection or measure immunity. Sanitation and disinfection of animal quarters, and of equipment used for preparation of animal food products, are often of critical importance in preventing infection and in controlling its spread.

Important zoonoses of domestic animals that are readily transmissible to man, for whom they have serious consequences, are tuberculosis, brucellosis, anthrax, taeniasis, trichinosis, echinococcosis, and rabies. Some of these can be successfully controlled at their animal source.

Tuberculosis in cattle is controlled by a rigorous routine of tuberculin testing and destruction of positive skin reactors. Further control of transmission of bovine tuberculosis to man has been achieved by routine pasteurization of milk.

Brucellosis is a disease of cattle, goats, and pigs. Infection is transmitted to man most frequently through raw milk and milk products and by contact with infected animals or their tissues. This disease is an occupational hazard to those who work with animals or meats (farmers, butchers, veterinarians) but it can also involve the milk-drinking public. Control has been achieved for the latter by milk pasteurization. Methods to control the animal reservoir include elimination of animals with infection demonstrated by serologic tests, immunization of young animals, and improved environmental sanitation.

Anthrax is a disease of cattle, sheep, horses, and goats, transmissible to man by contact or by the airborne route. The infectious bacterial agent is an aerobic, spore-forming bacillus. When the vegetative bacilli are exposed to air, in animal excreta or carcasses, they rapidly form spores capable of survival for many years. Contact with contaminated wool, hair, or hides of once-infected animals is one of the most frequent avenues of human anthrax infection. Control of the disease can be achieved by mass immunization of animals with anthrax vaccine, isolation of sick animals, and proper handling and disposal of infectious carcasses to prevent contamination of soil.

Taeniasis, or intestinal tapeworm infestation, is acquired by man through ingestion of beef or pork tissues containing encysted larval forms. The adult forms of these tapeworms live in man's intestinal tract, producing ova that are shed in feces. The domestic animal host becomes infected from grazing or rooting on ground polluted with human feces containing the ova, but for the animal infestation is systemic rather than intestinal. Control of this type of disease depends on recognizing infected animal meat by careful inspection for encysted larvae and breaking the chain of transmission from man to animals. Sanitary disposal of feces and sewage, and protection of animal feeding grounds, would eliminate these infections when combined with fully adequate methods for inspection and cooking of meats (Fig. 9–13, p. 223).

Trichinosis is maintained by man, hogs, and rodents as principal alternate hosts. Infestation is systemic in each host, involving encystment of larvae in striated muscle. The cysts are not visible on inspection of pork, so that control depends on thorough cooking or "curing" of all pork products. Protection of hogs from sources of infection is also essential to control. This includes elimination of raw garbage feeding (containing infectious scraps of pork) and extermination of rats and mice on the premises. It should be pointed out that some wild animals (bears, foxes, opossums, raccoons) are also hosts to this parasite. Game hunters and others who eat such meat run the risk of acquiring trichinosis unless adequate cooking is assured (Fig. 9–11, p. 221).

Echinococcosis is prevalent in some parts of the world in sheep, cattle, and pig reservoirs of systemic larval infection. Dogs and wild carnivores maintain the cycle of the parasite in nature, by eating infected flesh from which the adult worm develops in an intestinal infestation. Eggs from this source are shed in the feces of dogs and may be ingested by man as well as by grazing animals. Control of human disease centers largely on preventing dogs from playing their role in transmission. This includes strict personal hygiene on the part of persons who associate with dogs, particularly in endemic areas; deworming of dogs; adequate disposal of discarded flesh of infected, slaughtered animals so that dogs cannot find access to this source; and public education concerning the nature and transmission of this disease (Fig. 9–13, p. 223).

Immunization of dogs is an essential control on a number of infectious diseases that may afflict them, such as distemper, canine hepatitis, leptospirosis, and rabies. Of these, the latter is also extremely dangerous to man. The natural reservoir of rabies virus is in a number of wild animal species (foxes, wolves, raccoons, skunks, and bats; Fig. 11–6) and in dogs. Transmission among animals or to man is usually accomplished by the bite of a rabid animal. Rabies control depends on elimination of its source in sick animals and on protection of healthy dogs by immunization with rabies vaccine. Immunization of human beings is not recommended unless there is evidence of exposure to rabies.

When suspicion of rabies exists, biting dogs must be quarantined for a week to ten days and observed for symptoms of disease. They must be destroyed if rabies is proven, and efforts must be made to find other animals possibly exposed. These also must be quarantined (for three months) and vaccinated, or, if necessary, destroyed. Stray dogs should be elimi-

nated, and others immunized, with booster doses at regular intervals (every one to three years, depending on the vaccine). Domestic dogs should not be permitted to run freely, particularly in areas with endemic foci of rabies among wild animals.

Rabies control in wild animals is under supervision of federal and state fish and wildlife agencies. Hunting, trapping, and baiting methods are used to find and dispose of sick animals and to bring epidemics under control.

Rodent Control. Rats and mice are reservoirs for a number of infectious diseases that affect man. Chief among these are bubonic plague and murine typhus, both of which are transmitted by fleas to rodents or to man. Public health control of plague and other diseases associated with rodents includes a continuing program for extermination of rats by poisoning techniques and by rat proofing of buildings. Rat-infested ships arriving in port are fumigated with hydrocyanic gas, while their docking lines are shielded to prevent escape of rats by this route.

Shellfish Control. Shellfish harvested from sewage-polluted waters may be infectious reservoirs of enteric diseases, notably typhoid fever, other salmonelloses, and infectious hepatitis. Sanitary regulations for the shellfish industry are provided and enforced by local and state health departments, with cooperation of the Public Health Service.

Fig. 11–6. This captured bat has been submitted to a public health laboratory for examination for rabies. At least three varieties of bats are known to carry rabies. (Center for Disease Control, D.H.E.W., Atlanta, Ga.)

Control of Arthropod Vectors

Diseases transmitted solely by arthropod vectors may be controlled to the extent that it is possible to eliminate the insect reservoir or prevent its human contacts. Principal members of this group of diseases are yellow fever, epidemic typhus, and malaria (group 3 in Table 9–2, p. 213). Since man and the insect host are the only reservoirs for these diseases, significant control can be achieved by insect eradication programs, coupled with immunization or chemoprophylaxis as indicated.

Diseases transmitted by arthropods from several animal reservoirs are much more difficult to control. Complete eradication of the vector would be required to eliminate them. Widely based arthropodborne infections include viral encephalitis, many rickettsial diseases, leishmaniasis and trypanosomiasis (both protozoan), filariasis (the tissue roundworm), and bubonic plague. Man is not a part of the reservoir in the first two examples, which are maintained in a number of animal hosts, as are the other diseases mentioned. Arthropod control and protection of human beings from insect contacts are important, but other means of prevention must also be found.

Those diseases in which insects serve only as mechanical vectors do not depend on this means of transmission alone. Flies and other nonparasitic insects are often incriminated in transmission of such diseases as poliomyelitis, salmonellosis, and other bacterial infections, but these diseases may be transmitted by many other routes as well. Insect control in this case, therefore, would not lead to elimination of these diseases, although it may help to reduce their incidence.

Control of Insect Breeding. The most effective methods of eradicating insect species are those that prevent breeding, the techniques depending on the insect. Mosquitoes are among the most important vectors of infectious diseases and the most usual targets of insect eradication programs. They breed in stagnant water: in swamps, still pools, unused rain barrels, or any forgotten vessel that collects and holds water for long periods (Fig. 11–7). Mosquitoes can be eradicated or well controlled only if all these areas are found and drained or treated. Oily films

Fig. 11–7. Health workers engaged in mosquito control. Over large parts of Africa, there is so much malaria that present methods of vector control are inadequate to check the disease. (Reproduced from Rowe, David: The Forgotten People, *World Health*. June 1976, WHO photo by J. Mohr.)

spread on the surface of water prevent larvae from getting oxygen, or insecticides may be used, but draining and filling are most effective.

Sanitary disposal of garbage and sewage and protection of manure or compost remove some of the breeding places of flies. These and other insects, such as lice, bedbugs, and mites, that breed in and around human dwellings are best controlled by strict sanitation and effective personal hygiene, reinforced by use of chemical insecticides.

Insecticides. A number of chemical compounds kill insects on contact. These are used in sprays directed at insects but are not effective in large-scale control unless they knock out significant numbers. Residual insecticides are more effective in eradication because they can be applied to many surfaces where insects light or crawl. Insecticidal activity is maintained on these surfaces for many weeks or months. DDT is an insecticide of the residual type commonly used in mosquito control. It is also effective in killing lice. It can be used as a dust for clothing and skin, one of the methods effective in control of the louse vector of epidemic typhus, or as a spray for inner and outer walls of houses where mosquitoes rest. In control of malaria and other mosquito-borne diseases, entire communities are usually included in spraying programs. Other insecticides of this type include chlordane and dieldrin. All these compounds are hydrocarbons to which insects may develop resistance. When this happens, other types of chemicals, such as organophosphates (malathion), may be used as residual insecticides. An unfortunate ecologic side effect of the use of insecticides is their increasing concentration (biomagnification) to potentially toxic levels in the tissues of plants, animals, and, eventually, man.

Protection of Human Beings from Insects. Control of insect-borne diseases for which man is a reservoir includes careful screening of known infectious cases from the vector. All possible precautions must be taken during the infective period of yellow fever or malaria to screen patients from mosquitoes, for example. In the case of epidemic typhus, patients must be deloused and prevented from acquiring further infestation from others. Personal hygiene and cleanliness of clothing or bedding are effective in control of lice and other ectoparasites. Protective clothing is indicated when exposure to insects such as ticks or mites is anticipated, and frequent careful examination of the body should be made following such exposure. Above all, knowledge of the important vectors of infectious diseases should be sought by all who find themselves newly exposed to risk of infection.

Questions

1. What are the major environmental reservoirs of infectious disease in the public domain?
2. Briefly describe water purification methods.
3. What public health methods are used to control disease and reservoirs of infection?
4. What is the value of immunization to the individual? To the community? Why should immunization programs be maintained?
5. List diseases that have been successfully controlled with vaccines. What diseases defy control through the use of vaccines? Why?
6. Outline basic immunization programs for (a) children in the first two years of life; (b) children from two to six years of age; (c) teen-agers; (d) adults.

Additional Readings

*Benenson, Abram S. (ed.): *Control of Communicable Diseases in Man,* 12th ed. American Public Health Association, Washington, D.C., 1975.

Cranston, L.: Communicable Diseases and Immunization. *Can. Nurse,* **72**:34, Jan. 1976.

Editorial: Facing up to the Drinking Water Problem. *Am. J. Public Health,* **66**:635, July 1976.

Erickson, Suzanne: Epidemic. *Am. J. Nurs.,* **76**:1311, Aug. 1976.

Hubbert, W. T.; McCulloch, W. F.; and Schnurrenberger, P. R. (eds.): *Diseases Transmitted from Animals to Man,* 6th ed. Charles C Thomas, Publisher, Springfield, Ill., 1975.

*Available in paperback.

Marks, Geoffrey, and Beatty, W. K.: *Epidemics.* Charles Scribner's Sons, New York, 1976.

*Lilienfeld, Abraham M.: *Foundations of Epidemiology.* Oxford University Press, New York, 1976.

*Rosen, George: *Preventive Medicine in the United States 1900–1975.* Neale Watson Academic Publications, Inc., New York, 1975.

*Snow, John: *The Broad Street Pump,* in Roueché, Berton (ed.): *Curiosities of Medicine.* Berkley, Medallion Ed., New York, 1964.

Ventura, Jacqueline N.: The International Traveler's Health Guide. *Am. J. Nurs.,* **77**:968, June 1977.

Witte, John J.: Recent Advances in Public Health; Immunization. *Am. J. Public Health,* **64**:939, Oct. 1974.

*Available in paperback.

Part

Two

Microbial Diseases and Their Epidemiology

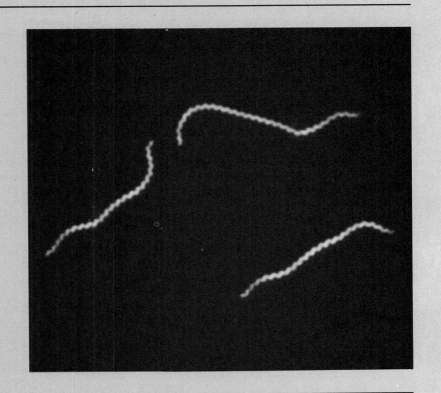

INTRODUCTION TO PART TWO

The preceding sections of this text have dealt with basic concepts concerning the nature of microbial diseases. The respective contributions of pathogenic microorganisms and the human host to the pathogenesis of infectious disease have been discussed, and epidemiologic principles involved in treatment, control, and prevention of injurious infection have been reviewed.

In Part Two specific infectious diseases of importance to man are presented in a manner designed to emphasize application of epidemiologic principles to their management. The sequence of approach is based on the practical considerations that confront those involved in patient care with each case of infectious disease: recognition of its important *clinical features,* establishment of the *laboratory diagnosis,* and institution of *appropriate control measures.* Each disease is discussed with respect to its clinical appearance and pathogenesis, the availability of laboratory methods for confirming the diagnosis (what specimens are of value, how and when they should be collected), and the most important aspects of its epidemiology. All necessary control measures are indicated in each case.

Some infectious diseases are easily spread by several routes of transmission and entry; others are self-limited but nonetheless capable of wide dissemination among the patient's contacts; some are particularly restricted by developmental and physiologic requirements of the infectious microorganism as well as by opportunities it may have for entry to or exit from the human body; and still others may be acquired by some fortuitous means but are not transmissible under ordinary circumstances. Measures for controlling further spread of infection must be pertinent to the demands of a specific situation and based not only on some knowledge of the infectious microorganism and its paths of transferral, but also on the probability of exposure of susceptible human hosts. Diseases encountered in the private home, or in rural situations, sometimes have a very different epidemiologic potential from those occurring in crowded urban districts or in hospitals.

With these considerations in mind, discussions of specific infectious diseases have been grouped primarily on the basis of the most common route of entry to the human host available to the microbial agent. They are arranged in four sections (IV through VII) dealing, respectively, with entry through the respiratory tract, the gastrointestinal route, skin and mucosal surfaces, and parenteral routes. Repeated emphasis is placed on the fact that the site of entry of an infecting organism does not always constitute the focus of developing infection, nor is it necessarily associated with the route of exit and transmission that may subsequently be available to the microorganism in question.

Within each section, secondary groupings of diseases are made by chapter, according to the nature of their causative agents (*viz.,* bacteria, viruses, fungi, animal parasites, and others). The general epidemiology of the four major groups, based on entry route, is summarized in a chapter introducing each section. These chapters (12, 17, 21, and 25) provide general recommendations for the nursing control of infectious diseases acquired through that route.

The Appendixes contain four tables summarizing important procedures for the immunologic diagnosis of infectious diseases, aseptic nursing precautions, and methods for the bacteriologic control of sterilizing equipment.

Section IV

Infectious Diseases Acquired Through Inhalation

12

Epidemiology of Infections Acquired Through the Respiratory Portal

General Principles

The human respiratory tract affords a site of entry and initial multiplication, though not necessarily *local* disease, to a large and highly diverse number of microbial agents. The three major groups represented are *bacteria, viruses,* and *fungi.*

Many of the agents classified together on the basis of their morphology and growth characteristics produce diseases of widely divergent types. Thus, among bacterial diseases, entities such as scarlet fever, whooping cough, and tuberculosis have few clinical relationships. One of the common signs of bacterial infection is the leukocytic response the body makes. This is usually polymorphonuclear in character and can be observed at the site of bacterial localizations or in white blood cell counts of circulating blood. In tuberculosis, however, the tissue response is quite different, resembling that induced by fungi, while in diphtheria and scarlet fever the production of bacterial toxins changes the entire clinical picture from one of mere infection to that of systemic toxicity.

Some of the viruses display even more clinical versatility, being associated with syndromes ranging from the mildest respiratory illness to severe disease in or beyond the pulmonary system. Furthermore, a particular clinical syndrome, such as "bronchitis," "pneumonia," or "aseptic meningitis," may be caused by any one of a number of different, unrelated viruses. Viral diseases are usually acute and self-limited. A characteristic feature of viral and rickettsial infections is the intracellular localizations of the parasites. They do not elicit large cellular responses from the body as a rule, and the leukocytic force that does meet them is of the mononuclear rather than the polymorphonuclear type.

Fungi that cause systemic disease are met by a fairly uniform type of local tissue reaction designed to wall them off and contain them through the activity of mononuclear cells of the reticuloendothelial system. The pathogenic fungi are relatively inert in the strange medium of human tissues, as they normally live a saprophytic life in the soil. However, they are often capable of survival and slow multiplication in the face of the best defenses the human body can offer, as are the "higher bacteria," which include the tubercle bacillus and actinomycetes. These organisms produce chronic diseases characterized by slowly progressive lesions. The cellular formations mustered by the body fight them every step of the way, but they are sometimes defeated in this effort by an overreactivity (hypersensitivity) of immunologic defense.

Although the infectious agents and diseases described in ensuing chapters of this section are quite diverse, it is important from the epidemiologic and nursing viewpoint to recognize that they have a common entry point in the respiratory tract. From this initial point of localization, even though it be transient, further extensions may occur in the respiratory tract or to other parts of the body. During the time respiratory localizations persist, the infectious agent may be present in mucous secretions of sputum and transmissible to others through nasal and oral discharges.

As a portal of entry, the respiratory tract is very vulnerable to microorganisms disseminated in air or carried into the mouth on fingers, objects, food, or water, but many defensive mechanisms operate effectively to prevent exogenous infection. The flushing action of saliva and nasal mucus, lymphoid tissue of the pharynx and nasopharynx, and the constant action of the "ciliary mucous escalator" of the tract all function to prevent airborne particles from reaching the lower bronchial tract or the lungs.

As a portal of exit, the respiratory tract has no equal in the ease and frequency with which it permits transfer of the microorganisms contained in sputum, saliva, or throat secretions to the outside environment. In ordinary speaking, or clearing of the throat, a minor spray may be ejected, but in coughing or sneezing not only is the spray projected farther but the secretions are nebulized, so that many more droplets and smaller ones are disseminated in the air (see the illustration on page 295). Smaller and lighter droplets may hang suspended for a time, settling slowly and drying to droplet nuclei which remain in the air or are swept back into it from dry surfaces, such as clothing, bed linens, dustcloths, or dry mops. When such droplets reach food or water, the organisms they contain may find an even wider dissemination.

Extension of infection from an initial site in the respiratory tract to systemic localizations may open other portals of exit to the microorganism. Thus, in smallpox the virus is present in discharges from open pustular skin lesions as well as in similar foci on the oral membranes, so that *transmission* may occur from both avenues. Smallpox infection is *acquired,* however, only through respiratory entry, not directly through the skin. In poliomyelitis, virus implantation first occurs in the pharyngeal membranes, and virus can be transmitted from them early in the disease, but later it is excreted primarily by the fecal route from sites of intestinal localization.

Successful transfer of infection to the respiratory tracts of other persons by the oral and nasal discharges of infected individuals often depends on the closeness of contacts, numbers of organisms inhaled, and the site of their localization. Some viral diseases, such as influenza and measles, are highly contagious, probably because their infectious agents are capable of quick and easy implantation in cells of the upper respiratory membranes, and also because they are transferred in fairly large numbers in pharyngeal secretions. Tuberculous infection may result from the inhalation of a single tubercle bacillus (or of a few in one droplet nucleus), but it must be deposited deep in the respiratory tract, beyond the ciliated epithelium. The same thing applies to bacterial spores (anthrax), to the agent of psittacosis, and to fungous spores, which do not implant in the upper respiratory tract and may be easily eliminated if they are not deeply inhaled. The ability of microorganisms to survive in air or on inanimate surfaces, and the duration and intimacy of contacts, are also factors to be considered in judging the communicability of infection.

Some noncommunicable diseases that are acquired by man through the respiratory route are not usually directly transmitted from person to person. This is true of a number of the zoonoses (see Chap. 9, p. 202), such as Q fever and psittacosis, described in Chapter 15. Anthrax, brucellosis, and tularemia are other examples, but these zoonoses are dealt with in later sections according to their more common routes of spread to man. The systemic mycoses represent another important group of infectious diseases that are not transmissible directly from man to man. Although these fungous diseases do not all have a respiratory entry, they are discussed as an entity in Chapter 16 of this section because their pathogenesis and epidemiology are quite similar.

Patient Care

It should be recognized that measures taken for the control of active infection acquired through or transferred from the respiratory tract must be based on existing knowledge of the infectious agent, with respect to its localizations and activities during different clinical stages of disease and its capacity for retaining infectivity in the outside environment.

Isolation of the Patient

The recommendations of the Center for Disease Control regarding "respiratory" or "strict" isolation (Table 12-1) are based on current epidemiologic information. The purpose of respiratory isolation is to prevent the transmission of organisms that are commonly spread by airborne *droplets* or *droplet nuclei,* or by *direct contact.* The diseases in this category are not usually spread by indirect contact with freshly contaminated articles. Strict isolation, however, is designed to prevent the transmission by *direct, indirect,* or *airborne routes* of diseases that are highly communicable.

It will be noted that many diseases requiring strict isolation are of respiratory origin whereas some (burn infections, skin infections, neonatal herpes) are easily acquired through direct or indirect contacts with the skin (hands are an important vector of these). Rabies is included here because the respiratory secretions of the infected human patient may be highly infectious for those caring for him. Conversely, noncommunicable respiratory infections such as the systemic mycoses and zoonoses do not, with the exception of inhalation anthrax, require isolation of the patient, but precautions must be taken with his respiratory or other secretions. In the case of pulmonary anthrax, strict isolation is recommended to prevent contamination of the hospital environment with the highly resistant spores of the bacterial agent.

The precautions suggested by the Center for Disease Control for maintaining either respiratory or strict isolation are printed on cards designed to be posted on the patient's door, as shown in Table 12-1. The reverse side of each card lists diseases requiring such isolation.

Respiratory Precautions

Precautions in managing respiratory infections are centered primarily on the control of the patient's infectious oral and nasal secretions, to prevent direct or airborne contact with others. The following general types of procedures may be indicated, depending on the nature of the disease:

1. The patient may be isolated to some degree, as discussed above.

2. The patient is instructed as to the importance of his own control of respiratory discharges. Adequate muffling of coughs and sneezes with paper handkerchiefs and the disposal of secretions in a container that can be incinerated are advised.

3. Masks are worn by all persons entering the room except those known to be immune to the patient's disease. The patient himself may be masked under circumstances that require removing him from the room for treatment or special examination.

4. Medical and nursing equipment is individualized for the patient insofar as possible. Concurrent disinfection or sterilization techniques are carried out with special attention to thermometers, inhalation therapy equipment, airways, laryngoscopes, emesis basins, and other items contaminated with infectious secretions or vomitus.

5. Dishes and eating utensils used by all patients should be washed by sterilizing techniques. The chief precaution indicated for patients with respiratory infections is the instruction of dietary personnel in careful hand washing after handling used trays. In some instances added emphasis can be given by placing the finished tray in a large paper or plastic bag before it is removed from the patient's room.

6. Laundry techniques should routinely assure disinfection of all linen and protection of clean laundry from contamination. Soiled linen should be handled with economy of motion to prevent shaking infectious organisms into the air with lint. As in the case of soiled dishes, laundry can be placed in marked bags to emphasize the hazard it represents. Blankets, mattresses, and pillows may also require special attention in many instances. Protection of mattresses and pillows with plastic covers or other impervious materials is helpful provided the technique for disinfection of their surfaces is adequate.

7. Clean gowns protect the clothing from contamination and for this reason may be advisable for

Table 12–1. Isolation Cards Recommended by the CDC

Card to Be Posted on Door of Patient's Room	Specified Diseases*
Respiratory Isolation *Visitors—Report to Nurses' Station Before Entering Room* 1. Private room — *necessary;* door must be kept closed 2. Gowns — not necessary 3. Masks — must be worn by any person entering room unless that person is not susceptible to the disease 4. Hands — must be washed on entering and leaving room 5. Articles — those contaminated with secretions must be disinfected	**Diseases Requiring Respiratory Isolation** Measles (rubeola) (14) Meningococcal meningitis (13) Meningococcemia (13) Mumps (14) Pertussis (whooping cough) (13) Rubella (German measles) (14) Tuberculosis, pulmonary (including tuberculosis of the respiratory tract) — suspected or sputum positive (smear) (13)
Strict Isolation *Visitors — Report to Nurses' Station Before Entering Room* 1. Private room — *necessary;* door must be kept closed 2. Gowns — must be worn by all persons entering room 3. Masks — must be worn by all persons entering room 4. Hands — must be washed on entering and leaving room 5. Articles — must be discarded or wrapped before being sent to central supply for disinfection or sterilization	**Diseases Requiring Strict Isolation** Anthrax (inhalation) (23) Burn wound, major, infected with *Staphylococcus aureus* or group A *Streptococcus* (27) Congenital rubella syndrome (14) Diphtheria (pharyngeal or cutaneous) (13) Disseminated neonatal *Herpesvirus hominis* (herpes simplex) (22) Herpes zoster, disseminated (14) Lassa fever (14) Marburg virus disease (14) Plague, pneumonic (13) Pneumonia, *Staphylococcus aureus* or group A *Streptococcus* (13) Rabies (27) Skin infection, major, infected with *Staphylococcus aureus* or group A *Streptococcus* (22) Smallpox (14) Vaccinia (generalized and progressive and eczema vaccinatum) (14) Varicella (chickenpox) (14)

*The number in parentheses indicates the chapter in this textbook where the disease is described.

those who come into close contact with the patient. However, gowns are not effective mechanisms for preventing spread of airborne infections and should not be viewed as such.

8. Additional precautions may be indicated for diseases that can be transmitted by feces, or by discharges from skin lesions, as well as by respiratory secretions.

9. Terminal disinfection-cleaning of the room or unit is often indicated, with special attention to hori-

zontal surfaces of furniture, bed frames, and floors, which hold settled accumulations of infectious droplets or droplet nuclei.

10. Hand washing comes first as well as last. The hands are one of the most important agents of infection transfer, and the easiest to control — simply by awareness of their role.

Questions

1. What defense mechanisms operate to prevent exogenous infection of the respiratory tract?
2. How are microorganisms transmitted from the respiratory tract to the outside environment?
3. What factors contribute to the successful transfer of respiratory tract infections among human beings?
4. What is the purpose of respiratory isolation?
5. List precautionary procedures that should be followed in caring for patients with respiratory infections.

Additional Readings

*Benenson, Abram S. (ed.): *Control of Communicable Diseases in Man,* 12th ed. American Public Health Association, Washington, D.C., 1975.

Editorial: What Needs to Be Done to Eradicate Tuberculosis. *Am. J. Public Health,* **62**:127, Feb. 1972.

Garner, Julia S., and Kaiser, A. B.: How Often Is Isolation Needed? *Am. J. Nurs.,* **72**:733, Apr. 1972.

Isolation Techniques for Use in Hospitals, 2nd ed. U.S. Department of Health, Education and Welfare, Public Health Service, Washington, D.C., 1975, DHEW Pub. No. 76-8314.

*Pierce, Alan K. (ed.): *Basics of RD, Vol. 1, 2, 3, Updated 1975.* American Lung Association, New York, 1975.

Tuberculosis Programs 1973. Tuberculosis Program Reports, U.S. Department of Health, Education and Welfare, Center for Disease Control, Atlanta, Ga., Dec. 1974.

*Available in paperback.

Bacterial Diseases Acquired Through the Respiratory Tract

13

The bacterial diseases that are acquired by way of the respiratory tract include some of the most important and serious infectious problems man has known. They are important because the respiratory route of access makes their control difficult and assures their widespread distribution in populations everywhere. This means that not only are low-grade but debilitating infections with some of these agents very common, but also the threat of epidemics of acute infections must be continuously controlled.

The diseases described in this chapter are diverse in their nature and symptomatology, but bacterial infections share a few common features. Some pathogenic bacteria are called *pyogenic* because the body responds to them with the defensive production of pus. Purulent exudates at the site of microbial localizations in the tissues are one of the most frequent signs of bacterial infection, whether it be pharyngitis, pneumonia, meningitis, or a focal abscess. When infection is severe, a general leukocytic response occurs, with elevated white blood cell counts (usually at least 15,000 cells per cubic millimeter). High fever (102° F [38.9° C] and above) and pain at the localizing site of infection (acute sore throat, chest pain, headache) are also common signs of developing bacterial disease. In some instances, disseminating bacterial toxins take a further systemic toll, as in diphtheria and some streptococcal infections.

Many acute bacterial diseases that were once accompanied by crippling or fatal effects (meningococcal meningitis, scarlet fever) can now be controlled by prompt, effective antibiotic therapy. Early clinical recognition and rapid laboratory confirmation of the diagnosis are essential to the choice of appropriate treatment and the control of active infection.

Unfortunately, effective vaccines are not available for many of these diseases, and, for these, prompt chemotherapy of the active case remains the best control measure. For diphtheria and whooping cough, however, the threat of epidemic disease has been eliminated by mass immunization, begun in childhood. The prevention of such diseases can be continued only through the maintenance of immunization programs for susceptible populations.

Upper Respiratory Infections (URI)

The Clinical Disease. Many common bacterial infections of the upper respiratory tract are superficial and nonspecific as to symptoms, pathology, and etiology. They may be acute, chronic, or recurrent. Febrile reactions are mild or do not occur. The basic picture is one of inflammation and hyperemia (congestion). Terms such as *pharyngitis, laryngitis, tonsillitis,* and *sinusitis* indicate the regional mucous membranes involved. Otitis media, or inflammation of the middle ear, results from extension of pharyngeal infection through the eustachian canal. Mucosal injury following viral infections, inhalation of toxic vapors or smoke, excessive dryness, pollen or dust allergies — all may permit unusual growth and activity of members of the indigenous bacterial flora. These secondary infections may be further characterized by the formation of purulent or serous exudates, localizing abscesses, and hyperplasia of tissues chronically involved (adenitis, adenoiditis, follicular tonsillitis). Persistent infections are frequently marked by hypersensitive responses to bacterial cells or their products. More specific symptoms and pathology may be produced by truly pathogenic invaders.

The Organisms. Infections of the upper respiratory mucosa may be caused by a variety of microorganisms. The most important agent of *bacterial* pharyngitis is the group A beta-hemolytic *Streptococcus* (p. 316). Although other potential pathogens such as *Haemophilus influenzae,* pneumococci, *Staphylococcus aureus,* Gram-negative enteric bacilli, and *Pseudomonas* species are often isolated from pharyngeal cultures, these bacteria are thought not to play a role in infection at that site. Rather they are involved in other upper respiratory infections such as otitis and sinusitis.

Laboratory Diagnosis. Smears and cultures may be unrewarding in distinguishing normal microbial flora of upper respiratory mucosa from those responsible for nonspecific infections. However, *cultures taken with thin swabs or wire loops* aimed at the principal lesions may reveal the causative microorganisms with some precision.

Care must be taken to avoid contaminating the swab or loop with saliva or the mucous secretions of uninvolved tissues. The collected material should be placed *immediately* in supportive "transport" medium, or streaked out on a blood agar plate, pending its delivery to the laboratory. Swabs should never be replaced in empty tubes where they may dry out before laboratory culture can be initiated, because fastidious streptococci and *Haemophilus* species may die, and even the hardier organisms (staphylococci or enteric bacilli) may be greatly reduced in numbers, so that their significance in culture is missed. In the laboratory, transport media are streaked on appropriate agar plates and incubated. Isolated colonies appearing on incubated plates are identified and, if indicated, tested in pure culture for antibiotic susceptibility. In general, pneumococci and other streptococci appearing on such plates are not routinely tested with antibiotics. With rare exceptions, pneumococci are uniformly susceptible to penicillin. Most streptococci are also, but in any case their association with clinical disease should first be confirmed, particularly if they are not beta-hemolytic strains (see Streptococcal Diseases, p. 316).

Serologic tests are not appropriate in nonspecific infections of the upper respiratory tract, because antibody levels induced by normal flora are not significantly higher than those seen in persons free of symptoms.

Epidemiology

I. Communicability of Infection

A. RESERVOIR, SOURCES, and TRANSMISSION ROUTES. The reservoir of these infections is man, and the source is usually endogenous. When man-to-man transmission is involved, the route is by airborne droplets or direct contact with secretions from the nose and throat of infected persons.

B. INCUBATION PERIOD. Unknown for endogenous infection; probably very short (one to two days) when transmissible.

C. COMMUNICABLE PERIOD. During time of active inflammation, excessive mucus production, or drainage of purulent exudates from nose and throat.

D. IMMUNITY. Specific immune responses either do not occur naturally or are not protective. The nonspecific resistance of normal, healthy tissues appears to constitute the major defense mechanism. Local injury to mucosal surfaces predisposes to infection of endogenous origin and increases susceptibility to exogenous infections should exposure occur.

II. Control of Active Infection

A. ISOLATION. Not warranted, but patient should avoid directly exposing those who may be unusually susceptible to infection (infants, aged people, surgical or obstetric patients).

B. PRECAUTIONS. Infected persons should practice careful personal hygiene: cover coughs and sneezes, collect nasal discharges in disposable tissues and discard these in closed paper or plastic bags for incineration, wash hands frequently and carefully.

C. TREATMENT. Nonspecific supportive measures designed to remove surface accumulations of exudates, promote blood supply, and reduce allergic reactivity of tissues, if any. Specific antibiotics are sometimes effective, but should be reserved until adequate clinical observation and laboratory studies have demonstrated their suitability in the individual case.

D. CONTROL OF CONTACTS

Case finding: Indicated only when risk of infection to exposed persons is great, as in hospital nurseries, operating rooms, or intensive-care units. Infectious personnel working in these areas should be removed from this duty until their infections are no longer communicable.

III. Prevention

IMMUNIZATION. Limited value. The use of autogenous bacterial vaccines (killed suspensions of bacteria isolated from the patient's lesions) and toxoids may be effective in reducing the hypersensitive response of tissues to these agents, provided a causal relationship exists between the organisms chosen for the vaccine and the active infection.

Bacterial Pneumonias

Pneumococcal Pneumonia

The Clinical Disease. The bacterial pneumonias, lobar or bronchial, are most frequently of pneumococcal origin. These are acute infectious diseases involving alveolar or bronchial structures of the lungs, or both. The onset is sudden and marked by chills, fever, chest pain, difficult breathing (dyspnea), and cough. The sputum is characteristically bright red or rusty with blood.

In *acute lobar pneumonia,* the alveolar spaces of an involved lobe fill with a fibrinous exudate containing red blood cells and polymorphonuclear leukocytes, and the lung becomes consolidated. Pneumococci are present in large numbers in the exudate; they may also be found in the bloodstream during early stages of disease, or persistently if infection is not arrested. These are extracellular, capsulated organisms. Their capsules prevent their ingestion by phagocytic cells and hence contribute to the virulence of pneumococci. The capsular material is antigenic, however, stimulating the production of antibodies that combine with it, change its surface properties, and thus promote phagocytic uptake of the bacterial cells. Once taken up by PMN's and macrophages, the pneumococci are readily killed.

Antibody production reaches a peak between the fifth and tenth days and may bring about dramatically sudden recovery of untreated cases by "crisis." Before the antibiotic era this was the usual resolution of recovered cases, often augmented by immunotherapy with a serum specific for the capsular antigen type of the causative strain. Today pneumococcal pneumonia is treated promptly with an antimicrobial agent, usually penicillin, which prevents or arrests lobar consolidation and terminates the disease rapidly. Pulmonary damage is not permanent.

Bronchopneumonia is more diffuse and less localized than the lobar variety, involving the bronchial tree rather than the alveoli.

Pneumococcal pneumonia is seldom a primary infection, but occurs as a secondary result of injury induced by other causes, e.g., virus infection, chemical irritation, allergic damage, alcoholism, impaired pulmonary circulation (as in chest surgery or cardiac disease). Pneumococci are members of the upper respiratory flora in 40 to 70 percent of normal persons. They may produce active infection if resistance is reduced by a predisposing cause, or they may be transmitted to others by asymptomatic infected persons.

The factors that predispose to pneumonia may similarly lead to extension of infection along mucosal surfaces of the upper respiratory tract. Pneumococcal sinusitis and otitis media may occur, and involvement of mastoid surfaces may lead to the further threat of pneumococcal meningitis. Pneumococci circulating in the bloodstream may reach the brain by this route as well, to localize on the meninges or within the brain.

Laboratory Diagnosis

I. Identification of the Organism (*Streptococcus pneumoniae*)

A. MICROSCOPIC CHARACTERISTICS. The pneumococcus is characteristically a Gram-positive diplococcus; it occurs in pairs or in short chains of pairs. The cocci are lancet shaped, the broad ends of a pair adjacent, the thinner ends pointing away from each other (Fig. 13–1).

B. CULTURE CHARACTERISTICS. Pneumococci are somewhat fastidious and do not survive well in competition with hardier organisms of the mixed throat flora present in expectorated sputum specimens. For this reason, sputum specimens submitted to the laboratory for confirmation of pneumococcal infection should be clearly identified as to purpose, so that appropriate culture techniques can be initiated.

Pneumococci are well supported by enriched culture media containing blood (see Fig. 4–3, p. 69) incubated in an atmosphere of 5 to 10 percent CO_2. The colonies growing on blood agar plates are surrounded by a zone of incomplete (alpha) hemolysis, usually accompanied by greening. In this respect they resemble alpha-hemo-

lytic streptococci, but can be differentiated from the latter by differences in colony morphology, some biochemical reactions to carbohydrates, and their greater sensitivity to bile salts and other surface-active agents.

C. SEROLOGIC METHODS OF IDENTIFICATION. Pneumococcal capsules are composed of an antigenic polysaccharide that stimulates antibody production. When anticapsular antibody is formed, it combines with the polysaccharide and changes the capsular properties so that phagocytosis proceeds normally. When the reaction of antibody with capsulated pneumococci is viewed under the microscope, the capsule appears to swell and its outer edge becomes sharply demarcated (Fig. 13–2). This reaction (often identified by the German word for swelling, *quellung*) is applied in the laboratory identification of pneumococci and in serologic recognition of their many different antigenic types.

About 82 serologic types of pneumococci have been distinguished on the basis of differences in the structure of their antigenic capsular polysaccharides, as detected by specific antiserums. Only a few of these types are associated with pneumococcal disease, types 1 to 8 accounting for most cases of adult pneumococcal pneumonia, while these plus types 14 and 19 are responsible for most cases in children.

The serologic identification of pneumococcal strains was a matter of clinical urgency before antimicrobial drugs were developed for therapeutic use because antiserums of homologous type were used in the treatment of pneumococcal infections. Passive immunization with antibody directed against the specific serologic type of

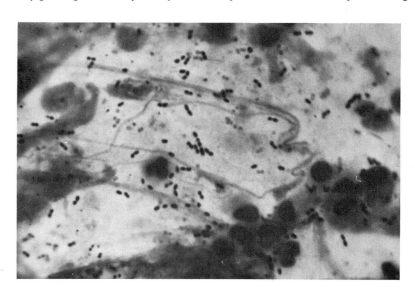

Fig. 13–1. Photomicrograph of *Streptococcus pneumoniae* in a sputum specimen. (1000×.) (Courtesy of Dr. Ramon Kranwinkel, Danbury Hospital, Danbury, Conn.)

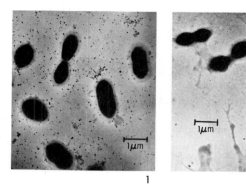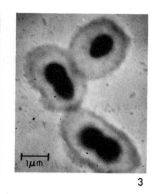

1   2   3

Fig. 13–2. Pneumococcus typing by capsular swelling reaction. *1. Streptococcus pneumoniae* type III were not exposed to pneumococcus antiserum. Their capsules are visible only as faint halos surrounding the bacteria. *2. S. pneumoniae* type III in type I serum. These organisms were exposed to a heterologous antipneumococcal serum prepared with type I organisms. The appearance of the capsules does not differ from those seen in the photograph to the left. *3. S. pneumoniae* type I in type I serum. Type I pneumococci were exposed to a homologous type I antipneumococccal serum. A strong "quellung" reaction has occurred as a result of specific antibody combination with capsular antigen. Note lancet shape of the organism. (Reproduced from Mudd, S.; Heinmets, F.; and Anderson, T. F.: The Pneumococcal Capsular Reaction, Studied with the Aid of the Electron Microscope. *J. Exp. Med.,* **78:**327, 1943.)

pneumococcus causing infection was of immediate value to the patient who had not yet developed his own active pneumococcus antibody. Chemotherapy has obviated the clinical need for pneumococcus typing, but the procedure remains valuable in tracing the epidemiology of pneumococcal disease.

II. Identification of Patient's Serum Antibodies. Impractical in view of cultural methods of diagnosis and the efficacy of prompt chemotherapy.

III. Specimens for Diagnosis (submitted, whenever possible, *before* antibiotic therapy is started.)

A. Blood samples drawn in the early febrile stage of disease may yield pneumococci promptly upon appropriate culture.

B. Sputum specimens for *smear* and *culture:* Gram-stained smears can be made and examined immediately for the presence of typical Gram-positive diplococci. Culture confirmation is usually possible within 24 to 48 hours.

C. Spinal fluid (if pneumococcal meningitis or other CNS involvement is suspected) for *smear* and *culture.*

Epidemiology

I. Communicability of Infection

A. RESERVOIR, SOURCES, AND TRANSMISSION ROUTES. Man is the reservoir; the organisms are present in respiratory secretions of asymptomatic carriers and patients with active infection. Transmission is by direct contact with infected persons, by droplet spread, and by indirect contact with infectious secretions on recently contaminated surfaces.

B. INCUBATION PERIOD. Unknown for infection of endogenous origin; probably very short (one to two days) for transmitted infection.

C. COMMUNICABLE PERIOD. Probably during time when pneumococci are present in large numbers in nasal and oral secretions. Prompt chemotherapy reduces these numbers sharply and shortens the communicable period (two to three days).

D. IMMUNITY. The nonspecific resistance of normal healthy tissues constitutes the major defense mechanism, holding most infections to an inapparent level. The immunity following an attack is type-specific; its duration depends on the level of antibody reached before antibiotic therapy eliminated the antigenic stimulus.

II. Control of Active Infection

A. ISOLATION. Not indicated; secretion precautions only.

B. PRECAUTIONS. During the communicable period, concurrent disinfection of respiratory secretions and of contaminated items. Terminal cleaning includes disinfection of furniture and floor.

C. TREATMENT. Penicillin is the usual drug of choice. Pneumococci are also susceptible to a wide range of antimicrobial drugs (erythromycin, cephalosporin, clindamycin). The route of administration (oral, intramuscular, or intravenous) depends on the severity of symptoms, localization of infection, characteristics of the drug, and response of the patient.

D. CONTROL OF CONTACTS. None indicated.

E. EPIDEMIC CONTROLS. Generally not necessary, unless outbreaks are threatened in closed populations with low resistance (pediatric wards, geriatric institutions, military hospitals, prisons). Vigorous sanitary and hygienic measures are indicated with chemoprophylaxis for the unusually susceptible. Epidemic outbreaks must be reported to the local health department; individual case reports are not required.

III. Prevention

IMMUNIZATION. Active immunization with bacterial vaccines or polysaccharide antigens is impractical except for persons at high risk (alcoholics or individuals with chronic diseases, for whom recently developed pneumococcal vaccines are now recommended).

Common Bacterial Pneumonias Other Than Pneumococcal

The Clinical Disease. About 20 percent of bacterial pneumonias are caused by organisms other than the pneumococcus. They are acute febrile infections that may localize in any segment of the bronchopulmonary system. They are often sequels to virus infections (influenza, viral pneumonia) or other injury and are caused most often by "opportunistic" bacteria that are, like the pneumococcus, normal inhabitants of the upper respiratory tract, e.g., *Staphylococcus aureus, Streptococcus pyogenes, Klebsiella pneumoniae,* and *Haemophilus influenzae.*

The pathogenesis of these pneumonias is similar to that caused by *S. pneumoniae.* They occur most frequently in persons whose normal resistance has been lowered in some way. The degree of permanent damage done to lung tissues varies, however, with the capacities of the causative organism to induce injury of parenchymal cells. Unlike the pneumococcus, some of these bacterial invaders may produce necrosis of local tissues or blood vessel walls through the action of exo- or endotoxins. The hemolytic toxins of *S. aureus* and *S. pyogenes,* and the endotoxins of *K. pneumoniae* (a Gram-negative bacillus), induce hemorrhagic consolidation and eventual necrosis of lung parenchyma if untreated. This damage is resolved with difficulty and is often replaced by permanent scar tissue (fibrosis).

K. pneumoniae and *H. influenzae* share with the pneumococcus the pathogenic feature of a polysaccharide capsule that protects them from phagocytosis. The capsular polysaccharides of these organisms are also antigenic and can be identified by serologic methods, including the capsular swelling test. There are 77 known antigenic types of *K. pneumoniae,* but only six of *H. influenzae,* the most pathogenic of the latter being type b. The incidence of these pneumonias is not high, but their morbidity and fatality depend on the patient's resistance and on the adequacy of chemotherapy. In *Klebsiella* pneumonia, pulmonary abscess formation, often associated with bronchial obstruction, is a common complication.

H. influenzae infection is seen most often in children and is usually caused by type b strains. It often begins in the upper respiratory tract as a pharyngitis, sinusitis, or epiglottitis. In the latter case, severe infection can lead to laryngeal obstruction, sometimes requiring tracheotomy to prevent asphyxia. Invasion of the lower respiratory tract, with development of *H. influenzae* pneumonia, occurs in adults as well as children, often as a sequel to virus infection. (The organism acquired its name because it was isolated frequently from cases of pneumonia that occurred during the 1918 viral influenza epidemic and was wrongly concluded to be the cause of the primary disease.) *H. influenzae* frequently invades the bloodstream, from upper or lower respiratory sites of heavy infection, and in children meningitis is a common result (see Secondary Bacterial Meningitis). *H. influenzae,* type b, is the most usual cause of meningitis in infants and very young children, probably because they have not yet developed antibodies as a result of colonization and subclinical infection.

Secondary bacterial pneumonias can also be caused by a variety of Gram-negative commensalistic bacteria, particularly those commonly associated with the intestinal tract or environmental sources, e.g., *E. coli, Proteus* and *Pseudomonas* species, and *Serratia marcescens.* These are almost always nosocomial infections seen in hospitalized patients, especially those with underlying pulmonary disease who are receiving inhalation therapy. Prior chemotherapy of these patients may have contributed to the pneumonia by modifying the normal flora of the superficial mucosae. Treatment with inappropriate antibiotics, such as penicillin, to which Gram-negative bacilli are resistant, results in a buildup of these organisms in the upper respiratory tract and increases the risk of further infection. In addition, the water in ventilator reservoir nebulizers and room humidifiers can easily become contaminated with *Pseudomonas,* and a number of pneumonia outbreaks linked to such sources have been reported.

Laboratory Diagnosis
I. The Organisms
A. STREPTOCOCCI—hemolytic strains of serologic group A (the primary human pathogens). Microscopically, these are Gram-positive cocci occurring in chains of medium to great length. In culture they are somewhat fastidious but are supported by blood-enriched media incubated aerobically or under increased CO_2 tension. Colonies on blood agar are surrounded by a zone of complete (beta) hemolysis. Serologic identification is based on differentiation of their somatic polysaccharide antigens. Numerous groups, designated by letters, are recognized but most of the strains associated with human disease fall into group A. (See the discussion of streptococcal diseases in this chapter.)

B. STAPHYLOCOCCI—microscopically, Gram-positive cocci in clustered arrangements, resembling grapes. These organisms grow readily in most simple nutrient media. Strains most frequently associated with human pathogenicity are hemolytic on blood agar, display a golden pigment, and produce a plasma-coagulating enzyme. When these characteristics are seen, strains are reported as *Staphylococcus aureus.* (See the discussion of staphylococcal diseases in Chap. 22.)

C. KLEBSIELLA PNEUMONIAE—microscopically, a Gram-negative plump, short bacillus with a large capsule. They are nonsporing and nonmotile. Klebsiellae grow with ease on common laboratory media, their colonies being quite mucoid because of capsular polysaccharide production. The latter are antigenic and can be identified by serologic methods such as the quellung reaction. There are many antigenic types but most strains that cause human respiratory infection fall into capsular types 1 and 2. (Other types are associated with urinary infections.)

D. HAEMOPHILUS INFLUENZAE—microscopically, a Gram-negative bacillus, slender but pleomorphic. Most are short and coccoid, but long filamentous rods may also be seen. This organism is nonmotile and nonsporulating. It requires enriched media and microaerophilic incubation for growth. *H. influenzae* produces antigenic capsular polysaccharides identifiable by the quellung technique. There are six antigenic types, lettered a through f.

E. GRAM-NEGATIVE AEROBIC BACILLI—most of the Gram-negative bacilli associated with the intestinal and upper respiratory membranes are aerobic, nonsporulating rods, indistinguishable from each other microscopically. The major genera include *Escherichia, Enterobacter, Proteus, Serratia,* and *Pseudomonas* (see Chap. 4, pp. 91–92). They are identified by genus and species on the basis of biochemical properties, particularly characteristic enzymic pathways in the utilization of carbohydrates. Except for *Escherichia,* members of these genera are also common in the environment (soil, water).

II. Identification of Patient's Serum Antibodies.
Except for streptococcal infection, serologic diagnosis is not routinely practiced.

III. Specimens for Diagnosis
A. Sputum specimens for smear and culture.
B. Blood specimens for culture.
C. Throat or nasopharyngeal swabs, in transport medium, may be useful for culture, particularly if concurrent URI exists.
D. Specimens should be collected for laboratory diagnosis before antibiotic therapy is started, whenever possible. Prompt transport to the laboratory is essential for successful isolation of fastidious pathogens, such as *Haemophilus* and streptococci.

Epidemiology. The epidemiology of these infections is similar to that of pneumococcal pneumonia. Specific immunity plays a minimal role, the best defense and preventive control being the maintenance of healthy, nonspecific resistance.

Secondary bacterial pneumonias occur sporadically, never in epidemic patterns. Their nursing care does not require isolation, but appropriate precautions should be taken with the patient's secretions. Concurrent disinfection of inhalation therapy equipment is of particular importance.

Specific antimicrobial therapy is based on the identification of the infectious agent and antibiotic susceptibility testing. Penicillin is the drug of choice for streptococcal infections and for staphylococci that do not produce penicillinase. Staphylococci that are penicillinase producers are generally susceptible to synthetic penicillins that are impervious to penicillinase, e.g., methicillin and oxacillin. For *H. influenzae* pneumonia or meningitis, ampicillin has been the drug of choice. However, a number of *H. influenzae* strains also produce a penicillinase that inactivates this antibiotic, and for these the drug of choice is chloramphenicol. *K. pneumoniae* is characteristically resistant to penicillin and is treated with either a cephalosporin or gentamicin, or a combination of these two drugs. These variations in microbial responses to antibiotics underline the necessity for adequate antibiotic testing in the laboratory.

Legionnaires' Disease

During the summer of 1976 an epidemic of respiratory illness swept through a group of people who had attended a state convention of American Legionnaires in Philadelphia. Before it ended, this epidemic took 29 lives and hospitalized some 170 persons of the 5000 or so who had gathered for the convention. The outstanding clinical symptoms were those of pneumonia, with chest pains, lung congestion, headache, dizziness, chills, and high fevers up to 107° F (41.7° C). As reports of illness and deaths mounted during the days and weeks that followed, it became obvious that the Legionnaires' convention, and one Philadelphia hotel in particular, had been the focal point, and identification of the agent of this disease became urgent.

At that time in the United States attention had been focused on the possibility that an epidemic of "swine flu" might be anticipated later in the year.

Swine flu is a type of viral influenza that occurs in domestic hogs but is rarely seen in humans, with one notable exception. In 1918 a deadly pandemic of influenza inundated human populations in many parts of the world, gathering a high mortality rate as it went. Although the viral agent was not identified at the time, later serologic evidence indicated strongly that swine influenza virus (or a closely similar one) had been responsible. Therefore, when a few human cases of swine flu, with one death, were identified in the early months of 1976, it was realized that the majority of the population was once again susceptible to this disease. In spite of considerable controversy, the federal government agreed to the recommendations of experts in the U.S. Public Health Service, and elsewhere, that a nationwide effort be made to develop a vaccine for swine influenza virus and to establish a mass immunization program against the possibility of a winter epidemic.

In this atmosphere, the outbreak of respiratory illness in Pennsylvania in late July and August of that same year immediately raised the question of influenza. Intensive laboratory studies definitely ruled out this possibility, however, along with other known viral or bacterial agents. Armies of epidemiologists and laboratory workers continued to sift for evidence of the nature of the disease. Chemical toxins as well as microbial agents were sought. Many samples of autopsied lung tissue and blood specimens were sent to the Public Health Service's Center for Disease Control (CDC) in Atlanta, as well as to the laboratories of other experts in the Pennsylvania State Health Department and elsewhere. Because of the similarity of the Legionnaires' symptoms to those of the rickettsial disease called Q fever, the CDC paid particular attention to this possibility in its range of techniques for isolation of the agent. These included inoculation of guinea pigs and embryonated chicken eggs. However, the Legionnaires' serum samples showed no antibodies for known rickettsial organisms. Finally, in the early weeks of 1977, bacteria-like organisms were seen in the yolk sacs of eggs inoculated with material from guinea pigs that had in turn been injected with lung tissue from an autopsied human victim. Serologic evidence that this organism had been the cause of the Legionnaires' epidemic was provided by the demonstration of antibodies against it in the serum of surviving individuals who had been ill, as well as in the stored serums taken earlier from these persons. A significant rise in antibody titer was seen in serum taken after recovery as compared with that drawn during the acute illness.

Efforts to identify the isolated organism continue. It is a bacillus, larger in size than a rickettsia — Gram-negative and rather pleomorphic. Although initially it would grow only in animal tissues, it was later cultivated on enriched bacteriologic media. It is extremely fastidious, however, requiring three to five days to form pinpoint colonies on an artificial medium.

Serologic tests conducted with this organism indicate that it may also have been the cause of other respiratory illnesses previously unidentified. Serum samples preserved after a similar, but much smaller, outbreak that occurred in 1965 in a Washington, D.C., hospital revealed significant levels of antibody for the unidentified bacterium. Many questions remain unanswered as yet, however. Nothing is known to date about the organism's origins, transmission, or pathogenicity. It appears to have a very low communicability, since there have been no secondary cases either in the Philadelphia outbreak or elsewhere, among close family relatives of the victims or among those who cared for them medically.

Some evidence has accumulated that erythromycin may be an effective drug against the Legionnaires' disease bacillus. This evidence has emerged from antibiotic susceptibility testing *in vitro,* from protection studies with embryonated eggs, from treatment studies with infected guinea pigs, and from a review of individual human case reports. The optimal antibiotic therapy of this disease has not yet been determined, however.

Sporadic cases and small clusters of cases of pneumonia apparently caused by the Legionnaires' disease agent have now been recognized in a number of states and in Europe. Many of these, as well as cases in the Philadelphia outbreak, have been characterized by severe pulmonary involvement, with extensive bilateral consolidation. A number of deaths occurred in older age groups or in persons compromised by underlying disease. This is an interesting example of an apparently "new" disease that has hitherto gone unnoticed in its sporadic appearances, partly at least because its microbial agent has been extremely difficult to isolate or identify, and partly perhaps because pulmonary deaths are not unexpected in some compromised patients. It required a large-scale outbreak, in an identifiable group of people, against a background of national alertness to another epidemic problem, to bring about coordinated efforts to reveal the microbial agent. Obviously, however, this organism has been restricted to date in its transmissibility. Further information on this point will be of great interest to the public health.

Primary Plague Pneumonia

The Clinical Disease. Plague is an endemic infection of rats and other wild rodents, maintained by a flea vector. When man is bitten by this flea and infected, he develops the bubonic form of plague (see Chap. 26). In late or terminal stages of bubonic plague, the invading organism (*Yersinia pestis*) disseminates to the lung, where it localizes and multiplies in large numbers. The sputum and respiratory secretions of such patients are extremely infectious and can transmit the organism directly to others. The respiratory entry of the infectious agent then results in primary plague pneumonia. This disease may occur in epidemic patterns among persons closely exposed to a plague source and to each other. It is a highly fatal infection, characterized by rigor, severe headache, generalized pain, difficult breathing, productive cough, and high fever. Sputum is watery, frothy, and easily projected in a droplet spray for several feet by the coughing patient. Prompt diagnosis and antibiotic therapy may be lifesaving.

The *laboratory diagnosis* and the *epidemiology* of plague are discussed in Chapter 26.

Meningitis

General Comments

The term *meningitis* signifies an acute inflammation of the meningeal membranes of the brain and spinal cord. It may be associated with a variety of nonspecific injuries, but it is also commonly caused by the localization of an infectious microorganism

Table 13–1. Important Microbial Agents of Meningitis

Type of Infection	Mode of Spread to Meninges	Organisms Most Commonly Involved
Bacterial meningitis		
Infants under 2 months	From intestinal tract or skin via bloodstream Direct from birth canal (rare)	*Escherichia coli* *Streptococcus,* group B
Children, 2 months to 5 years	From upper respiratory tract via bloodstream Direct invasion	*Haemophilus influenzae,* type b *Streptococcus pneumoniae* *Neisseria meningitidis*
Adults	From upper respiratory tract via bloodstream Direct invasion	*Neisseria meningitidis* *Streptococcus pneumoniae* *Haemophilus influenzae,* type b (occasional in oldest age group)
Any age	Direct invasion after head injury, neurosurgical or diagnostic procedure	Staphylococci *Streptococcus,* group A *Streptococcus pneumoniae* *Pseudomonas aeruginosa*
Tuberculous meningitis	From lesion in lung via bloodstream	*Mycobacterium tuberculosis*
Viral (aseptic) meningitis	From upper respiratory or intestinal tract via blood	Mumps virus Enteroviruses
Viral encephalitis	From upper respiratory or intestinal tract via blood	Mumps virus Enteroviruses Herpes virus
	Arthropod vector	Arboviruses
Fungal meningitis	From lesion in lung or other organ via blood	*Cryptococcus neoformans*

(bacterium, virus, or fungus) on the meninges (Table 13–1). The clinical as well as the epidemiologic pattern of infectious meningitis varies greatly with the type of organism involved.

Bacterial meningitis is clinically acute, characterized by an outpouring of purulent exudate over the membrane surfaces. The term *septic meningitis* is sometimes used with reference to this purulence. The exudate is largely composed of polymorphonuclear cells, which appear in great numbers in the spinal fluid. Epidemiologically, most types of bacterial meningitis are associated with respiratory routes of entry. The bacterial agents are often commensals of the respiratory mucosa. From this local site they may reach the bloodstream and infect other organs, with meningitis a secondary result of bacteremia. In some instances, direct penetration from the upper respiratory tract and invasion of the central nervous system may occur. Meningitis may also be secondary to head injury, or to surgical or diagnostic procedures involving the central nervous system. Spinal anesthesia, lumbar puncture, and pneumoencephalography may set the stage for opportunistic pathogens that invade directly.

The age and immune status of the host are correlated with the specific types of bacteria involved in meningeal infections. The major defense mechanism against bacteremia and the spread of infection to the central nervous system is thought to be the activity of opsonizing serum antibodies. In newborn infants less than two months of age, whose immune mechanisms are undeveloped, meningitis may be the result of colonization of the intestinal tract or skin by organisms encountered in the birth canal, with ensuing spread through the infant's bloodstream. Direct invasion by commensalistic bacteria of the birth canal may occur, but this is rare. The organisms most frequently involved in neonatal meningitis are group B streptococci and *E. coli,* both common members of the vaginal flora. In older children from two months to five years of age, meningitis results from bacteremia or direct extension with organisms colonizing the upper respiratory tract, the most frequent agent being *Haemophilus influenzae,* type b (sometimes *Streptococcus pneumoniae* or *Neisseria meningitidis*). In most adults, these are also the common modes of meningeal infection, with *N. meningitidis* (meningococcus) the usual agent in young adults and *S. pneumoniae* (pneumococcus) in the older age group. *H. influenzae* meningitis is occasionally seen in aging patients (over 65) who have lost their serum antibody for type b strains. Infections following head injuries, or those of iatrogenic origin, most often involve staphylococci, group A streptococci, pneumococci, and *Pseudomonas aeruginosa.*

The clinical picture of tuberculous meningitis is an exception to that of other bacterial infections of the meninges. This is a subacute or chronic progressive disease, infection being spread through the blood from a primary lesion in the lung. There is no purulent polymorphonuclear exudate, the local lesion, wherever it occurs, being a *granuloma* composed of lymphocytes and other mononuclear cells (see pp. 331–32).

Viral meningitis (sometimes called *aseptic meningitis* because it is nonbacterial and not purulent) is clinically a more benign disease, usually of short duration. Cellular reaction is less marked than in bacterial infections and is characterized by a predominance of lymphocytic cells. The viral agents of meningitis may come and go on either respiratory or oral routes of entry and transfer, reaching the meninges by way of the bloodstream, as do bacteria.

Viral *encephalitis,* or infection of brain tissue itself,

is a much more severe disease, involving cerebral dysfunction as well as all the manifestations of meningitis.

Fungal meningitis is very different clinically and epidemiologically from bacterial or viral meningeal disease. The fungal organisms are airborne and usually enter the body by the respiratory route, but they come from *soil* reservoirs. Man is not a reservoir for them and does not transmit them directly. Clinical infections induced by fungal agents are chronic, persistent, and slowly progressive. From primary lesions in the lungs or reticuloendothelial system, fungi may be disseminated to other organs or to the meninges. When meningitis occurs it is also chronic, progressive, and often fatal. The cellular reaction is predominantly lymphocytic.

Meningococcal Meningitis

The Clinical Disease. The only bacterial meningitis that can spread in epidemic form is that caused by *Neisseria meningitidis,* the meningococcus. This is an acute disease characterized by sudden onset of fever, severe headache, painful rigidity of the neck, nausea, and vomiting. Convulsions are often seen in children. Delirium or coma is frequent. The meninges are involved in an acute inflammatory reaction characterized by a purulent exudate. If the disease is not arrested before this exudate becomes abundant, the meninges later become thickened as inflammation is organized. Hemorrhage and thrombosis of small blood vessels contribute to the damage, which may be residual if treatment is delayed or ineffective.

Meningococci enter the body through the nasopharynx and localize there. They may remain there without any injury to tissues, or they may induce a local inflammation. In a small proportion of cases, meningococci enter the bloodstream. They may be filtered out and killed by the body's defensive mechanisms, or meningococcemia may ensue, with a possible distribution of the organisms to many foci of multiplication throughout the body, notably the skin, joints, lungs, adrenal glands, and central nervous system. Destruction of capillary walls by meningococci leads to small petechial hemorrhages in the skin and other affected tissues. Fulminating infections may lead to adrenal hemorrhage, circulatory collapse, and shock (Waterhouse-Friderichsen syndrome). The clinical appearance of meningococcemia is marked by fever, chills, acute malaise, petechial rash, and prostration. The disease may be rapidly fatal, with death occurring in a few hours, before meningitis can develop.

Although the usual route of spread to the meninges involves bacteremia, meningococci may penetrate through the thin bony lamina posterior and superior to the nasopharynx and invade the central nervous system directly.

Laboratory Diagnosis

I. Identification of the Organism (*Neisseria meningitidis*)

A. MICROSCOPIC CHARACTERISTICS. The organism is a Gram-negative diplococcus. Each member of a pair is characteristically flattened on its adjoining side and rounded on its outer edge, so that it resembles a kidney bean. In direct stained smears of purulent spinal fluid the diplococci are often seen in intracellular positions within polymorphonuclear phagocytes.

B. CULTURE CHARACTERISTICS. Meningococci belong to the genus *Neisseria*, which contains one other important pathogen (*Neisseria gonorrhoeae*). These organisms are aerobic, nonsporulating, nonmotile, and quite fastidious in their growth requirements. They grow best on blood-enriched media incubated microaerophilically, with added CO_2. Their colonies are transparent, glistening, and nonhemolytic. They are distinguished from gonococci by differences in their ability to produce acid from carbohydrates.

C. SEROLOGIC METHODS OF IDENTIFICATION. Like pneumococci, pathogenic meningococci possess capsules composed of antigenic polysaccharides that stimulate antibody production. Their recognition by capsular swelling or precipitation techniques may be of value in diagnosis, and in epidemiologic tracking of strains involved in outbreaks of meningitis. Counterimmunoelectrophoresis is the most useful application of the precipitation technique for the detection of capsular antigens in cerebrospinal fluid or serum (see Chap. 7, p. 174).

The capsular antigens fall into several serologic types. Groups A, B, and C were once the most prominent, group A being responsible for worldwide epidemics, while groups B and C were associated with sporadic or endemic disease. In recent years, group C has become more predominant, and new serotypes labeled X, Y, Z, 29-E, and W-135 have been identified. Of these, group Y has assumed clinical importance as the cause of small outbreaks of meningococcal disease. Group A is endemic in North Africa and still causes epidemics in some parts of the world, the most recent being a large one in Brazil.

II. Identification of Patient's Serum Antibodies. A rising titer of antibodies produced during the course of illness can be demonstrated by agglutination techniques, but this is seldom necessary.

III. Specimens for Diagnosis

A. Blood cultures drawn as soon as possible after onset of symptoms.

B. Spinal fluid samples for *immediate* reading of Gram-stained smears and initiation of culture. Serologic demonstration of capsular polysaccharide antigens in spinal fluid, if successful, provides a rapid, confirmed diagnosis. Cell counts and chemical analysis for glucose and protein are also essential.

C. Nasopharyngeal swabs, in transport medium, may detect carriers.

D. Petechial lesions on the skin may be scraped or punctured and submitted for smears and culture.

E. Blood samples to detect capsular antigens in serum by counterimmunoelectrophoresis.

IV. Special Considerations

A. Meningococcal meningitis is a medical emergency requiring prompt, accurate diagnosis and therapy. Permanent tissue damage or death may result from delay. Pretreatment specimens, aseptically collected in adequate quantity and promptly examined, are essential to laboratory diagnosis.

B. Spinal fluid must be analyzed by chemical and cytologic methods, as well as by culture techniques. Collection in three separate sterile tubes is recommended for speedy and efficient distribution to appropriate sections of the laboratory. The quantity per tube should be predetermined by consultation with the laboratory so that the requirements of its methods can be met. If only one tube is used for collection, it should be submitted *first* to the microbiology laboratory for aseptic withdrawal of an aliquot for smear and culture.

C. Any material submitted for culture of meningococci should be clearly identified for this purpose so that appropriate examinations may be made promptly by the laboratory.

D. Meningococci lyse rapidly in extravasated body fluids without nutrient. They are very sensitive to temperature changes. For these reasons specimens for meningococcus culture should be delivered *without delay* to the laboratory and placed in the hands of the bacteriologist, not in incubators, in refrigerators, or on unattended laboratory benches.

Epidemiology

I. Communicability of Infection

A. RESERVOIR, SOURCES, AND TRANSMISSION ROUTES. Man is the only known reservoir. The nasal and oral secretions of infected persons are the principal source. Epidemics are spread primarily by asymptomatic carriers rather than by infectious cases and have occurred most frequently in closed populations, such as military groups. Meningococci do not survive long in the environment; transmission is by direct contact or droplet spread. In recent years a rising incidence of meningococcal isolations from the genitourinary tract has been reported. Changing social attitudes and sexual habits are probably responsible for introducing these organisms to the urogenital tract and anus of both males and females. This may constitute a new infectious hazard for the newborn passing through a colonized birth canal. At least one case of fatal neonatal meningococcal meningitis has been reported, the mother's cervicovaginal cultures being positive for *N. meningitidis*.

B. INCUBATION PERIOD. The extremes are two and ten days, the average being three to four days.

C. COMMUNICABLE PERIOD. The disease is potentially transmissible while meningococci are present in respiratory discharges. Effective chemotherapy generally clears the nasopharyngeal focus within 24 to 48 hours.

D. IMMUNITY. Asymptomatic carriers greatly outnumber cases of infectious disease, indicating generally low susceptibility to clinical illness. Specific immunity following recovery from the disease is of uncertain importance in providing future protection. Capsular polysaccharide vaccines for artificial active immunization have been developed and are being used for high-risk groups, e.g., military recruits. Success with group A and group C vaccines has been particularly encouraging.

II. Control of Active Infection

A. ISOLATION. Respiratory isolation for at least 24 hours following initiation of specific therapy, or until the nasopharynx is culturally negative.

B. PRECAUTIONS. During the communicable period, concurrent disinfection of respiratory secretions and of contaminated items. Terminal cleaning should include disinfection of furniture and floor.

C. TREATMENT. Intravenous penicillin in high dosage is the regimen of choice for either meningitis or meningococcemia. Chloramphenicol is useful for those patients who are hypersensitive to penicillin or related drugs.

D. CONTROL OF CONTACTS

1. *Quarantine.* None.

2. *Prophylaxis.* Immunization with groups A and C polysaccharide vaccines appears to be a safe, effective method for providing artificial immunity and preventing disease caused by organisms of these two groups. The durability of artificially acquired immunity remains to be determined.

The choice of drugs for chemoprophylaxis depends on the susceptibility of the strain of *N. meningitidis* to which contacts have been exposed. The Center for Disease Control has reported that 23 percent of case isolates of meningococci in the United States are resistant to sulfonamides but susceptible to rifampin. The CDC therefore recommends that rifampin be used in the United States for family contacts of cases of meningococcal meningitis unless the infecting strain has been proven susceptible to sulfadiazine. In the latter case sulfadiazine would be the drug of choice for prophylaxis (see also Chap. 10, p. 268). Penicillin is ineffective in eliminating the carrier state and is not recommended for the treatment of contacts.

Local health departments should be consulted as to currently acceptable chemoprophylactic regimens.

3. *Case Finding.* Unnecessary.

E. EPIDEMIC CONTROLS. Individual cases must be reported to local health departments. Chemoprophylaxis of close contacts and surveillance of their health and living conditions limit spread of an outbreak by reducing the number of carriers and permitting early segregation of the clinically ill. Outbreaks occur most frequently in closed populations (military barracks, schools, camps, and institutions). Physical separation, increased ventilation of quarters, and vigorous insistence on good personal hygiene help to limit outbreaks.

III. Prevention

A. IMMUNIZATION. Not routine. Immunoprophylaxis may prove to be the best means of control in the future for selected groups at risk.

B. CONTROL OF RESERVOIRS, SOURCES, AND TRANSMISSION ROUTES. Public education concerning the source of infection, respiratory hygiene, avoidance of direct contacts, and hazards of crowding in living, working, or traveling conditions.

Secondary Bacterial Meningitis

The Clinical Disease. Acute bacterial meningitis may be secondary to respiratory tract infections caused by organisms of the indigenous flora. Entry of these organisms into the bloodstream from the site of active lesions in the upper or lower respiratory tract probably constitutes their chief route of access to the meninges. The latter event does not occur frequently, presumably because the blood's defensive barriers are effective in eliminating most bacteremic infections. Bacteria of enteric origin are also sometimes responsible for acute meningitis if they find a disseminating route from colonizing sites in the bowel. The clinical disease produced in these situations resembles meningococcal meningitis in every way, except that petechial lesions characteristic of meningococcal injury are usually not seen in other bacterial infections.

Fractures of the skull or other head injuries may create possibilities for direct invasion of the subarachnoid space by bacteria. Surgical and other procedures involving even minor trauma to the central nervous system may also open a pathway for invading microorganisms and may result in meningitis.

Laboratory Diagnosis
I. Identification of the Organisms. Differentiation of these meningitides requires bacteriologic recognition of their agents. The organisms most frequently involved are shown in Table 13–1. Some of these, notably *Haemophilus influenzae,* type b, and *Streptococcus pneumoniae* are also associated with bacterial pneumonias. A variety of other bacteria may also be responsible for secondary meningitis. These include a number of commensal or pathogenic bacilli of the intestinal tract, such as *E. coli,* species of *Proteus* and *Pseudomonas,* and *Salmonella.* The bacteriology of enteric bacilli is described in Chapter 18.

Tuberculous meningitis is an important disease that may occur as a sequel to pulmonary tuberculosis. This disease is discussed in the last section of this chapter.
II. Identification of Patient's Serum Antibodies. Serologic diagnosis is seldom helpful in these infections.

III. Specimens for Diagnosis
A. Blood for culture, particularly in the early stage of disease.
B. Spinal fluid for smear and culture.
C. Throat swabs or sputum specimens sometimes help in recognizing endogenous sources of meningeal infection.
IV. Special Considerations. Acute infectious meningitis is a medical emergency. See Special Considerations in the laboratory diagnosis of meningococcal meningitis, items A, B, and C.

Epidemiology
I. Communicability of Infection
A. RESERVOIR, SOURCES, AND TRANSMISSION ROUTES. Man is the reservoir. Endogenous infectious lesions in the respiratory tract or other organs are often an immediate source of meningeal infection. Nasal, oral, and intestinal discharges of infected persons may be a source for susceptible individuals. Transmission is by direct contact, by droplet spread, or by the fecal-oral route.

B. INCUBATION PERIOD. Not measurable in secondary meningitis.

C. COMMUNICABLE PERIOD. Organisms readily transmissible, but communicability depends on susceptibility of contacts.

D. IMMUNITY. General resistance to the pathogenic potential of these organisms is high, but children are often more susceptible than adults, particularly in the case of *H. influenzae.* This organism is one of the most common causes of meningitis in infants and children. Antibody response may lower susceptibility to severe *H. influenzae infection,* but in general the role of specific immunity to other organisms of this group is not significant.

II. Control of Active Infection
A. ISOLATION. Not necessary, provided the possibility of epidemic spread of meningococcal meningitis has been ruled out by prompt bacteriologic diagnosis.

B. PRECAUTIONS. Concurrent disinfection of respiratory secretions or other discharges as indicated by possible systemic foci of infection. Special care in disposal of urine and feces may be indicated, for example, if *Salmonella* or other pathogenic enteric bacillary agents of meningitis are localized primarily in the kidney or gallbladder.

C. Treatment. Antibiotic drugs are effective when given promptly on onset of symptoms. Speed is essential in identifying the causative organism and determining its susceptibility to antibiotics. Pending preliminary and confirmed information from the laboratory, antibiotics are given empirically to control the disease.

D. Control of Contacts. Generally not indicated except in the case of *Salmonella* infections.

E. Epidemic Controls. Generally not applicable. Case reports should be sent to local health departments.

III. Prevention

A. Immunization. None.

B. Avoidance of infectious contacts by susceptible persons, particularly children. Support and maintenance of healthy resistance.

Listeriosis

The Clinical Disease. Listeriosis is a disease of animals and man that has attracted new interest in recent years. It is a sporadic disease in man, or so it would seem because of the difficulty of recognizing it clinically or bacteriologically.

The organism has a widespread distribution among many animal hosts and may be acquired by man through inhalation, ingestion, venereal contacts, or congenital transfer. The probability is that the organism is readily transmissible, meets with resistance in the normal human adult, but lives commensalistically on his tissues until he becomes more susceptible to infection. Adults with underlying debilitating diseases (such as neoplasms), and particularly those who are maintained on steroid therapy, may develop listeriosis, manifest usually as an acute meningitis.

The onset of listerial meningitis is sudden, signaled by fever, chills, acute headache, stiff neck, and other signs of purulent bacterial meningitis. Delirium is common; coma or shock may follow. Bacterial infection can be confirmed by proper laboratory examination of spinal fluid.

The organisms may be found on the genital mucosae and can be transferred through venereal con-

tacts. The clinical result of venereal transfer among adults is not remarkable and may go undetected, but if conception occurs, the fetus may be severely involved. Pregnant women infected venereally may display mild febrile symptoms related to a light septicemic spread of the organisms. Infants acquire the infection *in utero*. The fetus may be aborted or stillborn. Those who survive the prenatal period may have a massive septicemia, manifest at birth by multiple infectious abscesses in internal organs and on external surfaces (granulomatosis infantiseptica); others may develop meningitis neonatally.

Laboratory Diagnosis

I. Identification of the Organism (*Listeria monocytogenes*)

A. Microscopic Characteristics. The organism is a Gram-positive, nonsporulating, motile, aerobic bacillus.

B. Culture Characteristics. Listeria is sometimes difficult to grow in primary isolation on artificial culture media from clinical specimens. When it does appear on blood agar plates among mixed flora of vaginal or other mucosal cultures, its small hemolytic colonies may be mistakenly interpreted as streptococci if not smeared. Microscopically, *Listeria* resembles nonpathogenic members of the coryneform group (diphtheroids), which are common commensals of mucosal and skin flora. Accurate bacteriologic diagnosis depends, therefore, on recognition of *Listeria* as a Gram-positive rod (not a coccus) displaying hemolytic properties and motility (unlike diphtheroids).

Cultures of spinal fluid and deep tissue abscesses from cases of listeriosis yield the organism in pure culture, sometimes with difficulty on initial attempts. Storage of specimens at refrigerator temperatures for days, or weeks, may lead to more successful recovery of *Listeria* upon subculture.

C. Serologic Methods for Identification. Not useful.

II. Identification of Patient's Serum Antibodies. Not routine.

III. Specimens for Diagnosis

A. Spinal fluid for smear and culture.

B. Blood for culture in the early stages of meningitis.

C. From postpartum mothers with infected infants, cultures of vaginal discharge, placenta, milk.

D. From infected infants, cultures of blood, urine, meconium, lesions.

IV. Special Laboratory Tests

A. Specimens for culture are divided in the laboratory for immediate culture and for repeated subculture after refrigerator storage.

B. Suspicious organisms isolated from cultures may be instilled in a rabbit's eye. *L. monocytogenes* produces a characteristic keratoconjunctivitis.

Epidemiology

I. Communicability of Infection. See initial discussion.

II. Control of Active Infection

A. ISOLATION. Since human resistance appears to be normally high, disease is not readily communicable, although the organisms are transmissible. Therefore, isolation of adult meningitis cases is not warranted. Infectious infants and mothers should be isolated until lesions are healed or discharges are bacteriologically negative for *Listeria.*

B. PRECAUTIONS. Concurrent disinfection of all items associated with care of infectious infants or the vaginal discharges of mothers. Disinfection of respiratory secretions of meningitis patients during the acute stage. Terminal disinfection of units.

C. TREATMENT. Tetracycline therapy is the usual choice; ampicillin and streptomycin in combination are also used.

D. CONTROL OF CONTACTS. Not indicated.

E. EPIDEMIC CONTROLS. None indicated. Cases should be reported to local health departments for the record, but this is not a feature of control.

III. Prevention

A. IMMUNIZATION. Immunizing agents are not available. (Protective immunity is apparently not acquired as a result of disease.)

B. CONTROL OF RESERVOIRS, SOURCES, AND TRANSMISSION ROUTES. Education of persons (especially pregnant women) continuously exposed to domestic animals, with particular regard to precautionary techniques for handling diseased animals.

Streptococcal Diseases

General Comments

The streptococci are a large and diverse family of microorganisms, among which several groups are of great clinical importance. Collectively they are associated with a wide variety of infectious diseases ranging from superficial infections of the skin or membranes to serious systemic diseases often characterized by the toxic effects of the cellular or extracellular products of the infecting strain. The classification of streptococci on the basis of their properties and the diseases they produce is discussed here.

The important routes of entry for streptococci are the respiratory tract, the skin and superficial membranes, and traumatized tissues. Streptococcal diseases associated with respiratory entry include *streptococcal sore throat* (pharyngitis-tonsillitis) and *scarlet fever.* These and the important *sequelae* of such infections caused by group A streptococci, e.g., rheumatic fever, glomerulonephritis, and erythema nodosum, are described in this section. Endocarditis is also included here because of its frequent association with streptococci found among the indigenous respiratory and oral flora.

Two other specific streptococcal disease entities, with a different entry route and epidemiology, are *erysipelas* and *puerperal fever.* Discussion of these will be found in Chapters 22 and 28, respectively, the former being a skin contact disease, the latter of iatrogenic origin. In addition to such entities, streptococci may be responsible for a variety of less specific infections, depending on their opportunity for localization. Diseases such as osteomyelitis, mastoiditis, lymphadenitis, and peritonitis may be of streptococcal origin. In skin diseases such as impetigo, or in wound infections, hemolytic streptococci are often found together with other bacterial pathogens, e.g., staphylococci.

Streptococcal Virulence. Little is known of the pathogenicity and virulence of streptococci other than those of group A. For these, such properties are related to cellular and extracellular substances that have a toxic effect on human cells (see also Chap. 6).

Capsular and Cell Wall Substances. The *capsule* of group A streptococci is an envelope of hyaluronic acid. This mucoid substance is chemically identical to the hyaluronic acid of human connective tissue. Therefore, during infection it does not elicit an antibody response, being treated as "self" by the immune system. Since the capsule cannot be coated with specific antibodies, as in the case of the pneumococcus, for example, there can be no enhancement of phagocytosis by opsonization. The capsule's chemical nature thus helps to protect the organism against the accelerated phagocytosis that occurs when opsonins coat bacterial surfaces.

Beneath the capsule, and extending into it as minute projections or fimbriae, is a component called *M protein*. This substance appears to play a role in the adherence of streptococci to epithelial cells of the upper respiratory membranes. In addition, it is actively antiphagocytic, preventing ingestion of the organism by phagocytes. These properties make the M protein an important virulence factor. It is also antigenic, conferring type specificity on group A strains.

The cell wall constituent of greatest serologic significance, especially among hemolytic strains, is a *carbohydrate*. The serologic classification of streptococci is based on the antigenic specificity of this cell wall carbohydrate, as first described by Dr. Rebecca Lancefield in the early 1930s. The Lancefield method recognizes antigenic differences in the carbohydrates of various strains, separating them into groups designated by letters (A through O). Further classification of group A strains into numbered types is then based on differences in their M protein antigens. Since the M protein is the important antiphagocytic factor, immunity to group A streptococcal infections depends on anti-M antibodies, and thus is type specific. For example, an individual host may be resistant to a strain of type 12 but not to type 4 streptococci.

Many group A strains also possess another antigenically distinct protein called *T protein*. Serologic typing of T proteins permits further differentiation of streptococcal strains responsible for outbreaks of infection.

Extracellular Products. The *hemolytic toxins* (*streptolysins*) of group A streptococci distinguish them from other species found more usually among commensal flora of the upper respiratory tract. The streptolysins of virulent strains completely destroy red blood cells in laboratory media. This property is referred to as *beta* hemolysis. (Strains of the viridans group, which occur most frequently in the throat, effect an incomplete hemolysis of blood cells termed *alpha* hemolysis. It is often accompanied by a green discoloration of the zone of incomplete lysis surrounding viridans colonies. Nonhemolytic streptococci are referred to as *indifferent,* or *gamma,* strains.)

One type of streptolysin of group A streptococci is antigenic (streptolysin *O*). Specific antibodies produced in response to its presence in the body during active streptococcal infection are called *antistreptolysins*. Laboratory detection of antistreptolysin O in a patient's serum indicates present or past infection with group A beta-hemolytic streptococci.

In addition to hemolytic toxins, virulent streptococci often produce the enzymes *hyaluronidase* and *streptokinase,* as well as a *leukocidin*. Hyaluronidase production accounts for some of the invasive properties of these organisms, assisting in their spread through local tissues by breaking down connective-tissue matrix. It is called a "spreading factor" on these grounds. Streptokinase activity assists in invasive spread because of its defibrinating effect on blood plasma, while leukocidin is destructive to phagocytes.

An *erythrogenic toxin* produced by some strains of beta streptococci is responsible for the skin rash of scarlet fever. The soluble toxin is distributed through the body via the bloodstream from a site of streptococcal infection in the throat. Its toxicity for blood vessels and cells in the skin results in the characteristic diffuse reddening of scarlatinal rash. This substance is antigenic and stimulates formation of an antibody capable of neutralizing its effects on dermal tissues.

Distribution of Serologic Groups of Streptococci. Most strains of streptococci of pathogenic importance to man fall into group A, while both human and animal pathogenicity is associated with group B. In recent years an increasing incidence of group B strains as colonizers of the female genital mucosa has been observed. This has had mounting significance as an infectious hazard for infants born of colonized mothers. Numerous instances of serious, often fatal, generalized infection or meningitis of the newborn have been caused by group B streptococci. In some cases the infected mothers have also developed grave postpartum group B disease (see Chap. 28).

Group C strains have been largely restricted to animal hosts, but group D contains organisms of human importance. Some of these are called entero-

Table 13–2. Classification of Streptococci and Their Diseases

Lancefield Group	Species	Hemolysis	Cellular Products	Extracellular Products	Clinical Diseases
A	*S. pyogenes*	Beta	Hyaluronic acid capsule Group A carbohydrate M protein T protein	Streptolysins Hyaluronidase Streptokinase Leukocidin Erythrogenic toxin	Pharyngitis-tonsillitis Skin infections (impetigo) Erysipelas Scarlet fever Puerperal fever Rheumatic fever Glomerulo-nephritis Endocarditis
B	*S. agalactiae*	Beta	Group B carbohydrate	Streptolysins	Neonatal sepsis and meningitis Puerperal fever
C	*S. equi*	Beta	Group C carbohydrate	Streptolysins	Endocarditis
D	Enterococci (*S. faecalis*)* Nonenterococci (*S. bovis*)*	Gamma	Group D carbohydrate–amino acid complex		Endocarditis Urinary tract infections
K	*S. salivarius*	Alpha	Group K carbohydrate	Glucan-like substance	Endocarditis Dental caries
H	*S. sanguis*	Alpha	Group H carbohydrate	Glucan-like substance	Endocarditis Dental caries
Not grouped	*S. mitis*	Alpha			Endocarditis Dental caries
Not grouped	*S. mutans*	Alpha		Glucan	Dental caries Endocarditis
Not grouped (anaerobic strains)	*Peptostreptococcus* species	Gamma			Brain abscess Lung abscess Puerperal fever

*Other species of group D streptococci occur that may be alpha, beta, or gamma hemolytic.

cocci, or fecal streptococci, because they are part of the normal flora of the intestinal tract. They have a pathogenic potential, however, being often associated with urinary tract or wound infections, as well as endocarditis. Nonenterococcal members of group D, i.e., strains that are not part of the normal enteric flora, may also cause endocarditis or urinary tract infection. Proper laboratory identification of group D strains is clinically important because of their differences in susceptibility to penicillin. Enterococci are highly resistant to penicillin and more difficult to treat than nonenterococci of group D, which are very susceptible to this antibiotic.

Other Lancefield groups are also sometimes involved in human disease. Some of the important properties of streptococci, and the human diseases with which they are associated, are shown in Table 13–2. Not all streptococci can be grouped by the Lancefield method, for some lack a group-specific carbohydrate in their cell walls. This is true of some of the alpha, or viridans, streptococci found normally on the respiratory or oral membranes. Anaerobic streptococci (peptostreptococci) of the normal intestinal tract and vagina cannot be grouped serologically for the same reason. These organisms are not primarily pathogenic but may cause opportunistic infections.

Serologic grouping and typing of streptococci are not essential to routine bacteriologic diagnosis because most strains can be distinguished on the basis of their biochemical properties, including their responses to antimicrobial agents (e.g., bile salts, bacitracin, penicillin). Serologic identification can be very useful, however, in epidemiologic tracking of sporadic or epidemic streptococcal disease. Since infection with group A strains may be followed by serious sequelae, e.g., rheumatic fever and acute glomerulonephritis, early recognition of these organisms, and effective chemotherapy, is very important.

Clinical Streptococcal Diseases

1. Streptococcal Sore Throat. Streptococcal infection of pharyngeal tissues is characterized by edema and reddening of posterior and anterior surfaces of the throat and soft palate. (They have a swollen, "beefy" look.) Petechial lesions are sometimes seen. When tonsils and other lymphoid tissues are involved, an acute exudative response occurs. Cervical lymph nodes may become tender and enlarged; fever and malaise are frequent.

2. Scarlet Fever. This disease results when (1) the causative strain involved in streptococcal sore throat produces *erythrogenic toxin,* and (2) the person in whom this occurs is *not* immune to the toxin. The

lesions in the throat appear together with the characteristic skin rash. The febrile reaction intensifies, nausea and vomiting may occur, and the white blood cell count rises. The rash is distributed to the neck and chest, the axillary and inguinal folds, and the soft skin of the inner sides of the arms and thighs. (Desquamation of these skin areas occurs during convalescence.) The tongue becomes swollen and reddened, and the papillae protrude, giving an effect described as "strawberry tongue."

Erythrogenic toxin is antigenic, stimulating production of specific antitoxin. There are three antigenically distinct types of this toxin. As antitoxin levels rise in circulating blood, the skin reaction begins to subside. The neutralization of toxin by antibody at the sites of rash forms the basis for an immunologically diagnostic skin test: the *Schultz-Charlton reaction.* Scarlatinal antitoxin injected intradermally into the center of a reddened area of skin produces visible blanching at the site of injection, within a few hours. Antitoxin levels produced during the course of disease are durable and protect the individual from the erythrogenic effects of future streptococcal throat infections caused by strains producing the same antigenic type of toxin. This probably accounts for the fact that many cases of streptococcal sore throat occur without development of scarlet fever. Also, different strains of streptococci produce varying amounts of the toxin. The incidence and severity of scarlet fever have been declining, but this is probably related not only to antitoxin immunity, but also to frequent therapeutic use of penicillin and other antibiotics to which beta streptococci are highly susceptible. Prompt therapy of initial throat infection effectively prevents development of scarlet fever by also aborting production of erythrogenic toxin by the infecting strain.

The Dick test is an immunologic method of demonstrating scarlatinal antitoxin immunity in healthy individuals. This test is similar to the Schick test used to determine diphtheria antitoxin immunity. In both instances, a small amount of the pertinent bacterial toxin is injected intradermally. If no antitoxin is available in the circulating blood to neutralize the toxin, local damage to skin at the injection site will occur in a few hours. If the individual possesses a protective level of antitoxin, however, the toxin will be neutralized, and the injection site will remain essentially normal in appearance.

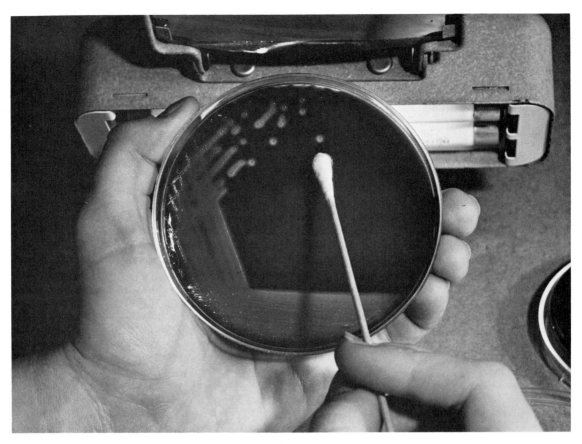

Fig. 13-3. Streptococcal colonies growing on a blood agar plate. There is a zone of complete (beta) hemolysis around each colony. (Reproduced from Maloy, L. B.: Lab Helps Prevent Strep Epidemic. *Health News,* December 1970. Courtesy of New York State Health Department, Albany, N.Y.)

Laboratory Diagnosis

I. Identification of the Organism (*Streptococcus pyogenes,* group A)

The microscopic, cultural, and serologic characteristics of virulent streptococci may be summarized as follows:

A. They are Gram-positive cocci that grow in chains (see Fig. 2-5, *B*, p. 23).

B. They are facultative aerobic organisms that grow best on blood-enriched media. Microaerophilic incubation (decreased atmospheric O_2 and increased CO_2 provided in a "candle jar") enhances growth. Isolated colonies on blood agar media are characteristically surrounded by zones of *beta hemolysis* (Fig. 13-3).

C. They may be serologically identified by grouping and typing methods when epidemiologic considerations warrant.

II. Identification of Patient's Serum Antibodies

A. Antibacterial (somatic) antibodies are type specific. They have no practical value in diagnosis of current infection because they arise late in disease.

B. Antistreptolysin titers are indicative of current or past pharyngeal infection with beta-hemolytic strains of streptococci. The titer (level of antibody) is a clue to the recency of infection.

C. The Schultz-Charlton skin test with scarlatinal *antitoxin* may have diagnostic value in recognition of scarlet fever rash.

D. The skin test with Dick *toxin* may be useful for determining antitoxic immunity. This test does not diagnose scarlet fever. It is applied when epidemiologic information is desired as to the history of previous infection with erythrogenic strains. Dick-negative persons are immune to the toxin of scarlet fever, but not to the bacterial infection of streptococcal sore throat.

III. Specimens for Diagnosis
A. Throat or nasopharyngeal swabs, in transport medium for culture.
B. Blood for culture.
C. Whole blood (or serum) for antistreptolysin titer. At least two samples should be submitted: one collected during the acute stage of illness (when the titer may be low), and another taken in the late or convalescent stage to determine whether the antibody level has increased.

IV. Special Laboratory Test
The fluorescent antibody technique is used for recognition of group A streptococci in respiratory secretions or in culture.

Epidemiology

I. Communicability of Infection
A. RESERVOIR, SOURCES, AND TRANSMISSION ROUTES. Man is the reservoir for group A streptococci. Asymptomatic carriers as well as patients in acute and convalescent stages of disease shed these organisms in their respiratory secretions. Transmission is by droplet spread, direct contact, and indirect contact with many temporarily contaminated environmental sources, including milk or food. Contaminated hands carry these organisms from handkerchiefs to clothing, skin, and food, as well as objects handled. Droplet spray contaminates many nearby surfaces also, including the faces of close contacts.

B. INCUBATION PERIOD. Rarely more than three days.

C. COMMUNICABLE PERIOD. Persists during the incubation period and throughout clinical illness or longer, unless adequately treated. The carrier state may last for many months, and untreated persons with exudative lesions are a constant source of virulent organisms. Adequate antibiotic therapy eliminates the communicability of disease, if not all streptococci, within 24 to 48 hours. Repeated cultures may be useful in marking the end of the communicable period.

D. IMMUNITY. Susceptibility to streptococcal sore throat and scarlet fever is general, although certain immunities develop as noted: antitoxin immunity to erythrogenic toxin prevents scarlet fever but not sore throat; antibacterial immunity is specific for the infecting type but does not prevent repeated infection with other types. Artificial immunization, active or passive, is not practical.

II. Control of Active Infection
A. ISOLATION. Usually not indicated except for streptococcal skin infections (Chap. 22). If warranted (e.g., a case of scarlet fever in a pediatric ward), the patient may be placed on respiratory isolation for 24 hours from the start of chemotherapy. Treatment should be continued until infection is eliminated. This varies with the severity of the individual case, but the average course of therapy is at least ten days. Infection control may be verified by repeated throat cultures.

B. PRECAUTIONS. Concurrent disinfection of respiratory discharges and all contaminated items. Terminal disinfectant-cleaning.

C. TREATMENT. Penicillin is the usual drug of choice. Streptococci are also susceptible to erythromycin and clindamycin.

D. CONTROL OF CONTACTS
1. *Quarantine.* Not indicated.
2. *Prophylaxis.* Immunization not available. Chemoprophylaxis may be indicated for unusually susceptible contacts (the very young, the very old or debilitated.)
3. *Case Finding.* Not indicated except in outbreaks.

E. EPIDEMIC CONTROLS. Local health departments require individual case reports of scarlet fever. For other streptococcal diseases, reports of epidemics are required, so that efforts can be made to find the source and mode of transmission. Serologic identification of the group and type of streptococcal strain involved in active infections is of importance in detection of sources. Suspected persons, milk, or food can then be checked by bacteriologic and serologic techniques. People involved in epidemic outbreaks are isolated and treated with antibiotics. Contaminated commercial supplies of food or milk are removed from the market. The latter sources may be pinpointed by examining the distribution of cases and their possible exposure to common transmission routes.

III. Prevention

A. Public education on the ease and hazards of transmission of streptococcal diseases; the need for personal hygienic controls and for prompt medical control of developing illness.

B. Control of food and milk handlers.

Inspection of dairy cows for mastitis or other infection.

Sanitation of dairy establishments.

Pasteurization of milk (boil or discard if evidence of streptococcal contamination exists).

C. Penicillin prophylaxis for children and others for whom risk of repeated infection is great. The group includes those who have had rheumatic fever or other diseases as an aftermath to streptococcal infection.

Sequelae of Group A Streptococcal Infections

Late manifestations or consequences of streptococcal throat infections include development of *rheumatic fever, acute hemorrhagic glomerulonephritis,* or *erythema nodosum* in a small proportion of patients. These diseases are neither infectious nor communicable. They are believed to develop as a consequence of delayed hypersensitive responses to streptococcal antigens.

1. Rheumatic Fever. This is the most important and frequent of the poststreptococcal diseases. Following upper respiratory tract infection with group A beta streptococci, about 1 to 3 percent of patients (the incidence is highest in childhood) develop rheumatic fever after an average interval of three weeks. Major clinical symptoms include fever, carditis, and polyarthritis (inflammatory involvement of joints, usually migratory in pattern). The severity of symptoms varies from the mild and scarcely noticeable to acute illness with irreversible cardiac damage. The lesions of the disease may include skin involvement (subcutaneous nodules), and central nervous system injury (chorea). Rheumatic fever may persist as a chronic disease for years, or it may be a recurrent acute or subacute problem following repeated streptococcal infections. Most serologic types of the group A strains have been demonstrated in the predisposing throat infections. When numerous cases of rheumatic fever are grouped as an aftermath of a streptococcal epidemic, epidemiologic studies can relate them to the specific infecting strain.

2. Acute Hemorrhagic Glomerulonephritis. Hemorrhagic lesions in the glomeruli may develop following streptococcal infection with certain strains. The clinical symptoms include hematuria, albuminuria, generalized edema, and hypertension. The average interval between infection and this type of sequel is shorter than for rheumatic fever. Glomerulonephritis is usually an acute disease from which most patients recover without residual damage, and recurrent episodes of this type are rare. Occasionally, symptoms of rheumatic fever may occur with those of glomerulonephritis, in a more serious and generalized hypersensitive response to streptococcal proteins or toxins.

3. Erythema Nodosum. This disease is not always specifically related to preceding streptococcal infection, but can appear as a recurring sequel to repeated bouts with hemolytic streptococci. It is characterized by the cropping of tender, red, subcutaneous nodules, distributed chiefly over the arms and legs, accompanied by fever and general malaise. This type of lesion is sometimes also seen in rheumatic fever, tuberculosis, and certain systemic fungous diseases, e.g., coccidioidomycosis. It is thought to be the result of hypersensitivity to the microbial antigens involved.

The laboratory diagnosis of these diseases is retrospective with regard to previous bacteriologic evidence of streptococcal infection. Current demonstration of significant titers of antistreptolysins in the patient's serum gives support to the clinical diagnosis.

The epidemiology of streptococcal sequelae does not include considerations of their *control,* since they are not communicable. The primary concern is *prevention,* particularly of rheumatic fever with its potential for permanent cardiac damage. Prompt, effective antibiotic therapy of active streptococcal disease, eliminates the organism and may prevent the distribution of sensitizing streptococcal proteins through the body, thus reducing the incidence of rheumatic fever. Penicillin prophylaxis for those who have had rheumatic fever can successfully prevent its recurrence.

Subacute Bacterial Endocarditis

Subacute bacterial endocarditis (SBE) is included in this section dealing with the streptococcal diseases because it is so often associated with streptococci indigenous to the respiratory and oral flora. Other commensalistic organisms of the normal mucosal surfaces may be involved in SBE, however, including various species of Gram-negative enteric bacilli.

This disease is not communicable. Its epidemiology reveals no relationship to entry of infectious pathogenic microorganisms by respiratory or other routes. SBE is an infection of *endogenous* origin, involving localization of organisms from indigenous sources at some site of previous injury on the endocardium, usually a valve surface. At least two factors appear to be essential to development of subacute bacterial endocarditis: (1) a preexisting defect of valvular endocardium (though this may not be clinically detectable before the onset of SBE), and (2) bacteremic distribution of microorganisms from a resident focus on a skin or mucosal surface disrupted by trauma, surgery, or some manipulation.

The majority of patients with SBE (about 75 percent) have a history of rheumatic fever as the most probable cause of preexisting cardiac damage. Congenital cardiac defects and valvular injuries induced by arteriosclerosis or syphilis may also set the scene for SBE.

Minor dental or throat surgery (tonsillectomy, adenoidectomy) is often considered to be the initiating cause of transient bacteremias that lead to SBE. Many other possibilities exist as well: nonsurgical manipulation of genitourinary tissues (catheterization, cystoscopy, prostatic massage), ulcerative erosions or perforating lesions of the bowel, or surgical transection of surfaces with a normal flora — all represent possible mechanisms by which commensalistic microorganisms may be introduced into the bloodstream. Ordinarily these bacteremias are terminated

rapidly, the organisms being incapable of localization in normal systemic human tissues. SBE develops if the microorganisms localize on damaged valvular surfaces and multiply there. The site may characteristically be involved in a fibrinous response to injury, and small aggregates of platelets in fibrin may be formed. Microorganisms can grow in this milieu protected from the action of phagocytes or other bactericidal components of blood. Their presence accelerates fibrinous tissue reaction and "vegetations" are built up around their proliferating growth. After such infection has been established, there is usually a steady and continuous bacteremia.

Increasing valvular dysfunction leads to symptoms of cardiac distress. If the situation is not arrested by antimicrobial treatment, the valve may be occluded by vegetative growth or fail to function because of loss of elasticity, with fatal results in either case. These friable vegetations break off in small pieces in the surging flow of blood propelled through the valve by ventricular contraction. Such infectious emboli then travel down the arterial vessels, finally lodging in and occluding vessels that are too small to pass them further. The seriousness of this situation depends on the nature of the tissue thus deprived of its blood supply and also on the capacity of the organisms to continue multiplying at the new site.

The laboratory diagnosis of SBE depends on successful isolation of the causative organism from blood. Since there is a fairly continuous release of organisms from the valvular lesions, cultures taken at any time are likely to be positive. However, only small numbers of organisms are usually present, and several days may be required for their growth and identification. Several samples of blood for culture should be drawn over the course of two or three days before antimicrobial therapy is instituted. Isolation of the same organism from a majority of them is strong evidence for a diagnosis of SBE.

Strains of alpha-hemolytic *Streptococcus* are most frequently isolated from blood cultures in SBE. Fecal streptococci (enterococci) are also common, while coliform bacilli and anaerobic organisms are less frequent.

It should be noted that modern methods of heart surgery, including the use of arterial catheters, open easy avenues for infection of damaged or prosthetic valves. Rigorous aseptic precautions are required to protect patients undergoing cardiac surgery or catheterization from directly implanted infection. Successful surgical techniques may be compromised or negated by microbial contamination that develops into infection. The organisms most frequently involved in this situation are staphylococci of the skin and mucosa. Some are benign coagulase-negative strains of the resident skin flora (*Staphylococcus epidermidis*), which generally induce subacute but troublesome problems; others are the virulent coagulase-positive pigmented variety (*Staphylococcus aureus*) capable of causing acute or overwhelming infection, particularly of traumatized or malfunctioning tissues. (See Chap. 28.)

The treatment of subacute bacterial endocarditis of streptococcal origin generally employs penicillin and streptomycin in combination when organisms of the alpha-hemolytic (viridans) group are involved. Penicillin-resistant enterococci may be treated with vancomycin or with ampicillin in combination with streptomycin or gentamicin. Antibiotic susceptibility testing of the isolated organisms is essential in indicating the choice of drugs. The physician may find it of even greater value to know what dosage is required to kill, rather than merely suppress, the organism. Quantitative tube dilution assays of the antibiotic are performed with the organism isolated from the patient, so that the dose, frequency, and route of administration of the drug may be adjusted to maintain a bactericidal level of antibiotic in the bloodstream and tissues. The patient's blood serum can also be assayed for its bactericidal activity against the causative agent as a further measure of the adequacy of therapy and of response to it. (See Chap. 4.)

The epidemiology of SBE points to the need for preventive measures (not for communicable infection controls): prevention of the causes of endocardial injury (rheumatic fever is the primary target), and control of transmission of infecting organisms from their sources. Chemoprophylaxis has helped to reduce the incidence of rheumatic fever, and it is also useful in decreasing the occurrence of bacteremia after surgery or other provocative procedures. In the latter application, antibiotics are given before, during, and after surgery, but this practice should take into consideration the susceptibility of the cardiac patient to the risk of infection as opposed to others for whom a transient bacteremia is not a major challenge. Antibiotic-induced changes in the normal flora and the emergence of resistant bacterial strains counterbalance the prophylactic use of antibiotics in nonselective situations.

Whooping Cough (Pertussis)

The Clinical Disease. Whooping cough is a highly communicable, acute respiratory disease, particularly prevalent in nonimmunized children. It begins with a catarrhal stage that resembles a cold, except that it persists for 10 to 12 days, the initial mild cough becoming more progressively irritating, and finally paroxysmal. At this time, the organism is extending on a path of descent from its first upper respiratory localizations downward along the mucosal surfaces of the larynx, trachea, and the bronchial tree. The epithelial tissue along the way is injured and a leukocytic exudate is formed. Pneumonia may develop if the deep structures of the lung are reached and the organisms multiply there in the interstitial tissues. Secondary opportunists may proliferate in alveolar exudates, but the pneumonia of pertussis is interstitial. Excessive mucus production in the bronchial tract may plug the lower airways. This is the cause of the paroxysm, which is characterized by a series of struggling coughs on a single inhalation ending with an involuntary "whoop" of inspiration through narrowed, mucus-filled passages. The mucus is clear but very tenacious and difficult to expel. Its obstruction of airways, together with thickening of interstitial tissue in the lung itself, reduces the amount of oxygen available for the blood. Convulsions may ensue if hypoxia continues. Antimicrobial drugs and immunotherapy may reduce the severity of the disease, but they do not change its course to any marked degree. The paroxysmal stage may remain acute for two to three weeks and persist for one or two months.

The mortality rate is high for infants: about 80 percent of the deaths occur in children under one year of age. Active immunization programs have greatly reduced the incidence of this disease in the past 20 years, but epidemic cycles continue every two to four years. Pertussis may occur in all seasons, but the epidemic incidence is highest in the winter and early spring.

Laboratory Diagnosis

I. Identification of the Organism (*Bordetella pertussis*)

A. MICROSCOPIC CHARACTERISTICS. The organism is a small Gram-negative bacillus closely resembling *Haemophilus*.

B. CULTURE CHARACTERISTICS. The pertussis bacillus requires complex media for its isolation and growth. Bordet-Gengou agar is the most satisfactory. It is a mixture of potato infusion, glycerol, blood, and agar. It may be used in a Petri dish as a "cough plate" held a few inches in front of the patient's mouth during his coughing or, preferably, it may be inoculated with a nasopharyngeal swab. Penicillin is added to the medium to inhibit growth of other respiratory flora. The pertussis bacillus, which is resistant to penicillin, may then grow as small transparent colonies surrounded by a zone of dark discoloration of the medium.

C. SEROLOGIC METHODS OF IDENTIFICATION. This organism possesses several components that are antigenic. The agglutination technique can be used to confirm the identity of isolated organisms.

II. Identification of Patient's Serum Antibodies. Agglutinins for *B. pertussis* do not rise to significant levels until the third or fourth week and are of little diagnostic value for this reason.

III. Specimens for Diagnosis

A. Nasopharyngeal swabs, in transport medium, for culture.

B. Cough plates are an alternative to swabs.

C. Blood for white cell counts. (Leukocytosis is striking [15,000 to 30,000], with a predominance of lymphocytes unusual in bacterial infections.)

Epidemiology

I. Communicability of Infection

A. RESERVOIR, SOURCES, AND TRANSMISSION ROUTES. Man is the only reservoir. Infectious respiratory secretions transmit the infection directly by droplet spread or indirectly by contamination of objects in the environment (including hands and handkerchiefs).

B. INCUBATION PERIOD. Not more than 21 days, usually seven to ten days.

C. COMMUNICABLE PERIOD. Communicability is highest in the late days of the incubation period and during the catarrhal stage. Familial contacts continue to run a high risk of infection until the paroxysmal stage begins to subside. In the schoolroom or out of doors, communicability is reduced by distance and by the susceptibility of the organism to effects of drying. For general control, the disease should be considered communicable for a period extending from the seventh day following exposure to the end of the third week of the paroxysmal stage.

D. IMMUNITY. Resistance to pertussis depends on specific immunity. The incidence of clinical disease and of inapparent or atypical infections is highest in childhood, and especially in preschool years. Active immunity acquired during an attack of the disease in childhood is durable, but occasional second attacks occur in exposed adults. Artificial active immunization is effective and should be started in infancy.

II. Control of Active Infection

A. ISOLATION. Children and other patients hospitalized with severe pertussis pneumonia must be isolated. Precautions are of most urgent importance in hospital nurseries and pediatric wards. At home, the sick child should be separated from susceptible children (particularly young infants) insofar as practical. Outdoor play for the afebrile patient is better than confinement in close quarters. The school-age patient must be kept at home for the duration of the recognized communicable period (usually about four weeks).

B. PRECAUTIONS. Concurrent disinfection of nasal and oral secretions, vomitus, and all articles contaminated with these discharges. Hospital units are cleaned and disinfected terminally. At home, frequent cleaning and disinfection of the sickroom should be practiced. The protection of susceptible children includes separating and boiling dishes used by the infectious patient.

C. TREATMENT. Antibiotic therapy is not markedly effective but may help to modify the most severe symptoms, prevent secondary pulmonary infection, and shorten the communicable period. Erythromycin or ampicillin is the drug of choice. In severe cases, human gamma globulin may be given to provide the support of passive antibody.

D. CONTROL OF CONTACTS

1. *Quarantine.* Nonimmune children *exposed* to whooping cough must be excluded from school or other public contacts with susceptible children for a

two-week period. This time corresponds with the incubation and communicability of the disease. Exceptions may be granted for public schools that have adequate surveillance programs for the detection of early illness in children. Younger children and infants must be guarded from infection with particular care.

2. *Prophylaxis.* Susceptible children and adults exposed to whooping cough are candidates for an injection of gamma globulin. *Passive* immunization following exposure may ameliorate the disease or prevent it. *Active* immunization with vaccine is not usually prophylactic for exposed individuals because antibody response is too slow to be protective during the time when the causative organism implants and multiplies on susceptible respiratory membranes.

3. *Case Finding.* The obvious case may be found and isolated easily. Atypical cases, or those that have been misdiagnosed, must also be found.

E. EPIDEMIC CONTROLS. Individual cases must be reported to local health departments. Beyond this, accurate case finding of unrecognized active infections is the key to controlling further spread.

III. Prevention

A. IMMUNIZATION. Artificial active immunization during infancy affords the best protection against whooping cough. Killed bacterial vaccines are available that effectively stimulate the production of protective antibody. Pertussis vaccine (usually combined with diphtheria and tetanus toxoids [DTP]) may be given to infants two or three months of age. Timing should be geared to seasonal and familial risk of infection. Immunization with currently available vaccine requires three injections at monthly intervals during the first year of life. A booster dose of the triple vaccine is advisable during the second year and another when school age is reached. Once the active immune process is underway, subsequent exposures to infection are minimized as to risk and result.

B. CONTROL OF RESERVOIR, SOURCES, AND TRANSMISSION ROUTES. The most effective control of the reservoir and sources of infection is accomplished by specific immunity actively acquired through the natural route of infection or, more safely, by artificial immunization programs.

Diphtheria

The Clinical Disease. Diphtheria does not occur in large epidemics today. It remains a potential threat, however, as evidenced by sporadic cases and small outbreaks that occur among persons who lack naturally or artificially acquired active immunity against the *toxic* components of infection. While the incidence of clinical diphtheria has sharply declined since the introduction of nearly universal artificial immunization, the *case fatality* rate of 5 to 10 percent has not changed significantly.

Diphtheria begins as an acute infection of the mucous membranes of the upper respiratory tract. The growing diphtheria bacilli produce a toxin that destroys the local epithelial cells and induces an inflammatory response at the site. The necrotic surface epithelium becomes enmeshed in a fibrinous and cellular exudate that lies like a membrane over the involved area. This *pseudomembrane* formation advances as the organisms continue to grow. It may cover tonsils and pharynx or extend downward to the larynx and trachea. Laryngeal diphtheria may cause death by suffocation, particularly in infants or young children. The pseudomembrane is tough and cannot be lifted without tearing the structures beneath, leaving an exposed, bleeding undersurface.

Patients often appear extremely toxic, even though fever is moderate (100° to 102° F [37.8° to 38.9° C]). Cervical lymph nodes become enlarged and tender, and there may be extensive edema of the neck. The organisms continue to proliferate in the pseudomembrane and to produce toxin, which is absorbed and distributed through the body. The toxin may cause severe injury to parenchymal cells of the liver, adrenals, kidneys, and (a more immediate threat) the myocardium. Cardiac deaths are frequent among patients succumbing in the stage of systemic toxic involvement. Cranial and peripheral nerves, both motor and sensory, may also be damaged, with resulting paralysis of ocular muscles, the soft palate, or the extremities, depending on the nerve affected.

Aside from this picture of classical diphtheria, the organism is sometimes associated with wound or skin infections. These superficial infections may also be marked by local membrane formation, but absorption of toxin with resulting systemic toxicity either does not occur or is very minor in effect. Diphtheritic skin or wound infections are common in tropical countries. The lesions may persist without healing for weeks or months.

The severity of diphtheria infections reflects the capacity of the organism to establish, multiply, and produce exotoxin. The speed of multiplication and the amount of toxin produced (and absorbed) are a measure of virulence. The most severe disease is produced by strains that elaborate large quantities of toxin in a short time.

Prompt *clinical* diagnosis of diphtheria is the most important factor in prevention of crippling or fatal toxicity. Immediate passive immunization with antitoxin provides circulating antibody to neutralize toxin *before* it can attach to cells in the myocardium or elsewhere and cause irreparable damage. Simultaneous therapy with an appropriate antibiotic then eliminates the infectious agent from the throat. The diagnosis is confirmed in the laboratory by demonstration of the organism in smears and cultures of the throat exudate.

Laboratory Diagnosis

I. Identification of the Organism (*Corynebacterium diphtheriae*) (*Klebs-Loeffler bacillus*). The diphtheria bacillus is one of several species belonging to the genus *Corynebacterium*, but it is the only one of pathogenic importance. The others are common commensals of the skin and mucous membranes, and the bacteriologist must distinguish these "diphtheroid" species from *C. diphtheriae* in cultures of material from the nose and throat.

A. MICROSCOPIC CHARACTERISTICS. The corynebacteria are Gram-positive bacilli, non-spore-forming, and nonmotile. *C. diphtheriae* is a slender, attenuated bacillus, with granular contents that stain irregularly and give it a beaded or barred appearance. Individual rods are often club-shaped. They lie in characteristic "V" and "Y" formations, or in palisaded groups (Fig. 13–4). Diphtheroids appear in this arrangement also, but they are usually shorter, thicker rods than diphtheria bacilli and have a coarser appearance. The distinction may be very difficult to make in direct smears of throat exudate, but the smear report is evaluated by the physician in the light of clinical findings. If these suggest diphtheria, the decision to give antitoxin is made at once, with or without bacteriologic confirmation.

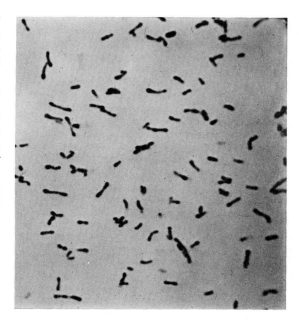

Fig. 13–4. *Corynebacterium diphtheriae,* the organism that causes diphtheria. Note the club shapes, beading, and V and Y figures. (Reprinted with permission of U.S. Public Health Service Audiovisual Facility, Atlanta, Ga.)

B. CULTURE CHARACTERISTICS. All corynebacteria grow aerobically on blood agar, but certain special media are useful in selection and differentiation of the diphtheria bacillus from other organisms present in throat cultures. Loeffler's medium is often used because it does not support the growth of streptococci and pneumococci, but offers enough nutrient for growth of *C. diphtheriae*. The addition of potassium tellurite to blood agar inhibits development of many throat organisms and also distinguishes colonies of *C. diphtheriae* from some other bacteria that are capable of growing in the presence of this chemical. The former take up the tellurite and react with it, the result being a gray to black coloration of *C. diphtheriae* colonies, which makes them stand out from others.

The morphology of *C. diphtheriae* is confirmed by Gram-stained smears, and the toxigenicity of the isolate is then tested by the Elek plate method (see Chaps. 4 and 7). Since the virulence of this organism depends on its toxigenicity, this *in vitro* method for demonstrating toxin production is equivalent to a virulence test in animals. If toxin precipitation by antitoxin is demonstrated in this test, the need for prompt immunotherapy is underscored.

The genetic mechanism of toxin production is explained by the fact that toxin-producing strains of *C. diphtheriae* are infected with a bacteriophage that directs the cellular manufacture of the toxin. The virus acts as a gene within the bacterial cell. Uninfected bacterial cells are incapable of toxin production, although they retain their ability to invade and multiply in human epithelial tissue.

II. Identification of Patient's Serum Antibodies. The demonstration of circulating antitoxin is not appropriate in the diagnosis of diphtheria, because active production of antibody in the sick individual occurs much too late to have diagnostic value. Healthy individuals may be tested for antitoxin immunity, however, by a skin test with diphtheria toxin. This is the *Schick test* described previously (Chap. 7). Intracutaneous injection of the toxin induces no local reaction in persons with protective levels of circulating antitoxin, but a Schick-positive reaction indicates an inadequate immunity and the need for artificial active immunization with diphtheria toxoid.

III. Specimens for Diagnosis. Material from the suspected lesion in the throat is submitted for smear and culture. This is the only type of specimen of diagnostic value, since diphtheria bacilli do not invade the body widely, nor do they enter the bloodstream. Material is best collected by running a thin swab or wire loop under the lifted edge of the pseudomembrane, for the organisms may be present in nearly pure culture in the active area beneath, but many saprophytic organisms may be living on the dead tissue at the surface of this membrane.

Epidemiology

I. Communicability of Infection

A. RESERVOIR, SOURCES, AND TRANSMISSION ROUTES. The reservoir is man, clinically or inapparently infected. Discharges from the normal respiratory membranes of carriers are a source of infection as well as those from active throat or skin lesions. The organism is transmitted in direct contacts, by droplet spread, and through indirect contacts with objects contaminated by respiratory secretions. Milk contaminated after pasteurization or raw milk may transmit infection.

B. INCUBATION PERIOD. Usually only a few days (two to five).

C. COMMUNICABLE PERIOD. This varies with the

persistence of bacilli in the lesions or on recovering mucous membranes. The carrier state may persist for many weeks following convalescence. Some health department regulations require continued quarantine of the convalescent until two or three negative nose-and-throat cultures have been obtained on consecutive days. Antibiotic therapy, or the surgical removal of tonsils and other harboring tissues when indicated, reduces the number of carriers in a shorter time, but these measures have not eliminated the problem completely.

D. IMMUNITY. Resistance to clinical disease is chiefly dependent on antitoxin immunity, passively or actively acquired, by natural or artificial means. An attack of disease usually confers a durable active immunity (the passive antibody acquired through antitoxin administration in the acute stage of infection does not persist for more than two to three weeks). Infants born of immune mothers acquire passive immunity congenitally, but this does not last more than five or six months. Artificial active immunization with diphtheria toxoid affords the most effective control.

II. Control of Active Infection

A. ISOLATION. Patients ill with diphtheria should be in strict isolation until bacteriologically negative throat-and-nose cultures can be demonstrated after antibiotic therapy has ceased. Two cultures from the nose and two from the throat, taken on consecutive days, are required as a basic rule. When cultures continue to be positive after clinical recovery is complete, a negative toxigenicity test on the latest organism recovered can release the patient from isolation. If the culture is toxigenic, a course of antibiotic therapy is repeated. In situations where laboratory evidence is difficult to obtain, isolation may be terminated 14 days after onset with fair assurance of safety.

B. PRECAUTIONS. Concurrent disinfection of all articles involved in respiratory transfer of infection. Terminal cleaning-disinfection of the room should be thorough.

C. TREATMENT. Antitoxin is given in a single dose of 20,000 to 80,000 units as indicated by the duration and severity of symptoms, after careful predetermination of hypersensitivity to horse serum proteins. Penicillin and erythromycin are the antibiotics of choice for antibacterial therapy, but they have no effect on diphtheria toxin or the toxic symptoms.

D. CONTROL OF CONTACTS

1. *Quarantine.* Intimate susceptible contacts of an active case should be held in *modified* quarantine (see Chap. 10) until nose and throat cultures are reported negative. Adult contacts who care for children or handle food should be removed from these occupations until bacteriologic evidence shows them not to be carriers. Contacts found to have inapparent infections should be placed on a course of antibiotic therapy.

2. *Prophylaxis.* Prophylactic use of antitoxin is indicated for nonimmune children under ten years of age who have been intimately exposed to the disease. These children should also be started on a course of active immunization with toxoid. The physician may determine the need for active or passive immunization of older children and adults exposed to the disease if he can obtain an accurate history of previous immunization. If not, the Schick test may be applied. Whether or not they require *passive* immunization, Schick-positive persons exposed to diphtheria should receive toxoid immunization, as a primary course or in a booster dose, depending on the history. Immunization may be particularly important for older people whose immunity has waned.

3. *Case Finding.* In communities where the population has been universally immunized, carriers and atypical cases are rare. When an unusual incidence of clinical diphtheria begins to appear in any community, however, a careful search for possible sources should be made.

E. EPIDEMIC CONTROLS. Immediate surveys are made to determine the general level of immunity in the population involved. Active immunization is conducted on the scale indicated, concentrating on the youngest children but extending to all age groups. The search for contacts and sources is intensified. Individual case reporting is required as a rule in this country and others. The international control of diphtheria requires recent immunization of persons traveling in or through endemic areas.

III. Prevention

A. IMMUNIZATION. Artificial active immunization with diphtheria on a mass scale has provided effective control of the human reservoir of this disease. It should be provided early in life, preferably at three to four months of age, and reinforced at intervals, at least through adolescence. In infancy, purified alum-precipitated toxoid is given together with tetanus toxoid and pertussis vaccine, in a combined preparation known as *DTP*. The initial series requires several injections at four-to-six-week intervals.

This preparation combining three immunizing agents may be used in an initial program for children up to the age of four years. Nonimmune children between the ages of 4 and 12 are given diphtheria and tetanus toxoid only, and this is feasible for nonimmune adults also. Booster toxoids for adults are adjusted in their preparation to avoid the complication of hypersensitive reactions to bacterial proteins encountered naturally or through previous immunizations.

It has become clear in recent years that artificial immunization begun in infancy must be reinforced at primary-school age and again during adolescence. During the past two decades an increased age incidence of clinical diphtheria has been observed in all countries where mass immunization of children has been practiced. This is because the organism tends to die away in immunized populations, so that there is no repeated, natural stimulus to immunity through inapparent infection. As artificially acquired immunity wanes, therefore, it is not reinforced by natural infection, and the adult population slowly becomes more susceptible. The introduction of carriers into such a population may lead to an outbreak involving the nonimmune of all ages, as reported from Holland during World War II and more recently from Denmark and Great Britain.

Adults who are unusually subject to the risk of exposure to diphtheria carriers or infectious patients should have a booster dose of the type of toxoid prepared for their age group. Diphtheria exposure is particularly likely for hospital personnel in contact with patients, for military personnel traveling to or stationed in endemic areas, and for professional travelers (such as airline flight personnel), teachers, and others in contact with many people from different parts of the world.

Future control of diphtheria depends on continuing education of people everywhere as to the necessity of adequate artificial active immunization.

Table 13–3. Mycobacteria in Infectious Disease

Disease	Species	Host(s)	Route of Entry
Tuberculosis	Mycobacterium tuberculosis	Man	Respiratory
	Mycobacterium bovis	Cattle and man	Alimentary (milk)
Pulmonary disease resembling tuberculosis (mycobacterioses)	Mycobacterium avium*	Fowl and man	Respiratory
	Mycobacterium intracellulare*	Man	
	Mycobacterium kansasii	Man	Environmental contacts? (water and soil)
	Mycobacterium szulgai	Man	
	Mycobacterium xenopi	Man	
Lymphadenitis (usually cervical)	Mycobacterium tuberculosis	Man	Respiratory
	Mycobacterium scrofulaceum	Fowl and man	Environmental contacts? (water and soil)
	Mycobacterium avium	Man	
	Mycobacterium intracellulare	Man	
Skin ulcerations	Mycobacterium ulcerans	Man	Environmental contacts?
	Mycobacterium marinum	Fish and man	Marine contacts
Leprosy (Hansen's disease)	Mycobacterium leprae	Man	Human contacts
Saprophytes: water, soil; human skin and mucosae	Mycobacterium smegmatis		
	Mycobacterium phlei		
	Mycobacterium gordonae		

*Mycobacterium avium and M. intracellulare are so closely related that they are often identified as the Mycobacterium avium complex.

Tuberculosis

Tuberculosis is a chronic infectious disease caused by one member of a large group of bacteria associated with numerous human and animal infections. These organisms are members of the genus *Mycobacterium,* a number of which may infect man, producing pulmonary disease similar to tuberculosis or more superficial infections of the skin or lymph nodes. Some of these are primarily pathogens of cattle, fowl, reptiles, or fish, whereas others are nonpathogenic saprophytes of soil and water that may also be found among the commensal flora of man's skin or superficial membranes.

The pathogenic mycobacteria are intracellular parasites that primarily infect macrophages (large mononuclear phagocytes). The great majority can be cultivated on artificial media and are therefore not obligate parasites. Some members of the group, notably the causative agent of leprosy, *Mycobacterium leprae,* cannot be cultivated but grow only in certain animals or in tissue cultures of macrophages. Some of these obligate intracellular parasites are associated with disease in rodents, but *M. leprae* is the only one of clinical importance to man.

Mycobacterium tuberculosis, the most usual cause of human tuberculosis, has a respiratory route of entry, as do other strains associated with pulmonary disease. Leprosy (Hansen's disease) is a human contact disease (see Chap. 22). Cutaneous mycobacterial infections are acquired by contact with infected animal or marine sources. The most important mycobacterial species are shown in Table 13–3, together with the diseases they cause.

Since the pulmonary disease caused by several mycobacterial species is clinically indistinguishable from that due to *M. tuberculosis,* the discussion of tuberculosis that follows refers to this organism, the host responses it induces, and its epidemiology.

The Clinical Disease. Tuberculosis is a chronic disease, characterized in its course by a continuing interplay between the durable invading organism and host mechanisms for resistance, the latter being greatly affected by development of hypersensitive responses. Specific immunity is primarily of the cell-mediated type, but nonspecific resistance is also extremely important.

Pathogenesis. Virulence of tubercle bacilli is associated with their capacity for establishment and multiplication in the human host and with the degree of their resistance to his defenses. After first entry of the bacilli into a susceptible host, production, progression, or healing of tuberculous lesions depends on the numbers of invading organisms and their success in multiplying locally.

When lesions are produced, they appear as one of two types, depending on local host reaction. *Exudative lesions* are formed by an acute inflammatory reaction, with engulfment of tubercle bacilli in a fluid exudate containing polymorphonuclear and monocytic cells. This type of lesion may subside and heal; progress and lead to necrosis of local tissue; or develop into a *tubercle.* This second type of lesion is called *productive* (i.e., producing new tissue) because it consists of cells (without exudate) organized in a fairly definite way around the offending organisms. It is also called a *granuloma* because it resembles a tumor and is composed of granulation tissue. In the central part there is a compact mass of large multinucleated cells (giant cells), formed by coalescence of mononuclear cells. Surrounding this is a zone of "epithelioid" cells, and the outer zone is composed of lymphocytes, monocytes, and fibroblasts. The tubercle bacilli reside in the center of the tubercle, principally intracellularly, within giant cells or mononuclear cells. Eventually the tubercle may become fibrous at its periphery and calcified within. The organisms may remain viable for years even under these conditions, or they may slowly dwindle and die. Often, the central area of the tubercle becomes soft and cheesy, in a process called *caseation* or *caseation necrosis.* Caseous tubercles sometimes break open, spilling their contents into surrounding

tissue and forming a cavity that then may heal by fibrosis and calcification.

Spread of the organisms through the body may occur in a number of ways. From exudative lesions or cavitating tubercles they may extend directly into adjacent tissue, enter the bloodstream or the lymphatic flow, or find their way along mucosal surfaces. The cavitation of a caseous tubercle in the lung, for example, may spill organisms into a bronchus. From here they may pass downward into the lung or be coughed up into the throat, swallowed, and passed into the gastrointestinal tract. When tubercle bacilli *initially* enter the body through respiratory or alimentary routes, they are spread from their original site through lymphatic ducts into regional lymph nodes.

Patterns of Disease. Two types of tuberculosis are recognized: *primary,* or first-infection type, and *reinfection* tuberculosis.

In *primary tuberculosis,* the initial lesion is of the acute exudative type from which a rapid spread occurs via lymphatics to regional nodes. The exudative lesion may quickly resorb and heal, usually with some residual fibrosis and scarring, but the involved lymph nodes become caseous. With time, the lymph nodes generally are calcified to some degree. Primary tuberculosis occurs most frequently in childhood, but may also be seen in adults who have not acquired infection previously. In first infections the primary lesion may occur in any part of the lung.

Reinfection tuberculosis is a more chronic disease, characterized by lesions of the productive type. Reinfection may be acquired from new exogenous sources, or it may be caused by organisms that survive in and extend from primary lesions (endogenous reinfection). Regional lymph nodes are seldom involved in reinfection. The prominent lesions of this type of disease almost always begin in the apex of a lung and progress downward. Tubercles form, become caseous, and cavitate; the organisms pass down the bronchial tree to new sites where tubercle formation again resists their spread.

Endogenous reinfection may occur many months or years after first infections appear to be healed. Fibrosis and calcification are not always sufficient to suppress and finally destroy the organism. Lowered resistance of the host may permit a breakthrough of organisms from a primary focus or a regional lymph node, with a resulting distribution and spread as described previously. Occasionally, organisms from a cavitating focus enter the blood or lymph stream in sufficient numbers to assure widespread distribution throughout the body. General infection then ensues, with tubercle formation in many organs. This distribution is called *miliary* tuberculosis because the tubercles are small and hard and resemble millet seed.

Resistance, Hypersensitivity, and Immunity. The speed and extent with which reinfection tuberculosis progresses are partly determined by the numbers of organisms released from productive foci and the route of spread available to them. Progress of the disease is also determined by the increasing capacity of the host to resist: that is, to localize bacilli and retard their growth or destroy them and thus limit their dissemination. The nonspecific resistance of the inflammatory process and of reticuloendothelial cells plays a primary role in first infection. Cellular immunity (CMI) also develops, increasing the reactivity of mononuclear cells. They become more adept at phagocytosis, and tubercle formation becomes more rapid. Although the CMI response inhibits the miltiplication of tubercle bacilli within mononuclear cells, it is unable to bring about destruction of all of the organisms. Tubercle bacilli within macrophages located in a tubercle somewhere in the lung or lymphatic system may remain in check for long periods but are capable of multiplying again if local conditions and lowered resistance permit. At the same time, delayed hypersensitivity to the proteins of the tubercle bacillus may lead to a destructive inflammatory reaction as new multiplication occurs. Resultant necrosis of adjacent tissues can then contribute to the spread of the organisms and progressive disease. Hypersensitivity can thus be involved in a new cycle of caseation, cavitation, and spread.

The tuberculin skin test for hypersensitivity to the proteins of tubercle bacilli becomes positive during the course of first infection and remains so. Methods for testing and the interpretation of reactions are described on pages 333–36.

Clinical Symptoms. The wide variety of symptoms in tuberculosis reflects the fact that the orga-nism may involve any organ to which circumstances permit its spread. The intensity of symptoms depends on the numbers and location of foci and the host response to them. There may be no symptoms at all in a first infection, or the primary course may be rapidly progressive and sometimes fatal. The signs of reinfection tuberculosis may also be minimal or prominent. Common symptoms include fever and malaise, easy fatigue, and loss of weight. In pulmonary disease there may be a productive cough yielding blood as well as sputum. Tuberculosis of kidney, liver, bone, genital organs, or other tissues may occur simultaneously by hematogenous spread, usually from a pulmonary focus, or one or another of these sites may be involved separately, with symptoms characteristic of the localization.

Tuberculous meningitis may occur as a result of bacteremia. It is more commonly seen in children following first infection than in adults. The symptoms are more chronic and slowly progressive than those of the acute bacterial meningitides (see Meningitis, this chapter), with lower cell counts that are predominantly lymphocytic. The clinical form of tuberculous meningitis is similar in some basic respects to that seen in extensions of fungous disease.

Laboratory Diagnosis

I. Identification of the Organism (*Mycobacterium tuberculosis*). The tubercle bacillus belongs to the genus *Mycobacterium,* whose most important members are shown in Table 13–3.

A. MICROSCOPIC CHARACTERISTICS. The mycobacteria stain with difficulty. Once they take an appropriate stain, however, they resist decolorization, even by acids, and are referred to as "acid-fast" bacilli because of this property. The Ziehl-Neelsen stain (see Chap. 4) is commonly used to identify them, the organisms appearing red against a blue background. Another stain, using the fluorescent dyes auramine and rhodamine, permits rapid detection of the organisms as bright objects against a dark background. This technique is used in many laboratories today.

Tubercle bacilli are slender straight or slightly curved rods, often granular or beaded in appearance. Like the corynebacteria (see Diphtheria, this chapter), they are often seen in palisade formations, or in "V" and "L" positions that suggest a snapped stick. (The colloquial phrase "red snapper" is often quite appropriate.) (Fig. 13–5.)

B. CULTURE CHARACTERISTICS. The growth rate of

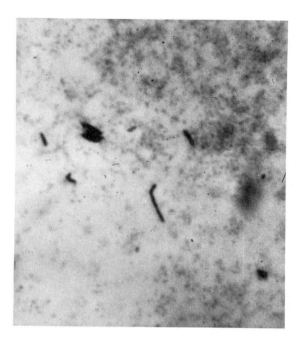

Fig. 13–5. A photomicrograph of sputum stained by the Ziehl-Neelsen method. The dark rods are tubercle bacilli, *Mycobacterium tuberculosis*. (1000×.) (Courtesy of Center for Disease Control, D.H.E.W., Atlanta, Ga.)

tubercle bacilli and most mycobacteria is much slower than that of other bacteria. Several days or weeks are required for primary isolation, with an average incubation time of three to six weeks on laboratory media of most practical value. They are aerobic, but their growth may be enhanced with increased atmospheric CO_2. Incubation is at 35° C.

A number of media are available for culture of tubercle bacilli. Some of these contain complex nutrients, such as egg, potato, and serum. Others are simpler media formulated to provide the factors best known to promote growth. Many laboratories use two or more different media to ensure successful cultures (Fig. 13–6).

Since tubercle bacilli may produce lesions in many parts of the body, many types of specimens can be appropriate for culture. Some of these, however, contain a number of commensalistic organisms indigenous to the region (sputum, voided urine, etc.). Many bacterial species will also grow on the media used for isolation of tubercle bacilli, and they grow much faster. Their multiplication on the media depletes it rapidly of nutrient,

and also makes the search for tubercle bacilli virtually impossible in the mixed culture. For these reasons it is necessary, prior to culture, to treat such specimens with chemicals that destroy bacteria other than *M. tuberculosis*. This takes advantage of the latter's greater resistance to alkalies, acids, quaternary ammonium compounds, and other chemical agents. The chemical treatment is carefully controlled to avoid undue suppression of tubercle bacilli themselves. Fluid specimens are then concentrated by centrifugation and the sediment is inoculated on appropriate culture media. Cultures are routinely examined every week for developing growth of *M. tuberculosis* (Fig. 13–7), or every day in cases of clinical urgency. Negative cultures are not discarded until they have been incubated for at least eight weeks.

The lipoid, hydrophobic nature of the surface of tubercle bacilli accounts for their heaped, clumped colonial growth on the surface of solid media. Certain compounds with a detergent or wetting action (for example, Tween 80) are used in liquid media to permit more dispersed and rapid growth. The waxy character of the surface of this organism contributes to its resistance to chemical and physical agents. Tubercle bacilli are not as readily killed by antiseptics or disinfectants as are other vegetative microorganisms. They are also more resistant to drying and are capable of surviving for long periods in dried sputum or other infectious discharges. For this reason, special care must be taken in selection of chemical disinfectants, in adjustment of their concentrations, and in timing of the disinfection period when tubercle bacilli are to be killed by chemical methods.

C. SEROLOGIC METHODS OF IDENTIFICATION. None.

II. Identification of Patient's Antibodies

A. CIRCULATING (HUMORAL) ANTIBODIES. These occur in the serum of tuberculous patients, are measurable by complement fixation and other tests, but have *no* diagnostic value.

B. SKIN TESTS. The *sensitizing* antibody of tuberculin allergy is associated with mononuclear cells wherever they occur in the body, including epidermal tissue. Skin tests using extracted protein antigens of tubercle bacilli can detect this antibody of hypersensitivity. When antigen is injected into skin it reacts with cell-attached antibody, and this interaction induces injury to hypersensitive cells at the local site of injection. Since the hypersensitivity of tuberculosis is long-lasting, a positive skin reaction indicates past infection that may or may not be currently active. (The possibility of current infection must then be determined by clinical findings, x-ray, and bacteriologic methods.) A negative tuberculin reaction is displayed by individuals who have never been

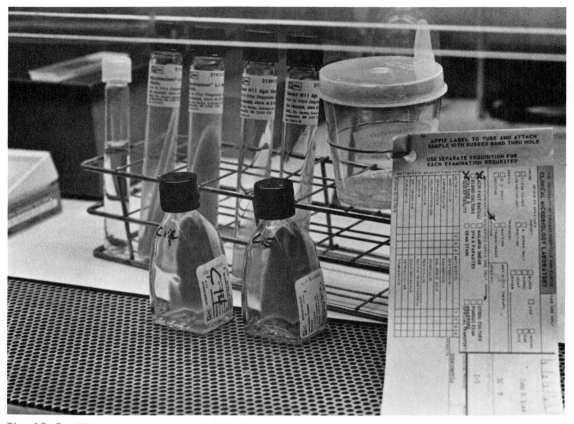

Fig. 13-6. This sputum specimen will be inoculated on several different kinds of culture media for growth of tubercle bacilli. The work is being done in a bacteriologic safety cabinet, the shield of which can be seen at the top of the photograph.

infected, by those who are in the preallergic early stage of first infection, and often by those with advanced terminal tuberculosis.

1. *Skin-Testing Antigens.* Old tuberculin (OT) is a concentrated filtrate of a broth culture of tubercle bacilli. It contains, in addition to the tuberculoproteins derived from the organism, a number of irrelevant constituents of the bacilli and of the medium that may induce nonspecific reactions in the skin. A purified protein derivative (PPD) is obtained by chemical fractionation and purification of OT. This purified antigen is preferred for skin testing.

2. *Skin Test Doses.* Hypersensitive persons may have severe local reactions to tuberculin. Introduction of the antigen into the skin may also induce renewed or increased reactivity at *focal* sites of infection elsewhere in the body. For this reason, caution must be exercised in skin testing. A series of skin test doses may be given, beginning with very dilute material. If no reaction occurs, repeated doses may be given, each in an increased concentration, until some hypersensitivity is displayed. If even a fully concentrated dose fails to elicit a skin response, the individual is considered tuberculin-negative.

3. *Skin-Testing Methods.* A number of techniques have been devised for introducing tuberculins to the cells of the skin. The goal of each is to detect hypersensitivity with accuracy, without inducing untoward reac-

tions. *Accuracy* can be increased by using purified antigens, as described. *Safety* is promoted by using minimal doses, or by applying them at superficial levels of skin, where reactivity is less intense and absorption of antigen at systemic foci less likely. Some of the techniques currently available are described below, together with their advantages and disadvantages:

The Vollmer patch test utilizes OT in a gauze strip (with lanolin) applied to the *surface* of the skin. This is the least accurate but the safest way to test for tuberculin hypersensitivity. The antigen is not pure, but the skin is not very absorptive, and reactions, when they occur, are seldom intense. It is a reasonable screening method for groups of children or adults who may be expected to display a high rate of infection.

The Heaf test utilizes OT for the *intracutaneous* introduction of antigen. A "gun" device is used to inject antigen from six spring-released needle points, which enter only the epidermal layer. The method is convenient and safe for mass surveys, but its accuracy is not quantitative.

The tine test employs dried OT on multiple metal tines held in a round plastic head. The tines are pressed against the skin for intracutaneous insertion of antigen. Each unit is used once and discarded. This method is similar to the Heaf test in advantages and limitations.

The Mantoux test requires the *intradermal* injection of antigen, which may be either OT or PPD. This is the most accurate method of tuberculin skin testing. An aliquot of 0.1 ml of quantitatively diluted antigen is introduced from a needle and syringe into a site just within the epidermal layer. Specificity is gained by using purified antigen; safety is controlled by adjusting the dose dilution and by making successive tests, if neces-

Fig. 13–7. Colonies of *Mycobacterium tuberculosis* growing on tubed, slanted agar. Note the heaped, dry, rough appearance of the growth.

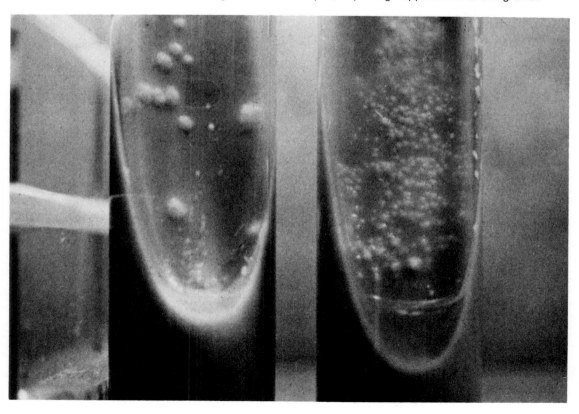

sary, with increasing concentrations. The necessity for individual syringes and needles is of some disadvantage in mass surveys. The choice of method depends largely on considerations of the epidemiologic background of the group or individual to be tested and of the risk of excessive reactivity.

4. *Reading and Interpreting Tuberculin Skin Tests.* The application, reading, and interpretation of skin tests are the physician's responsibility, because these tests involve clinical reactivity of diagnostic import. Nursing and laboratory personnel should be well informed, however, as to the nature of the materials used, the method of application, and the significance of results.

Tuberculin reactivity is of the delayed type, and of relatively long duration. Positive reactions require several hours to begin; they reach a maximum in one to three days. Skin tests should be read in 48 hours, or no later than 72 hours. They are interpreted as positive if an area of induration of at least 10-mm diameter can be felt at the injection site. Edema and erythema may also be seen, and intense reactions proceed to central necrosis. Strongly positive reactions may persist for several days, but weak ones disappear quickly after 72 hours. Nonspecific reactions may appear during the first 24 hours but do not persist to the second day.

Tuberculin reactivity begins within four to six weeks after infection and persists for long periods. It may be lost in the presence of overwhelming tuberculous infection, or during the course of such diseases as measles, sarcoid, and Hodgkin's (anergy). As pointed out previously, a positive test does not necessarily denote present infection, nor does it imply immunity. In the absence of *current* infection it may signify an increased resistance to superinfection. This expectation must be guarded, however, in view of the increased *reactivity* of hypersensitive persons, which may contribute to extension of focal infection.

III. Specimens for Bacteriologic Diagnosis

A. In *pulmonary* tuberculosis, sputum collections are the most appropriate specimens for smear and culture. If the patient is unable to produce sputum, gastric washings may be useful. In *disseminated,* or *miliary,* forms of the disease, any specimen that can represent a suspected focus of infection may be relevant: pleural fluid, synovial fluid, peritoneal fluid, or spinal fluid; urine, aspirated bone marrow, or exudate from lesions in bone, biopsied tissue, and other suspected material. Blood cultures are sometimes rewarding if blood is drawn during a period of active bacteremia, but they are less satisfactory than material obtained from a focal lesion. Stool speci-

mens have the least value and offer the greatest technical difficulties. They may be warranted only in rare instances of suspected involvement of intestinal lymphatic tissue, when this cannot be confirmed by other means.

B. *Sputum* should be collected in the morning when the patient first awakens. The pooling of sputum that occurs in the bronchi during sleep collects many more bacilli than can be found in casual specimens coughed up during the day. However, any specimen that appears to contain caseous particles should be submitted for culture. If the patient is nebulized in the early morning, a more representative specimen can often be produced. Propylene glycol should not be used in the nebulizer, however, since this substance can inhibit or kill tubercle bacilli.

Twenty-four-hour sputum collections are to be discouraged because of their heavy contamination with oral bacteria, which may multiply still further in specimens held at room temperature throughout the collection period.

C. *Urine.* A single collection of the first morning specimen is most satisfactory and presents fewer technical problems to the nurse and the laboratory than a 24-hour pooled specimen to which the same objections apply as with sputum collections.

D. The *precautions* described for the collection of sputum should be intelligently applied to the collection and transport of *any specimen,* fluid or solid. Collections are made in sterile containers *without* added preservative. They must be kept tightly capped. The patient should be instructed to avoid the collection of saliva and to handle the container hygienically. Containers must not be overfilled lest their contents leak through the cap. When the containers are ready for transport to the laboratory, the outer surfaces should be wiped clean with disinfectant, the cap again tested for fit, and the transporter instructed to carry the container in an upright position, within a paper or plastic bag.

E. *Smears* can be made directly from clinical specimens, or from their concentrates. A smear report indicating the presence of acid-fast bacilli must be considered presumptive evidence of tuberculosis, but it is important to remember that acid-fast saprophytic organisms may be found in sputum, urine, gastric washings, and similar specimens. Growth characteristics of the organisms must be identified in culture, or in inoculated animals, to confirm laboratory diagnosis.

F. *Cultures* are made directly or on centrifuged concentrates of fluid or other specimens that do not have a mixed, commensalistic flora. Sputum and other contaminated materials are first treated with a bactericidal agent that does not destroy all the tubercle bacilli as well. Inoculated media are incubated until mycobacterial growth is observed, or for a minimum of eight weeks. Identification of growth is made on the basis of microscopic and colonial appearance (Fig. 13–7), pigmentation, and a few important biochemical properties.

IV. Special Laboratory Tests

A. ANIMAL INOCULATIONS. Guinea pigs are highly susceptible to tuberculous infection. They may be inoculated subcutaneously with a portion of the clinical specimen that has been prepared for simultaneous culture. Animals are observed for signs of developing infection, tuberculin-tested in three to four weeks, and autopsied after six weeks. Infection is confirmed by smears and cultures of involved tissues. This method of laboratory diagnosis is expensive and hazardous, and should be reserved for those cases in which cultural diagnosis is difficult or inconclusive.

Guinea pigs, rabbits, and other animals may sometimes be used to establish the virulence of a strain of *Mycobacterium* isolated in culture. Animals vary in susceptibility to strains of *M. tuberculosis* from human, bovine, or other sources, a fact that may be of value in determining both origin and virulent properties of a pure culture. For example, rabbits and guinea pigs used together can distinguish virulent human and bovine strains. Both animal species develop severe infection with tubercle bacilli of bovine origin, but guinea pigs are also susceptible to human strains, while rabbits are relatively resistant to the latter.

B. ANTIBIOTIC SUSCEPTIBILITY TESTING is of particular value in determining resistance of strains from treated cases of tuberculosis. (See Treatment.)

Epidemiology

I. Communicability of Infection

A. RESERVOIR, SOURCES, AND TRANSMISSION ROUTES. Man is the most common reservoir of human tuberculosis, the source of infection being the respiratory secretions of persons with active pulmonary lesions. Transmission may occur by direct or indirect contact with patients with open lesions, but the most probable and usual route is by inhalation of airborne droplet nuclei. The closed conditions of familial, military, or institutional living may permit prolonged exposure to an active case if the latter is unrecognized. This type of situation most frequently contributes to spread of infection and the incidence of active cases.

Diseased cattle are a reservoir for the bovine strain, *Mycobacterium bovis*. The usual source of infection of man with this organism is raw milk (or other dairy product) from tuberculous cattle. The lesions acquired through gastrointestinal entry are usually extrapulmonary, and this type of infection is not directly transmitted by man. Bovine tuberculosis may also be acquired through inhalation of airborne organisms in and around barns housing diseased animals or by handling infectious animal products. Human pulmonary infection with bovine strains is communicable. The cattle reservoir has been largely controlled through continuous tuberculin testing of herds and pasteurization of milk.

The epidemiology of other pulmonary mycobacterioses is discussed for the group on page 341.

B. INCUBATION PERIOD. The period from known effective exposure to the appearance of a primary lesion of first infection is usually about four to six weeks. The time required for development of new lesions following reinfection varies with host resistance and with the source and numbers of organisms. The interval between first infection and reinfection tuberculosis may be of many years' duration.

C. COMMUNICABILITY. The communicable *period* persists for a given patient for as long as he discharges tubercle bacilli. The *degree* of communicability depends on the intensity of contamination of air with infectious droplets. This is influenced by the coughing habits of the patient, as well as by the dynamics of air circulation and fallout. Actual transmission to another person during the communicable period depends on inhalation of at least one infectious droplet nucleus. Inhaled tubercle bacilli must be carried far down the airways to a point in the lung where fixed mononuclear phagocytes can pick them up and their intracellular multiplication can begin. The sputum of an infected patient must carry fair numbers of bacilli to assure this process. It becomes a most dangerous source of communicable infection when it is sprayed into the air by forceful coughing or sneezing. In a closed environment, shared by a number of people, communicability of infection from an active case may be high.

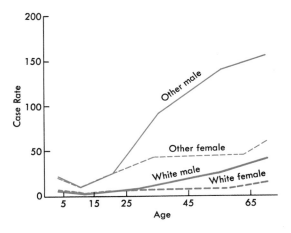

Fig. 13-8. Tuberculosis case rates for white and other races, by sex and age, United States, 1974. (Reproduced from *Tuberculosis in the United States*. August 1976, Bureau of State Services, Tuberculosis Control Division, Center for Disease Control, D.H.E.W., Atlanta, Ga.)

D. IMMUNITY. Susceptibility to infection is general but is influenced by factors of age, sex, race, nutrition, and general health. Children under three years of age are most susceptible, but the incidence of tuberculosis in older children has declined sharply with improved controls on bovine tuberculosis. In 1974 the U.S. Department of Health, Education, and

Fig.13-9. Distribution of tuberculosis cases in cities and other areas of the United States, 1974. (Reproduced from *Tuberculosis in the United States,* August 1976, Bureau of State Services, Tuberculosis Control Division, Center for Disease Control, D.H.E.W., Atlanta, Ga.)

Distribution of cases

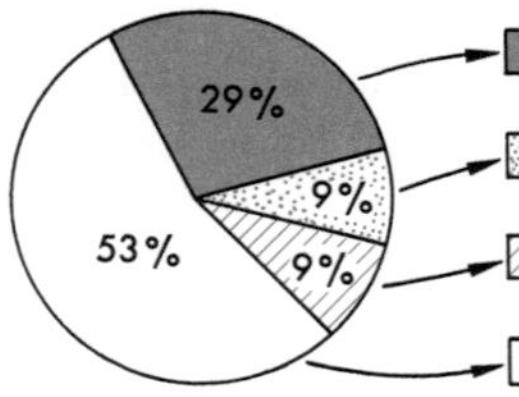

	Size of City (population)	New Cases	Case Rate
29%	500,000+	8,865	25.7
9%	250,000 to 500,000	2,773	21.3
9%	100,000 to 250,000	2,651	16.1
53%	All other areas	15,833	10.7
	United States	30,122	14.2

Welfare reported the incidence of new active cases to be lowest in children between the ages of 5 and 15, and rising steadily, sometimes sharply from age 20 to 65 and beyond (Fig. 13-8).

The white races appear to be generally less susceptible than blacks and American Indians. It is not certain whether the higher resistance of Caucasians is related to a longer history of endemic infection, with selection of resistant survivors, or to what extent differences in living conditions and nutrition are important. Differences observed in the incidence of tuberculosis in men and women may be a reflection of physiologic influences or of different degrees of exposure under varying living conditions.

In the United States the geographic distribution of new cases of tuberculosis is much greater in large cities than in small ones or in rural areas (Fig. 13-9). In the decade preceding 1974, the tuberculosis case rates in big cities was consistently at least double that for all other areas. This is related to the density of populations in the lowest socioeconomic groups as well as the concentration of the highest risk groups (nonwhites, especially males). The susceptibility of the latter group is further reflected in the case rates for the southern portion of the United States for the three-year period from 1972 to 1974 (Fig. 13-10). Another factor contributing to higher case rates in large American cities is the presence of legal or illegal aliens from areas of the world where tuberculosis is less well controlled. In 1976 there were 141 cases of active tuberculosis among legal aliens living in New York City, this group representing 6.5 percent of all new cases reported for the entire city in that year. This number represented a sharp increase in active cases in a group for which effective surveillance can be maintained (in 1975 there were 1007 inactive cases and 105 cases of active tuberculosis among legal aliens, compared with 962 inactive and 141 active cases in 1976). This significant rate of infection in a controlled segment of the population points up the problem of unrecognized and untreated active tuberculosis among unregistered, illegal aliens residing in New York or other cities.

A bacterial vaccine is available that confers a definite degree of increased resistance to tuberculosis. This vaccine is known as BCG, in recognition of the French workers who first suggested its possible value ("bacille de Calmette et Guérin"). The immu-

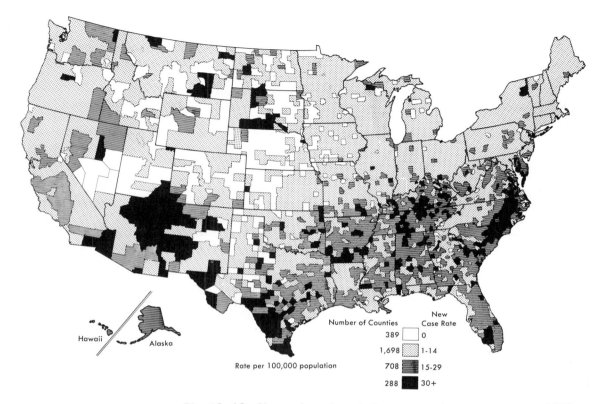

Number of Counties	New Case Rate
389	0
1,698	1-14
708	15-29
288	30+

Rate per 100,000 population

Fig. 13–10. New active tuberculosis case rates by county, average 1972–1974. (Reproduced from *Tuberculosis in the United States*. August 1976, Bureau of State Services, Tuberculosis Control Division, Center for Disease Control, D.H.E.W., Atlanta, Ga.)

nizing material is a living, attenuated strain of tubercle bacilli of bovine origin. The evidence has been considered convincing that BCG provides some protection against naturally acquired human tuberculosis. Accordingly, its use is recommended for tuberculin-negative individuals who run a high risk of continuing exposure to infection, a group that includes medical students, physicians, nurses, and laboratory personnel. Vaccinated persons become tuberculin-positive and display an increased resistance to new natural infection with virulent organisms.

II. Control of Active Infection

A. ISOLATION. The communicability of tuberculosis can be brought under control by specific antimicrobial therapy coupled with effective hygienic practices. Patients with active infection should be hospitalized for treatment, at least until sputum (or other discharge) is bacteriologically negative. During their stay in a hospital, patients are placed in respiratory isolation. When patients are treated at home, the principles of isolation should be observed while they remain in a communicable stage of disease. This is particularly important if there are young children or other highly susceptible individuals in the family. Public health nursing supervision should be provided, with instruction for all family members, as well as the patient, in the hygiene of tuberculosis prevention.

B. PRECAUTIONS. During the communicable period, care of hospitalized patients should be individualized, with concurrent disinfection of sputum and articles exposed to it. Since tuberculosis is chiefly an airborne infection, the most effective precautions are those that prevent tubercle bacilli from reaching the air in significant numbers.

The patient is taught to muffle coughing or sneezing with paper handkerchiefs, which are then carefully disposed of in paper bags for incineration. Effective masks are appropriate only for personnel, visitors, and others, under special circumstances in which there are face-to-face contacts, or coughing is uncontrolled. Uncooperative patients are sometimes a problem, but failure to cooperate may also be involuntary (as in throat or laryngeal examination).

Gowns are not effective in preventing spread of airborne infections, and their use is unnecessary from this point of view. For those in close contact with patients in a highly communicable stage of tuberculosis, gowns may be useful simply to protect their clothing from gross contamination.

The selection of good tuberculocidal disinfectant agents, and their proper use in routine disinfection of thermometers and other equipment, or for daily cleaning of floors and furniture, are essential precautions in the care of the hospitalized patient.

When the tuberculous patient has become sputum-negative on antimicrobial therapy, ordinary hygienic practices are sufficient for continuing control of infection.

C. TREATMENT. Most first infections heal with out recognition or treatment. Recognized, active cases of tuberculosis should be hospitalized, to prevent a chain of contact infections and to give the patient immediate treatment.

Currently, the most useful antituberculous drugs are isoniazid (INH), ethambutol, and rifampin. They are generally used in one or another combination, because drug resistance to any one of them may develop quickly in infecting strains of tubercle bacilli. Laboratory testing of drug susceptibility should be done on initial cultures of tubercle bacilli and repeated if the patient's sputum remains bacteriologically positive after six months or reverts after being negative. Changes in the therapeutic regimen may be necessary if laboratory studies demonstrate drug resistance.

Surgical techniques are also used in the treatment of tuberculosis. When anatomically and medically feasible, removal of diseased tissue, under cover of antimicrobial therapy, can prevent further systemic extension of infection. When systemic infection is extensive, or localized in vital organs, intensive antimicrobial therapy is used, but fatality rates in these situations remain high.

D. CONTROL OF CONTACTS

1. *Quarantine* of contacts is not indicated.

2. *Prophylaxis.* Immunoprophylactic use of BCG may be warranted for close contacts of active cases who are *tuberculin-negative.* Household and medical contacts might be included.

Chemoprophylactic use of isoniazid may be indicated for those contacts of active cases for whom risk of infection is particularly grave. This group includes *tuberculin-positive* individuals who can be expected to display unusual susceptibility to reinfection, for example, children under three years of age, diabetics, and patients on steroid therapy for other conditions (steroids reduce the inflammatory response that is so important in resistance to infection).

3. *Case Finding.* Chest x-ray and tuberculin test surveys of contacts are of particular value. In view of the recognized hazards of repeated x-irradiation, tuberculin skin testing is preferred for *continuing* surveillance of exposed negative reactors. This should be done on a semiannual basis for children, and at least annually for adults (particularly nursing, medical, and laboratory personnel). If negative skin reactors convert to a positive tuberculin response, chest x-ray and other intensive studies are useful in detecting early infection. Prompt chemotherapy may then abort development of serious disease.

E. EPIDEMIC CONTROLS. Case reports to local health departments are obligatory in most communities. This procedure permits epidemiologic analysis of the distribution of cases, so that unusual grouping and incidence of cases may be promptly recognized and an immediate search made for the origins of infection. International control calls for x-ray screening of immigrants.

III. Prevention

A. IMMUNIZATION. BCG vaccination provides some protection of specific value. When properly performed, BCG vaccination converts at least 90 percent of tuberculin-negative persons to a state of skin sensitivity. The protection afforded is only partial, and its duration remains undetermined. Mass vaccination with BCG is applicable only in situations where risk of infection and number of tuberculin-negative contacts of active disease are both high.

B. CONTROL OF RESERVOIRS, SOURCES, AND TRANSMISSION ROUTES. Methods for prevention of further infection may be summarized here as follows:

1. Find and treat active tuberculosis as rapidly as possible.

2. Find the source (or sources) of each active case, isolate the individual, and apply chemotherapeutic methods as indicated.

3. Continue and widen the search for bovine tuberculosis. Eliminate tuberculin-reactive cattle. Pasteurize milk.

4. Provide BCG vaccination for tuberculin-negative persons who are closely or continuously exposed to risk of infection.

5. Provide chemotherapy for tuberculin-positive *exposed* individuals for whom the risk of infection is great.

6. Conduct frequent community surveys by skin testing or roentgenographic methods.

7. Provide public health nursing supervision of active cases under treatment at home.

8. Provide continuing public education concerning the importance, origin, and control of tuberculosis.

Other Mycobacterioses. Mycobacteria other than *M. tuberculosis, M. bovis, M avium,* and *M. leprae* have a wide distribution in nature. Many of them are found in soil and water and could easily infect man or animals either through inhalation or ingestion. The epidemiology of the diseases caused by these organisms has not been well defined, however. Except for those that cause cutaneous lesions, there is no evidence that infection can be spread from person to person. There is a good deal of evidence, however, that those associated with human pulmonary disease are widely, though silently, distributed in the population. This is indicated by the fact that their protein extracts (equivalent to PPD from tubercle bacilli) produce skin reactions in many persons tested. Since even tuberculin-negative persons may react to extracts from strains other than *M. tuberculosis,* it would appear that this skin reactivity is related to specific subclinical infection with other strains, and not to hypersensitivity to the related tubercle bacillus.

It has been customary for microbiologists to refer to these strains as "atypical," "unclassified," or "anonymous" mycobacteria. This is because when they were first recognized as agents of pulmonary disease similar to tuberculosis, some 20 to 25 years ago, it was obvious that they are culturally and biochemically different from *M. tuberculosis.* Since then they have been classified into four groups (Runyon's groups I to IV) on the basis of their pigments, photosensitivity, and speed of growth.

The "atypical" mycobacteria are relatively resistant to the antituberculous drugs in common use. A high proportion are isoniazid resistant, but many can be successfully treated by using combinations of three or more drugs.

Table 13–4. Summary: Bacterial Diseases Acquired Through the Respiratory Tract

Clinical Disease	Causative Organism	Other Possible Entry Routes	Incubation Period	Communicable Period
I. Upper Respiratory Infections				
Pharyngitis	*Streptococcus, group A*		1–2 days	During time of active inflammation
Laryngitis (children)	*Haemophilus influenzae*			
Sinusitis Otitis	*Streptococcus pneumoniae* *Streptococcus group A* *Haemophilus influenzae* *Staphylococcus aureus* Gram-negative enteric bacilli			
II. Bacterial Pneumonias				
A. Pneumococcal pneumonia	*Streptococcus pneumoniae*		1–2 days	Probably during time sputum is bacteriologically positive
B. Other bacterial pneumonias	*Streptococcus pyogenes* *Staphylococcus aureus* *Klebsiella pneumoniae* *Haemophilus influenzae*		1–2 days	As above
C. Legionnaires' disease	Unclassified bacillus	Unknown	Unknown	Unknown
D. Primary plague pneumonia	*Yersinia pestis*	Bubonic plague is arthropod-borne (see Chap. 26)	3–4 days	As above
III. Bacterial Meningitis				
A. Meningococcal meningitis	*Neisseria meningitidis*		3–4 days	While nasopharynx is bacteriologically positive

Specimens Required	Laboratory Diagnosis	Immunization	Treatment	Nursing Care
Throat swab	Smear and culture	Limited value	Antibiotic choice depends on testing of isolated organisms	Principles of personal hygiene, support of healthy resistance
Discharge from site	Smear and culture			
Sputum Blood	Smear and culture Culture	None	Penicillin	Secretion precautions; concurrent disinfection of respiratory secretions
Throat and naso-pharyngeal swabs Sputum Blood	Smear and culture Smear and culture Culture	None	Antibiotic choice depends on testing of isolated organisms	As above
Lung tissue Pleural fluid Serum	Culture Culture Serology	None	Erythromycin	Concurrent dis-infection of respiratory secretions
Sputum Blood Serum	Smear and culture Smear and culture Serology	Killed vaccine	Sulfonamides Tetracyclines Streptomycin	Strict isolation
Spinal fluid Blood	Smear and culture Culture	None	Penicillin Rifampin (contacts or carriers)	Isolation and respiratory precautions for first day after therapy initiated

Table 13–4. *(Cont.)*

Clinical Disease	Causative Organism	Other Possible Entry Routes	Incubation Period	Communicable Period
B. Secondary bacterial meningitis	*Haemophilus influenzae* *Streptococcus pneumoniae* *Streptococcus pyogenes* *Staphylococcus aureus* Gram-negative enteric bacilli	Enteric bacteria possibly enter by fecal-oral route	Unknown	Respiratory transmission does not necessarily lead to meningeal infection
C. Listeriosis	*Listeria mono-cytogenes*	Venereal contact Congenital transfer Ingestion (animal sources)	Unknown	Man-to-man transfer unusual except for con-genital infection
IV. Streptococcal Diseases				
A. Strep sore throat	*Streptococcus pyogenes*		1–3 days	While throat secretions are bacteriologically positive
B. Scarlet fever	*Streptococcus pyogenes*		1–3 days	As **IV** A
C. *Sequelae of group A strep infections:* Rheumatic fever			3 weeks post-infection	Not communicable
Acute hemorrhagic glomerulo-nephritis			1–3 weeks postinfection	Not communicable
Erythema nodosum			Poststrep infection	Not communicable
D. Subacute bacterial endocarditis	*Streptococcus* species Gram-negative enteric bacilli and others		Unknown	Not communicable

Specimens Required	Laboratory Diagnosis	Immunization	Treatment	Nursing Care
Spinal fluid Blood	Smear and culture Culture	None	Antibiotic choice depends on testing of isolated organism	Concurrent disinfection of respiratory secretions; special care in case of enteric bacterial agents focused in kidney or gallbladder
Spinal fluid Blood Maternal vaginal discharge, placenta Infected infant's blood, lesions	Smear and culture Culture Smear and culture Smear and culture	None	Tetracyclines Ampicillin Streptomycin	Isolation of infants Concurrent disinfection in care of infants and infected mothers
Throat or nasopharyngeal swabs	Smear and culture	None	Penicillin	Concurrent disinfection of respiratory secretions
As **IV** A	As **IV** A (Schultz-Charlton skin test with scarlatinal antitoxin)	None	Penicillin	As **IV** A
Serum	Antistreptolysin titer	None	Penicillin	No special precautions
Serum Urine	Antistreptolysin titer Microscopic analysis (no culture)	None	Penicillin	No special precautions
Serum	Antistreptolysin titer	None	Penicillin	No special precautions
Blood	Culture	None	Penicillin + streptomycin*	No special precautions

*Or according to susceptibility of organism.

Table 13–4. *(Cont.)*

Clinical Disease	Causative Organism	Other Possible Entry Routes	Incubation Period	Communicable Period
V. Whooping Cough	*Bordetella pertussis*		7–10 days; not more than 21	Highest during late incubation and catarrhal stages
VI. Diphtheria	*Corynebacterium diphtheriae*		2–5 days	Carrier state may be persistent
VII. Tuberculosis	*Mycobacterium tuberculosis*	Gastrointestinal (bovine strains)	4–6 weeks for primary infection	While sputum is bacteriologically positive
	"Atypical" mycobacteria	Human, animal and environmental contacts		Not communicable person-person?

Specimens Required	Laboratory Diagnosis	Immunization	Treatment	Nursing Care
Nasopharyngeal swabs Cough plates Blood	Smear and culture Culture White blood cell count	Killed vaccine	Erythromycin Ampicillin	Hospital isolation (especially on pediatric service) Concurrent disinfection of nasal and oral secretions, vomitus, and contaminated articles
Swab of suspected lesion in throat	Smear and culture Elek test	Toxoid	Diphtheria antitoxin Penicillin Erythromycin	Strict isolation until bacteriologically negative; concurrent disinfection of nasal and oral secretions and contaminated articles
Sputum Gastric washings Body fluids Exudates from localized lesions Blood (rarely) Stools (rarely)	Smear and culture Smear and culture Smear and culture Smear and culture Culture Culture	BCG vaccine for persons at risk	Isoniazid Ethambutol Rifampin Surgery?	Isolation while sputum or other discharges are bacteriologically positive Concurrent disinfection of respiratory secretions and contaminated articles

Questions

1. What specimens are needed for the diagnosis of pneumonia?
2. List the organisms that cause pneumonia.
3. What is Legionnaires' disease? How was the causative organism isolated?
4. What are the important microbial agents of meningitis?
5. How are streptococci classified? Why is this classification useful?
6. What are the possible sequelae of group A streptococcal infections?
7. How can pertussis be prevented?
8. What factors contribute to the severity of diphtheria?
9. Why must a patient with diphtheria be given antitoxin immediately?
10. Why is the tubercle organism called "acid fast"?
11. How does primary tuberculosis differ from reinfection tuberculosis?
12. How does hypersensitivity contribute to tubercular disease?
13. What does a positive tuberculin test signify?
14. What is the communicable period for tuberculosis?

Additional Readings

Barham, Virginia Z.: Tuberculosis Care, 1971: Changing the Attitudes of Hospital Nurses. *Nurs. Outlook,* **8**:538, Aug. 1971.

Barkin, R. M.; Greer, C. C.; Schumacher, C. J.; and McIntosh, K.: *Haemophilus influenza* Meningitis. *Am. J. Dis. Child.,* **130**:1318, Dec. 1976.

Bassili, W. R., and Stewart, G. T.: Epidemiological Evaluation of Immunization and Other Factors in Control of Whooping Cough. *Lancet,* **1**:471, Feb. 28, 1976.

Crane, Lawrence R., and Lerner, M. A.: Gram-Negative Pneumonia in Hospitalized Patients. *Postgrad. Med.,* **58**:85, Sept. 1975.

Howe, Charles: Legionnaires' Disease: The Laboratory Search for a Baffling Killer. *Med. Lab. Observ.,* **9**:37, May 1977.

Hyslop, Newton E., and Swartz, M. N.: Bacterial Meningitis. *Postgrad. Med.,* **58**:120, Sept. 1975.

McClement, John H.: Tuberculosis: Chemotherapy and Treatment. *Postgrad. Med.,* **58**:97, Sept. 1975.

McGuckin, M.: Microbiological Studies, Tips for Assisting with Cultures of CSF and Other Body Fluids, Part 5. *Nursing '76,* **6**:17, Apr. 1976.

Mizer, Helen: Group B Streptococci in Neonatal Infection. *Am. J. Mat. Child Nurs.,* **3**:21, Jan.-Feb. 1978.

Myers, J. A.: Tapering Off of Tuberculosis Among the Elderly. *Am. J. Public Health,* **66**:1101, Nov. 1976.

O'Grady Roberta, and Dolan, Thomas: Whooping Cough in Infancy. *Am. J. Nurs.,* **76**:114, Jan. 1976.

Patterson, M. J.,and Hafeez, A. E.: Group B Streptococci in Human Disease. *Bacteriol. Rev.,* **40**:774, Sept. 1976.

Reichman, Lee B.: Tuberculosis Care: When and Where? *Ann. Intern. Med.,* **80**:402, Mar. 1974.

Taylor, Carol M.: Pneumococcal Pneumonia. Your Patient's Second Threat. *Nursing '76,* **6**:30, Mar. 1976.

Weg, John G.: Tuberculosis and the Generation Gap. *Am. J. Nurs.,***71**:495, Mar. 1971.

Wood, W. Barry: *From Miasmas to Molecules.* Columbia University Press, New York, 1961.

Youmans, G. P.: Relation Between Delayed Hypersensitivity and Immunity in Tuberculosis. *Am. Rev. Resp. Dis.,* **111**:109, 1975.

*Youmans, Guy P.; Paterson, P. Y.; and Sommers, H. M.: *The Biologic and Clinical Basis of Infectious Diseases.* W. B. Saunders Co., Philadelphia, 1975.

*Available in paperback.

Viral Diseases Acquired by the Respiratory Route

14

Viral diseases acquired by respiratory transmission and entry include some of the most contagious infections known. Influenza, for example, often occurs in rapidly spreading epidemic form, and the common viral diseases of childhood, such as mumps, chickenpox, and measles, are so highly communicable that they are not often escaped.

The diseases of this chapter are presented in two major sections: (1) those whose agents infect the upper or lower respiratory tract following respiratory entry; and (2) those that are systemic infections induced by agents entering by the respiratory route (these include mumps, measles, and the pox diseases).

There are a few systemic viral diseases that may have either alimentary or respiratory tract entry. These include poliomyelitis and Coxsackie viruses. Since the gastrointestinal route of spread and entry appears to be most common for these diseases, they are discussed in Chapter 19 with other viral infections of intestinal tract entry. It should be noted, however, that respiratory transmission is also possible and appropriate precautions should be applied.

Unlike many of the bacterial infections considered in Chapter 13, and particularly those of the upper respiratory tract where chronic bacterial entrenchments occur so frequently, the viral diseases described in this chapter are always acute and self-limited, although they may leave residual damage. They do not respond to antibiotic therapy, but are resolved by the body's defensive mechanisms. Specific immunity, artificial or natural, may play a large role notably in smallpox, chickenpox, measles, and mumps. Viral diseases of the respiratory tract itself, however, such as influenza or the common cold, induce only transient immunity. Artificial immunization is of limited value in controlling the latter infections, but it has been of major importance in controlling systemic viral diseases.

The easy and frequent communicability of these infections by the respiratory route accounts for their epidemic character and presents a difficult challenge in their nursing care, especially when this must be provided in hospitals. In some instances (e.g., smallpox or measles) even the most rigid techniques may not accomplish their purpose fully, but they must be maintained with careful attention to principle and detail for the greatest possible safety of all concerned. The general principles of respiratory isolation should be reviewed (see Chap. 12).

Viral Diseases of the Respiratory Tract

Influenza

The Clinical Disease. Influenza is an acute respiratory infection accompanied by fever, chills, and general malaise. Headache, muscular aches and pains, and a feeling of exhaustion that borders on prostration are common. The tissues of the upper respiratory tract are inflamed and have a bright red, mucoid appearance. Lymphoid tissue is often enlarged, but there is no purulent exudate. There is some coryza but it is not so prominent in influenza as it is in "head colds" or in bacterial infections of the upper respiratory tract. A dry, hacking cough is usual, and there may be laryngitis with hoarseness. The disease is usually self-limited, with fever for about three days and recovery in a week or so.

Complications. The chief complication of influenza is pneumonia, resulting from initial *viral* injury extending to the epithelium of the bronchial tract and alveoli, and frequently followed secondarily by *bacterial* invasion of these tissues. The most common secondary invaders are *Haemophilus influenzae,* staphylococci, streptococci, and pneumococci (see pp. 306–307).

When influenza occurs in epidemic form, it spreads rapidly through large segments of the communities involved. Under these circumstances many people with chronic debilitating diseases or other reason for lowered resistance are among those infected, and it is for this group that the risk of complication and death is greatest. The death rate is highest among elderly people with underlying cardiac, pulmonary, or renal diseases, and pregnant women, secondary bacterial pneumonia being the most serious complication. During epidemic years, the general mortality may greatly exceed the normal expectancy.

Central nervous system involvements may also occur as postinfluenza complications, although these are rare. The *Reye syndrome,* for example, is an acute encephalitis accompanied by fatty degeneration of the liver and other viscera. This syndrome has been associated with influenza B (see Laboratory Diagnosis, section I B), but may also follow other virus infections such as chickenpox. It is seen in children and young adolescents and has a high mortality rate (36 percent and 58 percent in two recent outbreaks). The incidence of the Reye syndrome is very low, however. In the United States following an outbreak of influenza B in 1973–1974, 379 cases of Reye syndrome were reported, with 55 in 1975 and 52 in 1976. The pathogenesis of this illness is not known.

Guillain-Barré syndrome (GBS), usually a self-limited but generalized paralysis, is another rare central nervous system disease linked with influenza (B and C) and possibly other virus infections. GBS received a great deal of attention in the latter part of 1976 as a result of the intensive surveillance of the national swine influenza vaccination program conducted in the United States in that year. Approximately ten of every one million persons who received swine influenza vaccine developed GBS (see Epidemiology, section III A). The mechanism by which viruses induce this syndrome is not clear.

Laboratory Diagnosis

I. Identification of the Organism. During epidemic periods influenza is usually diagnosed on a clinical basis, but sporadic cases can be identified only by laboratory findings. Mild or asymptomatic infections probably go unrecognized for the most part.

A. Influenza virus (Fig. 14–1) can be recovered from throat secretions by inoculating these into embryonated chicken eggs. The virus multiplies in the amniotic and allantoic fluids, which can be harvested from the eggs in three to four days and tested for virus activity (Fig. 14–2). One of the properties of this virus that is most easily recognized is its ability to agglutinate the red blood cells of several animal species. Influenza, like mumps and measles viruses, belongs to a group called "myxoviruses." The prefix *myxo* means slime and refers to an affinity for mucin. Red blood cells have a surface mucoprotein to which virus attaches, the result being agglutination of the cells, or hemagglutination. Since viruses other than influenza also have hemagglutinating properties, however, it is necessary to test the harvested egg fluids with specific influenza antiserums. If influenza virus is present, specific antibody combines with it and prevents hemagglutination from occurring. This is called an HI test, or hemagglutination-inhibition.

B. Serologic identification of influenza virus involves the recognition of antigenically distinct strains. There are three immunologic types of influenza virus, known

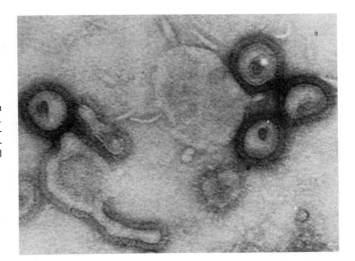

Fig. 14–1. Electron micrograph of Asian influenza virus, magnified 80,000×, shows the fine structure of the virus particles. (Courtesy of Lederle Laboratories, Division of American Cyanamid Co., Pearl River. N.Y.)

as A, B, and C. Type A strains are found among animals, e.g., swine, horses, and domestic fowl, as well as in the human population. Types B and C have been isolated from man only. The A and B strains have been of major importance in human epidemics, whereas C strains occur only sporadically.

Type C influenza is antigenically stable, but A and B viruses undergo frequent antigenic change. These antigenic variations are due to changes in viral components that occur as the virus passes through human populations. Consequently, the classification of influenza strains is based on differences in the antigenicity of such components as the nucleocapsid, envelope protein, hemagglutinin, and a surface enzyme called neuraminidase (the latter brings about the release of virus bound to red blood cells by hemagglutinin). The designation of strains indicates the antigenic type of nucleocapsid (A or B), the geographic origin of an isolate (e.g., Victoria, London, Hong Kong), the strain number, and the year of its isolation. Thus, A/England/42/72 indicates an A variant isolated during an epidemic in England in 1972, while B/HK/5/72 identifies the Hong Kong epidemic B strain of 1972. Simpler terms such as A/Victoria (isolated in Australia) are often used as a broader designation of a type and its origin. The "swine flu" variant isolated in the United States early in 1976 has been termed A/New Jersey (Hsw1N1), indicating its origin (Fort Dix, New Jersey), the antigenic identity of its hemagglutinin (H) with that of swine influenza virus (sw1), and its neuraminidase subtype (N1). The identification of such type-specific variants in epidemics is important for the preparation of immunologically effec-

tive vaccines so that future epidemics may be forestalled (see Epidemiology, section III A).

II. Identification of Patient's Serum Antibodies. Antibody production occurs during a clinical attack of influenza. Antibodies appear in the circulating blood in increasing concentration following onset, reaching a peak level in about two weeks and persisting for about four weeks. Thereafter the level gradually falls until a new infection is acquired.

The patient's antibodies can be identified by the complement-fixation technique, or by the hemagglutination-inhibition method described above, using laboratory-propagated virus. For diagnostic purposes, it is important that two specimens of serum be tested: one obtained during the *acute* stage of illness and one taken about two weeks later in the *convalescent* stage. The serologic diagnosis is made on the basis of a *rising titer* of antibody during the course of the disease, the changing level indicating an active immune process. This pairing of serums is necessary, because low levels of influenza antibody persist in normal persons, between infections, and do not signify current disease.

III. Specimens for Diagnosis

A. *Throat washings,* or garglings, for virus isolation. The patient's mouth should first be rinsed with a mild antiseptic, followed by water, to clear out saliva and many commensalistic mouth bacteria. A gargling solution should be obtained from the virus laboratory. The solution is a transport medium (buffered saline or broth) to be used as a throat rinse

Fig. 14–2. Fertile eggs are inoculated with seed virus, the first step in the production of many virus vaccines. After injection, the egg is sealed with collodion and incubated to permit maximum virus growth. (Courtesy of Eli Lilly and Company, Indianapolis, Ind.)

and returned to its container. Washings should be sent immediately to the virus laboratory, or frozen until their transport can be arranged.

B. An "acute" serum taken within a day or two of the onset of illness.

C. A "convalescent" serum taken two weeks after *onset.*

Note: The first serum specimen can be retained in the refrigerator until the second blood is drawn and the two sent to the virus laboratory as a pair. To prevent hemolysis in storage or transit, serum should be separated from whole blood and stored in a sterile tube using aseptic technique to prevent bacterial contamination and deterioration of the serum.

D. A brief clinical history and tentative diagnosis, specifying the dates of onset and of specimen collection, should accompany every patient's specimens.

Epidemiology

I. Communicability of Infection

A. Reservoir, Sources, and Transmission Routes. The reservoir for human influenza is man, although animal reservoirs (swine, horses, fowl) may be possible sources of new human variant strains. It is thought that genetic recombinations may occur among human and animal virus strains if they occasionally find their way from one species to the other. The frequency of such man-to-animal-to-man transmissions and their epidemiologic significance have not been firmly established, however.

Among infected human beings, transmission is by airborne droplets, by direct contact, and by indirect contamination of fomites with respiratory secretions. The virus probably does not survive long outside the body, so that contacts must be reasonably close.

B. Incubation Period. Very short, usually only one or two days.

C. Communicable Period. Probably does not extend past the acute febrile stage (three to four days).

D. Immunity. Specific active immunity confers resistance, but antibodies must be present at the site of virus entry to be protective. An effective concentration of antibody in respiratory secretions as well as in the circulating blood can protect against reinfection with the same strain for a year or two. However, resistance is type specific and depends on the individual's exposure to similar strains. Repeated infections with different influenza virus variants broaden the immunity. Protective immunity can also be obtained with influenza vaccines, the scope of the preventive effect depending on the antigenic types incorporated in the immunizing material.

In the absence of type-specific immunity, human beings are universally susceptible. The incidence of infections with respect to age groups reflects opportunity for previous exposure, and is influenced by social and environmental factors. In general the incidence is highest in schoolchildren, lower in children under six, and decreases progressively among adults of advancing age. When populations encounter new strains, or when old ones recur after some time, age incidence becomes more uniform. The disease is generally most severe, and mortality from complications highest, in elderly adults.

II. Control of Active Infection

A. ISOLATION. Patients may be isolated during the acute stage if their close contacts include those for whom the risk of infection might be serious.

During epidemics, patients admitted to the hospital with suspected influenza should be grouped together whenever possible. Respiratory or strict isolation may be considered, depending on the degree of risk to susceptible patients and the severity of the epidemic. Visitors should be limited, if not restricted altogether.

B. PRECAUTIONS. For hospitalized patients, nursing precautions should include concurrent disinfection of respiratory discharges and of items or equipment contaminated by these secretions during the acute stage. The patient should be taught handkerchief control of coughing and sneezing, whether he is at home or in the hospital.

C. TREATMENT. Amantadine is a drug that has been studied extensively in clinical trials for its prophylactic value in preventing influenza. It is a nontoxic drug that can be taken for several months (100 to 200 mg per day) without serious reaction. Clinical studies also indicate that it has therapeutic value in uncomplicated influenza. If given early in the course of illness (within 48 hours of onset), it can shorten the duration and relieve the severity of symptoms.

The usual antimicrobial drugs are ineffective in primary influenza and should be used only to treat secondary bacterial complications.

D. EPIDEMIC CONTROLS. Individual contact controls are not feasible in influenza, and sporadic cases are not reported unless specific information is available as to the laboratory diagnosis of type specificity. Health departments maintain surveillance programs to determine the prevailing types and the patterns of epidemics. This information is exchanged at state, national, and international levels so that vaccine manufacture may keep pace with epidemic expectancy. Appropriate vaccines must be available and administered as indicated, *before* epidemics get underway, because the disease develops and spreads rapidly, outstripping the immune response to active immunization.

During epidemics unnecessary public gatherings should be avoided, but the disruption of office, school, or other institutional functions is not warranted.

Hospitalization of influenza cases is indicated only for patients at high risk or those with complications. On admission, immunization should be recommended for high-risk patients who have not been previously vaccinated. Those who have been immunized should be considered for chemoprophylaxis with amantadine. Elective admissions should be curtailed and visiting discouraged. Isolation procedures may be instituted as indicated above. Hospital rules should include routine immunization of all personnel prior to the epidemic season.

III. Prevention

A. IMMUNIZATION. Mass immunization against influenza in periods between epidemics has not generally been considered feasible because of the limited duration of vaccine protection, and also because for healthy persons this is usually a mild disease. Efforts to prevent or control it have been aimed at protecting those at the greatest risk of developing serious complications and dying. Such persons, including chronically ill adults and children as well as people over age 65, should be vaccinated annually regardless of the amount of influenza in their communities. Vaccination is also advisable for all hospital personnel and for others in public services.

Vaccines in routine use are inactivated "whole-virus" or "split-virus" preparations. "Whole-virus" vaccines contain intact virus particles that have been propagated in chick embryos, purified, and inactivated with formalin. "Split-virus" preparations are made by disrupting the virus particles with organic solvents or detergents. The toxicity of such split products is reduced but they are also somewhat less immunogenic (they stimulate lower levels of antibody). In the United States, the Bureau of Biologics of the Food and Drug Administration regularly reviews the formulation of antigens in influenza vaccines, recommending new formulation with contemporary antigens as indicated. The vaccines of the late 1970s contain virus antigens referable to current epidemic strains, i.e., A/Victoria/75 and B/Hong Kong/72 in combination, and a single (monovalent) vaccine for A/New Jersey/76, the "swine flu" virus. The latter is also available in a bivalent vaccine, combined with A/Victoria.

The story of the 1976 "swine flu" vaccine program in the United States is of great interest. This program represents the first attempt ever made at mass immunization against influenza. The rationale for it lies in the history of influenza epidemics and in current knowledge of influenza virus antigen variation, or shift. Historically, influenza makes sudden pandemic (exceptionally wide) appearances characterized by high morbidity and a general rise in mortality, exceeding the normal expected death rate for the time and place where the epidemic occurs. The concept of excess deaths attributable to influenza or its complications has become an index for detecting epidemics and for measuring their extent and severity. Since the first isolation of the influenza virus, in 1933, retrospective serologic studies as well as ongoing studies of new outbreaks indicate that pandemics occur about once every ten years. These pandemics have always been caused by major antigenic mutants of the virus to which the population has been universally susceptible. In intermediate periods, minor antigenic variants appear every two to three years, causing smaller outbreaks. Perhaps the worst influenza epidemic in recorded history was that which occurred during the closing years of World War I, in 1917–1918. More than 15 million deaths occurred throughout the world before that pandemic had run its course. A return of that dreadful form of the disease has been a matter of major concern. Indirect but substantial serologic evidence indicates that it was caused by the swine influenza virus. The outbreak began in the midwestern United States and is presumed to have then established the infection in swine, from whom the first influenza virus of any variety was first isolated years later (1933). That virus had by then disappeared from the human population, leaving only a residual antibody record in previously infected persons, but a persistent form of active infection in swine.

Recurring major epidemics or pandemics were recorded in the late 1920s, 30s, 40s, 50s, and 60s, but the swine variant had never been seen again in human cases. Early in 1976, however, it was isolated at a military camp in New Jersey where it infected several hundred soldiers and caused one death. This was the first evidence of man-to-man transmission of an animal virus among people who had no contact with swine, and it sparked real concern that a wide dissemination of the swine variant might occur once again. The last previous pandemic had begun in 1968, originating in Hong Kong with a type A variant called A2 or "Asian" flu, and spreading rapidly around the world. In the United States it had caused 30 million cases, with an estimated 33,000 excess deaths. By 1976 the decade of the Hong Kong A2 variant was closing, leaving the population immune to it but susceptible to a new A variant that appeared to be emerging — a variant suspect of having once before devastated human populations.

Another cause for concern was related to the characteristics of still another, then current, influenza strain that had caused an interim epidemic of less major proportions than A2/HK/68. This strain was the type B variant that also originated in Hong Kong in 1972. The unusual thing about B/HK/72 is its ability to infect swine under natural conditions of man-to-swine contact. This raised the possibility that the type A swine virus and the type B Hong Kong strain might have infected swine simultaneously, undergone genetic recombination, and produced a new swine virus capable of infecting and reproducing in man.

The decision to attempt mass vaccination in the United States was thus based on several important considerations and speculations. It was a gamble, in a sense, undertaken to prevent millions of cases and thousands of deaths in the event that the A/New Jersey/76 swine variant should spread widely and rapidly. The effort was also predicated on the fact that influenza vaccines have a record of being 70 to 90 percent effective, rarely producing serious reactions. The largest objections to the program in the beginning were related to the disparity between its enormous cost and what seemed to some the relatively small risk that the new variant would spread and replace the Hong Kong type. Although several million people were immunized uneventfully, in the end the program was aborted before it could reach its full goal because of the unexpected appearance of the Guillain-Barré syndrome and its linkage to the vaccine (see p. 350). By that time also (late 1976 and early 1977) it had become apparent that the New Jersey strain had not spread beyond Fort Dix and that the risk of an epidemic had all but vanished.

The problem of GBS remains. Although the great majority (about 93 percent) of cases seen in 1976 were unrelated to influenza immunization, there was good epidemiologic evidence that the remainder were tied to the swine flu vaccine: the weekly incidence of cases of Guillain-Barré rose and fell with the number of vaccinations, whereas that of non-vaccine-related GBS cases remained steady; the GBS attack rate in the last quarter of 1976 was very much greater for vaccinated than for nonvaccinated persons; and the onset of GBS in vaccinated persons occurred between 10 and 28 days following vaccination. Although the estimated risk of vaccine-related GBS was quite small (1 case per 100,000 vaccinations), GBS is a serious illness. Fortunately, most patients recover completely, but 5 to 10 percent of GBS cases have some residual weakness, and about 5 percent die. The pathogenesis of the disease remains unknown. About 40 percent of nonvaccinated cases are associated with mild viral infections of the upper respiratory tract, but at least 23 percent give no history of previous infection, immunization, or other recognizable predisposing factor.

B. Chemoprophylaxis. Amantadine appears to be the first practical specific antiviral drug. It has been a successful prophylactic when given during an epidemic period, for it does not have an enduring effect. It is thought to prevent virus penetration of infected cells after attachment, but it protects only against A2 influenza (Asian) and not against type B virus. It is recommended as a prophylactic primarily when there is not enough time for vaccination to be

protective. In A2 epidemics, optimal control may be possible if vaccination is provided first and amantadine prophylaxis is then started.

The Common Cold and Other Acute Viral Respiratory Diseases

Clinical Diseases. The common cold and a number of other acute infections of the upper, and sometimes the lower, respiratory tract are caused by viruses. These infections are seldom marked by clinical features of individual diagnostic value, and their viral agents fall into diverse classifications. *Clinically,* two types of problem are recognized: nonfebrile illnesses with symptoms localized in the upper respiratory tract, and more severe involvements of lower respiratory structures, with constitutional symptoms, including fever, general malaise, and, sometimes, gastrointestinal disturbances. *Virologically,* differences are noted in the localization and type of cellular injury produced by different viruses, as well as their antigenic and biologic properties. *Epidemiologically,* these diseases are characterized by an easy distribution from the human reservoir by respiratory transmission routes, general susceptibility, and high incidence rates with varying degrees of immune response.

1. *The common cold* is a catarrhal infection of the nose and throat, characterized by the local discomfort of congestion and excessive mucus secretion, as well as general malaise and fatigue. Fever is rare and the duration is generally limited to a few days or a week. Secondary bacterial infections of involved mucous membranes are frequent, inducing sinusitis, laryngitis, tracheitis, bronchitis, otitis media, or combinations of these syndromes.

2. *Febrile viral diseases* of the respiratory tract generally involve either the underlying lymphoid tissue and the superficial membranes of the upper respiratory tract or the vulnerable tissues extending downward from larynx and trachea into the bronchial tree and lungs. Localizing signs are related to the nature and site of viral injury: e.g., lymphoid swelling; hoarseness and croup; the dry, irritated cough of laryngeal or tracheal involvement; the

productive cough of bronchitis or pneumonia; or deep pulmonary infection, which may lead to interstitial reactions or production of consolidating alveolar exudates.

> **Virologic Diagnosis.** The recognition and identification of the individual viral agents of these diseases are difficult and time-consuming at the practical level. Continuing study, however, has led to the association of numerous viruses with upper and lower respiratory diseases. Some of them belong to virus groups capable of diverse localizations in human tissues other than those of the respiratory tract. The diversity of names given to these organisms reflects the variety of situations in which they have been identified, as well as their biologic relationships. Table 14-1 lists some viruses associated with colds or more serious respiratory diseases and indicates their relationship to other epidemiologically or clinically distinct infections.

Epidemiology

Communicability of Infection. Man is the only reservoir for these viruses, so far as is known. They are transmitted by droplets spread on airborne routes, by direct contacts, or by indirect contamination of the immediate environment. Some may also be spread by the fecal-oral route.

The incubation period for the common cold is generally one to two days. The infectious agent is probably communicable for a day or so before symptoms begin, and throughout the course of active infection. In the more acute diseases, involving less accessible tissues, the incubation period is longer: from a few days to a week or more. The infections are communicable while active symptoms persist.

Susceptibility to these infections is general, although young children and infants are more frequently and seriously involved than adults. Specific immunity is induced by these diseases, but it is often short-lived. Reinfections are common but may be milder than preceding attacks because of lingering immunity.

Control and prevention of the common viral respiratory infections constitute a difficult problem, whose importance can be judged by the frequency of attack and the extent to which disability interferes with school or work. These viral infections often predispose to secondary bacterial complications that may become entrenched, chronic, and marked by hypersensitive responses. Personal and public attention to hygienic and sanitary measures that limit respiratory and oral spread of infection constitutes the chief control within families and groups. Antibiotics have no effect on viral infections and should be used only when bacterial complications arise. In those situations, antibiotics should be selected on the basis of laboratory recognition of significant bacterial strains, individually tested for susceptibility.

Systemic Viral Diseases Acquired by the Respiratory Route

Mumps

The Clinical Disease. Mumps is a familiar disease involving the parotid glands. The infection is usually generalized because viremia frequently occurs after the virus has multiplied in the parotid gland, or in the superficial epithelium of the upper respiratory tract. The virus may localize subsequently in other salivary glands; the testes or ovaries, especially in adolescents or adults; the thyroid gland; and occasionally the central nervous system.

Mumps begins typically as a parotitis, with sudden onset of fever and swelling and tenderness of one or both parotids, followed sometimes by enlargement of sublingual or submaxillary glands. Tissue damage is not great, but the swelling causes pain, particularly with the pressure of mouth and throat movements. The swelling reaches a peak in about two days and persists for a week to ten days, outlasting the fever. When other localizations occur, they generally follow the salivary gland inflammation in about a week, but they can occur simultaneously or even in the absence of parotitis. Tissue injury is not severe or permanent, except in the testis, which may be compressed by the swelling within its limiting membrane and later atrophied. If bilateral orchitis occurs (which is not usual), the damage may lead to sterility. Ovarian infection does not result in sterility, because the ovaries can swell without compression by a limiting anatomic sheath. Meningoencephalitis occurs in fewer than 10 percent of cases; fatality is rare.

Table 14–1. **Viruses Associated with Respiratory Diseases**

Virus	Isolation from	Associated Disease
Rhinoviruses (more than 100 are recognized)	Nasopharyngeal washings	Common colds (adults and children)
Adenoviruses (at least 30 types)	Normal and infected adenoids and tonsils Pharyngeal washings Infected conjunctiva Stools and rectal swabs in infected cases	Undifferentiated acute respiratory disease (ARD) Acute pharyngitis Acute conjunctivitis Pharyngoconjunctival fever Epidemic keratoconjunctivitis Bronchitis and bronchial pneumonia ⎱(children) Gastroenteritis ⎰
Parainfluenza viruses (4 antigenic groups, each with several virus types)	Nasal and pharyngeal washings Lung tissue	Mild or acute upper respiratory infection Laryngotracheobronchitis (croup) Bronchitis Pneumonia
Respiratory syncytial (RS virus)	Pharyngeal washings	Infant bronchiolitis and pneumonia Acute upper respiratory illness in childhood
Coxsackie viruses (enterovirus family) (2 antigenic groups, at least 29 types) (named for a town in New York State where first isolated)	Stool specimens Throat washings Spinal fluid	Common colds Herpangina (vesicular pharyngitis) Pleurodynia (epidemic, febrile myalgia, usually thoracic) Aseptic meningitis, sometimes with paresis resembling poliomyelitis Neonatal myocarditis
ECHO viruses (enterovirus family) (at least 31 antigenic types) (*E*ntero*c*ytopathogenic *h*uman *o*rphan virus = ECHO)	Stool specimens Throat washings Spinal fluid	Common colds Undifferentiated respiratory or enteric illness Aseptic meningitis, often with rash

Laboratory Diagnosis. Mumps parotitis is diagnosed clinically, without need for laboratory studies, but when other involvements occur without the characteristic initial syndrome, the virus laboratory may be needed to establish the diagnosis.

I. Identification of the Organism. Mumps virus can be propagated in chick embryos or in monkey kidney tissue culture. It can be identified as a *myxovirus* (see p. 350) by its hemagglutinating property. It is also characterized by its ability to lyse red blood cells, as well as agglutinate them. Serologic identification of virus is made by demonstrating specific neutralization of these properties by mumps antiserum.

II. Identification of Patient's Serum Antibodies. A serologic diagnosis of mumps can be made by demonstrating a rising titer of specific antibody following an attack of mumps. A sample of serum taken soon after onset is compared with another one drawn about two or three weeks later. The titer of the "convalescent" serum must be at least four times higher than that of the "acute" serum, to confirm a mumps diagnosis.

Hemagglutination-inhibition, complement-fixation,

and neutralization techniques are employed. The latter method involves neutralization by antibody of mumps virus infectivity for eggs or for tissue cultures.

III. Specimens for Diagnosis

A. Saliva, spinal fluid, or urine may be collected for virus isolation, as indicated by localization of symptoms, a few days after onset.

B. An "acute" serum taken within a day or two of onset.

C. A "convalescent" serum collected two to three weeks after onset.

D. A brief clinical history and tentative diagnosis with recorded dates of onset and specimen collection must accompany the specimens.

Epidemiology

I. Communicability of Infection

A. RESERVOIR, SOURCES, AND TRANSMISSION ROUTES. Mumps is a human disease, the virus source being the saliva of infected persons. It is transmitted by airborne droplets, by direct contacts, and by articles freshly contaminated by infectious saliva. The transfer must be rapid to be effective, apparently through close contacts.

B. INCUBATION PERIOD. May be as short as 12 days or as long as a month, but the average time is 18 days to three weeks.

C. COMMUNICABLE PERIOD. Mumps is most communicable at the time symptoms begin, but virus may be present in saliva from about four days prior to onset until nine days after the first swelling.

D. IMMUNITY. Mumps is generally a mild disease of childhood and confers a lasting immunity. Inapparent infections are frequent and also lead to antibody production as well as skin hypersensitivity. Mumps does not occur with as much regularity as other childhood virus infections, and consequently many adults may be susceptible. When risk of infection among adults is high (young adults in schools or military camps, hospital personnel, parents exposed to infected children), a skin test for hypersensitivity can be used to detect those who have had previous infection, and who are therefore probably immune. Susceptible individuals are skin-test negative and may be candidates for mumps vaccine.

II. Control of Active Infection

A. ISOLATION. The patient should be isolated (respiratory isolation, if hospitalized) from *new* contacts for as many days as indicated by the persistence of active symptoms of parotitis, usually seven to nine days. His contacts of the previous week presumably were exposed and should be watched for signs of developing infection.

B. PRECAUTIONS. Respiratory and oral secretions are infectious. Concurrent disinfection of dishes and other contaminated articles should be carried out. The hospitalized patient should have individualized care in isolation.

C. TREATMENT. There is no specific treatment for mumps, but mumps hyperimmune globulin (prepared from human convalescent sera) may lower the risk of orchitis if given immediately upon the onset of parotitis. Patients with immunodeficiencies who have been exposed to mumps should be considered especially for such prophylaxis.

D. CONTROL OF CONTACTS. The routine investigation or quarantine of contacts is not warranted, especially if they are children. On the other hand, surveillance of exposed, susceptible (skin test-negative) adults may help to prevent serious effects if passive immunoprophylaxis is given at the first sign of developing symptoms.

E. EPIDEMIC CONTROLS. When epidemics are underway, there are no recognized measures of practical value in control.

III. Prevention

A. ACTIVE IMMUNIZATION. Live attenuated vaccine is available either alone or in combination with rubella and measles live virus vaccines. It provides solid immunity and is safe for children from one year of age and older. The recommended time for mumps immunization is at one year of age, together with measles and rubella vaccines (MMR injection). Nonimmune older children and adolescents should receive the vaccine before they reach maturity, but this and other live virus preparations are contraindicated for immunosuppressed or immunodeficient individuals. As of 1977 mumps immunization is required by law in New York State, together with rubella, measles, polio, and diphtheria vaccinations.

Unless exempted for medical or religious reasons, parents must submit evidence of all these immunizations for their children before they can be admitted to schools in New York.

Measles (Rubeola)

The Clinical Disease. Measles is a common, but sometimes severe, infectious disease of childhood. It begins with fever and the sneezing, coughing, and conjunctival irritation so usual in colds or other upper respiratory infections, but the patient often appears sicker than a mere cold would warrant. In three or four days, the characteristic blotchy, dark-red rash appears on the face to confirm the impression that something more serious than a cold is developing. The skin rash, or exanthem, progresses quite dramatically within the next 24 to 48 hours, extending to the neck, the chest, and finally the entire trunk. In the meantime, systemic features of disease have intensified, fever reaching 104° to 105° F (40°–40.5°C), the cough deepening to an obvious symptom of bronchitis, the patient becoming progressively irritable and dyspneic. Physical examination of the mouth and throat reveals rashy lesions (enanthem) on the buccal mucosa and palate. These mucosal lesions, called Koplik's spots, often appear before the skin rash is visible and may provide the first grounds for clinical diagnosis.

In about a week or ten days the rash begins to lose its angry red look and slowly becomes brownish. Fever and respiratory symptoms subside rapidly, barring complications, and an uneventful recovery ensues. The symptoms of measles are referable to the activities of rubeola virus, which enters the body by the respiratory route, localizes, and multiplies in lymphoid tissue of the pharynx. After about ten days of incubation, as the catarrhal period begins, viremia occurs. Virus can be found not only in respiratory discharges, but also in conjunctival secretions and in the blood for at least two days after the rash appears. Superficial capillaries in the skin are injured either by virus or by virus-antibody interactions, and it is this pathology that leads to the rash. Koplik's spots on oral mucous membranes also reflect little focal areas of capillary damage and consequent formation of cellular exudate at the site. A generalized lymphoid tissue reaction also occurs in response to virus activity.

Both upper and lower respiratory tracts become more susceptible to secondary bacterial infection, and this is the most frequent complication of measles. Hemolytic streptococci, in particular, may superimpose new injury, causing a more severe bronchitis or bronchopneumonia, but this can usually be well controlled with antibiotic therapy. A rare, but more serious, complication arises if the virus enters the central nervous system (1 in 1000 cases). The encephalomyelitis that results is fatal in up to 30 percent of cases displaying this syndrome, and about 40 percent of survivors are left with some degree of brain damage.

Laboratory Diagnosis. Measles is usually diagnosed on the basis of its characteristic clinical appearance. When the rash is atypical, or Koplik's spots fail to appear, differentiation from other virus exanthems may be difficult, and in such cases laboratory confirmation of the diagnosis may be valuable.

I. Identification of the Organism. Measles virus can be propagated in tissue cultures of human cells, such as kidney and amnion cell cultures. It produces characteristic changes in these preparations, and its typical inclusion bodies can be seen in the nuclei of invaded cells. When harvested from tissue cultures, rubeola, which is a myxovirus, can be demonstrated to have hemagglutinating properties. Serologic identification of virus is obtained with measles antiserum, which specifically inhibits the hemagglutinating activity of rubeola or neutralizes its infectivity for human cell cultures.

II. Identification of Patient's Serum Antibodies. Specific measles antibody appears in the patient's serum during the course of disease, rising in titer from the acute to the convalescent period. These antibodies may persist for many years following an attack of measles and are sufficiently protective to make second attacks infrequent. In suspected cases of adult measles, serologic diagnosis depends particularly on comparison of antibody titer during the convalescent stage of disease with that of the acute period, which may reflect a titer lingering from previous encounters with the virus. At least a fourfold rise above this level must be demonstrated in the second sample to provide serologic evidence of current infection.

The patient's serum is tested for its ability to react with known strains of measles virus, by complement fixation, neutralization of infectivity, or inhibition of red cell agglutination.

III. Specimens for Diagnosis

A. Nasopharyngeal washings and blood samples can be submitted for *virus isolation* in the early acute phase, within 24 hours of the first appearance of rash. Blood should be defibrinated or anticoagulated; throat secretions should be placed in holding medium provided by the laboratory. These materials should be sent immediately for culture or frozen pending their transport.

B. An "acute" serum sample taken at the time the rash appears.

C. A "convalescent" serum collected two to three weeks after onset.

D. A brief history and tentative clinical diagnosis with recorded dates of onset and specimen collection must accompany the specimens.

Epidemiology

I. Communicability of Infection

A. RESERVOIR, SOURCES, AND TRANSMISSION ROUTES. Measles is one of the most readily transmissible of infectious diseases, with a human reservoir. It is spread by airborne droplets derived from respiratory discharges of infected persons, who may remain infectious from the time of onset of symptoms, throughout the catarrhal period, and for four or five days of the eruptive stage. Infectious droplets are transferred directly or indirectly by contamination of clothing or other articles with secretions from the nose and throat. Subsequent drying releases into the air droplet nuclei containing viable rubeola virus.

B. INCUBATION PERIOD. From exposure to the time of first symptoms of fever and "cold," usually about ten days. Development of the rash requires another three to four days.

C. COMMUNICABLE PERIOD. See I, A above.

D. IMMUNITY. Generally, a single attack of measles confers a lifelong protective immunity. Following a mild, or aborted, case, immunity may be less durable, and in these instances second attacks of unmodified measles may occur. Maternal antibody is transferred *in utero,* but congenitally acquired immunity does not persist in the newborn for more than a few months. Artificial passive immunization is sometimes of value in aborting disease, as discussed below with regard to control of contacts.

Measles is endemic in all habitable parts of the world and is most prevalent in winter and spring. The immune state of the population in any given area is the principal determining factor in the frequency of epidemics. Susceptibility is universal, but the disease is most common and widespread among children, so that about 90 percent of people reaching age 20 have acquired immunity through infection. In large urban communities the disease is endemic, but it appears in epidemic patterns of mild disease among the susceptible young or the newcomers about every second year. In rural or isolated areas, the intervals between outbreaks are longer, but the epidemic distribution includes people of all ages who have not been previously exposed, and the disease is often more severe.

II. Control of Active Infection

A. ISOLATION. The patient should be isolated for a period of seven days from the time the rash appears to minimize the risk of further exposure of susceptible contacts, particularly if these include infants or small children for whom the disease may be serious. The patient should be protected against secondary bacterial complications, particularly if hospitalized.

B. PRECAUTIONS. Nursing techniques are directed at control of respiratory spread. Concurrent disinfection of all articles contaminated by nose-and-throat secretions is indicated. Terminal disinfection is wise in view of the possible survival of virus within the protective protein coats of settled droplet nuclei.

When measles is prevalent or epidemic in the surrounding community, hospitals should be particularly careful to protect their very young or very sick patients from infection acquired through visitors or through new, unscreened admissions.

C. TREATMENT. There is no specific treatment for measles. The use of antimicrobial drugs is aimed at prevention of secondary bacterial infections (especially in patients with underlying debilities), or at treatment of complications as they arise.

D. CONTROL OF CONTACTS. Live attenuated vaccine protests against measles if given to nonimmune contacts on or before the day of exposure. If given later in the incubation period it may prevent severe illness, without producing adverse effects. If vaccine is contraindicated (immunodeficiency, immunosuppression, leukemia, tuberculosis, pregnancy), measles immunoglobulin (human) may be given in the first three days after exposure or even later. The earlier immunoglobulin is administered, the better the chance for preventing disease altogether.

E. EPIDEMIC CONTROLS. Case reporting is required by health departments, because it provides the best opportunity for prompt protection of vulnerable contacts. Where these are concentrated in institutions or isolated villages, public health measures include, as indicated by the local situation, isolation of the sick, prompt immunization programs aimed at all potentially susceptible individuals, and immunoprophylaxis for those who should not receive live virus.

III. Prevention

IMMUNIZATION. Live attenuated measles vaccine provides the most effective preventive control. Since the fall of 1966 when a measles eradication program was begun in the United States, the incidence of measles has decreased sharply. Recent surveys indicate that between 60 and 70 percent of children in the one-to-four-year age group are now immune to measles, as a result either of vaccination or of infection (Fig. 14–3). Sporadic epidemics now occur primarily among nonimmunized children.

The majority of children develop fever and rash following vaccination, but these symptoms are minimal and subside quickly. They can be reduced or avoided for children who can be expected to react adversely by the simultaneous injection of immune globulin at a different site.

The vaccine is given in a single dose. Infants should be immunized at about one year of age. Vaccination of older children is limited to those with no history of measles. The vaccine should not be given to persons with defective immune systems, active tuberculosis that is not being treated, leukemia, lymphoma, or severe allergy to eggs (in which the vaccine is prepared). Live virus vaccines should never be given to pregnant women. Institutionalized children with chronic pulmonary diseases are particular candidates for measles vaccination because the disease could be especially dangerous for them.

If a candidate for vaccination has already received measles immune serum globulin, a period of 8 to 12

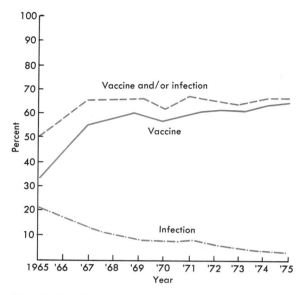

Fig. 14–3. Measles. Immunization status of one- to four-year-old children, 1965–1975. (Reproduced from *United States Immunization Survey*:1975. June 1976, Bureau of State Services, Immunization Division, Center for Disease Control, D.H.E.W., Atlanta, Ga.)

weeks should be allowed before the vaccine is given to avoid neutralizing the latter with the concentrated antibody present in the former.

German Measles (Rubella)

The Clinical Disease. German measles is one of the frequent but mild viral exanthems of childhood. Its chief importance lies in the effect it may have on infants born of mothers who contract the infection during early months of pregnancy.

Rubella virus probably enters the body through the respiratory tract, but is disseminated by the bloodstream. Symptoms begin with a mild upper respiratory disturbance, followed by enlargement of cervical lymph nodes, characteristically those of the back of the neck, and finally include the eruption of a generalized macular rash. The rash is distributed first on the face and head, spreads rapidly to the neck and trunk, but subsides in two or three days. Rapid recovery without complication is the rule, but in adults occasional complications result from virus localization in joints or the central nervous system.

When rubella occurs in pregnant women, transfer of the virus across the placenta to the fetus may cause serious and permanent damage to the developing embryonic tissues, particularly if they are infected during the stage of their earliest, most rapid formation. The danger of such effects diminishes after the fourth or fifth month of pregnancy. Stillbirths and spontaneous abortions may occur, and babies brought to term may display one or more serious anomalies, such as heart defects, liver and spleen enlargements with dysfunction, deafness, cataracts, mental retardation, or other brain damage.

Laboratory Diagnosis. Clinically, rubella must be distinguished from measles, scarlet fever, and a number of rash-producing virus infections.

For *virus isolation* in the early, acute stage of disease, throat washings, blood, urine, or stool specimens should be submitted to the laboratory.

For *serologic diagnosis,* hemagglutination-inhibition, neutralization, and complement-fixation tests of serum are available. The laboratory should be consulted by the physician requesting rubella serology. The choice of test and the timing of blood collection depend on the patient's age, exposure risk, clinical condition, and history of rubella vaccination.

Decisions concerning abortion often rest upon diagnostic laboratory tests. *It is essential that these be adequately planned and interpreted.*

Epidemiology

I. Communicability of Infection

A. RESERVOIR, SOURCES, AND TRANSMISSION ROUTES. So far as is known, man is the only reservoir for German measles. The nasopharyngeal secretions of infected persons contain the virus and transmit it when sprayed into the air. Airborne droplet nuclei and direct or indirect contacts are the usual transferral routes.

B. INCUBATION PERIOD. In adults, symptoms become apparent about 14 to 21 days after exposure. Average incubation time is 18 days.

C. COMMUNICABLE PERIOD. This disease is very communicable during the week before *and* the week after the appearance of the rash.

D. IMMUNITY. A single attack of German measles appears to provide a permanent immunity. However, in the absence of natural infection or active immunization, susceptibility is general. Infants may be protected for the first few months of life by maternal antibody acquired *in utero.*

In natural infection, rubella antibodies begin to appear in the serum soon after the rash subsides and rise to a peak within two or three weeks. Thereafter the persistence of antibody provides lifelong immunity. Demonstration of antibody by serum assay therefore constitutes an index of immunity that is more reliable than a history of rubella. A serum titer of 8 or higher (see Chap. 7, p. 169, for the definition of titer), in the absence of current or recent clinical symptoms of rubella, indicates immunity resulting from some previous exposure. When symptoms suggest current, active rubella infection, the diagnosis can be confirmed by demonstration of a four-fold rise in antibody titer in a "convalescent" serum compared with an "acute" serum, or of the presence of IgM antibody (primary response to active infection) as distinct from IgG antibody (secondary and persistent after previous infection [see Chap. 7, pp. 161 and 164]).

It is particularly important that the laboratory diagnosis be established for women in the first trimester of pregnancy who have symptoms suggestive of rubella, or who think they have been exposed to the disease. In many American states and elsewhere in the world, documentation of rubella in early pregnancy provides legally acceptable grounds for abortion, to prevent the birth of babies with congenital deformities. Conversely, demonstration of an unchanging rubella titer of 8 or more in an asymptomatic pregnant women exposed to rubella signifies that she is immune and not infected, and that abortion on this basis is not indicated.

II. Control of Active Infection

A. ISOLATION. German measles is such a mild disease that isolation or other controls are neither necessary nor desirable under ordinary circumstances. The situation is different, however, in the case of rubella contracted by pregnant women.

The epidemiology of congenital rubella is related to the fact that outbreaks of German measles can occur after childhood. In the spring of 1964 the United States experienced a large epidemic among young adults which resulted in congenital disaster for thousands of babies. Improved laboratory techniques made it possible to study many of those infants extensively. It was discovered that they remain infectious for many months after birth, the virus being recoverable from throat secretions or spinal fluid up to 18 months postnatally. It is thought that this continuing infection with rubella virus may account for the progression of symptoms observed in affected babies, especially those with central nervous system defects.

It is also important to recognize that congenitally infected infants are capable of transmitting virus to susceptible adults who care for them. The hazard is great only for young women in the first trimester of pregnancy. If these contract the disease, a 30 percent incidence of congenital infection can be expected in the first three months, weighted by an 80 percent record of infection in the initial four weeks of pregnancy. Furthermore, apparently normal babies, born of mothers who were infected, may shed virus, and inapparent infections of mothers may affect the fetus to the same degree as those characterized by rash. For all these reasons, it is considered advisable for susceptible pregnant women to avoid unnecessary contacts with newborn babies, especially during the first trimester.

B. PRECAUTIONS. None are indicated.

C. TREATMENT. No specific treatment for rubella is available. If the patient is pregnant, and the diagnosis of rubella is confirmed by the laboratory, therapeutic abortion is strongly advised. The use of gamma globulin as an alternative to abortion is not indicated because it does not prevent viremia in the mother or the transfer of virus in the blood to the fetus.

D. CONTROL OF CONTACTS. Susceptible women of childbearing age who are in contact with babies or others suspected of having rubella infection should be identified by serologic tests. If antibody cannot be demonstrated in their serum, they should be immunized with rubella vaccine, provided the restrictions described in the section below on prevention are noted.

E. EPIDEMIC CONTROLS. None that are effective.

III. Prevention

IMMUNIZATION. In 1969 a live attenuated rubella vaccine became available in the United States. A single dose of this vaccine is recommended for all children between age 1 and 12 years. It may also be given to adolescent or young adult females, but since it is a live vaccine the following restrictions must be carefully observed: (1) it should not be given to a pregnant female; (2) it should not be given to persons who already possess rubella antibody, as demonstrated serologically; and (3) it should not be given to any female who is likely to become pregnant within three months of vaccination. With the latter restriction in mind, rubella vaccine may be given to a susceptible woman during the postpartum period, when the possibility of an unrecognized pregnancy is remote.

The vaccine produces transient arthritis and arthralgia in about 30 percent of teen-age or older females, and this fact also should be weighed in the decision to vaccinate. Its use is contraindicated for persons whose immune mechanisms are in any degree abnormal.

Mass immunization of susceptible children in their prepubertal years holds the promise of rubella control in the future. Immunization surveys indicate that about 60 percent of children in the one-to-four-year age group are immune to rubella as a result of vaccination programs, while about 68 percent have immunity as a result of either vaccine or infection or both (see Fig. 14–4).

Chickenpox and Herpes Zoster

The Clinical Disease. Chickenpox (varicella) and herpes zoster (shingles) have been a part of the human scene for many centuries, but it is only in recent years that the relationship of their viral agents has become apparent. *Chickenpox* is a mild but highly communicable disease of childhood, characterized by fever, malaise, and an itching vesicular skin eruption. Epidemiologically and clinically, its chief importance lies in its resemblance to mild or modified smallpox, from which it must be carefully differentiated. *Zoster*, or shingles, bears little clinical resemblance to chickenpox but is apparently a re-

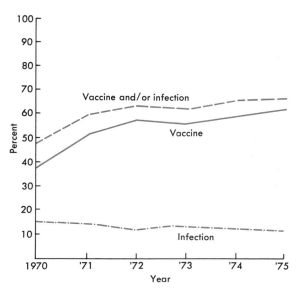

Fig. 14–4. Rubella. Immunization status of one- to four-year-old children, 1970–1975. (Reproduced from *United States Immunization Survey*: 1975. June 1976, Bureau of State Services, Immunization Division, Center for Disease Control, D.H.E.W., Atlanta, Ga.)

current manifestation of persistent infection with the same virus. It occurs sporadically in adults, who develop an incapacitating, painful inflammation of posterior nerve roots and ganglia, the path of the affected sensory nerves being marked by crops of vesicular eruptions of the skin of the area. These vesicles are morphologically identical with those of chickenpox, but the clinical appearance of zoster is quite distinct from the childhood disease.

Varicella virus is believed to enter the body through the respiratory mucosa and, after a period of viremia, to localize in the cells of the skin. The first symptoms of fever and malaise appear about 12 to 14 days after exposure. As skin localization occurs, the rash appears first on the trunk, then progresses to the face, limbs, and sometimes the mucosa of the mouth and throat. The lesions are often more abundant on covered than on exposed areas of the body, and they appear in successive crops for three to four days, so that different stages of pock development

may be seen at the same time. The pocks begin as papules, then develop into weepy vesicles, which become crusted as they begin to heal. The lesions resolve without scarring as a rule, and recovery is usually rapid and uneventful. The disease has an extremely low mortality rate and only occasionally may lead to dangerous complications such as viral pneumonia in adults or encephalitis in children.

Zoster also begins with fever and malaise as virus multiplication proceeds, and extreme tenderness develops along the dorsal nerve roots where virus localization has occurred. In a few days crops of vesicles appear on the skin area supplied by the sensory nerves of the affected root. The eruption is generally unilateral, appearing most commonly on the trunk along the lower rib line, sometimes on the shoulders, neck, or head.

Laboratory Diagnosis. Clinically, the appearance and sequence of distribution of skin lesions usually distinguish chickenpox from other "pox" diseases. When epidemiologic considerations raise doubts concerning the possibility of smallpox, however, laboratory identification of these viruses or of the patient's specific immune response becomes imperative.

Morphologically, varicella and smallpox viruses have much in common, but distinctions can be made in the location of inclusion bodies within infected host cells or in the cells of tissue cultures. *Positive identification of virus* can be made by cultivating vesicle fluid in tissue culture or chick embryos, and by confirming morphologic features of inclusion bodies or the damage they induce, with serologic tests using specific antibody. Similarly, *patient's serum antibodies* can be identified in tests with known strains of viruses.

Epidemiology
I. Communicability of Infection
A. RESERVOIR, SOURCES, AND TRANSMISSION ROUTES. Man appears to be the only susceptible host for chickenpox and zoster virus, the infected portion of the human population being the reservoir. Chickenpox is one of the most highly communicable and common epidemic diseases of childhood. It spreads rapidly, the chief source of infection being respiratory secretions or discharges from early vesic-

ular lesions of the mucous membranes or skin. Airborne droplets as well as direct and indirect contacts with infectious secretions carry infection into the respiratory tract of exposed contacts. Infection does not occur directly through the skin.

B. INCUBATION PERIOD. Two to three weeks are commonly required following exposure. The average case becomes symptomatic in about 14 to 16 days.

C. COMMUNICABLE PERIOD. Infectivity is greatest from a day or so before the rash appears to the fifth or sixth day of the eruption. Scabs are not infective, but the disease remains communicable while fresh crops of vesicles are appearing. Zoster is not readily communicable because the virus is not present in the respiratory tract, but this infection can be the source of varicella in children and has been known to initiate epidemics.

D. IMMUNITY. Infection with either varicella or zoster confers a long-lasting immunity to the respective diseases. Children who have recovered from experimentally induced zoster or varicella show a cross-immunity to both, but zoster may occur in adults who have had a natural varicella infection at an early age.

Susceptibility is universal among those who have not had the disease, but chickenpox in adults is often more severe than it is in children, and this has sometimes contributed to the difficulty of distinguishing it from smallpox. In temperate zones, chickenpox is most prevalent in the winter and spring, but zoster occurs sporadically, without seasonal incidence.

The viruses isolated from chickenpox and from zoster are either identical or share closely related antigens. The most workable theory concerning the relationship of their diseases is that chickenpox virus may sometimes infect nerve cells and remain latent and inactive within them for long periods. A stress or physiologic insult may later provide an opportunity for activation of the latent virus, with the characteristic result known as zoster.

II. Control of Active Infection

A. ISOLATION. Children should not be hospitalized with chickenpox unless complications or other problems require it. By the time complications appear, chickenpox is usually no longer communicable. Cases sometimes develop in patients already in the hospital, and in this situation every precaution must be taken to prevent spread to other susceptible persons. The patient should be strictly isolated, and susceptible contacts quarantined. It is particularly important to protect other patients who may be on steroid or other therapy that suppresses the immune mechanism. Children with leukemia are subject to severe or possibly fatal consequences if they contract chickenpox.

In the community, isolation of sick children is not effective in stopping spread of epidemics, but they should be kept out of school for the week of the eruptive stage, and away from very young susceptible contacts.

B. PRECAUTIONS. In the hospital, nursing care of chickenpox requires concurrent disinfection of objects contaminated by respiratory secretions or the discharges of fresh skin lesions. The patient's care should be individualized; gowns should be worn; and particular attention given to proper techniques for handling soiled bedclothing and linens. Terminal disinfection of the isolation room is recommended also.

C. TREATMENT. There is no specific treatment for chickenpox or varicella. Scratching of the very itchy eruption may lead to secondary bacterial infection of open vesicles. This can usually be controlled locally by ointments that reduce itching. It is important for the nurse to assist the patient in maintaining strict cleanliness of the skin, with careful attention to hands and fingernails. Hand and nail scrubbing is essential also for the nurse and physician who have been in close contact with the patient.

D. CONTROL OF CONTACTS. Quarantine of contacts is ordinarily not effective, and prophylactic measures are not indicated, except possibly for exposed susceptibles for whom chickenpox might be a serious disease (children or adults on antibody-suppressive drugs). Gamma globulin has been recommended in these occasional situations, in the hope of achieving modification of the infection.

Chickenpox is not a reportable disease except for adult cases occurring in areas where smallpox is not endemic.

III. Prevention.
There are no specific means of preventing chickenpox. Vaccines are not available and the therapeutic use of gamma globulin ordinarily is not justified, except for high-risk patients. Of

chief concern is the early distinction of chickenpox from smallpox in adolescents or adults.

Smallpox (Variola Major)

The Clinical Disease. Smallpox is a severe, acute infectious disease that has been recognized and feared for many centuries in all parts of the world. Recorded descriptions of its epidemic effects extend from the writings of Galen in the second century. Wherever it has been introduced into noninfected, susceptible populations, it has left a trail of suffering, disfigurement, and death. It has been brought under successful control by artificial immunization and virtually eliminated within the past decade by the

Fig. 14–5. The face of smallpox. (Reproduced from *World Health: Smallpox Target Zero*. October 1972. WHO Photo by L. Matlowsky.)

efforts of the World Health Organization. Intensive case finding and vaccination of direct and indirect contacts have eradicated the disease from lingering pockets of infection in India, Bangladesh, and Africa. Many countries have remained free of endemic smallpox through surveillance and control programs and strict monitoring of travelers arriving from previously infected areas. The requirement for continuing immunization programs has diminished and may eventually be terminated altogether.

Variola virus enters the susceptible host through the upper respiratory mucosa. During the incubation period, virus multiplication proceeds in lymphoid tissue, and bloodstream distribution occurs, with localization of virus in reticuloendothelial cells in many parts of the body. More virus is released from these cells to the circulating blood and localization then begins in the cells of the skin. Symptoms begin when the second stage of viremia becomes intense. The onset of fever and chills with intensifying malaise is usually sudden but may occur gradually. The skin rash begins to appear after one or several days of fever, erupting first on the face and then symmetrically on the arms, trunk, and legs (Fig. 14–5). The exanthem progresses through macular, papular, and vesicular stages to formation of pustules about the eighth or ninth day. Crusts harden on the pustules and fall off after another ten days or two weeks.

The nature and full development of the rash are an index of severity of disease and of the extent of the patient's immunity. A partially immune person, such as a vaccinated contact who has not yet developed full immunity or an individual whose immunity is waning between vaccinations, may develop fever and all the symptoms of the preeruptive stage, but experience little or no further progression of disease. In severe cases, the rash may become confluent, and even hemorrhagic. There is a direct correlation between death rates and the extent and character of the rash.

Recovery is signaled by drying of the pustular lesions. Successive crops of eruptions do not occur in smallpox, so that individual pocks all begin to heal at about the same time. As scabs fall away, healing is usually accompanied by scarring, the pitted scars fading gradually from pink to white.

Complications that may occur during smallpox infection include secondary bacterial invasion of skin lesions, sometimes with resultant septicemia. Patients who die in the first week of illness often show evidence of heart failure and pneumonia, as a result either of overwhelming virus infection or of secondary bacterial invasion. The majority of deaths occur in the second or third week, at the height of the pustular phase. There may be hemorrhages in the skin, conjunctiva, the oral mucosa, bowel, uterus, or vagina. Pregnant women with smallpox frequently abort, but if infection occurs near the the end of pregnancy the child may be brought to term. Infants may acquire the disease *in utero,* or at the time of birth if the mother's infection is still active.

A mild form of smallpox called *alastrim* is caused by a variant of the smallpox virus known as *variola minor.* The eruption is less profuse and malignant, symptoms are less profound, and the disease is seldom fatal. When this form of smallpox is transmitted to susceptible contacts, they develop a light disease. On the other hand, when mild smallpox is the result of modification of *variola major* in a partially immune individual, transmission to a susceptible person can give rise to typically severe smallpox if that contact possesses no immunity.

Laboratory Diagnosis. Most cases of smallpox can be diagnosed clinically on the basis of the appearance, development, and distribution of skin eruptions, together with a history of sudden febrile illness preceding exanthem. A history of possible contact must be sought, including the question of travel in parts of the world where smallpox is endemic or of recent contact with persons who reside in those areas. The clinical suspicion of smallpox calls for immediate preventive measures to be instituted for all contacts, even before laboratory confirmation can be obtained. Clinical diagnosis is most difficult in cases of modified variola major, or of alastrim, when pocks may appear atypical. Smallpox must also be distinguished from chickenpox, drug eruptions, meningococcemia, and other viral diseases accompanied by skin eruption.

The diagnostic virus laboratory can provide confirmation of clinical smallpox, or be of assistance in doubtful cases, by the following means:

I. Identification of the Organism

A. *By examination of stained smears* of material from skin lesions. The report is presumptive but can be obtained quickly; i.e., within an hour of the laboratory's receipt of suitable material. Smallpox viruses are rather large, being about one third the size of the smallest bacteria. They can be seen as aggregates, or inclusions called Guarnieri bodies, within the cytoplasm of infected cells. These bodies can be stained and visualized with the ordinary light microscope. Rapid diagnosis can also be made by examining vesicle fluid under the electron microscope for the presence of characteristic viral particles.

B. *By culture of appropriate material* in chick embryos. This method may provide confirmation of the diagnosis within three to five days. The material to be cultured includes *blood,* if it can be obtained during the febrile preeruptive stage, *scrapings* from the early papular lesions, *fluid* from vesicles or pustules, and *crusts.* Virus can usually be seen in material from the skin lesions and propagated in eggs. Smears of blood are not rewarding, but cultures may be successful.

C. *Serologic confirmation* of the nature of the virus harvested from chick embryo cultures is obtained by complement-fixation tests, or by demonstrating neutralization of virus infectivity for eggs by specific smallpox antiserum.

II. Identification of Patient's Serum Antibodies. Circulating antibody can be detected in patient's serum only after several days or a week of illness. Complement-fixation and hemagglutination-inhibition tests are generally employed with known strains of smallpox virus. Serologic diagnosis of smallpox requires demonstration of a rising titer of antibody during the course of disease, with a "convalescent" serum titer at least four times higher than the "acute" serum titer. The first specimen of blood for serologic testing should be drawn immediately upon appearance of the rash, the second one in two or three weeks.

Epidemiology

I. Communicability of Infection

A. RESERVOIR, SOURCES, AND TRANSMISSION ROUTES. Smallpox is a human disease, the reservoir and source of virus being the infected person. From the early stages of eruption, virus is present in oral and respiratory secretions, as a result of erosion of focal lesions in the mucous membranes. When vesicular skin lesions rupture, their discharges are also a source of infection. Transmission of smallpox does not occur through skin-to-skin contacts, but through introduction of virus into the respiratory tract of a susceptible contact. Direct contacts may easily convey virus, but indirect contacts and airborne droplet nuclei or dust may also transmit infection. Smallpox virus can survive drying for long periods, and scabs that have fallen off may remain infectious for years. This means that contacts need not be close, and that contamination of the air may result not only from the patient's respiratory secretions but also from dissemination of skin discharges, drying particles, and dust stirred into the air. The hands and clothing of hospital personnel, bedclothes and linens, bedside equipment, furniture, floor, walls, and curtains all may be contaminated by virus in a drying but viable state, scattered on air currents or deposited directly. The virus may also be transported for long distances to induce new outbreaks, as reported for cases occurring in England that were traced to cotton imported from an endemic area.

B. INCUBATION PERIOD. The virus requires from 7 to 16 days for multiplication before the first symptoms appear. The average time is 9 to 12 days from exposure to onset, and another three or four days to skin eruption.

C. COMMUNICABLE PERIOD. The patient must be considered infectious from onset of fever and other initial symptoms until all scabs have dropped away. Communicability is greatest during the early stages of the rash.

D. IMMUNITY. Resistance to smallpox depends on specific immunity, conferred either by an attack of the disease, by vaccination, or by congenital transfer of maternal antibody. The latter does not persist long after birth, and children may be vaccinated early in life for this reason. Following vaccination, antibodies can be demonstrated within 9 to 14 days, reach a maximum titer within two or three weeks, and remain demonstrable for a few years. Revaccination every three to five years maintains immunity at a protective level and has been considered adequate in nonendemic areas. In countries where smallpox persisted longest (Asia, Africa, South America), revaccination at more frequent intervals has been necessary. Until all danger of residual, undetected smallpox is past, it is important for people whose exposure risk is great to maintain an effective vaccination schedule. Such persons include customs and quarantine officers, international travelers, military personnel, and hospital employees.

II. Control of Active Infection

A. ISOLATION. Immediate consultation between the physician and appropriate public health authorities is mandatory. The strictest isolation and precautions are required for protection of nonimmune contacts, at home or in the hospital. Only persons whose vaccination is current can safely care for the patient, and even these persons should be promptly revaccinated at this time.

B. PRECAUTIONS. Careful techniques must be employed for concurrent disinfection of all articles associated with the patient, particularly if they must be taken from the area. Individualized equipment should be used whenever possible. Oral and nasal discharges are deposited in containers suitable for burning, and other disposable items also are incinerated. Steam pressure sterilization or boiling is used for nondisposable items when possible; otherwise thorough disinfection and cleaning procedures must be used. Clean gowns should be worn by all who enter the patient's room and carefully discarded so that clothing is not contaminated in the process. Nonimmune personnel or visitors may not be admitted to the patient's room, and even immune visitors should be discouraged during the communicable period because they may carry the infection out of the hospital to others.

Particular attention must be given to the handling and disposal of soiled bedclothes and bedding, because virus can be disseminated easily from such items in lint. Susceptible laundry workers may acquire smallpox through handling contaminated linen. Soiled linens, blankets, and gowns should be carefully folded without shaking, their contaminated sides turned inward, and placed in marked, sealed bags impermeable to moisture for transport to the laundry. If available laundry techniques cannot be relied upon to disinfect soiled linen, it should be placed loosely in a permeable bag and sterilized in the autoclave before washing.

Isolation and precautions are maintained until the patient displays no residual scabs. Terminal disinfection-cleaning should be thorough and includes sterilization of pillows, mattresses, and other bedding.

C. TREATMENT. There is no specific therapy for smallpox. Antibiotics are of value in preventing or controlling secondary bacterial infections during the pustular stage. Gamma globulin has prophylactic value during the incubation period.

D. CONTROL OF CONTACTS. All of the patient's contacts are promptly vaccinated, or revaccinated, or quarantined for a period of 16 days timed from the last exposure. These precautions apply to all persons who have lived or worked on the same premises as the patient or have had other close contacts. Non-immune newly vaccinated persons are kept under surveillance for a 16-day period. Persons with evidence of previously acquired immunity are carefully watched until the height of reaction to current vaccination has passed. Any contact who develops febrile illness during the period of quarantine or surveillance is promptly isolated until smallpox can be ruled out.

Passive as well as active immunization may be of value for nonimmune contacts. Gamma globulin is obtained from military personnel and others who have been recently vaccinated. It should be given together with vaccine as soon as possible after exposure.

Case finding is of ultimate importance in smallpox. The source of each patient's infection is of primary concern to all health authorities, who make a careful review of all associated cases with exanthematous or other suggestive symptoms.

E. EPIDEMIC CONTROLS. Case reporting of smallpox is required everywhere, under international law. If an epidemic threat occurs, local health authorities institute rigorous enforcement of quarantine on all contacts until evidence of successful vaccination is obtained. Surveillance is maintained after vaccination for a 16-day period. The public is informed of the necessity and nature of controls, and of availability of vaccine. Physicians and hospitals are supplied with fresh, potent vaccine, and public vaccination stations may be established as well. If the disease appears to be spreading, mass immunization of partial or whole populations within the affected area may be attempted.

International requirements for the control of smallpox include government notification of the World Health Organization as well as adjacent countries of new cases appearing in a nonendemic area. The International Sanitary Regulations of WHO provide special measures for vehicles arriving by any route from countries where smallpox persists. Strict regulations prevent the travel of infected persons.

III. Prevention

A. IMMUNIZATION. Active immunization by vaccination affords the most reliable means of controlling smallpox.

In the early part of the twentieth century, the annual case rate in the United States was over 40,000. By 1942 the number of annual cases had been reduced to less than 1000, and in the current decade only an occasional imported case has been reported, with few secondary transmissions.

The idea of protecting human beings from the effects of smallpox by injecting them with a small dose of material obtained from a mild case was first introduced about two thousand years ago by the Chinese. In our own time, the first safe application of this technique dates from 1798 with Jenner's use of cowpox virus for vaccination. Cowpox is a mild exanthematous disease of cattle caused by a virus immunologically related to variola. Jenner's success with cowpox virus led to its continuing use and the gradual decline of smallpox in populations of Western Europe and America. In the course of time another virus, called "vaccinia," emerged during laboratory propagation of cowpox and variola viruses. Its origin is not known, but it is also antigenically very similar to variola and even more attenuated for man. Vaccinia virus is now used for the preparation of immunizing vaccine for smallpox.

Vaccinia virus is propagated in calves and harvested from their skin lesions for vaccine manufacture. Calf lymph vaccine thus contains live attenuated virus. When introduced into human skin it produces a localized infection at the site of inoculation and stimulates production of antibodies capable of preventing development of smallpox infection. To accomplish this, vaccine must be fresh and potent at the time of inoculation. Under some circumstances it is capable of causing generalized skin or systemic infections.

The dangers of postvaccinal complications have called for reassessment of the smallpox vaccine and its application in mass programs. The complications include generalized vaccinia, usually in very young children with eczematous skin problems; postvaccinal encephalitis (about 1 per 100,000, but most likely in adults receiving primary vaccination or in very small babies); and progressive vaccinia in persons with impaired immune mechanisms. Such problems may be eliminated in the future by the preparation of inactivated as well as more attenuated vaccines. The former may be used to provide partial immunity on primary vaccination, the latter to continue and maintain immunity. In the United States, however, vaccination is now an elective rather than a routine procedure.

Where vaccination is practiced, it must be correctly applied and its effect properly interpreted if it is to achieve the desired result. The following factors are of greatest importance:

1. *Time and Frequency of Vaccination.* Vaccination must be successfully completed *before* exposure to smallpox if it is to guarantee protection.

Primary vaccination can be carried out in children in their second year of life. Severe reactions to the vaccine are seen most commonly in infants less than one year old; therefore, vaccination in the first year is not recommended.

Revaccination every three to five years is indicated in nonendemic countries; more frequently if the exposure risk is high and sustained.

Hospital personnel and others under special risk should maintain immunity through vaccination programs instituted upon employment and continued at regular intervals as indicated.

Contraindications to vaccination include skin diseases (such as eczema), agammaglobulinemia, and leukemia, particularly when these occur in young children. Because the vaccine is live, the risk of generalized vaccinia infection is higher in patients with preexisting skin lesions or suppressed mechanisms for immunity. Eczematous patients should not be permitted to come in contact with recently vaccinated persons, and the eyes of young children should be protected against secondary inoculation of virus from their own or others' fresh vaccinations. Primary vaccination should not be given to pregnant women. If danger of active smallpox is real, vaccination may be considered the lesser risk and is performed simultaneously with administration of vaccinia gamma globulin.

2. *Technique of Vaccination.* Vaccine is pressed into the most superficial layer of epidermis by a multiple pressure method. The site is prepared by wiping away excess skin oil with acetone or ether. When the skin is dry (important because these solvents may affect the virus), a drop of vaccine is placed on it and pressed in lightly by stroking through it several times with the side of a sterile needle. The needle should not draw blood or involve any area more than 3 mm wide in any direction. Excess vaccine is removed with a sterile gauze pad (carefully discarded into a container for incineration), and the site is left to dry *without* a protective covering. The preferred site of vaccination is the deltoid surface of the arm. Leg vaccination is not recommended because irritation of clothing can enlarge the lesion or encourage secondary bacterial infection.

3. *Interpretation of Reactions to Vaccine.* Vaccination sites are observed, and reactions recorded, at least twice (on the third and ninth days), because the time of development of the maximum reaction has a bearing on the immunologic interpretation of the reaction, as follows:

A *primary* reaction, or "take" (also called *vaccinia reaction*), occurs in a *nonimmune,* fully susceptible person. A local lesion is formed, but does not reach its maximum development before the seventh to ninth day. During the first three or four days there may be no visible reaction; then a papule appears, surrounded by a small area of erythema. The papule progresses toward the vesicular stage on the fifth or sixth day, the vesicle reaching full development on or about the ninth day. It becomes pustular thereafter, and a crust forms, dries, and drops away usually by the end of the second week. The small scar that is left is usually permanent. It will be noted that this sequence of events for the single lesion of vaccination in a nonimmune person is entirely similar to that which occurs as the multiple eruptions of smallpox develop in a susceptible individual.

An *accelerated* reaction (also called *vaccinoid*) occurs in an individual with *partial immunity.* The sequence of development is the same as for the primary reaction but occurs more rapidly, and the lesion is usually less conspicuous. A small vesicle is formed by the third to the seventh day, and the *total* sequence is completed by the tenth day. This muted reaction reflects the presence of residual antibody in the vaccinated individual which neutralizes some of the virus activity.

An *immediate* reaction (also called an *immune* or early reaction) should occur if the vaccinated individual is *fully immune* to smallpox. A small papule develops within three days and promptly disappears.

Errors in interpretation most often surround the immediate reaction. Unfortunately, this type of response also occurs if the vaccine is not potent, and even if the virus is dead. It may also occur if the individual is allergic to vaccine components other than virus, or if the skin is injured unduly by the vaccinating technique. Proof that the vaccine batch used was potent can only be obtained by observing its capacity to induce accelerated reactions in revaccinated persons or primary results in nonimmune, previously unvaccinated persons. If this proof cannot be obtained, or if there is any reason for doubt, individuals displaying immediate reactions should be revaccinated. An immediate reaction cannot be accepted as valid for a person without a history of an attack of smallpox or of previous vaccination.

Proper technique and potent vaccine applied together cannot fail to produce one of the three types of reactions. Most or all first vaccinations should result in the primary, vaccinia reaction. About 50 percent of persons revaccinated after more than ten years should develop accelerated reactions; the remainder show the primary response. Persons who maintain a vaccination schedule at regular intervals display immune or accelerated reactions.

A completely negative response to vaccine has no immunologic significance. The failure, or "no take," usually means inactive vaccine, or inadequate technique, or both. The vaccination should always be repeated in such case.

Infectious Mononucleosis

The Clinical Disease. This is an acute infectious disease characterized by fever, pharyngitis, tonsillitis, cervical lymphadenopathy, and in some instances disease of the spleen and liver (with jaundice). It is rarely fatal, but death has been associated with rupture of the spleen.

Laboratory Diagnosis

I. Identification of the Organism. The agent is a virus resembling a herpes type. It is called EB (Epstein-Barr) or HL (herpeslike) virus. It is associated not only with infectious mononucleosis but also with nasopharyngeal carcinomas and with a lymphoma known as Burkitt's tumor (indigenous in children of Central Africa).

II. Identification of Patient's Serum Antibodies. Antibodies against EB virus arise early in the acute stage of disease, reach high levels during convalescence, and persist for years. The indirect immunofluorescent technique is used to demonstrate them.

"Heterophil" antibodies, or sheep cell hemagglutinins, also develop during the course of infectious mononucleosis. The serologic test for identifying them has been in use for many years and is still the simplest diagnostic technique routinely available for this disease. Serum from some normal persons also agglutinates sheep red cells, but agglutinin absorption distinguishes heterophil antibody of mononucleosis. The *normal* antibody is absorbed when the serum is mixed with guinea pig tissue, but not when mixed with beef red cells. *Heterophil* antibody of mononucleosis is absorbed by beef erythrocytes but not appreciably by guinea pig tissues. Thus, two serum samples giving the same sheep cell hemagglutination titer can be differentiated as in the following examples:

| Heterophil Titer Before Absorption | Titer After Absorption with | | Diagnosis |
	Beef RBC	Guinea Pig Tissue Cells	
224	224	0	Normal serum
224	0	112	Infectious mononucleosis

Table 14–2. Summary: Viral Diseases Acquired by the Respiratory Route

Clinical Disease	Causative Organism	Other Possible Entry Routes	Incubation Period	Communicable Period	Specimens Required
Influenza	Influenza virus		1–2 days	Acute febrile stage (3–4 days)	Throat washings or garglings "Acute" and "convalescent" serums
Common cold	Rhinoviruses Adenoviruses Coxsackie viruses ECHO viruses	By fecal-oral route	1–2 days	During time of active symptoms	
Mumps	Mumps virus		18 days to 3 weeks	4 days prior to onset of symptoms until 9 days after first swelling	"Acute" and "convalescent" serums
Measles (rubeola)	Measles virus		About 10 days	From onset of symptoms through catarrhal period and for 4–5 days of eruptive stage	Blood serums Nasopharyngeal washings
German measles (rubella)	Rubella virus	Transfer of virus across placenta	14–21 days	Week prior and subsequent to appearance of rash	Blood serums Throat washings, blood, urine
Chickenpox and herpes zoster	Varicella virus		2–3 weeks	Greatest from a day before rash appears to 5th or 6th day of eruption	Throat washings Blood serums
Smallpox	Variola virus	Transfer of virus across placenta	7–16 days	Onset of fever until all scabs have dropped away; greatest during early stages of rash	Exudate from skin lesions Blood serums
Infectious mononucleosis	Epstein-Barr (EB) virus	Congenital?	2–6 weeks	Before onset and during clinical illness	Blood serum
Lassa fever	Lassa fever virus		7–14 days	3 weeks or more from onset	Throat washings Pleural fluid Blood Serum
Marburg virus disease	Marburg virus	Venereal	5–7 days	Weeks	Blood Urine Organs Serum

Laboratory Diagnosis	Immunization	Treatment	Nursing Management
Virus isolation Antibody titer	Killed vaccine	Amantadine	Isolation during acute stage if close contacts are high risks
Virologic diagnosis difficult and time-consuming at practical level	Killed vaccine for adenoviruses, use limited	None specific	Principles of personal hygiene, support of healthy resistance
Antibody titer	Live vaccine	None specific: gamma globulin if orchitis risk is great	Respiratory isolation; concurrent disinfection of dishes and contaminated articles; hospitalized patient isolated
Antibody titer Virus isolation	Live vaccine Gamma globulin for high-risk individuals	None specific	Respiratory isolation; concurrent disinfection of articles contaminated by nose-and-throat secretions; terminal disinfection
Antibody titer Virus isolation	Live vaccine	None specific	Respiratory isolation (to protect pregnant women)
Virus isolation to distinguish from smallpox virus Antibody titer	None	None specific	Strict isolation in hospital to protect susceptible children, infants, and adults
Smear; to seek Guarnieri bodies Virus isolation in chick embryo Antibody titer	Live vaccinia virus	None specific	Strict isolation; only personnel with current vaccination should care for patient; concurrent and terminal disinfection until patient no longer displays residual scabs
Heterophil agglutination test	None	None specific	Isolation not necessary; concurrent disinfection of oral and respiratory discharges
Virus isolation Antibody titer	None	None specific	Strict isolation; concurrent and terminal disinfection
Virus isolation Antibody titer	None	None specific	Strict isolation; concurrent and terminal disinfection

> *III. Specimens for Diagnosis.* Serum samples for heterophil agglutination.
> Blood counts show marked lymphocytosis with many abnormal lymphocytes.
> Chemical analysis of blood reveals dysfunction of liver parenchyma.

Epidemiology

I. Communicability of Infection. Man appears to be the reservoir of this disease, although studies of EB virus have shown that it also occurs among other primates. It is distributed among humans in many parts of the world. In some areas studied, half of the one-year-old children and 90 percent of children over four years of age and adults displayed EB antibody.

Possibly some infections are acquired congenitally, but the major routes of transmission are not clearly known. Clinically apparent infectious mononucleosis occurs most commonly among children and young adults. Among these, active disease appears to be transmissible by direct, intimate contacts; probably through the respiratory route. (Mononucleosis has been called the "kissing disease" because of this communicability.)

The incubation period among contacts appears to be two to six weeks with a communicable period beginning before symptoms appear and extending for the duration of clinical signs of infection in the oral and pharyngeal mucosa.

II. Control of Active Infection. It is not necessary to isolate cases of infectious mononucleosis, but nursing precautions should include concurrent disinfection of oral and respiratory discharges, or articles soiled by them.

There is no specific therapy.

Effective preventive or control measures must await better clarification of the natural patterns of this infection.

Lassa Fever

Lassa fever is an acute viral disease characterized by ulcers in the mouth and pharynx, lymphadenopathy with swelling of the face and neck, and conjunc-tivitis. In severe cases there may be pleural involvement, cardiac failure, renal shutdown, and encephalitis. The laboratory diagnosis is made by isolation of the virus from throat washings, pleural fluid, or blood. The patient's serum antibodies can be identified by complement fixation and neutralization techniques. Handling such specimens is hazardous for laboratory workers, who must take strict aseptic precautions during all phases of testing.

This disease was first described in 1969 in Nigeria. Since then it has been recognized in neighboring African countries, particularly Liberia and Sierra Leone. It has a high mortality rate (30 to 50 percent), but subclinical infections are common as revealed by serologic surveys. The reservoir is in wild rodents, with transmission to man through infected rodent saliva and urine. Man-to-man transmission then occurs, and infections among laboratory workers have occurred.

The incubation period is 7 to 14 days, and the disease is communicable during the early acute stage while virus is present in the throat. Human beings also excrete virus in the urine for periods up to three weeks or more. Susceptibility is general, but immunity follows infection.

Patients must be cared for in strict isolation until throat washings and urine are free of virus, or for at least three weeks. Concurrent disinfection must be practiced for the patient's excretions, blood, and for all objects with which he comes in contact. Terminal disinfection is a necessity. There is no specific treatment, immunization, or prophylaxis. Cases must be reported to local health authorities as rapidly as possible so that a search may be made for other undiagnosed cases, who are then isolated or quarantined.

Marburg Virus Disease

Marburg virus disease is another illness with a reservoir in an African animal, in this case the African green monkey. The first known cases were seen in Germany and Yugoslavia in 1967, where 31 people, seven of whom died, were infected through exposure to African green monkeys. This disease is not common and is not known to occur in clinically inapparent form. It has a sudden onset with fever, generalized pain in the arms and legs, headache, prostration, vomiting, and diarrhea. There is multiple involvement of liver, kidney, heart, central nervous system, and skin, where a maculopapular rash erupts. The virus can be identified in the laboratory by its effects on tissue cultures or experimental animals, and serologic tests are available.

The natural reservoir for this disease is unknown, except for its occurrence in monkeys. Man-to-man transmission is through airborne droplets as well as by venereal contact (the latter transmission has been reported as late as seven weeks following infection). Laboratory workers and other hospital personnel are at risk in handling blood or other specimens from the infected patient. Strict isolation and other control measures as described above for lassa fever must be followed.

Questions

1. What is influenza? What secondary organisms may complicate the disease? Why?
2. Who should be immunized against influenza?
3. What glands are affected in mumps?
4. When is active immunization for mumps indicated?
5. How may measles become clinically complicated?
6. What is the most effective preventive control measure for measles?
7. What is the clinical importance of German measles?
8. Why should pregnant women who are not immune to rubella avoid unnecessary contacts with newborn babies?
9. Why is smallpox vaccination no longer required for infants in the United States?
10. What is lassa fever? What is Marburg virus disease?

Additional Readings

Barnes, Asa: Infectious Mononucleosis. *Nurs. Dig.,* **4**:51, Jan. Feb. 1976.

Beveridge, W. I. B.: *Influenza: The Last Great Plague.* Neale Watson Academic Publications, Inc., New York, 1977.

Conrad, J. L.; Wallace, R.; and Witte, J. J.: The Epidemiological Rationale for the Failure to Eradicate Measles in the United States. *Am. J. Public Health,* **61**:2304, Nov. 1971.

Erickson, Suzanne: Epidemic. *Am. J. Nurs.,* **76**:1311, Aug. 1976.

Giesbrecht, Edith: Infectious Mononucleosis, the Kissing Disease. *Can. Nurse,* **68**:37, Feb. 1972.

Henderson, Donald A.: The Eradication of Smallpox. *Sci. Am.,* **235**:25, Oct. 1976.

*Hospital Practice: *Status Report on Influenza.* Hospital Practice Publishing Co., New York, 1977

Langer, William L: The Prevention of Smallpox Before Jenner. *Sci. Am.,* **234**:112, Jan. 1976.

Langmuir, Alexander D., and Schoenbaum, S. C.: The Epidemiology of Influenza. *Hosp. Prac.* **11**:49, Oct. 1976.

Millian, S. J., and Wegman, D.: Rubella Serology: Applications, Limitations, and Interpretations. *Am. J. Public Health,* **62**:171, Feb. 1972.

Pinneo, Lily, and Pinneo, Rose: Mystery Virus from Lassa. *Am. J. Nurs.,* **71**:1352, July 1971.

Reid, Winifred: Congenital Rubella. *Can. Nurse,* **67**:38, Jan. 1971.

Wright, W. J.: Lassa Fever. *Nurs. Times,* **72**:578, Apr. 15, 1976.

*Available in paperback.

15

Primary Atypical Pneumonia, Psittacosis, and Q Fever

The diseases discussed in this chapter are associated with distinctive organisms, each recognized as true bacteria. Their morphologic, physiologic, and epidemiologic characteristics are unusual, however. They are unrelated to each other, except for their common capacity to induce disease in man following an effective respiratory entry.

Primary Atypical Pneumonia

The Clinical Disease. The term *primary atypical pneumonia* refers to a clinical syndrome rather than to any one of the several infectious agents that may induce it. Clinically, the phrase implies an acute lower respiratory infection of primary origin. The disease is characterized by fever and other constitutional symptoms, and some degree of pulmonary infiltration, this being patchy, irregular, or diffuse, but seldom if ever consolidated. There is usually not a marked leukocytic response, the white blood cell count remaining within normal limits. This picture is "atypical" with respect to most bacterial infections, which characteristically induce exudative reactions in the lungs or bronchi.

Nonbacterial infectious pneumonias of the infiltrative type are caused by a number of different microbial agents. A large and heterogeneous group of *viruses* has been associated with a variety of respiratory infections, including primary pneumonitis (Table 14–1, p. 357). In addition to these, an unusual type of bacterium called *Mycoplasma pneumoniae* is characteristically associated with primary atypical pneumonia.

Mycoplasmas are unlike other bacteria in that they do not possess a rigid cell wall. Consequently, their protoplasmic structures are very plastic and pleomorphic (Fig. 15–1 *B*). Because of their plasticity and extremely small breadth, even the longest filaments of *Mycoplasma* are filterable. Like rickettsiae and viruses, they are intracellular parasites *in vivo,* but they are less dependent than these organisms, for they can grow on cell-free media *in vitro.* For many years mycoplasmas were classified as "pleuropneumonia-like" organisms (PPLO) because of their resemblance to the infectious agent of pleuropneumonia in cattle. They are now recognized as very prevalent in man, in both health and disease, as well as in many species of animals with which he has contact. *Mycoplasma* strains are found among the normal flora of the respiratory and genital membranes. They are also associated with inflammatory conditions of these tissues (cervicitis, urethritis, prostatitis). They have been identified in pus from deep tissue abscesses and associated with a variety of systemic illnesses. In experimental animals, *Mycoplasma* infection can induce a chronic arthritis, with lesions that closely resemble those of rheumatoid arthritis in man (Fig. 15–2).

Mycoplasma pneumoniae produces a spectrum of respiratory disease in man that varies from mild upper respiratory tract infections to severe bilateral pneumonia. Primary atypical pneumonia due to

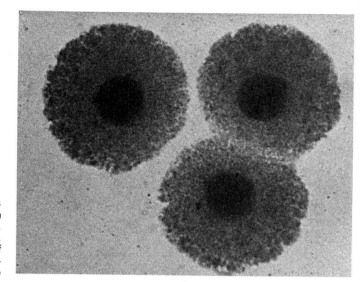

Fig. 15–1. *A. Mycoplasma* colonies growing on agar are typically dense in the center with a delicate perimeter, giving them a "fried egg" appearance. *B.* A scanning electron micrograph of *Mycoplasma pneumoniae.* These organisms lack a rigid cell wall and therefore appear quite plastic and pleomorphic. Note filamentous, bacillary, and coccal shapes. (*A* courtesy of Dr. Yvonne Faur and Mr. Martin Weisburd, Public Health Laboratories, Department of Health of the City of New York. *B* courtesy of Dr. Michael G. Gabridge, University of Illinois at Urbana.)

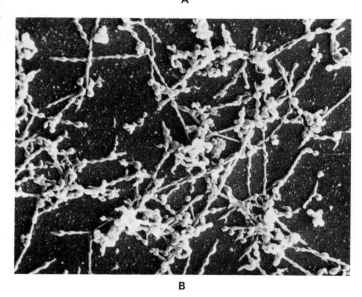

A

B

M. pneumoniae, however, is usually a rather mild, self-limited infection that occurs most often in children and young adults. The clinical picture is variable, with an insidious onset. Within about two weeks of exposure the symptoms begin with fever and chills, headache, malaise, and cough. The sputum is mucoid or purulent, but the cough is not always productive. Usually one lower lobe of the lung is involved in a patchy consolidation, but bilateral lower lobe involvement may also occur. Upper lobe pneumonia, pleural effusion, and pulmonary abscess are sometimes seen, but these are rare.

Fig. 15–2. Appearance of a mouse joint several months after injection of a strain of *Mycoplasma*. Note swelling and distortion similiar to those seen in human arthritis. (Reproduced from Cole, B. C.; Ward, J. R.; Jones, R. S.; and Cahill, J. F.: Arthritis of Mice Induced by *Mycoplasma arthritidis*. *Infect. Immun.*, **4**:344, 1971.)

Untreated primary atypical pneumonia usually resolves within two or three weeks, but antibiotic therapy with erythromycin or tetracycline can bring about prompt resolution of symptoms.

Laboratory Diagnosis

I. Identification of the Organism. (*Mycoplasma pneumoniae*)

MICROSCOPIC AND CULTURAL CHARACTERISTICS. Mycoplasmal pneumonia may be difficult to distinguish clinically from pneumonia due to other infectious agents, particularly viruses, and the laboratory diagnosis may require some time. Very often a presumptive diagnosis is made on the basis of clinical and x-ray findings and the patient's response to trial therapy with tetracycline or erythromycin. However, a laboratory diagnosis can be made with cultures and serologic methods.

Mycoplasmas require special enriched media for growth, and culture methods are not routinely available in every clinical laboratory. On agar media, the colonies are difficult to see without a lens because they are extremely small, and because the delicate, protoplastic cells extend through and within the medium rather than being heaped on the surface as are other bacterial colonies. When the agar surface is viewed through the low-power lens of a microscope, *Mycoplasma* colonies have a typical "fried egg" appearance, being dense in the center where filaments of growth extend downward into the agar and then spill over to the surface to form a delicate perimeter (Fig. 15–1, *A*). Colonies of *M. pneumoniae*

display only the dense "yolk." Microscopically, these organisms display many different forms: long filaments, bacillary and coccal shapes, disks, and granular bodies (Fig. 15–1, *B*).

II. Identification of Patient's Serum Antibodies. *Mycoplasma pneumoniae* was identified in the early 1960s as the cause of those cases of atypical pneumonia that are associated with development of "cold agglutinins" in patients' serum. This term refers to the fact that the serum agglutinates human erythrocytes in the cold, at 4° C, but not at 37° C (group O cells must be used). Such activity is not displayed by normal serum, but increases markedly in about 70 percent of cases of mycoplasmal pneumonia.

In addition to a rising titer of cold agglutinins, patients develop specific antibodies to *Mycoplasma pneumoniae*. These can be demonstrated by complement fixation and by the immunofluorescent staining technique.

III. Specimens for Diagnosis
A. Pharyngeal swabs and sputum for isolation of *M. pneumoniae*.
B. An "acute" serum taken soon after the onset of symptoms.
C. A "convalescent" serum taken about two weeks after the date of onset.

Epidemiology

I. Communicability of Infection

A. RESERVOIR, SOURCE, AND TRANSMISSION. The reservoir of mycoplasmal infection is man, its source being nasal and oral secretions of infected persons. Infection is transferred by droplet inhalation, or by direct and indirect contacts.

B. INCUBATION PERIOD. Usually about two weeks, but may be from 7 to 21 days.

C. COMMUNICABILITY is probably highest during the first week of illness.

D. IMMUNITY. Immunologic evidence, as well as the incidence of *M. pneumoniae* in normal persons and in patients with primary atypical pneumonia or febrile upper respiratory illness, indicates that this organism is widely disseminated. The incidence of disease in the general population is low, but the attack rate is higher among children and young adults. Epidemics occur most frequently among military and institutional populations. The incidence is highest in the fall and winter, but outbreaks within families or closed groups may occur throughout the year.

II. Control of Active Infection. Control of mycoplasmal pneumonia does not appear to hinge on specific preventive measures. The hospitalized patient does not require isolation, although suitable precautions should be taken with respiratory secretions. When an outbreak appears to be spreading, isolation of the sick may be useful in control.

M. pneumoniae is quite resistant to penicillin, but susceptible to erythromycin and the tetracycline antibiotics. The latter drugs are useful in severe cases of mycoplasmal infection, erythromycin being preferred for children because tetracyclines may cause staining of immature teeth. In mild cases of mycoplasmal infection, especially in the absence of a specific laboratory diagnosis, antibiotic therapy may be avoided because the disease is self-limited and also because it is difficult to distinguish clinically from viral infections for which antibiotics are useless.

III. Prevention. No immunizing or other specific preventive measures are available.

Psittacosis (Ornithosis, Parrot Fever)

The Clinical Disease. Psittacosis is a respiratory and systemic infectious disease of man transmitted by birds, such as parrots, parakeets, and lovebirds (known as psittacine birds), pigeons, and farmyard fowl. When the clinical disease of man was first recognized and associated with psittacine birds, it was called psittacosis to emphasize this relationship. Later it became obvious that a large variety of birds and domestic poultry may transmit the disease, and the more general term *ornithosis* was introduced to describe these infections when they occur in, or are acquired from, birds other than the psittacine variety.

Human infection may occur without clinical symptoms, or as a mild disease, but it may also be seen as a severe pneumonia. The infectious agent gains entry to the body through the respiratory tract. Localization occurs primarily in cells of the reticulo-endothelial system. After a period of multiplication, the organism is distributed through the body in the bloodstream, signaled by a sudden onset of fever, chills, headache, and malaise. The respiratory tract involvement may resemble influenza or primary atypical pneumonia. In mild cases, this may be the extent of symptoms, with recovery in about a week. Pulmonary infection may be severe, however, with a cough productive of mucopurulent sputum. Fatalities seldom occur with treatment, but when they do they are attributable to vascular damage. Rose spots may be seen in the skin as a result of capillary flushing, nosebleeds are common, and thrombophlebitis may occur, with threat of embolism. Early and adequate antibiotic therapy of recognized psittacosis usually assures recovery without circulatory or other complications.

Laboratory Diagnosis. The clinical diagnosis of psittacosis can be confirmed in the laboratory by isolation and identification of the agent from human cases, or suspected bird sources, and by serologic methods.

I. Identification of the Organism (*Chlamydia psittaci*). The agents of psittacosis and ornithosis are intracellular parasites that form a distinct biologic group. They have several properties in common with the microbial agents of two other very different kinds of human disease: lymphogranuloma venereum, a venereal disease, and trachoma, a serious eye infection (see Chap. 22). These organisms are classed together as bacteria in the order Chlamydiales, genus *Chlamydia* (see Chap. 2, p. 32, and Table 4–2, p. 98). They are Gram-negative, nonmotile organisms with cell walls of the bacterial type. Like other bacteria they multiply by binary fission (although through a more complicated developmental cycle), possess ribosomes, and are inhibited by many antibiotics. Unlike many bacteria, however, they are restricted to an intracellular life because they do not possess all of the cellular mechanisms required for independent metabolism.

Chlamydia psittaci can be readily cultivated in chick embryos, small laboratory animals, and tissue culture. It is identified by its growth in infected cells where it forms "inclusion bodies" (see p. 32) and by its reactivity with specific antiserum.

II. Identification of Patient's Serum Antibodies. Serologic diagnosis is made by testing two samples of the patient's serum, one taken during the early, acute phase of disease and another after at least two weeks. Antibodies can be demonstrated by complement-fixation and agglutination techniques, or by neutralization of infectivity of the agent for animal cells. However, antibody production may be delayed in the antibiotic-treated patient, and a two-week serum sample may not show a significant rise in titer over the "acute" serum in such cases. Repeated serologic testing done at three and four weeks after onset may confirm the diagnosis in treated cases.

III. Specimens for Diagnosis

A. Specimens of blood, throat washings, sputum, or vomitus are submitted for isolation of psittacosis agent. Suspected bird tissues also yield psittacosis agent.

B. "Acute" and "convalescent" serum samples are drawn at appropriate intervals, as described in II above, and submitted to the laboratory for serologic diagnosis.

C. A brief clinical history should accompany specimens for laboratory study. The tentative clinical diagnosis should be stated, together with a notation of the date of onset of symptoms and the dates on which specimens were collected.

Epidemiology

I. Communicability of Infection

A. RESERVOIR, SOURCES, AND TRANSMISSION ROUTES. The psittacosis agent is transmitted to man primarily from a reservoir in infected birds, but person-to-person transmission also occurs occasionally. Birds bred for sale as pets, such as parakeets, canaries, and thrushes, are frequently involved. Other important reservoirs for ornithosis are found among domestic fowl, particularly turkeys, ducks, and geese, and also pigeons and some water birds. Infection rates are high among workers in pet shops and aviaries, poultry farmers, pluckers, and processors. Infected birds may show symptoms such as diarrhea, droopiness, ruffled feathers; but often they do not appear to be ill. The infectious agent is present in the respiratory secretions of infected birds and also in droppings. It spreads rapidly among birds by contact and on air routes. The dust from dried, infected droppings is widely disseminated on air currents created by the motion of wings and feathers. Young birds may acquire the infection from their parents while still in the nest, and although the infection is often inapparent, they continue to excrete the agent for long periods.

Human infection is contracted through inhalation of infected droplets or dust particles. When direct man-to-man transmission occurs, the source of infection is the sputum of infected persons, the agent being spread in airborne droplets or possibly through indirect contacts with sputum-contaminated articles. Nurses and others who care for psittacosis patients are sometimes involved in outbreaks of the disease.

B. INCUBATION PERIOD. About ten days are required for symptoms to appear following exposure, the range being 4 to 15 days.

C. COMMUNICABLE PERIOD. Human patients are infectious during the acute stage of illness. Birds may be infectious for long periods.

D. IMMUNITY. An attack of the disease confers some protective immunity, but second attacks occur because of inadequate resistance. Psittacosis has a worldwide, nonseasonal distribution. The number of cases reported annually is directly related to the volume of traffic in birds. In the general population it is a sporadic disease, with occasional family outbreaks stemming from an infected pet bird.

II. Control of Active Infection. Isolation of the patient is important during the acute febrile stage of illness. Respiratory precautions should be observed, with concurrent disinfection of all discharges. Masks should be worn by nurses or others in close, continuous contact with patients whose coughing cannot be properly shielded (very young, very ill, or uncooperative persons). Terminal disinfection-cleaning should be thorough in view of this organism's capacity for survival in dust.

Treatment with tetracycline antibiotics is often effective, particularly when it is begun early and continued for several days after fever and other symptoms have subsided. Relapses occur when specific therapy is inadequate. Tetracyclines are also used in the treatment of sick birds.

Contacts should be investigated, but neither quarantine nor immunization is recommended. Identification of the source of infection in diseased birds is essential to control, not only to prevent further spread of human cases from a current source, but to prevent a widening flock epidemic. Infected pet birds are traced to their pet shop or aviary origin; domestic flocks are examined closely if they are involved. Microbiologic diagnosis is made at autopsy of infected birds, and appropriate flock controls are instituted. These may include mass dosing with tetracycline, elimination of sick birds and incineration of their bodies, and thorough cleaning and disinfection of premises, nests, and roosts.

To ensure effective application of control measures at the source of infection every human case must be reported to the local health authority.

III. Prevention. Effective immunizing vaccines are not available for prevention of psittacosis. Prevention is best achieved by regulations on the import of and traffic in psittacine birds, with quarantine of imports as indicated or of pet shops associated with cases. Improved methods for raising birds for pet sales and prophylactic feeding of tetracyclines have helped to prevent spread of infection in aviaries.

Q Fever

The Clinical Disease. Q fever was first recognized as a clinical entity in the mid-1930s in Australia. It was named by the letter Q to indicate "query," pending the demonstration of its agent, soon identified as a rickettsia. Q fever is an acute systemic disease with a sudden onset of fever and chills,

Fig. 15–3. An electron micrograph of *Coxiella burnetii*, the rickettsial agent of Q fever. (23,600×.) (Courtesy of Dr. R. L. Anacker. Electron micrograph by Mr. W. R. Brown, Rocky Mountain Laboratory, U.S.P.H.S., Hamilton, Mont.)

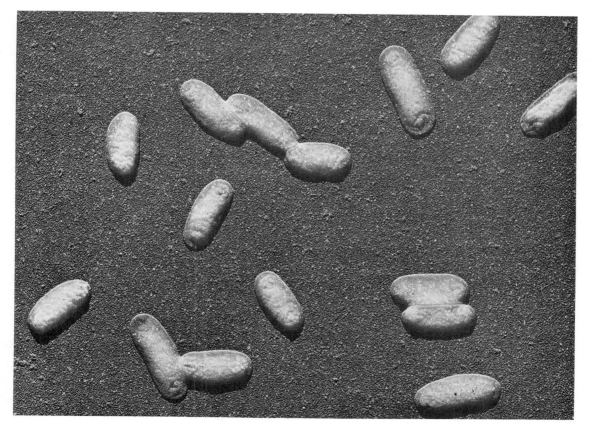

Table 15–1. **Summary: Primary Atypical Pneumonia, Psittacosis, Q Fever**

Clinical Disease	Causative Organisms	Other Possible Entry Routes	Incubation Period	Communicable Period	Specimens Required
Primary atypical pneumonia	*Mycoplasma pneumoniae*		7–21 days	During first week of illness	Pharyngeal swabs and sputum "Acute" and "convalescent" blood serums
Psittacosis	*Chlamydia psittaci*		4–15 days	Acute stage of illness, especially when coughing, not highly communicable	Blood, throat washings, sputum, vomitus "Acute" and "convalescent" blood serums
Q fever	*Coxiella burnetii*	Ingestion (milk), through conjunctiva and minor skin lesions	2–4 weeks	Ordinarily not communicable from man to man	Blood, spinal fluid, sputum, urine "Acute" and "convalescent" serums

myalgia, and severe headache. The upper respiratory tract is not commonly involved, but an interstitial pneumonia is frequent, with chest pain and cough. Nausea and vomiting often occur, and in some cases symptoms may be referable to hepatic or meningeal involvement or endocarditis. Chronic infection persisting for a few months has been noted, but usually recovery is complete in about three weeks.

Laboratory Diagnosis. Clinical and x-ray diagnosis reveals signs of viral (atypical) pneumonia, but these are not specific for Q fever. Laboratory studies such as blood counts and tests for cold agglutinins (negative in Q fever but positive in mycoplasmal pneumonia) are also nonspecific. Laboratory diagnosis is best provided by identification of the organism or the demonstration of specific antibodies in the patient's serum, or both.

I. Identification of the Organism. The causative agent, *Coxiella burnetii* (Fig. 15–3), is an obligate, intracellular parasite, like other rickettsiae. Important differences are noted for this organism as compared with other organisms of its type. Its morphology is irregular,

with forms large enough to be mistaken for usual bacteria, or so minute as to be barely visible in stained preparations examined by the light microscope. It is filterable and resistant to physical and chemical agents. It is the only rickettsia that does not produce a cutaneous rash and does not require an arthropod vector to infect man. The organism infects ticks, lice, mites, and other arthropods, and these probably have a role in its transmission among animals, but human infection with *Coxiella* can be acquired directly by inhalation, by ingestion in unpasteurized milk, through the conjunctiva, or through minor lesions in skin. This distinction from other rickettsiae, all of which are transmitted to man by insect vectors (see Chap. 26), together with the other differences noted, is the basis for the genus name *Coxiella* rather than *Rickettsia*.

II. Identification of Patient's Serum Antibodies. Tests for specific antibody in the patient's serum, early and late in the disease, are performed by complement-fixation and agglutination methods. Agglutinin titers are significantly high by the end of the second week; complement-fixing antibody titers, by the end of the fourth week.

Laboratory Diagnosis	Immunization	Treatment	Nursing Management
Culture Rising antibody titer	None	Erythromycin Tetracyclines	Concurrent disinfection of respiratory secretions
Isolation of organism Rising antibody titer	None	Tetracyclines Chloramphenicol	Isolation during acute febrile stage, respiratory precautions, masks for personnel if patient cannot control coughing; concurrent disinfection of all discharges; terminal disinfection
Isolation of organism in animal or chick embryo Rising antibody titer	Inactivated vaccine	Tetracyclines Chloramphenicol	Isolation not warranted

III. Specimens for Diagnosis

A. Specimens of blood taken during the febrile period, as well as sputum, urine, or spinal fluid (as indicated by symptomatology), are injected into small laboratory animals or chick embryos for isolation of the organism. Laboratory facilities for this work should include special safeguards against the high risk of infection of laboratory personnel and others who work in the same building.

B. Serologic diagnosis is simpler, safer, and preferred by the diagnostic laboratory. "Acute" and "convalescent" samples of serum, timed as indicated in II above, are submitted for testing. A brief clinical history should accompany these specimens, indicating the tentative diagnosis and the dates of onset and serum collection.

Epidemiology

I. Communicability of Infection

A. RESERVOIR, SOURCES, AND TRANSMISSION ROUTES. Domestic animals and ticks (possibly other arthropods as well) are the natural reservoirs of this infection, man being an incidental victim and unimportant to its natural maintenance. As a zoonosis, Q fever is typically an inapparent infection with a worldwide distribution in domestic livestock. Animals with inapparent infections may shed large numbers of the organisms in milk, urine, and feces, and in placental tissues and blood. Raw milk from infected animals is probably a source of infection for man, but the richest common source of the organisms is the dust of pasture soil and of barns housing infected animals. Inhalation of airborne rickettsiae is thought to be the most important route of transmission. Q fever may be an occupational hazard for workers in abattoirs, in plants where wool or hides are processed, and also in laboratories where the organism is handled.

Direct person-to-person transmission does not ordinarily occur, but Q fever has been reported as a hospital cross-infection and as a contact infection for nurses and physicians, particularly at postmortem examination. At least one case of transmission of Q fever by blood transfusion has been documented. In this case the blood donor was later shown to have been in the incubation period of his own infection, acquired from contact with goats, and was asymptomatic at the time the blood was donated. The recipient was a hospitalized surgical patient with no other possible contacts with sources of the organism. The diagnosis of Q fever was made retrospectively, for both the recipient and the donor, by serologic methods. Serum samples from the goats in question were also positive for Q fever.

B. INCUBATION PERIOD. From two to four weeks, with an average time of 18 to 21 days.

C. COMMUNICABLE PERIOD. Not ordinarily transmissible from man to man (see I, A above).

D. IMMUNITY. Susceptibility to Q fever appears to be general in areas where the organism is endemic in animals. Q fever is endemic in California and some other states where it occurs among farmers, dairy workers, veterinarians, and others who work with domestic animals.

II. Control of Active Infection

A. ISOLATION. Person-to-person transmission is unusual and does not warrant isolation.

B. PRECAUTIONS. Blood, sputum, feces, and urine should be disposed of by sterilizing methods. The organism can survive for months on clothing, on farm implements, in dust and dirt, and presumably on hospital equipment. Concurrent disinfection techniques should be practiced in the hospital unit where the Q fever patient is cared for, and terminal disinfection-cleaning is advisable as well.

C. TREATMENT. Tetracyclines are the drugs of choice. Chloramphenicol is also useful.

D. CONTROL OF CONTACTS. Human contacts are usually not infected, but a search should be made for the source of infection in animals, unpasteurized milk, or the laboratory.

E. EPIDEMIC CONTROLS. Case reports are indicated in areas where Q fever is endemic. Persons exposed in an outbreak are observed and promptly treated with antibiotics should symptoms occur.

III. Prevention

A. *Immunization* can be effected with an inactivated vaccine. This procedure is recommended for laboratory workers and others whose risk of infection is high.

B. Pasteurization of milk prevents dissemination of this infection.

C. Control depends on limiting spread of infection from animals, regulations on shipment of livestock, and vaccination of animals where economically feasible.

Questions

1. How does primary atypical pneumonia differ from most bacterial pneumonias?
2. How do mycoplasmas differ from other bacteria?
3. What clinical diseases are attributed to the mycoplasmas?
4. How does the psittacosis organism differ from other bacteria?
5. How can psittacosis be prevented?
6. What are the natural reservoirs for Q fever infections?

Additional Readings

Brown, G. L.: Part I, Infectious Diseases. 'Q' Fever. *Nurs. Mirror,* **141**:46, Oct. 16, 1975.

Cole, B. C.; Ward, J. R.; Jones, R. S.; and Cahill, J. F.: Chronic Proliferative Arthritis of Mice Induced by *Mycoplasma arthritidis. Infect. Immun.,* **4**:344, Oct. 1971.

*Jawetz, E.; Melnick, J.; and Adelberg, E. A.: *Review of Medical Microbiology,* 12th ed. Lange Medical Publications, Los Altos, Calif., 1976.

*Available in paperback.

Systemic Mycoses: Fungous Diseases Acquired Through Respiratory, Parenteral, and Endogenous Routes

16

Mycotic Infections: Clinical and Epidemiologic Distinctions

Clinically, mycotic diseases of man fall into three distinct patterns: *superficial infections* that involve only the outermost epithelial structures of the body (skin, hair, and nails), *systemic mycoses* that arise from fungal infections of deep tissues, and *subcutaneous mycoses.* Superficial fungous infections are usually chronic and difficult to treat, but are not a threat to general health because their causative agents do not invade deeper tissues and are not disseminated. Fungi responsible for systemic infections, on the other hand, can invade deeply and be disseminated widely in the body. They may remain inactive or be held quiescent by the body's defensive reactions, producing few if any symptoms of importance, or they may cause progressively advancing disease. Subcutaneous mycoses are generally acquired through trauma to the skin, their agents being introduced into deep dermal tissues. Although such infections may involve extensive subcutaneous areas and induce marked regional lymphadenopathy, generally they are not disseminated to internal organs.

Important *epidemiologic* distinctions can also be made for these three types of mycosis. Fungi of superficial infections have a natural reservoir in soil, man, and some animals and are directly transmissible, by close or indirect contacts between infected individuals. Most agents of both the deep and subcutaneous mycoses, however, appear to have a reservoir only in soil, where they live as saprophytes. They are capable of parasitism but are not directly transmissible from one person to another. The systemic mycoses are generally acquired through inhalation of fungal spores from a soil source, whereas subcutaneous fungal infections are acquired through injury to the skin, with introduction of spores into underlying tissues.

In view of these considerations, the superficial fungous infections are described in Section VI, together with other diseases acquired and transmitted through skin contacts (Chap. 22). Systemic and subcutaneous mycoses are grouped together here because, irrespective of their routes of entry, the basic patterns of their pathogenesis and epidemiology are closely similar.

Systemic and Subcutaneous Mycoses: Sources, Entry Routes, Pathogenesis, Epidemiology

Sources. There are many species of fungi associated with systemic and subcutaneous disease in man. Most of these are derived from an exogenous soil reservoir and are capable of causing a primary mycosis following an effective entry route into the healthy body. Others are more often associated with opportunistic infection in persons with underlying disease or an immunodeficiency. One important genus of yeastlike fungi, *Candida,* has several species that live commensally on human mucosal surfaces. From this endogenous source *Candida* species (particularly *C. albicans*) can induce serious infection, often in infants, aged people, or those debilitated by other diseases. Two bacterial diseases, nocardiosis and actinomycosis, have many features in common with the true mycoses. Their microbial agents are members of the Actinomycetales order of "higher" bacteria, which includes the mycobacteria (see pp. 95–96). One of these genera, *Actinomyces,* is a member of the commensal flora of the oral mucous membranes and the other, *Nocardia,* is commonly found in soil. Table 16–1 shows a classification of the primary systemic, subcutaneous, and opportunistic mycoses, distinguishing the similar diseases of bacterial origin and indicating source and entry routes.

Free-living fungi are not ubiquitously distributed

in nature; often they are concentrated in particular areas where soil conditions permit their growth. Some fungous diseases occur in widely scattered parts of the world, others in limited geographic regions. Little is known about the natural life of these fungi or about the factors that determine their geographic distribution. At least two pathogenic species, *Cryptococcus* and *Histoplasma,* are often associated with birds, such as pigeons, chickens, and starlings. They may be spread in bird droppings (which act as an enrichment medium) or be carried on their contaminated feathers and distributed widely on air currents (Fig. 9–8, p. 216). Bats may also be a reservoir for *Histoplasma.*

Entry Routes. In soil, these primitive plants produce many spores, which are capable of long survival with or without moisture, in earth, on wood or plant surfaces. When spore-laden dust is stirred up by feet, shovels, or bulldozers, or by wind currents, the airborne spores may be inhaled in sufficient numbers to reach the lungs and induce primary pulmonary mycoses. This is a common entry route for diseases such as histoplasmosis and coccidioidomycosis.

In other instances, fungous spores may be introduced subcutaneously, through minor skin injuries incurred by people working with contaminated soil, plants, or wood. Splinters and thorns are common agents of such injuries. The primary infection in such cases is extrapulmonary, as in sporotrichosis or chromomycosis. Although most subcutaneous mycoses do not disseminate to deep, systemic tissues, sporotrichosis sometimes does so. Rarely, sporotrichosis may occur as a primary pulmonary disease that can be misdiagnosed as tuberculosis.

Opportunistic fungal infections have increased with the widespread use of broad-spectrum antibiotics, corticosteroid and immunosuppressive therapy, and the self-administration of injectable drugs such as heroin. Open-heart surgery, indwelling intravenous catheters, hyperalimentation, and underlying malignancies (leukemia, lymphoma) are also associated with opportunistic mycotic infection. The distinction between primary and opportunistic mycoses is sometimes difficult, however. Diseases such as histoplasmosis and cryptococcosis, for example, which often appear as primary, have occurred with increasing incidence among patients on prolonged

Table 16–1. Classification of Systemic and Subcutaneous Mycoses

Type	Sources	Entry Routes	Primary Infection	Disease	Causative Organism(s)
Primary systemic mycoses	Exogenous	Respiratory or parenteral	Pulmonary or extrapulmonary	Histoplasmosis	*Histoplasma capsulatum*
				Coccidioido-mycosis	*Coccidioides immitis*
				Blastomycosis (North American)	*Blastomyces dermatitidis*
				Cryptococcosis	*Cryptococcus neoformans*
				Paracoccidioido-mycosis (South American blastomycosis)	*Paracoccidioides brasiliensis*
Subcutaneous mycoses	Exogenous	Parenteral	Extrapulmonary	Sporotrichosis	*Sporothrix schenckii*
				Chromomycosis	*Phialophora, Fonsecaea, Cladosporium* species
				Mycetoma (Madura foot)	*Madurella, Nocardia, Petriellidium* species, and others
Opportunistic mycoses	Endogenous	Skin or mucosae	Superficial or disseminated	Candidiasis	*Candida albicans* and other species
	Exogenous	Respiratory	Pulmonary	Aspergillosis	*Aspergillus fumigatus* and other species
	Exogenous	Respiratory or parenteral	Pulmonary or extrapulmonary	Mucormycosis	*Mucor, Rhizopus, Absidia,* and others
Bacterial diseases resembling mycoses	Exogenous	Respiratory	Pulmonary	Nocardiosis	*Nocardia asteroides* and other species
	Endogenous	Oral mucosa	Pulmonary or extrapulmonary	Actinomycosis	*Actinomyces israelii*

corticosteroid therapy. Conversely, although nearly all patients with aspergillosis or mucormycosis have some underlying predisposing disease, some cases may occur as primary infections. In healthy individuals, however, opportunistic fungi of low pathogenicity must gain entry in some extraordinary way (be inhaled in large numbers from a rich source in soil or be introduced traumatically) to establish primary disease.

Endogenous fungi such as *Candida* species similarly must have some opportunistic assistance to induce damaging infection in superficial or deep tissues. *Candida* strains live commensally on normal skin, and on oral, vaginal, and intestinal mucosae. Alteration of the normal bacterial flora by antibiotic therapy may permit their overgrowth with resultant damage to local tissues. Lowered host resistance, trauma, or other circumstances may contribute to further extension of *Candida* infection. Superficial candidiasis is much more frequent than systemic disease, but *Candida* may localize in lungs, meninges, the endocardium, or other tissues. The mycotic types of infection caused by bacteria of the genus *Actinomyces* often appear to be opportunistic also. *Actinomyces* may be found around carious teeth and in tonsillar crypts. Tooth extraction or other trauma to local tissues may permit entry to deeper structures of the face and neck, with a resultant cervicofacial infection known as "lumpy jaw." Pulmonary actinomycosis may result from aspiration of the organisms, and abdominal infection sometimes occurs if they are swallowed. A related genus, *Actinomadura,* is associated with the subcutaneous infection known as mycetoma, or Madura foot. Nocardiosis, however, another mycotic type of disease of bacterial origin, has an exogenous source in soil and appears most often as a pulmonary infection following respiratory entry.

Pathogenesis.   The typical tissue reaction to fungi reaching systemic positions is a chronic granuloma with necrosis or abscess formation. This is the same general type of lesion that develops in tuberculosis (see Chap. 13, p. 331), with certain differences characteristic of the proliferating organism or of its site of localization. Mononuclear, epithelioid, and giant cells surround and attempt to contain the organism, and this group is surrounded peripherally by dense accumulations of lymphocytes and fibroblasts. The different fungal agents germinate and reproduce in different ways. Under the pressure of human tissue opposition, proliferation may be halted or held to a minimum for long periods of time, or it may advance progressively as conditions permit. A necrotizing or suppurating granuloma may release fungi into adjacent tissues where new reaction to them occurs, or into lymphatic channels or blood vessels that distribute them widely in the body to other foci of localization.

In primary pulmonary mycoses resulting from inhalation of fungous spores, granulomatous lesions may be distributed through the lungs, involving any portion. The extent of disease varies. There may be a single, small lesion or multiple foci involving large areas of both lungs. Areas of consolidation can resemble bacterial pneumonia or extensive tumor formation. Cavities may be formed, or there may be miliary, fine or coarse nodular patterns. Progressive extension of infection within the lung may occur, or there may be hematogenous spread to other tissues (liver, spleen, kidneys, brain and meninges, bone, and skin) from a primary pulmonary site.

Skin and subcutaneous lesions sometimes develop by dissemination from systemic (usually pulmonary) foci, or they may represent the primary form of disease at the site of entry of the agent. Paracoccidioidomycosis (South American blastomycosis), for example, is a primary pulmonary mycosis with early lesions in the lung and later involvement of subcutaneous tissues, especially at the mucocutaneous borders of the mouth and nose, with extensive lymphadenopathy. It is possible, however, that such facial lesions may sometimes represent the primary site of entry of the fungus. In the subcutaneous mycoses, granulomas are formed in the deep tissues of the skin. Often these nodular lesions tend to break down and ulcerate, discharging through the surface. In sporotrichosis, minor injuries to the skin of the hands or arms admit the spores from thorns, splintered wood, or soil. The organism then slowly advances from the site of entry, where a subcutaneous nodule usually forms, along the course of the regional draining lymphatics. Disseminated forms of sporotrichosis occur but are not common. As pointed out previously, this disease may develop as a primary pulmonary infection, though rarely. The lesions of chromomycosis and mycetoma are characteristically limited to tissues of the extremity originally exposed to infection (usually the feet and legs), but they can be extensive and disabling.

Hypersensitivity, a manifestation of cell-mediated immunity, is the characteristic immunologic response to mycosis. This specifically altered reactivity to components of the invading fungus probably contributes in many instances to necrosis of granulomatous lesions and to progressive advance of infection, as it does in tuberculosis of the reinfection type (Chaps. 7 and 13). Humoral immunity of protective value does not develop in mycotic diseases. Circulating antibodies measurable by serologic methods do appear in many instances, but although they have limited diagnostic and prognostic value, they do not have an apparent role in resistance.

The clinical symptoms vary among these diseases, or for a particular mycosis, with the localization of the organism and the degree of nonspecific as well as hypersensitive reactivity evoked. Pulmonary mycosis often begins as a subacute respiratory infection, with low-grade fever, dyspnea, and nonproductive cough. As the disease progresses, these symptoms intensify, but the cough becomes productive of purulent sputum. Pain in the chest, loss of weight, fatigue, and night sweats may ensue. Disseminated disease produces symptoms referable to hypersensitivity and to the involvement of other organs.

The diagnosis is sometimes difficult because the clinical characteristics of systemic mycotic lesions are very similar to those of tuberculosis and other granulomatous diseases with protean symptomatology. Laboratory identification of the causative fungus in smears and cultures of sputum, body fluids, or pus from localized lesions usually affords a secure diagnosis.

The microscopic characteristics of many pathogenic fungi as they are seen in tissues, or in direct smears and wet mounts made from clinical specimens, are different from those that appear when the organisms grow saprophytically in culture. This difference in the structure of parasitic and saprophytic phases of the same fungus is of some epidemiologic interest, because infected human beings do not transmit the parasitic fungi directly to others. It is only when the parasitic forms have converted to a saprophytic type of growth outside of the body that spores or other structures with an infectious potential are produced.

Skin tests and serologic methods are sometimes helpful in establishing the presumptive diagnosis of mycotic disease, or in indicating the prognosis. Skin tests of the tuberculin type, using sterile, somewhat purified culture filtrates of the organisms, produce delayed reactions of hypersensitivity if they are positive. This may indicate either currently active or past (possibly healed) infection. However, skin testing usually has more epidemiologic and prognostic value than diagnostic usefulness. For example, in areas where histoplasmosis is endemic, more than half the population may have positive skin tests yet be clinically asymptomatic. Conversely, negative skin tests are not uncommon in the presence of clinically active histoplasmosis or coccidioidomycosis, and such findings cannot rule out the diagnosis. Prognostically, a reversion from a positive to a negative skin test often indicates failure of the immune mechanisms (*anergy,* the opposite of allergy or hypersensitivity). This may happen in the terminal stages of overwhelming, progressive infection, signaling a grave prognosis.

Among serologic methods, complement-fixation has some diagnostic value in histoplasmosis, blastomycosis, and coccidioidomycosis, although some degree of cross-reaction occurs in these infections. A rising positive titer (fourfold increase or more) probably has more value in raising the possibility of a fungous infection than in making a specific diagnosis. Similarly, a significant increase or decrease in the titer of sera drawn over a period of time may be prognostic, indicating the advance or wane of infection. As in the case of skin testing, however, a negative complement-fixation result is not conclusive evidence that active mycosis is not present.

Other serologic methods used in the diagnosis of mycotic disease include latex agglutination, fluorescent antibody techniques, and precipitin tests, such as tube precipitation, immunodiffusion (ID), and counterimmunoelectrophoresis (CIE). In general, fungal antigens used in these procedures tend to lack either specificity or sensitivity, or both, but ID and

CIE methods are promising in that they are quite sensitive in detecting specific antibodies.

Judicious coupling of skin tests with appropriate serologic methods can be helpful in the clinical diagnosis, but the sequence in which these tests are done may be important. In histoplasmosis, for example, the skin testing antigen itself may stimulate some antibody response and thus influence the results of serologic tests performed later. In this instance, therefore, it is recommended that blood specimens for the serologic diagnosis of histoplasmosis be drawn *before* a skin test is done and that the laboratory be informed as to this sequence. In coccidioidomycosis, the skin-testing antigen has no effect on antibody production, and in this case it makes no difference if the skin test precedes serologic testing.

Epidemiology. Systemic mycoses are frequently limited to particular geographic areas of the world. Within these areas a majority of people may acquire the fungous infection, but a very small minority of these develop serious progressive disease. Some mycoses, such as histoplasmosis and coccidioidomycosis, appear to be increasing in importance. This may be a reflection of improved diagnosis, an increased rate of exposure among the ever-larger traveling public, or a greater number of people with heightened susceptibility. Some of the factors that lower resistance to mycotic infection are (1) other persistent, debilitating disease (cancer, diabetes, tuberculosis, malnutrition); (2) widespread use of chemotherapeutic drugs and hormones that alter human metabolism and disturb balances among commensalistic flora of the body; (3) immunodeficiencies or the use of immunosuppressive drugs; (4) drug addiction; and (5) local acute or chronic injury that permits fungi to enter deep tissues (accidental or surgical trauma; ulcerative lesions of skin or oral or intestinal membranes; or the lesions of avitaminosis, x-irradiation).

The incubation period of systemic mycosis varies with the numbers of organisms introduced and the ease of their access to sites where they can proliferate. Fungi do not multiply rapidly, and as a general rule it requires at least ten days to two weeks for the first symptoms of infection to appear. In many instances, the incubation period may extend from three to five weeks, or it may be impossible to define.

Control of active mycotic infection does *not* require isolation of the patient or any special measures to protect his contacts. Nursing care of these diseases does include application of concurrent disinfection techniques to sputum or other discharges (as may be clinically appropriate), as well as to contaminated articles. Terminal disinfection-cleaning should be provided. These precautions are necessary to prevent possible conversion of the fungus from the parasitic to saprophytic forms, with resultant unusual concentrations of the latter.

When epidemics of airborne mycoses occur (coccidioidomycosis, histoplasmosis), recognition of their sources makes it possible to apply dust control measures or to eliminate heavy concentrations of fungus in limited areas (chicken houses, barns, starling roosts may be sources of *Histoplasma* infection). These procedures may be of preventive value when applied on a routine basis in endemic areas.

Treatment of deep fungous infections varies with the disease. Amphotericin B is an effective broad-spectrum antifungal agent used in the treatment of most systemic mycoses. It is not equally effective in all pulmonary infections, but it appears to be the most potent drug available for each of them. Unfortunately, its use is complicated by toxic and undesirable side effects, which include fever, nausea and vomiting, anemia, and renal failure. This toxicity of amphotericin creates problems in administering it, so that other drugs may be employed alternatively. In coccidioidomycosis and histoplasmosis, however, amphotericin is the only effective therapeutic agent currently available.

Other drugs such as 2-hydroxystilbamidine, 5-fluorocytosine (flucytosine), imidazoles (clotrimazole and miconazole), and hamycin have been used with varying degrees of success. Sporotrichosis is most commonly treated with potassium iodide. Selected cases of blastomycosis may respond to hydroxystilbamidine, but the drug of choice for this disease is amphotericin. Cryptococcosis and systemic candidiasis are often treated with flucytosine and amphotericin in combination. This combined use makes it possible to reduce the amount of amphotericin administered, thus decreasing its toxicity for the patient.

The antibiotics useful in bacterial infections have no value in fungal diseases. The bacterial diseases that simulate mycoses, however, can be successfully treated with antibiotics. Sulfonamides are the drugs of choice for nocardiosis, and penicillin is the most effective agent for actinomycosis.

Effective immunologic approaches to the prevention or treatment of the mycoses are not available. Recently, however, efforts have been made to treat patients with progressive, disseminated coccidioidomycosis experimentally with "transfer factor," a substance extractable from sensitized T lymphocytes (see Chap. 7, p. 166). This is an attempt to transfer cell-mediated immunity passively from a hypersensitive donor to patients whose immune mechanisms have been overwhelmed by advancing infection. Of the small number of patients thus treated (75), one-third have shown prompt, dramatic improvement, a third responded moderately or more slowly, and the remainder showed no response.

Primary Systemic Mycoses

Individual mycotic diseases are briefly described in outline form on the following pages. The preceding section should be consulted for general information when a particular disease is being studied. Specific details are inserted only when they represent an exception to the general features of mycosis.

Histoplasmosis

Causative Organism. *Histoplasma capsulatum.*

The Clinical Disease. Primary pulmonary infection is frequent in endemic areas, but usually asymptomatic. Small granulomas in the lung heal with calcification, as evidenced by chest x-ray and histoplasmin-positive and tuberculin-negative skin tests. (Calcified lung lesions cannot be distinguished from healed tuberculosis, or coccidioidomycosis, by x-ray alone.) With heavy exposure to airborne infection,

clinical pneumonia develops. Illness may be prolonged. It can subside with spontaneous healing or continue to disseminated disease in a small minority of cases. When dissemination occurs, the reticuloendothelial system is particularly involved, with localizations in lymph nodes, spleen, and liver. Granulomatous lesions may be found in many organs. Progressive histoplasmosis is usually fatal, its course being more rapid in children than in adults.

Incubation Period. When infection can be traced, about 5 to 18 days.

Diagnosis

I. Skin tests. A positive intradermal test with histoplasmin (a culture filtrate of the mycelial phase of the fungus) denotes either past or recent exposure to *H. capsulatum* or an antigenically related organism. The skin test may be negative in the late, disseminated stage of disease.

II. Laboratory. *Specimens* of sputum, urine, blood; pus aspirated from lesions; biopsies from bone marrow, liver, lymph nodes, skin.

In tissue sections, Histoplasma appears as an intracellular, single-celled yeastlike organism often packed within giant cells or macrophages (Fig. 16–1).

In culture the organism is *dimorphic:* at 37° C on enriched media simulating *in vivo* conditions the organism is yeastlike, as it is in tissues. At atmospheric temperature on ordinary media a saprophytic mold appears, with mycelium, microconidia, and characteristic spiny ("tuberculate") chlamydospores (macroconidia; Fig. 16–2).

Blood for serologic tests is useful. A rising complement-fixation titer is strong evidence of active disease.

Epidemiology. The organism is found in soil enriched with bird droppings, as in chicken coops, pigeon cotes, barns, and also in caves inhabited by bats. Endemic areas are found throughout the world, in temperate and tropical zones. In the United States, localized distribution in enriched soils occurs in many central and eastern states, including Tennessee, Maryland, Virginia, the District of Columbia, some parts of New York State, Ohio, Indiana, Missouri, and Arkansas. Infection is usually airborne, carried into the lung by inhalation.

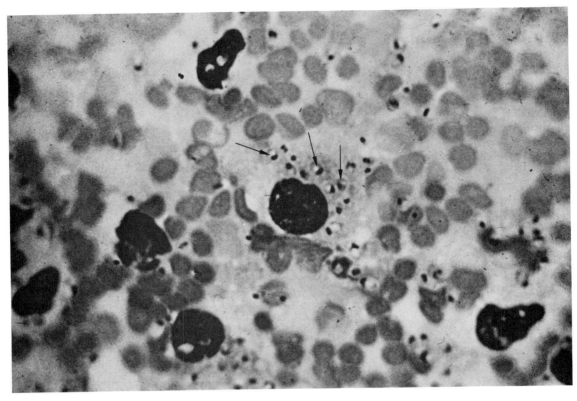

Fig. 16–1. Photomicrograph of *Histoplasma capsulatum* in tissue section. Small yeastlike, single-celled organisms are seen within macrophages. (1500×.) (Courtesy of Center for Disease Control, D.H.E.W., Atlanta, Ga.)

People of all ages are susceptible to histoplasmosis. The highest incidence of disseminated, fatal disease occurs in infants and in the aged, but in endemic areas up to 90 percent of the population may display evidence of exposure (positive skin tests) without signs of active infection. There are no significant racial or sex differences, except that males predominate among adult infected patients, probably because of greater exposure. Disseminated histoplasmosis is associated more than casually with leukemia, lymphoma, and Hodgkin's disease.

Histoplasmosis is often, but not strictly, a rural disease. People from urban areas who visit the country may be exposed to windblown spores, to contaminated garden soil fertilized with chicken manure, or to caves where bat guano (excrement) is abundant

(spelunkers run the risk of such exposure). Urban sources of exposure also exist, as in areas where starlings roost or in soil around old houses and other buildings inhabited by bats.

Amphotericin B is the drug of choice. A new antibiotic, saramycetin, has been found effective in clinical trials but is not commercially available.

Coccidioidomycosis

Causative Organism. *Coccidioides immitis.*

The Clinical Disease. This mycosis typically begins as a respiratory infection. The primary infection may be asymptomatic or it may resemble acute influenzal

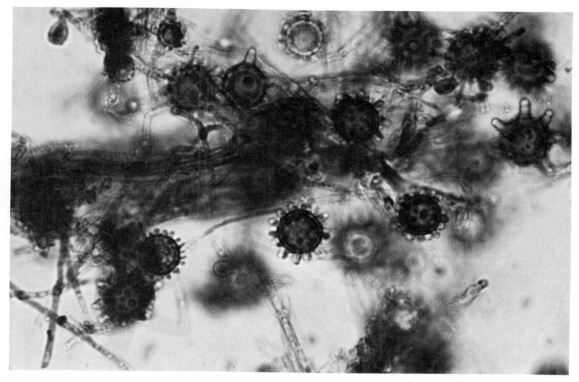

Fig. 16-2. Photomicrograph of the mycelial phase of *H. capsulatum* (stained with lactophenol cotton blue), showing large, spiny (tuberculate) chlamydospores. (800×.) (Reproduced from Hazen, E. L.; Gordon, M. A.; and Reed, F. C.: *Laboratory Identification of Pathogenic Fungi Simplified*, 3rd ed. 1970. Courtesy of Charles C Thomas, Publisher, Springfield, Ill.)

illness, with fever, chills, cough, and chest pain. These symptoms may subside without residual traces, or healing may be accompanied by fibrosis and calcification of granulomatous lesions in the chest. Rarely, the disease progresses to a disseminated form resembling tuberculosis, with lesions in lungs, bones, joints, subcutaneous tissues, skin, internal organs, brain, and meninges. With widespread dissemination, symptoms are acute; prostration and death may occur within a few weeks.

Incubation Period. For primary infection, between ten days and three weeks. Development of granuloma is often gradual and inapparent.

Diagnosis

I. Skin Tests. A positive intradermal test with coccidioidin is acquired in most cases of coccidioidomycosis and is relatively specific; that is, there are few or only mild reactions to histoplasmin or other fungal antigens. The test may remain positive for many years, even in recovered patients who have left endemic areas and have not been reexposed. In disseminated disease, however, the skin test may become negative.

II. Laboratory. *Specimens* of sputum; pus from aspirated lesions; biopsies; gastric washings; spinal fluid if indicated by symptoms.

Microscopic examination reveals thick-walled spheric structures (spherules) containing many en-

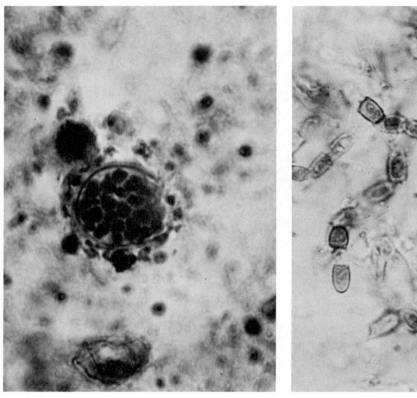

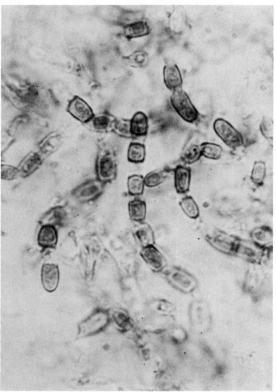

A

B

Fig. 16–3. *Coccidioides immitis. A.* Mature spherule seen in infected tissue. (800×.) *B.* Fragmenting arthrospores develop in cultures at room temperature. (1000×.) (Courtesy of Dr. John W. Rippon, The Pritzker School of Medicine, The University of Chicago, Chicago, Ill.)

dospores (Fig. 16–3, *A*). In tissue sections or biopsies of granulomas, the spherules may be seen within giant cells, or in the acellular areas of central necrosis.

In culture it is difficult to reproduce the *in vivo* conditions that encourage spherule production (the endospores resemble single yeast cells but do not reproduce by budding). At atmospheric temperature on ordinary media the saprophytic growth consists of an abundant aerial mycelium. The branching hyphae develop into chains of fragmenting arthrospores, which are easily disseminated and highly infectious (Fig. 16–3, *B*).

Blood for serologic tests is useful. Precipitins and complement-fixing antibodies appear in the course of symptomatic disease. The latter may be delayed for some weeks but persist for a few months. A rising titer of complement-fixing antibody indicates the probability of disseminated disease and a poor prognosis.

Epidemiology. Soil and spore-laden dust are sources of infection for man and a number of domestic animals. In the United States the disease is endemic in the southwest, particularly the San Joaquin Valley of southern California. Other endemic areas are found in parts of Central and South America.

C. immitis grows in soil, producing large numbers of arthrospores like those seen in cultures. These are easily disseminated on air currents. It has been demonstrated that in the hot summer the fungus disappears from the desert floor, persisting in rodent burrows where it is protected from solar heat and radiation. After the winter and spring rains, it spreads again over the desert floor.

The arthrospores of the culture and saprophytic phase are extremely infectious. In an unusual example of person-to-person transmission from a hospitalized patient with osteolytic coccidioidomycosis, six members of a medical staff were infected by inhaling arthrospores that had been produced on the patient's plaster cast. The tissue form of the fungus had converted to the spore-forming saprophytic form in the plaster.

People of all ages are susceptible, but in an endemic area the disease is usually a mild one of early childhood or frequent in adults who have migrated to the area. Disseminated fatal disease occurs most often among infected pregnant women and in dark-skinned males. Adult white females more frequently display allergic lesions (erythema nodosum, erythema multiforme), presumably because of an inherent resistance and ability to develop a rapid immunologic defense. Such allergic lesions are uncommon in dark-skinned men.

Amphotericin B is the only drug of currently established value in the treatment of coccidioidomycosis. It is less effective than in histoplasmosis and must be given for extended periods at maximum tolerable doses. Miconazole is a potentially useful drug for the treatment of this disease. Promising results have been reported in clinical trials of this imidazole.

Blastomycosis (North American)

Causative Organism. *Blastomyces dermatitidis.*

The Clinical Disease. Systemic blastomycosis usually begins as a primary pulmonary infection acquired through inhalation of spores. Initial symptoms are those of an upper respiratory viral infection, but progressive infection leads to severe pulmonary involvement and dissemination to subcutaneous tissues, bones, genital and visceral organs, and the central nervous system. Cutaneous blastomycosis may result from dissemination of systemic disease (highly fatal) or it may be a primary infection resulting from local entry of the organisms into the skin.

Incubation Period. Difficult to define; probably a few weeks.

Diagnosis

I. Skin Tests. Blastomycin is neither specific nor sensitive enough to be of diagnostic value and is no longer commercially available.

II. Laboratory. *Specimens* of sputum; pus and exudates from lesions; biopsy of skin granulomas.

Microscopic examination reveals the tissue form to be a large, thick-walled budding yeastlike organism (Fig. 16–4, *A*).

In culture the organism grows in the *in vivo* yeastlike form on blood agar incubated at 37° C (Fig. 16–4, *B*). At room temperature on ordinary media the saprophytic mold forms a white to tan colony with single-celled smooth conidia.

Blood for serologic tests may be useful, but cross-reactions with *Histoplasma* antigen can occur. Complement-fixation tests can be negative in active blastomycosis, but if positive they may have prognostic value in clinically proven blastomycosis. The immunodiffusion test is specific, with a sensitivity of 80 percent. Sera from early acute, convalescent, and postrecovery stages of disease should be drawn for the serologic diagnosis.

Epidemiology. Blastomycosis appears to be a relatively uncommon but severe disease. The reservoir is believed to be in soil, the mode of transmission via inhalation of spore-laden dust, but the ecologic niche of the organism has not been precisely determined. The disease occurs primarily in Canada and the United States. A number of reports indicate that it may also have a wide distribution in Africa. The infection also occurs in dogs, cats, and horses.

This is a sporadic disease that occurs among people of all ages, with a somewhat higher incidence in middle age. It is more frequent in males than females, but no occupational associations have been demonstrated.

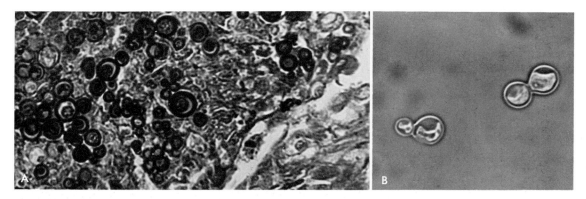

Fig. 16–4. *Blastomyces dermatitidis. A.* Large, spheric, thick-walled budding cells in human lung tissue. (400×.) *B.* Budding cells from growth on blood agar at 37°C. (800×.) (Reproduced from Hazen, E. L.; Gordon, M. A.; and Reed, F. C.: *Laboratory Identification of Pathogenic Fungi Simplified*, 3rd ed. 1970. Courtesy of Charles C Thomas, Publisher, Springfield, Ill.)

Amphotericin B is the most effective drug in severe cases, but 2-hydroxystilbamidine has had success in the therapy of mild forms of the disease. Clinical trials with the new antibiotic, saramycetin, have had good results, especially with dermal lesions.

Cryptococcosis

Causative Organism. *Cryptococcus neoformans.*

The Clinical Disease. Infection in man usually occurs through the respiratory tract, with resulting mild pulmonary infection frequently undiagnosed. The most common clinical extension of cryptococcosis in man is the slow development of a chronic meningitis. The course of cryptococcal meningitis may extend over several years, resembling degenerative central nervous system disease, syphilitic or tuberculous meningitis, or brain tumor. The organism may also gain entry to the body through the skin, or the mucosa of the upper respiratory or intestinal tracts, with isolated lesions in subcutaneous tissues, lymph nodes, tongue, muscles of the back, and other areas. (Oral entry is especially frequent in animals.) Lesions of this type may remain localized or extend to the brain and meninges.

Incubation Period. Not defined; meningitis follows usually upon inapparent pulmonary infection.

Laboratory Diagnosis. *Specimens* for wet mounts and culture include spinal fluid, sputum, and exudates from cutaneous lesions if they occur.

Microscopic examination of wet mounts of centrifuged spinal fluid or other specimens reveals a budding yeast surrounded by a large capsule. The latter can also be seen against a dark background, obtained by suspending the material to be examined in India ink (Fig. 16–5).

In culture the organism grows as a yeast on all media at temperatures ranging from 20° to 37°C. Capsule production can be demonstrated on special media or by mouse inoculation.

Serologic diagnosis can be made with good precision using the slide latex agglutination method (the laboratory must take steps to assure elimination from the patient's serum of a factor present in rheumatoid arthritis that also agglutinates the latex antigen). In CNS infection, cryptococcal antigen can be demonstrated serologically in spinal fluid.

Skin tests are not available.

Epidemiology. The organism grows saprophytically in the external environment. It can be isolated consistently from pigeon nests and droppings, and from soil in many parts of the world. Transmission is thought to occur through inhalation of contaminated

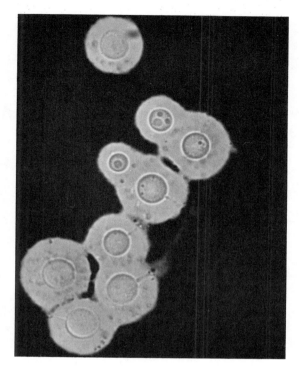

Fig. 16–5. India ink preparation of *Cryptococcus neoformans*, showing its large capsule. (1000×.) (Reproduced from Dykstra, Mark A.; Friedman, L.; and Murphy, J.: Capsule Size of *Cryptococcus neoformans*, Control and Relationship to Virulence. *Infect. Immun.*, **16**:129, 1977.)

dust, but development of disease depends on the intensity of exposure and the numbers of organisms reaching the lower respiratory tract.

Sporadic cases occur on a worldwide basis, in domestic and wild animals as well as in man. The disease is more frequent in human males than in females. Although there are no significant differences related to age, race, or occupation, the disease is associated more than randomly with leukemia, lymphosarcoma, Hodgkin's disease, diabetes, and prolonged steroid therapy. For this reason it is considered by some to be an opportunistic infection.

The prognosis is good for pulmonary and cutaneous forms of the disease, but disseminated or meningeal cryptococcosis may be fatal. Flucytosine therapy is effective in the former types of infection, but the

organism becomes resistant to this drug during treatment. Low doses of amphotericin given together with flucytosine appear to decrease the frequency of resistance to the latter, and the combination of these two drugs has an additive effect on susceptible strains of *Cryptococcus*. This combined use of amphotericin and flucytosine is now the treatment of choice for cryptococcal meningitis and is recommended for other severe forms of the infection as well.

Paracoccidioidomycosis (South American Blastomycosis)

Causative Organism. *Paracoccidioides brasiliensis.*

The Clinical Disease. This disease is endemic in South America, particularly Brazil. It is often called *paracoccidioidal granuloma* because the tissue form of the organism resembles *Coccidioides*. It produces a chronic granulomatous disease of the lung, the mucocutaneous membranes of the face (Fig. 16-6),

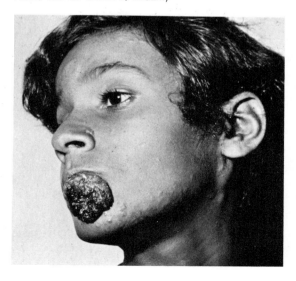

Fig. 16–6. Mucocutaneous facial lesion caused by the fungus *Paracoccidioides brasiliensis*. Note also the enlargement of cervical lymph nodes. (Courtesy of Dr. José Lisbôa Miranda, Universidade Gama Filho, Rio de Janeiro, Brazil.)

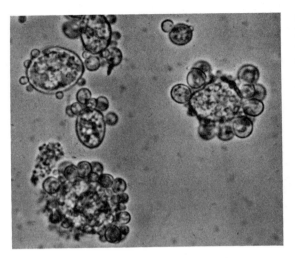

Fig. 16–7. *Paracoccidioides brasiliensis.* Photomicrograph of yeast cells showing multiple budding (wet mount made from a culture grown on blood agar at 37°C). (2500×.)

oral membranes, lymph nodes, and visceral organs. Cervical lymphadenopathy may resemble that of Hodgkin's disease. Disseminated infection is highly fatal; localized disease may persist for years if untreated.

Incubation Period. A few weeks.

> **Diagnosis**
> *Skin Tests.* Limited diagnostic value but useful epidemiologically in detecting early cases and in delimiting endemic areas.
> *Laboratory. Specimens* of pus from lesions, biopsies. Sputum if pulmonary disease occurs.
> *Microscopically,* the tissue form of the organism is a large, thick-walled budding yeast resembling *Blastomyces dermatitidis,* except that many cells show *multiple* budding.
> *In culture,* the tissue form grows at 37° C on blood agar (Fig. 16–7). At room temperature on ordinary media the organism is a saprophytic mold.
> *Serologic diagnosis* is made by immunodiffusion and complement-fixation tests.

Epidemiology. The organism is thought to be associated with vegetation, wood, or soil. The infection is primarily acquired through inhalation, but it is possible that it may sometimes be introduced through the skin or oral membranes, presumably by direct contact with contaminated vegetative material. The incidence of this mycosis is highest among young male adults whose manual labor brings them in contact with the fungus, but the disease is sporadic rather than epidemic.

This mycosis is uniformly fatal if not treated. Sulfonamides are the drugs of choice in pulmonary and chronic disseminated forms of the disease. These drugs control but do not eliminate entrenched infection. Amphotericin B is recommended for patients who do not respond to sulfonamides, for those who relapse after treatment with the latter, and for those with advanced disease.

Subcutaneous Mycoses

Like the systemic mycotic diseases, those with a parenteral entry route are not transmissible from man to man. Subcutaneous mycoses tend to be chronic and slowly progressive, but they usually do not disseminate to systemic tissues. Three important diseases of this type are described briefly here.

Sporotrichosis

Causative Organism. *Sporothrix schenckii.*

The Clinical Disease. This is a chronic infection usually initiated by entry of the organism through the skin of an extremity (frequently a hand or arm) and characterized by development of nodular lesions in subcutaneous tissues along the path of the regional lymphatic drainage. The nodules frequently soften, ulcerate at the surface, and discharge this pus. The clinical picture is highly characteristic. Dissemination of the organism to other parts of the body occurs rarely.

Incubation Period. The initial lesion may be seen within three weeks, or it may require up to three months to appear.

> **Laboratory Diagnosis**
> *Specimens.* Pus or biopsy from subcutaneous lesions for smear and culture.

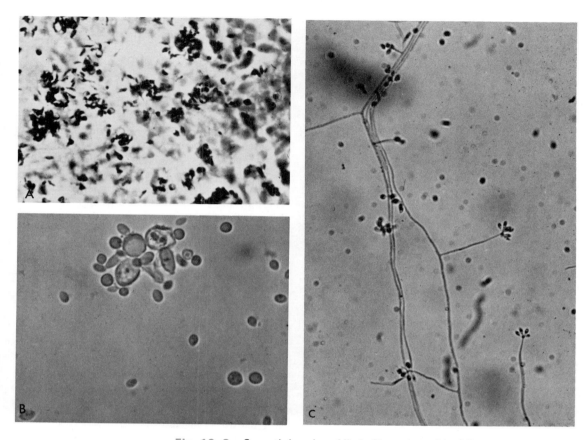

Fig. 16–8. *Sporothrix schenckii. A.* Cigar-shaped budding (yeast) cells in a histologic section of a nodular lesion. (400×.) *B.* Yeastlike cells taken from a culture grown on blood agar at 37°C. (800×.) *C.* A microscopic view of the mycelial growth obtained at room temperature (note the delicate petallike spores). (400×.) (*C* reproduced from Rippon, John W.: *Medical Mycology: The Pathogenic Fungi and the Pathogenic Actinomycetes.* W. B. Saunders Co., Philadelphia, 1974.)

Microscopic examination of pus does not always reveal the tissue phase (Fig. 16–8, *A*) of the organism, which is a small, Gram-positive, cigar-shaped, budding cell.

In culture the yeastlike tissue form grows on enriched media at 37° C (Fig. 16–8, *B*). At room temperature on simple media the organism is a mold, often darkly pigmented. Delicate hyphae on the aerial mycelium carry small ovoid spores in petallike clusters (Fig. 16–8, *C*).

Epidemiology. The organism grows in soil, on wood, and on vegetation. Infection is a sporadic occupational hazard of gardeners and farmers. An epidemic was reported from South Africa, involving about 3000 gold-mine workers commonly infected by contaminated mine timbers. The disease has been reported from all parts of the world and has been seen in animals as well as man.

The cutaneous lymphatic form of the disease responds well to treatment with potassium iodide. This therapy is less satisfactory in disseminated infection. Amphotericin B is given in the latter instance.

In the nursing care of sporotrichosis, special attention should be given to the disposal of contaminated dressings. These may be infectious for persons with open skin lesions.

Chromomycosis (Verrucous Dermatitis)

Causative Organism. Several species of darkly pigmented fungi, notably of the *Cladosporium, Fonsecaea,* and *Phialophora* genera.

The Clinical Disease. This is a chronic, slowly progressive, granulomatous infection of the skin and lymphatics. Warty cutaneous nodules develop very slowly. They become prominent vegetations that do not always ulcerate. The lesions are located as a rule on an exposed extremity and are almost always unilateral. The feet and legs are generally involved, possibly because they are frequently exposed to the soil source of the organism. Surgical excision of early lesions and treatment with amphotericin B or flucytosine help to limit the slow progression of the disease. The infection does not become generalized.

Incubation Period. Unknown, but probably requires weeks or months.

> **Laboratory Diagnosis**
> *Specimens.* The crusts of lesions or biopsy material.
> *Microscopic examination* of the tissue reveals dark-brown, thick-walled septate bodies about the size of a leukocyte. These structures reproduce by splitting rather than budding. The several fungi that cause this disease produce the same form in tissues. *In culture* these fungus species produce dark-brown or black mold colonies at room temperature on simple media. They do not grow at 37° C.

Epidemiology. The organisms probably exist in wood, soil, or vegetation and are introduced through traumatized tissue. Sporadic cases occur in many parts of the world, but this is primarily a disease of rural tropical regions (the West Indies, Central and South America, the Orient). It is seen frequently in male laborers.

Mycetoma (Madura Foot, Maduromycosis)

Causative Organism. Numerous soil fungi, including *Madurella, Petriellidium* (or *Allescheria*), *Phialophora,* and *Acremonium* species (also the "higher" bacteria, *Nocardia* and *Actinomadura*).

The Clinical Disease. This is a chronic infection of subcutaneous tissues, characterized by swelling and nodule and abscess formation, with suppurative extension through sinus tracts or fistulae. The foot is commonly involved, especially in tropical areas where people go without shoes. Deep extension of the infection involves muscles and bones. The affected area swells to a globular shape, losing its structural form. Sulfonamide or antibiotic therapy is of value if the infecting organism is one of the "higher" bacteria of the genera *Actinomadura* or *Nocardia.* The true fungi that cause this disease do not yield to these drugs, but may respond to amphotericin therapy. Intractable disease may require amputation to save the patient from death by secondary bacterial infection.

Incubation Period. The first lesions may require months to develop; progression takes place over a period of years.

> **Laboratory Diagnosis**
> *Specimens.* Draining pus from sinus tracts or biopsy material.
> *Microscopic examination* and *culture* may reveal bacillary forms or fungi. Culture techniques should include methods appropriate for the growth of either.

Epidemiology. Mycetomas are seen most frequently in tropical and semitropical areas where people acquire the infection through the injured skin of bare feet. The disease is seen occasionally in scattered areas of the United States, and in Mexico, South America, Africa, and Asia.

Opportunistic Mycoses

Many and varied fungi may be involved in the opportunistic mycoses that are frequently seen in the host already compromised by inadequate resistance or underlying disease. The organisms often originate in environmental sources to which many are exposed, but they lack pathogenic qualities under ordinary circumstances. A few yeasts or yeastlike fungi have endogenous sources in the human body, living as commensals among the normal flora of the skin and superficial membranes, but, of these, species of the genus *Candida* are most common and most notorious for their opportunism. Some of the outstanding features of candidiasis are described here.

Two of the fungous diseases of exogenous origin that are ordinarily seen in the compromised host are also discussed briefly in this section. These are aspergillosis and mucormycosis. The latter is a general term referring to infection caused by various members of the order *Mucorales*, subdivision *Zygomycotina*, notably *Mucor*, *Rhizopus*, and *Absidia* (see Chap. 2, p. 42).

Candidiasis

Causative Organism. *Candida albicans* (also *C. tropicalis, C. stellatoidea, C. parapsilosis, C. guillermondii,* and others).

The Clinical Disease. Candidiasis is usually a superficial mycosis of skin or mucous membranes, where the organism is ordinarily a member of the normal flora. A number of factors involving injury to or unusual susceptibility of tissues may permit local entry of the organism or development of systemic progressive disease. These factors are stated below under Epidemiology.

Several types of clinical candidiasis may be described briefly as follows:

Oral candidiasis (thrush) is particularly frequent in infants. Loosely adherent, white, pseudomembranous patches of fungous growth occur on the tongue and buccal mucosa (Fig. 16–9).

Vaginal infection is common in pregnancy, diabetic women, and those on antibiotic therapy. It involves vulvar surfaces as well as the vagina, producing irritation, intense pruritus, and discharge.

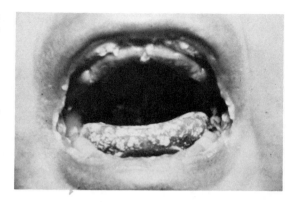

Fig. 16–9. Oral candidiasis (thrush) is frequently observed in infants. (From the collection of the Mycology Laboratory, Columbia University College of Physicians and Surgeons, New York, N.Y. Courtesy of Dr. Margarita Silva-Hutner.)

Cutaneous candidiasis often involves smooth skin of the intertriginous folds, particularly in axillary, inguinal, and inframammary regions. The interdigital folds of the feet and hands may also be involved. The cutaneous form may also be quite generalized. These areas become reddened and weepy with exudate from small vesicles or pustules.

Nail infections, onychia and paronychia, are common. Swellings develop around or in the nailbed, and the nails become thick, hard and deeply grooved.

Systemic candidiasis may begin as bronchial, pulmonary, or renal infection, usually in persons debilitated by coexisting disease. Dissemination of the organisms through the bloodstream may lead to grave extensions of infection to the meninges or the brain, and sometimes the endocardium.

Incubation Period. Infant thrush develops within two to five days of exposure. In other forms of infection the incubation period is variable and difficult to ascertain.

Laboratory Diagnosis
Specimens. Scrapings or pus from surface lesions, sputum, stools, spinal fluid, as indicated by clinical symptoms.

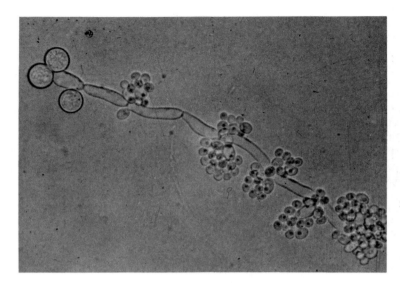

Fig. 16–10. Slide culture of *Candida albicans*. (800×.) Note the pseudomycelium, clusters of small blastospores, and three large terminal chlamydospores. (Courtesy of Dr. Coy D. Smith, University of Kentucky College of Medicine, Lexington, Ky.)

Microscopic examination of clinical specimens in wet mounts or stained smears reveals an oval, budding yeast. Very often the pseudomycelium produced by these organisms in culture is also seen in the tissues.

In culture, the organism grows well on simple media both at room temperature and at 37° C. Under both conditions it is a yeastlike fungus, with a pseudomycelium produced by elongation and pinching off of germinating cells. When grown on media deficient in nutrient, *Candida albicans* produces clusters of thick-walled chlamydospores (resting spores). This feature is often used to distinguish this species from other members of the *Candida* genus that are not frequently associated with mycosis (Fig. 16–10).

Serologic methods are not useful in the diagnosis of superficial candidiasis, but in systemic disease a number of tests have value. These include agglutination, latex agglutination, immunodiffusion, and counterimmunoelectrophoresis.

Epidemiology. Factors that commonly predispose to *Candida* infections include the following:

1. Antibiotic therapy, which disturbs the normal bacterial flora that holds *Candida* in check under ordinary circumstances.

2. Debilitating organic disease, particularly diabetes.

3. Chronic infectious disease, such as tuberculosis.

4. Malignant tumors and leukemia.

5. Nutritional deficiencies.

6. Continuing maceration of tissues caused by frequent or prolonged immersion of the hands in water; or excessive perspiration in the frictional, intertriginous folds (axillary, inguinal, inframammary, interdigital).

7. Narcotics addiction. Dirty needles may introduce *Candida* and other organisms that settle into deep tissues.

8. Open-heart and other surgery requiring venous catheterization may open a path for exogenous or endogenous spread of *Candida*.

Increased susceptibility to oral and dermal *Candida* infections is also seen in newborn infants, particularly premature babies and others with underlying organic problems. *Thrush,* as these superficial infections are often called, is frequently communicable in the nursery situation (Fig. 16–9). It may be transmitted to babies by nursing mothers with breast infection, and it may also appear as a nursery outbreak arising from a common environmental source, such as inadequately sterilized bottle nipples.

Control. In the control of active infection, isolation is necessary only in the case of infected newborn babies in nurseries. When nursery epidemics occur, every effort must be made to find the source of infection. (Bottle nipples should be considered as the first possibility.) Concurrent disinfection and terminal cleaning techniques must be carefully maintained. In these situations, swabs can be used to collect material from the mouths of all new babies entering the nursery. Swabs can be sent to the laboratory, in sterile saline or in transport medium, for microscopic examination, and for culture. If *Candida albicans* is demonstrated, these babies should be segregated whether or not they display active thrush. The laboratory can provide smear reports on the day the specimen is received, with culture results following on the second or third day.

Treatment. Some forms of candidiasis, especially superficial infections, may clear without therapy. Vulvovaginitis in the pregnant woman, for example, usually improves following delivery. In superficial candidiasis, control of underlying factors or disease is important and may obviate the need for further treatment, e.g., good diabetic therapy or judicious changes in or discontinuance of antibiotic therapy.

Superficial lesions are usually treated topically. The antibiotic nystatin may be used as a cream or an ointment for skin infections or as a solution for oral and vaginal douches. Topical clotrimazole and miconazole are also of demonstrated value.

Systemic candidiasis is a serious disease with a poor prognosis if untreated. Amphotericin B must be used in these cases, alone or in combination with flucytosine.

The prophylactic use of nystatin has been recommended for premature infants to protect these particularly vulnerable patients from oral or disseminated candidiasis.

Aspergillosis

Causative Organism. *Aspergillus fumigatus.* Other strains may include *A. flavus, A. niger, A. clavatus,* and *A. terreus.*

The Clinical Disease. Aspergillosis is a granulomatous disease of the lung or other tissues. Pulmonary disease may be necrotizing, with cavitation and hematogenous spread to other organs. Other forms of aspergillosis include otomycosis, sinusitis, and mycetoma. Pulmonary disease with this fungus is often secondary to tuberculosis, silicosis, tumor, immunodeficiency, or steroid therapy.

Colonization of a dilated bronchus or pulmonary cavity may occur, with formation of a fungus ball or tightly matted mass of mycelial fragments. The latter form of infection, called aspergilloma, rarely produces symptoms because it is merely a saprophytic growth in the cavity, with no tissue invasion. However, hemoptysis (spitting of blood) may be serious enough to require surgical intervention.

Allergic aspergillosis may occur in healthy persons who are frequently exposed to the spores in an environmental source (soil, compost, wood chips). Inhalation of spores and colonization of the respiratory tract without invasive disease may so sensitize the individual that each subsequent exposure may be followed by asthmatic symptoms.

Incubation Period. A few days to weeks following exposure.

Laboratory Diagnosis
Specimens. Sputum, bronchial aspirates, draining pus, or other material, depending on suspected localization.

Microscopic examination and *culture* reveal the typical structures of this fungus, which grows as a mold at room temperature and above.

Serologic methods and skin tests may be of little or no diagnostic value in invasive aspergillosis because the immune mechanisms of such patients have been undermined by predisposing disease. In aspergilloma and allergic aspergillosis, skin tests and serologic techniques (immunodiffusion, counterimmunoelectrophoresis) can be diagnostically useful.

Epidemiology. Healthy persons have a high degree of resistance to this fungus, found ubiquitously in the external environment.

Invasive pulmonary aspergillosis has an extremely poor prognosis. Amphotericin B is required, but the underlying disease often itself has a poor prognosis as well. Any immunosuppressive therapy in use should be reduced or discontinued in the face of superimposed mycosis.

Bronchial colonization without invasion is treated

supportively, particularly with measures designed to improve pulmonary function and drainage.

Mucormycosis

Other mycoses of this type include infections caused by members of the *Zygomycotina* subdivision of fungi, formerly the Phycomycetes. These infections are variously called mucormycosis, phycomycosis, and zygomycosis. The latter term refers to the characteristic formation of a zygospore in the sexual reproduction of fungi of this group (see Chap. 2, p. 41).

Causative Organisms. Species of *Mucor, Rhizopus,* and *Absidia.*

The Clinical Disease. Mucormycosis is characterized by massive invasion of blood vessels (with formation of emboli), pulmonary lesions, infection of orbital and nasal sinus tissues, gastrointestinal ulcerations, and terminal central nervous system involvement. Disseminated infection is usually rapidly fatal, even before the laboratory diagnosis can be made.

Laboratory Diagnosis

The fungi can be recovered readily from appropriate clinical specimens, growing rapidly as coarse, wooly molds at 37° C. However, the mere isolation of *Mucor, Rhizopus,* or *Absidia* species from sputum does not constitute a diagnosis of mucormycosis, since patients with bronchiectasis may cough up fungous spores, especially after exposure to a dusty environment. If pulmonary invasion has occurred, the sputum will be repeatedly positive for the fungus species involved, but cultures will be negative after a few days if the colonization was transitory or noninvasive. Curetted tissue for direct microscopic examination is most important for confirming the diagnosis, for hyphal segments of the fungus can then be visualized, actively growing among the tissue cells.

In making the diagnosis it is important for the laboratory to report these organisms even though they may often be mere contaminants. The physician may then weigh the significance of the laboratory report with the clinical findings in recognizing an invasive mucormycosis.

Epidemiology. These fungi are common environmental saprophytes to which healthy individuals

have a high degree of resistance. There are a few members of the group associated with primary mycoses, seen largely in tropical areas, but most cases are opportunistic in patients with predisposing illnesses, such as diabetes, malignancy, or immunodeficiency. Therapeutic control of the underlying disease constitutes the best control, coupled with administration of amphotericin B.

Bacterial Diseases Resembling Mycoses

The actinomycetes are filamentous organisms with fungus-like characteristics such as branching hyphae. Their principal morphologic features place them among the true bacteria, however: small size, cell wall composition, primitive organization of nuclear material, fragmentation into bacillary or coccal elements, response to antibiotics, resistance to antifungal agents. The diseases they cause resemble the true mycoses in many respects and they are discussed here for this reason.

The classification of the order Actinomycetales is shown in Table 4-1 (pp. 95–96). This order contains four important families, including the mycobacteria, discussed in Chapter 13 as agents of tuberculosis or similar diseases. The streptomycetes constitute a second major group. These are soil organisms from which many useful antibiotics have been derived. The other two families are the Actinomycetaceae and Nocardiaceae, each containing a genus of pathogenic importance: *Actinomyces* and *Nocardia,* respectively. The latter are soil organisms associated with pulmonary disease in man (nocardiosis) and also with some types of mycetoma, described previously (p. 400). Species of *Actinomyces* have an endogenous source among the commensalistic flora of the human mouth. (A fifth family, the Thermomonosporaceae, also contains a pathogenic genus, *Actinomadura,* associated with mycetoma (p. 400). This organism had been classified, until recently, as a species of *Nocardia.*)

Nocardiosis

Causative Organism. *Nocardia asteroides, Nocardia brasiliensis.*

The Clinical Disease. Systemic nocardiosis is a chronic pulmonary disease simulating tuberculosis clinically and bacteriologically. The organisms are variably acid-fast filamentous bacilli. Unlike *M. tuberculosis,* they are matted, branching, and display many coccoid forms as well. Dissemination of the organism from pulmonary lesions results in scattered abscesses in subcutaneous tissues, the peritoneum, and the brain. The disease has a high mortality rate unless diagnosed before widespread metastases have occurred.

Incubation Period. Not known.

Laboratory Diagnosis

Specimens for smear and culture include sputum, spinal fluid, exudates, biopsies. Examination of clinical specimens may reveal pigmented flecks or *granules.*

Microscopically, granules appear as tangled masses of filamentous, branching bacillary forms. The organisms are Gram positive and acid fast (variable).

In culture the organism grows slowly on ordinary media at room temperature or 37° C. It is *aerobic.* The colony is often waxy and folded. Orange, yellow, and red pigments are produced by different strains. (See also the laboratory diagnosis of actinomycosis, p. 410, and Fig. 16–11.)

Epidemiology. Contaminated soil or dust is the source of infection. Airborne organisms are presumably the source of pulmonary disease; local tissue entry through minor wounds of skin is the origin of mycetoma. This disease is sporadic in all parts of the world.

Nocardiosis may be a primary disease or an opportunistic one in the host compromised by malignancy or immunodeficiency. The prognosis depends on early diagnosis, before metastasis to the brain can occur. Surgical incision of abscesses with drainage of pus is important whenever possible. Treatment with sulfonamides for prolonged periods is helpful if instituted early. In disseminated cases, sulfonamides are combined with an antibiotic, such as ampicillin, erythromycin, or minocycline.

Actinomycosis

Causative Organism. *Actinomyces israelii* (human oral mucosa).

The Clinical Disease. Actinomycosis is a chronic granulomatous and suppurative disease of endogenous origin. The localization of primary infection depends on access of the organisms to deep tissues from their original sites on normal upper respiratory

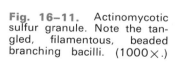

Fig. 16–11. Actinomycotic sulfur granule. Note the tangled, filamentous, beaded branching bacilli. (1000×.)

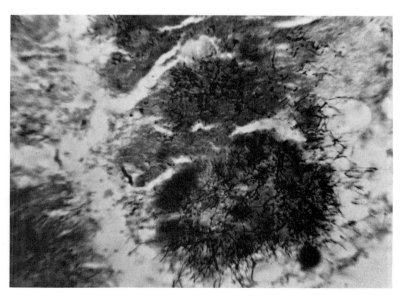

Table 16-2. Summary: Mycotic Infections

Clinical Disease	Causative Organisms	Other Possible Entry Routes	Incubation Period	Communicable Period
Primary (Pulmonary) Systemic Mycoses				
Histoplasmosis	*Histoplasma capsulatum*	Parenteral	5–18 days	Not usually communicable from man to man
Coccidioidomycosis	*Coccidioides immitis*	Parenteral	10 days to 3 weeks	Same as above
Blastomycosis (North American)	*Blastomyces dermatitidis*	Parenteral	Probably a few weeks	Same as above
Cryptococcosis	*Cryptococcus neoformans*	Parenteral	Not defined	Same as above
Paracoccidioido-mycosis (South American blasto-mycosis)	*Paracoccidioides brasiliensis*	Intestinal Parenteral?	A few weeks or undefined	Same as above
Subcutaneous Mycoses				
Sporotrichosis	*Sporothrix schenckii*	Parenteral	3 weeks to 3 months	Not usually communicable from man to man

Specimens Required	Laboratory Diagnosis	Immunization	Treatment	Nursing Management
Sputum, pus, bone marrow, spinal fluid	Wet mounts Cultures	None	Amphotericin B	Isolation not necessary; concurrent disinfection of contaminated articles prevents conversion of fungus to saprophytic form
Biopsies of liver, spleen, lung, lymph nodes	Fungal stains for tissues Cultures			
Skin test	Positive for histoplasmin			
Blood serum	Serologic tests			
Sputum, pus from lesions, gastric washings; spinal fluid	Wet mounts Cultures	None	Amphotericin B Miconazole	Same as above
Skin test	Positive for coccidioidin			
Blood serum	Serologic tests			
Sputum, pus and exudates	Wet mounts Cultures	None	Amphotericin B Hydroxystilbamidine	Same as above
Biopsies of skin granulomas	Fungal stains for tissues Cultures			
Blood serum	Serologic tests			
Sputum, spinal fluid, exudates from lesions	India ink mounts Cultures	None	Flucytosine Amphotericin B	Same as above
Blood serum} Spinal fluid }	Serologic tests			
Pus from lesions, biopsies, sputum	Wet mounts Cultures	None	Sulfonamides Amphotericin B	Same as above
Blood serum	Serologic tests			
Pus or biopsy of subcutaneous lesions	Wet mounts Fungal stains for tissues Cultures	None	Potassium iodide Amphotericin B	Isolation not necessary; concurrent disinfection of discharges and contaminated articles prevents conversion of fungus to saprophytic form

Table 16–2. Summary: Mycotic Infections (*Cont.*)

Clinical Disease	Causative Organisms	Other Possible Entry Routes	Incubation Period	Communicable Period
Chromomycosis	Dark-pigmented fungi: *Cladosporium, Fonsecaea, Phialophora*	Parenteral	Unknown	Same as above
Mycetoma	*Madurella Petriellidium Phialophora Acremonium* Bacteria: *Nocardia Actinomadura*	Parenteral	First lesions may take months to develop	Same as above
Opportunistic Mycoses				
Candidiasis	*Candida albicans* and other species	Parenteral Endogenous Skin or mucosa	Infants (thrush), 3–5 days	During time of lesions
Aspergillosis	*Aspergillus fumigatus, A. flavus,* et al.		Few days to weeks	Not communicable
Mucormycosis	*Mucor Rhizopus Absidia*		1–2 days	Not communicable
Bacterial Diseases Resembling Mycoses				
Mycetoma (see Subcutaneous Mycoses)				
Nocardiosis	*Nocardia asteroides Nocardia brasiliensis*	Parenteral	Unknown	Not usually communicable
Actinomycosis	*Actinomyces israelii*	Endogenous	May require days or months	Same as above

Specimens Required	Laboratory Diagnosis	Immunization	Treatment	Nursing Management
Crusts of lesions Biopsy material	Same as above	None	Amphotericin B Flucytosine	Same as above
Draining pus or biopsy material	Same as above	None	Amphotericin B for fungi, Sulfonamides or antibiotics for bacterial infection	Same as above
Scrapings or pus from surface lesions, sputum, stools, spinal fluid, as indicated Blood serum	Wet mounts Smears Cultures Serologic tests	None	Nystatin Imidazoles Amphotericin B Flucytosine	Isolation of infants with thrush; concurrent disinfection and terminal cleaning techniques must be carefully maintained; find source of infection in nursery
Sputum, bronchial aspirates, pus, blood Allergy: skin tests Blood serum	Wet mounts Cultures Serologic tests	None	Amphotericin B	Isolation not necessary; concurrent disinfection
Curretted tissue; sputum, pus from areas involved	Wet mounts Cultures	None	Amphotericin B	Same as above
Sputum, spinal fluid exudates, biopsies	Smear and culture	None	Sulfonamides	Isolation not necessary; concurrent disinfection
Sputum, pus from draining sinus tracts, biopsies as indicated	Smear and culture	None	Penicillin Tetracycline	Same as above

membranes. Tooth extraction or other injury to the mouth or jaw may lead to cervicofacial infections. Aspiration of the organism into the lungs may result in primary pulmonary disease, while gastrointestinal entry may permit primary infection of the abdominal wall. Abdominal actinomycosis can also extend to secondary pulmonary involvement. The lesions are granulomas that frequently break down to suppurative abscesses. Multiple sinus tracts are formed, which drain off pus to adjacent tissues or to the surface. The drainage contains so-called "sulfur granules," which are actually tangled masses of the infectious organism.

Incubation Period. Not easily defined; may require days or months after trauma permits tissue penetration.

Laboratory Diagnosis

Specimens. Sputum, pus from draining sinus tracts, or biopsy as indicated by clinical symptoms.

Macroscopic examination of pus should reveal tiny hard "sulfur granules," about the size of a pinhead. Their presence may give the drainage a gritty quality. On *microscopic* examination the granules are seen to be composed of matted, tangled bacillary filaments (Fig. 16–11). These organisms are Gram positive and nonacid fast.

In culture, Actinomyces species are *anaerobic* to microaerophilic. Culture techniques must include aerobic incubation of suitable media to rule out *Nocardia. Actinomyces* species require enriched media incubated at 37° C. The filamentous bacilli display true branching.

Epidemiology. These organisms have a natural reservoir in man and also in cattle. The organisms are presumably transmissible, but *disease* is not communicable. This type of disease is not frequent in man, but occurs sporadically in all parts of the world, the frequency being highest in adult males.

As in nocardiosis, surgical incision and drainage of abscesses are important to treatment. Actinomycosis can be successfully treated with penicillin, but the drug must be given for prolonged periods (six months or longer). Tetracycline is a useful alternative for patients who are sensitive to penicillin.

Questions

1. How are the mycotic diseases of man classified clinically?
2. Why has the incidence of opportunistic fungal infections increased?
3. Why is the diagnosis of systemic mycoses sometimes difficult?
4. Briefly describe the nursing care in active mycotic infections.
5. Describe oral and vaginal candidiasis.
6. What factors commonly predispose to *Candida* infection?
7. When is patient isolation necessary in *Candida* infection?

Additional Readings

Busey, John F.: Modern Concepts in the Diagnosis and Management of the Pulmonary Mycoses. *Clinical Notes on Respiratory Diseases,* **14**:3, Spring 1976. Published by American Thoracic Society, Medical Section of American Lung Association.

Emmons, Chester, W.; Binford, C. H.; Utz, J. P.; and Kwon-Chung, K. J.: *Medical Mycology,* 3rd ed. Lea & Febiger, Philadelphia, 1977.

Gauder, John P.: Cryptococcal Cellulitis. *J. Am. Med. Assoc.,* **237**:672, Feb. 14, 1977.

Hazen, Elizabeth, L.; Gordon, M. A.; and Reed, F. C.: *Laboratory Identification of Pathogenic Fungi Simplified,* 3rd ed. Charles C Thomas, Publisher, Springfield, Ill., 1970.

Rippon, John W.: *Medical Mycology: The Pathogenic Fungi and the Pathogenic Actinomycetes.* W. B. Saunders, Co., Philadelphia, 1974.

Section V

Infectious Diseases Acquired Through Ingestion

17

Epidemiology of Infections Acquired Through the Alimentary Portal

General Principles

The gastrointestinal tract is the site of entry of many varieties of microorganisms, carried in with food and water, by hands, or by objects placed in the mouth. Many organisms do not survive their trip through the alimentary canal, being killed by the stomach's acidity or crowded out by microbial competitors of the commensalistic intestinal flora. Many that do survive are unable to implant, and these are simply eliminated. The mechanisms associated with the alimentary portal of entry are discussed in some detail in Chapter 6.

Three types of organisms primarily responsible for the orally acquired diseases described in this section are *bacteria, viruses,* and *animal parasites.* Most of the *bacterial* diseases are caused by members of the group known as "enteric bacteria." They are Gram-negative, aerobic, motile bacilli similar in morphology and in many biochemical respects to the large group of coliform bacilli that normally inhabit the human or animal intestinal tract. These diseases are described in Chapter 18 together with a few caused by bacteria of other families. The *viral* diseases of enteric origin are induced by agents classified as "enteroviruses," the most important being poliomyelitis virus. The agent of infectious hepatitis also appears to have many of the epidemiologic features of an enterovirus. These diseases are described in Chapter 19. The *animal parasites* that gain

oral entry to the body represent every major class of the pathogenic species: *Protozoa, Nematoda* (roundworms), *Cestoda* (tapeworms), and *Trematoda* (the liver, intestinal, and lung flukes). The diseases associated with ingestion of infective parasitic forms are reviewed in Chapter 20.

The Intestinal Tract As a Portal of Exit. Important aspects of the epidemiology of these diseases relate to the gastrointestinal tract as a *portal of exit* as well as an entry site, and to infective human feces as a transmitting agent for pathogenic microorganisms. Patient management must be based on both considerations, as are preventive measures. Most diseases acquired by the alimentary route are transmitted from that source as well. Exceptions to this rule are found mainly among parasitic diseases induced by parasites lodged in parenteral tissues from which they cannot escape (trichinosis, echinococcosis, cysticercosis), and a few systemic bacterial infections in which ingested organisms invade the tissues from the bowel but do not remain implanted in the latter site (brucellosis, bovine tuberculosis, tularemia).

Transmission of infection from the intestinal tract may be accomplished in several ways: (1) by direct simple fecal-oral transfer from one person to another, soiled hands being the agent; (2) by indirect contacts with fecally contaminated objects placed in the mouth; (3) indirectly through contamination of food or water; (4) indirectly from fecally polluted soil; and (5) by a circuitous route through animal or plant hosts that not only infect man but may be infected by human fecal sources of pathogenic organisms. The communicable bacterial diseases, as well as the viral and protozoan infections with which we are concerned here, are transmissible from person to person by one of the first three methods. This is true also of pinworm infection, but other animal parasites are not directly communicable. The eggs of two nematode species (*Trichuris* and *Ascaris*) require a period of maturation in soil before they become infective, but this indirect route of transfer is much less involved than that required by other helminths that undergo cyclic development in different host species.

Direct fecal-oral transmission is always a matter of poor or inadequate personal hygiene. Improper toilet habits and failures in hand washing after defecation or before eating are chiefly at fault. Diseases spread in this way are often seen in outbreaks among familial or close institutional contacts. In families, young children are the most frequent disseminators of microorganisms, and this is very difficult to control even in the best regulated households with high sanitary and hygienic standards. The situation is also difficult in institutions that care for children or for persons who are mentally incompetent. Overcrowding contributes to disease transmission by contact, as does negligent staff control of sanitary facilities. The threat of epidemics through contact transmissions is limited to the number of susceptible persons who can be exposed, but it is also influenced by the numbers of persons whose infection is unrecognized and the extent to which they can spread disease before they are detected. A number of enteric diseases are transmissible before symptoms become apparent, a notable example being poliomyelitis. When this situation is coupled with generally unsanitary standards of living within a community, poor socioeconomic conditions, the malnutrition of poverty, and a climate that encourages spread, infection fans out rapidly through the susceptibles, and those who develop disease provide the visible framework of the epidemic. The persistence of inapparent, asymptomatic carriers within a group constitutes another silent source of infection for susceptible contacts. Infant members of any population are always the most vulnerable, even to organisms that

live as harmless commensals in adults. The modern hospital nursery can provide a serious threat of epidemic infection, spread rapidly by contacts that for adults would be casual or insignificant. The best control of infections derived from the intestinal tract, and easily spread by direct transmission, begins with sanitary toilet practices scrupulously followed by adults and taught to children from the earliest possible time. Thorough hand washing is essential to control of all directly transmissible diseases, but particularly those of enteric origin.

Indirect contact with fecally contaminated objects usually transfers infection to the hands, and from there to new objects, to food, or directly to the mouth. Personal hygiene prevents transfers which begin with soiled hands and end by soiling the hands of others. Mechanical vectors, especially flies, play a role in this, bringing infection from their breeding places in fecal deposits. Public and personal measures are both required to control the potential spread of disease by flies and other insects.

Indirect transmission through contaminated food or water provides a serious public health hazard when large groups of people are exposed to the same source of infection. Food is usually contaminated by persons with inapparent or disregarded infections. People with colds, sore throats, mild diarrhea, pimples, boils, or skin "rashes" may pass infection on to others for whom they prepare or serve food.

Control of such sources of food contamination is not easy. It can be achieved only when general education and community understanding support the efforts of public health authorities to find and remove sources of infection of common food supplies (Chap. 11). Food handlers should develop an awareness of their possible role in transmitting disease and a corresponding sense of responsibility.

Protection of water supplies is also achieved only by continuous supervision and guidance of public habits with the support of an educated community. Much is taken for granted, today, in countries that have achieved a good measure of public sanitary controls. One has only to look at areas of the world where such measures have not been well established (or glance back a few decades at the history of infectious diseases) to realize that safety is a matter of continual effective control of sewage wastes and protection of water from contamination. Water is a ubiquitous medium, necessary to and utilized by all forms of life. Without control of its pollution, it can be a source of explosive, widely disseminated epidemics.

Indirect disease transmission from contaminated soil is limited to those pathogenic microorganisms that can establish a reservoir in this medium. These include the fungi that induce systemic mycoses (Chap. 16), larval forms of animal parasites that mature in soil and enter the human body through the skin (see hookworm disease, Chap. 24), and those intestinal roundworms of man whose ova require development under conditions offered in soil. The latter (*Trichuris* and *Ascaris*) are appropriate for discussion in this section (Chap. 20) because their epidemiology involves their transmission to soil through infective feces and back to human beings through the oral route of entry.

Tapeworm and trematode diseases of man are for the most part not directly transmissible to other persons, but man transmits many of them indirectly, through environmental reservoirs, to animal hosts that continue the parasite's cycle. Human excreta containing parasite ova, when permitted to contaminate soil or water, are a source of infection for animals or aquatic species. Man acquires these parasites by ingesting them in an infective stage developed in the animal host.

Control of diseases perpetuated by human contamination of soil or water begins with sanitary disposal of feces, either through sewage systems, well-constructed septic tanks, or chemically disinfected privies. In areas of the world where sanitary standards are not well controlled, it is useful to apply additional measures designed to interrupt the chain of transfer when this involves animal hosts. Domestic animals that serve as intermediate or alternate hosts can be protected from infection from human sources. When animals such as rats or snails are involved in the chain they can be eliminated, or at least held in control. Continuing educational campaigns are required to inform the public of the nature and sources of these diseases, and of the necessity for good sanitation, care in choosing meats and other foods, and adequate cooking methods.

Patient Care

The care of patients with infections acquired through and transmitted from the gastrointestinal route primarily requires control of spread through infectious feces. Some or all of the precautions outlined below may be indicated.

Isolation of the Patient

Complete isolation of patients is generally not required except for children, among whom direct fecal-oral transmission is difficult to prevent. Pediatric patients admitted with the same infection may be assigned to the same room; otherwise, infected children should be segregated in private rooms. Infected nursery babies must also be cared for in separate rooms, an isolette being inadequate for isolation.

Adults with enteric infection do not require private room isolation, provided all necessary precautions are taken in their care (see below). Exceptions include adult patients with behavioral problems or those who have fecal incontinence.

The recommendations of the Center for Disease Control concerning the care of patients with enteric diseases are shown in Table 17-1.

Enteric Precautions

1. Proper hand washing, gown and glove techniques, and excretion precautions are necessary to control cross-infection.

2. Patients must be instructed and supervised in hand-washing techniques, particularly with reference to use of a shared toilet. Sharing can be permitted only if each patient understands that hands must be washed before and after using toilet facilities.

3. Bedpan collections may be emptied directly into toilets or hoppers if the local sewage system is adequate. If not, bedpan contents must first be chemically disinfected (see Appendix III).

4. Bedpans, enema equipment, rectal tubes, and similar items should be individualized for infectious patients. Bedpans should be sterilized between each use, by either steam or boiling water.

Table 17–1. Enteric Precautions Recommended by the CDC

Card to Be Posted on Door of Patient's Room

Enteric Precautions

Visitors — Report to Nurses' Station Before Entering Room
1. Private room — *necessary for children only*.
2. Gowns — must be worn by all persons having direct contact with patient.
3. Masks — not necessary.
4. Hands — must be washed on entering and leaving room.
5. Gloves — must be worn by all persons having direct contact with patient or with articles contaminated with fecal material.
6. Articles — special precautions necessary for articles contaminated with urine and feces. Articles must be disinfected or discarded.

Diseases Requiring Enteric Precautions*

Cholera (18)
Diarrhea, acute illness with suspected infectious etiology
Enterocolitis, staphylococcal (18)
Gastroenteritis caused by
   Enteropathogenic or enterotoxic *Escherichia coli* (18)
   Salmonella species (18)
   Shigella species (18)
   Yersinia enterocolitica (18)
Hepatitis, viral, type A, B, or unspecified (19)
Typhoid fever (*Salmonella typhi*) (18)

*The number in parentheses indicates the chapter in this textbook where the disease is described.

5. Autoclave sterilization is indicated for durable equipment. Rectal thermometers and other heat-sensitive items must either be discarded and incinerated, or carefully disinfected.

6. Dressings and tissues should be placed in a leakproof plastic or paper bag. This bag should be closed securely and placed in a wastebasket in the patient's room. The wastebasket should be lined with an impervious bag, sealed on removal from the room, and incinerated without being opened.

7. The patient's used linens should be placed in a laundry bag tagged for precautionary use. Bags that are soluble in hot water are preferred because they can be placed unopened in hospital washing machines. Insoluble bags must be opened in the laundry room and their contents dumped into washing machines without sorting. Such bags must then be either washed or discarded.

8. Mattresses and pillows should be covered with an impervious plastic that can be cleaned with germicidal detergent. For terminal disinfection this covering should be removed with the linens and laundered.

9. Dishes, utensils, and trays should be placed, after each meal, in an impervious plastic bag tagged for precautionary handling. Leftover food should be wrapped and discarded; liquids can be poured into a drain or toilet. Everyone who handles these dishes should be alert to the need for careful hand washing. Disposable dishes and utensils may be useful in many situations.

In the following chapters, the discussion of each disease includes the precautions required in patient care.

Questions

1. What factors must be considered in the nursing management of active intestinal infections?
2. How can infections be transmitted from the intestinal tract?
3. What enteric precautions are recommended by the CDC?

Additional Readings

*Center for Disease Control: *Botulism in the United States.* 1899–1973 Handbook for Epidemiologists, Clinicians and Laboratory Workers. Issued June 1974.

Editorial: Facing Up to the Drinking Water Problem. *Am. J. Public Health,* **66**:635, July 1976.

* *Isolation Techniques for Use in Hospitals,* 2nd ed. U.S. Department of Health, Education, and Welfare, Public Health Service, Washington, D.C., 1975. (DHEW Pub. No. 76-8314).

Keusch, Gerald: Bacterial Diarrheas. *Am. J. Nurs.,* **73**:1028, June 1973.

Worldwide Spread of Infections with *Yersinia enterocolitica. WHO Chronicle,* **30**:494, 1976.

Zaki, M. H.; Miller, G. S.; McLaughlin, M. C.; and Weinberg, S. B.: A Progressive Approach to the Problem of Foodborne Infections. *Am. J. Public Health,* **67**:44, Jan. 1977.

*Available in paperback.

Bacterial Diseases Acquired Through the Alimentary Route

18

The bacterial diseases acquired through the alimentary route fall into three major categories: (1) active infections of the gastrointestinal tract caused by the ingestion of viable bacteria, usually in contaminated food or water, with resulting acute gastroenteritis; (2) intoxications caused by the ingestion of preformed toxins produced by bacteria growing in contaminated foods; and (3) systemic infections caused by the dissemination of ingested microorganisms from intestinal sites to internal tissues and organs.

Infectious gastroenteritis may be caused by a number of bacterial agents. The basic clinical syndrome, common to most of these infections, is abdominal distress and diarrhea, often with nausea and vomiting. Clinical distinctions, except in the case of shigellosis and cholera, are usually difficult or impossible, and specific diagnosis depends on the laboratory identification of the organism in the patient's feces or vomitus, or in suspect food or water. For a number of reasons a specific diagnosis may not be made at all in many instances: these infections are usually self-limited and characterized by rapid, complete recovery; many cases of gastroenteritis are probably caused by viruses, whose isolation and identification are time consuming and expensive; a variety of noxious chemical agents, other than bacterial toxins, may induce similar symptoms (such agents include medications, allergens, poisonous plant and animal substances). However, when the cause is infectious and the disease takes an epidemic form, because it is being transmitted directly from infected individuals or indirectly from contaminated food or water sources, every effort is made to establish the definitive diagnosis so that the source of the agent may be found and controlled.

The term *bacterial food poisoning* traditionally refers to those diseases of bacterial origin that develop abruptly after ingestion of contaminated food. They often occur as outbreaks among many persons who shared the food, their symptoms developing simultaneously or within close time periods. Most frequently the illness is an acute gastroenteritis caused by the ingested microorganisms themselves or by a preformed bacterial toxin, as in the case of staphylococcal food poisoning. Botulism is a notable exception, this being a systemic disease caused by a powerful, preformed neurotoxin that affects the central nervous system.

The important bacterial systemic diseases that are

commonly acquired through the alimentary route include typhoid fever and related *Salmonella* infections, and brucellosis. The infectious agents involved find their way from local sites of multiplication in the intestinal tract to visceral organs, usually by lymphatic and hematogenous routes.

Public control of all the diseases to be discussed in this chapter depends on the protection of food and water supplies, and on the sanitary disposal of sewage. The same principles apply in hospitals. Food must be protected from environmental contamination and from handlers who may be carriers of infection. Enteric precautions must be followed carefully in the management of patients, in most cases.

Acute Gastrointestinal Infections

The most prominent symptom of most of these infections is diarrhea, often accompanied by crampy abdominal pain, nausea, and vomiting. Mild fever, chills, and headache may be present, but these symptoms of infection are not invariable. Many bacterial species have been incriminated in acute gastroenteritis, especially as it relates to food poisoning. The most important of these are discussed here.

Salmonellosis

The *Salmonella* genus is remarkable because of its size, wide distribution in nature, and nearly uniform pathogenicity of its many species for man, animals, and birds. Some species are more frequently associated with one type of host than with another, but all of them have pathogenicity for man. *Salmonella typhi,* the agent of typhoid fever, is the only one of these species that appears to be restricted to a human reservoir.

Morphologically the salmonellae are indistinguishable from those members of the normal intestinal flora collectively called "Gram-negative enteric bacilli." The latter include *Escherichia coli* and species of *Proteus, Providencia, Klebsiella, Enterobacter,* and *Citrobacter*. All of these, and the salmonellae, are Gram-negative, facultatively anaerobic, non-spore-forming rods, many of them motile. They are distinguished from each other by differences in biochemical properties (enzymatic pathways in the utilization of carbohydrates such as lactose, dextrose, sucrose, and others, for example), and by recognition of their specific antigenic components. The biochemical characteristics of the salmonellae are fairly uniform, but their antigenic classification differentiates a large number of species (more than 1800 different serotypes). In recent years, the taxonomy of the group has been simplified by the recognition of three principal species: *Salmonella typhi, Salmonella choleraesuis,* and *Salmonella enteritidis*. There is only one serotype in each of the first two of these species, while all the rest are designated as varieties of *S. enteritidis*. The system for naming the latter employs the postscript *var,* meaning variety, following the principal species designation, for example, *S. enteritidis var paratyphi A, S. enteritidis var typhimurium,* and so on.

Salmonella infections are most often acquired by ingestion of food or water contaminated with these bacilli derived from a fecal source. Soiled hands frequently play a role in implanting these organisms in food, or directly into the mouth. Meat and other products of infected animals, eggs from infected birds, and shellfish are also often involved in the far-flung distribution of salmonellosis. Inapparent human infection is a common source of sporadic or epidemic disease until it is recognized.

Salmonella infections may be one of three major types, and, sometimes, a combination of these: *enteric fever, bacteremia,* or *gastroenteritis*. The first two are systemic infections resulting from distribution of organisms through the blood following intestinal tract entry, while the third (and most common) is intestinally localized infection. All species of *Salmonella* have the potential to produce any of these clinical situations, but certain strains are characteristically associated with either systemic or intestinal infection. Typhoid fever is the prototype of severe enteric fever and is classically associated with *Salmonella typhi*. Paratyphoid fever is a milder form of enteric fever, most frequently caused by *S. enteritidis var paratyphi A* or *B*. Bacteremia is a more fulminant systemic invasion, with a wide dissemination of focal abscesses. *S. choleraesuis* is characteristically associated with this type of disease. The enteric fevers and bacteremia are discussed in the last section of this chapter, under Infectious Systemic Diseases.

The largest number of *Salmonella* strains (varieties of *S. enteritidis*) are agents of gastroenteritis. While they are an enormously diverse group antigenically, their clinical and epidemiologic potential is very similar.

Gastroenteritis (Salmonella Food Poisoning). The symptoms of this infection begin from 8 to 48 hours following ingestion of the organisms, with sudden onset of abdominal pain, nausea, vomiting, and diarrhea, often accompanied by fever and chills. In severe cases, the patient may be prostrate. The signs are those of severe irritation of the intestinal mucous membranes, perhaps from release of endotoxic components of the cells when large numbers are ingested, multiply rapidly, and die. The organisms do not enter the bloodstream ordinarily (although occasionally they do so, some time after the enteritis has subsided), but can be recovered from stools and vomitus.

Occasionally more than one strain of *Salmonella* may be found in a patient's stool, and sometimes a case that begins as a gastroenteritis may develop into enteric fever.

Laboratory Diagnosis

I. Identification of the Organism. The microscopic characteristics of the salmonellae were described above. These organisms grow with relative ease in laboratory culture, but their isolation from the mixture of bacteria present in stools presents some difficulty. This problem is overcome by selective agar media that suppress growth of coliform and other enteric organisms but permit *Salmonella* species to grow. The incorporation of lactose in these media, together with an indicator that responds with a color change when acid is produced, serves to differentiate the colonies that grow. *Salmonella* species, with rare exceptions, are indifferent to lactose and do not break it down into acid end products. Their colonies are therefore colorless in appearance on differential media. Coliform bacilli, on the other hand, characteristically ferment lactose. Their colonies take on the indicator color as the pH changes with breakdown of the sugar to acid products. This color differentiation makes it possible to select suspicious, colorless colonies and inoculate them as pure cultures into other media necessary for their identification. Certain other organisms normally found in stool specimens, such as *Pro-*

teus and *Pseudomonas* species, are also lactose-negative and must be differentiated from *Salmonella*, by recognition of a combination of biochemical characteristics (fermentative or oxidative enzymes for other carbohydrates, utilization of certain amino acids, and other metabolic properties).

Serologic identification of isolated organisms biochemically typical of *Salmonella* is made with specific antiserums, by simple slide agglutination techniques. The somatic antigens of *Salmonella* fall into several groups. Antiserums containing group-specific antibodies identify the isolated strain as a member of a *Salmonella* group. Species identification of the strain then depends on recognition of individual flagellar antigens, using antisera that contain specific flagellar antibodies. (See Chap. 7.)

II. Identification of Patient's Serum Antibodies. The *Widal test* is a method for identifying agglutinating antibodies that arise in the patient's serum during *systemic Salmonella* infection. It is useful in typhoid and paratyphoid fevers, and in *Salmonella* bacteremia, when the presence of the organisms in parenteral tissues stimulates antibody formation. In gastroenteritis without bloodstream invasion, however, there is usually no antibody response, and the Widal test is not helpful.

> *III. Specimens for Diagnosis.*
>
> *In gastroenteritis,* repeated stool specimens should be submitted for culture during the acute stage. Culture of vomitus is sometimes useful, if stool results are negative. When positive stool results are obtained, cultures should be repeated a week or so after the patient has returned to normal, to check on the possibility of lingering infection and establishment of the intestinal carrier state. Other cultures and serologic testing are not indicated unless signs of enteric fever or bacteremia develop. (See page 436.)

Epidemiology

I. Communicability of Infection

A. Reservoir, Sources, and Transmission Routes. Man is the only reservoir for *Salmonella typhi* and the *primary* reservoir for species that cause paratyphoid fever. Other strains have a worldwide distribution in domestic and wild animals, some reptiles such as turtles, and poultry. The source of infection is the feces of infected animals or persons, and the environmental reservoir is usually either (1) food derived from an infected animal (meats, eggs, milk and other dairy products), (2) food contaminated during storage by the feces of infected animals, especially rodents, (3) food contaminated by an infected person during its processing or preparation (meats, eggs, milk), (4) water contaminated by infectious sewage, (5) food obtained from contaminated water (shellfish), or (6) direct or indirect contact with persons having clinical or inapparent infection. When infection is derived from any of the first five of these sources, the result is likely to be an outbreak among persons sharing the contaminated food or water supply. The numbers of cases resulting from contact with an infected individual depend on the nature and closeness of contacts and on the activities of the infected person, with particular respect to food handling for family or public consumption.

B. INCUBATION PERIOD. Gastroenteritis usually begins in about eight to ten hours. It may be delayed up to 48 hours, depending on the numbers of organisms ingested and the degree of multiplication in the intestinal tract.

C. COMMUNICABLE PERIOD. Communicability continues throughout active infection, and for as long as salmonellae are shed in excreta. Adults generally do not become permanent intestinal carriers, but babies and young children may shed the organisms for months.

D. IMMUNITY. Gastrointestinal infection produces little or no immunity and attacks may be experienced after each exposure.

II. Control of Active Infection.

A. ISOLATION. Children must be isolated in private rooms when cared for in hospitals, but this is not necessary for cooperative adults. Patients sharing toilet facilities must be instructed in hand-washing techniques.

B. PRECAUTIONS. These include concurrent disinfection of all infectious excreta and of objects contaminated by them (bedpans, rectal thermometers, enema equipment, clothing). Disinfection techniques should be used frequently for toilet seats and other bathroom facilities, especially when these are shared. Individual bedpans should be provided for bedridden patients. Gowns and gloves should be worn by all persons in direct contact with the patient and by those who must handle contaminated articles. Hand washing must be practiced scrupulously by the patient and by personnel in attendance.

These precautions should be continued until the patient's stool cultures appear to be free of salmonellae. This assurance is usually based on three negative cultures of stools obtained on successive days during convalescence. If any culture is positive, supervision is continued for one month's time. Patients must be excluded from food handling while culture positive or until three negative stool cultures can be obtained.

C. TREATMENT. Supportive and symptomatic therapy is of greatest value in *Salmonella* gastroenteritis. Antimicrobial agents are not useful and may even prolong the period of fecal excretion of the organisms.

D. CONTROL OF CONTACTS. Food handlers are the particular targets of efforts to prevent spread of *Salmonella* infection. The close contacts of infected patients are examined in a search for unreported cases or inapparent carriers.

E. EPIDEMIC CONTROLS. Individual case reports in epidemics of *Salmonella* gastroenteritis guide the efforts of epidemiologists to find active cases and carriers, as well as food or water sources. Special situations may arise in hospitals, as in the case of a *Salmonella*-contaminated dye administered by gastric intubation to patients undergoing studies of abdominal tumors. Another hospital outbreak arose from the practice of serving raw eggs in milk shakes, or other form, to nutritionally debilitated patients. In this instance, eggs marketed by local poultry farmers were the source of infection.

These incidents point up the necessity for continuing awareness of the many living and environmental reservoirs of *Salmonella* infection, and for controls on obvious sources. When outbreaks occur, controls are tightened, with exclusion of suspected foods from the market; recommendations for pasteurization, boiling, or exclusion of milk; thorough cooking of eggs, poultry, and meats; and chlorination or boiling of water, depending on epidemiologic evidence.

III. Prevention

A. IMMUNIZATION. Bacterial vaccines do not confer protection against *Salmonella* gastroenteritis, but are of value in preventing typhoid fever (See pages 437–38.)

B. CONTROL OF RESERVOIRS, SOURCES, AND

TRANSMISSION ROUTES. Many factors involved in prevention of salmonellosis may be summarized as follows:

1. Public water supplies must be protected, purified, and chlorinated (see Chap. 11). Private water supplies must also be protected. When doubt exists concerning the safety of water for drinking or washing, it should be either boiled or disinfected with chlorine or iodine preparations. Disinfectants are available for individual use for this purpose.

2. Adequate disposal of sewage is essential.

3. Meats, eggs, and milk must be protected by sanitary processing. Infection should be eliminated at the animal source, insofar as possible. Stored foods should be safeguarded from possible contamination by rodent feces.

4. Shellfish should be protected by control of sewage disposal in waters from which they are harvested.

5. Milk should be pasteurized; eggs, meats, and shellfish should be cooked when questions concerning their purity arise.

6. Food handlers and methods for preparing foods in public places should be under continuing supervision by local health departments.

7. Flies should be controlled by adequate screening and judicious use of insecticides at their breeding places or entry routes.

8. Carriers must be supervised and restricted from food handling.

9. Food handlers and the public must be continuously educated as to the sources and transmission of *Salmonella* infections.

Shigellosis (Bacillary Dysentery)

The Clinical Disease. Shigellosis, or bacillary dysentery, is an acute intestinal disease characterized by diarrhea with abdominal cramps, vomiting, and fever. The liquid stools often contain blood and mucus and are passed with difficulty because of rectal spasm (tenesmus). The large bowel is inflamed. Damage to the mucous membrane leads to ulceration, bleeding, and formation of a "pseudomembranous" exudate on ulcerated areas. The infection is usually self-limited, but when it subsides the intestinal ulcers heal by granulation and are filled in with scar tissue. Uncomplicated recovery is

the usual rule, although the clinical severity varies with the causative strain, the age of the patient, and his or her general condition. Ordinarily the case fatality rate is less than 1 percent, but in epidemics in tropical countries where living conditions and sanitation are poor, the disease may be rapidly fatal to as many as 20 percent of cases.

The intestinal injury that results from *Shigella* infection is thought to be caused by the toxicity of somatic antigens (endotoxins), released from the organisms as they die and lyse. One strain, *Shigella dysenteriae,* also secretes an exotoxin that adds to the damage and leads to more serious disease, marked by a general intoxication. Unlike salmonellae, which may invade the bloodstream and cause enteric fever, shigellae do not enter parenteral tissues but remain in the bowel. The fever that often accompanies shigellosis is probably due to some absorption of toxic products and to dehydrating effects of the diarrhea.

Laboratory Diagnosis
I. Identification of the Organism

A. MICROSCOPIC CHARACTERISTICS. The shigellae closely resemble other enteric bacteria in appearance. They are Gram-negative, nonsporulating, nonmotile, facultatively anaerobic bacilli.

B. CULTURAL CHARACTERISTICS. Methods for isolation of *Shigella* species from stool specimens are very similar to those described for salmonellae. The chief problem is to distinguish them from nonpathogenic enteric bacilli that abound in feces. This can be done by using selective and differential agar media that contain inhibitors for the coliform group, and lactose, with an indicator. *Shigella* species are indifferent to lactose (one ferments it very slowly), and therefore, their colonies are colorless on these plates, in contrast to lactose-fermenting coliform bacilli, which take on the indicator color. Colorless colonies are subcultured for further identification. They are distinguished from *Salmonella* and other species by lack of motility, by the fact that those carbohydrates that they do utilize usually are fermented without gas production, and by other biochemical parameters.

C. SEROLOGIC IDENTIFICATION. Final identification of isolated *Shigella* species is accomplished by slide agglutination tests of the growth mixed with specific *Shigella* antisera. The genus contains four antigenic groups: Group A (*S. dysenteriae*), group B (*S. flexneri*), group C (*S. boydii*), and group D (*S. sonnei*). Some of these groups contain more than one strain, but those named are prototype species.

II. Identification of Patient's Serum Antibodies. Circulating antibodies (agglutinins) appear in the blood during shigellosis but are not correlated with recovery from infection or with protective resistance to second attacks. It is seldom practical to test for these antibodies as a diagnostic procedure because they are frequently found in normal persons who have had previous infection, and because bacteriologic diagnosis is generally more reliable.

III. Specimens for Diagnosis. Bowel contents only are suitable for culture of *Shigella* species. Specimens should be received by the laboratory in the freshest possible condition, because these organisms are undergoing autolysis in the bowel during the symptomatic stage of disease, and the survivors may be neither numerous nor metabolically active. They are in competition for nutrient with other bacteria present, and for this reason, also, fresh material should be cultured promptly.

Organisms are often concentrated in flecks of mucus and blood present in diarrheic stools, and these may profitably be selected for culture.

Rectal swabs have a clinical and nursing advantage and yield good bacteriologic results. If the patient is examined by sigmoidoscopy, cultures may be taken from the ulcerative lesions visualized through the instrument.

If serologic diagnosis is attempted, serum samples should be taken during acute and convalescent stages, separated by at least ten days, so that a rising titer of antibody may be recognized if it occurs.

Epidemiology

I. Communicability of Infection

A. RESERVOIR, SOURCES, AND TRANSMISSION ROUTES. The reservoir for shigellae is the human intestinal tract, and infected feces are always the source. Transmission may occur by the direct fecal-oral route, or indirectly through contaminated objects, hands, or mechanical vectors, but contamination of water or food enables the widest distribution.

B. INCUBATION PERIOD. Usually short, within one to four days, but not more than seven days.

C. COMMUNICABLE PERIOD. For as long as the organisms are present in feces. The carrier state does not develop so frequently in shigellosis as it does in salmonellosis. Most convalescents do not shed the bacilli for more than two or three weeks, but occasional persons remain chronically infected and may suffer relapses.

D. IMMUNITY. Specific antibodies develop during the course of disease but are not protective against new infections. People of all ages are susceptible, but shigellosis is often more serious in children and in debilitated adults. It has a worldwide distribution. Epidemics are more frequent and severe in tropical and other areas where nutrition is poor, living conditions are marked by poverty, and sanitation is inadequate. Outbreaks are often associated with institutional living, particularly in orphanages, convalescent homes for children, mental hospitals, and jails.

II. Control of Active Infection.

A. ISOLATION. Children must be isolated during the acute stage of illness. A private hospital room is not necessary for adults, provided precaution techniques are carefully carried out.

B. PRECAUTIONS. Concurrent disinfection of feces and contaminated objects (bedpans, toilet seats, bed linens, rectal thermometers, sigmoidoscopes, enema equipment, clothing) must be thorough. Feces may be disposed of directly into the sewage system if this is adequate; if not, fecal matter should first be disinfected chemically. Gowns should be worn by all who come in contact with the patient, and scrupulous attention should be given to hand washing, both by the patient and by personnel.

C. TREATMENT. Many *Shigella* strains have become resistant to a number of antimicrobial drugs that were once effective for treatment, e.g., the sulfonamides. For this reason, antibiotic susceptibility testing should be done to determine the most effective agent for individual isolated strains. Ampicillin, the tetracyclines, trimethoprim-sulfamethoxazole, or chloramphenicol may be useful for some strains. Effective antimicrobial drugs eliminate shigellae quickly from the feces, abating fever and diarrhea within 12 to 18 hours after treatment is initiated. Antidiarrheal drugs, on the other hand, may prolong fever, diarrhea, and excretion of shigellae, and these agents are contraindicated in shigellosis.

The most important supportive therapy is control of dehydration, with fluid and electrolyte replacement.

D. CONTROL OF CONTACTS. As in the case of salmonellosis, the control of food handlers is important in this disease. When the contacts of active cases include food handlers, the latter are excluded from this occupation until they are proved to be bacteriologically negative by fecal culture. A search is made among contacts for other cases of active shigellosis, atypical cases, and carriers.

E. EPIDEMIC CONTROLS. All cases must be reported so that the source of infection may be found in the population or in the environment. Investigations are centered on common water, food, and milk supplies.

III. Prevention

A. IMMUNIZATION. Effective vaccines are not available.

B. The prevention of shigellosis is based on control of its human reservoir, and sanitary control of its environmental sources, through chlorination of water, adequate sewage disposal, fly control, and protection of food, water, and milk from human or mechanical vectors.

Cholera

The Clinical Disease. Cholera is an acute diarrheal disease caused by an exotoxin (enterotoxin) produced by an organism called *Vibrio cholerae*. Following their ingestion in fecally contaminated water, the organisms colonize the small bowel, secreting enterotoxin that binds rapidly to the intestinal epithelial cells. Its effect is to activate an enzyme within mucosal cells, which in turn increases the production of an organic phosphate (cyclic adenosine monophosphate) by the epithelium. The rising concentration of this compound markedly alters the permeability of the mucosa (Fig. 18-1), resulting in an outpouring of fluid and salts. There is a profuse diarrhea, with "rice water" stools containing mucus and mucosal epithelial cells. Onset is sudden, and accompanied by nausea, vomiting, abdominal pain, and severe dehydration. Collapse, shock, and death follow if the patient is not promptly and continuously rehydrated until infection subsides. This disease may occur in explosive epidemic patterns in susceptible populations, with death rates as high as 75 percent. In endemic areas the fatality rate varies from 5 to 15 percent.

Cholera is strictly an intestinal infection. The organisms do not reach the bloodstream or establish systemic foci. Consequently, the disease is ordinarily self-limited. Recovery, when it occurs, is usually

Fig. 18–1. Scanning electron micrograph of adrenal cells. *A.* Untreated control cells. *B.* Cells treated with cholera toxin. Note that treated cells appear to be rounded up and thickened, with a marked change in the cell membrane surface that affects their permeability. (Reproduced from Donta, Sam: Toxigenic *Escherichia coli* and Diarrheal Disease. *Clin. Med.*, **85**:21–24, 1978.)

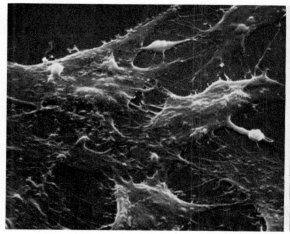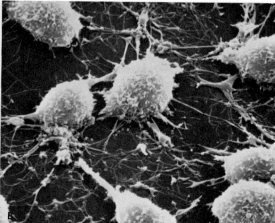

quite rapid, with no signs of residual damage. Convalescents may shed vibrios in feces for two or three weeks, but the carrier state does not appear to be permanent.

Laboratory Diagnosis

I. Identification of the Organism (*Vibrio cholerae*). *Microscopically,* vibrios are small Gram-negative, aerobic, nonsporulating bacilli, curved in the shape of a comma. They have a single polar flagellum, which gives them very rapid motility.

Culturally, vibrios grow on ordinary laboratory media but, in addition, a selective medium, thiosulfate-citrate-bile salt-sucrose agar (TCBS) is recommended for their isolation. Two of the individual metabolic features by which they are recognized are their ability to grow in media with a strongly alkaline pH (8.5 to 9.5) and their capacity to produce indole and reduce nitrate in peptone media. The indicator system used to detect the latter reaction produces a red color, and the method is therefore called the "cholera red test."

Serologically, vibrios can be identified by slide agglutination tests, using specific antiserum.

II. Identification of Patient's Serum Antibodies. Specific agglutinating antibodies against the vibrios appear in the patient's serum during cholera infection (the reasons for this are not clear, since cholera is apparently not a systemic infection). They do not provide long-term protection against reinfection, but their presence provides a basis for serologic diagnosis of current disease. A significant increase in antibody titer must be demonstrated in the convalescent stage as compared with the acute stage of disease. Antibodies directed against the enterotoxin may also be detected in serum.

III. Specimens for Diagnosis. Stool specimens or vomitus should be sent for bacteriologic culture.

An "acute" and a "convalescent" pair of serum samples may be submitted for serologic diagnosis. The time interval between collection of these samples should be at least seven, preferably ten, days.

Epidemiology

I. Communicability of Infection

A. RESERVOIR, SOURCES, AND TRANSMISSION ROUTES. Cholera is a human disease, transmitted by infected persons in feces or vomitus. Contaminated water supplies are the usual vehicle for epidemic spread. Food is not ordinarily involved in epidemics, but may be the source of sporadic cases in endemic areas where mild cases of cholera continue. In the latter situation, water is less important as a transmitting agent than direct contacts of the household variety. Hands, utensils, clothing, and flies may contaminate food or carry infection directly to the mouth.

B. INCUBATION PERIOD. Usually a matter of two to three days, but in rapidly spreading cholera, many cases begin within a few hours of exposure.

C. COMMUNICABLE PERIOD. Cholera is transmissible so long as vibrios are present in vomitus or feces. They usually persist in the intestinal tract for several weeks after recovery.

D. IMMUNITY. Active immunity results from an attack of disease or from artificial immunization with bacterial vaccine, but the latter is less effective than previous illness in preventing subsequent disease.

Both circulating immunoglobulins and local secretory IgA are produced by the body. These antibodies act on the bacteria themselves, but antibodies to the enterotoxin are also produced. Secretory IgA developed in the small bowel is thought to function by preventing attachment of the vibrios to intestinal epithelial cells. The role of antitoxic antibodies is not clear. If they act by intercepting enterotoxin before it binds to mucosal cells, they might be overcome, if present in limited amount, by large quantities of enterotoxin produced during active infection.

Human susceptibility to cholera is not consistent. Cholera has been endemic in India and other parts of Asia for the past century and a half and has caused several major pandemics. The latest one (1961–1977), however, has not yet reached the western hemisphere. In endemic areas, the ratio of persons with asymptomatic infections to those with clinical disease varies from 4:1 to 10:1, depending on the infecting strain. In these areas cholera is predominantly a disease of children, but when it spreads to previously uninvolved sections, the attack rates are at least as high in adults as in children.

II. Control of Active Infection

A. ISOLATION. Cholera patients should be cared for with strict enteric precautions. Isolation is not necessary, but effective hand washing and scrupulous cleanliness are basic to the management of cholera wards in hospitals.

B. PRECAUTIONS. These include concurrent disinfection of feces, vomitus, and contaminated equipment and bedding. Chemical disinfection of feces must be carried out when adequate sewage systems are not available. Personnel in contact with the patients must wear clean gowns and practice careful hand washing.

C. TREATMENT. Adequate rehydration of the patient with restoration of electrolyte balance is most important. Tetracycline antibiotics may dramatically reduce the duration and volume of diarrhea as well as speed the elimination of vibrios from the feces. Trimethoprim-sulfamethoxazole and chloramphenicol are useful, but somewhat less effective than tetracycline.

D. CONTROL OF CONTACTS. Exposed contacts are kept under surveillance for several days. If fecal cultures are positive for vibrios, long periods of observation may be required. A search is made for other cases of cholera, and for possible sources in common water supplies or among household contacts.

A prolonged gallbladder carrier state may develop in up to 3 percent of convalescents, particularly adults, but the role of such carriers in transmission is not clear. Surveillance of such individuals is maintained whenever feasible.

Vaccination of contacts is not warranted, since it cannot prevent current infection within the short incubation period. Passive immunization is not available.

E. EPIDEMIC CONTROLS. Case reporting is required by the International Sanitary Regulations of WHO. Measures described above are expanded to the entire community when epidemics threaten. When the general water supply cannot be immediately purified, drinking water must be boiled. Cases are isolated, and supervisory controls are tightened on members of households where cholera has occurred, and on public eating places and food supplies.

III. Prevention. Cholera prevention depends on continuing application of the principles of control outlined above. Immunization has a short-term value only. In endemic areas cholera vaccine must be given repeatedly to persons who run the risk of frequent exposure. Most authorities, including WHO and the U.S. Public Health Service, do not recommend the routine use of vaccine in prevention of cholera. However, individuals traveling to epidemic areas should have a valid vaccination certificate (obtained within six months) to facilitate entry to other countries that may require it. (The United States no longer requires such a certificate for entry.) Personal and public sanitary measures remain the best protection.

Vibrio parahaemolyticus

Vibrio parahaemolyticus is a microorganism related to *V. cholerae* that has been receiving increasing recognition as an agent of bacterial diarrhea. This vibrio was first isolated in an outbreak of food poisoning in Japan in 1951 and is now the most common cause of bacterial gastroenteritis in that country. In recent years, *V. parahaemolyticus* has been documented as the cause of several outbreaks in the United States, particularly along the East Coast. Fish or shellfish are most commonly involved, raw fish or "sashimi" in Japan, cooked seafood (steamed crabs) in the United States. The organism's survival in frozen shrimp has also been described. Vibrio gastroenteritis is primarily a disease of the summer months, when there are large numbers of vibrios in coastal seawaters and when bacterial multiplication in unrefrigerated foods is encouraged.

V. parahaemolyticus food poisoning appears to be an infectious rather than a toxic disease. Symptoms usually begin about 12 hours after ingestion of contaminated food, but the incubation period ranges from 2 to 48 hours. The disease may be clinically indistinguishable from *Salmonella* gastroenteritis, with abdominal pain, diarrhea, nausea, and vomiting. Mild fever, chills, and headache are also characteristic. The diagnosis must be established bacteriologically by isolating *V. parahaemolyticus* from the feces or vomitus of the patient. These vibrios are not found in the feces of asymptomatic persons, and they appear to be pathogenic only for man. They can be cultured on ordinary media to which 1 to 3 percent sodium chloride has been added and on TCBS agar (see page 424). Enteropathogenic strains are hemolytic on blood agar.

Vibrio gastroenteritis is not a serious disease as a rule. Recovery is usually complete within two to five days. The very low fatality rate is associated with infection in aged, debilitated patients.

Enteropathogenic Escherichia coli

Certain strains of *Escherichia coli* have long been known to cause severe or lethal diarrhea in newborn domestic animals, such as piglets and calves. Since the late 1940s, when a number of outbreaks of severe diarrhea occurred among human infants and were associated with certain serologic types of *E. coli,* the pathogenicity of these normally commensal enteric bacilli has been recognized. They have been shown also to be a major cause of "traveler's diarrhea," a frequent problem among persons visiting other countries.

Early studies of enteropathogenic *E. coli* indicated that these strains belonged to 12 to 15 antigenically distinct groups, but the mechanism of disease production was not known. Recent investigations have shown that at least two different mechanisms may be involved: mucosal invasion similar to that seen in shigellosis (enteroinvasive) or secretion of an enterotoxin that produces a diarrheal effect similar to that described for cholera (enterotoxigenic). Mucosal invasion may be related to surface properties of certain *E. coli* strains that allow them to adhere initially to the intestinal epithelium. Enterotoxin production is determined by the presence of a plasmid (extrachromosomal DNA fragment; see Chap. 3) that can be transferred from one *E. coli* strain to another by sexual conjugation. This plasmid coding of enterotoxin is similar to the role of plasmids in the transfer of antibiotic resistance and may explain the relatively high incidence of multiple antibiotic resistance among enteropathogenic *E. coli*. The fact that transferable plasmids are responsible for enterotoxin production may also explain the observation that all strains of the same serotype do not necessarily produce diarrheal disease. Those that do not are thought either to have lost the enterotoxin-producing plasmid or never to have acquired it, whereas those that are pathogenic have accepted and maintained it.

Not all of the classic serologic types of *E. coli* that were implicated in infantile diarrhea in the 1940s and 1950s have been found to produce enterotoxin, and other serologic types that *do* produce enterotoxin have now been identified. At the present time a reevaluation of the relationship between antigenic markers and disease production is in progress. Those diseases caused by enteropathogenic strains are dis-

tinct from *E. coli* infections of parenteral tissues, particularly of the urinary tract (Chap. 28). Parenteral infections due to *E. coli* occur most often in hospitalized patients, compromised by underlying problems related to surgery, catheterization, prolonged chemotherapy, or immunosuppressive treatment.

Epidemic Diarrhea in Nurseries. Epidemic diarrhea is a clinical syndrome classically seen in hospital nurseries for the newborn. The epidemic pattern can be alarming, with swift spread from one baby to another. Clinically, there is severe, watery diarrhea inducing dehydration and shock. This is a prostrating syndrome not seen in the general community, but in hospital nurseries it has a high fatality rate. It is not related to the sporadic cases or community epidemics that sometimes result from spread of infectious agents by poor sanitation but occurs instead among babies, healthy or sick, who are under hospital supervision. The incidence is higher in urban hospital nurseries than in rural communities.

Diarrhea is most probably the result of colonization of the small bowel by enteropathogenic strains of *E. coli* that produce enterotoxin. The action of the latter is similar to that of the enterotoxin produced in cholera. The incidence is higher among bottle-fed as compared with breast-fed babies, possibly because of the protective effects on the latter of secretory IgA ingested with colostrum.

Laboratory Diagnosis

I. Identification of Organisms. A number of species of *E. coli* are of importance in epidemic diarrhea in infants. These may have distinct antigenic components recognizable by serologic techniques using specific antiserums. When a recognized group is found, it is reported by numbers assigned to its somatic (or O) antigens, e.g., O111, O55, O119.

The demonstration of enterotoxin production by these strains requires experimental animals or tissue culture. These are difficult and time-consuming techniques not available in most clinical laboratories.

II. Identification of Patient's Serum Antibodies. It is not always possible to make a serologic diagnosis of these infections, for several reasons: the antibody-forming mechanism is not well developed in babies; these diarrheal infections may be too fulminating, or too short lived, to permit antibody response; serologic testing depends on the use of a battery of specific *E. coli* antigens, not all of which may be readily available. Nevertheless, many pathogenic serotypes of *E. coli* have been confirmed as causative agents in individual outbreaks by recognition of rising titers of specific antibodies in infants' serums.

III. Specimens for Diagnosis. Stool specimens should be submitted for bacteriologic culture. The laboratory should be alerted to the possibility of an epidemic of infant diarrhea so that appropriate methods may be used to distinguish enteropathogenic serotypes of *E. coli* from others normally present in feces.

Serum samples are not often of value in routine diagnostic testing, for reasons described in II above. They may sometimes be submitted for special testing, or for referral to large diagnostic centers. Serum collections should always be paired, an "acute" and a "convalescent" specimen being drawn at onset of symptoms and ten days to two weeks later.

Epidemiology

I. COMMUNICABILITY OF INFECTION.

A. *Reservoir, Sources, and Transmission Routes.* The adults who surround babies in hospitals are the probable reservoirs of these infections: mothers, physicians, nurses, and other attendants. Since the organisms involved are often members of adult intestinal flora, the source of infection is probably fecal. Transfer may be direct from contaminated hands of adults, or indirect through contamination of environmental surfaces or objects to which hospitalized babies are commonly exposed, e.g., weighing scales, and the contaminated dust of ill-kept nurseries. Formulae may be an indirect source of infection if they are contaminated by personnel during their preparation, improperly sterilized or pasteurized, stored in inadequate refrigerators, or saved for another feeding after a first try. Pathogenic serotypes of *E. coli* have been found on the bassinet linen and clothing of babies, in nursery dust and dirt, in solutions and formulae, on instruments and weighing scales, and in cultures taken from attendant personnel — their feces, hands, and clothing. These organisms, and others of similar origin, may also be present in large numbers on the mother's perineum or as contaminants of her birth canal at time of delivery, if she is not properly prepared. Breast-fed babies may acquire their infections at feeding times, from the mother's soiled hands, breasts, or sheets. In premature nurseries, the humidity chambers of incubators are often a colonizing site for *Pseudomonas* strains.

B. *Incubation Period.* This is often difficult to determine, but incubation of organisms infectious for babies probably does not require more than two to four days, with a maximum estimate of three weeks.

C. *Communicable Period.* Epidemics of diarrheal disease are communicable among nursery infants for as long as symptomatic babies remain in the area, the environmental source persists, or the adult carrier of infection remains undetected.

D. *Immunity.* Babies are not immune to this type of infection. They are most susceptible during the first year of life. Epidemics are most common in nurseries for the newborn or for older babies sick with other diseases. Premature babies are even more susceptible. Congenital immunity appears to be lacking, although it is possible that sIgA in mother's milk may have some protective effect for breast-fed babies. This may be offset if the baby acquires infection through breast feeding.

II. CONTROL OF ACTIVE INFECTION. Symptomatic babies should be immediately isolated, and "infection security" must be tightened in the nursery. Cleaning procedures may have to be revised, disinfection practices reviewed, and personnel instructed in the principles of cross-infection prevention. It is imperative that hands be carefully washed, not only after the personal toilet, but after each infant's diaper change.

Case reports should be made when two cases have appeared concurrently within a nursery, for this may be the start of an epidemic whose source should be found without delay. The isolation of babies under these circumstances requires the establishment of two completely separated units: a "contaminated" and a "clean" nursery, with different facilities and personnel for each. There should be no interconnecting links between them. When the safety of the clean nursery for new babies remains in doubt, it may be necessary to close the maternity service until the source of the infection is found or the problem is solved by vigorous, thorough disinfection-cleaning. Newborn infants may not be admitted to the contaminated nursery. This should remain closed until sick babies recover and are discharged. Contact infants in the closed area should remain under close observation for at least two weeks from the time of their last exposure to an infected baby, although they are not necessarily kept in the hospital throughout this period if they appear to be well and progressing normally. While in the hospital these babies should have individualized care. When the last contact or infected baby has left the contaminated nursery, it may be reopened after thorough, detailed disinfection-cleaning.

Other steps indicated by epidemic emergencies include the following:

A. Call consultative meetings of the hospital's infection control committee to evaluate the problem and the efficacy of measures taken to combat it. Local health departments may also be of epidemiologic assistance.

B. Survey mothers and obstetric and nursery personnel for symptoms of illness or for bacteriologic evidence relating them to infant cases.

C. Survey babies discharged from the hospital within the preceding weeks for evidence of a missed case, or a late-developing infection.

D. Review methods for formula preparation.

E. Review methods for nursery or general hospital sanitation.

F. Review nursing and medical techniques in the delivery room, obstetric service, and nursery.

G. Take stool cultures from sick or exposed babies, mothers, and personnel to determine the infectious agent of the epidemic, if possible.

H. Establish firmer standards for asepsis wherever weaknesses are noted.

Successful therapy depends on prompt replacement of lost fluids and electrolytes. Antimicrobial therapy is based on clinical and microbiologic evidence of its efficacy and relevance. However, the antibiotic resistance of many strains of *E. coli* derived from the adult population is a serious problem and may hamper efforts to achieve specific therapeutic control of the situation.

III. PREVENTION. The general principles followed in prevention of serious cross-infections in nurseries fall under two major headings:

A. *Asepsis.* Control of the general environment of the nursery should be rigid and detailed. Safety of formula preparations is particularly vital. Nursery laundry must be clean and sterilized. Cleaning techniques should include routine use of effective disinfectants, properly chosen for their purpose. Humidity chambers and inner walls of incubators should receive particular attention: interior surfaces should be disinfected daily; chambers, as indicated by periodic bacteriologic testing. (Consult the manufacturer's instructions for methods of disinfecting humidity chambers or plastic walls.) Examining tables, weighing scales, and equipment shared by infants must be disinfected between uses. Individual sterile pads should be used for babies placed on surfaces in common use.

B. *Personal Contacts.* Good floor planning and traffic control are essential. Separate, disconnected facilities should be provided for newborns, premature infants, and older babies. Newborn and premature units should offer individualized equipment for each infant. Isolation facilities must be available for segregation of sick or quarantined babies. Infants born of mothers with any recognized infection should be quarantined until clinical and microbiologic evidence warrants release. Babies admitted from other hospitals or from the community are quarantined pending evaluation.

Nursery personnel should not do double duty in other areas of the hospital, and access to the nursery area should be denied for unauthorized visitors. All who enter are required to wear a clean gown and scrub their hands if they are to handle babies.

On the obstetric service, mothers who breast-feed their babies are carefully instructed in hygienic technique. Babies should not be breast-fed or visit with mothers who have infections.

Effective techniques for routine bacteriologic surveillance of the nursery can prevent the buildup of unsuspected environmental reservoirs of infection. Prompt recognition and reporting of the earliest signs of infant infection prevent wide dissemination from an unsuspected human reservoir.

Dysentery Caused by *E. coli*. As previously indicated, it has become apparent in recent years that some *E. coli* strains can produce a disease resembling shigellosis. The clinical syndrome is dysentery, with abdominal pain and bloody diarrhea. Several outbreaks in both infants and adults have occurred. The isolated *E. coli* strains, when studied in experimental animals, have been shown to have the ability to penetrate epithelial cells, multiply, and produce disruptive effects, with necrosis and ulceration, accompanied by an intense inflammatory reaction. These capabilities are not possessed by enterotoxin-producing strains, but the mechanisms of epithelial invasion have not been clarified.

Traveler's Diarrhea. "Traveler's diarrhea," or "turista," is often a troublesome problem. It occurs in people of all ages visiting in tropical or semitropical countries and is characterized by transient diarrhea and abdominal cramping. Until recently, numerous efforts to identify specific microbial causes have failed in most instances. New evidence now suggests that enterotoxin-producing strains of *E. coli* may be responsible for an important number of cases. Further clarification may be possible when simpler methods are developed for laboratory identification and characterization of the enterotoxin.

Yersinia enterocolitica

Yersinia enterocolitica is a Gram-negative bacillus of animal origin that has assumed increasing importance as a cause of severe enterocolitis in humans. Infections due to this organism have been spreading in various parts of the world since the early 1960s. Prior to that time, yersiniosis had been known as an agent of epidemic illness in domestic and wild animals, such as pigs, sheep, and rabbits. The first human cases were diagnosed in northern Europe, in 1963, but they have now been recognized in numerous countries, including the United States.

In man, the commonest form of illness is acute enterocolitis. It affects people of all ages, but about two-thirds of the cases are in children under seven years of age. The symptoms of diarrhea, abdominal pain, and sometimes fever are like those caused by salmonellae, shigellae, or pathogenic strains of *E. coli*. Occasionally, however, the organisms may infect the mesenteric lymph nodes, producing symptoms suggestive of acute appendicitis. In older compromised patients, bacteremia may develop, with diverse symptoms of systemic infection. Even previously healthy persons have developed life-threatening disease due to disseminated yersiniosis, with focal abscesses in the liver, spleen, kidneys, or other viscera. About 20 percent of cases develop erythema nodosum or arthritis following an initial bout of diarrhea and fever. The mortality rate remains undefined, but these infections are often severe.

The diagnosis of *Y. enterocolitica* infections depends on isolation and identification of the organism from feces, in uncomplicated enterocolitis. Surgery for suspected appendicitis may reveal mesenteric lymphadenitis, and culture of excised lymph nodes then establishes the diagnosis. In disseminated infection, blood or urine cultures may be diagnostic, and, if necessary, biopsy of involved tissues may yield the organism. *Y. enterocolitica* grows well on enriched media, but on the selective and differential media normally used for stool cultures it grows best at room temperature rather than 37° C. Such media are generally incubated first at 37° C to rule out salmonellae, shigellae, or enteropathogenic *E. coli,* then held at room temperature for another 24 hours. Newly developing colonies are then tested for the characteristic biochemical features of *Y. enterocolitica*. They possess many of the properties of other Gram-negative enteric bacilli, but can be differentiated by a number of tests.

The serologic diagnosis of yersiniosis involves testing the patient's serum with a battery of strain-specific antiserums, a difficult matter on a routine basis. There are about 30 known antigenic types of *Y. enterocolitica,* but they lack a common group antigen. For this reason, only laboratories with a complete collection of strains of known serotypes and the corresponding antiserums can perform definitive serologic testing.

The antimicrobial drugs that may be used in the treatment of *Y. enterocolitica* infections include gentamicin, kanamycin, and chloramphenicol. The organism is resistant to penicillin and ampicillin.

The organism appears to be widespread in nature. It has been isolated from water, fruits, and vegetables as well as from a number of animal species. A number of small mammals appear to be healthy carriers. The epidemiology of human infections has not been completely defined, but most evidence points to primary entry through the digestive tract from environmental or animal sources. Entry on a respiratory pathway may also be possible, but it has been assumed that the organism follows the usual transmission routes of other enterobacteria. In one recent outbreak among schoolchildren in a New York county, foodborne transmission was documented, the source being chocolate milk supplied to the school cafeterias by a local dairy. Of 218 affected children, 33 were hospitalized for suspected appendicitis, and 13 had appendectomies before the nature of the illness was clarified. Among the operated children mesenteric adenopathy and inflammation of the terminal ileum were frequent observations, but the removed appendices were normal or showed lymphoid hyperplasia. Cultures of a number of sick children and of representative samples of the suspected milk, coupled with epidemiologic surveys, eventually revealed the source and cause of the infection. Milk had previously been suspected but not proven to be the vehicle of outbreaks of *Y. enterocolitica* in England and Canada.

Clostridium perfringens

Clostridium perfringens is a Gram-positive, anaerobic, spore-forming bacillus, responsible for about 3 percent of reported outbreaks of food poisoning, and for about 11 percent of individual reported cases.

There are numerous species of *Clostridium,* but those of pathogenic importance fall into three groups according to the types of diseases they produce: (1) *C. perfringens* and others that characteristically cause a variety of tissue infections associated with traumatic wounds (the typical clinical pattern is that of gas gangrene, discussed in Chapter 27; the association of *C. perfringens* with food poisoning is discussed briefly below); (2) *C. tetani,* the causative agent of tetanus, a disease resulting from the production of a potent exotoxin by this organism following its traumatic introduction into the tissues (see Chap. 27); and (3) *C. botulinum,* the agent of botulism, a food poisoning caused by ingestion of a powerful, preformed neurotoxin formed by the organism as it grows in contaminated food. The latter disease is described later in this chapter (pp. 432–34).

The pathogenic clostridia are noted for their production of exotoxins that can induce serious injury to human tissues. *C. perfringens* produces four major toxins and is classified into five types, A through E, on the basis of its production of these potent substances. Type A is primarily responsible for diseases in man, and types B, C, D, and E cause a variety of diseases in animals. All of these types occur naturally in the intestinal tracts of animals, and type A can be found normally among man's intestinal flora.

Clostridial food poisoning is associated with the ingestion of foods contaminated with *C. perfringens* of type A. Outbreaks are usually linked to products such as gravy that provide an anaerobic environment for the organism's growth. The organism produces heat-resistant spores that may survive the temperatures at which contaminated meat is cooked. If the meat, or its gravy, is then allowed to stand at room temperature, the spores germinate and the growing bacilli produce toxin. Food poisoning probably results as much from the ingestion of preformed toxin as from organisms continuing to multiply in the bowel.

Symptoms usually appear within 8 or 10 to 24 hours after eating contaminated food. The disease is mild as a rule, with crampy abdominal pain and diarrhea the only symptoms. Nausea, vomiting, and fever are seldom prominent. The illness is brief, symptoms subsiding usually within 24 hours. Chemotherapy is not warranted, and treatment is supportive only.

Individual cases probably go undiagnosed in many instances because of the mild nature and short duration of this type of food poisoning. When outbreaks occur, the diagnosis is established by identifying the organism in appropriate anaerobic cultures of the incriminated food. Fecal cultures may be useful if *C. perfringens* is isolated from affected persons in higher-than-normal numbers.

A rare type of clostridial food poisoning, called enteritis necroticans, is caused by type C strains of *C. perfringens.* This is an acute illness characterized by severe abdominal pain, diarrhea, vomiting, and prostration. Shock may ensue and death may be rapid as a result of diffuse, necrotizing enteritis. Sporadic cases of this disease have been reported, but outbreaks have been limited to New Guinea, where they are associated with occasions for "pig feasting" and the consumption of inadequately cooked pork.

Clostridial food poisoning can be prevented by

avoiding long periods in which cooked foods are held at mild temperatures. If foods must stand for several hours before they are served, they should be kept at very warm or very cool temperatures (above 60° C or below 5° C).

Bacillus cereus

Bacillus cereus is another Gram-positive, spore-forming bacillus, but unlike the clostridia it is an aerobic organism. It has been incriminated in several outbreaks of food poisoning. The incubation period is about ten hours, after which symptoms of abdominal pain, profuse watery diarrhea with rectal spasm (tenesmus), and nausea occur. Fever is rare, and the illness persists only for about 12 hours. Experimental evidence indicates that large numbers of the organism must be ingested to produce disease. *B. cereus* produces a toxin, but its role in food poisoning is not clear.

Foods most likely to be contaminated with *B. cereus* include grain and vegetable dishes. These should not be allowed to remain at room temperature for long periods after cooking, since spores may germinate and the bacilli multiply rapidly.

Staphylococcal Enterocolitis

Staphylococcal enterocolitis is the result of infection of the bowel wall with *Staphylococcus aureus*. This is not a food poisoning and is a different entity from the gastroenteritis caused by staphylococcal enterotoxin discussed in the next section of this chapter. *Staphylococcus aureus* is a normal, though minor, member of man's intestinal flora. It may become abnormally abundant in the bowel as a result of the suppression of other fecal organisms by antimicrobial drugs. When this happens, staphylococci may actively infect the intestinal epithelium, producing mucosal ulceration. In severe cases there may be necrosis, with formation of a pseudomembrane. The symptoms include moderate to severe diarrhea. The stools are watery and may contain blood and mucus. Fever, abdominal pain, nausea, and vomiting are usual, and the white blood cell count is elevated. Loss of fluid and electrolytes may

have serious consequences leading to shock. The diagnosis is supported by the demonstration of large numbers of staphylococci in cultures and Gram-stained smears of feces.

Staphylococcal enterocolitis is a sporadic disease seen most often in compromised hospital patients. Many are surgical patients who have received large amounts of antimicrobial drugs to suppress the intestinal flora prior to gastrointestinal or genitourinary tract surgery. However, only a minority of patients thus treated develop this syndrome. The first step in treatment is to stop the antibiotics being used (chloramphenicol, tetracyclines, and neomycin have been implicated most frequently) and to restore fluid and electrolyte balance. Oral vancomycin is effective in treating staphylococcal enterocolitis, but to avoid the risk of bacteremia and systemic infection, intravenous therapy with a penicillinase-resistant penicillin or cephalosporin is generally recommended (see Chap. 10).

An exotoxin-producing *Clostridium*, *C. difficile*, has recently been implicated in an almost identical syndrome termed "antibiotic-associated colitis."

Diseases Caused by Ingestion of Preformed Bacterial Toxins

Food poisoning caused by preformed bacterial toxins falls into two clinical patterns: acute gastroenteritis, as induced by staphylococcal enterotoxin, and the sytemic intoxication typical of botulism.

Staphylococcal Enterotoxin Poisoning

Acute food poisoning due to staphylococcal enterotoxin is one of the principal forms of bacterial enteric poisoning in the United States. This illness is an intoxication induced by preformed toxin produced by staphylococci multiplying actively in food before it is consumed. Symptoms begin within one to six hours, usually about two to four hours after the contaminated food is eaten. The onset is abrupt, and often violent, with nausea and vomiting, cramps, and diarrhea. The patient may be prostrate, with a depressed temperature and blood pressure. Severe as these symptoms are, they pass rather quickly, usually within a day or so, as the toxin is eliminated and the insulted intestinal mucosa recovers.

The *diagnosis* of staphylococcal intoxication can be made tentatively from clinical symptoms, supported by isolation of staphylococci in significantly large numbers from the suspected food if it is still available for testing. The presence of many staphylococci in vomitus is also significant, but their isolation from fecal specimens has little meaning, because these organisms may be found among normal intestinal flora. When staphylococci are isolated from food or stomach contents, they may be tested for enterotoxin production or typed with bacteriophages (see p. 520).

The *source* of staphylococcal contamination of food is usually a person with an infected lesion on hands, arms, or face, although these organisms may also reside on healthy skin and respiratory mucosa. A great variety of foods may support growth of staphylococci, but those most often involved are custards and pastries made with cream, salads and salad dressings, and meats that have been sliced or ground. When staphylococci are implanted into foods they require a few hours to grow and elaborate the toxin. Growth and multiplication occur most rapidly in warm rooms. Cooking kills the organisms, but if toxin has already been formed it may survive, for it is stable to heat. Refrigeration does not kill the organisms or the toxin, but inhibits microbial growth and thus prevents enterotoxin formation. For this reason, meats and foods prepared in advance of meals must be kept under refrigeration until they are eaten or cooked.

Control of staphylococcal food poisoning depends on the recognition of cases (outbreaks must be reported to health departments), the food source, and the infected human reservoir. Bacteriologic cultures are made of suspected food and of handlers (from lesions, or from the nose), and also from patients' vomitus. Strains of staphylococci isolated from these sources are then compared as to phage type, enterotoxin production, and other properties to relate them to the case and its source. Control of food handlers includes education in personal hygiene and supervi-

sion in maintaining sanitary techniques for handling, preparing, and refrigerating foods. Persons with obvious skin infections should not be permitted to handle foods until the condition has been treated and cured. In many areas the public health laws require refrigeration of foods prepared for public consumption.

Botulism

This acute and often fatal disease results from ingestion of a toxin preformed in foods by the anaerobic bacterial species *Clostridium botulinum*. Unlike other forms of bacterial food poisoning, botulism is a systemic intoxication. The powerful toxin is absorbed from the intestinal tract and causes motor paralysis of cranial and peripheral nerves, effects that are sometimes irreversible and fatal. Symptoms sometimes, but not always, begin with gastrointestinal irritation, vomiting, and diarrhea. Often, the first signs are related to neurotoxin activity: headache and dizziness, ocular disturbances (double vision), with paralysis of oculomotor nerves. Neuromuscular involvement of the intestinal wall may lead to severe constipation; death usually results from respiratory paralysis or cardiac failure. An incubation period of *12 to 36 hours,* rarely longer, reflects the time required for the toxin to be absorbed from the intestinal tract, disseminated through the body, and attached to peripheral nerves. The amount of toxin ingested may also affect the length of the incubation period, although this is a very powerful substance, capable of exerting effects in very small doses. Patients who recover do so slowly, as toxin is inactivated and eliminated, and the inhibitory effects at the myoneural junctions slowly regress. They may require artificial respiration for some time. Immunotherapy with *botulinal* antitoxin is sometimes lifesaving.

The *diagnosis* of botulism can often be made on clinical grounds but requires bacteriologic confirmation and identification of its source. The organism may be isolated from the suspected food, if it is available for testing, but the best proof lies in demonstrating (1) that the food is toxic for mice, and (2) that this toxicity can be specifically neutralized by

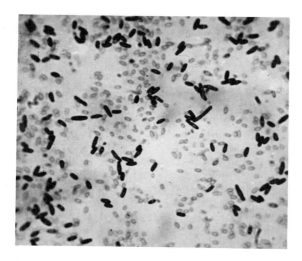

Fig. 18-2. *Clostridium botulinum*. Vegetative bacilli (darkly stained) and many free spores. The latter are resistant to staining, appearing as pale, empty rings or ovals in this photomicrograph. (1000×.) Special techniques are required to demonstrate the thickness of bacterial spore walls and the density of their contents (see also Fig. 2-11). (Reproduced from Dolman, C. E.: Botulism. *Am. J. Nurs.*, **64**:119, Sept. 1964.)

botulinal antitoxin. When canned foods are involved, related lots or batches are tested.

C. botulinum is an anaerobic, spore-forming, motile, Gram-positive bacillus, normally found in soil and in the intestinal tract of animals and fish (Fig. 18-2). Its spores may contaminate vegetables and other foods. When these are preserved by canning techniques, spores may survive inadequate processing and germinate later under the anaerobic conditions provided in a can or jar. The vegetative bacilli then proliferate and elaborate toxin. Canned foods are frequently eaten with little or no subsequent cooking, and under these circumstances active toxin is ingested. Boiling of foods for at least ten minutes will inactivate the toxin. The chief danger lies in the fact that these foods may display little or no change in appearance or odor or at least none related to the presence of toxin, so that the urgent necessity for boiling or discarding them may not be apparent. The foods most commonly involved are home-canned vegetables, olives, mushrooms, corn, spinach, and string beans. Commercial canning processes have

largely overcome the problem, but outbreaks have been related to commercially packed tuna fish, smoked fish, and potato soup.

Control of botulism hinges on safe measures for processing foods (especially those likely to harbor *Cl. botulinum*), prompt recognition of the disease and its source, and recall from the market of suspected foods. Cases must be reported so that a search for the source can be made without delay. Persons known to have eaten the incriminated food may be treated prophylactically with botulinal antitoxin. This procedure is most effective if antitoxin reaches the bloodstream before the toxin does so. There are seven types of *C. botulinum*, each with its own distinct antigenic type of toxin (A through G). It is essential that prophylactic antiserum contain antibodies specific to the type involved in the disease, if this can be determined, or that polyvalent antiserum, containing several type-specific antibodies, be used. Types A, B, and E are most prevalent in human disease. Type G has been found in soil but has not been known to cause botulism.

Government controls now provide for inspection of commercially canned foods, although occasionally new problems arise. It is vitally important that housewives be educated in safe methods for preparing home-canned foods. Spores of *Cl. botulinum* may survive long periods of boiling, especially in neutral or alkaline foods, so that unless the food has an acid reaction it cannot be safely processed except in a pressure cooker. If there is the slightest reason for doubt, home-canned nonacid foods should be boiled for at least ten minutes before they are served or even tasted by the cook. The contents of swollen or damaged cans (Fig. 18-3) should never be eaten.

Table 18-1 shows some of the important distinctions that can be made for bacterial food poisoning caused by the ingestion of infectious microorganisms and that resulting from consumption of preformed bacterial toxins.

Sudden Infant Death (SID) Syndrome. Recently, evidence has accumulated indicating that botulism may be responsible for some cases of sudden infant death, sometimes referred to as "crib death." SID has been a baffling syndrome, most likely to occur in

Fig. 18-3. The swelling of this can is due to the activity of microbial contaminants within. Its contents should not be eaten (Courtesy of U.S. Food and Drug Administration, Washington, D.C.)

infants up to three months of age. Because its victims are apparently normal, healthy babies often found dead in their cribs, death was generally attributed to accidental suffocation in bedclothes. This has been shown to be seldom, if ever, the true cause of death, but the basic etiology has remained undocumented. A number of cases of infant botulism have been recognized, however, and a small percentage of these have died. The age pattern matches that of SID, as do the symptoms of respiratory paralysis. It has been postulated that spores of *C. botulinum* reaching the digestive tracts of infants, from unknown sources (they are common in soil and can be carried by dust), may germinate there. This does not happen in the adult bowel, with its teeming microbial flora, but conditions in the infant may permit it. Infants may

then be highly susceptible to minute quantities of botulinal toxin produced by the multiplying bacilli. Studies now underway have yielded culture isolates of *C. botulinum* in a few cases of SID and traces of the toxin in others. Further investigations may elucidate the epidemiology of infant botulism and its possible role in SID.

Infectious Systemic Diseases

The Enteric Fevers (Typhoid and Paratyphoid Fever)

Typhoid Fever. Typhoid fever is a systemic bacterial disease of man caused by *Salmonella typhi*. Like other salmonelloses, this infection is transmitted by the fecal-oral route via contaminated food and water. Unlike most others, however, it does not lead to an acute gastroenteritis but to involvement of parenteral sites.

When *S. typhi* is ingested in contaminated food or water, it first localizes in the deep lymphoid tissue of the intestinal wall (Peyer's patches). The lymph flow carries the organisms into the thoracic duct, and from there they reach the bloodstream. They may be widely disseminated, but localize most usually in the liver and gallbladder. The kidneys may also be infected, as well as the spleen, bone marrow, or lungs. In liver and kidneys there is focal necrosis of parenchymal cells at the site of colonization; lymphoid tissue becomes enlarged and more cellular. The gallbladder is always infected and is a major source of organisms that appear in the stool during the course of the disease and often continuing into convalescence. Localization in the lungs produces a bronchitis or pneumonitis. The cellular response is mononuclear rather than polymorphonuclear, and the total number of white blood cells in the circulating blood is often depressed.

The symptoms of typhoid fever include a gradual, insidious onset, with fever rising in steps to an average height of 104° F (40° C), with a relatively slow pulse. There is often abdominal tenderness, but constipation is more frequent than diarrhea. Rose spots often appear on the trunk during the first or second week, together with symptoms of pulmonary involvement. In severe cases, the patient may become

Table 18–1. Summary of Bacterial Food Poisoning

Disease	Agent	Incubation Period	Communicable Person to Person	Symptoms	Control
Infectious gas-troenteritis	*Salmonella* sp.	8–48 hr	Yes	Fever, N, V, D,* for several days.	Prevent contamination of food, water; control carriers, food handlers; cook and store foods at proper temperatures
	Shigella sp.		Yes		
	Vibrio sp.		No		
	Y. entero-colitica		Yes		
	C. perfringens†		No		
	B. cereus		No		
Staphylococcal intoxication	Preformed entero-toxin of staphy-lococci	1–6 hr	No	Acute abdominal cramps N, V, D* 1 or 2 days; subnormal temperature	Prevent contamination and multiplication of staph in food; control food handlers
Botulism	Preformed neuro-toxin of *C. bot-ulinum*	12–36 hr	No	Central nervous system intoxication; double vision; flaccid paralysis	Prevent contamination and germination of spores in canned foods; reheat canned foods before serving (boil 10 min); anti-toxin for all who have eaten suspected food

*N, V, D = nausea, vomiting, diarrhea.
†Preformed toxin involved as well as active infection.

delirious or stuporous (the Greek word for stupor is *typhos*).

The organisms may be recovered from the blood during the first week to ten days when bacteremia and dissemination occur. Stool cultures may be positive from the onset, although the greatest numbers of organisms are found after two or more weeks. They may remain positive, even long after the convalescence. Urine cultures may yield the organisms in the second and third weeks, and continue to do so for long periods thereafter. The organisms may be found in bone marrow, bile drainage, or sputum if localization in bone, the gallbladder, or the lungs has occurred.

The immune response to typhoid fever begins in the early days of infection, and recovery is associated with a rising titer of antibodies during the second and third weeks. Then organisms can no longer be found in the bloodstream, the temperature gradually returns to normal levels, and the convalescent stage ensues. Foci of typhoid bacilli may nonetheless persist in the gallbladder, spleen, or kidney within mononuclear cells, out of reach of circulating antibody. From such sites they may continue to be shed into the intestinal tract (by way of the common bile duct when the gallbladder is involved) or excreted in urine. The patient remains asymptomatic, but he may become a persistent carrier.

Paratyphoid Fever. This is a milder disease than typhoid fever. It usually has a more rapid onset, with fever, involvement of lymphoid tissue in the intestine, and spleen enlargement. In more severe cases there may be rose spots and other symptoms similar to typhoid. The duration of symptoms may be from one to three weeks. *S. enteritidis var paratyphi A* and *B,* or closely related serotypes, are the causative organisms.

Salmonella Bacteremia

Salmonella gastroenteritis is sometimes followed by bloodstream invasion, with scattered localization of organisms and the production of focal abscesses. Although many *Salmonella* strains are potentially capable of such invasion, some species, e.g., *S. choleraesuis* and *S. enteritidis,* are more likely to be associated with severe illness. There may be perineal and pelvic abscesses, meningitis, endocarditis, pneumonia, pyelonephritis, arthritis, or osteomyelitis. Such infections with salmonellae are often associated with other chronic disease or defective host defenses. Patients with sickle cell anemia, for example, appear to be more susceptible, as are those with metastatic cancer. Peripheral vascular grafts are also subject to colonization by salmonellae, presumably from a low-level source of infection in the bowel, with periodic subclinical bacteremia. Smaller numbers of organisms may be able to colonize foreign bodies in the vascular system than would be required for intact normal blood vessels.

Laboratory Diagnosis

I. Identification of the Organism. See pages 418, 419.

II. Identification of Patient's Serum Antibodies. The *Widal* test is useful in demonstrating antibody titers during systemic *Salmonella* infection. This is an agglutination method in which known *Salmonella* O (somatic) and H (flagellar) antigens are mixed separately with patient's serum, the latter being diluted serially to find its maximum capacity (titer) for agglutinating the antigen. A *rising* titer of agglutinins occurring from the onset through the course of the disease signifies active infection. A single finding of typhoid or paratyphoid agglutinins, in only one serum sample, may not distinguish antibody persisting from a previous infection, from vaccination, or during the carrier state.

III. Specimens for Diagnosis.

A. Blood cultures should be taken frequently, especially during the first week of disease. Bone marrow cultures may also be useful in demonstrating bacteremic spread.

B. Stool specimens should be submitted for culture at frequent intervals, particularly after the first week of disease.

C. Urine specimens may be positive for *Salmonella* by culture, usually in the second and third weeks of infection.

D. Bile drainage may be collected for culture, during convalescence or later, to check for persistence of the carrier state, and to demonstrate biliary tract localization of the organisms.

E. Sputum, pus from abscesses, synovial fluid, spinal fluid, and similar collections as indicated by the clinical signs of infection may yield the organisms in culture.

F. An "acute" serum taken during the first week of disease, as close to onset as possible, should be submitted for the Widal or "febrile agglutinins" test.

G. A "convalescent" serum should be collected after ten days to two weeks for a second Widal titer to demonstrate a rising antibody response to active infection.

Epidemiology

I. Communicability of Infection

A. Reservoir, Sources, and Transmission Routes. As mentioned previously (p. 418), man is the only reservoir for *S.typhi,* and a primary one for the *S. paratyphi* serotypes. Family contacts may be transient carriers. The carrier state is most common among persons who acquire their infection during middle age and is especially frequent among females. One of the most famous typhoid carriers was a woman nicknamed "Typhoid Mary" (her name was Mary Mallon). She is known to have been responsible for ten outbreaks of typhoid fever, involving 51 cases and three deaths, through her activities as a food handler.

B. INCUBATION PERIOD. For typhoid fever, usually one to two weeks, sometimes three weeks; for paratyphoid fever, about one to ten days; for bacteremia, difficult to define.

C. COMMUNICABLE PERIOD. While the organisms persist in excreta, usually from the first week through convalescence. About 10 percent of typhoid fever convalescents shed bacilli for varying periods up to three months, and about 2 to 5 percent become permanent carriers. The carrier rate is variable in paratyphoid fever and other systemic infections.

D. IMMUNITY. Human populations show a general susceptibility to *Salmonella* infections, with more severe reactions among the very young and the very old. Exposure to systemic infection generally results in immunity, but the degree and duration are variable. Second attacks occur but are not usual after age 30 or 40.

II. *Control of Active Infection*

A. ISOLATION. A private room is not required (except for children) but enteric isolation procedures must be followed. Typhoid fever patients should not be cared for at home unless adequate sanitary conditions exist and good nursing care is available. Their rooms should be screened from flies and kept scrupulously clean. They may not be released to resume general activities, especially if these include food handling, until stool and urine cultures appear to be safely free of salmonellae. This assurance is generally based on three negative cultures of stool (and urine, if indicated) obtained on successive days during convalescence. If any culture is positive, supervision is continued for one month's time. If three negative cultures can then be obtained, the patient may be released from supervision; if not, this re-

quirement may be continued by health authorities for as long as the situation appears to demand.

B. PRECAUTIONS. These include concurrent disinfection of all infectious excreta, and of objects contaminated by them (bedpans, bedlinens, rectal thermometers, enema equipment, clothing). When community sewage disposal systems are adequate, toilet facilities may be used directly, or for bedpan contents. Disinfection techniques should be used frequently for toilet seats and other bathroom facilities, especially when these are shared. In hospitals, individual bedpans or toilets should be provided for patients with enteric infections, or arrangements should be made for sterilization of all bedpans. When good sewage disposal is not available, preliminary treatment of bedpan contents with chemical disinfectants is necessary. Gowns should be worn, and hand washing should be practiced with scrupulous care, by the patient as well as by personnel.

C. TREATMENT. Chloramphenicol is the drug of choice for typhoid fever and other systemic infections. Antibiotic susceptibility testing should be performed with the patient's strain. If it is resistant to chloramphenicol, ampicillin or trimethoprim-sulfamethoxazole may be of value.

D. CONTROL OF CONTACTS. Close contacts are excluded from food handling until cultures prove them to be free of inapparent infection. Household and medical personnel caring for active cases may be actively immunized with vaccine.

The source of every case of typhoid or other enteric fever is investigated. The possibilities of unreported cases or asymptomatic carriers are explored, together with the question of contaminated food or water as a source of common exposure in outbreaks.

E. EPIDEMIC CONTROLS. Individual case reports are required by local health authorities. Suspected food is discarded; milk must be either boiled, pasteurized, or discarded until proven safe. Suspected water supplies must be treated with chlorine or iodine or boiled before use. Vaccine is not recommended during outbreaks because it creates difficulties in the serologic diagnosis of suspected illness.

III. *Prevention*

A. IMMUNIZATION. Vaccine is available for typhoid fever only. In a primary series, it is given in two injections, several weeks apart. A single booster dose is given once in three years. Current recommendations indicate vaccination of persons subject to unusual exposure, through either occupation or travel. Vaccine is also used in areas of high endemicity, or in institutions where sanitation cannot be maintained.

Vaccine elicits the production of agglutinating antibodies, which appear in the circulating blood and persist for some time. When serum from patients with febrile disease of suspected *Salmonella* origin is tested for typhoid agglutinins, a positive titer must be interpreted in the light of a history of vaccination, or of repeated natural exposure to, or active infection with, these organisms.

B. CONTROL OF RESERVOIRS, SOURCES, AND TRANSMISSION ROUTES. Prevention of typhoid fever depends on adequate protection of water supplies (travelers in endemic areas should take care to use water that is boiled, chlorinated, or disinfected with iodine tablets); sanitary disposal of human excreta; pasteurization of milk and dairy products; fly control; sanitary supervision of food processing in public eating places, with special attention to foods eaten raw and to the provision and use of hand-washing facilities; and limitations on the collection and marketing of shellfish from clean waters.

The identification and continued supervision of typhoid carriers is an essential preventive measure. If the carrier state persists for a year, carriers must continue under restriction of occupation until six consecutive negative cultures of feces and urine have been taken at intervals of one month. Carriers sometimes become impatient and uncooperative under surveillance, and this necessitates authentication of specimens yielding negative cultures. Antibiotic therapy with ampicillin, if prolonged, may end the carrier state. If this fails, removal of the gallbladder may be considered.

Patients convalescing from enteric fevers must be warned of the significance of the carrier state, if they remain culture-positive. They should be given special instruction in personal hygiene, in hand washing after use of the toilet and before meals, and in the sanitation of excreta when adequate public sewers are not available. They must be excluded from food handling until satisfactorily demonstrated to be free of continuing infection.

Food handlers and the public must be continuously educated as to the sources and transmission of typhoid and other *Salmonella* infections.

Brucellosis

The Clinical Disease. Brucellosis is a systemic infection, characterized by *variability* — of entry route, of tissue localizations, of onset and symptoms. It is really a disease associated with animals. Called Bang's disease when it affects cattle, it also occurs in hogs, sheep, goats, horses, and other domestic or wild animals. It is sometimes transmitted to man through infectious tissues or milk. The ingestion of contaminated dairy products affords one of the major entry routes of this infection, but the organisms also may enter the body through skin contacts or on inhalation.

Whatever their portal of entry, the organisms usually find their way through lymphatic channels to regional lymph nodes, the thoracic duct, and the bloodstream, with eventual dissemination to many organs. They are localized most frequently on tissues of the reticuloendothelial system, being taken up by mononuclear phagocytes of tissues such as spleen, liver, bone marrow, and lymph nodes. The *cellular* reaction around them is granulomatous, as noted so often in chronic infectious diseases. The nodules frequently undergo central necrosis, and if they break down, extension of the infection results. Cholecystitis, bone marrow invasion, and meningitis sometimes occur. The *immune* reaction is one of hypersensitivity, as evidenced by skin sensitivity to endotoxins extracted from the bacterial cells.

The symptoms of brucellosis reflect this dissemination and localization of the organisms with hypersensitive tissue responses. The onset is usually insidious, with malaise and weakness, assorted muscle aches and pains, and an intermittent fever that arises late in the day but falls during the night, the patient being drenched with sweat at that time. With abrupt, acute onset, fever and chills are more pronounced, though characteristically intermittent, and there may be severe headache, backache and exhaustion. There may be gastrointestinal symptoms, enlarged lymph nodes and spleen, hepatitis with jaundice, and mental depression. These generalized symptoms may subside for several weeks, but the disease often becomes chronic, with persisting or extending lesions. Endocarditis or pyelonephritis may develop, endometritis may occur in women, and various forms of nervous system disease may be seen.

Gradual recovery may occur, often with residual allergy or damage to bones and other tissues; or infection may become latent and asymptomatic, with recurrent chills and fever occurring at times over the years as old lesions are reactivated and produce symptoms.

Laboratory Diagnosis

I. Identification of the Organisms. There are three major species of the *Brucella* genus: *Brucella abortus* (the primary reservoir is in cattle), *Brucella melitensis* (goats are the chief reservoir), and *Brucella suis* (swine are the principal hosts). Any of these strains may infect man or animals other than the principal host. *B. suis* is the species most commonly isolated in the United States at present.

Microscopically, brucellae are very small Gram-negative coccobacilli, nonmotile and non-spore forming. They are much smaller and more delicate in appearance than the enteric bacilli of the human intestinal tract.

Culturally, these organisms are quite fastidious. They require enriched culture media, and some will not grow unless additional atmospheric CO_2 is provided during 37° C incubation. They are differentiated from other Gram-negative bacilli and from each other by a few biochemical reactions and by serologic agglutination with strain-specific *Brucella* antisera.

II. Identification of Patient's Serum Antibodies. Ag-glutinating antibodies appear in the circulating blood during early stages of infection and persist for some time, but are not always in high titer during later, chronic stages of disease. *Opsonins* also develop, as recognized by tests that demonstrate enhanced phagocytosis of brucellae in the presence of patient's serum (opsonocytophagic test, Chap. 7). *Precipitating* antibodies can be recognized during the acute stage and may be diagnostic at that time, but disappear when infection subsides or becomes latent. The evanescent nature of the immune response with respect to humoral antibodies makes serologic diagnosis difficult, particularly during chronic, persistent stages. The results of these tests must be interpreted in the light of the patient's history with regard to time of possible exposure, onset, and current stage of symptoms.

Hypersensitive antibodies can be demonstrated by skin tests, using an extract of *Brucella* endotoxins. This product is called "brucellergen" and elicits skin reactions of the delayed type in individuals who have experienced *Brucella* infection. Some limitations on interpretation exist in this case also, because a positive test does not necessarily indicate current *active* infection, and a negative result may be obtained during acute infection, before the hypersensitive response has begun.

III. Specimens for Diagnosis

A. Repeated blood cultures are useful during early stages of acute infection, or during febrile periods of chronic disease. Cultures of bone marrow may yield the organism also.

B. Biopsies of lymph nodes, spleen, liver, or other tissues apparently involved in infection; body fluids, such as spinal fluid; or sections of nodules in tissues removed by surgery may be bacteriologically diagnostic.

C. Serum samples for agglutination, precipitation, or opsonocytophagic tests may be submitted. Only the first of these is performed on a routine basis, and the results of each test must be interpreted in the light of difficulties in serologic diagnosis discussed above.

Epidemiology

I. Communicability of Infection

A. Reservoirs, Sources, and Transmission Routes. From domestic animal reservoirs, brucellosis is transmitted to man through direct contacts with animal tissues (placentas, aborted fetuses, slaughterhouse blood and meat), inhalation of contaminated dust of barns or abattoirs, or ingestion of unpasteurized dairy products from infected animals. It is rarely transmitted from person to person, even when the organisms are present in discharges from the infected human case.

B. Incubation Period. When the onset is slow and insidious, the incubation period is difficult to determine, but may be estimated at two or three weeks or several months. In cases with an acute onset, incubation may require from one to three weeks.

C. Communicable Period. Brucellosis is not ordinarily communicable among human beings.

D. Immunity. The duration and effectiveness of active immunity resulting from disease are uncertain. Reinfections may occur if exposure is massive or strains possess particular virulence. Judging by the numbers of people who can be infected through

Table 18-2. Summary: Bacterial Diseases Acquired Through the Alimentary Route

Clinical Disease	Causative Organism	Other Possible Entry Routes	Incubation Period	Communicable Period
Acute Gastrointestinal Infections				
Salmonellosis	*Salmonella en-teritidis* (many serotypes)		8–10 hours, up to 48 hours	For duration of active infection and as long as salmonellae are shed
Shigellosis (bacillary dysentery)	*Shigella dysenteriae* *Shigella flexneri* *Shigella boydii* *Shigella sonnei*		1–4 days, not more than 7 days	As long as organisms are in feces
Cholera	*Vibrio cholerae*		2–3 days; few hours during outbreak of cholera	While organisms are present in vomitus and feces; may persist in intestinal tract for several weeks after recovery

ingestion of contaminated milk, it would appear that a majority of human beings are susceptible to *Brucella* infection. Pasteurizing milk prevents this form of exposure on any large scale, and brucellosis remains primarily an occupational hazard for farmers, veterinarians, and slaughterhouse workers exposed to infected animals. Artificial active immunization is not available for humans, but a bacterial vaccine is used for calves in endemic areas.

II. Control of Active Infection

A. *Isolation* of the patient with brucellosis is not necessary, and infection precautions are followed only as normally indicated for safe disposal of excreta and secretions.

B. *Treatment* of diagnosed *Brucella* infection with antimicrobial drugs generally must be prolonged to eradicate organisms from protected intracellular sites in granulomatous nodules, and relapses may be fre-

Specimens Required	Laboratory Diagnosis	Immunization	Treatment	Nursing Management
Stool specimens Vomitus	Culture Culture	None	None (unless enteric fever develops)	Enteric isolation procedures; concurrent disinfection of all excreta and contaminated objects; individual toilets or bedpans; chemical disinfection of excreta if sewage disposal is inadequate; gowns; *hand washing* by patient and personnel
Bowel contents Rectal swabs Material taken from ulcerative lesions during sigmoidoscopy	Culture Culture Culture	None	Ampicillin Tetracyclines Trimethoprim-sulfamethoxazole	Same as above
Stools Vomitus	Culture	Killed vaccine	Fluid and electrolyte replacement	Same as above
Acute and convalescent serums	Agglutinin titers		Tetracyclines	

quent. Drugs of choice are the tetracyclines and streptomycin, often administered in combination.

C. *Contact controls* are applicable only to infected animal reservoirs, or to detection of sources of infection in unpasteurized dairy products from infected herds. Suspected animals are tested for skin hypersensitivity and for agglutinating serum antibodies. Reactive animals are slaughtered.

D. *Epidemic controls* depend on good case report-ing of human infections. These may be mild or unrecognized because of difficulties in bacteriologic or serologic diagnosis. When brucellosis is suspected or confirmed, local health agencies can provide effective investigation of sources and set up appropriate controls on infected animals or their products.

Table 18–2. *(Cont.)*

Clinical Disease	Causative Organism	Other Possible Entry Routes	Incubation Period	Communicable Period
Epidemic diarrhea in nurseries	Enteropathogenic *Escherichia coli*		2–4 days; maximum of 3 weeks	As long as babies remain symptomatic and adult carrier or environmental source remains undetected
Gastroenteritis	*Vibrio parahaemolyticus*		8–48 hr	Not communicable from man to man
	Yersinia enterocolitica	Respiratory?	As above	As long as organisms are in feces
	Clostridium perfringens		As above	Not communicable
	Bacillus cereus		As above	Not communicable
Enterocolitis	*Staphylococcus aureus*		Undefined	While organisms are numerous in feces

Diseases Caused by Ingestion of Preformed Toxin

Staphylococcal intoxication	Staphylococcal enterotoxin		1–6 hours	Not communicable from person to person
Botulism	*Clostridium botulinum* toxin		12–36 hours	Not communicable from person to person

Specimens Required	Laboratory Diagnosis	Immunization	Treatment	Nursing Management
Stools from sick and exposed babies Stools from mothers and personnel	Culture Virus isolation (to exclude viral etiology) Culture	None	Fluid and electrolyte replacement Antibiotic choice depends on testing of isolated organism	Isolation of symptomatic babies; set up clean and contaminated nurseries; consult infection control committee; review techniques in obstetric service and nursery; survey mothers and personnel for symptoms of illness; survey recently discharged babies
Stool specimens, Vomitus	Cultures	None	None	Enteric isolation procedures
As above Lymph nodes	Cultures Cultures	None	Streptomycin Tetracyclines	Same as above
Suspect food Stools	Cultures Cultures	None	None	Supportive
Suspect food Stools	Cultures Cultures	None	None	Supportive
Stools Blood if bacteremia	Smears and cultures Cultures	None	Vancomycin Methicillin Cephalosporin	Enteric isolation procedures
Vomitus Suspected food (if available)	Smear and culture Culture for staph Phage typing Toxicity test	None	None	Supportive
Suspected food Serum	Smear and culture Toxicity for mice Toxicity for mice	None	Antitoxin	Supportive

Table 18-2. *(Cont.)*

Clinical Disease	Causative Organism	Other Possible Entry Routes	Incubation Period	Communicable Period
Infectious Systemic Diseases				
Typhoid fever	*Salmonella typhi*		1-2 weeks, sometimes 3 weeks	Through active infection, and as long as salmonellae are shed in excreta; 10% of convalescents shed bacilli up to 3 months, 2-5% become permanent carriers
Paratyphoid fever	*Salmonella enteritidis var paratyphi* A or B		1-10 days	Through active infection and for as long as salmonellae are shed in excreta; carrier rate is variable
Salmonella septicemia	*Salmonella choleraesuis*		Same as above	Same as above
Brucellosis	*Brucella abortus* *Brucella melitensis* *Brucella suis*	Inhalation Skin contacts	1-3 weeks	Not ordinarily communicable from person to person

Specimens Required	Laboratory Diagnosis	Immunization	Treatment	Nursing Management
Blood cultures	Culture	Heat-killed vaccine	Chloramphenicol	Enteric isolation procedures; concurrent disinfection of all contaminated objects; scrupulous attention to *hand washing*
Stool specimens	Culture		Ampicillin	
Urine specimens	Culture		Trimethoprim-sulfamethoxazole	
Bile drainage				
Sputum				
Pus from abscesses	Smear and culture			
Synovial fluid				
Spinal fluid (as appropriate clinically)				
Acute and convalescent blood serums	Widal titer			
Same as above	Same as above	Same as above	Same as above	Enteric isolation procedures
Same as above	Same as above	Same as above	Same as above	Enteric isolation procedures
Blood cultures	Culture	None available for humans	Tetracyclines	Isolation not necessary; safe disposal of excreta and secretions
Bone marrow			Streptomycin	
Biopsies of lymph nodes, spleen, liver; spinal fluid	Smear and culture			
Serum samples	Agglutinin titers			

III. Prevention. Prevention of brucellosis depends on recognition of its sources in animals; limitation of its spread among them; and control of occupational hazards, infected meats, milk, and other animal foods.

Vaccines for human use are not available. They are sometimes used for the immunization of calves, but only in areas where the disease is prevalent.

Questions

1. Name the three clinical categories of bacterial diseases acquired through the alimentary route.
2. What is "bacterial food poisoning"?
3. What is the most prominent symptom of acute gastrointestinal infections?
4. What is the causative organism of bacillary dysentery?
5. Why is it necessary to send a stool culture to the laboratory without delay?
6. How does botulism differ from other bacterial food poisoning?
7. What conditions are necessary for *C. botulinum* to produce toxin?
8. How can botulism be controlled?
9. What is the causative organism of epidemic diarrhea in hospital nurseries?
10. What is the probable reservoir of infection for epidemic diarrhea?
11. What organism has been implicated in sudden infant death?
12. Where do typhoid bacilli localize? What implications does this have for public health?
13. Describe the effects of *Vibrio cholerae* on the human intestines.
14. How is brucellosis transmitted to man?
15. What control measures are necessary for active brucellosis?

Additional Readings

Darland, Gary; Ewing, W. H.; and Davis, B. R.: *The Biochemical Characteristics of* Yersinia enterocolitica *and* Yersinia pseudotuberculosis. U.S. Department of Health, Education and Welfare, Public Health Service (D.H.E.W. Pub. No. (CDC) 75-8294). Washington, D.C., Jan. 1975.

*Hailey, Arthur: *The Final Diagnosis.* Bantam Books, New York, 1970.

Hirschhorn, Norbert, and Greenough, W. B.: Cholera. *Sci. Am.,* **225**:15, Aug. 1971.

Jordan, C. M.; Powell, K. E.; Corothers, T. E.; and Murray, R. J.: Salmonellosis Among Restaurant Patrons: The Incisive Role of a Meat Slicer. *Am. J. Public Health,* **63**:982, Nov. 1973.

*McGrew, Roderick E.: *Russia and the Cholera, 1823-1832.* University of Wisconsin Press, Madison, 1965.

McGuckin, M.: Microbiologic Studies, Part 4. What You Should Know About Collecting Stool Specimens, *Nursing '76,* **6**:22, Mar. 1976.

*Roueché, Berton: A Game of Wild Indians and Family Reunion in *Eleven Blue Men.* Berkley Medallion Editions, New York, 1953.

Saslow, Milton; Nitzkin, Joel L.; Feldman, Ronald; Boine, William; Pfeiffer, Kenneth; and Pearson, Margaret: Typhoid Fever, Public Health Aspects. *Am. J. Public Health,* **65**:1184, Nov. 1975.

Storlie, Frances: Ann: A Child with Botulism Poisoning. *Nurs. Clin. North Am.,* **6**:563, Sept. 1971.

Taylor, Andrew: Botulism and Its Control. *Am. J. Nurs.,* **73**:1380, Aug. 1973.

Williams, E.: Infectious Diseases, Brucellosis, Part 4. *Nurs. Mirror,* **141**:45, Dec. 25, 1975.

Youmans, Guy P.; Paterson, P. Y.; and Sommers, H. M.: *The Biologic and Clinical Basis of Infectious Diseases.* W. B. Saunders Co., Philadelphia, 1975.

*Available in paperback.

Viral Diseases Acquired Through the Alimentary Route

19

A number of important viral diseases of man are acquired through the fecal-oral route of transmission. Some of the viral agents involved characteristically localize in the intestinal tract following entry but are then disseminated and exert their most pathogenic effects in other tissues of the body. These include the enteroviruses and the agents of viral hepatitis.

There is also a group of viruses that appear to be transient in the intestinal tract as well as in respiratory secretions and that have been difficult to correlate with clinically obvious disease. Their morphology and physicochemical properties are sufficiently uniform to classify them as one group among the RNA viruses, but their name — *reoviruses* — is an acronym revealing uncertainty as to their clinical significance: *r*espiratory *e*nteric *o*rphan viruses. The term *orphan* in this connection denotes a virus unassociated with a recognizable disease entity. Reoviruses (Fig. 19–1) are often encountered in the feces of infants and young children. The frequency of their isolation from stools (they are encountered in

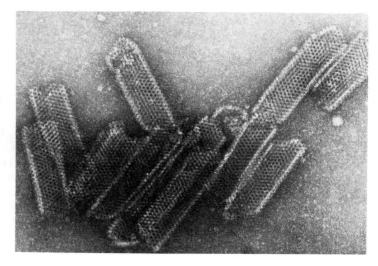

Fig. 19–1. Electron micrograph of reovirus-like particles. (160,000×.) (Courtesy of Dr. Teruo Kimura, Osaka City Institute of Public Health and Environmental Sciences, Osaka, Japan.)

respiratory secretions also) and the prevalence of reovirus antibodies in man suggest that human infection with these viruses is common. Conclusive evidence linking them to clinically recognizable diseases remains lacking, although a reovirus-like agent called *rotavirus,* has recently been associated with acute diarrhea in children. Rotaviruses (named for their spheric shape) are most prevalent during the winter months and are thought to be a major cause of "winter" gastroenteritis in children from six months to two years of age. They have not been successfully cultivated in tissue culture systems, however, and the diagnosis generally must be made by direct electron microscopy of fecal filtrates in which the virus is observed. The technical difficulty and expense of such a procedure have made it difficult to characterize these viruses and to monitor their frequency as disease agents.

In this chapter we shall be concerned with the principal viral diseases transmitted through the alimentary route, and with their clinical and epidemiologic patterns.

Enterovirus Diseases

The enteroviruses are members of a group called *picornaviruses* because they are the smallest of the RNA viruses (*pico* = very small; *rna* indicates the type of nucleic acid they possess). These are spheric viruses, having cubic symmetry, lacking an envelope, and measuring only 20 to 30 nm in diameter (see Table 2–3, p. 36). Rhinoviruses, associated with common colds, constitute a second subgroup of the picornaviruses (Table 14–1).

There are three "families" of enteroviruses: (1) the agents of *poliomyelitis,* (2) *Coxsackie* viruses, and (3) a group known as *ECHO* viruses. (The latter term is derived from the descriptive phrase "*Entero*Cy-topathogenic *H*uman *O*rphan" viruses, originally describing the fact that these organisms were isolated from human enteric sources, produced injury to human cells in tissue culture, but were "orphans" because they were not then recognized as agents of clinical disease.)

All of the enteroviruses are widely distributed in the human population, and all are far more commonly associated with asymptomatic infection of the intestinal tract than they are with clinically apparent disease. When they induce injury, their clinical and epidemiologic patterns are varied. The upper respiratory tract may be involved in their primary entry, localization, and subsequent transmission, but they usually implant more permanently in the bowel. From the latter site they may be disseminated to parenteral tissues of the infected host, causing cell damage and host reaction, and they may also be transmitted, by the fecal-oral route, to the oropharynx of neighboring human hosts.

The clinical patterns of diseases produced by enterovirus infection differ in the nature and tissue focus of lesions produced, and in severity of damage. They range from simple upper respiratory syndromes, such as the common cold, to serious illnesses, including paralytic poliomyelitis, myocarditis, and meningitis. Some Coxsackie and ECHO virus types have been associated only with diseases of the respiratory tract, but others are known to produce several kinds of systemic infections. Conversely, a particular clinical entity (meningitis, bronchitis, pneumonitis) may be caused by any one of several different enteroviruses. The clinical classification of these viruses is, therefore, not very satisfactory, except for their common relationship to the alimentary route of entry, implantation, and transmission. Virologically, they have some physical and chemical properties in common. They are small viruses, in the range of 17 to 28 nm in size; their architecture is similar, as revealed by electron micrography; and their nucleic acid structures are closely related. Antigenically, the three families of poliomyelitis, Coxsackie, and ECHO viruses represent distinct groups. The poliovirus group contains three major types, while the other two contain a much larger number each.

Poliomyelitis

The Clinical Disease. Poliovirus infection is usually inapparent. As a disease, it may take one of three forms, or a combination of these merging together. (1) Most commonly, poliomyelitis is a mild, *nonparalytic* illness characterized as a gastrointestinal disturbance, with fever, nausea and vomiting, headache, sore throat, and drowsiness. The course is short, with recovery in a few days. This syndrome can be diagnosed as an abortive poliomyelitis only by laboratory identification of the virus or of specific antibody response. (2) Patients with nonparalytic poliomyelitis may develop symptoms of meningitis, with stiffness and pain in the neck and back. This is an *aseptic meningitis,* typical of the kind produced by many viruses (see Chap. 13, p. 311). These symptoms may continue up to about ten days and be followed by rapid, complete recovery. The diagnosis again depends on laboratory findings, although clinical diagnosis may be based on the prevalence of typical poliomyelitis among the patient's contacts. (3) Paralytic poliomyelitis may develop without preliminary warning, but more usually follows the initial, mild form of febrile illness. Viral damage to lower motor neurons results in a flaccid paralysis, usually asymmetric and commonly involving lower extremities, although this varies with the site of virus localization. In bulbar polio, there may be brainstem invasion with paralysis of respiratory muscles or spasm and incoordination of nonparalyzed muscles.

Poliovirus enters the body through the mouth, transmitted by fecally contaminated objects or hands or by infectious pharyngeal secretions. It implants in the lymphoid tissue of the pharynx (tonsils) and the intestinal wall (Peyer's patches) and multiplies in these sites. Virus can be isolated both from the throat and from feces during the incubation period and in the early days of illness, but after a week it is difficult to recover from the throat. It is excreted in feces for a number of weeks and is often identified in stools of persons with inapparent infection. Virus can be isolated from the blood of patients with nonparalytic polio, and from others in the preparalytic stage of central nervous system disease.

Circulating viruses probably cross the blood-brain barrier through transcapillary diffusion. They multiply within the neurons, causing variable degrees of injury. The anterior horn cells of the spinal cord are frequently but not exclusively the sites of intracellular localization and multiplication. Posterior horn cells and sensory ganglia may be involved, as well as intermediate gray matter. (The term *poliomyelitis* means that the gray matter [*polio-*] of the spinal cord [*-myel-*] is inflamed [*-itis*].) Damage to these cells may be severe and may proceed rapidly to complete destruction. Cells that are not destroyed may later recover their function. Inflammation appears at the site where neurons are attacked, with focal collections of mononuclear cells and some polymorphonuclears. The virus does not affect peripheral nerves and muscles directly, but these undergo changes reflecting motor neuron damage. Muscle involvement usually reaches its peak quickly, within a few days. Recovery from paralytic polio is slower than in most acute infections, requiring several months, and residual damage often remains crippling for a lifetime.

Laboratory Diagnosis

I. Identification of the Organism. Poliovirus (Fig. 19-2) can be propagated in cells of human or monkey tissue cultures. It can be identified by complement-fixation tests with specific antisera or, more usually, by antibody neutralization of its infectivity for tissue cultures. There are three antigenic types of poliovirus, numbered 1, 2, and 3.

II. Identification of Patient's Serum Antibodies. Antibodies to poliovirus appear early in the disease, whether it is abortive or paralytic. They may also be detected in the blood of persons with inapparent infection, vaccinated persons, and contacts of the latter; therefore, serologic diagnosis of active poliomyelitis requires paired serum samples. The first serum should be collected as soon after onset as possible, the second within three to four weeks of the start of symptoms. Serums are tested with the three antigenic types of poliovirus by neutralization or complement-fixation techniques. The titer of the second sample must be at least four times greater than that of the first to be considered diagnostically significant of active disease.

III. Specimens for Diagnosis

A. *Throat washings* collected within a few days of onset of symptoms may be submitted for virus isolation. The patient gargles with a broth medium supplied by the laboratory. This washing should be kept frozen until it can be transported to a diagnostic virus laboratory.

B. *Spinal fluid* should be collected for virus isolation in cases of aseptic meningitis or paralytic disease and frozen until cultured.

C. *Fecal specimens* or rectal swabs in transport broth may be collected at any time during the course of illness or convalescence, but the optimum time for virus recovery from stools is during the first two

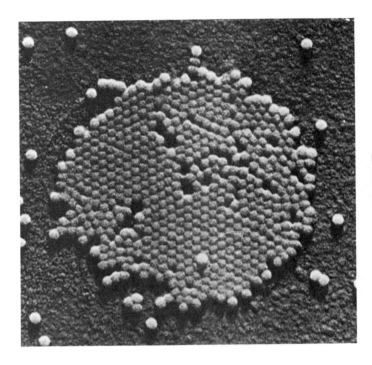

Fig. 19–2. Electron micrograph of the poliomyelitis virus. (148,000×.) (Courtesy of Lederle Laboratories, Division of American Cyanamid Co., Pearl River, N.Y.)

weeks. These samples also should be frozen pending virus culture.

D. *Acute and convalescent serums.* See II above.

Epidemiology

I. Communicability of Infection

A. RESERVOIR, SOURCES, AND TRANSMISSION ROUTES. Man is the only known reservoir of poliovirus infection. Respiratory secretions and fecal contamination are sources of infection during the incubation period and for a short time after onset. Contacts must be fairly close and direct, since infection rates are highest in familial households. When good standards of sanitation and hygiene prevail, direct and indirect oral contacts probably account for transmission of virus, but in areas where living conditions are poor, spread of disease from fecal sources is important. The latter type of spread is enhanced by a warm climate and is greatest during the summer season of temperate climate zones. During epidemic periods, poliovirus has been recovered from flies, roaches, and sewage, but the role played by contaminated vectors, food, or water does not appear to be prominent. Direct human contacts are the most usual factor in transmission.

B. INCUBATION PERIOD. Usually between one and two weeks, but may range from three days to four or five weeks.

C. COMMUNICABLE PERIOD. Virus is present in throat secretions and in feces before the onset of illness. It usually persists in the throat for about a week, and in feces for varying periods, up to about six weeks.

D. IMMUNITY. Type-specific antibodies induced both by clinically inapparent and by recognizable infection confer a durable resistance to reinfection. When second attacks occur, they are related to infection with a different type of poliovirus. Congenital passive immunity is acquired by infants *in utero,* but does not persist for more than about six months, neonatally. Artificial active immunization with poliovirus vaccines provides immunity to all three types of virus. This immunity is durable, provided vaccine is given in adequate initial and booster doses.

Human populations all over the world are susceptible at all ages to poliomyelitis. Many adults have acquired immunity, but those who have not been exposed or vaccinated remain susceptible, as do most children. In areas where the majority of children become immune early in life because of repeated exposures under poor and unsanitary living conditions, poliovirus is maintained in only a small portion of the population. In temperate zones, and among people who maintain good hygienic standards, there may be a high percentage of susceptibles of all ages, subject to an epidemic pattern of spread if poliovirus is introduced. Following such epidemics, the virus is spread in a slow fashion, restricted by sanitary controls, until a new generation grows to the age when it participates most actively in transmission (the preschool years). When strains of high virulence are spread under conditions that promote rapid dissemination (warm weather influences human contacts, environmental reservoirs, and vectors), and the proportion of susceptibles is high, a new epidemic breaks out. The majority of infections in epidemics are inapparent, with about 1 percent of those infected actually displaying clinical disease.

II. Control of Infection

A. ISOLATION. When patients suspected of poliomyelitis are hospitalized, it is not necessary to isolate them, but they should be placed on excretion precautions for the duration of their hospital stay. If this exceeds six weeks, the need for continued precautions should be reevaluated since virus shedding has not been documented beyond that time. Isolation at home is also unwarranted since it has no value for previously exposed contacts, but new exposures should be avoided.

B. PRECAUTIONS. During the first week of illness, respiratory precautions should be followed as well as concurrent disinfection techniques for fecal excreta or articles exposed to fecal contamination. When adequate sewage disposal systems are not available, bedpan collections must be chemically disinfected before discard. Terminal disinfection-cleaning is necessary.

C. TREATMENT. There is no specific antimicrobial therapy for poliomyelitis. Passive immunotherapy appears to have no value after symptoms have begun.

D. CONTROL OF CONTACTS. *Quarantine* of contacts is not justified in view of the usual wide spread of virus before active cases appear. In individual situations, it may be effective to limit activities of family members to prevent their contacts with persons not previously exposed.

Case finding among the contacts of the sick individual is important to assure their treatment and prevent further spread.

E. EPIDEMIC CONTROLS. Case reporting is required so that community protection from epidemic polio can be provided without delay. Since the development of effective vaccines, mass immunization has produced a striking reduction in the incidence of paralytic poliomyelitis. Paralytic cases now occur chiefly in pre-school-age children who have not been vaccinated or have not received adequate doses of vaccine. These are usually children whose families live under the poorest socioeconomic conditions.

When the incidence of reported cases indicates the possibility of an outbreak, mass immunization for the involved community is practiced as rapidly as possible. Oral vaccines can be administered to large numbers of people in a very short time (Fig. 19-3).

III. Prevention

A. IMMUNIZATION. Two kinds of poliovaccine are available. Both provide durable active immunity.

The Salk vaccine contains formalin-inactivated viruses of the three types of polioviruses. It is given in a series of four injections, the first three doses at six-week intervals, and the last one after about six months. This series may be given at any age, from early infancy on. Booster doses must be given every two to three years. The killed virus is disseminated through the body and stimulates production of circulating antibody, which can then neutralize the activity of infective viruses if they reach the blood stream from intestinal sites. Use of inactivated poliovirus vaccine (IPV) does not prevent intestinal infection, or the excretion of virus in the feces of infected persons, but it prevents development of paralytic disease if live virus reaches the bloodstream.

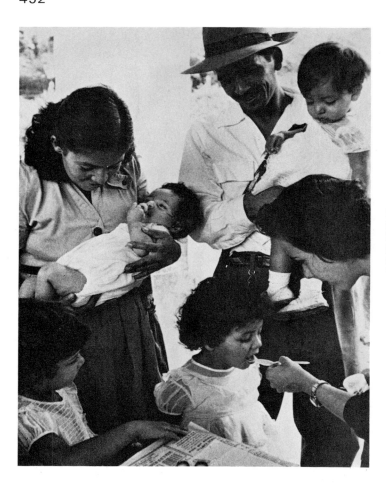

Fig. 19–3. Parents are watching their children receive oral poliomyelitis vaccine at a village center in Colombia, South America. (Reproduced from *World Health,* July 1976. WHO photo by H. Page.)

Oral poliovirus vaccines (OPV), also called *Sabin* vaccines, contain live attenuated strains of all three antigenic types. The live viruses in OPV multiply in intestinal sites, inducing resistance of the alimentary tract to later infection through production of secretory antibodies (sIgA). They also stimulate the production of humoral antibodies that prevent CNS localization of infective virus. For primary immunization with trivalent oral vaccines (T-OPV), infants should receive the first dose at six weeks to three months of age when they are given their first DPT inoculation (see Table 11–1, p. 282). A second dose is fed after an interval of six to eight weeks, and a third given 6 to 12 months later. At about 20 months of

age a fourth dose can be administered (Fig. 19-3), with a final booster for all immunized children at the time they enter school. Primary immunization for older children and teen-agers follows a similar schedule. Adults are not routinely immunized in the United States unless the risk of exposure is increased. Reinforcement immunization should be given to persons of all ages who are directly exposed to active cases or are planning travel to areas where polio is endemic.

Mass immunization with oral poliovirus vaccines has reduced the incidence of paralytic poliomyelitis to extremely low levels. In the United States there were only 14 cases reported in the year 1974, three of these associated with the vaccine itself. About one in several million vaccine recipients develops paralytic disease within a month of taking OPV. Although this frequency is negligible, it constitutes the main reason OPV is not recommended for adults in countries like the United States where the risk of natural exposure to poliomyelitis is very low.

Unfortunately, as the incidence of poliomyelitis has decreased, the public stimulus to maintain effective levels of immunity has also waned. If immunization programs decline, the numbers of susceptibles will again increase, raising the possibility of new epidemics. For the individual victim of paralytic poliomyelitis, the consequences of failure to provide vaccine can be tragically enduring (Fig. 19–4). Furthermore, the declining incidence of poliomyelitis has resulted in loss of clinical experience with a disease in which adequate supportive care may be essential for the patient's survival. A rising case fatality rate in the United States for children under five years of age (12.1 percent in the period 1969–1974 as compared with 4.1 percent in 1960–1962) may reflect inadequate clinical support, arising from inexperience with the complications of poliomyelitis.

The availability of a highly effective vaccine makes it possible to prevent such problems.

B. OTHER PREVENTIVE MEASURES. If epidemics of poliomyelitis appear again in the community, all susceptibles should be protected. Booster immunization may be indicated for prophylaxis. Febrile conditions should be treated with particular care, because undue exertion may promote viremia and CNS involvement in polio infections. Tonsillectomies or other nose, throat, and oral surgery should be avoided for nonimmunized children during seasons in which poliovirus infections are frequent because if children are harboring virus in lymphoid tissue of the oropharynx such operations may release the organism to the bloodstream or directly to adjacent peripheral nerves, and thence to the central nervous system.

Coxsackie Virus Diseases

Clinical Syndromes. The Coxsackie viruses were first recognized in a study of an outbreak of a disease that simulated paralytic poliomyelitis but was milder

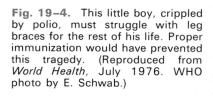

Fig. 19–4. This little boy, crippled by polio, must struggle with leg braces for the rest of his life. Proper immunization would have prevented this tragedy. (Reproduced from *World Health*, July 1976. WHO photo by E. Schwab.)

and did not leave residual paralysis. Clinically, this syndrome was one of aseptic meningitis with some muscle weakness, or paresis. In the laboratory, virus isolated from these cases was shown to be distinct from poliovirus in its antigenic and infective properties, and so it was named for the town (Coxsackie, New York) in which it had caused an epidemic. Since that time (1948), Coxsackie viruses have been found to have a worldwide distribution and to be associated with a number of acute illnesses, some relatively mild and of short duration, others of more serious consequence.

Two antigenic groups of Coxsackie viruses, A and B, are recognized, each containing several serologic types. They are similar to the ECHO viruses in their physical properties, as well as in the patterns of clinical illness produced in man and in their epidemiology. Firm diagnosis of their clinical diseases depends, therefore, on laboratory isolation and characterization of the viral agent in each case. By such means, certain agents and serologic types have been recognized as more prevalent causes of specific syndromes, as follows:

Aseptic Meningitis and Encephalitis. These are among the most common manifestations of Coxsackie virus infection, caused by serotypes of both groups, A and B. Aseptic meningitis caused by Coxsackie viruses cannot be distinguished clinically from that induced by other viral agents. Encephalitic infection may produce paralysis like that seen in poliomyelitis, but this is usually less extensive and more transient when caused by Coxsackie viruses.

Herpangina. This is an acute brief illness characterized by fever, sore throat, difficulty in swallowing, vomiting, and abdominal pain. The pharynx typically shows many vesicular lesions. Recovery occurs within two to four days. This is often a summer illness of children, associated with group A Coxsackie viruses, but older persons may also experience it. Aseptic meningitis has been reported as a further extension of infection, but this is not common.

Pleurodynia. Fever and acute muscle pain in the chest wall characterize this disease. Malaise, headache, and other ill-defined symptoms occur but disappear in a week or two, with complete recovery.

Group B Coxsackie viruses are the chief cause of this syndrome, which sometimes occurs in epidemic form, most often in the summer and fall.

Respiratory Infections. Some group A viruses may produce the rhinopharyngitis typical of the common cold (see Chap. 14). Some strains may also induce nodular lesions of the lymphatic tissue of the pharynx, the syndrome being similar to herpangina but milder. Both group A and B Coxsackie viruses have been associated with primary atypical pneumonia.

Neonatal Myocarditis. Neonatal myocarditis or encephalomyocarditis is associated with Coxsackie group B viruses, but it is not known how babies acquire this infection. During the first week of life, lethargy, diarrhea, vomiting, and feeding problems, sometimes with fever, may occur. These symptoms may disappear or there may be rapid development of cardiac and respiratory distress, the baby being cyanotic and dyspneic, with evidence of myocarditis. The disease may be fatal but infants who recover do not display residual damage.

Other clinical syndromes sometimes associated with Coxsackie viruses include hepatitis, acute conjunctivitis, gastroenteritis, and febrile illnesses that are accompanied by rashes of the face, neck, and trunk.

Laboratory Diagnosis

I. Identification of the Organism. These viruses can be propagated in tissue cultures or in suckling mice and recognized by distinct pathogenic effects. A number of antigenic types in both groups A and B have been identified by specific antisera that neutralize infectivity of the viruses or induce complement fixation when combined with viral antigen.

II. Identification of Patient's Serum Antibodies. Rising titers of neutralizing antibodies appear in patients' serums during the course of these viral infections. Their identification with known strains of viruses provides serologic confirmation of Coxsackie disease.

III. Specimens for Diagnosis. Throat or nasopharyngeal washings and stool specimens are submitted for viral isolation. Spinal fluid may yield the virus in aseptic meningitis, and fecal specimens are positive in this instance also.

"Acute" and "convalescent" serum samples are

useful for serologic diagnosis. The first serum collection is made as soon as possible after onset; the second, in about two weeks.

Epidemiology

I. Communicability of Infection. Coxsackie viruses appear to have a human reservoir only. Their frequent presence in throat washings or feces and the easy spread of infections such as herpangina among familial contacts suggest transmission by fecal-oral or oral-oral direct or indirect contacts. The incubation period is short (a few days) and the communicable period is probably limited to the acute stage of illness. Encounters with these viruses appear to be frequent, resulting in formation of antibodies that persist for years. Many adults possess antibodies to a wide variety of antigenic types.

II. Control of Active Infection. The question of isolation and precautionary care of these infections is often academic, because virologic diagnosis is seldom complete until long after the patient's recovery, and clinical diagnosis is difficult or inconclusive. It has been recommended that cases of aseptic meningitis, particularly if they resemble poliomyelitis, be placed on enteric isolation precautions during the febrile stage. Maternity or nursery patients who develop illness suggestive of Coxsackie B virus infection should be placed under appropriate precautions at once. Attempts to control contacts are not practical, but family members or medical personnel with suspected viral infections should be excluded from visiting maternity and nursery units.

ECHO Virus Diseases

Clinical Syndromes. ECHO viruses are frequently isolated from the human intestinal tract but are less often documented as causative agents of disease. Diagnosis is entirely dependent on laboratory tests in which virus is isolated and identified, or serum antibodies are detected in significant titer, or both.

Diseases that ECHO viruses can produce are similar to those caused by Coxsackie viruses. They range from common colds to febrile respiratory illnesses, gastroenteritis, and aseptic meningitis. The latter syndrome may include muscle weakness and spasm and must be differentiated from poliomyelitis. This type of disease in young children is often accompanied by a rash.

Laboratory Diagnosis. ECHO viruses are best cultivated in monkey kidney tissues. Their isolation from throat washings or feces (spinal fluid, in aseptic meningitis) affords the best diagnosis. More than 30 antigenic types of ECHO viruses have been recognized. It is not uncommon to find more than one type in a given specimen. The size of this group makes serologic diagnosis difficult under ordinary circumstances, for the patient's serum must be tested against the entire battery, often by more than one serologic method.

Epidemiology. The ECHO group appears to have a worldwide distribution. Infections occur frequently during infancy and childhood, and the viruses have a rapid spread among intimate contacts. ECHO infections probably occur far more widely in the community than is recognized on the basis of incidence seen in hospitalized patients, but there are no effective means of controlling community spread. The protection of young infants from older children or adults with acute febrile illnesses is always advisable.

Viral Hepatitis

Viral hepatitis is a major cause of acute liver disease in the United States and other parts of the world, particularly the developing countries. It is caused by at least two separate agents, hepatitis A virus (HAV) and hepatitis B virus (HBV). Clinically the various syndromes produced by HAV and HBV are indistinguishable, but these viruses are serologically distinct and differ in their epidemiologic patterns. Some serologic evidence is emerging that still another type of hepatitis virus (hepatitis C?) may exist and possibly others.

Characteristics of the Viruses

Progress in understanding the nature of viral hepatitis has been slow because of difficulties in cultivating the viruses *in vitro* or *in vivo*. In recent years, however, advances have been made with the development of new specific serologic tests, successful infection of certain nonhuman primates (marmosets and chimpanzees) with hepatitis A virus, and the discovery of a surface antigen of hepatitis B virus in the blood of many human beings. Both HAV and HBV remain unclassified because there is still no convenient and practical means of cultivating them, but a number of their characteristics have been determined by electron microscopy, serologic techniques, and physicochemical assays.

Hepatitis A Virus. HAV is a small, 27-nm particle morphologically similar to picornaviruses and a number of other small, spheric viruses (Fig. 19–5). It can be demonstrated in the feces of patients with acute hepatitis A by using specific HAV antiserum to produce antigen-antibody aggregates that can be seen with the electron microscope. This technique, called immune electron microscopy, or IEM, has also been used to demonstrate clumping of the fecal particles by acute and convalescent serums from patients with acute hepatitis A. It is not yet known whether the particle is the complete virion, but its association with hepatitis A has been well documented by the patterns of its appearance in the feces of infected as compared with uninfected patients. Fecal shedding occurs late in the incubation period and in the early days of illness, corresponding with the time of known fecal infectivity, but the particles have not been seen in stools from uninfected persons or those with other viral infections of the liver.

Hepatitis A, that is, the disease caused by HAV, has been known previously as infectious hepatitis or epidemic hepatitis. Its primary transmission is by the fecal-oral route, and therefore its spread tends to occur in an epidemic pattern. It may also be transmitted by parenteral injection, as shown in volunteers, but serologic studies indicate that it is rarely spread by blood transfusion. Contaminated needles and syringes may play a role among drug addicts, but for practical purposes the natural spread of hepatitis A appears to be from person to person. The route may be direct or by way of contaminated food or water. The incubation period is relatively short, with abrupt onset of symptoms.

Hepatitis B Virus. HBV can infect chimpanzees and rhesus monkeys but has not been successfully propagated in other experimental animals or in tissue cultures. It remains infective for humans for months in serum stored at 4° C and for years if frozen at −20° C. It is resistant to heat, but its infectivity for volunteers is destroyed by heating to 60° C for ten hours, or to 98° C for two minutes. Ether, formalin, activated glutaraldehyde, and 5 percent sodium hypochlorite also inactivate the virus.

The most important discovery with respect to

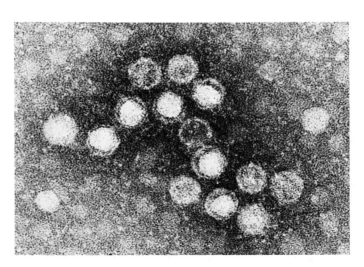

Fig. 19–5. Electron micrograph of hepatitis virus A grown in marmoset liver. (320,000×.) (Courtesy of Dr. Bohdan Wolanski, Merck Sharp & Dohme Research Laboratories, West Point, Pa.)

HBV was the finding of the so-called Australia antigen in human blood. This antigen, first noted in the serum of an Australian aborigine, was later found in the blood of many apparently healthy people as well as convalescents of viral hepatitis. For a time it was referred to as "hepatitis-associated antigen," or HAA, but it is now called "hepatitis B surface antigen," or HB_sAg, because it is known to be derived from the envelope of the hepatitis B virion. Earlier studies of this antigen led to the recognition that it occurs commonly in chronic HBV infection. HB_sAg may be found in the blood for many months or years, but is not necessarily associated with active clinical disease.

As demonstrated by electron microscopy, several structures carrying virus-specific antigens have been found in the blood of patients infected with HBV. A 42-nm spheric particle discovered by Dane, and called the Dane particle, carries HB_sAg on its surface. The core of the Dane particle, measuring about 20 nm, contains another antigenic component called the core antigen, or HB_cAg. The core also contains DNA and an enzyme (DNA polymerase) that catalyzes the formation of new DNA in viral replication.

During active infection, HBV core components are synthesized within the nuclei of liver cells, while the surface or coat antigens form in the cytoplasm of the hepatocytes. Completed viral cores pass from the nucleus into the cytoplasm where they are surrounded by surface antigen. The fully assembled Dane particles may then be released into the bloodstream. The coat antigen, HB_sAg, is produced in excess of the core antigen and may be found in the blood as an unassembled empty particle, lacking DNA and DNA polymerase. It is this antigen that is often found in the blood in chronic HBV infection, sometimes appearing spheric, with a 20-nm diameter, sometimes tubular, in lengths up to 200 nm. The surface of these HB_sAg particles is antigenically complex, and particles of different subtypes can be recognized serologically.

Hepatitis B has often been called serum hepatitis, or homologous serum jaundice, because of its frequent transmission by parenteral injection of blood or blood products. Other routes of spread are now known to be possible, however, for HB_sAg has been found, not only in blood, but also in feces, urine, bile, saliva, tears, semen, spinal fluid, vaginal secre-

tions, breast milk, and cord blood. It is possible that nonparenteral transmission may be common, but the exact mode of such spread is not yet known. The incubation period of hepatitis B is generally much longer than that required for HAV, and the onset is more gradual. The epidemiology of hepatitis B is discussed more fully in Chapter 28, with other infections commonly acquired through medical and surgical procedures. The most prominent differences between HAV and HBV are shown in Table 19–1.

The Clinical Syndromes. Clinically, infection with hepatitis A or B virus may result in a spectrum of syndromes. Often it begins as many other infections do, with fever, malaise, vomiting, anorexia, and discomfort. These symptoms subside after a few days and are succeeded by jaundice. The liver may be enlarged and tender, and bile may appear in the urine. Injury appears to be restricted to the liver, the degree of destruction to parenchymal cells determining the severity and mortality of the disease. In the most fulminating cases, which are unusual, death may occur within about ten days; in more slowly advancing cases death occurs in one or two months. The fatality rate is low, however, and the majority of cases recover, though residual liver damage may take some time to repair. The disease in children is mild and probably occurs more frequently than is recognized. The infection is more severe in adults with underlying liver disorders and in pregnant women.

Clinically inapparent infection is more common than symptomatic disease, the proportion being estimated at 50 to 90 percent.

Laboratory Diagnosis

I. Identification of the Organism. The agents of viral hepatitis have not been successfully cultivated in tissue culture or in experimental animals other than primates. Immune electron microscopy is not practical, or available, on a routine basis for the identification of HAV, and the diagnosis of hepatitis A rests with laboratory tests for liver function, serologic tests to rule out other viral infections (e.g., yellow fever, malaria, infectious mononucleosis), clinical findings, and epidemiologic information.

Hepatitis B is diagnosed in the same way, but in this case routine serologic tests for identification of HBV surface antigen, HB_sAg, are available. These include radioimmunoassay, complement-fixation, counter-

Table 19–1. Some Important Features of Viral Hepatitis A and Viral Hepatitis B

	Viral Hepatitis A	Viral Hepatitis B
Synonym	Infectious hepatitis	Serum hepatitis
Principal sources	Food, water	Blood products
Principal transmission route	Fecal-oral (possibly parenteral, respiratory)	Parenteral injection (possibly oral)
Detectable entity	Virus particle in liver, feces	HB_sAg (surface antigen) and HB_cAg (core antigen) in blood, saliva, other body fluids
Incubation period	15–50 days (average 30)	60–175 days (average 12–14 weeks)
Onset of symptoms	Acute	Gradual
Communicable period	About 2 weeks before to a few days after jaundice	Blood may be infective for months to years
Immunity following infection	For HAV only	For HBV only
Passive immunization	Effective	Poor
Precautions	Enteric	Enteric
Laboratory diagnosis	No isolation techniques	No isolation techniques
	No specific serologic tests available routinely; immune electron microscopy useful in research	Serologic tests for HB_sAg, Hb_s and HB_c antibody

immunoelectrophoresis, and immunodiffusion. Usually HB_sAg is detectable in the blood about four weeks (range, one to seven weeks) before the onset of hepatitis and remains so for one to six weeks. It cannot be found in the blood of all patients with hepatitis B, however, and a negative test does not rule out HBV infection.

II. Identification of Patient's Serum Antibodies. Serologic diagnosis of hepatitis A is not yet possible, since the virus cannot be propagated conveniently and the antigen is not available.

A fourfold or greater rise in anti-HB_s titer following active hepatitis is diagnostically useful for HBV infection. Radioimmunoassay is available for this purpose, as well as passive hemagglutination techniques. Routine tests for anti-HB_c are being developed. Antibody to HB_cAg can be detected from the time of onset of hepatitis B, and a rising titer to this antigen should therefore be helpful in diagnosis.

III. Specimens for Diagnosis. Until convenient methods for virus isolation are developed, the diagnosis is made on clinical and epidemiologic grounds, supported by tests for abnormal liver function and serologic exclusion of other viral diseases.

Acute and convalescent serums may have limited value as outlined in II, above. Blood for serologic tests for HBV surface antigen (HB_sAg) should be collected.

Epidemiology of Hepatitis A

I. Communicability of Infection

A. RESERVOIR, SOURCES, AND TRANSMISSION ROUTES. Man and some nonhuman primates are the reservoir. There are three general sources of orally transmitted infection: (1) water supplies contaminated by human sewage; (2) food such as shellfish from naturally contaminated sources, especially if not cooked at temperatures that inactivate the virus; and (3) food prepared or handled by infectious persons who practice poor hygiene. Cooking in the latter instance is irrelevant if the food is handled just before serving.

Intimate contact with infected persons may result in transmission, and within families or institutions it may lead to epidemics. The most explosive epidemics are likely to originate from drinking water, particularly during summer months under conditions of camp life.

Parenteral transmission is possible through injection of contaminated blood, either by transfusion or through the use of contaminated medical, dental, or tattooing instruments. Blood-sucking insects have been suspected of transmitting hepatitis, but current evidence makes this seem to be an unlikely route of transfer.

B. INCUBATION PERIOD. Varies from 15 to 50 days, with an average of one month.

C. COMMUNICABLE PERIOD. Studies in human volunteers and epidemiologic evidence indicate that the disease is communicable from the latter half of the incubation period and for a few days after onset of jaundice. Recent studies on fecal excretion in humans using immune electron microscopy, however, suggest that the person who already has symptoms of hepatitis A is unlikely to transmit the infection to others.

D. IMMUNITY. Immunity is widespread in the adult population as a result of childhood infection. Second attacks of clinically apparent infectious jaundice are rare. Early infections are probably largely inapparent, or so mild as to escape recognition. The persistence of antibodies in adult serum and their concentration in preparations of gamma globulin make it possible to provide passive immunoprophylaxis for contacts.

II. Control of Active Infection

A. ISOLATION. Hospitalization of uncomplicated cases is discouraged since it exposes other patients and medical personnel to the risks of infection. Isolation at home is unwarranted, since family contacts have been fully exposed before the patient's symptoms begin. In the hospital, a private room is not required except for children or uncooperative adults.

B. PRECAUTIONS. Enteric precautions are emphasized in the care of viral hepatitis patients. Concurrent disinfection techniques should be applied to the disposal of feces, urine, and blood.

If they are not disposable, needles, syringes, and other instruments soiled with the blood of hepatitis patients must be handled carefully before sterilization. Suitable sterilization techniques are (1) autoclaving for 30 minutes at 121° C; (2) dry heat for one hour at 180° C or for two hours at 170° C; or (3) boiling in water for 30 minutes. Sterile disposable needles and syringes provide a convenient solution to this problem.

C. TREATMENT. There is no specific therapy for infectious hepatitis.

D. CONTROL OF CONTACTS. Quarantine of contacts is not indicated, but case finding is important so that close contacts may receive the protection of gamma globulin. Passive antibodies provide protection for about two months if given within the first week after exposure. Later immunization may also be effective, and if exposure is prolonged, a second dose may be indicated.

E. EPIDEMIC CONTROLS. Case reports are advisable to protect contacts and to prevent possible epidemic outbreaks. A search is made for the source of infection when the incidence of cases is unusually great. Mass immunization with gamma globulin is sometimes indicated when institutional outbreaks occur. Sanitary and hygienic standards are reviewed and strengthened as indicated, with particular regard to sewage disposal and the protection of food and water supplies. When outbreaks are associated with hospitals or physicians' offices, a careful review of sterilizing practices is indicated, giving special attention to syringes, needles, and small instruments. The source of blood or blood products given by injection is also investigated.

III. Prevention.

Positive control of hepatitis A awaits the development of a vaccine. If and when convenient animal models of infection are available and the virus has been fully characterized, an effective vaccine may become a possibility.

In the meantime, passive immunization is useful for travelers, medical personnel, and others under risk of exposure. When given to exposed contacts, early in the incubation period, gamma globulin further increases the percentage of asymptomatic infections. Although passive immunity prevents clinical disease from developing, a state of active immunity builds up as a result of the lingering presence of virus.

With regard to preventive measures in food handling, it must be remembered that peak fecal excretion and therefore maximum infectivity occur prior to the onset of symptoms. Since an infected food handler would typically be asymptomatic when most

Table 19–2. Summary: Viral Diseases Acquired Through the Alimentary Route

Clinical Diseases	Causative Organism	Other Possible Entry Routes	Incubation Period	Communicable Period
Poliomyelitis	Poliomyelitis virus	Respiratory tract	Usually 1–2 weeks, may range from 3 days to 4–5 weeks	Prior to onset of symptoms and during infection
Coxsackie virus diseases	Coxsackie viruses	Respiratory tract	3–5 days	During acute, febrile stage of illness
ECHO virus diseases	ECHO viruses	Respiratory tract	Unknown	During acute, febrile stage of illness
Viral hepatitis	Hepatitis A virus	Parenteral route	15–50 days	During incubation period and possibly through first week of symptoms

infectious, the routines of good hygiene and sanitary food processing are essential to the prevention of HAV transmission.

Preventive measures applicable to blood bank practices are described in the discussion of hepatitis B, Chapter 28, page 625.

Questions

1. What are reoviruses? Rotaviruses? Of what clinical importance are they?
2. Name the three "families" of enteroviruses.
3. What cells of the CNS are most susceptible to the poliovirus?
4. How does oral polio vaccine (OPV) protect an individual from acquiring poliomyelitis?
5. What are the most common manifestations of Coxsackie virus infections?
6. What two organisms cause viral hepatitis?
7. How can hepatitis A be controlled immunologically?

Specimens Required	Laboratory Diagnosis	Immunization	Treatment	Nursing Management
Throat washing Spinal fluid Rectal swab in transport broth	Virus isolation	Salk vaccine (killed) Sabin vaccine (live)	None specific	Enteric isolation procedures for duration of hospitalization
Acute and convalescent serum	Rising antibody titer			
Throat and nasopharyngeal washings Stool specimens Spinal fluid	Virus isolation	None	None specific	Enteric isolation during febrile stage; special attention to maternity and nursery patients
Acute and convalescent serum	Rising titer			
Throat washings, feces, spinal fluid	Virus isolation	None	None specific	See above
No microbiologic tests available Blood serum	None Serologic tests to rule out other viral infections (For hepatitis B, serologic tests for HB_sAg, or HB_s and HB_c antibody)	Passive only	None specific	Enteric isolation procedures

Additional Readings

Baranowski, Karen; Greene, H.; and Lamont, T.: Viral Hepatitis: How To Reduce Its Threat to the Patient and Others (Including You). *Nursing '76,* **6**:31, May 1976.

Erickson, Suzanne: Epidemic. *Am. J. Nurs.,* **76**:1311, Aug. 1976.

Hoeprich, Paul D. (ed.): *Infectious Diseases: A Modern Treatise of Infectious Processes,* 2nd ed. Harper & Row, Hagerstown, Md., 1977, Chaps. 67, 68, 84, 112, 115, and 126.

Krugman, Saul: Hepatitis, Current Status of Etiology and Prevention. *Hosp. Prac.,* **10**:39, Nov. 1975.

Melnick, Joseph; Dreesman, G. R.; and Hollinger, F. B.: Viral Hepatitis. *Sci. Am.,* **237**:44, July 1977.

Nahmias, J.; Gomez-Barreto, J.; Kohl, S.; Oleski, J.; and Flax, F.: New Microbial Agents in Diarrhea. *Hosp. Prac.,* **11**:75, Mar. 1976.

Paul, John R.: *A History of Poliomyelitis.* Yale University Press, New Haven, Conn., 1972.

Segal, Herbert E.; Llewellyn, Craig H.; Irwin, Gilbert; Bancroft, William H.; Boe, Gerard P.; and Balaban, Donald: Hepatitis B Antigen and Antibody in the U.S. Army. Prevalence in Health Care Personnel. *Am. J. Public Health,* **66**:667, July 1976.

The General Nature of Parasitic Diseases

Animal parasites capable of living in or on human tissues usually induce some degree of noticeable injury at sites of localization in the body. There are some exceptions to this general rule. There are a few protozoans that are commensal inhabitants of the human bowel, and some helminths (both roundworms and tapeworms) infest the intestinal tract as saprophytes, but for the most part these organisms are truly parasitic. They are dependent on one or more hosts to maintain them in nature, producing damage to the host as a result of parasitic activities. Some of the injury is simply due to their size. Even small protozoans, and immature forms of helminths, although microscopic, can induce structural damage within tissues. Coupled with this is the fact that they are living, metabolizing organisms, competing for nutrient with the body's own cells. Many utilize body materials and add injury by excreting metabolic end products toxic to cells around them, or to distant tissues if disseminated.

Epidemiology

Animal parasites gain entry to the human body through one of three major routes: the *alimentary tract, skin or mucosa,* or the direct *parenteral route.* The majority are taken in through the mouth, being derived from the feces of other infected persons or eaten in infected animal or aquatic plant foods. These are the diseases described in this chapter.

The control and prevention of parasitic diseases are based on knowledge of their communicability, and an understanding of the developmental requirements of parasites and their sometimes prolonged life cycles. The life cycles of animal parasites, and their communicability, are described in Chapter 9.

The parasitic organisms that establish disease following oral entry represent all the major taxonomic groups (protozoans, nematodes, cestodes, and trematodes) and three of the four types of life cycle development displayed by animal parasites:

Type 1, in which the cycle is maintained in one definitive host, and infection transfer occurs either directly from fecal sources or indirectly from fecally contaminated soil where parasite maturation occurs. This group includes the intestinal protozoans and nematodes, such as *Entamoeba, Enterobius, Trichuris,* and *Ascaris.*

Type 2, with a cycle maintained by alternate hosts. *Trichinella spiralis* is the only parasite of importance to man that displays this type of cycle. Its development in each host follows the same pattern, from adult maturation in the intestinal tract following ingestion of larvae encysted in the flesh of an infected animal to the production and dissemination of larvae through the tissues of the new host. Pork is the infected animal source for man.

Type 4, in which the cycle is maintained by a definitive host, together with one or more intermediate hosts. This group includes the *intestinal* cestodes, acquired by ingestion of the larvae-infected flesh of animals (beef, pork, or fish); the *tissue* cestode, *Echinococcus,* acquired by ingestion of the ovum derived from the intestinal tract of a definitive host (usually the dog); and some trematodes whose larval forms are encysted in marine animals or plants that are eaten (the liver and lung flukes).

Most parasites that enter by the alimentary route localize in the intestinal tract, but a few produce parenteral disease by invasion of adjacent tissues, or by disseminating larvae to systemic localizations.

Since 1965 there have been reports of rare cases (about 50 from various parts of the world) of fatal meningoencephalitis in humans caused by free-living amebae. Two genera, *Naegleria* and *Acanthamoeba,* have been incriminated in this form of invasive disease. Most of the patients involved are thought to have acquired the infection while swimming or diving in warm water (fresh or brackish), the organisms gaining entry through the nasal membranes. Although such infections are unusual, they illustrate the invasive potential that ordinarily nonparasitic microorganisms may display if given an opportunity to gain effective entry to the body.

The control of patients with active parasitic infestations never requires their isolation, but suitable precautions must be taken with feces or other excretions in which infective forms may leave the body. When infection can be acquired by direct transfer, as in amebiasis, or in pinworm infestation, measures appropriate to the individual situation are taken. For example, people with active amebiasis should not be involved in food handling until the disease has been treated, because amebae may be transmitted by this route. Controlling the spread of pinworm requires more stringent precautions, because the ova are present on perianal skin, not merely in feces, and are easily transferred by contaminated hands to many surfaces from which they can find their way into the mouth.

The source of any parasitic infestation should be sought, whether it is an infected human contact or an animal food, so that further spread of disease can be prevented. Preventive measures are based on interrupting the chain of transfer of parasites to other hosts in which they can develop. Sanitary disposal of feces and urine, sewage control, and protection of drinking-water supplies are important in any event; animal farming should include measures that prevent infection of animals raised for meat, and meat inspection provides further preventive control. Education of the public in the nature of these diseases, and the necessity for careful selection of government-inspected meats and for their adequate cooking are the final links in the chain of prevention in some cases. In others it rests on public understanding of the role played by certain fish and aquatic plants that are eaten raw or without thorough cooking.

Intestinal Infestation

Protozoans and helminths that parasitize the bowel produce varying degrees of surface injury to the intestinal lining. The most invasive protozoan,

Entamoeba histolytica, actually erodes, penetrating into intestinal submucosa and producing undermined, ulcerative lesions. One ciliated protozoan *Balantidium coli,* also induces ulcerative lesions, but these are usually not deeply erosive. Some species of amebae and flagellates live as saprophytes in the bowel, inducing no injury unless their numbers become excessive, and in this case their motility, activity, or accumulating metabolites may be irritating to lining cells.

Helminth infestation of the bowel is usually not serious per se, unless the worm actually parasitizes the tissues from that site, or its larval forms migrate beyond the bowel. The *hookworm* (acquired through larval penetration of the skin; see Chap. 24) is a notable example of a helminth adult that parasitizes the body from an intestinal site, attaching itself to the mucosa with a cutting mouth through which it ingests the host's blood. This situation is debilitating and can lead eventually to severe anemia if many worms are involved, or if it continues unchecked. In trichinosis and ascariasis, larval migrations into parenteral tissues occur, and these, rather than intestinal infestation, lead to the serious aspects of these diseases.

Many intestinal roundworms and tapeworms live saprophytically in the bowel. Some attach with hooklets and suckers and irritate the mucosa in this way, but derive nutrient from bowel contents, not from living tissue. Some have no means of attachment, but their activities or products may be locally injurious. If they are numerous and large enough (some species of tapeworm may reach up to 20 ft, or 6 meters, in length), they can create intestinal obstruction. The small intestinal roundworms that are not equipped for holding on, for example the pinworm (*Enterobius*) or whipworm (*Trichuris*), often find their way, or are pushed, into the cul-de-sac of the appendix. This seldom causes difficulty, but may be an incidental finding in appendices that have been surgically removed for some other reason.

Invasive Infestation

When animal parasites gain entrance to parenteral tissues, either from intestinal sites or directly through skin (by way of insect vectors or their own penetra-tive capacity), the body attempts to deal with them as it does with any foreign object, marshaling phagocytic cells and fibrinous exudates to contain them. If this succeeds, scar tissue formation and calcium deposit complete the job. The resulting degree of impairment of local tissue function depends on the extent of infestation; it may be minor and unnoticeable, or permanently handicapping. The total effect on health depends on the nature of tissue involved and the extent to which it contributes to the body's vital functions.

Immune response to antigenic components of animal parasites or their products is usually one of cell-mediated immunity and hypersensitivity. This response may occur when antigenic products of intestinal parasites are absorbed from the bowel, but the allergic reaction is most marked in parasitic diseases that are characterized by parenteral localizations of the parasite, in either adult or immature stages. Eosinophilic polymorphonuclear cells, which frequently play a role in allergic responses, are characteristically seen in increased numbers in circulating blood in several parasitic diseases. Hypersensitivity can sometimes be demonstrated by the use of skin-testing antigens extracted from parasites. These antigens evoke skin reactions of the delayed type of hypersensitivity. Humoral immunity is not prominent in the parasitic diseases. Circulating antibodies are generally not evoked by intestinal parasites but may be produced in response to those that invade parenteral tissues. Detection of humoral antibodies by serologic techniques can be of diagnostic value, therefore, in some invasive infestations. Artificial immunization procedures have not been developed, however, because of the generally poor humoral immunity induced by animal parasites and because of the antigenic complexity of these organisms.

Laboratory Diagnosis

Identification of Parasites. Parasitic infestations can seldom be diagnosed on clinical grounds alone. They almost always require laboratory confirmation by microscopic examination of appropriate specimens for ova, larvae, or adult forms of the suspected parasite. Stool specimens are relevant when intestinal infestations by protozoa or helminths are suspected. Small, selected portions of feces may be suspended in saline or water for direct examination as wet-mount cover slip preparations. The numbers of ova or other developmental stages of the parasite may be few in proportion to the quantity of feces passed at any one time. It is often necessary, for this reason, to reduce the bulk of fecal material, using concentration techniques that increase the density of parasitic forms in the residue.

When parenteral infestations are suspected, specimens of urine, sputum, or blood may be relevant, depending on the nature of the parasite in question and clinical indication of tissue localization. In some instances, parenteral lodgments of parasitic forms do not permit their migration into excreta, secretions from superficial tissues, or the bloodstream, and in such cases biopsies of involved tissue may reveal the organisms.

The various forms of animal parasites that may be present in such clinical specimens are usually easily identified by microscopic examination. The ova, larvae, and adult forms of helminthic parasites are large, readily visualized (even without staining, as a rule), and identified by characteristic morphologic features.

Protozoan parasites are also recognizable by microscopic techniques, but they are smaller and sometimes difficult to find in clinical specimens. A few protozoans can be cultivated on artificial laboratory media (e.g., *Entamoeba histolytica, Trichomonas, Leishmania,* or *Trypanosoma*) or propagated in laboratory animals (*Toxoplasma*), but these techniques are not always available on a routine basis.

Serology. When laboratory techniques for demonstrating the parasite itself are unrewarding, serologic methods may be a useful adjunct in diagnosis. This is particularly true for invasive parasitic diseases that evoke a humoral antibody response measurable by sensitive serologic tests. The specificity and sensitivity of current methods for serologic diagnosis have improved the reliability of such testing for a number of parenteral diseases, such as invasive amebiasis, toxoplasmosis, trichinosis, and others. Some of the most useful techniques include indirect hemagglutination (IHA), immunodiffusion (ID), indirect fluorescent antibody (IFA), and complement-fixation (CF).

Intestinal infestations by parasites generally do not stimulate an appreciable immune response, and these diseases cannot as a rule be diagnosed serologically.

Treatment

The treatment of parasitic diseases is often difficult because many antiparasitic drugs, while effective against the parasite, are also toxic for the host. Since the goal must be either complete eradication of the parasite or reduction in its numbers within the infected host, the physician must first be certain of the nature of the parasite, the location and relative intensity of the infestation, and the amount of damage done or threatened. These factors must be weighed against the difficulties inherent in the planned treatment. The physical condition of the patient, particularly one who has been compromised by other diseases, may contraindicate the use of some drugs. Successful chemotherapy requires that the chosen drug exert a maximal effect on the parasite with minimum toxicity for the host. In some instances it may be necessary to treat with a combination of agents, each effective against a different form of the parasite if more than one developmental stage is involved in the infection. The patient's care may also require supportive measures, as well as effective procedures to prevent reinfection.

Treatment of protozoan diseases usually must be aimed at complete elimination of the organisms, since protozoa multiply within the host much as bacteria do. The helminths, however, do not reproduce and increase their numbers in this sense, and for this reason antihelminth therapy may be designed to relieve the patient by reducing the density of worm infestation, rather than by eliminating the last worm.

In recent years, a number of broad-spectrum drugs (metronidazole, levamisole, pyrantel, and others) have been developed for the treatment of intestinal parasitic diseases, both protozoan and helminthic. These are less toxic than some of the older drugs and are well tolerated in most instances. The most serious problems in the chemotherapy of parasitic diseases are encountered in invasive infestations, particularly those that produce chronic, inflammatory lesions (filariasis, schistosomiasis, the trematode diseases, and others). Tissue entrenchments with extensive fibrosis may render chemotherapy virtually useless in such cases, and surgical intervention or supportive treatment may be the only recourse. The problem of drug resistance may also arise, as with some strains of the malaria parasite which now fail to respond to the drugs of routine choice and must be treated with alternative medications.

Intestinal Protozoan Diseases

Giardiasis

Organism. *Giardia lamblia.*

A flagellated protozoan (subphylum Mastigophora) with two developmental stages: a *trophozoite,* which has a tumbling, vibrant motility imparted by several pairs of flagellae located on one side of the pear-shaped body, a sucking disc that helps it to resist peristalsis, and a pair of symmetrically located nuclei which look like spectacles. It is about twice the size of a red blood cell in length, and half this in diameter. The *cyst* is nonmotile, is oval in shape, and has four nuclei when mature. Flagellar protoplasts can be seen within the cytoplasm.

Life Cycle (Type 1A; Fig. 9–9, p. ▮▮▮)
I. Definitive Host. Man is the reservoir.

II. Intermediate Host. None.

III. Human Infection
A. SOURCE OF INFECTION. Infected human feces.
B. INFECTIVE FORM. Cysts from contaminated food and water.
C. LOCALIZATION AND DEVELOPMENT. These organisms reproduce in the upper small intestine. They attach to epithelial cells, but do not invade intestinal tissue.
D. EXIT ROUTE. Feces. Trophozoites and cysts may both be found.

IV. Cyclic Development Outside of Human Host. Probably none.

Characteristics of Disease. This organism ordinarily lives saprophytically in the duodenum and jeju-

num, producing no symptoms. It may cause symptoms when it reaches sufficient numbers to involve large surface areas of the upper intestine. There are chronic diarrhea, dehydration, excessive mucus secretion, and flatulence. The stools are pale in color, possibly because of interference with fat absorption.

Diagnostic Methods. Identification of large numbers of trophozoites or cysts, or both, in feces.

Special Features of Control
I. Treatment. The drug of choice is metronidazole (Flagyl). This drug produces few adverse effects (headache, nausea, and vomiting may occur but are uncommon), but it is teratogenic for experimental animals and therefore contraindicated for pregnant women.

Quinacrine hydrochloride (Atabrine) is also effective, but frequently produces headache, dizziness, and vomiting.

II. Prevalence and Prevention. Giardiasis has a worldwide distribution but is more prevalent in areas of poor sanitation. A high incidence is sometimes associated with contaminated tap water. In the early 1970s, numerous tour groups traveling to Russia experienced a high incidence of giardiasis (23 percent of more than 1400 persons) associated epidemiologically with drinking tap water in Leningrad. Children are more frequently infected than adults. In the United States the carrier rate varies from 1.5 to 20 percent, depending on the age group and community surveyed.

Hospitalized infants and children with giardiasis should be isolated. Enteric precautions should be followed for infected adults.

Prevention depends on the same hygienic measures applicable to other diseases transmitted by the fecal-oral route. Individual sources of infection and vehicles of transmission should be sought, especially when outbreaks occur among families, in institutions, or among travelers.

Balantidiasis

Organism. *Balantidium coli.*
A ciliated protozoan (subphylum Ciliophora) with two developmental stages, a *trophozoite* and a *cyst.*

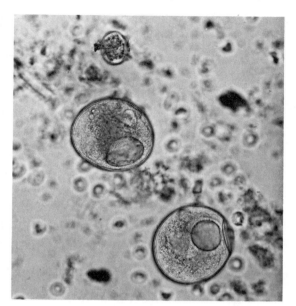

Fig. 20–1. Cysts of *Balantidium coli* with large macronuclei visible. (500×.) (Courtesy of Dr. R. H. Cypress, New York State College of Veterinary Medicine, Cornell University, Ithaca, N.Y.)

The trophozoite has a rapid motility imparted by the beating of cilia that cover its outer surface. Both forms possess two nuclei, one quite large (macronucleus) and the other small (micronucleus). This organism is the largest of the parasitic protozoa, measuring 50 to 200 μm in length and 40 to 70 μm in breadth (Fig. 20–1).

Life Cycle (Type 1A; Fig. 9–9, p. ▮▮▮)

I. Definitive Host. Swine and rats; man an occasional victim.

II. Intermediate Host. None.

III. Human Infection

A. SOURCE OF INFECTION. Infected human or swine feces (the latter exposure occurs on farms or in abattoirs).

B. INFECTIVE FORM. Cysts from indirect, environmental sources, or less frequently via direct human contacts. Human epidemics have developed under conditions of poor environmental sanitation or low standards of personal hygiene.

C. LOCALIZATION AND DEVELOPMENT. Motile trophozoites pass into the large bowel and burrow into mucosal surfaces, producing shallow ulcers. These seldom become undermined or eroded as is characteristic in amebic dysentery caused by *Entamoeba histolytica*. Invasive penetration into parenteral tissues is rare.

D. EXIT ROUTE. Feces. Trophozoites are found in active diarrheal infections; cysts are the usual finding in chronic infections.

IV. Cyclic Development Outside of Human Host

A. IN ENVIRONMENT. Trophozoites pass into more resistant cyst stage.

B. IN HOSTS OTHER THAN MAN. The two forms are the same as those seen in man.

Characteristics of Human Disease. Balantidia produce ulcers or subsurface abscesses in the mucosa and submucosa of the large intestine. Infection may be acute, chronic, or asymptomatic. In acute cases, severe dysentery with bloody, mucoid stools may develop. Chronic cases may display diarrhea alternating with constipation. Extraintestinal extensions have been reported producing cases of balantidial peritonitis, pyelonephritis, and vaginitis, but these are uncommon.

Diagnostic Methods. Microscopic examination of feces may demonstrate trophozoites or cysts.

Special Features of Control

I. Treatment. Antibiotic (tetracycline) and quinine (Diodoquine) drugs are effective in eliminating this parasite. Metronidazole (Flagyl) and paromomycin (Humatin) are alternative choices.

II. Prevalence and Prevention. The incidence of balantidiasis in man is low, although it has a worldwide distribution. Outbreaks have been traced to fecally contaminated water, but sporadic transmission is by soiled hands, food, or flies. Swine and man are the chief reservoirs of infection. Prevention depends on good sanitation and hygiene, as well as identification of any source of common infection.

Isolation of infected patients is not necessary, but

they should be instructed in the principles of personal hygiene.

Invasive Protozoan Diseases

Amebiasis (Amebic Dysentery)

Organism. *Entamoeba histolytica.*

An amebic protozoan (subphylum Sarcodina) with two developmental stages: the *trophozoite* or vegetative form, and a *cyst* stage.

The trophozoite (18 to 20 μm in diameter) is two and one-half to three times the size of an erythrocyte. It has an active ameboid motion and is capable of ingesting red blood cells when it parasitizes the human body (see Fig. 2–17, p. 39). The trophozoite has one nucleus and multiplies either by fission or by encystation. It can be cultivated through several generations on artificial laboratory media if these are properly enriched.

In human infection, *encystation* of the trophozoite occurs in the cecum and colon. The trophozoite rounds up, extrudes undigested food particles, shrinks in size, and secretes a cyst wall. Young cysts at first have one nucleus, but as they mature two consecutive mitotic divisions take place, producing four nuclei in each fully ripe cyst.

When viable mature cysts are swallowed, they pass unchanged through the stomach. When the acid environment around them becomes neutral or slightly alkaline in the small intestine, *excystation* takes place. The cyst wall weakens and the multinucleate ameba within squeezes out. Another mitotic division of the four nuclei ultimately leads to the formation of eight small trophozoites, or amebulae, each with one nucleus in its cytoplasm.

The amebulae do not colonize in the small intestine but are carried along into the cecum, where they may make contact with the mucosa and become lodged in the glandular crypts. Once they begin to feed and grow, the amebulae develop into mature trophozoites, and the developmental cycle is complete.

When examined microscopically, *E. histolytica* trophozoites and cysts are distinguished from those of other intestinal amebae by the details of their nuclear structure, cytoplasmic inclusions, and motility (in the case of trophozoites).

Life Cycle (Type 1A; Fig. 9–9, p. 219)

I. Definitive Host. Man (the only reservoir).

II. Intermediate Host. None.

III. Human Infection

A. SOURCE OF INFECTION. Infected human feces. Can be spread by direct fecal-oral contact, or through contaminated food or water. Water may be a source of epidemics.

B. INFECTIVE FORM. Cysts, which can survive in environmental sources.

C. LOCALIZATION AND DEVELOPMENT. Trophozoites parasitize the large bowel. They enter the cystic stage when fecal material becomes more solid and unfavorable for the trophozoites. Cystic forms may persist and be responsible for an asymptomatic carrier state. Erosion of capillaries in intestinal lesions may lead to dissemination of the organisms to the liver, lung, brain, or other organs.

D. EXIT ROUTE. Feces. Trophozoites are found in active intestinal infections; cysts in chronic intestinal disease or the carrier state. When other localizations occur, either form may be found in pus from the localized lesion, sometimes in sputum if pulmonary abscess location permits bronchial discharge of pus.

IV. Cyclic Development Outside of Human Host

A. IN ENVIRONMENT. Trophozoites may pass into the cyst stage, but are easily destroyed by chemical or physical pressures before this occurs. Cysts may survive for some time but do not evolve into trophozoites until ingested.

B. IN HOSTS OTHER THAN MAN. Does not establish disease in other hosts.

Characteristics of Human Disease. Amebiasis may be acute, chronic, or asymptomatic. Amebic trophozoites invade the wall of the large bowel, feeding on red blood cells. They may erode the local area, creating ulcerative lesions that can extend deeply through the submucosa under the entry point, with undermining of the mucosal surface. In acute dysentery, there may be profuse bloody diarrhea, with mucus in the stool but few pus cells. Chronic amebic dysentery is characterized by intermittent mucoid diarrhea, sometimes with blood and some pus cells, or there may be only occasional discomfort with mild diarrhea or constipation. When extension to parenteral sites occurs, abscesses form at the focus of infection in the liver, lung, brain, or elsewhere. The organism may extend to new sites if these abscesses rupture.

Diagnostic Methods. Microscopic examination of feces is indicated when intestinal infection is suspected. Trophozoites remain active for a very short time only, and for this reason stool specimens should be kept warm during their transport to the laboratory. Cold, casual specimens of stool are ordinarily not satisfactory for the diagnosis of amebiasis. When chronic or asymptomatic amebiasis is suspected, mild purgatives are sometimes useful in increasing the concentration of cysts in stool specimens and facilitating laboratory diagnosis.

Aspirated pus from parenteral lesions, or sputum, may reveal the organisms when systemic localizations are suspected.

Culture techniques are sometimes useful when microscopic examination fails to reveal the amebae or cysts.

Serologic tests are useful in the diagnosis of all forms of amebiasis except the cyst carrier state. They are particularly useful in extraintestinal infections. The most sensitive methods (93 to 97 percent positive in parenteral amebiasis) include indirect hemagglutination, immunodiffusion, and indirect fluorescent antibody tests.

Special Features of Control

I. Treatment. Metronidazole (Flagyl) is the drug of choice for both intestinal and extraintestinal forms of amebiasis, but is contraindicated for pregnant women. This drug is somewhat more effective against trophozoites than cysts but has eliminated the previous need for two agents in treating intestinal amebiasis: one slowly absorbed drug effective against cysts in the bowel, and another for the trophozoite. Emetine in combination with a tetracycline or chloroquine was formerly the preferred regimen and remains a useful alternative treatment.

Surgical aspiration may be necessary for the treatment of abscesses, but chemotherapy should be given first to inhibit dissemination of the infection during the surgical procedure.

II. Prevalence and Prevention. E. histolytica infection is prevalent in many parts of the world, but is associated particularly with areas of poor sanitation, such as the tropics. However, even usually well-controlled water supplies can become contaminated, as in a large Chicago outbreak that claimed about 100 lives, during the World's Fair of the 1930s. Even in temperate zones, the incidence of amebiasis is estimated at about 5 percent, while that of asymptomatic infection is thought to be much higher.

Patients with active amebiasis are not isolated. However, they as well as asymptomatic cyst passers should be removed from food-handling occupations until freed of the organism. Sources of infection among the patient's contacts or in his environment should be found if possible, using laboratory techniques for examination.

Preventive measures on a public scale involve sanitary controls on feces and sewage disposal, the protection of water supplies, and health agency supervision of public restaurants.

Toxoplasmosis

Organism (*Toxoplasma gondii*) **and Life Cycle (Type 4;** Fig. 9–13, p. 223)

An intracellular protozoan belonging to the *Sporozoa,* classified in the order *Coccidia.* Like other sporozoa, *T. gondii* has both sexual and asexual phases in its life cycle, the former taking place in the intestinal epithelium of cats. Forms called *oocysts* are shed in cat feces. These undergo further development in soil, becoming infective for man and other animals in about one to three days. When this form is ingested, the trophozoite develops. The latter is a crescent-shaped organism, 4 to 8 μm in length by 2 to 3 μm in width. It can invade and multiply in cells of the reticuloendothelial system, blood vessels, and visceral organs. Cats, which appear to be the principal reservoir and definitive host of the oocystic form acquire the infection by eating small infected animals (mice) and birds. Man and other warm-blooded animals (dogs, cattle, sheep, rodents, chickens) are intermediate hosts, harboring the invasive stage in parenteral tissues.

Characteristics of Human Disease. Human toxoplasmosis is acquired through either oral or congenital routes of transmission. Orally transmitted disease has a source either in cat feces or in the inadequately prepared meat of infected animals. Congenital toxoplasmosis is transmitted via the placenta from an infected mother who is usually asymptomatic.

The primary infection, orally acquired, seldom induces serious illness and may often go unrecognized, being marked only by fatigue and muscle pains. More acute disease is rare, with a range of symptoms from fever and lymphadenopathy to generalized muscle involvement or cerebral infection. The symptoms are usually nonspecific, resembling those of aseptic meningitis, hepatitis, pneumonia, or myocarditis, depending on localization of the parasite.

Congenital toxoplasmosis is much more severe, sometimes resulting in death of the fetus. Surviving infants may display liver disease, splenomegaly, or brain damage. Abnormalities include hydrocephaly, microcephaly, jaundice, or chorioretinitis. Cerebral calcifications contribute to mental retardation.

Diagnostic Methods. The organisms can sometimes be demonstrated in body fluids or biopsied tissues, but diagnosis depends more frequently on serologic tests. Indirect hemagglutination and indirect fluorescent antibody techniques are most useful.

Special Features of Control

I. Treatment. Sulfonamides combined with pyrimethamine are used for treatment but are not fully satisfactory. They are more effective in acute than in chronic forms of toxoplasmosis.

II. Prevalence and Prevention. Infection is common in man and animals, but clinical disease is uncommon. Preventive measures are of importance primarily for pregnant women who may transmit crippling or fatal disease to their fetuses. They should be alert to the possibilities of acquiring infection through handling or eating raw meat, or through contacts with cats and their litterboxes. Pregnant women who have had close contacts with cats should be followed serologically from the early days of pregnancy. If they display a rising titer of *Toxoplasma* antibody during this period, with evidence of newly acquired infection, decisions may have to be made concerning the advisability of therapeutic abortion.

Pneumocystosis

Organism (*Pneumocystis carinii*) and Life Cycle (Unknown)

Pneumocystis carinii is an organism of uncertain classification believed to be a sporozoan parasite and the causative agent of interstitial plasma cell pneumonia in infants. In sputum and smears from the lung, it appears as an extracellular cyst, rounded or oval with a diameter of 6 to 9 μm, containing up to eight nucleated bodies and enclosed by a viscous wall. This structure is thought to be a stage in the sexual reproduction of the parasite, like that of the *Sporozoa,* but the life cycle of *P. carinii* is not known. Similar structures have been seen in the lungs of many animals, but proof of their epidemiologic significance is lacking.

Characteristics of Human Disease. Pneumocystosis is an acute and often fatal disease in premature infants, occurring from three to six months after birth. It has also been seen as an opportunistic infection in older children and adults with immunodeficiencies or drug-suppressed immune mechanisms. In premature and other compromised infants there is an unusual infiltration of plasma cells; thickening of the alveolar walls; and a frothy, alveolar exudate filled with parasites and host cells. At autopsy the lungs are characteristically heavy and airless.

Diagnostic Methods. The diagnosis is made by microscopic demonstration of the organisms in biopsied lung tissue, aspirated tracheobronchial mucus, or occasionally sputum. Cultural and serologic methods are not satisfactory.

Special Features of Control. Pentamidine is the drug of choice, alone or in combination with sulfadiazine.

Pneumocystosis has been recognized in England and other countries of Europe, Asia, Australia, and North America. Epidemics have been seen in hospital nurseries and institutions, where the organism may be endemic as well. The route of transmission is not known, but nursery outbreaks suggest a direct spread, perhaps from asymptomatic adults via respiratory or oral membranes. The isolation of recognized cases is recommended, or at least their segregation from immunodeficient or premature babies.

Intestinal Nematode (Roundworm) Diseases

Enterobiasis

Organism. *Enterobius vermicularis* (pinworm).

The smallest of the parasitic intestinal roundworms, females measuring about 13 mm in length ($\frac{1}{2}$ to $\frac{3}{4}$ in.), the male about 5 mm. The gravid female (Fig. 20–2, *A*) produces eggs containing well-developed larvae, many of them motile within their shells. The ova are broadly oval, are flattened on one side, and measure about $60 \times 30\ \mu$m (Fig. 20–2, *B*); compare this with the red blood cell's 7-μm diameter.

Life Cycle (Type 1A; Fig. 9–9, p. 219)

I. Definitive Host. Man.

II. Intermediate Host. None.

III. Human Infection

A. SOURCES OF INFECTION. Ova on the perianal skin of infected persons, rarely in feces. Direct or indirect transfer, via clothing, bedding, toys, or other objects, and sometimes food.

B. INFECTIVE FORM. Ingested ovum.

C. LOCALIZATION AND DEVELOPMENT. Following ingestion of ova, the larvae hatch and develop into adults in the intestinal tract. After copulation, the female usually migrates to the anus where she produces thousands of eggs. The adults die, but reinfection from the ova is common, as well as transfer of infection.

D. EXIT ROUTE. Anus and perianal region.

IV. Cyclic Development Outside of Human Host

A. *Environmental* development of the embryonated ova occurs within a few hours, usually while eggs are still clinging to the skin. These ova are relatively resistant to drying and to disinfectants and may remain infective in dust, or on surfaces, for many days, although most die within two to three days.

B. *Hosts other than man* are not involved in the cycle.

Characteristics of Human Disease. Intestinal infestation with adult worms produces few, if any, symptoms. Perianal migrations of the female and oviposition in that region cause an intense local pruritus. This is often particularly irritating at night, inducing restlessness, insomnia, and incessant scratching. Children are frequently infected by other youngsters, and can transfer the infestation to an entire household.

Diagnostic Methods. Pinworm ova may be seen by microscopic examination of Scotch tape applied to the perianal skin. The eggs may also be present in feces, or in material scraped from beneath the fingernails. Adult worms, females and sometimes males, can also be found on the skin of the perianal region.

Special Features of Control

I. Treatment. The major problem in the treatment of pinworm infestation has been prevention of reinfection, difficult to accomplish even with major hygienic efforts. Drugs previously used (e.g., piperazine) required several doses each day for several days, and this course sometimes had to be repeated. The advent of newer anthelminthic drugs, such as pyrvinium pamoate (Povan) and pyrantel pamoate (Antiminth) has improved this picture considerably. These drugs have had a high degree of success (95 to 100 percent cure rates) with administration of a single oral dose. They are relatively nontoxic, palatable, and usually well tolerated.

When pinworms become entrenched in families or institutions, it is now feasible to deworm entire groups two or three times a year.

II. Prevalence and Prevention. Enterobiasis has a worldwide distribution with high infection rates in some areas. It is the most common helminth infection in the United States, with 5 to 10 percent of the population involved. Prevalence is highest among children and their mothers. Because of its easy transmissibility, pinworm infection is usually a familial or institutional infection.

Chemotherapy should be combined with vigorous methods of household disinfection, involving bathrooms, bed linens, bedroom dust. Careful personal

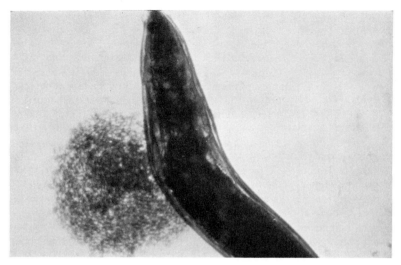

A

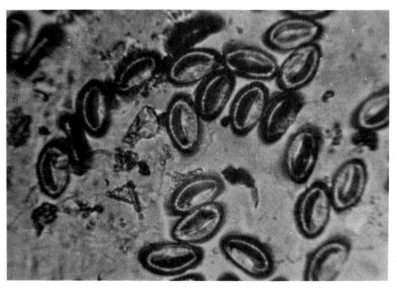

B

Fig. 20–2. *Enterobius vermicularis* (pinworm). *A.* A female pinworm extruding a cloud of eggs. (20×.) *B.* Pinworm eggs (500×) are characteristically deposited on the perianal skin. Each female discharges an average of about 11,000 eggs. (Courtesy of Mr. Hugo Terner, Montefiore Hospital and Medical Center, Bronx, N.Y.)

hygiene is also important, particularly with regard to cleanliness of hands and fingernails.

Trichuriasis

Organism. *Trichuris trichiura* (whipworm).

A slender little roundworm shaped like a whip, the posterior end being thickened, the anterior slim and drawn out. It is about twice the size of the pinworm, measuring up to 5 cm, the male a little shorter.

The characteristic ovum is barrel shaped, thick shelled, with a bulging protuberance at each pole (Fig. 20–3).

Life Cycle (Type 1B; Fig. 9–10, p. 220)
I. Definitive Host. Man.

II. Intermediate Host. None.

Fig. 20–3. A nonembryonated egg of the whipworm, *Trichuris trichiura.* The barrel-shaped ovum with a transparent plug at each end is characteristic of this species. (2500×.) (Courtesy of Mr. Hugo Terner, Montefiore Hospital and Medical Center, Bronx, N.Y.)

III. Human Infection
A. SOURCE OF INFECTION. Soil in which eggs from human feces have matured.

B. INFECTIVE FORM. Embryonated egg.

C. LOCALIZATION AND DEVELOPMENT. After ingestion, the shell of the ovum is digested away; the larva emerges and develops into an adult in the small intestine. Adults attach to the wall of the cecum. Each female can produce five or six thousand eggs per day.

D. EXIT FROM BODY. Eggs discharged in feces.

IV. Cyclic Development Outside of Human Host
IN ENVIRONMENT. Ova require 10 to 21 days in soil for further development of embryo before they are infective. Other living hosts not involved.

Characteristics of Disease. Light infestations produce few, if any, symptoms. Massive infestations may occur in children, with worm attachments extending far down the colon. Inflammation of the mucosa, mucous diarrhea, systemic toxicity, and anemia may result in severe untreated cases. These symptoms resemble those of hookworm disease. Adult infestations are usually asymptomatic.

Diagnostic Methods. Demonstration of *Trichuris* ova in feces, by microscopic examination.

Special Features of Control

I. Treatment. The firm attachment of adult worms to the intestinal wall makes trichuriasis difficult to treat. Mebendazole (Vermox) is a new broad-spectrum anthelminthic that is now the drug of choice. It is given in two daily doses for a period of three days and appears both safe and effective. Thiabendazole is an alternative but has some unpleasant effects, such as nausea, vomiting, and dizziness.

II. Prevalence and Prevention. The whipworm has a wide distribution but is especially prevalent in warm, moist climates where sanitation is poor and crowding contributes to ease of transmission.

Prevention requires good personal hygiene, sanitary disposal of feces, and careful washing of vegetables or other foods that could be contaminated with infective eggs from the soil.

Ascariasis

Organism. *Ascaris lumbricoides.*

The largest of the roundworms, cylindric, tapered, measuring 20 to 30 cm in length.

The ovum is large (50 × 75 μm), spheric to ovoid, with a very thick, rough outer shell (Fig. 20–4). Some infertile eggs, with a thinner shell, are produced also.

Life Cycle (Type 1B; Fig. 9–10, p. 220)
I. Definitive Host. Man.

II. Intermediate Host. None.

III. Human Infection
A. Source of Infection. Soil in which eggs from human feces have matured.
B. Infective Form. Embryonated egg.
C. Localization and Development. After ingestion, larvae hatch from the ova in the duodenum. They penetrate the wall of the small intestine, reach the portal circulation, and are carried through the right side of the heart into the lungs. They may remain in the lungs for several days (Fig. 20–5), then penetrate the pulmonary capillaries, reach the alveoli, and work their way up the bronchial tree. They are coughed up through the epiglottis, swallowed,

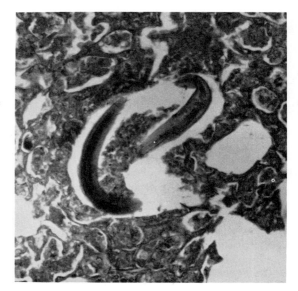

Fig. 20–5. *Ascaris lumbricoides* larvae in lung tissue. (800×.)

and on rearrival in the small intestine develop into adult males and females.

D. Exit from Body. Ova are discharged in feces. (Each female can produce approximately 200,000 eggs per day.)

IV. Cyclic Development Outside of Human Host
In Environment. Ova require two to three weeks of incubation in the soil for maturation of the embryo before they are infective. Other living hosts not involved.

Characteristics of Disease. *Intestinal* infestation may be mild and asymptomatic, or produce abdominal pain, vomiting, diarrhea, indigestion. Masses of worms can cause intestinal obstruction, and systemic toxicities may occur.

Pulmonary infestation with migrating larvae may induce a severe pneumonitis, sometimes with secondary bacterial complications. Fever, spasms of coughing, and asthmatic breathing are characteristic, and allergic reactions indicate hypersensitivity to the ascarids. These may include urticarial rashes and eosinophilia.

Fig. 20–4. The ovum of *Ascaris lumbricoides.* (730×.)

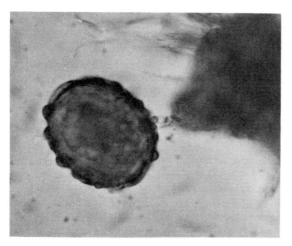

Larvae sometimes reach the general circulation and are lodged in other tissues, producing severe symptoms if they reach such organs as the kidney or brain.

> **Diagnostic Methods.** *Ascaris* pneumonitis is diagnosed clinically and by radiologic techniques. Occasionally, larvae may be seen in sputum. Later, the recognition of ova in feces during intestinal infestation confirms the diagnosis.

Special Features of Control

I. Treatment. Piperazine salts are used to treat intestinal ascariasis. Pyrantel pamoate, levamisole, and thiabendazole are alternative drugs. Intestinal obstruction caused by adult ascarids may necessitate intubation or surgery. There is no specific therapy for migrating larvae.

II. Prevalence and Prevention. Ascariasis is a common, worldwide infection. In moist tropical areas, the prevalence may exceed 50 percent of a population. Infection is heaviest and most frequent in young children of early or preschool age. Control depends on preventing contamination of soil with infective feces and education of children and adults in techniques of sanitation and hygiene.

Invasive Nematode (Roundworm) Diseases

Trichinosis

Organism. *Trichinella spiralis.*

A minute worm, smaller than the pinworm, just visible to the unaided eye. The female is *viviparous;* that is, she deposits viable larvae rather than eggs.

Life Cycle (Type 2; Fig. 9–11, p. 221)

I. Definitive Hosts. Swine, rats, and wild animals are alternate hosts. Man becomes an alternate host by eating infected meat, but does not maintain the parasite in nature.

II. Intermediate Host. None.

III. Human Infection

A. SOURCE OF INFECTION. Meat of infected animal (pork, bear meat).

B. INFECTIVE FORM. Viable larvae encysted in muscle tissue of animal.

C. LOCALIZATION AND DEVELOPMENT. After being swallowed, the larvae are set free from their encystment during gastric digestion of the meat. They pass into the duodenum and develop into male and female adults within four or five days. The females burrow into the mucosa of the villi, or even into submucosal layers or mesenteric lymph nodes, and begin to deposit larvae, which reach the lymphatic and blood circulation and are distributed through the body. Larvae may lodge temporarily in a variety of tissues from which they reenter the circulation. When they reach striated muscle tissue, they become encysted. Muscles most frequently involved are the diaphragm, those of the thoracic and abdominal walls, the tongue, biceps, and deltoid. Larvae cannot develop into adults unless they subsequently are ingested by an alternate host.

D. EXIT FROM HUMAN BODY. None, not communicable by man.

IV. Cyclic Development in Nature

A. IN ENVIRONMENT. None.

B. IN HOSTS OTHER THAN MAN. Carnivorous animals maintain this parasite. Swine are infected from garbage scraps containing infected meat. Adult and larval trichinae both develop in one host, but each worm requires two hosts to maintain itself.

Characteristics of Human Disease. Trichinosis often appears as a small, sporadic outbreak among family or other groups who have gathered around the same infected, undercooked roast pig or bear roast. (Sausages and other products may also contain infective larvae.) The severity of resulting disease depends on the number of viable larvae eaten. Trichinosis may be a mild infestation, or a serious and threatening disease. Initially there may be some gastrointestinal discomfort as the larvae invade the intestinal mucosa and develop into adults. During larval migration there are fever, often a characteristic periorbital edema, and then in a few days muscle pain, chills, weakness, and sometimes prostration. Eosinophilia is usually marked. There may be widespread injury due to localization of larvae. Respiratory distress and myocardial involvement constitute the most threatening events and may cause death. Recovery begins when larval migration ceases, the time varying between three and eight weeks. Residual handicap to muscle function depends on the numbers and location of encysted larvae. These are slowly calcified, but may remain viable for many months.

> **Diagnostic Methods.** Visualization of the parasite in the laboratory requires examination of muscle tissue taken by biopsy (Fig. 20–6). Bits of muscle may be sectioned and stained by histologic techniques, or they can be digested by proteolytic enzymes and examined microscopically for freed, motile larvae.
>
> A skin-testing antigen is available that induces a delayed type of hypersensitive skin reaction. Hypersensitivity to trichina antigen persists for many years following an attack of trichinosis, so that a positive skin reaction is not of itself indicative of current disease.
>
> Circulating antibodies appear in the patient's serum after two or three weeks and provide a useful basis for serologic diagnosis. Precipitin, complement-fixation, and fluorescent antibody tests are commonly used. Bentonite flocculation is a sensitive, specific method often applied in trichinosis. The antigen is prepared in this case by adsorbing it onto bentonite particles. When the patient's serum is added, antibody, if present, combines with the antigen, producing a fluffy flocculation of the bentonite that is readily visible.

Special Features of Control

I. Treatment. There are no specific drugs for the treatment of trichinosis. Thiabendazole is a new drug that has had some success in destroying larvae after muscle penetration, during the acute stage of myositis, but its evaluation is incomplete. Side effects of thiabendazole include nausea, vomiting, and dizziness. It is given together with corticosteroids, which reduce the inflammatory response and thus help to relieve the symptoms caused by larval encystment.

II. Prevalence and Prevention. Trichinosis has a wide distribution, but its prevalence depends on local practices with respect to preparing and eating pork. Control and prevention depend on eliminating sources of infection for hogs (feeding cooked garbage, eliminating infested animals detected by skin testing, destroying rats), processing meat by adequate cold storage or freezing, and final thorough cooking.

Fig. 20–6. *Trichinella spiralis* larva in a preparation of digested muscle tissue. (128×.) The patient from whom this muscle biopsy was obtained had eaten infested bear meat while on a hunting trip. The meat was cooked over a campfire and was not heated sufficiently to kill the viable larvae with which the bear was infested.

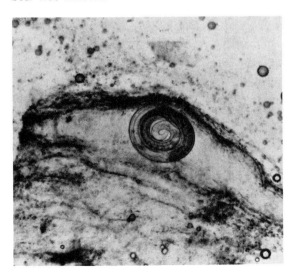

Toxocariasis (Visceral Larva Migrans)

Organisms (*Toxocara canis* and *Toxocara cati*) **and Life Cycle (Type IB)**

Toxocariasis, also known as visceral larva migrans, is a disease of young children caused by the larvae of animal ascarids. *Toxocara canis,* a canine ascarid, and *Toxocara cati,* its feline counterpart, normally infest dogs and cats, which maintain their host-to-soil-to-host life cycles in nature. In animals, this nematode infection is like that of ascariasis in man (see Fig. 9–10, p. 220). Man is an accidental host to whom the parasite is not fully adapted, and when human infection occurs, the life cycle is aborted in the stage of larval migration through viscera.

Man is infected by ingestion of *Toxocara* eggs produced by adult worms residing in the intestinal tract of a dog or cat, passed in animal feces, and matured in the soil. After ingestion, the embryonated eggs hatch in the intestine. Larvae penetrate the wall of the bowel and migrate to liver and lungs in the blood and lymph systems. The immature larvae may wander from the lungs to other tissues, continuing their migrations for years. They do not mature or make their way back to the intestinal tract for adult development, as do the larvae of *Ascaris lumbricoides.*

Visceral larva migrans (VLM) is most common in very young children who have close contact with dogs and cats, and also in those with the habit known as pica (eating unnatural foods such as soil, clay, starch, chalk, ashes). Animals are also infected by ingesting eggs from the soil. Even more important, however, a pregnant female dog can transmit infection to her fetuses through the migration of immature larvae into their tissues.

Characteristics of Human Disease. VLM is usually a self-limited, benign disease, most patients showing complete recovery over a period of years. The severity of disease depends on the intensity of the infestation, however. Clinically, it is characterized by fever, hepatomegaly, and pulmonary symptoms such as coughing and wheezing. The white blood cell count is often elevated, and there is a marked eosinophilia (80 to 90 percent). Endophthalmitis, caused by larval migration to the eye, is sometimes seen, and cerebral or myocardial symptoms occur occasionally.

Diagnostic Methods. Positive serologic tests using *Toxocara* and *Ascaris* antigens (indirect hemagglutination and flocculation tests are performed) suggest the diagnosis but are not conclusive. A definitive diagnosis can be made only by liver biopsy and demonstration of the larvae in the tissue sections. If the worms are few in number, this procedure may not succeed in establishing the diagnosis.

Special Features of Control. *Treatment* of VLM is difficult and should be instituted only in severe cases. Diethylcarbamazine (Hetrazan) or thiabendazole is used, but must be given in multiple daily doses for 30 days or until drug toxicity is manifested. Corticosteroids may relieve severe symptoms and are recommended as the first choice, before anthelminthic drugs are tried.

Prevention is best achieved by controlling contact of children with dogs and cats, particularly with regard to animal feces in play areas. Children's sandboxes, for example, are attractive to cats and should be covered when not in use. Animals should be adequately dewormed, and children should be discouraged from pica. All family members should be taught the importance of hand washing after handling soil and before eating.

Intestinal Cestode (Tapeworm) Diseases

Taeniasis

Organisms. *Taenia saginata* (beef tapeworm), *Taenia solium* (pork tapeworm).

The adult parasites are closely similar morphologically, differing in details of structure of the scolex (*T. solium* possesses a crown of hooklets [Fig. 20–7], *T. saginata* is unarmed) and internal features of the proglottids (Fig. 20–8). Adults are seldom less than 2 meters in length and may reach 6 m (Fig. 2–23, p. 45). The ova of the two species are morphologically identical. They are spheric (35 to 40 μm in diameter), have a striated outer shell, and the embryo within possesses hooklets.

Life Cycle (Type 4; Fig. 9–13, p. 223)
  I. Definitive Host. Man.

  II. Intermediate Host. Cattle (*T. saginata*) and swine (*T. solium*). Man can also serve as an intermediate host for *T. solium*.

  III. Human Infection
  A. SOURCE OF INFECTION. Infected, inadequately cooked flesh of beef or pork.

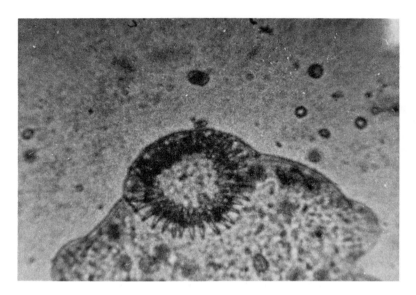

Fig. 20–7. The scolex of *Taenia solium*. Note the crown of hooklets and the muscular suckers. (100×.) (Courtesy of Mr. Hugo Terner, Montefiore Hospital and Medical Center, Bronx, N.Y.)

B. INFECTIVE FORM. Viable larvae encysted in muscle tissue of animals.

C. LOCALIZATION AND DEVELOPMENT. Following ingestion, larvae are excysted in the stomach and pass into the small intestine where the scolex attaches to the intestinal wall, developing into a mature worm within 5 to 12 weeks. The hermaphroditic proglottids produce many eggs.

D. EXIT FROM HUMAN BODY. Eggs are discharged in the feces. Gravid proglottids break off easily and may also be passed in the stool.

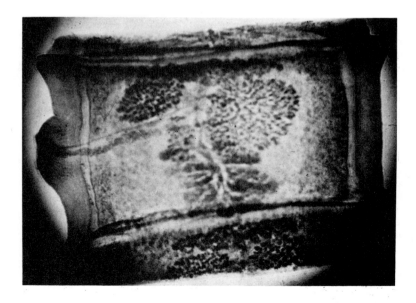

Fig. 20–8. A proglottid of *Taenia solium*. Note the central uterine canal and its few bilateral branches. Mature ova pass from the uterus along the genital canal (seen extending to the left in the photograph) and escape through the prominent lateral genital pore. (8×.)

IV. Cyclic Development Outside of Human Host
Animal hosts ingest eggs derived from infective human feces if this is deposited on soil where animals graze. When the eggs reach the animal's intestinal tract, the larvae are hatched, penetrate the intestinal wall, and are distributed through the body by the bloodstream. They localize and encyst in skeletal muscle, in forms referred to as *bladder worms,* or *cysticerci.* In the case of *Taenia solium,* this invasive form of infestation may also occur in humans (see Cysticercosis, p. 481).

Characteristics of Human Disease. Intestinal infestation with either of these species is relatively mild. There may be diarrhea, epigastric distress, vomiting, and weight loss in persistent infections. Anemia and eosinophilia sometimes occur, but not commonly. Although these worms are quite long, they rarely cause intestinal obstruction. The majority of patients with *Taenia* infections display no symptoms. *T. solium* has a dangerous potential, however, since it can cause the invasive, systemic form of disease called cysticercosis, as discussed later.

Diagnostic Methods. Laboratory diagnosis of intestinal taeniasis is provided by gross recognition of proglottids passed in feces, or microscopic identification of *Taenia* ova.

Special Features of Control
I. Treatment. The scolex must be successfully eliminated in treatment, or a new adult worm will be generated. Niclosamide, given in a single dose, is currently the drug of choice. It is not commercially available, but can be obtained through the Center for Disease Control in Atlanta, Georgia. Paromomycin (Humatin) is the recommended alternative. These drugs are more effective than the older medications used for tapeworms, but they are not without toxicity, causing nausea and abdominal pain. In treating *T. solium* infection, it is important to avoid vomiting, because gravid proglottids and eggs may be brought into the stomach by reverse peristalsis. If this happens, the eggs may hatch in the upper small intestine, penetrate, and cause cysticercosis.

II. Prevalence and Prevention. These tapeworm infections are prevalent wherever beef or pork is eaten raw or lightly cooked. The incidence is highest in eastern Europe, Africa, and Asia, as well as some tropical countries of the western hemisphere. *T. saginata* is more common than *T. solium,* and the latter is rare in the United States and Canada.

Control and prevention of *Taenia* infestations depend on sanitary disposal of human feces and protection of animal feeding grounds. Government meat inspectors reject grossly infected meats in abattoirs (bladder worms can be seen and felt in the tissues), but thorough cooking provides the final safeguard.

Diphyllobothriasis

Organism. *Diphyllobothrium latum* (fish tapeworm).
This tapeworm, like the *Taenia* species, can reach great lengths. The structure of the scolex differs from that of the taeniae (Fig. 20–9), the proglottids are very much broader than long, and the ovum is larger (about $50 \times 75 \,\mu m$). The ovum is broadly oval and has a rather thin shell with a lidded opening (operculum) at one end. The embryo within has hooklets.

Life Cycle (Type 4; Fig. 9–13, p. 223)
I. Definitive Host. Man and many animals (cats, dogs, bears, foxes, walruses, etc.).

Fig. 20–9. Scolex of *Diphyllobothrium latum.* (25×.) Note the long, deep muscular cleft by which this tapeworm attaches itself to the intestinal wall. Compare with the scolex of *Taenia solium* (Fig. 20–7). (Courtesy of Dr. Kenneth Phifer, Rockville, Md.)

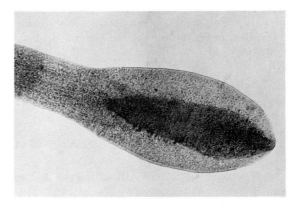

II. Intermediate Hosts. Copepods (aquatic arthropods) and fish.

III. Human Infection

A. SOURCES OF INFECTION. The flesh of infected freshwater fish.

B. INFECTIVE FORMS. Viable larvae in fish tissues.

C. LOCALIZATION AND DEVELOPMENT. After they are swallowed, larvae pass through the stomach into the small intestine. The scolex attaches to the wall and develops into a mature, egg-producing worm.

D. EXIT FROM HUMAN BODY. Eggs are discharged in feces. These are not infective for humans.

IV. Cyclic Development Outside of Human Host

Aquatic hosts ingest the embryos derived from eggs passed in human feces which is deposited in water or contaminates water from draining sewage. When the ova reach fresh water, their ciliated embryos are released and swim about until ingested by the copepod, the first intermediate host. The parasite undergoes some development within the arthropod for two or three weeks. If the latter is then eaten by a fish, the larval form develops further and migrates into the fish's muscles. Fish-eating marine or land animals, including man, then acquire the intestinal infestation by eating the fish raw. Man can avoid the problem by cooking fish thoroughly, or by freezing it for 24 hours at $-10°$ C.

Characteristics of Human Disease. The fish tapeworm does not ordinarily produce serious intestinal symptoms unless it becomes obstructive. Systemic intoxication sometimes results from absorbed metabolic wastes; and in some instances there is a severe primary anemia, produced because the worm in the intestine deprives the host of the vitamin B_{12} contained in his food and thus interferes with red blood cell formation.

> **Diagnostic Methods.** The characteristic eggs are identified by microscopic examination of feces. Proglottids and segments of worms are sometimes passed as well.

Special Features of Control

I. Treatment. As in taeniasis, if treatment does not release the scolex from its point of attachment and eliminate it, a new worm will be generated.

Treatment is the same as for the taeniae, with niclosamide given in a single dose.

II. Prevalence and Prevention. Diphyllobothriasis is endemic in northern and central Europe, parts of the Near and Far East, and on this continent, in Canada and the Great Lakes region, and Florida. These are all areas where freshwater fish are an important food, and where rural systems for sewage disposal may fail to protect fresh water from contamination. Infected human beings are the primary source of fish infestation in endemic areas. Control and prevention require protection of fresh water from human fecal pollution, as well as education of housewives in preparation of fish and the dangers of sampling partially cooked fish foods.

Hymenolepiasis

Organisms. *Hymenolepis nana* (dwarf tapeworm), *Hymenolepis diminuta* (rat tapeworm).

These are the smallest of the intestinal tapeworms, measuring only about 2.5 to 5 cm in length.

Hymenolepis nana infections are common, particularly in the southern United States. Its life cycle (type 1A, Fig. 9–9, p. 219) is unusual, for it is the only tapeworm that does not require an intermediate host. The adult worm lives in the human intestinal tract, producing ova that are discharged in the feces and are directly infective for others if swallowed. Ova remaining in the bowel may also hatch into larval forms within intestinal villi and continue development into adult forms within the bowel lumen (Fig. 20–10). This means that a single host may harbor the parasite indefinitely, as well as pass it along to others. Dwarf tapeworm infestation is generally mild until large numbers of adult worms are formed, causing intestinal irritation and systemic intoxication from absorption of worm products. Treatment with niclosamide or paromomycin is the same as for other tapeworms. Control requires breaking the chain of transfer among human contacts by improved sanitation and hygiene.

Hymenolepis diminuta infests man infrequently. It is primarily an intestinal infestation of rats, kept going by fleas that ingest ova passed in rat feces.

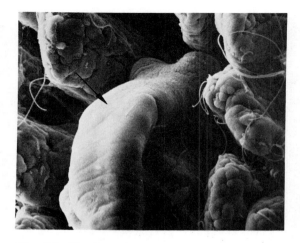

Fig. 20–10. Scanning electron micrograph of an adult tapeworm, *Hymenolepis nana* (arrow), attached by its scolex to the base of the intestinal villi in a mouse. (410×.) (Reproduced from Davis, C. P., and Savage, D. C.: Effect of Penicillin on the Succession, Attachment, and Morphology of Segmented, Filamentous Microbes in the Murine Small Bowel. *Infect. Immun.*, **13**:180, 1976.)

(Fleas are scavengers as well as bloodsucking insects.) The larval stage of development occurs in the arthropod host, which is then ingested by the rat host it parasitizes, and the intestinal cycle begins again (type 4). Human intestinal infection results from accidental swallowing of an infected flea or other arthropod. This situation can arise under squalid, rat-infested living conditions. Rat extermination, relief of poverty, and improvement of living conditions reduce the possibility of human exposure to this parasite.

Invasive Cestode (Tapeworm) Diseases

Cysticercosis

Human cysticercosis may result from ingestion of eggs of *Taenia solium* derived from the feces of a person with an intestinal infestation with this worm. The eggs hatch in the small intestine, and the larvae penetrate through the intestinal wall and are carried by the bloodstream into striated muscle tissue, subcutaneous loci, and sometimes to vital organs, where development of a cysticercus may have severe consequences. Autoinfection is also possible if eggs from an intestinal infestation reach the stomach or upper small intestine by reverse peristalsis. When human infection of this type occurs, no further development of the parasite is possible, for it finds no exit route by which it can reach another host.

The symptoms of cysticercosis are referrable to the localizations of the bladder worms. They have been reported most frequently in subcutaneous tissues, but also in the eye, brain, muscles, and visceral organs. Diagnosis requires excision of a cyst and microscopic examination. There is no specific chemotherapy for this disease, and surgery is the only effective treatment.

Echinococcosis (Hydatid Cyst)

Organism. *Echinococcus granulosus.*

A small tapeworm having only four segments: a scolex, an immature proglottid, one maturing and one gravid proglottid (Fig. 20–11, *A*). The ovum is indistinguishable from that of the *Taenia* species.

Life Cycle (Type 4; Fig. 9–13, p. 223)

I. Definitive Host. Domestic dogs and wild canines.

II. Intermediate Hosts. Sheep, cattle, swine, and sometimes man.

III. Human Infection

A. SOURCE OF INFECTION. Feces from infected animals, usually domestic dogs.

B. INFECTIVE FORM. Egg derived from adult worm living in canine intestine.

C. LOCALIZATION AND DEVELOPMENT. After being swallowed, the eggs hatch in the duodenum; the embryos migrate through the intestinal wall, enter the portal bloodstream, and eventually lodge in capillary filter beds, in the liver and various other tissues. A fluid-filled sac, or *hydatid cyst,* forms at sites of tissue localization (Fig. 20–11, *B*). Within the cysts many immature scolices develop but cannot mature (Fig. 20–11, *C*). The cysts enlarge, creating pressure and architectural damage to tissues. They

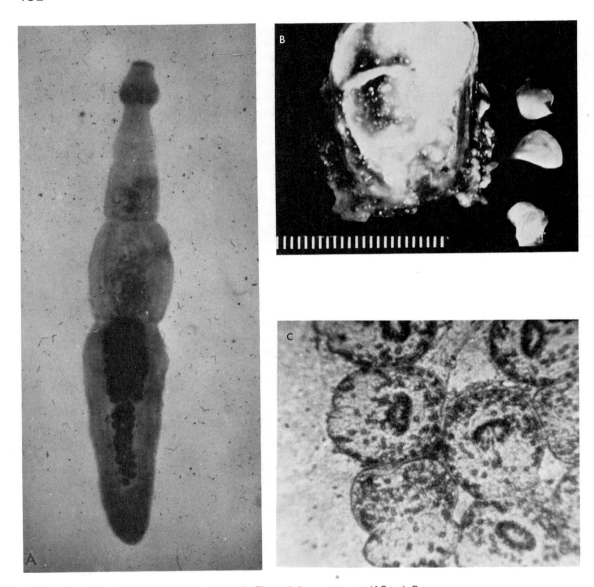

Fig. 20–11. *Echinococcus granulosus. A.* The adult tapeworm. (10×.) *B.* Cysts removed from the liver of a human case (note the numerous daughter cysts). *C.* Immature scolices in hydatid cyst fluid. (100×.) (*C* courtesy of Mr. Hugo Terner, Montefiore Hospital and Medical Center, Bronx, N.Y.)

may rupture and extend their contents to adjacent areas, where new cysts are formed.

D. EXIT ROUTE FROM THE HUMAN BODY. None, not communicable by man.

IV. Cyclic Development in Nature. Dogs and sheep (or other domestic animals raised for meat) ordinarily maintain this parasite. Sheep ingest the ova when they graze in pastures contaminated by infective dog feces, and they develop the larval form of tissue infestation described above for man. In rural areas, dogs usually have access to discarded portions of slaughtered animals. They ingest the larvae in these tissues and develop an intestinal infestation, which completes the cycle.

Characteristics of Human Disease. Echinococcosis is usually quite serious, although symptoms depend on the size and location of the cysts. They sometimes cause death, but they may also persist throughout life, producing little difficulty.

> **Diagnostic Methods.** Diagnosis is usually made on the basis of clinical and radiologic findings. Laboratory identification of the cysts and scolices within them can be made only after their surgical removal. If cysts rupture and discharge their contents to an orifice (bowel, urinary tract, bronchial tree), examination of feces, urine, or sputum may reveal immature scolices.
>
> Skin tests and serologic methods (immunoelectrophoresis, indirect hemagglutination, bentonite or latex flocculation) can help to establish the diagnosis.

Special Features of Control

I. Treatment. There is no specific chemotherapy for this disease. Surgical techniques are useful if the cysts lie in operable sites, and if they have not extended too widely. Cautious surgical dissection is important, because cysts are easily nicked or broken and may spill their infectious contents into the tissues of the surgical field. Preoperative injection of the cyst with alcohol, formalin, or aqueous iodine is recommended to kill scolices prior to surgery. A recently developed technique for freezing the operative site and then injecting a 0.5 percent solution of silver nitrate has been reported effective for killing the scolices.

II. Prevalence and Prevention. Echinococcosis is most common in countries where grazing animals are herded by dogs that also have close contact with humans: South America, the Mediterranean countries of Europe and Africa, and Australia. The disease is rarely acquired in the United States.

Control and prevention depend on breaking the chain of transfer, particularly from sheep or cattle to dogs. The disease declines when strict controls keep dogs away from slaughtered animal tissues. Human disease can be prevented by good household and personal hygiene, particularly when dogs are free-roaming members of the household in an endemic area.

Invasive Trematode Diseases (Liver, Intestinal, and Lung Flukes)

The trematode diseases described here are caused by a small group of animal parasites with many biologic features in common. They share the same group of definitive hosts (man, and some domesticated animals, such as dogs, cats, sheep, and swine) that acquire them by the common route of ingestion; and their cyclic development in intermediate hosts is similar, although this may differ in some details. They differ from blood flukes (*Schistosoma* species) in that their adult forms are hermaphroditic, while schistosomal flukes are bisexual. The latter live in such close association, however, that the distinction is not extremely important. The more useful epidemiologic distinction lies in the fact that larval forms of *Schistosoma* find their way into the body by penetrating exposed skin or mucosa from their environmental sources (see Chap. 24), while the flukes discussed here have an alimentary route of entry.

Life Cycle (Type 4; Fig. 9–13, p. 223)

I. Definitive Hosts. Man, dogs, cats, sheep, and swine.

II. Intermediate Hosts. Two are usual:
Snails are the first intermediate host.
Marine animals (crayfish, crabs, or fish) or aquatic plants (water chestnuts and others) act as second intermediate hosts. These are ingested by man and other land animals.

III. Human Infection

A. SOURCE OF INFECTION. Infective marine life eaten raw or partially cooked.

B. INFECTIVE FORM. Larvae (metacercariae) encysted in marine animals or plants.

C. LOCALIZATION AND DEVELOPMENT. After they are eaten, the encysted larvae are freed, by digestion, of surrounding plant or animal tissue. *Fasciolopsis buski,* the intestinal fluke, fastens to the intestinal wall and develops into an adult at that site. Larvae of other species of this group penetrate the intestinal wall and travel through body cavities or ducts to reach their respective sites of choice. The liver flukes (*Clonorchis* and *Fasciola*) usually settle down in the bile ducts within the liver, while the lung fluke (*Paragonimus*) finds its way from the abdominal cavity through the diaphragm into the pleural cavity, the lungs, or, finally, the bronchioles. The development of larvae into adult flukes is completed in these locations, usually after several weeks, and the hermaphroditic adult worms begin to produce eggs. Depending on the location of the adult worm, ova may find their way directly into feces (*Fasciolopsis*), down the bile ducts into the bowel and the feces (*Clonorchis* and *Fasciola*), or up the bronchial tree with sputum (*Paragonimus*). Some eggs may be deflected from these paths and be caught in surrounding parenchymal tissues, where defensive host reactions embroil them in cellular exudates that hold the ova but disrupt local tissue functions (as the adult worm may also do).

D. EXIT ROUTES FROM HUMAN BODY. Ova of liver and intestinal flukes exit through fecal discharges. Lung fluke ova exit in sputum. Other sites are possible if adults locate in aberrant positions. These ova are not infective for human beings.

IV. Cyclic Development Outside Human Host

A. IN ENVIRONMENT. The ova of these parasites must reach water to continue their developmental cycles. Ciliated embryos (miracidia) escape from their shells and swim about until they are taken up by snails.

B. IN HOSTS OTHER THAN MAN. Snails are the *first* intermediate host for each of these species (also for the schistosomal flukes). The parasites go through maturation stages in snails and emerge again as free-swimming forms called cercariae. These larval forms then encyst again in a second aquatic host — fish, crustaceans, or water plants — until they are ingested by a mammalian host that offers conditions that are suitable for their maturation as adults. The second intermediate host differs for each of these species.

Clonorchiasis

The *Clonorchis* adult lodges in the bile ducts. Symptoms may result from obstruction of the ducts, with jaundice or liver disease due to short migrations of the ova into hepatic tissue, with accompanying cellular reactions. The liver may become enlarged, tender, and cirrhotic. Symptoms may persist for many years. Satisfactorily curative drugs have not been developed for this infestation, but chloroquine phosphate has a suppressive action on the parasite. The disease is acquired through ingestion of larvae encysted in fish. Control requires sanitary disposal of human feces so that the ova of human infections do not reach water where snail and fish hosts may perpetuate the cycle. Thorough cooking of fish is another measure necessary in endemic areas (the Far East and Southeast Asia, including Hawaii, Korea, and Vietnam).

Fascioliasis

This disease occurs in sheep- and cattle-raising countries around the world, including the United States. The adult worm, *Fasciola hepatica,* develops from larvae ingested with infected water plants (such as watercress). It inhabits the bile ducts and produces symptoms referable to obstructive jaundice. Emetine treatment or surgical removal of the adult worm may be effective. Sheep and cattle are the principal animal reservoir. Control depends on elimination of infected animals, snail eradication, and sanitary protection of water greens cultivated for animal or human consumption.

Fasciolopsiasis

The adult trematode, *Fasciolopsis buski,* infests the intestinal tract, producing local irritation with resultant diarrhea, vomiting, anorexia, and other gastrointestinal discomfort. Absorption of toxic products by the worm induces systemic reactions and allergy. The face becomes edematous; the abdomen may be distended and painful with ascites, as well as with allergic edema of the wall. Treatment with hexylresorcinol is effective in clearing the worms from their intestinal attachments, and recovery is good if the accompanying intoxication has not been too damaging. This is primarily a disease of the Orient. Human beings and some of their domestic animals maintain the parasite. Fecal contamination of water leads to the parasite's further development in snails and eventual encystment on water plants, particularly the popular water chestnut, which is often eaten raw. Control can be achieved by sanitary disposal of feces, or its disinfection before use as a fertilizer; destruction of snail hosts in areas where water plants are harvested; or preliminary treatment of such vegetables by a few seconds' dip in boiling water.

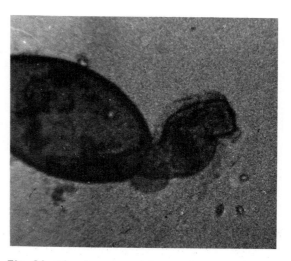

Fig. 20–12. An ovum of *Paragonimus westermani* with an open operculum through which the embryo is emerging. (1000×.) (Courtesy of Mr. Hugo Terner, Montefiore Hospital and Medical Center, Bronx, N.Y.)

Paragonimiasis

The adult lung fluke, *Paragonimus westermani*, infests pulmonary tissues. The ova may be coughed out in sputum, or swallowed and discharged in feces. When they reach water and a snail host, larval development proceeds (Fig. 20–12). Cercariae emerging from snails find their way into crayfish or crabs, and encyst in these second intermediate hosts. Man and other animals (dogs, cats, wild carnivores) that eat these infected crustaceans perpetuate the parasite. The human infestation can be treated with bithional or chloroquine phosphate, the former being the drug of choice. It can be prevented by adequate cooking of crustacean foods. In areas of the world where this disease is endemic (the Far East, parts of India, Africa, and South America), control of the snail host is important.

Table 20–1.   Summary: Parasitic Diseases Acquired Through the Alimentary Route

Clinical Disease	Causative Organisms	Other Possible Entry Routes	Incubation Period	Communicable Period
Intestinal Protozoan Diseases				
Giardiasis	*Giardia lamblia*		1–4 weeks	Throughout infection
Balantidiasis	*Balantidium coli*		Unknown	Same as above
Invasive Protozoan Diseases				
Amebiasis	*Entamoeba histolytica*		Usually 3–4 weeks	Same as above
Toxoplasmosis	*Toxoplasma gondii*	Via placenta	Unknown	Probably no man-to-man transmission except congenital
Pneumocystosis	*Pneumocystis carinii*	Respiratory	1–2 months	Unknown
Intestinal Nematode Diseases				
Enterobiasis	*Enterobius vermicularis*		3–6 week life cycle	Throughout infection
Trichiuriasis	*Trichuris trichiura*		Long, variable	Not communicable from man to man, ova require period of maturation in soil
Ascariasis	*Ascaris lumbricoides*		Several months	Same as above

Specimens Required	Laboratory Diagnosis	Immunization	Treatment	Nursing Management
Stools	Microscopic examination for trophozoites or cysts	None	Metronidazole Quinacrine	Isolation of infants and children; enteric precautions for adults
Stools	Same as above	None	Tetracyclines Quinine Metronidazole Paromomycin	No isolation; personal hygiene
Warm stools Blood serum	Same as above Serology	None	Metronidazole Emetine and tetracycline	No isolation; control food handlers; sanitary disposal of feces
Body fluids Biopsy Blood serum	Microscopic examination for trophozoites Serology	None	Sulfonamides with pyrimethamine	No isolation; instruct pregnant women about source in cats, raw meat
Lung biopsy Aspirated mucus Sputum	Microscopic examination for cysts	None	Pentamidine and sulfadiazine	Isolate known cases from premature or immunodeficient babies
Scotch tape of perianal area	Ova or adults	None	Pyrvinium pamoate Pyrantel pamoate	In home: vigorous disinfection involving bathroom, bed linens, bedroom dust, hands, fingernails; in hospital: concurrent disinfection of fecally contaminated objects, careful *hand washing*
Stools	Ova	None	Mebendazole	Sanitary disposal of feces
Sputum Stools	Larvae Ova or adults	None	Piperazine salts Pyrantel pamoate	Same as above

Table 20–1. (*Cont.*)

Clinical Disease	Causative Organisms	Other Possible Entry Routes	Incubation Period	Communicable Period
Invasive Nematode Diseases				
Trichinosis	*Trichinella spiralis*		2–28 days	Not communicable from man to man
Toxocariasis (VLM)	*Toxocara canis* *Toxocara cati*		Probably weeks or months	Same as above
Intestinal Cestode Diseases				
Taeniasis	*Taenia saginata* *Taenia solium* (see Cysticercosis below)		8–10 weeks	Same as above
Diphyllobothriasis	*Diphyllobothrium latum*		3–6 weeks	Same as above
Hymenolepiasis	*Hymenolepis nana* *Hymenolepis diminuta*		Unknown	*H. nana* communicable while parasite persists in intestines *H. diminuta* not communicable from man to man
Invasive Cestode Diseases				
Cysticercosis	*Taenia solium*			As long as *T. solium* persists in intestines
Echinococcosis	*Echinococcus granulosus*		Long, months to years	Not communicable from man to man
Invasive Trematode Diseases				
Clonorchiasis	*Clonorchis sinensis*		Unknown	Same as above
Fascioliasis	*Fasciola hepatica*			Same as above
Fasciolopsiasis	*Fasciolopsis buski*		About 2 months	Same as above
Paragonimiasis	*Paragonimus westermani*		Unknown	Same as above

Specimens Required	Laboratory Diagnosis	Immunization	Treatment	Nursing Management
Muscle biopsy Blood serum	Larvae Serology	None	Thiabendazole Corticosteroids	No special precautions necessary
Liver biopsy Blood serum	Larvae Serology	None	Corticosteroids	Same as above
Stools	Ova or proglottids	None	Niclosamide Paromomycin	Sanitary disposal of feces (see Cysticercosis below)
Stools	Ova or proglottids	None	Same as above	Sanitary disposal of feces
Stools	Ova	None	Same as above	*H. nana:* sanitary disposal of feces, careful hand washing before eating and after defecation; *H. diminuta:* no special precautions necessary
Cyst excision	Larvae	None	Surgery	No special precautions
Cyst excision Blood serum	Scolices Serology	None	Surgery	Same as above
Stools	Ova	None	Chloroquine	Sanitary disposal of feces
Stools	Ova	None	Emetine	Same as above
Stools	Ova	None	Hexylresorcinol	Same as above
Sputum Stools	Ova Ova	None	Bithionol	Same as above

Questions

1. What is the pathophysiology of amebiasis?
2. What are the two stages of development of *Entamoeba histolytica*?
3. Describe the development of *Ascaris lumbricoides* after ingestion of the egg.
4. What are the definitive hosts of *Trichinella spiralis*?
5. How does man become an alternate host for *Trichinella spiralis*?
6. Why must treatment for tapeworms successfully eliminate the scolex?

Additional Readings

Brown, Harold W.: *Basic Clinical Parasitology,* 4th ed. Appleton-Century-Crofts, New York, 1975.
Diefenbach, W. C. L.: Intestinal Parasites; Common and Becoming More So. *Consultant,* **16**:47, Jan. 1976.
Eveland, L.; Kenney, M.; and Yermakov, V.: The Value of Routine Screening for Intestinal Parasites. *Am. J. Public Health,* **65**:1326, Dec. 1975.
Faust, Ernest C.; Beaver, P. C.; and Jung, R. C.: *Animal Agents and Vectors of Human Disease,* 4th ed. Lea & Febiger, Philadelphia, 1975.
Imperato, Pascal J.; Shookhoff, Howard B.; Marr, John S.; Friedman, S.; and Hiva, C. L.: Parasitic Infections in New York City. *N.Y. State J. Med.,* **77**:50, Jan. 1977.
Katz, Michael: Three Serious Parasitic Infections Often Missed in Clinical Practice. *Postgrad. Med.,* **58**:149, Sept. 1975.
Mizer, Helen E.: The Tapeworm and the Noodle. *Am. J. Nurs.,* **63**:102, July 1963.
Potter, Morris E.; Kruse, Mary B.; Matthews, Muriel A.; Hill, Raymond O.; and Martin, Russell J.: A Sausage Associated Outbreak of Trichinosis in Illinois. *Am. J. Public Health,* **66**:1194, Dec. 1976.
*Roueché, Berton: A Pig from New Jersey. In *Eleven Blue Men.* Berkley Medallion Editions, New York, 1953.
Spencer, Francis M., and Monroe, L. S.: *The Color Atlas of Intestinal Parasites,* rev. ed. Charles C Thomas, Publisher, Springfield, Ill., 1975.

*Available in paperback.

Section VI

Infectious Diseases Acquired Through Skin and Mucosa

21

Epidemiology of Infections Acquired Through Skin and Mucosa

General Principles

Unbroken skin and mucous membranes offer a very effective barrier to most microorganisms, for reasons previously described (Chap. 7). Under normal circumstances, however, there is a good deal of wear and tear on these surfaces, and although they may appear to be intact, minor breaks may frequently occur. Such openings can afford a route of entry to the body for a number of microorganisms whose pathogenic properties permit them to establish themselves, multiply, and induce tissue reaction and injury. In some instances organisms introduced through the skin or mucosa remain localized there, the disease process affecting only superficial tissues, but in many cases wide dissemination of the pathogen may occur, and infection becomes systemic. Microorganisms capable of establishing human disease through entry portals in superficial tissues represent all the major groups of pathogens. Epidemiologically it is important to distinguish between those that can be transmitted through minor breaks in "intact" skin and those that usually must be introduced through penetrative or deep injury. The latter group includes the arthropod-borne infectious diseases as well as those associated with accidental or surgical injury, discussed in the chapters of Section VII.

The pathogenic organisms that can enter through "intact" skin are usually transmitted to man through simple, direct contacts with a source of infection. The most important and largest group of the contact diseases of man have a human reservoir, their sources are infected persons, and they are transmitted directly through human contacts. They include the venereal diseases, staphylococcal and streptococcal infections of the skin, leprosy, a few spirochetal diseases, superficial fungous infections, and bacterial and viral diseases of the conjunctiva and other ocular structures. These diseases associated with *human contacts* are discussed in Chapter 22.

Infectious diseases acquired through *animal contacts* include anthrax and tularemia. These can be transmitted to man through other routes as well, but cutaneous entry is frequent and important, and they should be recognized as contact zoonoses, as described in Chapter 23. Some other infectious diseases such as brucellosis, leptospirosis, and ringworm infections may also be contracted through direct contacts with animals, but in these instances we have emphasized an epidemiologic route of spread and entry that is either more frequent or involves larger numbers of people. Brucellosis, for example, can be transmitted in unpasteurized milk from infected cows, involving many more people than those in direct contact with the infectious source. The fungi associated with ringworm diseases often infect dogs, cats, and domestic farm animals and can be transmitted by them to man, but they are frequently and easily spread through human contacts, as discussed in Chapter 22.

492

Leptospirosis is an enzootic spirochetal infection of wild animals, dogs, and rodents. Man seldom acquires the disease directly from these animals, but he is infected through contacts with *intermediate environmental sources* contaminated by organisms shed through the urinary tract of infected animals. Soil and water may also be the source of certain parasitic diseases, notably hookworm and schistosomiasis. Hookworm larvae require a period of maturation in soil, after being passed in human feces, and are then capable of penetrating intact human skin to begin their cycle of development in a new host. *Schistosoma* larvae (or cercariae, as they are called) also enter the body by wriggling through the skin from their source in infested water. This parasite is maintained in its life cycle by man and by intermediate snail hosts. Man contaminates water with ova shed in feces or urine, and the snail liberates cercariae that have undergone development in its body. The skin itself is merely the entry point for these infectious organisms and is not importantly involved in disease; although sometimes parasitic larvae find themselves unable to progress beyond the skin by way of lymphatic or blood routes. Local tissue reactions may block them, or they may be physiologically incapable of further penetration. Their presence in skin is irritating, however, and induces a dermatitis called "ground itch" in the case of hookworm species, or "swimmer's itch" when *Schistosoma* are involved. These diseases that man acquires through skin contacts with environmental sources of the organism are described in Chapter 24.

Infectious lesions of the skin and mucosal surfaces represent the portal of exit and transmission of the human contact diseases, but man generally does not transmit the disease he acquires from animal contacts. He is an accidental victim of these and does not contribute to their maintenance in nature. In the case of the parasitic diseases, the organisms are perpetuated by human infections, but they are transmitted through and acquired from indirect environmental sources only.

Patient Care

The nursing management of contact diseases is based on a knowledge of their transmissibility from skin lesions and the route of entry their infectious agents may follow on transfer to another individual. Communicability varies greatly in this group, and the necessity for precautionary techniques is correspondingly diverse. General recommendations for the nursing management of diseases transmissible through skin or mucosal contacts are summarized below.

Isolation of the Patient

In many instances, the patient may be placed in partial or strict isolation, depending on his infection and its communicability. *Strict isolation* is indicated for staphylococcal and group A streptococcal infections, inhalation anthrax, and disseminated herpesvirus infections, because these are highly contagious and may take epidemic form (see Table 12-1, p. 299). *Partial isolation* should be provided for skin infections other than those caused by the organisms mentioned above.

Patients with venereal diseases are not isolated when cared for in hospitals. Under other circumstances they are instructed to isolate themselves from sexual contacts until they have been successfully treated. Diseases of animal origin, such as tularemia and leptospirosis, generally do not require isolation, nor is it indicated for hookworm infection or schistosomiasis.

Precautions

The recommendations of the Center for Disease Control regarding wound and skin precautions are shown in Table 21-1. Some of the infections cited are discussed in the last section of this text (VII) because they are wounds or systemic diseases acquired through parenteral rather than contact routes.

The following procedures are of particular importance in the hospital care of patients:

1. Gowns must be worn by all who attend the patient directly.

2. Masks may be necessary when dressings are being changed, but are generally not indicated under other circumstances.

3. Hand washing is essential, particularly after

Table 21–1. Wound and Skin Precautions Recommended by the CDC

Card to Be Posted on Door of Patient's Room

Wound and Skin Precautions

Visitors — Report to Nurses' Station Before Entering Room

1. Private Room — desirable.
2. Gowns — must be worn by all persons having direct contact with patient.
3. Masks — not necessary except during dressing changes.
4. Hands — must be washed on entering and leaving room.
5. Gloves — must be worn by all persons having direct contact with infected area.
6. Articles — special precautions necessary for instruments, dressings, and linen.

Diseases Requiring Wound and Skin Precautions*

Burns that are infected, except those infected with *Staphylococcus aureus* or group A *Streptococcus* that are not covered or not adequately contained by dressings (see Strict Isolation) (27)

Gas gangrene (due to *Clostridium perfringens*) (27)

Herpes zoster, localized (14)

Melioidosis, extrapulmonary with draining sinuses (27)

Plague, bubonic (26)

Puerperal sepsis — group A *Streptococcus,* vaginal discharge (28)

Wound and skin infections that are not covered by dressings or that have copious purulent drainage that is not contained by dressings, except those infected with *Staphylococcus aureus* or group A *Streptococcus,* which require Strict Isolation (22, 27)

Wound and skin infections that are covered by dressings and the discharge is adequately contained, including those infected with *Staphylococcus aureus* or group A *Streptococcus.* Minor wound infections, such as stitch abscesses, need only Secretion Precautions.

*The number in parentheses indicates the chapter in this textbook where the infection is described.

direct contact with the patient. Hands should be washed before and after dressing changes.

4. Concurrent disinfection varies in detail but should include incineration of contaminated dressings or mucosal discharges (in paper bags).

5. Soiled bedclothing and linens should be packaged for the laundry in individual marked bags, to ensure extra care in transport and handling prior to actual washing. The routine laundry technique should assure adequate disinfection of all laundry, with the exception of that contaminated by patho-genic sporeformers (such as the anthrax bacillus or clostridial species; see 6).

6. Linens, instruments, and equipment contaminated with spore-bearing organisms should be sterilized *prior* to their inclusion with materials to be handled by routine procedures. Depending on the nature of the material, sterilization may be accomplished by autoclaving, gas sterilization, or incineration.

7. Concurrent techniques for the cleaning and/or disinfection of units in which patients with infec-

tious diseases are treated should be performed with particular care and attention to detail.

8. The use of special precautionary techniques (other than those listed above) must be based on the individual case, the stage of infection, the site of the infectious agent, and the epidemiology of the disease in question.

Questions

1. List the general recommendations for the nursing management of diseases transmissible through skin or mucosal contacts.
2. For what diseases in this group is isolation generally not indicated?
3. For what diseases in this group is *complete* isolation indicated?

Additional Readings

Huxall, Linda K.: VD, The Equal Opportunity Disease; Part II: The "Social Diseases." *J. Obstet. Gynecol. Neo. Nurs.,* **4**:16, Jan.–Feb. 1975.

Morton, Robert S., and Harris, J. R.: *Recent Advances in Sexually Transmitted Diseases.* Longman Inc., New York, 1975.

Quirk, Barbara: VD, The Equal Opportunity Disease, Part I: This Is VD Too? *J. Obstet. Gynecol. Neo. Nurs.,* **4**:13, Jan.–Feb. 1975.

*Rosebury, Theodore: *Microbes and Morals, the Strange Story of Venereal Disease.* Ballantine, New York, 1976.

Sigerist, Henry E.: *Civilization and Disease.* Cornell University Press, Ithaca, N.Y., 1943.

Vandermeer, Daniel C.: Meet the VD Epidemiologist. *Am. J. Nurs.,* **71**:722, Apr. 1971.

*Available in paperback.

22

Infectious Diseases Acquired Through Skin and Mucosa, Transmitted by Human Contacts

Note: Diseases acquired through respiratory and oral mucosa are discussed in Sections IV and V.

Diseases acquired through penetrative or other injury to skin or mucosa are discussed in Section VII.

The major source of microorganisms that can infect the skin and superficial mucosae appears to be the infected skin or mucosal lesions of other persons. A few of the organisms involved, such as staphylococci and streptococci, may have a reservoir in the normal respiratory tract, but the epidemiology of these infections characteristically indicates transmission from one person to another through direct skin-to-skin contact. Fomites may play an indirect role in the spread of such diseases on occasion, but they are generally not a major factor except in closed populations. Prevention and control are therefore primarily dependent on principles of personal hygiene, as well as early diagnosis and adequate treatment. Effective vaccines and immunoprophylactic measures have proved difficult to develop and are not as yet available for control of the contact diseases.

Sexually Transmitted (Venereal) Diseases

Literally, the term *venereal* means "given or received through love," being derived from the name of the Roman goddess of love, Venus. Clinically, it refers to those infectious diseases that are transmitted only, or most commonly, through intimate sexual contacts.

The microbial agents of the sexually transmitted diseases (STD) are limited to and transmitted by human hosts. They do not survive in the environment or in other hosts under normal circumstances. They can attach to, colonize, and even penetrate mucosal surfaces but probably not normal intact skin, their usual entry routes being the genital or oropharyngeal surfaces.

Ordinarily, venereal diseases occur among adults and adolescents, but children can acquire infections through sexual or other close contacts with adults, and infants may be infected at birth or *in utero*. Infection from environmental sources is not common, but accidental transfer from freshly contaminated objects, infectious discharges, or transfused blood is occasionally reported, particularly in medical or laboratory personnel, or infants.

A variety of potentially pathogenic organisms can be transferred by venereal contacts. Among these, five are consistently associated with disease in adults: the spirochete of syphilis, the bacterial agents of gonorrhea, chancroid, and granuloma inguinale; and the chlamydial agent of lymphogranuloma venereum. Most of these agents cause acute or chronic reactions in tissues adjacent to the site of entry. In some instances, microorganisms transmitted through the genital route produce infections of little or no consequence in adults, but intrauterine infection of infants born of infected mothers may be severe or fatal.

Vaccines capable of providing effective immunity have not been developed against the agents of venereally acquired diseases, but efforts to achieve this kind of control continue. In the meantime, prevention remains an unsolved problem. Prompt diagnosis and adequate treatment of each discovered case afford the only external control possible. This in turn depends on the efficacy of community-wide efforts to inform the public of the nature and dangers of untreated STD. Physicians and laboratories are requested to make prompt reports of diagnosed cases, whereupon local health agencies deploy trained advisors to interview patients, find the sources of their infections if possible, and provide for treatment of all contacts. Such treatment is usually offered without charge in hospital or health department clinics, at government expense.

In the following discussions of individual STD, other special features of their epidemiology are pointed out when significant.

Syphilis

The Clinical Disease. In adults, untreated syphilis progresses through three stages extending over many years, with long intervening periods of latency. The causative spirochetes (*Treponema pallidum*) usually enter the body through intact mucous membranes but sometimes penetrate through breaks in skin at the site of their deposit. During the *primary* stage of disease, they multiply locally, producing a superficial lesion called a *chancre*. Some may also reach the draining lymph nodes of the area and spread to the bloodstream. The chancre usually requires about three to six weeks to develop. It is a painless ulcer with a hard, firm base, located usually on the penis in the male, and on the labia, vaginal wall, or cervix of the female. Depending on the mode of infection, the chancre may appear at other sites, such as the rectum, lip, or palm of the hand, but only one lesion is formed. The regional lymph nodes are often enlarged but not tender. The chancre heals spontaneously, even if not treated, but without therapy systemic infection continues.

Following disappearance of the chancre, the lesions of the *secondary stage* develop from two to four weeks later (usually within three months of the time of infection). Secondary syphilis commonly involves the skin, with lesions on the trunk, extremities, genitalia, palms, and soles of the feet. The rash may be macular, papular, or pustular, the lesions representing foci of disseminated spirochetes. Patchy lesions may also appear on the genital or oropharyngeal mucosa, or in the anal canal. These mucous patches (like the primary chancre) contain many treponemes and are highly infectious. Generalized lymphadenopathy is common in this stage of disease, and there may be other manifestations, such as arthritis, or involvement of the liver, spleen, eye, or central nervous system.

As with primary syphilis, the lesions of secondary disease disappear with or without treatment. If the syphilitic patient is untreated, or if therapy is inadequate, another latent, asymptomatic period ensues. In the early part of the latent stage, up to about two years from the time of infection, syphilis remains potentially infectious if secondary lesions recur. After two years, the late latent stage of the disease is not communicable by sexual contact but may be transmitted to a fetus by an infected mother.

Latency may continue for life, but about one-third of untreated cases advance to the *tertiary stage* of disease, or *late syphilis*. The onset of this stage may be as early as three years or as late as 30 years following the primary infection, but in most cases at least ten years elapse before tertiary symptoms develop. Late syphilis is characterized by the formation of soft, granulomatous lesions called *gummas*, which may occur in a variety of tissues: skin, bones, blood vessels, liver, central nervous system. Spirochetes may be difficult to demonstrate in these lesions, which probably represent a hypersensitive host response to the organism or its products. The location, number, and clinical effects of gummatous lesions are quite varied, producing symptoms like those of many other chronic diseases (syphilis has been called "the great imitator"). Manifestations may be benign, or there may be cardiovascular disease (aortic aneurysm is not uncommon), CNS involvement, or a combination of these. Some of the symptoms of neurosyphilis include *paresis* (incomplete paralysis), *tabes dorsalis* (also called *locomotor ataxia*, a shuffling walk induced by sensory tract degenerations in the cord), deafness, blindness, and progressive mental deterioration. With vigorous campaigns for the discovery and treatment of early syphilis, the tertiary disease has become uncommon in the United States and elsewhere.

Congenital Syphilis. Untreated syphilis in pregnant women may produce congenital disease of the fetus. The spirochete crosses the placental barrier and infects fetal tissues. Miscarriages and stillbirths occur, but the greatest tragedy involves live infants brought to term with generalized (including CNS)

lesions. Sometimes there are no obvious signs of syphilis at birth, but months or years later the congenital case develops blindness, deafness, or insanity. The superficial lesions of congenital syphilis are as infectious as those of the disease acquired in any other way. Syphilitic babies may transfer disease to adults who fondle and kiss them, and conversely, normal babies may be infected in this way by persons in primary or secondary stages of infection.

Laboratory Diagnosis. Syphilis may be diagnosed clinically from the appearance of characteristic lesions and a history of sexual contact with an infected case, but this disease is so protean in its symptomatology and so serious in its consequences that laboratory confirmation is always sought.

I. Identification of the Organism (*Treponema pallidum*)

A. Microscopic Characteristics. Treponemal spirochetes are slender spirals with about 8 to 15 coils evenly spaced at 1-μm distances, so that the entire length is from 8 to 15 μm. They are very thin (about 0.2 μm in width), and their cell walls are rather membranous and flexible. They are actively motile, with a rapid corkscrew rotation on their long axis. Their bodies also may curve in graceful, slowly undulating motions.

Treponemes do not take ordinary stains, but can be visualized in tissue sections treated with silver nitrate. Metallic silver precipitates on the surface of spirochetes, making them appear black against a light background of tissue (Fig. 22–1).

They can also be visualized in wet preparations of material taken from chancres if the microscope is adjusted for *dark-field* illumination, and this method is commonly used to make a rapid, presumptive diagnosis

Fig. 22–1. *Treponema pallidum* in a tissue section treated with silver nitrate. (Reproduced from *Health News,* January 1971. Courtesy of New York State Health Department, Albany, N.Y.)

of syphilis. The microscope technique is described in Chapter 4 (p. 65). The photograph on page 293 illustrates the appearance of a spirochete (a leptospire) when examined by the dark-field technique. Note that it is brightly illuminated against a very dark background, so that its coiled morphology can be clearly seen. The spacing of the coils and differences in motility help to distinguish treponemes from other spirochetes.

Treponema pallidum has not been successfully cultivated on artificial laboratory media, nor does it grow in embryonated chick eggs or in tissue cultures. It must be diagnosed on the basis of microscopic characteristics, in material obtained from active lesions, or by serologic testing of the patient's serum.

II. Identification of Patient's Serum Antibodies. Two types of antibodies develop during the course of syphilitic infection, and both can be detected in the circulating blood. One type is called reagin; the other is specific treponemal antibody.

REAGIN. Reagin is a nontreponemal antibody that, in the test tube, reacts with lipid antigens extracted from beef heart (cardiolipin) and other mammalian tissue, combining with them to fix complement or produce flocculation. Although the antigen is not derived from the spirochete, it is thought to be closely similar to lipid material formed in the body as a result of tissue destruction during active infection or to resemble some antigenic component of the organism. In any case, the antibody formed in response to a lipid that appears during the disease process gives positive reactions in the Kolmer complement-fixation test and in several flocculation tests. The one used most frequently today is the VDRL slide flocculation test (developed by the Venereal Disease Research Laboratory of the U.S. Public Health Service, hence its name). Current modifications of flocculation or agglutination techniques include such rapid reagin tests as Unheated Serum Reagin (USR) and Rapid Plasma Reagin (RPR), used for screening large numbers of patients' serums. These are the common and routine serologic tests for syphilis (STS). They all detect reagin, but these rapid tests are quite sensitive and, if reactive, should be confirmed by the VDRL or other nontreponemal serologic method. Any of these nonspecific methods (including the VDRL) may give false-positive tests for syphilis, particularly in other infectious diseases such as malaria, tuberculosis, leprosy, infectious mononucleosis, and viral hepatitis, as well as other spirochetal infections. Reagins also appear in spinal fluid during syphilitic infections of the central nervous system. In this case there is little likelihood of obtaining a false-positive test, but a negative result merely excludes CNS syphilis. It does not rule out syphilitic infections elsewhere in the body.

SPECIFIC TREPONEMAL ANTIBODY. Specific humoral antibodies against *T. pallidum* also arise during the course of the disease. A number of serologic tests for these have been developed and are particularly useful when reagin tests are suspected of being falsely positive. The first of the treponemal tests was the TPI, or *Treponema pallidum immobilization test,* utilizing live spirochetes cultivated in the testes of living rabbits. When the spirochetes are mixed with serum from a syphilitic patient, their surfaces are coated by antibody and they become immobilized, whereas in the presence of normal serum, lacking specific antibody, they remain typically active. This test is quite expensive, requiring colonies of live rabbits, and has been largely replaced by less difficult confirmatory tests, such as immunofluorescent and microhemagglutination procedures.

The most reliable fluorescent antibody test is called the FTA-ABS (*fluorescent treponemal antibody-absorbed*) technique. This is an indirect fluorescent test in which the patient's serum is brought in contact with lyophilized (freeze-dried) *T. pallidum* that has been fixed to a glass slide. This mixture is then overlaid with a fluorescein-tagged conjugate of antihuman globulin. If the patient's serum contains specific treponemal antibodies, the organisms are coated with fluorescein and can be seen under an ultraviolet microscope (see Chap. 7, p. 172). If the patient's serum has no antibody, there will be no coating and the spirochetes will not be visible. Preliminary treatment of the test serum with a related treponemal strain absorbs out group antibodies for commensalistic treponemes and makes the FTA test still more specific for *T. pallidum.*

Microhemagglutination tests for *T. pallidum* (MHA-TP) are the most convenient specific confirmatory tests now available. These employ sheep red blood cells sensitized with an antigen obtained from *T. pallidum* by ultrasonification. When the red cell–antigen complex is mixed with antitreponemal antibody, visible hemagglutination occurs (the term *micro-* refers to the small volume of blood required for this test). Commercial availability of the antigen and the simplicity of the test procedure are among the advantages of the MHA-TP.

The three confirmatory procedures (TPI, FTA-ABS, and MHA-TP) all have some disadvantages in the serologic diagnosis of syphilis. For example, all give positive results with nonvenereal treponemal diseases such as yaws, pinta, and bejel (see p. 526). This makes it impossible to diagnose latent syphilis in patients exposed to other treponematoses who present no clinical evidence of syphilis. Also, both the FTA-ABS and MHA-TP tests can infrequently give rise to false-positive results in certain patients. The fluorescent method may be positive in some autoimmune diseases, such as rheumatoid arthritis or systemic lupus erythematosus; the hemagglutination test, in similar connective tissue disease syndromes and in infectious mononucleosis.

It should also be noted that the specific treponemal tests remain reactive for many years following syphilitic infection whether the disease is treated or not. Therefore, these tests do not provide information regarding the clinical activity of syphilis, nor are they useful in following response to treatment. Reagin tests, on the other hand, may give rising or falling titers with changes in the status of disease. A rising titer may indicate recent infection, reinfection, or relapse — that is, changes in the activity of disease. A fall in titer signifies adequate treatment, and an unchanging low titer indicates inactivity of infection.

III. Specimens for Diagnosis

A. Tissue fluid expressed from primary or secondary superficial lesions is submitted for dark-field examination. Arrangements must be made with the laboratory for immediate examination of this material, because spirochetes die rapidly when they leave the body and cannot be identified with certainty when they are dead. Patients presenting with primary chancres should be questioned as to any self-treatment they may have attempted. Ointments, salves, and other home remedies do not influence the lesion, but may inactivate spirochetes at the surface, so that the dark-field examination is unsuccessful. In such instances, the lesion should be carefully cleansed with a bland soap or detergent and water, and the patient instructed to return in a few hours, or the following day, having used nothing but water to clean the area. Repeated dark-field examination may then be successful. If this approach creates a risk of losing control of the patient or his contacts, prior to treatment, the physician must weigh the value of a confirmed laboratory diagnosis in deciding whether or not to initiate treatment on clinical evidence alone, without a positive dark-field result. Spirochetes disappear from the superficial lesions within a few hours of antibiotic therapy, and there is usually no later opportunity to demonstrate them in the treated case. Presumptive serologic evidence can be obtained rapidly with the VDRL or other screening test, and if this is positive the diagnostic problem is solved. However, the primary lesion not infrequently appears a week or more in advance of detectable levels of reagin, so that a negative STS is not conclusive at this stage.

B. Late syphilis can be diagnosed by histologic examination of gummatous lesions removed by surgery or biopsied. Silver impregnation techniques are used to demonstrate the organisms in tissues.

C. Serologic tests for reagin in serum (VDRL) become positive between the third and sixth weeks of infection, and between the fourth and eighth weeks in spinal fluid from cases with CNS involvement. These antibodies may persist in low titer for months or years, even in treated cases, but in latent syphilis a small percentage of patients become reagin-negative.

D. Serologic tests for treponemal antibodies (TPI, FTA-ABS, MHA-TP) in the serum remain positive for many years, even in treated cases.

IV. Diagnostic Problems. Four particular problems should be noted in the serologic diagnosis of syphilis:

1. The patient with a history of treated syphilis who displays both reagin and treponemal antibody some years later has not necessarily been reinfected. However, a drop in reagin titer over the years, followed by a new rise, does suggest either reinfection or reactivation of an inadequately treated earlier infection.

2. An asymptomatic patient with latent syphilis may be reagin negative. In such cases, the specific treponemal tests are valuable, for they will be positive.

3. If a reactive reagin test constitutes the only presumptive evidence of syphilis and the primary stage is suspected, the FTA-ABS should be used in preference to the less sensitive MHA-TP, which may be negative in early primary syphilis.

4. If a reactive reagin test cannot be confirmed by any specific treponemal technique and is unsupported by clinical or historic evidence of syphilis, it is considered to be a "biologic false-positive" reaction.

Epidemiology

I. Communicability of Infection

A. RESERVOIR, SOURCES, AND TRANSMISSION ROUTES: see pages 496–97.

B. INCUBATION PERIOD. Average time from exposure to development of primary lesion is about three weeks, but this period may be as short as ten days.

C. COMMUNICABLE PERIOD. Untreated syphilis may be communicable, on the basis of intermittently active infection, throughout the two to four years of the primary and secondary stages, and during the first years of latency. With adequate treatment, infectivity ceases after about one day.

D. IMMUNITY. There is no natural resistance to syphilis among human beings of any age, sex, or race, but infection slowly leads to some immunity against reinfection, probably because of the development of specific treponemal antibodies. (Reagins do not play a protective role.) When syphilis is treated adequately during its early stages, eradication of infection returns the individual to a state of full susceptibility.

II. Control of Active Infection

A. ISOLATION. None. Antibiotic therapy eliminates the threat of communicability within 24 hours. To avoid reinfection, patients should not engage in sexual intercourse with previous partners who have not been treated.

B. PRECAUTIONS. Nursing care of patients with active superficial lesions includes concurrent disinfection of discharges from open lesions and of objects freshly contaminated through direct contacts.

C. TREATMENT. *Treponema pallidum* is highly susceptible to penicillin. Long-acting formulations of this drug, such as benzathine penicillin G, are customarily used to treat early or latent syphilis. The drug is given in two intramuscular doses of 2.4 million units each, separated by an interval of two weeks. If there is any question of the patient's return for the second treatment, a larger dose may be given initially. Patients with hypersensitivity to penicillin may be treated with other antibiotics, such as tetracycline or erythromycin, but these drugs are not as treponemicidal as penicillin, and a watch must be kept for remittent symptoms of active infection.

If syphilitic pregnant women are treated during the first weeks of pregnancy, congenital disease of the fetus can be avoided. After the eighteenth week, *in utero* treatment of the baby is indicated, as well as specific therapy for the mother.

The Jarisch-Herxheimer reaction is a febrile response to specific therapy that may occur in patients in the secondary or later stages of syphilis being treated for the first time. During the preceding weeks or months of spirochetal multiplication and dissemination throughout the body, hypersensitivity develops to antigenic components of these organisms, and many tissues may contain cell-fixed antibodies to these antigens. The sudden killing effect of penicillin or other treponemicidal drugs on numerous spirochetes localized throughout the body releases their antigens and elicits a generalized allergic tissue response characterized by fever and chills. The reaction usually occurs within 12 hours following the first dose of penicillin and subsides within the next 24 hours. It is not likely to occur when syphilis is treated in its primary stage, before hypersensitivity is well developed.

D. CONTROL OF CONTACTS. Prophylactic methods appear to be quite unreliable for the control of syphilis. The use of condoms and diaphragms, soaps and antiseptics has little practical value. Prostitution and other forms of sexual promiscuity assure the continuation of this disease, which has been steadily on the rise during the last few years, after an initial decline following the institution of antibiotic therapy. Case finding through investigation of contacts of known syphilitic infections remains the most effective form of control, through its offer of treatment for otherwise undetected cases. Tracing contacts of a known case is a job for public health nurses or epidemiologists trained in interview techniques and in the clinical implications of syphilis in its various stages. When a primary case is involved, the contacts of the previous three months should be investigated. A case of secondary or early latent syphilis may have spread the disease for six months to a year, while the discovery of a case of tertiary syphilis, or one of long latency, raises the possibility that marital contacts and children may have become infected. When syphilis is not diagnosed until it has reached the tertiary stage, it may be unrewarding to search for contacts beyond the household. With the diagnosis of congenital syphilis, all family members and household or other significant contacts must be investigated. Sources of infection should be sought in advanced as well as currently communicable stages of disease.

E. EPIDEMIC CONTROLS. All state health departments require individual case reports so that efforts can be made to find and treat contacts who represent sources of infection. Efforts to detect and control syphilis include compulsory serologic tests for applicants for marriage licenses, pregnant women, military personnel, and others. Personnel health services in hospitals and most business corporations also require that new employees have a serologic test for syphilis (STS).

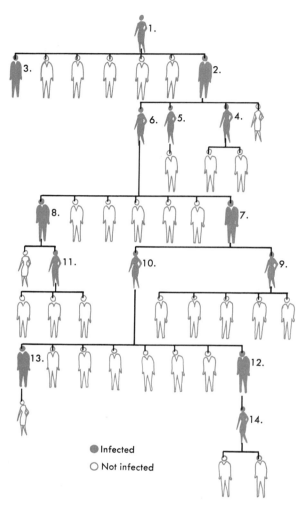

● Infected

○ Not infected

Fig. 22–2. A typical syphilis outbreak. Note how the chain of infection spread from one infected individual. (Reproduced from *Today's VD Control Problem 1975.* American Social Health Association, Palo Alto, Calif., 1975.)

In spite of countrywide and international measures to control it, the disease has spread in new epidemic patterns, particularly among adolescents. Syphilis can be widely disseminated from a single active case. Each new transfer has the same potential for fanning out widely, particularly among the sexually promiscuous (Fig. 22–2). The comfortable

thought that it can now be cured promptly, with relative ease, by penicillin has perhaps been misleading. The early cure of infection is essential in controlling its spread, but it leaves little or no residual resistance to reinfection. Furthermore, primary infection may go unnoticed (particularly in women) and therefore untreated, being spread from such silent sources to large numbers of contacts, including the children congenitally infected. Renewed and intensified efforts are being made by health agencies to maintain control through case finding and treatment.

III. Prevention. The prevention of this and other venereal diseases depends on the efforts of an informed public, armed with a basic knowledge of sex and health, to promote the social and physical welfare of its young people, in particular. Working together, the general public, its public health services, and the medical community could reduce the problem. More secure control and prevention await the development of an effective vaccine. This in turn depends on new methods for the cultivation of the spirochete, *in vitro* or in a practical animal model.

Gonorrhea

Gonorrhea has been known as a human disease from time immemorial, one of its earliest descriptions appearing in an Egyptian papyrus dating from 3500 B.C. The written records of every developing civilization, from China to Mesopotamia to early Rome, contain descriptions and prescriptions relating to a disease known for its characteristic urethral discharge or "issue" and the strictures or "strangury" that so often followed. It was Hippocrates, in the third century B.C., who defined its mode of transmission, listing it among the consequences of the "excesses of the pleasure of Venus." The misnomer "gonorrhea" came from the Roman physician Galen in the second century A.D. Galen mistakenly assumed that the male urethral discharge represented an involuntary loss of semen and coined the term (*gonos* = seed; *rhoia* = flow), which has remained in the medical language ever since. In the welter of vernacular and medical terms that have been applied to this disease, perhaps the phrase "the Great Sterilizer" holds the strongest implications of its consequences when untreated.

The Clinical Disease. Gonorrhea is a venereal disease of worldwide distribution that has reached alarming epidemic proportions in the past decade (Fig. 22–3). It is a genitourinary tract infection of adults that may proceed to disseminated systemic disease. In young female children it may appear as a vulvovaginitis, and in newborn infants infected at birth it may be manifest as a severe ophthalmic infection.

Adult Gonorrhea. In the *adult male,* gonorrhea begins as an acute urethritis, manifest by a thick, purulent discharge. If untreated, it may extend to an epididymitis or prostatitis. Urethral scarring may produce strictures requiring surgical relief. Not uncommonly, systemic involvements such as arthritis, endocarditis, or meningitis may develop (see Disseminated Gonococcal Infection) in untreated cases.

In the adult female, initial localization of infection may be in the cervix, urethra, or anal canal. It is often mild or asymptomatic in the early stage, producing no discharge or one that goes unnoticed. If not treated, infection may extend upward into the fallopian tubes and on into the pelvic peritoneum. The resultant salpingitis and pelvic inflammatory disease (PID) may heal by scarring when the suppurative process subsides, and this may lead to obstruction of the tubes and sterility. Silent, residual infection in untreated or unsuspected cases, particularly in women, is most responsible for continuing spread of the disease to new contacts.

Pharyngeal and *anorectal* infections have become increasingly common among heterosexual as well as homosexual adults of both sexes. Pharyngeal infection is seldom symptomatic, but may present as an exudative pharyngitis. Anorectal gonorrhea can produce a marked proctitis, with rectal discharge, tenesmus, and bloody diarrhea. More commonly it is asymptomatic or causes only mild rectal burning or pruritus.

Gonorrhea is primarily an infection of columnar and transitional epithelium. On entry into the body, the causative organism (*Neisseria gonorrhoeae,* commonly called the *gonococcus*) attaches to mucosal cells by means of pili and surface antigens present on

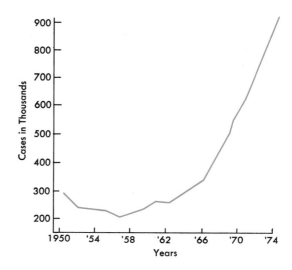

Fig. 22–3. Reported cases of gonorrhea in the United States, fiscal years 1950–1974. (Courtesy of Center for Disease Control, Venereal Disease Division, D.H.E.W., Atlanta, Ga.)

the cell wall. Following attachment, gonococci reach the subepithelial connective tissue by penetrating the intercellular epithelial spaces. Sometimes they are engulfed by epithelial cells (see Fig. 6–2, page 125), but more usually they are taken up in large numbers by polymorphonuclear leukocytes as part of an inflammatory response to their presence. Spread of gonococci to adjacent tissues is probably facilitated by their multiplication in the film of secretions that covers the genitourinary surfaces. Studies in women indicate that spread into the endometrial cavity and fallopian tubes is increased among those using an IUD (intrauterine device) for contraception and decreased in women using oral contraceptives. Menstruation also appears to facilitate spread to the upper genital tract, perhaps as a result of cervical dilatation. Gonococci attach to spermatozoa and thus could be carried on sperm into the upper tract.

Secretory IgA antibody to gonococci appears in vaginal and urethral secretions in gonorrhea, but its role is not yet clear. It may inhibit gonococcal attachment to mucosal cells or promote opsonization by PMN's. Conversely, gonococci produce an enzyme that can inactivate IgA and may thus perhaps protect themselves. Humoral antibodies, of both IgM and IgG types, also appear in genitourinary as well as disseminated gonorrhea.

If infection with *N. gonorrhoeae* is to occur, the organisms must come into contact with susceptible epithelium. For initial infection in adults, the only available surfaces are the urethra, endocervix, rectum, pharynx, and conjunctiva, since the organism cannot attach to and survive on cornified squamous epithelium. Coitus, fellatio, cunnilingus, and anal intercourse provide the routes of transmission to all but the latter of these sites. Transfer of gonococci to the eyes may be accomplished by unwashed hands contaminated with an infectious genital discharge.

Gonorrhea in Female Children.
The vaginal epithelium of prepubescent girls is not cornified, so that the entire vagina is susceptible to gonococcal infection. The genital labia may also be involved, and in severe cases extension into the urethra and bladder can occur. These mucous membranes are inflamed, and there is a profuse leukorrheal discharge. Sexual or other intimate contacts with infected adults appear to be responsible for most cases. There are some reports of transmission from mothers to infant daughters in the course of routine baby care, and a few cases may have resulted from sharing a bed with an infected parent. However, the question of child abuse should be considered in all such cases.

Gonococcal Ophthalmia of the Newborn.
Ophthalmia neonatorum may be a serious consequence of untreated, or undetected, maternal infection. The eyes of a newborn baby become infected at the time of delivery, during passage through the birth canal, if the mother has active gonorrhea. If the baby's eyes are not promptly treated, infection begins as an acute conjunctivitis, which rapidly involves the cornea and commonly ends in blindness (Fig. 22–4, *A*). This problem can be readily avoided by treating the eyes of all newborn babies prophylactically with silver nitrate solution (1 percent), a procedure required by the law of most states.

Intrauterine Infection.
A common form of neonatal gonococcal infection may be the "gonococcal amniotic infection syndrome." Infection of the chorioamnion may lead to orogastric colonization of the newborn. A study of infected neonates has indicated an increased morbidity among them, as compared with infants free of gonococci. There was a greater incidence of prematurity, evidence of chorioamnionitis, prolonged rupture of fetal membranes, and sepsis in the infant associated with maternal peripartum fever.

Gonococcal Conjunctivitis.
This infection may be acquired by people of all ages, usually through the transfer of genital infection by contaminated hands (Fig. 22–4, *B*). After an incubation period of one to three days, the conjunctiva becomes intensely red and swollen, and there is a profuse purulent discharge. Prompt treatment brings about full recovery, but neglect can lead to corneal ulceration, panophthalmitis, and eventual blindness.

Disseminated Gonococcal Infection (DGI).
In about 1 to 3 percent of patients with gonorrhea, disseminated infection occurs through the entry of gonococci into the bloodstream from a site of localized infection. This is more common in women than men, occurring most frequently during pregnancy or menstruation. DGI may follow asymptomatic genital, pharyngeal, or rectal infection as well as more severe local disease.

There are two stages of DGI. In the first there is a bacteremia, followed by the appearance of skin lesions and multiple joint involvement (polyarthralgia or polyarthritis). If DGI is untreated during this stage, the second phase may begin. Bacteremia ceases and polyarthritis disappears, but the infection localizes in a single joint. This is called the septic joint stage, because the affected joint develops a purulent effusion in which gonococci are usually present. There is a cellular infiltration of the synovium, which may lead to its destruction as granulation tissue is formed. Without treatment, this stage progresses to immobilization of the joint, with ankylosis and fibrosis.

Meningitis, endocarditis, myocarditis, and pericarditis are all possible complications of DGI, particularly in cases where bacteremia persists.

Laboratory Diagnosis.
The control of gonorrhea depends on its recognition and treatment, not only in obvious cases but in asymptomatic infections and con-

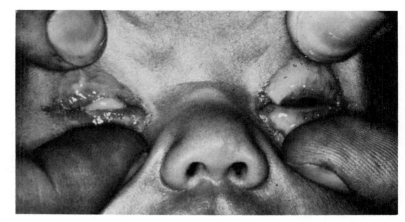

A

Fig. 22–4. *A.* Ophthalmia neonatorum in a newborn. Silver nitrate and antibiotics are used prophylactically to prevent this complication of maternal gonococcal infection. *B.* The inflammation in this eye is caused by *Neisseria gonorrhoeae.* Compare the possible routes of transmission that have led to eye involvement in *A* and *B.* (*A* courtesy of Dr. H. Hunter Handsfield, Department of Medicine, U.S.P.H.S. Hospital, Seattle, Wash.; *B* courtesy of Center for Disease Control, D.H.E.W., Atlanta, Ga.)

B

tacts of these. It is therefore essential that laboratory confirmation be obtained whenever possible. In recent years the practice of taking cultures on a mass scale, particularly from women, who are often asymptomatic carriers of infection, has become a major facet of a vigorous program to achieve better control. Serologic diagnosis is less satisfactory as a method for screening a population for infection, for reasons given in II below.

I. *Identification of the Organism* (*Neisseria gonorrhoeae*)

A. MICROSCOPIC CHARACTERISTICS. The gonococcus is a Gram-negative, kidney bean-shaped diplococcus,

morphologically indistinguishable from the meningococcus (see Chap. 13, p. 312). In stained smears of exudates, gonococci are characteristically found in intracellular positions within polymorphonuclear leukocytes (Fig. 22–5). They cannot be identified on the basis of morphology alone, however. Other Gram-negative organisms that normally inhabit mucosal surfaces, or that may be associated with infections, may look very much like gonococci.

With this in mind, the laboratory sends a guarded report of the findings in a smear of exudate. The language is descriptive, but avoids a commitment as to

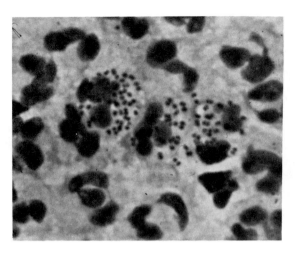

Fig. 22–5. *Neisseria gonorrhoeae* in a urethral smear. Note the intracellular position of the kidney-bean-shaped diplococci. (870×.) (Courtesy of Center for Disease Control, D.H.E.W., Atlanta, Ga.)

identification. When suspicious organisms are found, the report may read, "Gram-negative intracellular diplococci resembling gonococci were seen."

B. CULTURE CHARACTERISTICS. The presumptive diagnosis of gonorrhea is confirmed by culture techniques. *N. gonorrhoeae* is a fastidious organism, requiring enriched media containing hemoglobin, plasma, and yeast components. The best known of these media were developed at the Center for Disease Control of the Public Health Service, in Atlanta, Georgia (Thayer-Martin medium and its many modifications), and in the laboratories of the New York City Department of Health (NYC medium). In addition to factors required for the growth of gonococci, they contain a battery of antibiotics that do not affect *N. gonorrhoeae* but inhibit the growth of other organisms commonly found as commensals on the mucosal surfaces to be cultured. This permits gonococci, if they are present, to grow in relatively pure culture, free of competition or interference from other less fastidious organisms, and simplifies their isolation.

The requirement for CO_2 is particularly important on primary isolation of the gonococcus. In addition, this organism is extremely sensitive to the effects of excessive temperatures and drying. For these reasons, the success of cultures depends on the care with which specimens are collected and the speed of their transport to the laboratory. Preferably, a suitable growth agar medium is

inoculated directly and taken to the laboratory immediately, so that it can be placed under CO_2 incubation without delay. If transport to the laboratory is to be delayed for some hours, a growth-transport agar medium system should be used. Such systems provide a growth agar and also make provision for the necessary CO_2 atmosphere. Those commercially available are of two types: some utilize a small vial or bottle containing a slant of agar, the container being pregassed with a mixture of CO_2 and air; others in current use provide a bicarbonate tablet for inclusion in the system. The tablet may be placed in a sealable plastic bag together with the culture plate. Moisture from the agar medium is sufficient to activate the bicarbonate tablet, which gives off CO_2 within the bag, providing initial growth stimulus for the gonococci. Some systems utilize a culture dish that contains a well for the bicarbonate tablet. The growth medium in the dish is streaked with the specimen, a tablet placed in the well, and the dish closed firmly, then placed in a plastic bag that can be sealed. Whenever possible, cultures initiated in such transport systems should be incubated at 35° C during the time they are being held for transport to the laboratory.

Specimens taken directly to the laboratory may be streaked out there on a suitable growth medium and incubated in a candle jar or CO_2 incubator. After 24 to 48 hours of incubation at 35° C, colonies appearing on the culture plates are tested for the presence of oxidase, an enzyme possessed by all *Neisseria* and some other microorganisms. Oxidase-positive colonies are then smeared, Gram-stained, and examined microscopically. Those with typical morphology may then be distinguished from other *Neisseria,* including the meningococcus, by their reactions in carbohydrate media and confirmed as *N. gonorrhoeae.* Cultures from urogenital sources, i.e., cervix, vagina, or urethra (the latter from either sex), are generally reported as presumptively positive, without confirmatory techniques, if they meet the criteria of a positive oxidase reaction and typical morphology. Colonies grown from cultures of the anus or oropharyngeal sites, however, must be fully differentiated from those of *N. meningitidis* or other *Neisseria* that may also be isolated from such sources.

C. SEROLOGIC METHODS FOR IDENTIFICATION. The fluorescent antibody (FA) technique is sometimes of value in identifying an isolated strain, or in locating organisms in synovial fluid, skin lesions, spinal fluid, or even blood in cases of DGI. In some instances, the direct FA test on such material may be more helpful than culture in the diagnosis of DGI, particularly for patients who have had prior antibiotic treatment.

II. Identification of Patient's Serum Antibodies. Humoral antibodies arise during the course of gonococcal infection, but they are not protective and do not prevent reinfection. New infections may occur repeatedly and may be superimposed on a chronic, untreated one. Although numerous serologic techniques are available for detection of gonococcal antibodies (complement-fixation, immunofluorescence, precipitation, flocculation, latex agglutination, microhemagglutination), their diagnostic value is limited, because they do not reliably distinguish between currently infected individuals and those who had a past infection that was adequately treated. Lingering antibody in the latter situation is difficult to differentiate from antibody recently produced in response to a new infection. This limitation is particularly handicapping in screening large numbers of people for serologic evidence of new urogenital infection. Also, antibody levels may be low or absent in persons with genital infection only.

The chief diagnostic value of serologic tests is in the confirmation of chronic or disseminated gonococcal infection, particularly if cultures, which may be difficult to obtain or negative in such cases, fail to support clinical suspicion. All positive serologic results must be carefully weighed, however, against the clinical evidence for old or new infection.

III. Specimens for Diagnosis. Exudates from urogenital or other mucosal membranes for smear and/or culture.

In the male, Gram-stained smears of urogenital exudates are examined microscopically. If smears are positive for typical Gram-negative diplococci, cultures are unnecessary. Male patients are treated at once on the basis of positive smears. If smears are negative but there is clinical evidence or suspicion of gonorrhea, cultures are initiated. Exudates from the anus or pharynx should be cultured in any case, since smears may fail to reveal the organisms at these sites or distinguish morphologically similar organisms that may be present there normally. The usual presence of commensalistic *Neisseria* species in the throat, particularly, and the possibility of encountering *N. meningitidis* there require the cultural differentiation of suspicious isolates and confirmatory techniques for identification of *N. gonorrhoeae.*

In the female, exudates from all sources are routinely cultured. Smears may fail to reveal gonococci, especially in asymptomatic female infections, and are not recommended except to confirm clinically obvious symptoms. Smears from females can be especially difficult to interpret because organisms morphologically similar to gonococci are often found normally on genital membranes. Cervical discharges are most likely to yield *N. gonorrhoeae,* but vaginal, urethral, anal, or pharyngeal cultures may also be positive. Isolates from rectal and throat cultures must be distinguished from *N. meningitidis* or other *Neisseria* species and culturally confirmed as gonococci.

If joint involvement is suspected in chronic gonorrhea, aspirated synovial fluid should be submitted for culture.

Cultures of *blood* or *spinal fluid* may yield the organism if systemic dissemination has occurred.

In gonococcal ophthalmia, pus from the conjunctiva is collected.

For proper collection and transport of specimens to the laboratory see I B above.

Epidemiology

I. Communicability of Infection

A. RESERVOIR, SOURCES, AND TRANSMISSION ROUTES. Gonorrhea is a human disease. Exudates from the involved mucous membranes of infected persons are the source of infection, transmitted by direct contact, usually sexual.

B. INCUBATION PERIOD. Usually three to nine days, except in the case of ophthalmia neonatorum, which develops within 36 to 48 hours.

C. COMMUNICABLE PERIOD. If specific antibiotic therapy is given, symptoms and communicability subside within a day or so. Without therapy, adult gonorrhea may remain communicable for months or years. The disease in children generally does not persist for more than about six months. Conjunctival membranes should be considered infectious until all discharge has ceased.

D. IMMUNITY Effective natural immunity to gonorrhea does not develop as the result of an attack, and artificial immunization is not available. Susceptibility to this disease is universal without respect to age, sex, or race, and reinfection may readily occur.

II. Control of Active Infection

A. ISOLATION. Hospitalized infants and children should be isolated for 24 hours following the administration of antibiotics. No isolation is necessary for adults, but they should be instructed to avoid sexual contacts with previous partners who have not been treated.

B. PRECAUTIONS. Concurrent disinfection of exudates from lesions or articles soiled by them. Nurses and physicians should be aware of the susceptibility of the conjunctiva to gonococcal infection. Routine terminal cleaning of the unit is satisfactory.

C. TREATMENT. The gonococcus has a tendency to develop resistance to antibiotics used against it. When the sulfonamides were first discovered in the 1930s, they were effective for gonorrhea, but by the middle of World War II gonococcal resistance to them was prevalent. Penicillin was then introduced and became the most effective therapeutic choice for all forms of gonorrhea. By 1957, however, a fair percentage of gonococcal strains were resistant to ten times the amount of penicillin that had previously killed them easily. The percentage of resistant strains continued to rise until in 1969, in the United States, nearly 65 percent required extremely high doses of penicillin to kill them. This relative resistance to penicillin was paralleled by a rise in resistance to tetracyclines and was most prominent in strains isolated in Southeast Asia during the Vietnam War. Military personnel returning to the West Coast from Vietnam probably introduced these newly resistant strains, which then became distributed throughout the United States. Since that time, gonococcal resistance to antibiotics has leveled off, largely because very high-dose antibiotic therapy for gonorrhea was instituted then and has been maintained generally. This has reduced the number of treatment failures and the percentage of treated patients who harbor resistant gonococci, so that these strains are not perpetuated efficiently in the population.

The antibiotic resistance of gonococci is genetically determined. Single or multiple gene mutations may bring about resistance to different antibiotics. Gonococci are known to have plasmids (see Chap. 3, p. 61) that are unrelated to antibiotic resistance, but recently a new one has been acquired that has given rise to a fresh problem in penicillin resistance. This plasmid codes for the production, in the gonococcal cell, of a penicillinase typical of the kind produced by enteric bacteria that bear R plasmids (see Chap. 10, p. 263). The gonococcal penicillinase plasmid can be transferred between gonococci and also to *Escherichia coli* or other organisms by transformation or by conjugation. A rising incidence of gonococcal strains possessing this plasmid has been observed since late 1975, when they first appeared in the Philippines, then England, and, in early 1976, in the United States. It has been inferred that the penicillinase plasmid has been introduced into multiple gonococcal strains in nature, possibly first by conjugation with another species of bacteria, followed by exchange between strains of *N. gonorrhoeae*.

Gonococci that bear this plasmid exhibit absolute resistance to penicillin. In all cases, patients from whom such strains have been isolated have failed to respond to standard high-dose therapy with penicillin. Even doubling the dose has also resulted in treatment failure. These cases have been clustered in the Philippines, where 30 to 40 percent of gonorrhea in some cities is due to penicillinase-producing *N. gonorrhoeae* (PPNG). This prevalence is associated with American military personnel in the Philippines, among whom 50 percent of gonococcal infections are caused by PPNG. It is estimated that several hundred new cases of gonorrhea due to PPNG occur monthly in that area among the military. The number of isolates of PPNG strains has been increasing throughout the Far East, as well as in the United Kingdom and the United States. Most evidence indicates that they originated in the Philippines. By the spring of 1977, in the United States 150 isolates had been confirmed from 21 states. Some of these strains were related to travelers recently returned from the Philippines, but local outbreaks have occurred without any demonstrable connection to travelers from Asia.

Most strains of PPNG are multiply resistant to the penicillins, including ampicillin, carbenicillin, and mecillinam. The semisynthetic penicillins that are resistant to beta-lactamase (penicillinase), such as methicillin, oxacillin, cloxacillin, and nafcillin, also resist this enzyme produced by PPNG, but these drugs display too little activity against gonococci to be clinically useful. Some other classes of antibiotics such as cephalosporins are effective, but spectinomycin is the drug of choice for gonorrhea caused by PPNG. Treatment with this drug in a single intramuscular dose of 2 gm has been very effective to date. Clinical studies of the problems presented by PPNG continue, with an effort to develop further strategies for the control of gonorrhea caused by these strains. The Center for Disease Control has recommended that all patients for whom penicillin therapy fails be cultured for PPNG and that all isolates of the latter be reported. Patients in whom infection with PPNG has been confirmed should be recultured after successful spectinomycin therapy, and every effort should be made to trace and treat their sexual partners.

The standard therapy for uncomplicated urogenital gonorrhea still indicates penicillin as the drug of choice. Aqueous procaine penicillin G is given intramuscularly in two injections of 2.4 million units each, injected at different sites on one visit. This dose is immediately preceded by 1 gm of probenecid given by mouth. The latter drug delays the urinary excretion of penicillin, assuring that high blood levels of the antibiotic can be reached and maintained. Patients who are allergic to penicillin or probenecid can be treated with spectinomycin, tetracyclines, or erythromycin. All sexual contacts of patients in whom gonorrhea has been diagnosed and treated should be identified and treated as quickly as possible.

It is not uncommon for syphilis and gonorrhea to exist simultaneously. The penicillin schedule recommended for gonorrhea is also effective against incubating syphilis, but all patients with gonorrhea should have a serologic test for syphilis also. If the test is negative, no further serologic follow-up is necessary, provided penicillin was given. Patients treated with other drugs, however, should have another serologic test for syphilis in three to four months and, if it is positive, should receive additional therapy for the coincident infection.

Disseminated gonococcal infections are treated initially with intravenous injections of penicillin, followed by oral ampicillin. Meningitis and endocarditis, however, require at least ten million units of penicillin per day, given intravenously for ten days for meningitis, three weeks for endocarditis. Penicillin-allergic patients with such threatening infections may be treated with high doses of a cephalosporin or chloramphenicol.

D. CONTROL OF CONTACTS. Rapid case finding to locate and treat the source of gonococcal infection is essential for control of this disease. This requires trained interviewers, whose goal is to identify those sexual partners with whom the patient has had contact within ten days of onset of his (her) own infection. Full curative doses of antibiotics are given to such contacts, regardless of symptoms, and serologic tests for syphilis are also done.

All infants born to infected mothers are monitored for symptoms of developing infection. Orogastric and rectal cultures should be taken. Conjunctival cultures are not indicated if adequate prophylaxis was employed at birth to prevent ophthalmia neonatorum.

III. Prevention. Social forms of prophylaxis can be effective, but mutual sexual restraints are often unacceptable in today's society. The condom provides protection against the transmission of some venereal diseases but may not be tolerated, and washing is ineffective. Some vaginal contraceptives have an antigonococcal effect, but their value in prophylaxis has not been established. The use of antibiotics in low doses immediately after exposure may reduce the risk of infection, but it also increases the possibility that if infection is acquired, the strain will become antibiotic resistant.

A continuing search is being made for gonococcal antigens that may be of immunogenic value in a vaccine. Experimental infections in chimpanzees and human volunteers have provided hope of future success in the development of an effective vaccine.

Chancroid

The Clinical Disease. This venereal disease is relatively widespread, but is particularly well known to military personnel in tropical areas. It is also a disease of seaports and overcrowded cities. The chancroid is the initial lesion that occurs on the genitalia: a *soft* chancre with ragged edges. Unlike the hard chancre of syphilis, the ulcer is swollen and painful, as are regional lymph nodes. The chancroid and the nodes become suppurative and necrotic. Pus-filled inguinal nodes may break down and suppurate to the surface. These "buboes" exude the causative bacteria, which may then be inoculated into new sites. Extragenital lesions may be seen on other mucosal and skin surfaces (tongue, lip, breast, umbilicus).

Laboratory Diagnosis. *Haemophilus ducreyi* is the responsible agent. This is a small, Gram-negative bacillus, nonmotile and non-spore-forming. It is classified in the same genus as *Haemophilus influenzae* (see Chap. 13, p. 307), but it is culturally more fastidious, requiring heavy blood enrichment as well as microaerophilic incubation. It can be visualized in Gram-stained smears of exudates from the chancroid or from buboes as chaining rows of bacilli.

The differential diagnosis of chancroid includes dark-field examinations of exudate and serologic testing to rule out syphilis.

Special Features of Control. Sulfonamides are the drugs of choice. Tetracyclines and other antibiotics are also effective, but they can mask concurrent syphilis; therefore, careful diagnostic techniques for the latter disease must be used.

Chancroid has a short incubation period of three to five days and is communicable through infectious discharges of the genital lesions and buboes.

Measures for the control and prevention of this disease are similar to those described for syphilis and gonorrhea.

Granuloma Inguinale

The Clinical Disease. Granuloma inguinale is a chronic venereal disease chiefly of tropical areas but also found in temperate zones. A small nodule appears initially at the site of entry of the organism on the genital mucosa. When the nodule ulcerates, it may be mistaken for the hard chancre of syphilis, or the soft one of chancroid. As disease progresses, the ulcer spreads peripherally and may superficially involve a large area of the mucous membrane.

Laboratory Diagnosis. The causative organism is *Calymmatobacterium granulomatis*. Formerly known as *Donovania granulomatis* or, simply as "Donovan bodies," these organisms appear in biopsies of the lesions as short, Gram-negative bacilli crowded into mononuclear cells. They have been cultivated in embryonated eggs, and also on complex artificial media.

Special Features of Control. Tetracyclines are the drugs of choice for this disease. Streptomycin is a useful alternative.

This disease occurs more frequently among males than females. Many male patients are homosexuals, whereas the infection is uncommon in prostitutes or the heterosexual partners of infected cases.

Lymphogranuloma Venereum

The Clinical Disease. This disease has a worldwide distribution, but is most common in tropical or semitropical areas. In the United States it is endemic in southern regions bordering the Gulf of Mexico. It is a venereal infection to which all ages and races are susceptible, but is most common in sexually active young adults, including homosexuals. Lymphogranuloma venereum (LGV) begins with a small, painless primary lesion on the genital mucosa marking the site of entry. This may be a papule or an ulcerated lesion. Clinical evidence of systemic infection begins in about two weeks, with enlargement of the regional lymph nodes which may proceed to suppurate and drain, and with fever, general malaise, and sometimes involvement of membranous surfaces of the joints, eye, or brain. Chronic infection within pelvic lymphatics may be of long standing, eventually inducing blockade, with resultant elephantiasis of genital tissues, and strictures.

Laboratory Diagnosis. The agent of this disease is *Chlamydia trachomatis*. Its morphologic and biologic properties relate it to the agent of trachoma (p. 515), and it is a member of the same genus as the organism causing psittacosis (Chap. 15, p. 379).

With special staining techniques this organism can be visualized within the mononuclear cells of exudates obtained from draining lymph nodes, or in tissue biopsies. It can also be cultivated in yolk sacs of embryonated eggs, in cell cultures, or in mice. In preparations harvested from these sources it is identified by microscopic morphology and by serologic identification of its antigenic properties.

Antibodies produced during the course of LGV infection can be identified by complement-fixation or microimmunofluorescence tests, in serologic confirmation of the clinical diagnosis. The specific hypersensitivity that develops in this disease can also be recognized by the use of a skin-testing antigen, prepared from the organism cultivated in a chick embryo. This is known as the "Frei antigen," or the "Frei test," which elicits a reaction of the delayed hypersensitive type in infected persons. The antigen is shared by other microorganisms of the psittacosis-lymphogranuloma-trachoma group, which means that the Frei test is not specific for LGV alone but may also be positive in persons infected with related organisms.

Special Features of Control. *Specific treatment* with tetracycline for two to four weeks results in prompt relief of fever and pain, but enlarged lymph nodes require several weeks, or even months, to heal. Aspiration of lymph nodes and surgical relief of elephantiasis or strictures may be required. Sulfonamides and chloramphenicol are alternative drugs of choice.

The incubation period may require from one to three or four weeks. The primary lesion usually appears within 7 to 12 days, but, if it fails to develop, the first sign of infection may be an inguinal bubo appearing about a month after exposure. Untreated LGV may be communicable for months or years. With treatment the lesions are slowly healed.

Resistance to reinfection develops during the course of this disease, but the infecting organism may remain latent and viable within tissues, unaffected by antibody production.

LGV is primarily a venereal infection, but it is also sometimes transmitted to children through contacts with freshly contaminated environmental sources. Control and prevention depend on detection and treatment of sexual contacts, as outlined for syphilis and gonorrhea.

Other Venereal Infections

Genital transfer of infection may involve quite a number of organisms that ordinarily may produce localized symptoms related to the genitourinary tract of adults. In males, urethritis not associated with the gonococcus may be caused by several agents that can be transmitted venereally. In females, these organisms may cause a low-grade vaginitis or cervicitis. Most of these infections do not constitute a serious threat to the health of adults, although they may be responsible for reproductive failures. The most serious implications, however, may involve infants infected congenitally or at birth.

Nongonococcal Urethritis (NGU). Urethritis, particularly in males, that cannot be associated with gonococcal infection has come to be known as nongonococcal urethritis, or NGU. This is a clinical entity characterized by generally low-grade symptoms of pyuria, or urethral exudate with dysuria. It has been principally associated with two microorganisms, *Chlamydia trachomatis* and *Ureaplasma urealyticum* (a small strain of mycoplasma). Others less commonly involved include *Candida albicans* (see Chap. 16, p. 401), herpes simplex virus type 2, and *Trichomonas vaginalis,* discussed below.

Chlamydia trachomatis. Organisms classified in this chlamydial species are variously associated with several diseases, notably trachoma, inclusion conjunctivitis, and sexually transmitted entities such as lymphogranuloma venereum, NGU in males, and cervicitis in females. Although they belong to one species, those responsible for different clinical entities are serologically distinct, identifiable as specific immunotypes, as they are called.

Recognition of *C. trachomatis* as a cause of NGU requires demonstration of characteristic inclusion bodies (see Chap. 2, p. 32) in the urethral discharge, isolation of the agent in embryonated eggs and tissue cultures, and serologic identification of the organism as well as specific antibodies in the patient's serum. The same diagnostic procedures are used to identify chlamydial infection of the cervix, present in most female sex partners of men with chlamydial NGU but seldom found in the sexual contacts of men with *Chlamydia*-negative NGU. The chlamydial immunotypes that occur in these venereal infections in adults are the same as those associated with inclusion conjunctivitis of the newborn (p. 516) and can be traced from infants to the cervical secretions of infected mothers.

Ureaplasma urealyticum. This organism is a type of mycoplasma commonly found on the genital mucosa. (It was formerly called T-strain mycoplasma because it forms tiny colonies on agar media, but has been renamed to emphasize its unusual ability to split urea.) The causative role of ureaplasmas in NGU has not been as well established as that of the chlamydiae because they do not elicit an antibody response during infection, and also because they can often be isolated from men without urethritis. However, they are significantly prevalent in men with *Chlamydia*-negative NGU as compared with those having proven chlamydial infection. Their persistence in urine cultures is also associated with persistence or recurrence of NGU. These and other urogenital mycoplasmas (*M. hominis*) have been incriminated in various reproductive failures, such as infertility, premature births, and spontaneous abortions, but definitive proof of their pathogenicity has not been fully established.

The incidence of NGU is generally correlated with sexual activity. It occurs most often in young men, between ages 16 and 30 years, and the onset often occurs within two to three weeks following coitus with a new sex partner. Both chlamydial and mycoplasmal infections respond well to antibiotic therapy, but without treatment NGU may be a persistent annoyance for months and can lead to acute epididymitis. Both types of infection are treated with tetracyclines given orally for a period of at least one week. The only known means of preventing these infections is through the practice of good sex hygiene, as discussed for gonorrhea.

Trichomoniasis. The flagellated protozoan *Trichomonas vaginalis* (Fig. 22–6) may cause a particularly troublesome local infection in women. The organism is often found as a commensal of the genital mucosa. In men it lives on the urethral surfaces or in the prostate and seminal vesicles, where it almost never produces any symptoms. When transferred to women it may induce a chronic vaginitis characterized by a profuse malodorous discharge. Other bacteria (lactobacilli) that normally live in the vagina, maintaining the acidity of the membrane, are displaced by trichomonads. The latter feed on bacteria and also on leukocytes marshaled in response to their invasion. The membranes become superficially inflamed, and there is usually an intense pruritus of the vagina and vulva. Trichomoniasis is easily diagnosed by microscopic examination of vaginal secretions, or of seminal discharge from the male. The organisms are often present in urine as well. The actively motile parasite can be seen in unstained wet mounts. It can be cultured, also, on a medium similar to that used for amebic protozoans (see Chap. 20, p. 465). Trichomoniasis can be treated effectively

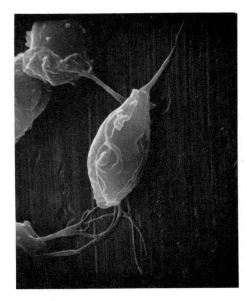

Fig. 22–6. A scanning electron micrograph of the flagellated protozoan, *Trichomonas vaginalis*. (5000×.) (Courtesy of Dr. William G. Barnes, Veterans Administration Hospital, Kansas City, Mo.)

with metronidazole given orally, but concurrent treatment of sexual contacts is also necessary to prevent reinfection. In women, restoration of the normally acid pH of the vagina is also important and is accomplished with mild douches that have a cleansing action as well.

Herpes Genitalis. Herpes simplex viruses (HSV) are responsible for a spectrum of human contact diseases, discussed more fully on page 527. Herpes genitalis is a distinct clinical entity usually associated with HSV type 2. It is venereally transmitted among adults, in whom it may appear as a primary or recurrent disease. The characteristic lesion is a vesicle filled with clear fluid, surrounded by an area of inflammation, edema, and congestion. These vesicular lesions may break down, becoming ulcerative and painful. In males, herpetic vesicles usually occur on the glans penis, the prepuce, or the penile shaft. In women, the principal site of genital herpes is the cervix, but lesions may also occur in the vagina or on the vulva. The distribution of lesions varies in both

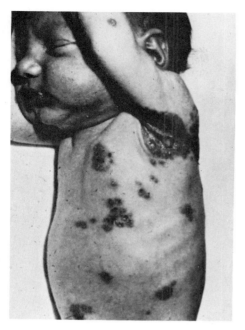

Fig. 22–7. An infant with congenital herpesvirus infection acquired from its venereally infected mother. (Courtesy of Center for Disease Control, D.H.E.W., Atlanta, Ga.)

sexes, however, depending on sex practices. Severe to fatal infection may be transmitted to the newborn infant, at the time of birth, from an infected mother's genitalia (Fig. 22–7). No specific treatment is available.

The clinical diagnosis of genital herpes is confirmed by demonstration of HSV type 2 in vesicular fluid, using tissue culture techniques. Microscopic examination of epithelial cells at the margin of a vesicle may also reveal intranuclear inclusion bodies, typical of this virus. Humoral antibodies to HSV can be detected by a variety of serologic techniques and may provide a serologic diagnosis. Studies of antibody levels in women with cervical cancer have led to the implication of HSV 2 as a possible agent of cervical tumors. Women with antibodies for HSV 2 appear to have a higher potential for developing cervical cancer than those without antibodies and, conversely, women with such tumors have a higher

incidence of HSV 2 antibodies than those who are free of HSV 2 infection.

Congenital and Neonatal Diseases Resulting from Venereal Infections. Some of the organisms transferred among adults during sexual intercourse may have little consequence except for a developing fetus or an infant born of an infected mother, but for these the results may be disastrous.

Intrauterine infections fall into two epidemiologic categories. Those acquired by extension of *nonvenereal* maternal infection include such diseases as congenital rubella syndrome, listeriosis (granulomatosis infantiseptica), and toxoplasmosis. Respiratory or alimentary routes of entry are involved in the initial infection of the mother in these instances. The second category includes those acquired by extension of maternal *venereal* disease, the most notable example being congenital syphilis. Gonococcal infection of the chorioamnion also falls in this group and may be more common than previously recognized.

Neonatal diseases acquired at birth from venereally infected mothers include gonococcal ophthalmia neonatorum, inclusion conjunctivitis of chlamydial origin, and herpes simplex virus infection. In each of these instances, the infant is infected during passage through the birth canal by exposure to infectious organisms on the mother's vaginal or vulval mucosa. The maternal infection may have been asymptomatic, but in any case unrecognized and untreated. Neonatal infections with other organisms belonging to the mother's indigenous mucosal flora are not usual, but may be due to faulty preparation for delivery or to breaks in aseptic technique.

The postnatal infectious diseases of infants are generally acquired through the usual routes of transfer from infected adult carriers, or from environmental sources. In some instances, however, infections acquired congenitally or at birth may not be immediately apparent. This is often the case in congenital syphilis, for example, infants appearing healthy at birth, with signs of infection developing only about the third week of life. Recognition of the nature and origin of infant infections is thus of great epidemiologic importance, requiring careful assessment of the need for treatment of the mother and other adults involved.

Brief descriptions of the infectious diseases of

Table 22–1.   Summary of Infectious Diseases of Infants

	Diseases	Organisms	Chapter Reference
Congenital infections	Syphilis	*Treponema pallidum*	22
	Gonococcal amniotic infection syndrome	*Neisseria gonorrhoeae*	22
	Listeriosis (granulomatosis infantiseptica)	*Listeria monocytogenes*	13
	Rubella syndrome	Rubella virus (German measles)	14
	Toxoplasmosis	*Toxoplasma gondii*	20
Infections acquired at birth	Gonococcal ophthalmia neonatorum	*Neisseria gonorrhoeae*	22
	Herpesvirus infections	Herpes simplex virus	22
	Inclusion conjunctivitis	*Chlamydia trachomatis*	22
	Pneumonia Septicemia Meningitis	Group B *Streptococci*	14
Neonatal infections	Epidemic diarrhea	Pathogenic serotypes of *E. coli*	18
	Myocarditis Encephalomyocarditis	Coxsackie virus	14, 19
	Skin infections (pemphigus neonatorum)	*Staphylococcus aureus*	22
	Pneumocystosis	*Pneumocystis carinii*	20

infants are included in the discussion of each disease, and they are summarized in Table 22–1.

Infectious Diseases of the Eye

Infectious eye diseases acquired from infected persons fall into two groups, those of *bacterial* etiology (including chlamydial infections), and those caused by *viral* agents.

Numerous strains of pathogenic *bacteria* are associated with eye infections, but the clinical result is similar in each case: there is an acute conjunctivitis, with production of a purulent exudate. Bacteria generally do not invade the epithelial cells of the conjunctiva or surrounding tissues, but multiply in extracellular positions where they may be taken up by phagocytes marshaled in the exudate. The clinical and pathologic features of *chlamydial* or *viral* infection, on the other hand, are related to invasion of the conjunctival epithelium, or cells of other ocular tissues, and to the intracellular activities of the infectious agent. In most of these invasive infections, initial acute conjunctivitis may be succeeded by a chronic stage of slow destruction with scarring which can lead to blindness.

Acute Bacterial Conjunctivitis

The Clinical Picture. Bacterial infections of the eye begin with symptoms of local irritation, tearing, and congestion of small blood vessels of the conjunctiva over the cornea and under the eyelids. As infection proceeds, the eyelids become swollen and reddened, and a reactive, cellular exudate is formed. The affected eye (both or only one may be involved) becomes sensitive to light and feels hot and painful to the patient. These symptoms vary in degree. They are more severe in patients who have been heavily infected, or in those who have little resistance. Neglect may lead to chronicity with more serious involvement of deep tissues. As a rule, these infections respond well to treatment and heal without scarring or other residual damage. Because vascular injection is often a marked feature of bacterial conjunctivitis, the popular word for this type of infection is "pink-eye."

Laboratory Diagnosis. Microscopic examination of the exudate from eye infections caused by bacteria usually reveals the organism, and culture techniques provide final identification. The agents identified most commonly are strains of pneumococci or species of *Haemophilus* and *Moraxella.* Other organisms associated with the respiratory membranes, such as streptococci or staphylococci, may also be responsible for conjunctivitis. It should be remembered, also, that gonococci from infected genital membranes can establish themselves on the conjunctival mucosa and produce severe infection in adults as well as in babies born of mothers with gonorrhea (see p. 504).

Epidemiology. These infections are most common among young children, or others with lowered resistance. They are spread from infectious throat or conjunctival discharges and may pass rapidly from one small child to another in kindergartens, nursery schools, or playgrounds. They often yield to local antibiotic therapy, but systemic dosage is sometimes required. Healing is usually rapid, without residual damage if treated early. The best control lies in prompt therapy and careful attention to hygienic care of normal eyes. Children should be taught as early as possible to respect their eyes, and to avoid rubbing them, especially with dirty hands. Parents and nurses both should be alert to the child who rubs his eyes to an unusual extent, for this may be the first symptoms of a developing infection.

Chlamydial Diseases of the Eye

Trachoma. This is a serious communicable disease of the eye prevalent in hot, dry areas of the world, particularly among people who live under the overcrowded and unsanitary conditions of poverty. It is widespread in Mediterranean countries, parts of Africa, the Middle East, Asia, and South America. In the United States it is a public health problem on Indian reservations of the Southwest.

Trachoma is caused by *Chlamydia trachomatis,* an intracellular bacterial agent. Different serologic types of this organism are associated with inclusion conjunctivitis and pneumonia of the newborn, chlamydial urethritis in adults, and lymphogranuloma venereum (see p. 510). These microorganisms are obligate intracellular parasites that can be seen as inclusion bodies within the cytoplasm of invaded cells. In trachoma and inclusion conjunctivitis they are found within epithelial cells of the conjunctiva.

Trachoma begins as an acute inflammatory conjunctivitis, but is later characterized by excessive accumulations of tissue cells in tumorlike formations. Lymphocytes accumulating in and beneath the conjunctiva form lymphoid "follicles." This hyperplasia usually involves the conjunctiva of the upper lid and may extend under the mucosa covering the upper half of the cornea. The cornea becomes vascularized, and as affected cells die they are filled in by fibroblastic scar tissue. The corneal epithelium is not infected, but becomes keratinized as a result of the tissue reaction, and this, together with scarring, not only deforms the eye but threatens blindness because of corneal opacity. Secondary bacterial infection is common and inflicts further damage on the cornea, usually with total loss of vision.

The laboratory diagnosis is made by microscopic demonstration of the typical intracytoplasmic inclusion bodies in epithelial cells of conjunctival scrapings. Special tissue stains or immunofluorescent staining is used. The agent may be cultivated in embryonated eggs or tissue cultures and identified serologically. Identification of specific antibodies in the patient's serum can be made with immunofluorescent methods, and secretory IgA antibodies can be detected in tears. It is sometimes difficult to distinguish the agent from that of inclusion conjunctivitis by serotyping, but the clinical picture and epidemiology of the two diseases are quite different.

Epidemiology. Trachoma is spread by direct or indirect contacts with conjunctival discharges from infected persons. It can be treated successfully if discovered early, so that prompt diagnosis constitutes the best means of control. Tetracyclines are the drugs of choice, but erythromycin and the sulfonamides are also effective. Children are more frequently affected than adults and must be diagnosed and treated before the disease has advanced to the point of scar formation if their recovery is to be complete. In areas where trachoma is prevalent, mass treatment programs with topical antibiotics are coupled with efforts to improve sanitary standards.

Inclusion Conjunctivitis. This is an acute conjunctivitis of newborn babies resulting from maternal venereal infection, as described previously. The organism, *C. trachomatis,* is rarely transmitted to adults from infected babies' eyes, but is sometimes associated with "swimming pool conjunctivitis" acquired by children and adults from nonchlorinated swimming water contaminated from genital sources.

The disease is less severe than trachoma, but if untreated it may persist in a chronic stage for several months. The incubation period is from 5 to 14 days (in the newborn this distinguishes it from gonococcal ophthalmia, which usually begins within 36 hours of birth). Infant conjunctivitis is characterized by acute inflammation, particularly of the lower lids. Purulent exudate is produced, but there is no lymphoid follicle formation or scarring. In older children and adults, there is lymphoid hyperplasia, as in trachoma, but it is most marked on the lower lids and resolves without scarring. Unlike trachoma, there is no involvement of the cornea in inclusion conjunctivitis.

Confirmation of the diagnosis can be obtained by laboratory demonstration of inclusion bodies within epithelial cells scraped gently from the conjunctival surface of the lower lid, also by culture methods and serologic techniques. The infection responds rapidly to tetracycline, erythromycin, or sulfonamide therapy. The newborn disease is difficult to prevent, since the mother's infection is usually inapparent, and neither penicillin nor silver nitrate instillation in babies' eyes affects the microorganism. Outbreaks associated with swimming pools are prevented by adequate chlorination of the water.

Viral Diseases of the Eye

Acute conjunctivitis is associated with at least three viruses: adenovirus (type 8), herpes simplex virus (type 1), and an enterovirus (type 70).

Epidemic Keratoconjunctivitis (Shipyard Eye). One of the *adenoviruses* (serologic type 8) is responsible for this acute conjunctivitis. The disease is usually accompanied by symptoms of general infection, with low-grade fever, headache, and edema of the tissues around the eyes. The conjunctivae are inflamed, and the cornea may become keratinized and opaque, although this is seldom permanent. Most patients recover completely within a few weeks. The disease is sometimes seen in small outbreaks occurring in shipyards and industrial plants. Eyes are frequently subject to trauma in such plants. They may become infected when they are treated in dispensaries or clinics where strict asepsis is not maintained. Surgical instruments and other objects can be contaminated by infectious conjunctival discharges and serve as a common source for spreading infection. This virus may also be spread by infectious nasal secretions.

There is no specific therapy for adenovirus conjunctivitis. Control depends primarily on industrial safety measures that protect the eyes, as well as careful technique in dispensaries. Sterilized instruments, general cleanliness, and well-scrubbed hands are essential in industrial clinics.

Herpetic Conjunctivitis. *Herpes simplex virus type 1* (see p. 527) may also infect the conjunctival membrane or the cornea. Infections of the eye may accompany herpetic lesions in the mouth and throat or may be confined to ocular tissues. They are sometimes superficial and quickly resolved, but more usually offer a serious threat of blindness. The cornea may be ulcerated and scarred, or keratinized. Antiviral drugs (such as iododeoxyuridine) are useful in suppressing the organism in superficial cells. Deep infections often require ophthalmic surgery. There are no specific control measures applicable to herpes simplex virus, for this is an organism that is constantly associated with man, and the majority of infections are latent or of minor importance.

Epidemic Hemorrhagic Conjunctivitis. An *enterovirus* (type 70) is associated with epidemics of a hemorrhagic conjunctivitis that has been observed in many African, Asian, and European countries. The infection is characterized by pain, swollen eyelids, injection of the conjunctivae, and subconjunctival hemorrhage. The hemorrhages vary in size, commonly involving both eyes. Systemic symptoms of an upper respiratory infection sometimes occur also. The infection resolves spontaneously in about one to two weeks and is not influenced by antibiotics. This disease, like other types of conjunctivitis, is transmitted by direct or indirect contact with the discharge from infected eyes, possibly also by infectious droplets from those with respiratory infection. Epidemics are most frequent in families, or in situations marked by overcrowding and poor hygiene.

Each type of viral conjunctivitis described above can be diagnosed by laboratory demonstration of the virus. Microscopy may reveal it in conjunctival scrapings, and it can be cultivated in tissue cultures or embryonated eggs, then identified serologically. Identification of a rising titer of specific antibody in the patient's blood confirms the diagnosis.

Infectious Diseases Involving or Entering Through Skin

Staphylococcal Infections

Staphylococci are among the most constant members of the resident flora of man's superficial tissues. There are two types, both of which live commensalistically on skin and mucous membranes: one type has very few of the properties required for pathogenicity (i.e., it is *relatively* avirulent) and produces disease only if given unusual opportunities for entry into the tissues of persons whose defensive mecha-

nisms are not in good working order; the other type has several virulent properties and probably always has a potential for pathogenicity, but is normally held in check by the many physiologic conditions that make for good general health and for the integrity of the superficial barrier tissues.

Current classification of the staphylococci assigns the species name *Staphylococcus epidermidis* to those strains that fail to display properties associated with virulence *when tested in the laboratory*. Those strains that do possess properties that contribute to virulence are given the species name *Staphylococcus aureus*. It must be emphasized that these distinctions are made in the laboratory when strains are isolated from clinical specimens and tested for certain characteristics *in vitro*. The laboratory's report should never be construed as an assignment of the pathogenic role played by a particular strain in an individual situation. This must be done on the basis of the clinical evidence, supported by laboratory studies and sometimes by epidemiologic data as well. Nevertheless, the greater potential for pathogenicity possessed by strains of *S. aureus* should always be recognized.

Diseases produced by staphylococci cover a very wide spectrum, from the simplest localized infection of the skin, or focal infections of deep tissues, to septicemia, disseminating disease, and death. The source of invading organisms is sometimes endogenous, resident staphylococci entering deep tissues through broken skin or mucosal barriers or multiplying vigorously at some local site because physiologic barriers have broken down. Probably more often the source of infection is purulent discharge from the active lesions of another person or the asymptomatic nasal carrier of a virulent strain. Any portal of entry may be involved, but the gastrointestinal tract is ordinarily not susceptible to staphylococcal infection from exogenous sources. The bowel mucosa may be seriously injured by overgrowth of endogenous staphylococci when antibiotic therapy or some other factor has reduced the numbers of Gram-negative enteric bacteria and upset the normal balance of microbial competitions in the bowel; and it may be acutely irritated by ingested staphylococcal *enterotoxin,* as described in the discussion of bacterial food poisoning in Chapter 18. With this exception, staphylococci may enter through any route if local circumstances permit. Skin is not only a frequent route, but the most common site of localized staphylococcal infection. For this reason, as well as for convenience, staphylococcal diseases are grouped here as *superficial* or *deep-tissue* infections.

1. Superficial Staphylococcal Infections. The typical lesion produced in response to staphylococcal infection is an abscess which may range in size and severity from the small but annoying **pimple,** to larger and more painful furuncles (boils) and carbuncles. *Furuncles* arise from abscesses localized in hair follicles. They are uncommon on the scalp, but may occur on many other hairy surfaces of the body, frequent sites being the axillae, the back of the neck, and the buttocks. (Abscesses arising in the follicles of the eyelashes are called *sties.*) *Carbuncles* are deep-seated abscesses of subcutaneous tissues. They are large, bumpy, and painful because there may be several closely associated, or intercommunicating, pockets of pus. These infections and others, such as staphylococcal *paronychia* (inflammation of soft tissues around nails) and *decubitus ulcers* (bedsores), usually arise at sites of minor but continuously irritating injuries to the skin. Such injuries may be caused by the constant pressure of clothing, the macerating effect of heat and moisture, or both of these physical effects coupled with lowered physiologic resistance. Bedridden patients who do not receive proper nursing care probably suffer most from these infections, which add their toll to the underlying debility that confines them to bed.

Acne is another type of skin infection commonly associated with staphylococci and other bacterial species that live commensally on skin. This is a problem of adolescence primarily associated with physiologic changes related to hormone output. Dietary factors may also play a role, as well as hypersensitivity to components of persistently infecting organisms. Acne is characterized by the recurrent appearance of crops of pimples on the face, often developing in such density that the epithelium cannot recover and is replaced by scar tissue. It can be controlled by skin soaps that reduce normal bacterial flora, judicious hormone therapy, and careful use of antibiotics. Tetracyclines applied topically may be helpful. In more severe cases, an oral antibiotic such as clindamycin may be given.

Staphylococcal skin infections are particularly serious in young children and infants who have few defenses against virulent strains. *Impetigo* (from the Latin word *impetare* meaning to attack) is the clinical term given to the contagious form of skin disease sometimes displayed by children. It is characterized by the formation of vesicles on exposed parts of the body. These are rather large "blisters" with an underlying base of inflammatory tissue. The serous fluid that fills them contains staphylococci, often in association with streptococci. The vesicles rupture and then crust over, healing later with some scarring. Fluid from ruptured vesicles is infectious for susceptible young contacts, and infection may spread in epidemic fashion among nursery-school or backyard playmates. It may spread in even more rapid patterns through a newborn nursery where susceptibility is greatest. (Impetigo of the newborn is sometimes called **pemphigus neonatorum** and should not be confused with noninfectious pemphigus of adults.)

Two other staphylococcal infections that affect children primarily, but also adults, are toxic epidermal necrolysis (TEN) and the scalded skin syndrome (SSS). These are characterized by the formation of large superficial blisters and separation of the upper layers of skin, which may peel off in "sheets." The staphylococcal toxins that produce these dramatic lesions are currently under investigation.

2. Deep-Tissue Infections Caused by Staphylococci. Within the tissues staphylococci may localize anywhere, depending on the route of entry and their accessibility to the bloodstream. From the upper respiratory tract they may extend to sinuses or the middle ear and mastoid, or downward to the bronchial tract and pulmonary beds. Staphylococcal pneumonia is one of the most serious bacterial pneumonias, because the organism and its toxins produce hemorrhagic, necrotizing lesions in the lung parenchyma. Within the upper respiratory tract itself, entrenched staphylococcal infection can be a stubborn and difficult problem, particularly when the causative strain is resistant to antibiotics.

Staphylococci sometimes cause puerperal fever, gaining access to the pelvic tract from the superficial genital mucosa during childbirth. Streptococci are associated with this type of disease more frequently (see Chap. 28).

Staphylococci may gain access to the bloodstream through skin wounds or abscesses and be distributed to various sites. They are capable of growing in any tissue. If local defenses are not adequate, abscesses can form in bone marrow (**osteomyelitis**), in the covering tissue that sheathes bone (**periostitis**), in liver or spleen, the kidney pelvis (**pyelonephritis**), or brain (**meningitis** or **encephalitis**). Depending on the site of entry, focal lesions develop in one or more of these sites; but if septicemia becomes fulminating and miliary abscesses develop, the result may be fatal.

Generalized infections are more frequent and dangerous in those of highest susceptibility: the very young or old, and those with problems that open the door to superimposed infection. The latter include surgical and obstetric patients, people with respiratory diseases, patients with renal or bladder dysfunctions (especially those that require catheterization), diabetics and others with endocrine disorders that affect resistance, and cardiac patients requiring arterial catheterization or open-heart surgery. Staphylococci (or other organisms) introduced to the endocardium directly by the latter techniques may result in a rapidly fatal **endocarditis,** or septicemia with generalized infection. Even a minor abscess in the skin may be the source of infection disseminated to systemic tissues if the fibrous capsule encasing it is broken by the pressure of squeezing, so that microorganisms within the lesion are released to surrounding tissues and to adjacent capillaries. It is particularly dangerous to squeeze pimples and boils that appear on the face, because the draining lymphatics and blood capillaries may carry released staphylococci directly into areas contiguous with the brain.

Laboratory Diagnosis

I. Identification of the Organism (*Staphylococcus aureus*) (*Staphylococcus epidermidis*)

A. MICROSCOPIC CHARACTERISTICS. Species of *Staphylococcus* cannot be distinguished by microscopic examination. They are Gram-positive cocci, typically arranged in grapelike clusters, nonmotile, and nonspore-forming.

B. CULTURE CHARACTERISTICS. Strains of *S. aureus* grow well under aerobic or anaerobic conditions on ordinary laboratory media. On blood agar they characteristically possess the *golden-yellow pigment* from which

their name was derived. They are often *hemolytic,* and virulent strains associated with lesions usually produce a substance called *coagulase* that causes clotting of plasma (see Chap. 6, p. 138).

Strains of *S. epidermidis* grow under the same conditions, but they possess a *white pigment* (they used to be called *S. albus* because of this). They are generally *nonhemolytic* and *coagulase-negative.*

These findings represent the general rule but variations may occur. Occasional strains of *S. epidermidis* may be hemolytic, and strains of *S. aureus* may be nonhemolytic or coagulase negative. Variants of either type may be associated with human lesions, sometimes with minor superficial infections of surgical wounds (e.g., stitch abscesses), or with chronic, low-grade, but threatening systemic infections occurring most frequently as an aftermath of cardiac surgery or catheterization.

C. SEROLOGIC METHODS OF IDENTIFICATION. Serologic methods are generally not applicable because the staphylococci are a large and heterogeneous group antigenically. The tools most useful in recognizing strains of common epidemiologic importance in hospital or community cross-infections are phage typing and antibiotic susceptibility testing (see Special Laboratory Tests below).

II. Identification of Patient's Serum Antibodies. The serum of most normal adults contains antibodies against numerous staphylococcal products and components: hemolysins, coagulase, leukocidin, hyaluronidase, and others. It is difficult, therefore, to base diagnosis on serum titers of these antibodies or to correlate them with increased resistance to infection.

III. Specimens for Diagnosis. Pus or fluid from any lesion in any part of the body may be submitted for smear and culture. Blood, spinal fluid, pleural, abdominal, or synovial fluid each may be appropriate for culture depending on clinical circumstances.

Stained smears of the exudate are examined for the presence of clusters of organisms resembling staphylococci. Culture reports indicate pigment, hemolysis, and coagulase activity, or they may read simply "*S. aureus*" or "*S. epidermidis,*" implying the combination of properties described above.

IV. Special Laboratory Tests. When it is suspected that cross-infections have been incurred from a single source of infection, two types of laboratory information may support or confirm this:

A. "*Antibiograms*" of staphylococci isolated from a series of infected patients may be compared. An anti-

Fig. 22-8. Phage typing of staphylococci. The surface of this agar plate was seeded with staphylococci, then inoculated with a series of numbered bacteriophages, each placed in a separate area. This strain of *Staphylococcus* was resistant to most of the phages tested, but was lysed by five of them, as evidenced by clear zones at the sites where phages 6, 47, 53, 81, and 83 were placed. (Courtesy of Center for Disease Control, D.H.E.W., Atlanta, Ga.)

biogram is the pattern of susceptibility and resistance shown by a given strain to a battery of antibiotics. When all isolated strains have similar or identical patterns, this constitutes presumptive evidence of their epidemiologic relationship.

B. *Phage types* of strains isolated from a series of infections may be compared. Coagulase-positive strains of *Staphylococcus* are generally sensitive to bacteriophages, each being lysed by a particular phage or group of phages. Staphylococcal bacteriophages are given numbers, and the type of a strain of *Staphylococcus* is indicated by the phages to which it is susceptible (Fig. 22-8). Thus, *S. aureus* of phage type 80/81 is a strain that is lysed by each of the two phages indicated by number. This has little significance of itself, but if many strains isolated from numbers of patients involved in an outbreak of staphylococcal infection all are reported from the laboratory as being of phage type 80/81 (or some other type), this means that the epidemic probably arose from one strain disseminated from a common source that should be identified and eliminated if possible. The strain may be spread by an asymptomatic nasal carrier, or from active lesions of an infected person. Some phage types appear to spread more rapidly than others, and to acquire drug resistance with greater speed than others.

Epidemiology

I. Communicability of Infection

A. RESERVOIR, SOURCES, AND TRANSMISSION ROUTES. Staphylococcal infections have their primary reservoir in man. They are spread from colonizing sources on normal human membranes (usually in the anterior part of the nose) or in open furuncles, draining sinuses, tonsillar abscesses, and similar lesions. They may be transmitted by direct personal contacts or indirectly through contaminated hands and objects. Airborne transfers may be important for viable organisms in dust, or in droplet nuclei from respiratory sources. In hospitals, dirty dry mops and dustcloths may contribute to their spread, as well as soiled bed linens, dressings, and unwashed hands of personnel.

B. INCUBATION PERIOD. This varies, and depends on the route of entry, the resistance of the infected individual, and the numbers of virulent organisms introduced. In surgical wounds, infection may appear within a day or two. When organisms are introduced directly into the bloodstream, symptoms may ensue within hours. Infections of normal "intact" skin, such as impetigo, boils, or other abscesses, usually require from four to ten days to appear.

C. COMMUNICABLE PERIOD. Staphylococcal infection is transmissible throughout the time purulent discharges drain from open lesions or the nasal carrier state persists.

D. IMMUNITY. Resistance to staphylococcal infections is based primarily on nonspecific mechanisms. Antibody titers are not consistent and seem to have little relationship to immunity against reinfection. Susceptibility is fairly general, but those at greatest risk are the newborn, those debilitated by chronic diseases or age, surgical patients, and those who must undergo protracted medical or surgical therapy for underlying organic diseases.

II. Control of Active Infection

A. ISOLATION. It is seldom necessary or feasible to isolate the patient being treated *at home*, but he should be kept away from infants or family members who are ill and debilitated. In the *hospital,* prompt isolation of all patients with staphylococcal infections is indicated, but the degree of isolation warranted may vary. In the newborn nursery, isolation should be complete and rigidly maintained (see Chap. 18, p. 427–28). On medical units, all cases of staphylococcal pneumonia should be placed in strict isolation, but for other infections the type of isolation depends on the degree of risk offered by draining lesions (see Tables 12–1 and 21–1, pp. 299 and 494).

B. PRECAUTIONS. Dressings from discharging lesions should be placed in a paper or plastic bag, sealed and burned. Bed linens and clothing should be carefully placed in tightly closed bags marked for care in transport and handling prior to laundering. If laundering routine cannot assure disinfection, sheets and bedclothes should be autoclaved or washed in boiling water with detergent. Concurrent disinfection must be carried out for all items and equipment in the patient's room that might be contaminated by infectious discharges or by drying droplet nuclei disseminated from the patient's secretions or exudates. A clean gown should be worn by each person who enters the room, including visitors. Upon leaving, each person discards the gown in a hamper provided near the hand-washing facility and carefully washes his hands with disinfectant liquid soap, drying them with paper towels. Terminal cleaning-disinfection must be thorough.

C. TREATMENT. Staphylococci have been notorious for their ability to become resistant to antibiotics within a short time of their introduction into clinical use, and this complicates the specific therapy of staphylococcal diseases. Because of the marked variability of staphylococci in their responses to useful antimicrobial drugs, and their ability to develop resistance rapidly, those responsible for serious infections, particularly, must be isolated and tested for antibiotic susceptibility. The results of laboratory testing serve as a guide to the choice of antimicrobial therapy coupled with appropriate supportive measures. Medical or surgical methods designed to promote drainage are used for localized skin infections, often with topical application of antibiotics. Systemic use of antibiotics is generally reserved for parenteral infection or the threat of extension from local sites.

Antibiotic resistance in *Staphylococcus aureus* is determined by chromosomal determinants or by extrachromosomal, self-duplicating genetic elements called plasmids (see Chap. 3, p. 61). The latter are transferred between strains by bacteriophage transduction or by transformation. Each plasmid codes for resistance to one antimicrobial agent or to several, and a given strain of *S. aureus* may possess several different plasmids or many copies of the same one. Thus one strain may display multiple antibiotic resistance, and this may be passed along to other strains through plasmid transfer. The appearance and persistence of plasmid-coded antibiotic resistance in the bacterial population are encouraged by the continued clinical use of particular antibiotics in the human population. In hospitals, which represent relatively closed situations where certain antibiotics are used repeatedly, the emergence of multiply resistant strains of staphylococci (and other organisms, such as enteric bacteria) offers serious difficulties in the treatment of threatening infections.

In the case of the penicillin resistance of *S. aureus,* genetic coding may involve production of the enzyme *penicillinase.* The genetic control of this enzyme also resides in a plasmid that can be transmitted to other staphylococci by transduction or transformation. Penicillinase acts on the beta-lactam portion of the penicillin molecule (see Fig. 10–12, p. 263) and is therefore often called *beta-lactamase.* Some semisynthetic penicillins, such as methicillin, nafcillin, oxacillin, and cloxacillin, are not destroyed by penicillinase, and these drugs may be clinically useful against staphylococci that produce the enzyme. They are not as active as penicillin, however, against Gram-positive bacteria. Some cephalosporin antimicrobics (see Fig. 10–13, p. 266) are similarly resistant to staphylococcal beta-lactamase and may be useful substitutes for penicillin, but these drugs are also less efficient therapeutically.

Some strains of *S. aureus* are now resistant to methicillin and other lactamase-resistant penicillins, and to the cephalosporins. Collectively referred to as methicillin resistant, such strains display multiple resistance to such useful drugs as tetracycline, erythromycin, lincomycin, kanamycin, and chloramphenicol. Generally they are susceptible only to gentamicin, vancomycin, clindamycin, and a few other, more toxic, agents. This resistance is independent of penicillinase production, but its precise mechanism has not been fully identified. Until recently, the frequency of methicillin-resistant staphylococci was greater in European hospitals than in the United States. They remain relatively infrequent in this country but have been responsible for some hospital epidemics among particularly vulnerable patients, such as those in burn units and nurseries.

The choice of specific antimicrobics for the treatment of staphylococcal infections must thus be made carefully and weighed against the seriousness of the clinical problem. In view of the rapid emergence of resistance among strains of S. aureus, hospitals sometimes restrict the use of an antistaphylococcal drug to treatment of only the most serious infections. For a valuable new drug, this practice can extend the period of its clinical usefulness.

D. CONTROL OF STAPHYLOCOCCAL INFECTIONS IN HOSPITALS. The control of hospital infections is based on an understanding of the three major reasons that hospitals represent the most dangerous focus for staphylococci:

1. Staphylococci rapidly become drug resistant. When constantly exposed to antibiotics, resistant strains emerge and susceptible ones are destroyed, as discussed above. Prevention and control of this situation depend on (a) vigorous therapy of staphylococcal infections with adequate doses of effective antibiotics for a time sufficient to eliminate the causative strain, and (b) restriction of antistaphylococcal drugs for use only in the most seriously ill patients. Indiscriminate use of antibiotics for minor infections or for general prophylaxis is discouraged.

2. Persons who run the greatest risk of serious staphylococcal infections are concentrated in hospitals and should have the greatest protection. Prompt identification and isolation of cases should be accompanied by efforts to find and eliminate sources so that others will not also be involved. High-risk patients (the newborn, those who have undergone prolonged surgery or are under continuing vascular or urinary catheterization) should be protected from infection by rigid aseptic techniques, using protective isolation (see Table 25–1, p. 560).

3. Patients and hospital personnel alike become carriers of drug-resistant strains after very short periods of exposure to hospital life. Both may be a source of organisms for exposed susceptible patients, but personnel constitute the most frequent, permanent, and mobile sources. When demonstrated to be active shedders of virulent strains that are antibiotic resistant and of phage types associated with the hospital's most common sporadic or epidemic infections, such personnel should be removed from work involving patient contacts, particularly the high-risk group.

When epidemic outbreaks occur in nurseries, surgical units, or elsewhere in the hospital, cultures are taken from infected patients, from the noses and lesions of contact personnel, and from environmental sources that might constitute a reservoir for persistent organisms. Antibiograms and phage types are obtained for all isolates of S. aureus, and the data are analyzed for a common source. Infected patients are isolated, aseptic techniques are reviewed and tightened, and personnel harboring the epidemic strain are removed from the area. Usually they are placed on antibiotic therapy, but this is not always effective in eliminating a carrier strain or in ensuring its permanent displacement.

Strict principles of personal hygiene should be applied to patients and practiced by personnel in all hospital situations, particularly when threats of staphylococcal outbreaks arise. Recurring infections often can be controlled by routine, judicious use of disinfectant skin soaps; brief daily ultraviolet exposures for carriers; laundry disinfection of clothing; and careful disinfectant-cleaning of the immediate environment (including bathroom facilities, bedside furniture, and personal toilet items).

III. Prevention

A. IMMUNIZATION. Vaccines prepared from killed cultures of staphylococci and toxoids containing several antigenic but inactivated toxic products of staphylococci are often used in treatment of recurrent furunculosis. Their success as immunizing agents has been difficult to establish, for they fail as often as they seem to improve the patient's resistance. They may be effective together with vigorous methods of skin and environmental asepsis.

B. CONTROL OF THE NASAL RESERVOIR. For patients with recurring staphylococcal skin infections, the nasal reservoir can be suppressed by topical application of antibiotic ointments. Gentamicin ointment is preferred, since few strains of S. aureus are resistant to this drug. It must be applied several times daily for a period of at least three months. This reduces the numbers and frequency of isolation of staphylococci from the nose, skin, and immediate environment of carriers, but prescribed treatment must be maintained to avoid recurrence of infection.

The principle of bacterial interference has also been applied to this problem. A strain of *S. aureus* originally carried in the nose can be replaced with a presumably less virulent one. This approach has been used for families with recurrent skin lesions, the original strain being eliminated first by topical ointments and oral antibiotics, then replaced by implanting a new, avirulent strain in the nose. Implantation of a nonvirulent strain has also been used to control epidemics of staphylococcal disease in newborn nurseries. Since newborns have not yet been colonized by bacteria, implantation can be made directly, in the nose and umbilicus. This method of control is reserved for use as a last resort, however, for "avirulent" strains are known to be capable of causing severe infections. Also, an implanted strain will be eliminated by antimicrobial therapy given later.

C. Prevention of staphylococcal infections in hospitals requires coordinated policy and the cooperation of all responsible for the safety of patients (see Chap. 10, pp. 254–59).

Erysipelas

The Clinical Disease. Erysipelas is a skin infection caused by strains of beta-hemolytic streptococci of group A. Systemic infection also occurs, with bacteremia, fever, malaise, and marked leukocytosis. The skin lesions occur most frequently on the face and legs, beginning at an inconspicuous portal of entry of the organism in the skin (perhaps rubbed into an abrasion or a hair follicle). The lesion is tender and red and spreads rapidly, having a raised advancing margin. On the face, erysipelas often spreads bilaterally from the nose across the cheeks in a "butterfly" pattern. This highly contagious infection is more severe in patients with debilitating conditions.

The streptococcal diseases and the role of streptococci in skin infections are discussed in Chapter 13. A few details pertinent to the laboratory diagnosis and epidemiology of erysipelas are mentioned here.

Laboratory Diagnosis. Specimens for smear and culture should be taken from the raised peripheral border of lesions. Blood cultures may yield the organisms if taken during febrile periods.

Epidemiology. Erysipelas occurs most frequently in aged people and in infants. It is characterized by recurrences in the same individual, possibly because of hypersensitivity or reinfection from endogenous sources. Its epidemiology is similar to that of other streptococcal diseases described in Chapter 13. Hospitalized patients should be isolated and strictest precautions should be followed, especially if the newborn nursery is involved (see Table 12–1).

Leprosy (Hansen's Disease)

The Clinical Disease. Leprosy is a chronic, very slowly progressive disease. It requires many years to develop and is perhaps the least communicable of infectious diseases. When untreated its progressive development leads to disfigurements that have been greeted for centuries with horror and fear. No social ostracism could be more complete, and no physical exclusion more dreadful than that to which lepers have been subjected throughout human history (Fig. 22–9). In recent years, medical knowledge has shed light and objectivity on the problem, relieving some pressure on the leprous patient.

Two forms of leprosy are recognized. Both require many years to develop, and both may be present in the same patient. One is a form characterized by granulomatous nodules, called *lepromas,* in the skin, mucous membranes (especially those of the upper respiratory tract), and some visceral organs. This nodular form is often spoken of as *lepromatous,* or *cutaneous,* leprosy (Fig. 22–10, *A*). The other form involves development of lesions around peripheral nerves, which leads to sensory damage, anesthesia, and atrophy of muscle, skin, and bone, particularly in the extremities. This *tuberculoid,* or *neural,* form of leprosy may lead to mutilating injuries of hands or feet that leave the patient open to secondary infections resulting in loss of fingers and toes, nasal cartilage, and other affected tissues.

Fig. 22–9. The stave church in Borgund, Norway. The door on the far right was reserved for lepers. (Courtesy of the Royal Norwegian Embassy Information Service, New York, N.Y.)

Laboratory Diagnosis

I. Identification of the Organism (*Mycobacterium leprae*). The agent of leprosy is an acid-fast bacillus, classified among the "higher" bacteria in the same genus as the tubercle bacillus (see Chap. 13, p. 330). Unlike the latter, *M. leprae* (originally called Hansen's bacillus) is an obligate intracellular organism. It has never been satisfactorily cultivated in cell-free media or in tissue culture. In recent years efforts to establish leprosy in experimental animals have succeeded, and the disease can now be studied in the mouse and possibly the armadillo.

When seen in the tissues of infected human beings or mice, aggregates of *M. leprae* resemble tubercle bacilli, lying in parallel bundles within cells of granulomatous tissue, particularly epithelioid and other mononuclear cells. Injected into the footpads of mice, the organism produces local granulomatous lesions. Subsequent inva-

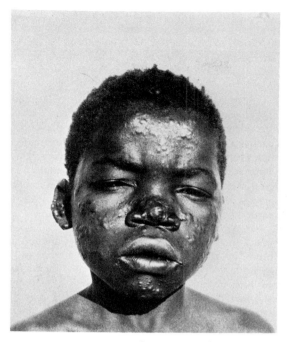
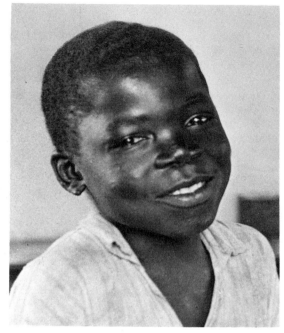

Fig. 22–10. *A.* An early case of cutaneous (lepromatous) leprosy. *B.* The same child following treatment with antileprosy drugs. (Courtesy of American Leprosy Missions, Inc., New York, N.Y.)

sion of striated muscle and destruction of dermal and peripheral nerve trunks in the mouse (feet, ears, nose, and tail) produce deformities like those seen in infected humans.

II. Identification of Patient's Serum Antibodies. Reliable serologic tests for leprosy have yet to be developed. A skin-testing antigen, *lepromin,* has some value in determining the prognosis, since it can distinguish the two forms of leprosy. The antigen is an extract of leprous tissue that produces hypersensitive skin reactions in patients with the neural type of leprosy, but the test is negative in lepromatous cases.

It should be remembered that sera from leprous patients often give false-positive results in reagin tests for syphilis (see p. 499).

III. Specimens for Diagnosis. Scrapings or biopsies of skin lesions are submitted for microscopic examination for intracellular acid-fast bacilli. The nasal mucosa is a preferred area for sampling, or the active periphery of a nodule cut through its epidermal layer and scraped along the inner edge of the cut for cells representing the central portion of the granuloma. Acid-fast bacilli are seen within mononuclear cells.

Cultures are attempted with this material to establish the fact that the organism cannot be propagated and that it is not *M. tuberculosis.*

Epidemiology

I. Communicability of Infection

A. RESERVOIR, SOURCES, AND TRANSMISSION ROUTES. This disease has a human reservoir only. Organisms present in discharges of infectious lesions are thought to be transmitted through the skin or mucous membranes of susceptible contacts. Infants and young children are the most susceptible, but their contacts with infectious parents or other adults must be long and intimate if they are to contract leprosy.

B. INCUBATION PERIOD. Very long periods of exposure are required, the average time for appearance of symptoms being three to five years. Infants heavily exposed from birth have been known to develop the disease in as short a time as seven months or, more usually, one to two years.

C. COMMUNICABLE PERIOD. Leprosy is considered infectious if bacilli can be demonstrated in skin or mucosal lesions. Usually many more organisms are demonstrable in lepromatous nodules than in the tuberculoid, neural lesions.

D. IMMUNITY. Human susceptibility to leprosy is universal but greatest during childhood. Immunity is of the hypersensitive type, as indicated by reactivity to lepromin. Patients with the more progressive lepromatous type of infection have no resistance and are lepromin negative, whereas tuberculoid patients as well as infected but asymptomatic persons are lepromin positive. The long incubation period and the closeness of contact required for transmission of the disease indicate a low communicability; increasing resistance with age is reflected in the diminished communicability of leprosy in older children and adults.

Leprosy is distributed primarily in tropical and semitropical areas of the world. There are an estimated 11 million total cases, half of which are found in China and India. The African continent has the next highest incidence, and there is a spotty distribution of foci in Mediterranean Europe, Hawaii, the Caribbean, and the southern Gulf States in the United States.

II. Control of Active Infection

A. ISOLATION. It is no longer believed that lepers should be rigidly isolated. During active stages of disease they are best treated in leprosaria, such as the National Leprosy Hospital in Carville, Louisiana, where experienced methods of medical treatment and control offer hope for suppression of clinical symptoms and a return to community living. Many patients with inactive lesions are treated at home.

B. PRECAUTIONS. Nursing care is very similar to that of the tuberculous patient. Respiratory and skin discharges should be disposed of by incineration. Bed linens and clothing should be carefully handled

and bagged for transport to the laundry. If the patient is in an active lepromatous stage, a gown is advisable to protect the clothing of the nurse or others who are in close contact, and hand washing is vital. Terminal disinfection-cleaning should be thorough.

C. TREATMENT. The sulfone drugs (diaminodiphenylsulfone, promin, and diasone) have proved effective in arresting active disease and in preventing reactivation of old lesions. Antituberculous drugs, such as streptomycin and rifampin, also have value. Other antibiotics are used to control secondary bacterial infection. The sulfones are given in gradually increased doses until a maintenance plateau is reached and continued for three or more years (Fig. 22-10, B).

D. CONTROL OF CONTACTS. The chief method of control involves case finding, especially among family contacts of the newly diagnosed individual case. Newborn infants are separated from leprous parents, but older children who have already been exposed are not segregated from their families. All active cases are reported and registered so that they can be periodically examined and referred for treatment when necessary.

III. Prevention. The appearance of new cases can be markedly reduced by protecting or segregating children born into families where leprosy exists. In endemic areas, survey and treatment clinics and health education of the public have helped to control the disease and to alleviate both suffering and fear.

Yaws

The Clinical Disease. Yaws is one of several nonvenereal diseases caused by spirochetes of the treponemal group. (*Bejel* and *pinta* are related syndromes.) It is an acute infection that becomes chronic and relapsing if not treated. The primary lesion, or "mother yaw," appears in three to six weeks at the site of skin exposure and inoculation. This lesion is a papule that enlarges into a papilloma, or tumorlike growth of epithelial cells. Mild symptoms of systemic involvement occur, and after some weeks successive crops of papules appear on the skin, persisting for months. There may also be destructive lesions in bone. The disease is marked by periods of latency and reactivation of progressive lesions that can be quite disfiguring. The superficial lesions are distributed on arms and legs, palms and soles, but may also occur on the trunk and on oral and nasal mucosa.

> **Laboratory Diagnosis.** Clinical diagnosis is based on the appearance and distribution of lesions and is supported by laboratory demonstration of spirochetes in the eroding and ulcerating papillomas. Dark-field examination or special staining of exudates is necessary. The causative organism is *Treponema pertenue.*
>
> Sera of patients with yaws or other treponematoses give positive results in serologic tests for syphilis.

Epidemiology. The reservoir for this disease is man. Exudates of skin lesions contain the organisms, which are transmitted by direct contact or indirectly by contaminated objects. Flies may play a role in transmission, but this is probably less frequent than personal contact.

Yaws is a disease of tropical and semitropical countries around the world. It occurs most frequently in children, and in areas where living conditions are poorest. The disease responds to a single intramuscular injection of penicillin in a long-lasting form, so that early diagnosis and prompt treatment constitute the best hope for control. Concerted efforts are being made through the World Health Organization to eliminate yaws in its endemic foci in many countries, through case finding and mass treatment programs, and improvement of standards of living.

Herpes Simplex Virus

Herpesvirus Infections. Herpes simplex virus (HSV) belongs to the group of viruses that includes varicella (chickenpox) and herpes zoster agents. It produces vesicular eruptions of the skin or mucous membranes similar to those of zoster (see Chap. 14), but it differs antigenically and in its epidemiologic patterns of spread. HSV often establishes a lifelong association with infected human beings. It may involve various mucous membranes (oral, pharyngeal, conjunctival, genital) or mucocutaneous areas in recurrent infections interspersed with long periods of asymptomatic latency.

In the wide spectrum of herpetic infections, most clinical syndromes are mild. One of the most common is the recurrent *herpes labialis* of adults. This is familiar to everyone as the "cold sore" or "fever blister" that crops up again and again in some individuals at a mucocutaneous margin of the lip. Primary acute *stomatitis,* an inflammation of the oral membranes, occurs in small children, and *herpes genitalis* is a sexually transmitted disease of adults (see p. 512).

More severe forms of herpetic disease occur less frequently. These include *keratoconjunctivitis* (see p. 516). chronic *eczema,* and *meningoencephalitis.* Generalized *neonatal infection,* acquired at birth from exposure to maternal genital lesions, is perhaps the most disastrous form of herpetic disease and is often fatal (Fig. 22–7). Ordinarily the newborn child is protected from critical infection during the first months of life by maternal antibody produced in the course of low-grade infections experienced by the mother. If the mother acquires her first genital infection during pregnancy, however, and the child is born before sufficient protective antibody can be transferred placentally, neonatal infection may ensue. The virus is widely disseminated in infant tissues, producing necrotizing lesions in the skin and mucosa, liver, and brain.

Two serologic types of HSV are recognized, displaying biologic as well as antigenic differences. HSV type 1 is characteristically associated with most nongenital infections, while HSV type 2 is the cause of genital lesions and neonatal disease. The latter has also been recovered from cases of aseptic meningitis. These viruses can be distinguished by different effects produced in infected tissue cultures and embryonated eggs.

> **Laboratory Diagnosis.** HSV can be isolated from skin and mucosal lesions or from spinal fluid in central nervous system infections and identified serologically. Serologic diagnosis can be made by demonstrating a rising titer of neutralizing antibodies.

Epidemiology. Man is the reservoir. HSV may be present in saliva, respiratory secretions, conjunctival discharges, and on the genital mucosa. In males, the principal reservoir may be semen. The virus is transmitted by direct personal contacts of all kinds.

Herpes labialis is the most common result of the reactivation of latent virus infection. Fever blisters and cold sores reappear on the face or lips, the vesicular lesions usually crusting in a few days, then healing spontaneously. Reactivated lesions may also involve other areas of skin, elsewhere on the body. This reactivated infection is usually the result of trauma (sunburn is a common cause), hormonal or other physiologic changes (e.g., menstruation), or concurrent infectious disease (respiratory infection, bacterial meningitis, malaria). Occasionally, reactivation may lead to serious CNS disease, such as meningoencephalitis, but the latter occurs most often as a primary herpetic infection, acquired through direct contact with the virus in the saliva of carriers.

Nurses and others who handle infants with skin lesions, or patients with chronic eczematoid disease, may be infected through direct contact. Extreme care should be used in such nursing care, with strict observance of appropriate precautions (see Tables 12-1 and 21-1). Newborn infants with disseminated herpetic disease must be cared for in complete isolation, precautions including the use of gowns, gloves, and masks.

Older patients with herpetic lesions should not be permitted to come in contact with newborn infants. Such exposures should also be prevented for burn patients or others with chronic skin diseases. Patients on immunosuppressive therapy, or those with immunodeficiencies, must also be protected from the risk of generalized herpetic disease.

There is no specific treatment for most forms of HSV infection other than acute conjunctivitis. In the latter, iododeoxyuridine and adenine arabinoside have some clinical value in modifying the course of infection. The only general measures for control lie in good personal hygiene and avoidance of contact with obviously infected persons.

Superficial Mycoses

Ringworm Infections. Superficial mycotic diseases of hair, skin, or nails are referred to collectively as ringworm infections. The clinical term for ringworm is *tinea* (derived from the Latin word for "worm"). In medical usage, this term is modified by another Latin word indicating the part of the body affected, e.g., *tinea capitis* (ringworm of the scalp), *tinea corporis* (ringworm of the body), *tinea pedis* (ringworm of the feet), and *tinea unguium* (ringworm of the nails). Many other more specific anatomic terms may be used for fungous infections of particular areas of the body, such as *tinea barbae* (the bearded area of face and neck are involved) or *tinea cruris* (inguinal folds are infected).

These "dermatomycoses" are caused by members of a group of fungi called *dermatophytes* because they are capable of invading skin or its keratinized appendages, but do not infect systemic tissues. There are three important genera of dermatophytes (*Epidermophyton, Microsporum,* and *Trichophyton*), described briefly below under Laboratory Diagnosis. A particular species of dermatophyte may cause a variety of clinical lesions in different areas of the body, and, conversely, different fungi may cause similar clinical manifestations.

Epidemiologically these infections have much in common. Some of the causative fungi have a reservoir in both man and animals, some in man only, but all are directly transmissible from person to person, unlike the systemic fungi (see Chap. 16). The route of transmission may be through direct contact with infected persons (or animals) or from indirect sources of environmental contamination derived from active cases. Methods for control and treatment of these infections are very similar and are described in outline form at the end of this section (see Table 22-2), following a brief description of the four major clinical forms of dermatomycosis and the causative fungi.

I. Tinea Capitis (Ringworm of the Scalp). This infection usually begins with the appearance of a small scaling papule which is red and itchy. Numerous papules may appear and spread peripherally. The hair of the affected area becomes brittle and is easily broken off, leaving patches of baldness (alopecia). Sometimes underlying tissue becomes inflamed and ulcerated, this type of lesion being called a *kerion*. The hair shafts themselves are infected with

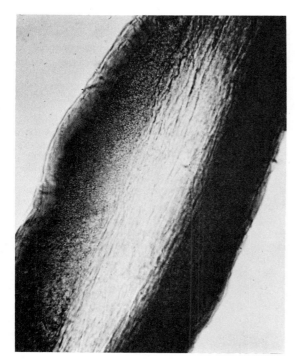

Fig. 22-11. Ectothrix infection of hair. Growth of *Microsporum audouini* forms a sheath around the hair shaft. (Reproduced from Rippon, John W.: *Medical Mycology: The Pathogenic Fungi and the Pathogenic Actinomycetes*. W. B. Saunders Co., Philadelphia, 1974.)

fungus (Fig. 22-11). Some of the fungi (*Microsporum* species) are fluorescent in ultraviolet light. This can be an aid to diagnosis: that is, the suspected area of the scalp can be examined under an ultraviolet ("Wood's") lamp, and fluorescing hairs can be removed with forceps for laboratory examination (Fig. 22-12).

Some *Trichophyton* species create a lesion known as "black dot" ringworm, the effect being created by dark broken stumps of hair shafts sticking up in an area of alopecia and scaling. These fungi do not fluoresce under the Wood's lamp.

Tinea favosa is a clinical lesion of the scalp caused by a particular species of *Trichophyton* (*T. schoenleinii*). Cup-shaped yellowish crusts, called *scutula*,

are formed. The lesion is deep and may heal by scarring, with permanent destruction of hair follicles.

2. Tinea Corporis (Ringworm of the Body). This term refers to fungous infections of the glabrous (smooth, nonhairy) skin anywhere on the body. The lesions are characteristically flat, spreading, and ring shaped. The periphery of the lesion is always the most active, being raised, erythematous, sometimes vesicular and weeping, sometimes dry and scaling. Moist lesions may form crusts as they heal. The infection advances peripherally, leaving a central area of healing, normal skin in many instances.

3. Tinea Pedis (Ringworm of the Feet). "Athlete's foot" is one of the most common and widespread superficial fungous infections. It may often appear as nothing more than a scaling or cracking of the skin, particularly between the toes; but it may become chronic and severe, with vesicle formation and maceration of tissues. Scaling may extend over the heels and soles of the feet, sometimes with vesicular eruptions or acute excematoid reactions. The tissues of the sole may be undermined by ulcerations, with resulting cellulitis, lymphangitis, and systemic reaction. This form of *tinea pedis* can be chronically disabling. The hands may also become infected, and allergic skin reactions to fungal products may appear on many parts of the body, but particularly arms and legs.

4. Tinea Unguium (Ringworm of the Nails). Nails of the hands or feet may become involved in chronic infections caused by dermatophytic fungi. (Infections with *Candida albicans* are also common, but have a somewhat different epidemiology, being of endogenous origin). The affected nails become discolored, thickened, brittle, and often deeply grooved or pitted. Caseous epidermal debris accumulates under the nail, and the top may separate. Eventually the whole nail may be destroyed if the fungus invades the entire plate. Unlike *Candida* or staphylococcal infections, there is usually no paronychial involvement.

Laboratory Diagnosis. The characteristic features of the three genera of dermatophytes may be summarized as follows:

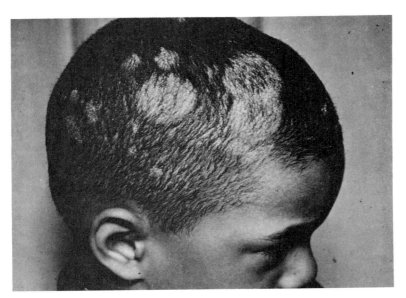

Fig. 22–12. Tinea capitis, or ringworm of the scalp, often occurs in childhood. The ultraviolet Wood's lamp might be used to examine this child's lesions. Fluorescence of infected hairs would be indicative of *Microsporum* infection. If no fluorescence can be seen, the infectious agent is probably a *Trichophyton*. The diagnosis must be confirmed by culture of infected hairs. (Reproduced from Rippon, John W.: *Medical Mycology: The Pathogenic Fungi and the Pathogenic Actinomycetes*. W. B. Saunders Co., Philadelphia, 1974.)

Epidermophyton. The colony is velvety and greenish-yellow, with radiating furrows. Microscopically, macroconidia identify the fungus. These are large, club-shaped, septate, smooth bodies usually born in clusters on the hyphae. The only important species is *E. floccosum.*

Microsporum. Colonies are velvety, woolly, or powdery, depending on the species. They also vary in color, one of the important species being light gray to brown: the other two, bright orange or yellow. Pigmentation is seen best on the underside of the colonies. The macroconidia are spindle-shaped, septate, thick-walled, and rough surfaced (Fig. 22–13). The three important species are *M. audouini, M. canis,* and *M. gypseum.*

Trichophyton. There are many species, displaying a variety of colony forms and pigments. On microscopic examination, these fungi are seen to produce many microconidia but often few macroconidia. The latter are thin, pencil-shaped bodies, multiseptate, thin-walled, and smooth (Fig. 22-14). Some important species are *T. mentagrophytes, T. rubrum, T. tonsurans, T. violaceum,* and *T. schoenleinii.*

Most of these fungi grow slowly in the laboratory, requiring two to four weeks for maturation and identification of characteristic reproductive structures. They are strict aerobes that grow best at room temperature (22° to 28° C) rather than at body temperature.

Specimens for Diagnosis. Scrapings of involved areas of skin, vesicular fluid, exudates, or hairs may be submitted for microscopic examination and culture. In general, it is best to take material from the active margins of advancing lesions, scraping deeply under the lip of crateriform lesions or into the inflamed base of vesicles or papules. Hairs and skin scrapings may be examined under the microscope directly, using a 10 percent hydroxide solution to clear cellular debris. When infected hair is examined, fungous spores may be seen invading the shaft itself (*endothrix*) or clinging to its external surfaces (*ectothrix*). Skin scrapings usually reveal branching hyphal fragments without spores.

These materials may be satisfactorily cultured on a variety of simple media, the most common of which is Sabouraud's glucose agar, usually containing antibiotics to inhibit growth of contaminating bacteria.

Treatment. Superficial mycotic infections are treated locally with a variety of fungistatic or fungicidal solutions, ointments, lotions, or dusting powders. Some of the imidazole compounds have been found clinically useful in topical applications. These include miconazole, econazole, and clotrimazole. In

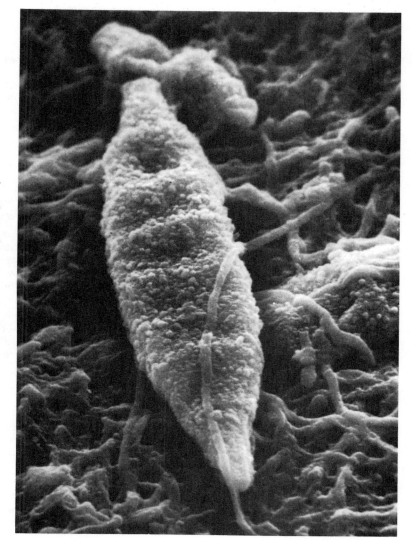

Fig. 22–13. Scanning electron micrograph of *Microsporum gypseum*. (800×.) This is a macroconidium lying among hyphae of the mycelial mat. Macroconidia of this genus are spindle-shaped, septate, and have rough, thick walls. Compare with Figure 22–14. (Courtesy of William G. Barnes, Ph.D., John D. Arnold, M.D., and Arthur Berger, Department of Pathology, General Hospital and Medical Center and Harry S. Truman Research Laboratory, Kansas City, Mo.)

clinical usage the imidazoles are active against superficial *Candida albicans* infections as well as the dermatophytes.

Systemic therapy of dermatophyte infections is presently limited to oral griseofulvin. Miconazole is an investigational new drug being evaluated for treatment of the systemic mycoses. It appears to be effective against the dermatophytes as well when given intravenously, but it is not yet available for oral therapy.

Table 22–2. Epidemiology of Superficial Mycoses

Disease	Causative Fungus	Reservoir	Sources	Transmission	Incubation Period
Tinea capitis	*Microsporum* sp. *Trichophyton* sp.	Man Dogs Cats Cattle	Lesions Combs Toilet articles Barber's tools Headrests of upholstered seats in theaters, trains, etc.	Direct or indirect contacts	10 to 14 days
Tinea favosa	*T. schoenleinii*	Man	Crusts from scutula	Direct or indirect contacts (favored by crowding and filth)	10 to 14 days
Tinea corporis	*Epidermophyton* *Microsporum* sp. *Trichophyton* sp.	Man and animals	Lesions Clothing Floors Shower stalls	Direct or indirect contacts	10 to 14 days
Tinea pedis	*Epidermophyton* *Trichophyton* sp.	Man	Lesions Shoes and socks Floors Shower stalls	Direct or indirect contacts	10 to 14 days
Tinea unguium	*Trichophyton* sp.	Man	Lesions	Person-to-person transmission unusual, but may extend to other nails of infected individual	Source and time of exposure usually unknown

Communicable Period	Special* Treatment	Precautions	Other Controls
While lesions remain active	Daily shampoo Fungicidal ointment Protect with skullcap At night, pull out infected hairs and reapply ointment Griseofulvin by mouth, especially for *M. audouini* infections X-ray epilation if indicated	Keep head covered with cotton cap that can be sterilized frequently by boiling or autoclaving	Investigate household for other infected cases Look for infection among pet animals or farm animal contacts
While lesions remain active	Epilation usually essential Other therapy as above	As above	More common in central and southeast Europe than in U.S.A.; seen in Russian and Polish immigrant refugees who are detained and treated before entry
While lesions remain active	Keep skin very clean by soap-and-water bathing Clean away crusts and scabs, if any, and apply fungicidal ointments Griseofulvin by mouth	Infected persons should not frequent public pools and gymnasiums Clothing, especially underclothes, should be disinfected if in contact with lesions	Look for source of infection among human or animal contacts
While lesions remain active	As above Keep feet clean and dry; expose to air as much as possible	Cotton socks or hose should be worn and disinfected between uses, preferably by boiling	Infected persons should not frequent public pools and gymnasiums; the care of such places includes daily disinfectant cleaning
While lesions remain active but little communicability for others	Nail is kept closely pared and filed Fungicidal soaks and ointments topically Griseofulvin by mouth X-radiation sometimes necessary but recurrences are frequent	As above if toenails are involved When fingernails are infected, gloves should be worn only to maintain contact with fungicides; otherwise nails should be kept clean and dry	As above if toenails are involved

*See text discussion.

Clinical Disease	Causative Organism	Other Possible Entry Routes	Incubation Period	Communicable Period
1. Sexually Transmitted (Venereal) Diseases				
Syphilis	*Treponema pallidum*	Parenteral Congenital	10 to 21 days	Probably communicable during 2–4 years of primary and secondary stages, and during first years of latency if untreated
Gonorrhea	*Neisseria gonorrhoeae*	Conjunctiva (newborn or adults) Chorioamnion	3–9 days; in ophthalmia neonatorum 36–48 hours	In adults, communicable for years if not treated; conjunctival membranes are infectious until discharges cease
Chancroid	*Haemophilus ducreyi*		3–5 days	Communicable through infectious discharges of genital lesions and buboes
Granuloma inguinale	*Calymmatobacterium granulomatis*		Unknown	While infective agent is found in lesions
Lymphogranuloma venereum	*Chlamydia trachomatis*		1–4 weeks	Untreated LGV may be communicable for months or years
II. Eye Diseases				
Acute bacterial conjunctivitis	Streptococci Staphylococci Pneumococci *Haemophilus* species Gonococci	Respiratory	Short, 24–72 hours	During active infection
Trachoma	*Chlamydia trachomatis*		About 5–12 days	During active infection

Specimens Required	Laboratory Diagnosis	Immunization	Treatment	Nursing Management
Tissue fluid aspirations Gummatous lesion biopsy Blood serum	Dark-field examination for spirochete Silver impregnation demonstrates spirochete Positive for reagin (VDRL) Positive for treponemal antibodies (TPI, FTA-ABS, MHA-TP)	None	Penicillin Tetracyclines Erythromycin	Isolation not necessary (antibiotic therapy eliminates threat of communicability in 24 hr); concurrent disinfection of discharges from open lesions and of freshly contaminated objects; *hand washing*
Exudate from lesions Blood, joint fluid in DGI	Smear, males Culture, females and smear-negative males Culture, direct FA	None	Penicillin Spectinomycin for PPNG cases	Hospitalized infants and children isolated 24 hr following start of antibiotics; concurrent disinfection; *hand washing*
Exudate from lesions	Smear and culture	None	Sulfonamides Tetracyclines	Concurrent disinfection of exudates from lesions and contaminated articles
Biopsy of lesions	Isolation of organism in chick embryos	None	Tetracyclines Streptomycin	Same as above
Exudate from lesions Tissue biopsy Blood serum Skin test	Smear and isolation in embryonated eggs Rising titer of antibodies Frei test positive	None	Tetracyclines Sulfonamides Chloramphenicol	Same as above
Exudate or discharge from eye	Smear and culture	None	Antibiotic ophthalmic ointment Sulfonamide ophthalmic ointment	Concurrent disinfection of exudate and articles contaminated by exudate
Epithelial cells scraped gently from eyelids Blood scrum	Smear for cytoplasmic inclusion bodies Tissue culture, eggs Rising titer of antibodies	None	Oral tetracycline Ophthalmic ointment	Same as above

535

Table 22–3. *(Cont.)*

Clinical Disease	Causative Organism	Other Possible Entry Routes	Incubation Period	Communicable Period
Inclusion conjunctivitis (newborn)	*Chlamydia trachomatis*	Contaminated swimming-pool water	Unknown	During active infection
Keratoconjunctivitis	Adenovirus (type 8) Herpes simplex virus (type 1)	Respiratory	Unknown	During active infection
Hemorrhagic conjunctivitis	Enterovirus (type 70)			

III. Skin Contact Diseases

Staphylococcal infections	*Staphylococcus aureus*	Respiratory Parenteral	Variable; in surgical wounds 1–2 days; impetigo, boils, 4–10 days	During time of purulent discharge of open lesion; and as long as nasal carrier state persists
Erysipelas	Beta-hemolytic streptococci of group A	Respiratory	Unknown, probably about 2 days	During active infection, no longer communicable 24 hours after institution of antibiotic therapy
Leprosy	*Mycobacterium leprae*		3–5 years	During time bacilli can be demonstrated in skin or mucosal lesions

Specimens Required	Laboratory Diagnosis	Immunization	Treatment	Nursing Management
Epithelial cells scraped gently from eyelids	Smear for cytoplasmic inclusion bodies Tissue culture, eggs	None	Tetracycline Erythromycin Sulfonamides	Same as above; isolate infected babies in nurseries
Blood serum	Rising titer of antibodies	None	Nonspecific except iododeoxyuridine for herpes	Careful aseptic techniques in dispensaries and industrial clinics
Exudate from lesions Blood Spinal fluid Pleural fluid Abdominal fluid Synovial fluid	Smear and culture Antibiograms Phage typing	Killed vaccines and toxoids	Antibiotics chosen on basis of susceptibility	Hospital isolation; very stringent aseptic techniques including burning of dressings, autoclaving of linen if laundry cannot assure disinfection; concurrent disinfection of all articles possibly contaminated by infectious discharges; gown and *hand washing* with disinfectant soap; terminal disinfection.
Scrapings from raised peripheral border of lesions Blood	Smear and culture Culture	None	Penicillin Erythromycin	Hospital isolation; concurrent disinfection of discharges and dressings from lesions; thorough terminal disinfection
Scrapings or biopsies of skin lesions Skin tests	Acid-fast smears Culture to rule out *M. tuberculosis* Lepromin test prognostic; + in neural form only	None	Sulfones Antituberculous drugs	Rigid isolation not necessary; respiratory and skin discharges incinerated; bed linens and clothing bagged for transport to laundry; gown and *hand washing;* thorough terminal disinfection-cleaning

Table 22–3. *(Cont.)*

Clinical Disease	Causative Organism	Other Possible Entry Routes	Incubation Period	Communicable Period
Yaws	*Treponema pertenue*		3–6 weeks	During time of active infection and while skin lesions are present
Herpes simplex infection	Herpes simplex virus, type 1 and type 2	Respiratory	Up to 2 weeks	During active infection

Fig. 22–14. Photomicrograph of a macroconidium of a *Trichophyton* species. (1400×.) Compare its morphology with the macroconidium of *M. gypseum* shown in Figure 22–13. (Courtesy of Dr. John W. Rippon, The Pritzker School of Medicine, University of Chicago, Chicago, Ill.)

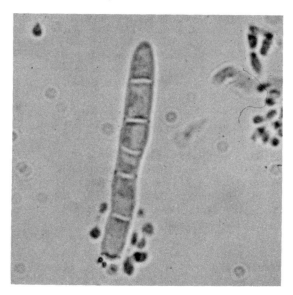

Specimens Required	Laboratory Diagnosis	Immunization	Treatment	Nursing Management
Exudates and scrapings from papillomas Blood serum	Dark-field examination Special staining Positive reagin test for syphilis	None	Penicillin	Concurrent disinfection of discharges and contaminated articles
Exudates from lesions Spinal fluid	Virus isolation HSV, 1, nongenital lesions HSV 2, genital and neonatal lesions Characteristic inclusions in stained cells from lesions	None	Nonspecific (iododeoxyuridine or adenine arabinoside for eye infections)	Careful personal hygiene; avoidance of contact with obviously infected persons; strict isolation of infected newborns
Blood serum	Rising titer of antibodies			

Questions

1. Why is it not usual for venereal diseases to be spread by means other than sexual contact?
2. What factors are involved in the control of venereal diseases?
3. What are the possible clinical consequences of untreated syphilis?
4. What is congenital syphilis?
5. What is the communicable period for syphilis? What nursing precautions are required for each of the stages of syphilis?
6. When is treatment for the syphilitic pregnant woman effective?
7. Name some of the routine serologic tests for syphilis (STS). Why is it necessary for a positive STS to be confirmed? How do confirmatory tests differ from routine screening tests?
8. What is the communicable period for gonorrhea? What nursing precautions are indicated for this disease?
9. What are the possible clinical consequences of untreated gonorrhea?
10. How are infants infected with *N. gonorrhoeae?* Young children?
11. What is the basis for and significance of penicillin resistance in gonococci?
12. Outline the steps required to make a laboratory diagnosis of gonorrhea in the adult male and the adult female.
13. Why is it important to transport cultures for gonorrhea to the laboratory promptly? Describe a growth-transport agar medium system and indicate the reasons for its value.
14. How can gonorrhea be prevented and controlled?
15. What are the two epidemiologic categories of intrauterine infections? Name some infectious diseases of infants that are acquired at birth.
16. What is trachoma? Inclusion conjunctivitis?
17. What are the significance and value of phage typing in staphylococcal infection?
18. Why are hospitals a dangerous focus for staphylococcal infections?

19. What are the two clinical forms of leprosy and how do they differ?
20. What nursing precautions are required in the care of a patient with leprosy?
21. Briefly describe the epidemiology of herpes simplex virus infections.
22. How does the epidemiology of the superficial mycoses differ from that of systemic mycotic infections?

Additional Readings

Atwater, John B.: Adapting the Venereal Disease Clinic to Today's Problem. *Am. J. Public Health,* **64**:433, May, 1974.

Brown, Marie Scott: Syphilis and Gonorrhea: An Update for Nurses in Ambulatory Settings. *Nursing '76,* **6**:71, Jan. 1976.

Brown, William J.: Acquired Syphilis, Drugs and Blood Tests. *Am. J. Nurs.,* **71**:713, Apr. 1971.

Caldwell, Joseph G.: Congenital Syphilis: A Nonvenereal Disease. *Am. J. Nurs.,* **71**:1768, Sept. 1971.

Elliot, Hazel, and Ryz, J.: *Venereal Diseases, Treatment and Nursing.* Williams & Wilkins Co., Baltimore, 1972.

Faur, Yvonne C.; Weisburd, M. H.; Wilson, M. E.; and May, P. S.: A New Medium for the Isolation of Pathogenic *Neisseria. Health Lab. Sci.,* **10**:44, Apr. 1973.

Gingrich, Roger; Drusin, L. M.; and Kavey, R. W.: Early Congenital Syphilis. *N.Y. State J. Med.,* **76**:283, Feb. 1976.

Gould, Donald: The Scourge of Venus. *World Health,* **26**:8, Nov. 1976.

Hoeprich, Paul D. (ed.): *Infectious Diseases: A Modern Treatise of Infectious Process,* 2nd ed. Harper & Row, Hagerstown, Md., 1977.

Kaslow, R. A.; Dixon, R. E.; Martin, S. M.; Mallison, G. F.; Goldman, D. A.; Lindsey, J. O.; Rhame, F. S.; and Bennett, J. V.: Staphylococcal Disease Related to Hospital Nursery Bathing Practices—A Nationwide Epidemiologic Investigation. *Pediatrics,* **51**:418, 1973.

Lenz, Philomene E.: Women, the Unwitting Carriers of Gonorrhea. *Am. J. Nurs.,* **71**:716, Apr. 1971.

Litt, I. F.; Edberg, S. C.; and Finberg, L.: Gonorrhea in Children and Adolescents, a Current Review. *J. Pediatr.,* **85**:595, Nov. 1974.

McCracken, G. H.: Infections in the Newborn. Neonatal Septicemia and Meningitis. *Hosp. Prac.,* **11**:89, Jan. 1976.

Mizer, Helen E.: The Staphylococcus Problem Versus the Hexachlorophene Dilemma in Hospital Nurseries. *J. Obstet. Gynecol. Neo. Nurs.,* **2**:31, Mar.–Apr. 1973.

Morton, Barbara: *Venereal Disease: A Guide for Nurses and Counselors.* Little, Brown and Co., Boston, 1976.

Rafferty, K. A.: Herpes Viruses and Cancer. *Sci. Am.,* **229**:26, Oct. 1973.

Roberts, Richard B. (ed.): *The Gonococcus.* John Wiley & Sons, New York, 1977.

*Roueché, Berton: A Lonely Road. In *Eleven Blue Men.* Berkley Medallion Editions, New York, 1953.

Sansarricqu, H.: Research into the Battle Against Leprosy. *World Health,* **26**:9, July 1976.

*Youmans, Guy P.; Paterson, P. Y.; and Sommers, H. M.: *The Biologic and Clinical Basis of Infectious Diseases.* W. B. Saunders Co., Philadelphia, 1975.

*Available in paperback.

Infectious Diseases Acquired Through Skin and Mucosa, Transmitted by Animal Contacts

23

Anthrax and tularemia are two important zoonoses transmissible from animals to human beings through direct contacts. Each may also be transmitted through indirect contact with animal products, or through entry routes other than skin and mucosa (respiratory, alimentary, and, in the case of tularemia, arthropod-borne transmissions may occur). The cutaneous route is perhaps the most direct and frequent and is emphasized here for that reason.

Other diseases such as brucellosis, leptospirosis, and ringworm (superficial mycoses) may also be acquired through the skin via contacts with infected animals, but in these instances alternate entry routes are more important. Brucellosis is discussed in Chapter 18 as an alimentary route infection. Leptospirosis is acquired most often from environmental sources, although its reservoir is in animals (Chap. 24). Ringworm is usually transmissible between infected human beings and is discussed in Chapter 22.

Anthrax

The Clinical Disease. The bacillary agent of anthrax is an aerobic sporeformer that is primarily associated with severe disease of domestic animals, notably sheep, cattle, and horses. The resistant spore of this organism is capable of surviving for years in a dried condition in soil, and in hides, hair, and wool of infected animals. Human anthrax has been an occupational disease among people who handle animal products for industrial processing. It also represents a hazard for farmers and veterinarians who work directly with infected animals. In recent years, this disease has become an infrequent problem in the United States. Domestic animal herds have been largely freed of the infection and imported animal products are sterilized either before processing or before final distribution for sale. Some effective controls have been instituted for handling raw products that would be damaged by sterilization, and immunizing agents are now available for protection of exposed workers.

Human anthrax usually begins as an infection of the skin, with formation of a lesion known as "malignant pustule." The bacilli or spores enter the skin, probably through minor abrasions at the point of pressure contacts with the hide or other product being handled. Within a day or so a papule appears at the site of entry, rapidly develops into a vesicle, becomes pustular, and progresses to necrosis. This necrotic ulcer, or "eschar," is sometimes underlaid by hard, swollen tissues. As the spores germinate, capsulated vegetative bacilli multiply in the wound, and a gelatinous edema affects the surrounding area. The bacilli spread through adjacent lymph channels into the bloodstream, multiplying freely. The overwhelming septicemia that results, if untreated, leads to shock and death within a few days. Similar disaster may follow the inhalation of spores or their ingestion.

541

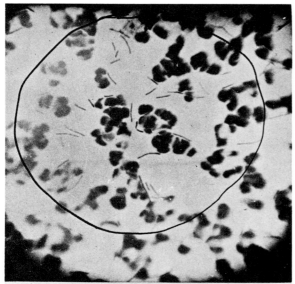

Fig. 23–1. This photomicrograph of *Bacillus anthracis* was annotated and initialed by Louis Pasteur in 1885. Pasteur confirmed Robert Koch's identification of the organism as the causative agent of anthrax in animals and went on to develop a method for vaccinating sheep and cattle against the disease. (Courtesy of the Pasteur Institute, Paris, France.)

Laboratory Diagnosis

I. Identification of the Organism (*Bacillus anthracis*) (Fig. 23–1)

A. MICROSCOPIC CHARACTERISTICS. The genus *Bacillus* includes large Gram-positive, spore-forming, aerobic rods. Many species of this genus are saprophytes, such as *B. subtilis* or *B. megatherium* which live in soil, dust, vegetation, or water and are seldom capable of surviving in human tissues or injuring them. *B. anthracis* is the only member of the group that possesses properties that are always pathogenic for man and animals. They can often be distinguished from other species of the genus by the fact that they are *encapsulated* and are *nonmotile*. They line up in chains, with individual ends squared off at regular intervals.

B. CULTURE CHARACTERISTICS. *B. anthracis* grows readily on blood agar incubated at 37° C under aerobic conditions. The colonies are nonhemolytic and dull gray. They sometimes have irregular margins but some variants are smooth and discrete with a "ground glass" appearance. Short outgrowths sometimes extend from the colony edge giving a "Medusa head" appearance. Colonies are mucoid and tenacious when touched with an inoculating needle.

Immunofluorescent staining can be used to identify the bacteria in smears made from cultures, or in direct smears of exudate.

II. Identification of Patient's Antibodies. Serologic diagnosis can be made by demonstrating a rising titer of antibodies in the patient's serum, using the indirect hemagglutination technique, immunodiffusion, or complement-fixation.

III. Specimens for Diagnosis

A. Exudate from the cutaneous lesion is submitted for smear, culture, and immunofluorescent staining.

B. Sputum specimens may reveal the organisms by smear or culture when inhalation anthrax is suspected.

C. Blood cultures should be collected as soon as the nature of the cutaneous lesion is suspected.

D. Acute and convalescent serums for serologic diagnosis.

Epidemiology

I. Communicability of Infection

A. RESERVOIR, SOURCES, AND TRANSMISSION ROUTES. The reservoir of anthrax infection is in a number of domestic animals such as cattle, sheep, horses, pigs, or goats. Infected animals die of anthrax, but their tissues may harbor spores for very long periods. Pasture soil also becomes contaminated and may remain so for years, with infection being spread among animals, from carcasses or from soil, by biting flies or other insects and by vultures. There is no effective method known for disinfecting the soil of these areas where animal anthrax has been prevalent, and this environmental contamination remains a possible source of infection for both animals and man. Anthrax may be transmitted to man from a great variety of animal products, including bone meal, shaving brush bristles, hair or wool used in textile industries, and hides processed for leather goods production. Spores from such sources may be inhaled or ingested (in contaminated meat) or enter through the skin.

The popularity of handcrafts and the importation of such items as goatskin drums, alpaca wool, and goat's hair have led to a few recent cases in the United States, each beginning as a cutaneous infection. One death resulted from dissemination of the disease before it was recognized clinically.

B. INCUBATION PERIOD. Usually less than a week.

C. COMMUNICABLE PERIOD. Anthrax has not been reported as transmissible from man to man, but the possibility of transfer of pulmonary infection exists. The infectivity of animal or soil sources may persist for years.

D. IMMUNITY. Animals and human beings that survive an attack of the disease are apparently resistant to reinfection. Natural resistance to anthrax varies among animal species and has not been determined with certainty for man, although serologic studies indicate that subclinical infections do occur among exposed persons. Persistent antibodies in such individuals may have a protective effect.

The infection is endemic in agricultural regions, but may fluctuate in its locations with changing soil and climatic conditions, or through introduction of the organism by vectors or in animal feeds. The incidence of human anthrax similarly varies in different parts of the world. The annual world total is between 20,000 and 100,000 cases, but in the United States the figure is now usually less than one hundred in a decade, sporadic cases being associated with occupational hazards.

II. Control of Active Infection

A. ISOLATION. Hospitalized patients with cutaneous lesions should be placed under "wound and skin isolation" (see Table 21–1, p. 494). Cases of inhalation anthrax, or patients with disseminated pulmonary infection, should be cared for in strict isolation (Table 12–1, p. 299). These precautions are advisable to prevent contamination of the environment with resistant spores. They should be continued until skin lesions or sputum is bacteriologically negative.

B. PRECAUTIONS. The organisms multiplying in human tissues are in the vegetative state. When discharged to the environment in exudates from cutaneous lesions, or in sputum, they convert rapidly to spores, and these can be destroyed only by incineration or steam sterilization. Contaminated dressings should be placed in paper bags and burned. Bed linens should be autoclaved before laundering. Nursing and medical equipment should be individualized for the patient insofar as possible. Disposable items are placed in bags for incineration: heat-stable items can be steam-sterilized; heat-sensitive equipment can be sterilized with gas (ETO). Gown technique is essential to prevent contamination of the clothing of those who come in close contact with the patient, and hand-washing facilities must be provided in the unit. Terminal disinfection-cleaning of the patient area is also important.

C. TREATMENT. Because anthrax can develop quickly into a fulminating septicemia, early diagnosis and treatment are essential. The drug of choice is penicillin, but the bacillus is also quite susceptible to most other clinically useful antibiotics. The mortality rate in treated anthrax is essentially zero, as compared with a 5 to 50 percent fatality among untreated cases, or those diagnosed too late.

D. CONTROL OF CONTACTS. A case report of anthrax leads to an immediate search for the source of infection and the prompt diagnosis and treatment of any other human cases exposed to and infected by the same source. Outbreaks are associated either with industrial workers or with the people of farming areas directly involved in the raising, handling, or slaughter of animals. Handicrafters and hobbyists working with animal leathers, hairs, and wools may be at risk if these items are acquired from areas where infection is endemic in soil or animals.

III. Prevention

A. IMMUNIZATION. A vaccine is available for artificial active immunization of animals and of persons exposed to high occupational risks. The material is a cell-free antigen obtained from anthrax exudate.

B. Control of natural reservoirs of anthrax requires special precautions with dead or dying animals to prevent contamination of others or of the environment. Carcasses must be cremated or buried in deep lime pits. In areas where anthrax is prevalent, healthy animals and their human handlers should be vaccinated annually.

Animal products for commercial processing are sterilized insofar as is practical. Bone meal is autoclaved before incorporation into animal feeds; fibers used in brushes are sterilized. Hair and wool can be washed with soap and exposed to formaldehyde. Safety measures for industrial workers include the provision of protective clothing and gloves, as well as immunization programs. Dust control in high-risk industrial areas may be necessary.

Table 23–1. Summary: Infectious Diseases Acquired Through Skin and Mucosa, Transmitted by Animal Contacts

Clinical Disease	Causative Organism	Other Possible Entry Routes	Incubation Period	Communicable Period
Anthrax	*Bacillus anthracis*	Respiratory Gastrointestinal	Less than 1 week	Not communicable from man to man
Tularemia	*Francisella tularensis*	Respiratory Gastrointestinal Parenteral	1–10 days	Not transmissible from man to man

Tularemia

The Clinical Disease. Tularemia is an infectious disease of wild animals, hares and rabbits being among the chief sources of human disease. The organism may find any one of several portals of entry: the skin or mucous membranes, including the conjunctivae; the respiratory route, gastrointestinal tract, or parenteral route if the organism is injected by the bite of infective insect vectors.

In some instances infection is limited to tissues at the portal of entry, with formation of a local ulcer and involvement of regional lymph glands. Cutaneous infection is frequent among those who handle or skin infected animals. The lesion takes an "ulceroglandular" form, with an ulcer at the site of entry, usually on the hands, arms, or face, and regional lymphadenopathy. Ocular infection may also be primary, with a papule forming on the eyelid. Conjunctivitis and swelling of adjacent lymphoid tissue occur as well in this "oculoglandular" form of dis-

Specimens Required	Laboratory Diagnosis	Immunization	Treatment	Nursing Management
Exudate from lesions Sputum Blood Acute and convalescent serums	Smear and culture Immunofluorescence Culture Rising titer of antibodies	Cell-free vaccine	Penicillin	Hospital isolation until lesions are bacteriologically negative (*prevents contamination of environment with spores*); contaminated dressings and disposable items burned; bed linens autoclaved before laundering; steam sterilization of heat-stable items; gas sterilization for heat-sensitive items; gown technique and *hand washing* essential; terminal disinfection-cleaning important
Exudates from ulcers, aspirated material from suppurating lymph nodes Blood Sputum Acute and convalescent serums	Smear and culture Rising titer of antibodies	Killed vaccines	Streptomycin Tetracyclines Chloramphenicol	Isolation not required; concurrent disinfection of discharges, dressings burned, syringes and needles handled with care and sterilized or burned immediately after use, *hand washing* important

ease. Primary tularemic pneumonia may result from inhalation of the bacilli. When they are ingested, necrotizing lesions may be formed in the mouth, pharynx, or gastrointestinal mucosa. Submaxillary, cervical, or mesenteric lymph nodes are involved, depending on the site of colonization. Occasionally, tularemia may begin without localizing signs, developing as a febrile systemic infection.

Whatever the site of initial localization, enlargement of regional lymph nodes may lead to their suppuration, dissemination of the organisms to various tissues of the body, and the formation of granulomatous nodules in systemic foci. These nodules may also break down, and the organisms may be distributed still further in a progressive disease pattern. Fatal septicemia and death may ensue in untreated cases, but early diagnosis and treatment remove this threat.

Laboratory Diagnosis

I. Identification of the Organism (*Francisella tularensis*)

A. MICROSCOPIC CHARACTERISTICS. The organism is a short Gram-negative bacillus, nonmotile, nonspore forming, aerobic or microaerophilic. Morphologically it is similar to *Yersinia pestis,* the agent of bubonic plague (see Chap. 26), with which it was formerly classified in the genus *Pasteurella.*

B. CULTURE CHARACTERISTICS. *F. tularensis* grows with some difficulty in laboratory media, requiring blood or tissue enrichments and added concentrations of cystine. Incubation at 37° C for two to three days yields minute colonies, appearing as transparent drops on the surface of blood agar plates.

The organism is identified and distinguished from *Pasteurella* and *Yersinia* species by biochemical methods, animal inoculations (this is hazardous, and should be done only with strict precautions for animals and their handlers), and serologic reactions of the isolate with specific antiserum (agglutination or precipitation techniques are used).

II. Identification of Patient's Serum Antibodies. A rising titer of agglutinating and precipitating antibodies occurs during the course of active infection. These may be identified with strains of *F. tularensis* or their extracted antigens. Demonstration of an increasing level of antibody is essential to serologic diagnosis of this disease, especially in persons who have been previously exposed and have a persisting titer of residual antibody.

III. Specimens for Diagnosis. Exudates from cutaneous or mucosal ulcers, material aspirated from suppurating lymph nodes, blood, or sputum may be appropriate for culture, as indicated by clinical symptoms of localizing infection. These materials should be collected, transported, and handled with strict aseptic technique, and full awareness of the infectivity of these organisms. Laboratory infections have been frequent, particularly when experimental animals are used in virulence studies.

Preliminary reports on stained smears made from these specimens (except blood samples) may sometimes tentatively confirm the clinical diagnosis, but in subacute or chronic infections the organisms may be extremely difficult to see in smears or to propagate in culture. Serologic diagnosis may be most useful in such cases.

At least two samples of the patient's serum are required for serologic diagnosis: one taken as soon as possible after onset of symptoms, and one taken about two weeks later, so that a rise in titer can be recognized if it occurs.

Epidemiology

I. Communicability of Infection. In nature tularemia is transmitted among animal hosts by several biting arthropods, such as deerflies, wood ticks, and rabbit ticks, and these insects may also transmit the disease to human beings. Man is an accidental host for the infectious organism, and does not transmit it to others. The *incubation period* following exposure to or ingestion of infected animal meat (or contaminated drinking water) is usually about three days, with a possible range of from one to ten days. Animal meats may remain infective, even though frozen, for periods up to three years.

Immunity follows recovery from active infection and is usually durable and solidly protective. Without acquired immunity, there is no natural resistance to tularemia.

II. Control and Prevention. There is no necessity for isolation of patients with active infection, since this is not a communicable disease. However, mucosal or cutaneous discharges should be disinfected, to avoid accidental transfer of infection from these materials, which can sometimes be teeming with the infecting organism. Dressings should be burned;

syringes and needles used to collect blood samples should be handled with particular care and promptly sterilized or incinerated.

Tularemia can be effectively treated with any of a number of antibiotics, particularly streptomycin, chloramphenicol, or tetracyclines. Cure is more rapid with early treatment, but specific therapy is continued for several days after fever and other symptoms have subsided to avoid the establishment of chronic, low-grade, persistent infection.

The source of infection is sought in each case, with a view to preventing further incidence from a lingering reservoir, such as remaining portions of game meat stored in the refrigerator. The sale or shipment of infected animals or meats is a matter of concern to public health authorities, who enforce controls on this possible mode of spread. Campaigns are conducted to educate the public, especially hunters, concerning the sources of infection and the techniques of prevention: wearing clothing protective against insect vectors, wearing rubber gloves when dressing wild game, thoroughly cooking such meat, and boiling water obtained in areas where infected animals may have contaminated it.

Immunizing vaccines are available but their use in the United States is restricted to those who must run a high risk of infection, notably laboratory workers.

Questions

1. Why has the incidence of anthrax decreased in the United States?
2. What is the reservoir for anthrax infection?
3. What type of organism is found in the genus *Bacillus?*
4. Why is it necessary to incinerate or steam-sterilize when disposing of articles contaminated with *B. anthracis?*
5. What animals are reservoirs for *Francisella tularensis?*
6. What precautions are necessary when caring for a patient with tularemia?

Additional Readings

Hubbert, William T.; McCulloch, W. F.; and Schnurrenberger, P. R. (ed.):*Diseases Transmitted from Animals to Man,* 6th ed. Charles C Thomas, Publisher, Springfield, Ill., 1975.
Parrish, Henry M.: Animal-Man Relationships in Today's Environment. *Am. J. Public Health,* **63**:199, Mar. 1973.

24

Infectious Diseases Acquired Through Skin and Mucosa, Transmitted from Intermediate Environmental Sources

The infective agents of the diseases described in this chapter have little in common except the epidemiologic importance of their intermediate sources in water or soil and their common route of entry through "intact" human skin or mucosal surfaces. The spirochetal agent of leptospirosis has a primary reservoir in infected animals; the other three organisms are animal parasites. Hookworm and *Strongyloides* species are intestinal roundworms maintained in man, with intermediate periods of maturation in soil (life cycle type 1), and the schistosomes are trematodes (blood flukes) that display a cycle of development in man or animals (definitive hosts) and in aquatic intermediate hosts (life cycle type 4). The infective form of each of these orgnisms is found in water or soil to which man may be exposed with some frequency, and each of them is capable of penetrating skin and establishing human disease by this route.

Control and prevention of these diseases depend in part on protecting man from contaminated water and soil, but perhaps more important on protecting the environment from sources of infection in primary human or animal reservoirs. It is also important to note the epidemiologic implications of the routing of these organisms in nature. They are not directly transmissible from man to man, but in the case of the animal parasites man contributes to their perpetuation. These are really parenteral, or systemic, infections, which exert major effects through deep-tissue localizations. The skin itself is seldom involved in any but minor skirmishes with the penetrating organism. The epidemiologic point is that these diseases can be acquired through simple environmental contacts, without noticeable injury to the skin.

Leptospirosis (Weil's Disease)

The Clinical Disease. Leptospirosis is an acute infection characterized by fever, chills, headache, malaise, and often by jaundice. Localization of the organisms occurs chiefly in the liver and kidneys, producing necrosis of tissue and dysfunction of the affected organ. In addition to jaundice, there may be hemorrhage, hemolytic anemia, and azotemia. The organism may be disseminated to other areas also, such as skin, muscle, or central nervous system. Leptospiral meningitis is clinically of the benign aseptic type (see Chap. 13, p. 311). The duration of this acute illness is from one to three weeks, with possible recurrences. Mortality is low, the outcome usually reflecting severity of damage to hepatic or renal tissues.

Laboratory Diagnosis

I. Identification of the Organism (*Leptospira interrogans,* formerly classed as *L. icterohaemorrhagiae*). *Leptospira* pathogenic for man include at least eight species, distributed among animal hosts in different parts of the world. They are morphologically indistinguishable but display antigenic differences as well as some variety in degree and severity of clinical disease.

Leptospira species are very tightly coiled spirochetes. They are about the same length as the treponemal agent of syphilis but their spirals are much finer, being only about 0.1 to 0.2 µm wide (see illustration on p. 293). One or both ends of the organism are often bent into a hook. These spirochetes have a very active corkscrew rotation and are so delicate that in dark-field preparations this combination of factors sometimes creates the illusion that they are chains of tiny cocci in rapid undulant motion. Like the treponemes, they do not stain well but can be impregnated with silver.

Leptospira can be propagated in chick embryos, or in a liquid or semisolid medium containing serum enriched with peptones, incubated at 30° C.

II. Identification of Patient's Serum Antibodies. Agglutinating antibodies develop during leptospiral infection but do not reach a peak level until five to eight weeks after the onset of infection. Antibodies can also be demonstrated by complement-fixation techniques. The antigens used in these tests are derived from leptospirae grown in broth media or chick eggs.

III. Specimens for Diagnosis. Blood, urine, or spinal fluid may be submitted for dark-field examination or for culture, the choice of specimen depending on clinical symptoms. These specimens may be inoculated into culture media, chick embryos, guinea pigs, or hamsters. The latter animals are very susceptible to *Leptospira* infection and die in a few days with jaundice and extensive hemorrhagic disease. Spirochetes can be demonstrated in large numbers in their tissues, blood, or urine.

Urine submitted for laboratory studies should be of a neutral or slightly alkaline pH, because leptospirae are extremely sensitive to acid and may be inactivated before the specimen reaches the laboratory. Preliminary alkalinization of the patient affords the best method for assuring survival of organisms pending their examination and culture from urine.

Epidemiology

I. Communicability of Infection

A. RESERVOIR, SOURCES, AND TRANSMISSION ROUTES. The reservoir of leptospirosis is in domestic and wild animals, including rats and other rodents. Infected animals excrete large numbers of spirochetes in their urine and may contaminate water used for drinking, swimming, or cultivating crops. People whose occupations may take them into contaminated water include workers in sugar cane and rice fields, sewer men and miners, processors in fish plants, and soldiers dug into trenches or foxholes. The disease may also be acquired by people in direct contact with infected animals, such as farmers, veterinarians, and abattoir workers. The route of entry is probably through the skin, or conjunctival, oral, or alimentary tract mucosa.

B. INCUBATION PERIOD. The average time required for development of symptoms following exposure is ten days, with a range from 4 to 19 days.

C. COMMUNICABLE PERIOD. Leptospirosis is not transmitted directly from person to person. Man is an accidental host for this organism.

D. IMMUNITY. Infection confers a long-lasting protective immunity, but otherwise human susceptibility appears to be universal.

II. Control and Prevention. Isolation and special precautionary measures are not necessary in the care of these patients. Treatment is largely supportive, although penicillin and other antibiotics may have some therapeutic effect in the early stages of infection. Antimicrobial drugs may not eliminate the organisms.

Control measures consist of a search for the source of infected water, programs for extermination of rats and other rodents, and use of protective clothing by workers exposed to infection. Contaminated waters should be closed to swimmers, and domestic animals should not be permitted access to such watering places. Vaccines are available but often elicit only a poor antibody response. They contain leptospiral strains prevalent in the local area and are administered when the risk of infection appears to be high.

Diseases Caused by Animal Parasites

The nature and epidemiology of the parasitic diseases are discussed at some length in Chapter 20, together with a review of the major types of life

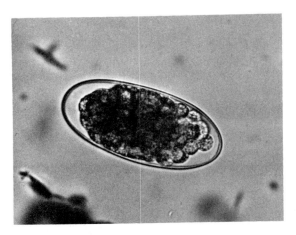

Fig. 24–1. Photomicrograph of a hookworm ovum. (1000×.)

cycles displayed by the animal parasites (see also Chap. 9). The diseases considered here are presented in outline form, describing cyclic development of each parasite in the human or other living host and its environmental requirements; characteristics of human infestation; diagnostic methods; and special features of control. Reference should be made in each case to the general discussions elsewhere in the text, as indicated above.

Hookworm Disease

Organisms. *Necator americanus* ("American hookworm"), *Ancylostoma duodenale* ("Old World" hookworm), *Ancylostoma braziliense* (agent of "creeping eruption"). The small adult roundworms have mouths equipped with chitinous plates, or "teeth," by which they attach themselves to intestinal mucosa (Fig. 5–5, p. 112).

Hookworm ova measure about 40 to 60 μm. They are oval in shape with rounded ends and have a thin outer shell. Normally, when they are passed promptly in the feces, the developing embryo has not yet reached a differentiated stage but appears cellular (Fig. 24–1).

Life Cycle (Type 1B; see Fig. 9–10, p. 220)
  I. Definitive Host. Man.

II. Intermediate Host. None.

III. Human Infection
A. SOURCE OF INFECTION. Soil in which larvae have matured after hatching from eggs derived from human feces.

B. INFECTIVE FORM. Larvae that penetrate exposed skin.

C. LOCALIZATION AND DEVELOPMENT. Larvae enter through the skin, reach lymphatics and the bloodstream, and are carried to the lungs. They migrate into alveoli, up the bronchial tree to the trachea, epiglottis, and pharynx, are swallowed, and reach the small intestine. They attach to the intestinal mucosa, develop into male and female adults, copulate, and produce eggs.

D. EXIT FROM BODY. Hookworm eggs are discharged in feces.

IV. Cyclic Development Outside of Human Host
IN ENVIRONMENT. In moist shaded, warm soil eggs hatch in one or two days (Fig. 24–2), and the emerging larvae feed on soil bacteria and organic matter. Larvae go through developmental changes, and in about a week reach a stage that is infective for

Fig. 24–2. Photomicrograph of a hookworm larva emerging from an ovum. (500×.) (Courtesy of Mr. Hugo Terner, Montefiore Hospital and Medical Center, Bronx, N.Y.)

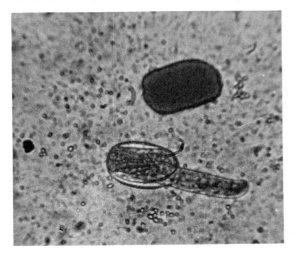

man. This stage may remain viable in soil for several weeks.

Characteristics of Human Disease. Hookworm infestation produces a chronic disease with symptoms related to the blood-sucking activities of the intestinal worms. It is estimated that a single hookworm may remove nearly 0.1 ml of blood per day. When this loss is multiplied by thousands of worms and by all the days of a long infestation, the debilitating effects of this disease may be well understood. Severe anemia of the microcytic, hypochromic type is a characteristic result. Children afflicted with this disease are often malnourished to begin with, and their infestation constitutes an added physical burden, leading to symptoms of weakness, fatigue, pulmonary disability on exertion, physical and even mental retardation.

Light infestations may produce few, if any, symptoms, but the infested individual may nonetheless perpetuate the disease if eggs discharged in feces have an opportunity to mature in soil. Treatment may be indicated for this reason alone.

In the United States, hookworm disease is caused by species of the genus *Necator,* whereas *Ancylostoma* infestations are common in Europe, Southeast Asia, and the Far East.

Dogs and cats are infested with a species of hookworm, *Ancylostoma braziliense,* that has a similar cycle in the animals' intestine, with larval maturation in soil. These larvae may also penetrate human skin but are incapable of further migration or development in the human body. They may remain alive in epidermal tissues, creating serpentine tunnels and producing intense local tissue reactions of erythema, cellular infiltration, induration, and sometimes vesicular eruptions. This form of epidermal hookworm infestation is known as "creeping eruption," or "ground itch."

Diagnostic Methods. Diagnosis of intestinal hookworm disease is established by microscopic identification of hookworm ova in feces. Quantitative estimation of the daily number of eggs being produced is sometimes useful to the physician who might decide not to treat a lightly infected patient if he is unlikely, by virtue of his living conditions and sanitary habits, to perpetuate the organism's life cycle.

Creeping eruption of hookworm origin is diagnosed on the basis of its clinical appearance and a history of exposure to domestic animals who may harbor the intestinal worm.

Special Features of Control

I. Treatment. Hookworm disease is treated with pyrantel pamoate (Antiminth, the drug of choice) or bephenium hydroxynaphthoate (Alcopara). Alternative drugs are tetrachlorethylene and thiabendazole. This specific therapy should be supplemented by supportive treatment of the accompanying anemia. A balanced diet is also an essential part of treatment.

Creeping eruption is treated locally with thiabendazole cream, ethyl chloride spray, or carbon dioxide snow to arrest the progress of larvae in the skin. Anthelminthic drugs may be given by mouth when there are many lesions present.

II. Prevention. Prevention of hookworm disease is primarily a matter of establishing and maintaining sanitary systems for disposal of feces so that the parasite cannot continue its cycle in soil. Campaigns to promote the wearing of shoes in rural areas where the organism is entrenched in the human population constitute a stopgap measure only. The disease can be eliminated by the installation and maintenance of adequate methods for feces disposal, in privies or latrines that can be chemically disinfected, or in sewerage systems.

Strongyloidiasis

Organism. *Strongyloides stercoralis.* The adult worms are smaller than hookworms and do not possess cutting teeth or plates for attachment.

Life Cycle (Type 1B; see Fig. 9–10, p. 220)
  I. Definitive Host. Man.

  II. Intermediate Host. None.

  III. Human Infection
  A. SOURCE OF INFECTION. Soil in which larvae have matured after discharge in human feces.

B. INFECTIVE FORM. Larvae that penetrate exposed skin.

C. LOCALIZATION AND DEVELOPMENT. Identical with those of hookworm, except that the ova usually hatch in the bowel, before they are discharged in feces.

D. EXIT FROM BODY. Larval forms are passed in the stool.

IV. Cyclic Development Outside of Human Host
IN ENVIRONMENT. The larvae mature in soil and may live freely for several generations. The infective form is closely similar to that of the hookworm, can penetrate human skin, migrate through the body, reach the lungs and then the intestinal tract.

V. Autoinfection. An unusual feature of this parasite is that larval maturation may occur within the infected bowel. Infective larvae may then penetrate the wall of the colon or rectum, enter the tissues, and migrate to lungs, bronchial tree, and intestinal tract, establishing a new generation of adults. Infestation may be continued in this way for many years, with frequent recurrences of symptoms.

Characteristics of Human Disease. The symptoms of strongyloidiasis vary according to the density of infection. Migration of larvae through the skin may induce local pruritus; their pulmonary passage sometimes induces pneumonitis or bronchial irritation; but the chief symptoms are related to the burrowing of developing larvae and adults in the villi of the duodenum and small intestine. Epigastric pain, nausea, vomiting, weight loss, weakness, and disturbances in bowel function are frequent symptoms, sometimes suggesting peptic ulcer. Allergic skin reactions and eosinophilia also occur with frequency. Light infestations may produce mild variations of these symptoms, or none at all.

Diagnostic Methods. Microscopic examination of the stool usually reveals viable, active *Strongyloides* larvae. These must be distinguished morphologically from hookworm larvae, which occasionally hatch in the bowel before they are discharged in feces. When strongyloidiasis is suspected but cannot be confirmed by stool examination, aspirated duodenal drainage may reveal larvae.

Special Features of Control

I. Treatment. The drug of choice is thiabendazole, the best alternative being pyrvinium pamoate. These drugs may cause nausea, vomiting, and dizziness, and patients must be followed closely while under treatment.

II. Prevention. Measures for control and prevention of *Strongyloides* infestations are the same as those required for hookworm disease. The most effective and permanent control is provided by installation of sanitary measures for disposal of human feces. Shoes are essential in endemic areas where the primary source of infection has not been controlled.

Schistosomiasis

Organisms. *Schistosoma mansoni* (occurs in the Caribbean, South America, Africa), *Schistosoma haematobium* (Africa, Middle East, India), *Schistosoma japonicum* (Far East, Philippines). The small adult worms are bisexual, but live in close association (Fig. 24–3, *A*). The ova are quite large, measuring about 50 µm in diameter and from 100 to 175 µm in length, depending upon the species. They have a transparent shell which in two species is equipped with a prominent spine (laterally placed on the *S. mansoni* egg [Fig. 24–3, *B*], terminal on the ovum of *S. haematobium*). The ovum of the third species (*S. japonicum*) has a very small, short curved hook on one side. When the embryo within the shell is fully developed, it is ciliated and, upon being hatched, can swim about freely. The hatched embryo is called a *miracidium* (Fig. 24–3, *C*)

Life Cycle (Type 4; see Fig. 9–13, p. 223)
I. Definitive Host. Man and domestic or wild animals.

II. Intermediate Host. Snails.

III. Human Infection
A. SOURCE OF INFECTION. Water infested with larval forms (cercariae) liberated from the intermediate snail host.

B. INFECTIVE FORM. Free-swimming larvae ("fork-tailed cercariae").

C. Localization and Development. Larvae penetrate the skin of persons swimming, wading, or working in infested water. They enter the bloodstream, are carried to the liver, mature to adult forms, and migrate through hepatic vessels or veins of the abdominal cavity. *S. mansoni* and *S. japonicum* remain in hepatic or mesenteric vessels, but *S. haematobium* usually finds its way into the venous complex of the pelvis. Deposited eggs find their way out of venules into the lumen of hepatic ducts, or pass more directly into the bowel or urinary bladder, depending on the localization of the adult worm. Many ova may be lodged in tissues adjacent to the site of their deposition and be immobilized there by local cellular reactions.

D. Exit from the Body. The ova of *S. mansoni* and *S. japonicum* may be discharged with feces; those of *S. haematobium* usually emerge into the

Fig. 24–3. Four stages in the life cycle of *Schistosoma mansoni. A.* Male and female adults. The slender female is interlocked with the larger male. (6×.) *B. Schistosoma mansoni* egg from a human case of schistosomiasis. (150×.) (*B* courtesy of Mr. Hugo Terner, Montefiore Hospital and Medical Center, Bronx, N.Y.) *C.* A miracidium hatched from the egg and now free-swimming. (240×.) Note the fine rim of cilia on the outer side of the cell membrane. *D.* Fork-tailed cercariae. (150×.) This form emerges following cyclic development of the miracidium in a snail host, swims about freely in water, and infects man by penetrating his skin.

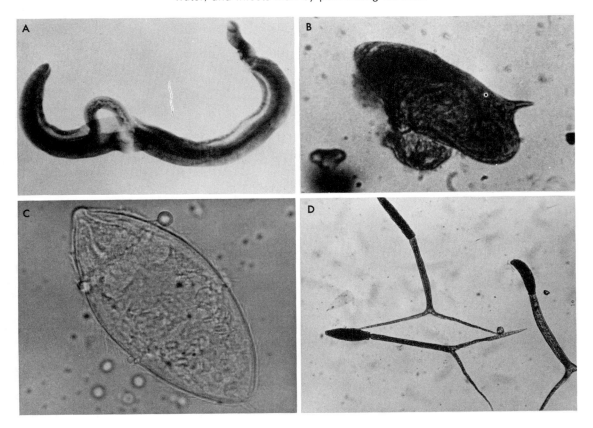

bladder and are excreted with urine, but may also find their way into the colon and be passed in the stool.

IV. Cyclic Development Outside of Human Host

A. IN ENVIRONMENT. *Schistosoma* ova must reach water to begin their maturation. The ciliated miracidia (Fig. 24–3, *C*) break out of their enclosing shells and swim about in water until they are ingested by snails, which serve as intermediate hosts for further development.

B. IN SNAIL HOSTS. The parasites undergo several morphologic changes in snail tissues, emerging finally as microscopic larval forms that also can swim freely. The larvae are called *cercariae* and are characterized by having forked tails, their anterior ends being thickened and elongated (Fig. 24–3, *D*). Glands in the anterior section of the larva produce enzymes that assist in penetration of human skin, once the parasite has become attached.

Characteristics of Human Disease. The symptoms of schistosomiasis are related to the localization of the adult worm and the local damage inflicted on adjacent tissues by migration of ova. Migration of the bisexual adults, which cling together in their passage through venules, may cause rupture of small vessels, but chronic injury is induced by ova wandering in parenchymal tissues. Polymorphonuclear exudates collect around the eggs, and abscess formation results. When ova lodge in the lumen of hepatic ducts or in the bowel, small ulcers may form around them. The eggs are sometimes dislodged and carried by the bloodstream into the lungs, where they may obstruct capillaries and give rise to abscesses in the pulmonary bed. In the liver they may obstruct the portal circulation, and this may lead to enlargement of the spleen. From the venous complex in the pelvis, eggs may reach the bladder, genital organs, or colon. Further migration of adult worms through connecting venules may lead to deposition of eggs in

Table 24–1. Summary: Infectious Diseases Acquired Through Skin and Mucosa, Transmitted from Intermediate Environmental Sources

Clinical Disease	Causative Organism	Other Possible Entry Routes	Incubation Period	Communicable Period
Leptospirosis	*Leptospira interrogans*		4–19 days	Not communicable from man to man
Hookworm disease	*Necator americanus* *Ancylostoma duodenale* *Ancylostoma braziliense*		Variable; ova are seen in stools 6 weeks after initial infection	Not communicable from man to man
Strongyloidiasis	*Strongyloides stercoralis*		About 17 days	Not communicable from man to man
Schistosomiasis	*Schistosoma mansoni* *Schistosoma haematobium* *Schistosoma japonicum*		4–6 weeks	Not communicable from man to man

other parts of the body, such as the skin, conjunctiva, or spinal cord and brain.

During the course of active infection and egg deposition, allergic reactions characterize the disease. Fever, urticaria, and eosinophilia may be intense. When the eggs trapped in parenchymal tissues eventually die, the surrounding reaction becomes granulomatous in character; pseudotubercles are formed, with foreign-body giant cells surrounding and engulfing the material of the dead ova. The eggs may be removed entirely, or they may become calcified and surrounded by scar tissue. Extensive scarring may disrupt normal tissue architecture and interfere with its function. Obstructions of blood vessels and bile ducts result from the body's efforts to heal the lesions induced by *Schistosoma* ova.

Diagnostic Methods. Diagnosis of schistosomiasis is confirmed by demonstration of characteristic eggs in feces, urine, or biopsied tissue. Since dead eggs may be excreted as well as live ones, the viability of the embryo should be confirmed by observation of movement, or by encouraging hatching of the miracidium in a water-diluted specimen incubated at 35° to 37° C. The demonstration of viable eggs indicates the presence of active adults and the necessity for treatment.

Special Features of Control

I. Treatment. Schistosomiasis has been treated with antimony compounds in the past, but because of their toxicity they have been largely superseded by several nonmetallic drugs, such as niridazole, hycanthone, and oxamniquine. These drugs vary in their effects on different *Schistosoma* species and must be selected accordingly. They are not without side effects and are contraindicated when schistosomiasis has already led to hepatic, renal, or cardiac disease. The nonmetallic drugs have also been shown

Specimens Required	Laboratory Diagnosis	Immunization	Treatment	Nursing Management
Blood Urine Spinal fluid	Culture Animal inoculation	Killed vaccines	Penicillin and other antibiotics suppress but do not eliminate the organism	No special precautionary measures indicated
Stools	Identification of ova		Pyrantel pamoate Bephenium	Sanitary disposal of feces
Stools	Identification of viable, active strongyloid larvae		Thiabendazole Pyrvinium pamoate	Sanitary disposal of feces
Stools Urine	Identification of ova		Niridazole Hycanthone Oxamniquine Antimonials	Sanitary disposal of feces

to have teratogenic activity and therefore should not be given to pregnant women.

II. Prevention. Major control of schistosomiasis can be achieved only by sanitary disposal of human excreta, which prevents infection of the snail host and perpetuation of the parasite's life cycle. Snail eradication and land reclamation programs afford some relief but cannot eliminate the disease while human infection and unsanitary practices persist.

Questions

1. What kind of organisms are the leptospirae?
2. What is the reservoir of leptospirosis?
3. What people are most likely to develop leptospirosis?
4. What measures are taken to control this disease?
5. What is the source of infection for hookworm disease?
6. How is the diagnosis of hookworm disease made?
7. Briefly describe the life cycle of the schistosomes.

Additional Readings

Brown, Harold W.: *Basic Clinical Parasitology,* 4th ed. Appleton-Century-Crofts, New York, 1975.
Faust, Ernest C; Beaver, P. C.; and Jung, R. C.: *Animal Agents and Vectors of Human Disease,* 4th ed. Lea & Febiger, Philadelphia, 1975.
*Roueché, Berton: A Swim in the Nile. In *Annals of Epidemiology.* Little, Brown & Co., Boston, 1967.
Trainer, Daniel O.: Wildlife as Monitors of Disease. *Am. J. Public Health,* **63**:201, Mar. 1973.

*Available in paperback.

Section VII
Infectious Diseases Acquired Parenterally

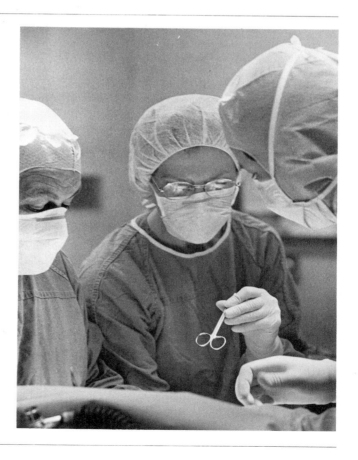

25

Epidemiology of Parenterally Acquired Infections

General Principles

Two important epidemiologic points may be made about the diseases presented in this section: (1) the infectious microorganisms are introduced directly into the tissues as a result of injury to (or inefficacy of) the normal skin or mucosal barriers, and (2) normally these infections are not directly transmissible among healthy persons. In some cases direct transmission is not possible, as in diseases requiring a sequential arthropod vector, and in others it requires unusual circumstances or impaired resistance.

The diseases of this section are arranged in three chapters according to the means by which infection is introduced. The major infectious diseases acquired through bites of arthropods are described in Chapter 26. They are divided into four groups with regard to the nature of their causative agents (bacterial, rickettsial, viral, and parasitic organisms). Chapter 27 includes the infections of deep tissues that can be acquired through accidental injury to skin or mucosal membranes, arranged in three groups according to the nature of the injury — those inflicted by the bites of animals, "street wounds," and burns.

Infections that may arise as a result of medical or surgical techniques are placed in Chapter 28. These include diseases transmitted through blood transfusions, contaminated needles and syringes, or catheterization procedures, and surgical wound infections. Iatrogenic infections (acquired through medical treatment) and those of nosocomial origin (initiated within a medical facility) constitute some of the most serious hazards of hospitalization for vulnerable patients, and their prevention is the basis for many of the cardinal principles of patient care.

In considering the parenteral route of entry for infection, it is important to recognize the reservoir or source of the infectious agent if effective barriers are to be established. The prevention of arthropod-borne diseases may require a combination of measures, such as insect control, immunizing vaccines, or chemoprophylaxis for those at risk (as in malaria prevention). In the case of accidental injuries from animal bites or "street wounds," contamination of exposed traumatized tissues may lead to serious infections such as rabies, tetanus, or gas gangrene. Such possibilities must be anticipated in the management of accidental injuries, and a knowledge of the source and epidemiology of these diseases is essential to the choice of appropriate preventive measures. The care of hospitalized patients who are particularly susceptible to infection — surgical patients; newborns; those with immunodeficient diseases or those undergoing immunosuppressive therapy; patients with malignancies, diabetes, or other underlying disorders — often requires special protective measures. These are based on an understanding of the degree to which the patient's resistance mechanisms have been impaired, as well as on effective techniques to achieve an appropriately aseptic environment.

It is among the latter group of patients that direct transfer of otherwise noncommunicable diseases becomes of greatest importance. The risk of cross-infection, from personnel who carry infectious microorganisms from one patient to another or from intermediate environmental sources, is always present and must be minimized in modern hospital practice. Above all, those responsible for patient care should be aware that their own minor infections (a sore throat, a febrile "cold," diarrhea, an infectious skin lesion) can become a disaster for the susceptible patient and may even be translated into an epidemic in closed hospital populations. The fact that many of the bacteria causing iatrogenic or nosocomial infections are antibiotic resistant, and that these are often carried by hospital personnel, adds serious emphasis to the need for effective infection control in hospitals.

Patient Care

The nursing management of parenterally acquired diseases is varied with respect to the precautions required to control active infection and, as pointed out above, requires a knowledge of the routes and agents of transfer of each disease.

Isolation of the Patient

The need for isolation depends as always on the transmissibility of the infectious microorganism as well as the susceptibility of those at risk. In the case of the arthropod-borne diseases, the nature and habits of the insect vector determine the necessity for isolation or screening of the patient in an insect-transmissible stage of infection. Patients with *tick-borne* diseases (for example, Rocky Moutain spotted fever) need not be isolated because ticks are not ubiquitous. They are encountered only in thickly vegetated areas, among thick weeds or bushy undergrowth; therefore, the infected human being who is sick and confined to bed is not a source of infection for this insect. Similarly, cases of diseases that are transmitted by *rat fleas* (bubonic plague) do not require isolation provided there is no problem regarding rat control in the immediate area. On the other hand, it is extremely important to isolate cases of yellow fever or malaria when patients remain in

an area where the *mosquito vectors* of these diseases are abundant. Isolation need not involve their segregation from other people, but patients should be housed in well-screened rooms and be protected at night by bed netting. Without such precautions, mosquitoes that feed on infected persons may spread the disease through a widening area, sometimes in epidemic patterns. *Louse-borne diseases* such as epidemic typhus fever and relapsing fever present a different kind of problem. Here again isolation is not necessary provided the patient has been completely deloused. Lice are ectoparasites that breed and live on the human body, and in clothing and bedding. Under conditions of crowding and filth they pass back and forth readily from one person to another and can spread the diseases they transmit very rapidly. In typhus fever zones, louse-infested contacts of patients should be quarantined until they have been deloused.

The isolation of persons with other parenterally acquired infections depends on the organism involved and its accessibility to patients with open or unhealed wounds, or to infants and others who are highly susceptible. Reference should be made in each case to the requirements for isolation and nursing precautions stated for each disease. General recommendations for the nursing management of parenterally acquired diseases include the following:

1. *Complete* isolation is indicated for patients with wound infections caused by staphylococci or group A streptococci (see Table 12–1, p. 229). This is particularly important for patients on surgical, maternity, or pediatric services. Newborn babies must not have contact with mothers who have active infections of any kind. Nursery isolation is indicated for babies whose mothers have infections at the time of delivery and for infants who develop signs of infectious illness postnatally.

Rabies patients must also be cared for in strict isolation because of the danger of the direct transmission of the virus through oral secretions.

2. *Partial* isolation may be necessary for patients with acute wound infections other than those mentioned above (see Wound and Skin Precautions, Table 21–1, p. 494). This includes gas gangrene cases and maternity patients with puerperal sepsis.

3. *No isolation* is necessary for patients with arthropod-borne diseases, but patients are screened

Table 25–1.   Protective Isolation (Procedures Recommended by the CDC)

Card to Be Posted on Door of Patient's Room

Protective Isolation

Visitors — Report to Nurses' Station Before Entering Room

1. Private Room — *necessary:* door must be kept closed.
2. Gowns — must be worn by all persons entering room.
3. Masks — must be worn by all persons entering room.
4. Hands — must be washed on entering and leaving room.
5. Gloves — must be worn by all persons having direct contact with patient.
6. Articles — see manual "Isolation Techniques for Use in Hospitals"*

Conditions That May Require Protective Isolation†

Agranulocytosis
Dermatitis, noninfected vesicular, bullous, or eczematous disease when severe
   and extensive
Extensive, noninfected burns in certain patients
Lymphomas and leukemia in certain patients (especially in the late stages of
   Hodgkin's disease and acute leukemia)

*2nd edition, 1975, U.S. Department of Health, Education, and Welfare, Public Health Service, Center for Disease Control, Atlanta, Georgia, Superintendent of Documents No. 017-023-00094-2.
† *Ibid.*

from mosquito and biting fly vectors in endemic areas. Adequate delousing is essential to the control of typhus fever and relapsing fever.

4. *Protective isolation* may be required for immunodeficient patients or others at special risk as outlined in Table 25–1.

Precautions

Nursing procedures vary greatly among the parenterally acquired infectious diseases, but the following general guidelines are applicable:

1. Special nursing precautions are not indicated in the management of insect-borne rickettsial, viral, and parasitic diseases, but in the case of bubonic plague, nurses and all other attendants must take stringent care to protect other patients and themselves if the pneumonic phase of plague develops.

2. Precautions in the management of wound in-

fections include the use of gowns for those who attend the patient directly, particularly at the time dressings are changed. Thorough hand washing is essential before and after wounds are examined or treated. Gloves provide a safety feature and are necessary on occasion. Clean hands are one of the most important factors in preventing spread of infection to environmental reservoirs or to other individuals (including oneself).

3. Concurrent disinfection varies in detail but includes incineration of contaminated dressings, mucosal discharges, or any material obtained from an infectious lesion. Objects contaminated by infectious discharges must be sterilized or carefully disinfected. Terminal disinfection-cleaning of the patient's unit should be thorough.

4. Soiled clothing and bed linen should be sent to the laundry in marked bags to be handled with care prior to washing.

5. Linens, instruments, and equipment contami-

nated with spore-bearing clostridia from cases of gas gangrene should be sterilized before they are cleaned by routine procedures. Disposable items may be incinerated, others autoclaved or placed in an ethylene oxide sterilizer.

6. Routine sterilization procedures for the preparation of surgical or other supplies should be reviewed at frequent intervals and checked by reliable methods.

7. Hospital areas involved in the care and treatment of patients under special risk of infection must be kept in scrupulously clean condition and protected from avenues of cross-contamination. These critical areas include operating rooms, recovery rooms, delivery rooms, nurseries, maternity, pediatric, and surgical services, and intensive-care units. Patients for whom the risk of superimposed infection is greatest include newborn infants; burn cases; all surgical patients, but especially those undergoing prolonged surgery, such as open-heart procedures, extensive bowel resection or pelvic evisceration, renal surgery or dialysis, vascular shunts, organ transplants, or the placement of orthopedic prostheses; medical patients with endocrine disorders, notably diabetes; and patients whose resistance mechanisms are suppressed, because of metabolic defects resulting in agammaglobulinemia or as a consequence of prolonged steroid therapy.

Strict asepsis should always be maintained for these patients throughout the critical periods of their illness, concurrent disinfection techniques should be practiced with attention to every detail so that environmental reservoirs of infection cannot be established, and the sterilization of supplies for these areas should be subjected to critical review at frequent intervals. Such reviews should be conducted by hospital infection control committees, supported by data supplied by the microbiology laboratory.

8. The use of special precautionary techniques must be decided on an individual basis in consideration of the nature of the infection and its probable routes of communicability.

Questions

1. What conditions may require protective isolation?
2. What infections require complete isolation?
3. What precautions are taken in the management of wound infections?

Additional Readings

Herban, Nancy J.: Nursing Care of Patients with Tropical Diseases. *Nurs. Clin. North Am.,* **5**:157, Mar. 1970.

Maki, Dennis G.; Goldman, D. A.; and Rhame, F. S.: Infection Control in Intravenous Therapy. *Ann. Intern. Med.,* **79**:867, Dec. 1973.

Ungvarski, Peter, J.: Parenteral Therapy. *Am. J. Nurs.,* **74**:1974, Dec. 1976.

Wilson, Jacqueline A.: Infection Control in Intravenous Therapy. *Heart Lung,* **5**:430, May–June 1976.

26 Infectious Diseases Acquired from Arthropod Vectors

A large group of human or animal diseases is perpetuated in nature by arthropod vectors that transmit infectious agents to mammalian hosts and play a vital role in the developmental cycle of the microorganisms. The microbial agents of this group fall in each of the major classes of pathogenic microorganisms except the fungi. The human diseases are presented here in groups according to the etiologic agent.

The epidemiology of these infections varies and holds different implications for man. He may be an accidental host and victim of parasites maintained primarily by animals and arthropods, or the primary or only mammalian reservoir, sustaining the parasite sequentially with an insect host. These diseases are usually acquired by man only from arthropod vectors and are not directly transmissible from one person to another. Exceptions to this rule may be noted in the case of plague. Bubonic plague is normally transmitted to man through the bite of the rat flea, but it may also be acquired through direct contact with infected animal tissues or even by droplet transmission. As described previously (Chap. 13, p. 309), the pneumonic phase of plague in human cases is directly transmissible from person to person.

Arthropods and their role as hosts of infectious disease are described in Chapter 9, and methods for their control are discussed in Chapter 11.

Bacterial Diseases Transmitted by Arthropods

Plague

The Clinical Disease. Plague is a virulent disease that has been a threat to human societies for centuries. During the course of human history, epidemics of the Black Death have devastated populations in various parts of the world. Some of the most ravaging epidemics occurred during the Middle Ages in the crowded cities of Europe, India, and Asia, sometimes killing as many as two-thirds of a given population. The last pandemic began in Hong Kong in 1894. As it spread through large segments of the world it left an estimated 43 million people dead. It was during that outbreak that the causative organism was discovered and the epidemiology of the disease began to be understood. Since then plague has come under control in most parts of the world and is rarely seen in epidemic form today. It remains endemic in the western United States and in other areas of the world where the wild rodent population harbors the organism, but human infection is limited to sporadic cases involving unusual exposure to rodents or their fleas. Public health controls have prevented further spread from such cases.

The plague bacillus can produce two principal forms of human illness: *bubonic* and *pneumonic* plague. Other clinical phases, such as septicemic or meningeal plague, sometimes occur.

The *bubonic* form results from the bite of a flea that is carrying the organisms it acquired by previously biting an infected rodent. The injected bacilli are carried along the lymphatic channels into regional lymph nodes, the thoracic duct, and the bloodstream, which disseminates them throughout the body. The most usual sites of localization are in the spleen, liver, and lungs. Symptoms begin suddenly, with fever, chills, severe headache, and exhaustion. There may be vomiting and diarrhea, and the patient often becomes delirious. Damage to lymph nodes creates hemorrhagic inflammation, swelling, and pain. These swellings are called "buboes" and occur most frequently in axilla or groin. They may become necrotic and suppurate to the surface. Hemorrhagic lesions in other parts of the body may also become necrotic. In the skin these necrosing areas appear very dark, hence the term *Black Death*. If septicemia ensues, there may be destructive lesions in every part of the body, with pleural or peritoneal effusions, pericarditis, or even meningitis. This form of plague is not directly transmissible to other persons, but may be carried to others by fleas.

The *pneumonic* form of plague results when organisms localize in the lungs of the victim of bubonic plague. This form is highly contagious, being spread by the coughing of the infected patient. Cases of *primary* plague pneumonia may then occur among people in contact with the patient (see Chap. 13).

Laboratory Diagnosis. The clinical diagnosis of plague is sometimes difficult until characteristic buboes appear. In the early stages the symptoms may resemble those of other diseases, so that laboratory diagnosis may be of critical importance.

I. Identification of the Organism (*Yersinia pestis,* formerly *Pasteurella pestis*)

A. MICROSCOPIC CHARACTERISTICS. The plague bacillus is a short Gram-negative rod, nonmotile and non-spore forming. The organisms characteristically show bipolar staining, which gives them the appearance of safety pins, with solid ends and an unstained space between. They can be positively identified by immunofluorescent techniques applied to smears of exudates.

B. CULTURE CHARACTERISTICS. *Y. pestis* is an aerobic organism that grows best at 30° C, on media enriched with blood or tissue fluids. The colonies are gray and viscous. They are distinguished from other species of *Yersinia,* or from *Pasteurella,* by their biochemical behavior and virulence for animals. Other useful methods for identification of the isolated organisms include typing with specific bacteriophage and serologic agglutination tests.

II. Identification of Patient's Serum Antibodies. Agglutinating, precipitating, and complement-fixing antibodies arise during the course of infection. Agglutination techniques are most commonly used to establish a serologic diagnosis by demonstrating at least a fourfold rise in antibody titer during the course of disease.

III. Specimens for Diagnosis. Blood samples should be collected repeatedly for culture.

Pus aspirated from buboes, paracentesis fluids, spinal fluid, or sputum each may be appropriate for smear and culture if localizing symptoms suggest their value.

Blood should be collected for *serologic testing* as soon as possible after the onset of symptoms, and again after two or three weeks.

Epidemiology
I. Communicability of Infection

A. RESERVOIR, SOURCES, AND TRANSMISSION ROUTES. Man is an accidental host for the plague bacillus, which has a primary reservoir in rats and wild rodents (squirrels, field mice, and others). Rodent fleas transmit the infection between animals. When the flea ingests organisms in a blood meal taken from an infected rodent, the organisms multiply and obstruct its feeding passage. The next time the flea bites, the new meal cannot be swallowed, so that the blood eddies over the obstructing mass of bacilli and is regurgitated back into the bite wound of the current victim, infecting him with its burden of organisms. Fleas are most likely to bite human beings who happen to be around when the rodent host dies, leaving the insect parasite in search of a new source of food.

Plague can also be acquired through direct contact with infected animal tissues, and accidental infections have been reported among laboratory workers. In a recent case of primary cervical node infection, droplet transmission from an infected pet cat has been strongly implicated. Such incidents are rare but point up the infectious potential of the organism. The most hazardous situations arise in cases of pharyngeal or pneumonic infection, transmissible on an airborne route by inhalation of droplets from infected patients. This can occur whether the pneumonic phase is primary or secondary to bubonic plague.

Plague in man is relatively rare in the United States today, but there has been a gradual increase in the number of cases seen in the past decade: 8 in 1965, 13 in 1970, 20 in 1975 as compared with 3 in 1950 and 1 in 1940. This incidence is associated with wild animal reservoirs and has been restricted to southwestern parts of the country. Urban plague has been controlled throughout the world, but sylvatic (wild rodent) plague remains endemic in many areas, e.g., South America, parts of Africa, the Near East, Indonesia, and Asia, as well as the western American states.

B. INCUBATION PERIOD. Two to six days. Average time for development of pneumonic plague is three to four days.

C. COMMUNICABLE PERIOD. Rat fleas may remain infected for long periods (days, weeks, months). Bubonic plague is not directly transmissible among people, but pneumonic plague is highly communicable until infection is successfully cleared by antibiotic therapy.

D. IMMUNITY. Recovery from plague is usually accompanied by a solid, durable immunity. This acquired immunity offers the only natural resistance to the human disease.

II. Control of Active Infection

A. ISOLATION. The patient with bubonic plague, if hospitalized, should be treated immediately for flea infestation but does not require isolation provided the clinical risk of pneumonic plague is averted by successful antibiotic therapy.

Patients with pneumonic plague must be cared for in the strictest isolation, with full precautions against airborne spread of infection. Such isolation must be maintained until at least the second day following the start of antibiotic therapy or until the patient responds to treatment clinically.

B. PRECAUTIONS. Nurses and physicians attending pneumonic plague patients must observe stringent precautions, to protect themselves and to avoid spread to others. Hoods, face masks with goggles, gloves, and overall protective clothing may be necessary. Prophylactic doses of antibiotics or sulfonamides offer further protection for personnel. The patient's excretions, purulent discharges, sputum, contaminated equipment, and bed linens must be concurrently disinfected. Terminal disinfection-cleaning must be thorough. Dead bodies must be handled with strict asepsis.

C. TREATMENT. Antibiotic or sulfonamide therapy is rapidly effective if instituted early. In pneumonic plague prognosis is best when treatment is begun within 24 hours of onset. Streptomycin and tetracyclines are drugs of choice and are often given in combination. Chloramphenicol is an effective alternative.

D. CONTROL OF CONTACTS. The immediate concern when cases of plague are reported is protection of contacts and prevention of spread by rodent fleas, or by airborne routes if pneumonic plague develops. Contacts of active cases are quarantined, dusted with insecticide powders, and observed during a six-day period for signs of infection. Antibiotic therapy is instituted with the first evidence of febrile illness. Where these procedures are not feasible, chemoprophylactic drugs are administered to contacts daily throughout the six-day incubation period.

III. Prevention

A. IMMUNIZATION. Killed vaccines are available that provide protective immunity for at least a one-year period. Artificial immunization is recommended for persons at high risk. Laboratory workers, medical personnel in areas where infection is endemic in the rodent population, travelers, and others whose occupation may expose them to the animal source of infection constitute the group that should be vaccinated.

B. CONTROL OF RESERVOIRS. The most effective methods for preventing plague are centered on destruction of rats and their fleas. Antirat campaigns attempt to eliminate unsanitary conditions that encourage rat breeding. When rodent controls are applied on a mass scale, simultaneous measures to eliminate rat fleas are necessary to prevent incursions by hungry insects into the adjacent human population.

Public health measures also include the rat proofing of buildings in sea- and airports, quarantine of ships from areas where plague has been reported, fumigation of such vessels as indicated, and frequent surveys of the native rodent population for evidence of plague infection. With such evidence, efforts are intensified to eradicate the animal reservoir and arthropod vector.

Relapsing Fever

The Clinical Disease. Relapsing fever is a spirochetal infection characterized by repeated episodes of fever that last two or three days and alternate with afebrile intervals of similar duration. This syndrome of relapse and remission may continue for two or three weeks. Symptoms are related to the presence of spirochetes in the bloodstream. Fever, chills, headache, and malaise mark the release of organisms from localizing sites in spleen, liver, kidneys, and other visceral tissues. There may be three to ten such relapses, of decreasing severity, before symptoms of active infection subside. In epidemic situations the mortality rate may be high (50 percent) among debilitated victims of this disease, but ordinarily it is in the range of 2 to 10 percent.

Laboratory Diagnosis

I. Identification of the Organism (*Borrelia recurrentis*). The spirochete of relapsing fever is a long, irregularly coiled, spiraled organism (twice the dimen-

sion of species of *Treponema* or *Leptospira*). The coils are haphazardly spaced. The organism rotates in the characteristic corkscrew fashion of spirochetes, but is more flexible than other species in its twisting and lashing motions. Unlike other spirochetes, species of *Borrelia* can be stained with bacteriologic dyes, such as methylene blue, Giemsa's and Wright's stains, and the safranin of the Gram stain (they are Gram-negative).

Borrelia can be cultivated in liquid media enriched with blood or serum, or in chick embryos, mice, and rats.

II. Identification of Patient's Serum Antibodies. Agglutinating and complement-fixing antibodies arise during the course of infection, but their identification is difficult because cultivated spirochetes used to identify them possess unstable antigenic components. These organisms also appear to undergo antigenic variation during the course of human infection, possibly as a result of interactions with antibody produced *in vivo*. The multiplication of new variants may account for the relapsing character of the infection, remissions occurring as specific antibodies for each variant reach a neutralizing level.

III. Specimens for Diagnosis. Blood samples should be collected repeatedly during the febrile periods for examination by stained smear, dark-field microscopy, culture, and animal inoculation. *Serologic tests* with spirochetal antigens may be inconclusive for reasons outlined in II above.

Epidemiology

I. Communicability of Infection. This disease has two epidemiologic patterns of spread. (1) Human epidemics arise when man and one of his common ectoparasites, the body louse, become reservoirs for the organism. In this situation the louse transmits disease from one person to another, the rapidity of spread being favored by overcrowded conditions among people who are unable to maintain hygienic standards or adequate nutrition. Wars, famines, floods, and other disasters favor louse-borne epidemics, now restricted to parts of Asia, Africa, and South America. (2) Under natural circumstances, this infection has a reservoir in wild rodents (squirrels, prairie dogs, and others) and is transmitted among them, or to man, by ticks. The endemic, tick-borne disease occurs sporadically in man in many parts of the world, including the western part of the United

States, Mediterranean Europe and Africa, the Near East, and South America.

The *incubation period* following the bite of a tick or louse varies from three days to about two weeks, with an average time of one week.

The disease is *not communicable* from man to man. Ticks may remain infective throughout their lifetime of many years. Lice become infective within a few days of ingestion of blood from infected persons and remain so within their short life-span of five to six weeks.

Immunity following an attack of relapsing fever does not persist for more than one or two years. Acquired immunity provides the only form of resistance.

II. Control and Prevention. Patients with active infection do not require isolation if free of insect parasites. Control of this disease depends on elimination of lice as the source of epidemic infection or suppression of ticks in areas where the disease is endemic among native rodents. Human infection can be treated with tetracyclines or chloramphenicol.

Bartonellosis (Oroya Fever and Verruga Peruana)

The arthropod vector of this infection is restricted to areas in mountains of northwestern South America. Bartonellosis constitutes an excellent example of a geographically limited bacterial infection. The organism (*Bartonella bacilliformis*) is a Gram-negative, small but highly pleomorphic, motile bacterium and is itself restricted to the human reservoir. It parasitizes red blood cells, producing an acute febrile disease marked by a severe, sometimes fatal anemia (*Oroya fever*). After a period of quiescence, an eruptive stage (*verruga peruana*) develops and runs a prolonged course. The lesions are granulomatous nodules that are highly vascular and resemble hemangiomas. The organisms within the nodules parasitize endothelial cells of the blood vessels. The disease responds to antibiotic therapy, although the organisms are not completely eliminated. Specific immunity prevents recurrence of Oroya fever, but verrugous disease may continue.

Bartonella infection is important in areas of Peru, Ecuador, and Colombia where the sandfly vector is active. Control requires suppression of the vector with insecticides.

Rickettsial Diseases Transmitted by Arthropods

General Epidemiology

The rickettsiae were described in Chapters 2 and 4 as coccobacillary, nonmotile, extremely small bacteria. They are obligate intracellular organisms (Fig. 26-1) that parasitize arthropods, notably ticks and mites, lice and fleas. These insects in turn parasitize animals, such as rodents, rabbits, and dogs. Together the insect and animal hosts maintain the reservoir and cycle of rickettsial infection. In most instances, man is an occasional, accidental victim of infection transmitted to him by an infective arthropod. The most outstanding exception to this general epidemiologic pattern is the form of typhus fever known as *epidemic typhus*. In this case, man himself is the only known reservoir for the rickettsial agent. When infected human beings are parasitized by lice, these insects acquire infection and transmit it from person to person. Rapid epidemic spread may occur among masses of people in overcrowded, unhygienic conditions.

Q fever is another exception to the usual route of transfer of rickettsial infection. The infective agent, *Coxiella burnetii,* is a rickettsia but displays certain biologic differences from other organisms of the group. It is maintained in nature by ticks, which transmit it to domestic animals. The human disease is acquired as a respiratory infection transmitted on airborne routes from sources associated with infected animals (Chap. 15). Q fever is the only human rickettsiosis not transferred by an infective arthropod.

Clinical Rickettsiosis

The arthropod-borne rickettsial infections are clinically and epidemiologically similar. In the human body these organisms parasitize endothelial cells and smooth-muscle fibers of small blood vessels. After introduction into the skin, they may be disseminated through the bloodstream, localizing in

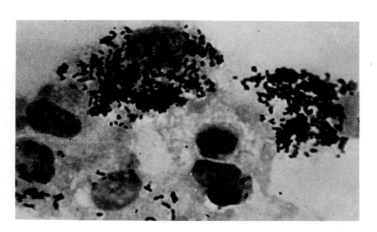

Fig. 26–1. Photomicrograph showing rickettsiae that have multiplied within macrophages in cell culture. (Reproduced from Beaman, Lovelle, and Wisseman, C. L., Jr.: Mechanisms of Immunity in Typhus Infections. VI. Differential Opsonizing and Neutralizing Action of Human Typhus Rickettsia-Specific Cytophilic Antibodies in Cultures of Human Macrophages. *Infect. Immun.,* **14**:1071, Oct. 1976.)

many peripheral vessels, as well as those of the heart or brain. The incubation period ranges from a few days to about three weeks, with an average time of two weeks. Clinical symptoms include fever, headache, and a typical skin rash. Depending on the virulence of the organism, or severity of infection, there may be prostration, delirium or stupor, myocarditis, or peripheral vascular collapse. These diseases respond well to antibiotic therapy, although antimicrobial drugs merely suppress the organisms and do not eliminate them from the body. Antibodies arise during the course of infection and the immune response leads to final recovery, usually without residual tissue damage. The rickettsial agent of epidemic typhus sometimes persists in its human reservoir for many years following recovery from the initial disease. These infected persons display no symptoms until some change in individual resistance permits a recrudescence of infection. This recurring form of typhus is called *Brill's disease.*

The rickettsial diseases are associated with particular types of arthropod vectors. They can be classified in three groups on the basis of insect transmission and of certain clinical features, as shown in Table 26–1.

Typhus Fever Group

Epidemic Typhus Fever. This is a severe disease with a high mortality rate (6 to 30 percent). During the Middle Ages it swept in epidemic form through armies and among the huddled refugee victims of wars. The louse is an ectoparasite of man that thrives when human conditions are at their worst. Unsanitary, crowded, malnourished people living in concentration camps, prisons, poorhouses, or slums are the usual hosts for body and head lice. In central and eastern Europe, Africa, Asia, Mexico, and South America, these same human beings also became endemic reservoirs for the rickettsial parasite, transmitting it to each other by way of their lice. The infected louse excretes the organism in its feces, defecating on the skin at the time it bites. When the bite is scratched, organisms are rubbed into the minor wound, penetrating to the tissues below the skin. It is small comfort to the human victim that the louse also dies of this infection.

Special features of the control of active typhus infection center on early treatment of the patient; delousing him, his clothing, surroundings, and contacts; and protecting medical personnel who care for him. Isolation of the patient is not necessary provided he is protected from further infestation by lice. Tetracycline or chloramphenicol is given as specific treatment. In epidemic areas, vaccines and sometimes antimicrobial drugs are used prophylactically for those who run a high risk of infection. Vigorous efforts are made to eliminate the louse from the resident population. DDT dusting is highly effective in delousing the skin, clothing, and bedding. Lindane and malathion are also useful.

Table 26-1. Classification of the Arthropod-Borne Rickettsial Diseases

Diseases	Insect Vector	Animal Host	Causative Organism
Typhus Fever Group (Flea- and Louse-Borne)			
Epidemic typhus	Louse	Man only	*Rickettsia prowazeki*
Endemic typhus (also called murine typhus)	Flea	Rat	*R. typhi*
Spotted Fever Group (Tick-Borne)			
Rocky Mountain spotted fever	Ticks	Rabbits Field mice Dogs (and ticks)	*R. rickettsi*
Boutonneuse fever	Ticks	Rabbits Field mice Dogs (and ticks)	*R. conori*
Mite-Borne Rickettsial Diseases			
Scrub typhus (tsutsugamushi fever)	Mite	Wild rodents	*R. tsutsugamushi*
Rickettsialpox	Mite	House mouse	*R. akari*

Lice breed on their human hosts, leaving eggs (nits) on the skin or hair, and must be carefully removed. Bathing, shampooing, and frequent washing of clothes are highly effective in discouraging louse infestation.

Typhus *vaccines* have proved to be of great value in the *prevention* of this disease. They are recommended for medical personnel in areas of risk, as well as for travelers or military units entering such regions.

Brill's Disease. This modified form of typhus may occur in an individual who has recovered from a previous attack of the disease. Following recovery, the organisms remain latent for many years, the disease recrudescing only when some condition of stress shifts the patient's balance of resistance. If this happens in an area where human louse infestation is widespread, the recrudescing case may be the starting point of a new epidemic of unmodified typhus.

Endemic Typhus Fever. Also called murine typhus, this is a sporadic disease occurring in many parts of the world where people are exposed to infected rats and their fleas. It was once an urban disease associated with granaries and other areas in cities or sea-

ports where rats congregate and breed. In the United States, continuing antirat campaigns have decreased the incidence of this disease to less than 50 cases per year. Murine typhus is clinically milder than epidemic typhus but runs a similar course, conferring active immunity on those who recover. The fatality rate is about 2 percent.

Spotted Fever Group

Rocky Mountain Spotted Fever. This disease was first recognized in patients who acquired their infection in the Rocky Mountain area. Since then it has been recognized as most prevalent in the southeastern part of the United States, particularly the Piedmont area. The disease also occurs in the western provinces of Canada, and in Mexico and South America.

The tick reservoir differs in various regions. The dog tick (*Dermacentor variabilis*) maintains the infection in the east; the wood tick (*Dermacentor andersoni*), in northwestern sections. Animals (rabbits, field mice, dogs) parasitized by ticks help to maintain the infection in nature. Ticks themselves are an important reservoir, because they pass the organism along, transovarially, to the eggs of succeeding generations.

Special features in the control of rickettsial spotted fever include antibiotic therapy of the patient and prompt removal and destruction of ticks from his body, if any remain at the time he becomes ill. Isolation and special precautions are not necessary. Prevention of this disease is primarily a matter of protection of individuals who may be exposed to ticks in endemic areas (hunters and campers, sheep herders, people engaged in land clearance or forest protection work). Vaccines reduce the chance of infection or modify disease. Protective clothing is important, but persons at risk learn to examine themselves closely for ticks and to remove them promptly. Dogs should also be examined frequently and carefully, so that ticks can be removed before they have embedded their biting mouthparts in the skin. It is not practical to attempt eradication of ticks, except in limited areas where the land has been cleared. Various insecticides, such as chlordane or lindane, control the vectors under these circumstances.

Boutonneuse Fever. This rickettsial spotted fever is endemic in Mediterranean countries. Similar diseases occur in other parts of the world, notably Australia, India, and Siberia. Each of these is associated with its own tick reservoir, as well as animal hosts parasitized by the arthropod.

Mite-Borne Rickettsial Diseases

Scrub Typhus (*Tsutsugamushi Fever*). Scrub typhus is a rickettsial infection limited to the eastern part of the world. Mites and wild rodents that maintain the natural infection are found in southeast Asia, Australia, many Pacific islands, and some sections of China, Japan, and Korea. There are numerous antigenic types of the rickettsial species. The immunity resulting from disease is not cross-protective, so that second and third attacks are not uncommon in resident populations. Military operations in these regions during the last three decades were marked by epidemics occurring among susceptible men newly exposed to mite infestation.

Scrub typhus and other mite-borne rickettsial infections are characterized by a primary lesion, or "eschar," that forms at the skin site where an infective mite attached and inoculated the organisms. Fever, headache, and rash follow, persisting about two weeks unless treated with antibiotics. Control of scrub typhus depends on protection against mites, together with chemoprophylaxis or vaccines that prevent disease even though infection may occur. Military measures include stripping vegetation from camp sites, insecticide spraying, rodent controls, impregnation of clothing and bedding with miticidal chemicals, and the use of insect repellents on the skin.

Rickettsialpox. This disease appears to be limited to some cities of Russia and of eastern and central United States (New York, Boston, Hartford, Philadelphia, Cleveland). It is a mild disease, probably often undiagnosed. It is carried by a mite that shares it with mice under natural circumstances. The human disease begins with an eschar at the site of the mite's attachment to the skin and proceeds as a febrile illness with a rash that lasts about a week. Rickettsialpox may appear in small outbreaks involving people in apartment houses or tenements infested with mice that carry the infection and maintain it with mites, the latter also parasitizing the human tenants. Control is readily accomplished by clearing basements and incinerator closets of trash that encourages mouse breeding. Insecticides are also helpful. No cases have been reported in the United States for several years.

Laboratory Diagnosis of Rickettsial Diseases

I. Identification of Organisms. Microscopic and cultural identification of rickettsiae is technically difficult, and hazardous for the laboratory worker. Biopsies of skin lesions may be stained by special techniques to reveal the organisms in endothelial cells. Specimens of whole blood (or macerated blood clots) can be inoculated into embryonated chick eggs, guinea pigs, or mice, to propagate the infectious agents, but these techniques are usually available only in public health reference laboratories.

II. Identification of Patient's Serum Antibodies. Specific antibodies arise during the course of rickettsial infections. These can be identified by complement-fixing or agglutination techniques using rickettsial antigens, but the latter are difficult to prepare and expensive to buy. These tests are usually deferred, therefore, until cross-reacting bacterial antigens have been tried.

Certain strains of *Proteus* bacilli (Gram-negative enteric bacteria) have been found to possess antigens in common with many rickettsiae, and therefore rickettsial antibodies in patients' serums are capable of agglutinating those related *Proteus* species. This circumstance forms the basis for the *Weil-Felix test*. The rise in antibody titer that occurs during infection can be detected with *Proteus* agglutination, three strains being used to differentiate the different kinds of rickettsial disease.

III. Specimens for Diagnosis. The patient's serum can be submitted to the routine microbiology or serology laboratory for the Weil-Felix test. At least two samples should be collected, one in the early stages of illness and another in about ten days to two weeks, to detect a rising level of antibody. It should be noted that if antibiotic therapy is given early in disease there may be little stimulus for production of antibody, and titers may be low and inconclusive.

If isolation and identification of rickettsiae are deemed necessary and desirable, two specimens of blood should be collected as soon as possible after onset of symptoms. The blood should be taken by aseptic technique and divided between two sterile tubes. One sample should be defibrinated, the other allowed to clot. The tubes can then be packed in an unbreakable container, placed in turn in an insulated shipping box containing ice to keep the blood cool, and forwarded to a reference laboratory equipped for rickettsial work.

Control and Prevention

I. Isolation. These diseases are not directly communicable from man to man. The first step required in management of cases is removal and destruction of insects on the patient's body. Continuing measures must be taken to protect patients and personnel from insects, particularly in louse-borne typhus. When this is assured, isolation is not necessary.

II. Precautions. The blood of patients with rickettsial diseases may be infectious. Syringes and needles used for blood collection should be handled with caution and sterilized promptly.

III. Treatment. Tetracyclines and chloramphenicol are the drugs used in rickettsial infections. Because drugs suppress but do not eliminate the organisms, the patient's immune mechanism is important to full recovery. Antibodies may fail to rise to protective levels when antimicrobial drugs are given early, and when drugs are withdrawn, relapses may occur. To prevent this, a second course of antibiotic is started a week after the end of the first course.

IV. Chemoprophylaxis. Chloramphenicol has been used prophylactically to control scrub typhus in endemic areas. It does not avert infection, but prevents development of disease if continued for a month.

V. Immunity. An attack of rickettsial disease usually confers permanent immunity against the specific infecting strain. This depends on judicious use of antibiotics.

VI. Immunization. Formalinized vaccines are available containing rickettsiae cultivated in chick embryos. They induce an immunity that remains effective for a few months to a year. Persons exposed to risk of infection should receive booster doses at regular intervals as indicated by the local situation.

Viral Diseases Transmitted by Arthropods

Arthropod-borne viruses are often called *arboviruses* to distinguish them from others that are transmissible from man to man. The arboviruses characteristically require arthropod hosts, multiply in these insects without damaging them, and are transmitted by their bites to man or other animals, including birds and reptiles. In some instances, the arthropod is itself a reservoir, passing a virus along to its own succeeding generations, but some arboviruses must cycle sequentially between mammalian reservoirs and the insect host.

The arbovirus group includes a large number of viruses, usually classified in several groups on the basis of antigenic relationships, insect vectors, and the types of disease they cause in man (Table 26–2). Mosquitoes, biting flies, and ticks are the major types of arthropods that play host to viruses. The biologic requirements of an insect species may sharply limit geographic distribution of the virus or the human disease associated with it, as we have seen in the case of bartonellosis.

The arbovirus diseases are presented here in two groups based on clinical distinctions: (1) systemic infections caused by *viscerotropic* viruses (those with a predilection for cells of visceral organs), and (2) primary infections of the brain and spinal cord caused by *neurotropic* viruses (those with an affinity for neural cells). The characteristic clinical syndrome of the latter group of diseases is *encephalitis* (plural, *encephalitides*).

Systemic Arbovirus Diseases

Yellow Fever

The Clinical Disease. Yellow fever is an acute infectious disease with symptoms that may be mild or extremely severe. The fatality rate is low (less than 5 percent) among people who live in tropical areas where the disease is endemic, but may be as high as 30 to 40 percent among susceptible people not previously exposed. At one time epidemic yellow fever was one of the great plagues of urban areas of South America, the Caribbean, seaports of the Atlantic and Gulf coasts in the United States, and those of Western Europe connected by trade vessels with the American focus of yellow fever. The epidemic disease came under control in 1900 when the Yellow Fever Commission under Dr. Walter Reed studied the disease in Cuba and demonstrated the causative filterable agent in the blood of patients and in a mosquito vector (species *Aedes aegypti*). Mosquito eradication programs promptly eliminated the disease from the large *urban* centers in the Americas. A few years later it became apparent that a sylvatic or *jungle* form of yellow fever is entrenched in mosquito vectors of the tropical rain forests of South America and Africa. In these areas, forest mosquitoes transmit virus to animals and to human beings, native populations displaying mild or inapparent infections. When infected human beings bring the virus into an area where the *Aedes* mosquito is abundant, this species may quickly pick up infection and pass it along to the resident human population. If the latter contains many susceptible people who have never before been exposed to the virus, the result may be a devastating epidemic. This means that

continuous control depends not only on vigilant efforts to keep the *Aedes* vector suppressed but also on active immunization of persons entering, residing in, or leaving areas where jungle yellow fever is endemic. Fortunately, effective vaccines are available to assist in this control.

When yellow fever virus is injected into human skin through the bite of an infective mosquito, the organism spreads to the lymph nodes, enters the bloodstream, and localizes in the liver, spleen, kidney, and bone marrow. After a few days, symptoms begin with fever, chills, headache, and malaise. Foci of virus multiplication in the viscera become necrotic. The liver parenchyma is often hardest hit, symptoms of jaundice being prominent (hence *yellow* fever). In severe cases the gastric mucosa may be involved, hemorrhage from the stomach producing black vomitus. Death may result if hepatic and renal damage is excessive. Patients who recover do so completely, without residual damage.

Laboratory Diagnosis. Virus is present in the bloodstream during the first three to five days of illness. It can be recovered and identified by inoculating mice with serum collected from a suspected case.

As disease progresses, neutralizing antibodies appear in the patient's serum in rising titer. They are identified by their specific capacity to neutralize the infectivity for mice of known virulent yellow fever strains. It is essential to demonstrate the increasing level of antibody that characterizes the course of active disease by comparing the titer of serum collected during the acute phase with that of a sample drawn after a week or ten days of illness.

Epidemiology

I. COMMUNICABILITY OF INFECTION

A. Reservoirs, Sources, and Transmission Routes. In urban areas, man and the *Aedes aegypti* mosquito are both reservoirs.

In jungle areas, mosquitoes maintain the natural infection with animals, transmitting it to man also if he enters the circle.

B. Incubation Period. Less than a week; as a rule three to six days.

C. Communicable Period. Virus is present in the bloodstream just before symptoms begin and for the next few days. During this time man is infective for the mosquito. Following the mosquito's blood meal, the virus requires about 10 to 12 days to develop and multiply before the mosquito becomes infectious. Mosquitoes remain infected while they live but do not pass virus along to their progeny.

D. Immunity. During infection antibodies are produced rapidly, and following recovery provide permanent immunity. Repeated exposure to virus in an endemic area provides protective immunity, as evidenced by a high incidence of inapparent infections. Congenital immunity is possessed by infants of immune mothers and is protective for the first few months of life.

II. Control of Active Infection

A. Isolation. It is not necessary to isolate patients, but it is essential that they be screened from mosquitoes, especially during the first few days of illness and viremia.

B. Precautions. The patients' living quarters and all those nearby are immediately sprayed with insecticides.

C. There is no specific antimicrobial therapy.

D. Control of Contacts. Nonimmune persons in the immediate vicinity of a new case of yellow fever are vaccinated promptly, to prevent their possible infection by local mosquitoes that may have newly acquired virus from the patient. A search is made for the origin of each new case, and a campaign is conducted against mosquito species that might carry and transfer the infection.

E. Epidemic Controls. Yellow fever is a matter of international concern. All governments work through the World Health Organization to report the disease and control it.

Urban yellow fever is fought by attacking the mosquito on a communitywide basis, using residual insecticide sprays. *Aedes aegypti* breeding places are searched out and treated with compounds that kill mosquito larvae. At the same time, yellow fever vaccine is given to all who live within the area ranged by infective species.

Jungle yellow fever cannot be controlled by an attack on the vector, because its breeding places are too numerous and inaccessible. Vaccination is the principal defense for those who live in endemic areas or travel into them. Public health authorities maintain surveillance programs in these regions, investigating fatal cases of human illness, searching forests

for monkeys that might have died of the disease, and conducting immunologic surveys of captured animals (especially primates) to determine the range of virus infection.

International control of yellow fever includes quarantining animals shipped from endemic regions, inspection and spraying of aircraft or other vessels arriving from infected areas, and requirements for vaccination of travelers. Vaccination certificates are valid for ten years. Persons without evidence of vaccination who have come from or passed through endemic zones are quarantined on arrival in countries where the *Aedes* vector exists. Quarantine periods are based on the expectation that persons incubating yellow fever should develop symptoms within 10 to 12 days of their last exposure to infective vectors.

III. Prevention

A. Immunization. A live attenuated strain of yellow fever virus is used for artificial active immunization. This variant, called the 17D strain, is highly antigenic but nonpathogenic for man. A single subcutaneous injection stimulates good levels of antibodies within a week to ten days, and provides a solid immunity for several years. Routine use of vaccine is desirable for persons exposed to the jungle type of endemic disease. In other areas, vaccine is used only under threat of an *Aedes*-carried urban epidemic.

B. Continuing control and surveillance programs are necessary to prevent recurrence of yellow fever as an epidemic disease.

Other Arbovirus Fevers. Many other systemic virus infections are carried by arthropods in different parts of the world. Some are mosquito borne, others are carried by ticks or biting flies. A group of these diseases is characterized, like yellow fever, by hemorrhagic lesions in mucosal areas (enanthems), but petechial or purpuric skin lesions also occur (exanthems). These hemorrhagic fevers, as they are called, occur in scattered areas (Crimea, southeast Asia and the Pacific Islands, Russia and Siberia, South America), usually limited by the range of the arthropod host harboring the virus.

Some arboviruses do not produce hemorrhagic injury, but characteristically elicit a maculopapular rash like that of measles. The fever lasts for a few days and is accompanied by lymph node involvement, joint and muscle pains, pain in the eyes, and conjunctivitis. Viruses of this group do not localize in the liver or other vital organs, and these fevers are rarely fatal. They are distributed geographically according to the range of their vectors.

When viruses are introduced to new arthropod hosts that are able to accommodate them, they may cause outbreaks of disease and become endemic in the geographic area of the new vector. In the Americas, *dengue* is a well-known example of an arbovirus disease introduced into the Caribbean area via the *Aedes aegypti* mosquito vector of the New World. Dengue (also known as "breakbone fever" because of the severe muscle and joint pain experienced by many patients) was first recognized in the Caribbean in 1963, when it caused several large outbreaks. It is now endemic in that area and in South America and is seen with increasing frequency among residents as well as vacationing travelers. The latter are warned by the U.S. Public Health Service to take precautions to avoid mosquito bites and to report any acute febrile illness they may experience within two weeks of returning home. This relatively mild virus disease is also endemic in parts of Asia, the southwest Pacific, and West Africa.

Other examples of arbovirus fevers include *Colorado tick fever,* seen in western areas of the United States, and *sandfly fever.* Several viruses of the latter group are vectored by biting flies of the genus *Phlebotomus.* (This name and the general term *phlebotomine* refer to the fact that sandflies are blood-sucking insects.) The diseases of this group have a scattered distribution in the Americas, as well as hot, dry sectors of Europe, Africa, and Asia, as shown in Table 26–2.

Arbovirus Encephalitides

There is a rather large group of encephalitides associated with arboviruses. Most are carried by mosquitoes, but there is also a small group transmitted by ticks. Birds and possibly many other animals (rodents, bats, reptiles, and domestic animals such as horses and sheep) appear to play a role in maintenance of these infections in nature, with the assistance of ectoparasites.

In the United States only mosquito-borne encephalitides are known, the tick-borne variety being restricted to spotty areas of northern and central Europe and Russia. The clinical diseases recognized in the United States and other parts of this hemisphere are *Eastern equine encephalitis* (EEE), *Western equine encephalitis* (WEE), *Venezuelan equine encephalitis* (VEE), and *St. Louis encephalitis* (SLE). (In the Far East, Japanese B and Murray Valley encephalitis are arbovirus infections carried by mosquitoes.)

Eastern Equine, Western Equine, Venezuelan Equine, and St. Louis Encephalitis

The Clinical Disease. These are acute, self-limited diseases characterized by inflammation of the brain, spinal cord, and meninges. They may occur as mild cases of aseptic meningitis, or they may be quite severe, with sudden onset of high fever, chills, nausea, and malaise with localizing symptoms of meningitis and encephalitis occurring within the next day or two. Pain and stiffness of the neck, drowsiness or stupor, and disorientation are frequent. In severe cases, there may be aphasia, convulsions, and coma. Spastic paralysis may occur. There may be sequelae, such as impaired mental function, blindness, deafness, or epilepsy, and paralysis may also be residual. The severity and mortality rate of these "American" encephalitides are not so great as those seen in Japanese B encephalitis, but they take their worst toll among infants and young children.

Laboratory Diagnosis. The viruses of these diseases can seldom be isolated from the blood of patients, because viremia usually occurs before onset of symptoms. By the time illness becomes apparent, virus is localized within the central nervous system. It is seldom identified except in brain tissue taken at autopsy. These viruses can be cultivated in mice and embryonated eggs, then identified by neutralizing or complement-fixing antibodies.

Serologic diagnosis can be obtained by identifying the patient's serum antibodies, using neutralization, complement-fixation, and hemagglutination-inhibition techniques. A rising titer of antibody must be demonstrated by comparing a sample of serum drawn in the acute stage of illness with another taken three to four weeks later.

Table 26–2. Classification of Some Arthropod-Borne Virus Diseases*

Virus Group	Name of Virus	Vector	Disease in Man	Where Found
Togaviruses				
Alphaviruses (group A)	Eastern equine	Mosquito	Encephalitis	Americas
	Venezuelan equine	Mosquito	Encephalitis	South America, Mexico, U.S.A.
	Western equine	Mosquito	Encephalitis	Americas
Flaviviruses (group B)	Dengue	Mosquito	Fever, rash, hemorrhagic fever	Africa, Asia, Pacific, South America, Caribbean islands, Australia, New Guinea
	Japanese (B)	Mosquito	Encephalitis	Asia, Pacific
	Louping ill	Tick	Encephalitis	Great Britain
	Murray Valley	Mosquito	Encephalitis	Australia, New Guinea
	Russian spring-summer	Tick	Encephalitis	Europe, Asia
	St. Louis	Mosquito	Encephalitis	Americas, Jamaica
	Yellow fever	Mosquito	Hemorrhagic fever	Africa, South and Central America
Bunyaviruses (group C)				
California group	California	Mosquito	Encephalitis	U.S.A., Canada
Sandfly fever group	Several	Sandfly (*Phlebotomus*)	Fever	South and Central America, Europe, Africa, Asia
Orbiviruses				
Ungrouped	Colorado tick fever	Tick	Fever	U.S.A.
Rhabdoviruses				
Vesicular stomatitis group	Vesicular stomatitis	Sandfly	Fever	North and Central America

*Partial list, adapted from Benenson, Abram S. (ed.): *Control of Communicable Diseases in Man,* 12th ed. A.P.H.A., Washington, D.C., 1975, pp. 17–20.

When outbreaks of these encephalitides occur, epidemiologists search for mosquito and animal reservoirs. Virus may be isolated from these sources and tested with patient's serum as well as with known antisera.

Epidemiology

I. COMMUNICABILITY OF INFECTION

A. *Reservoir, Sources, and Transmission Routes.* The identity of year-round reservoirs for these viruses is not known, but wild birds, rodents, and reptiles are thought to be important in the annual cycle. During the summer mosquito season, birds are the source of infection for mosquitoes, and the latter transfer it to man. The equine viruses are also transferred to horses, and large epizootics may occur among these animals. Neither horses nor men are a likely source of infection for mosquitoes, because the viremic stage is brief, and virus is not heavily concentrated in the blood. The virus of St. Louis encephalitis is not known to infect horses. A number of mosquito vectors may be involved. The most important of them belong to the genus *Culex*.

B. Incubation Period. Varies from four days to three weeks, with an average time of 10 to 15 days.

C. Communicable Period. These diseases are not transmissible from person to person, and man probably does not infect the mosquito. Viremia persists in birds for several days, and the mosquito remains infective throughout its life.

D. Immunity. In areas where these diseases occur frequently, most adults are immune because of frequent inapparent infection. Active disease results in solid immunity against the infecting virus strain. Infants, young children, and aging people are among the most susceptible because their immunity is incomplete or waning. See Table 26-2 for the geographic distribution of these encephalitides.

II. CONTROL OF ACTIVE INFECTION. Isolation and special precautions are not indicated for patients with these encephalitides. They must be clinically differentiated from others that may represent extension of systemic infection (poliomyelitis, mumps, rabies, postinfectious viral encephalitides, as well as those of bacterial or mycotic origin). There is no specific treatment.

Control and prevention of these infections depend on recognition of human cases, as well as related disease in horses or birds, and elimination of vector mosquitoes.

Vaccines are generally not available. For investigational purposes, an inactivated EEE vaccine can be obtained from the Center for Disease Control, U.S. Public Health Service, for those who are continuously at high risk of infection (laboratory workers and epidemiologists). These may also be passively protected, following known exposure, with specific immune serum from human or animal sources.

Parasitic Diseases Transmitted by Arthropods

Most arthropod-borne parasitic diseases are caused by protozoan species. There are three important genera of protozoa transmitted by arthropods to man and animals, each genus containing several species: *Plasmodium,* the agent of malaria; *Leishmania,* which is associated with three types of disease; and *Trypanosoma,* the cause of African sleeping sickness and of a South American infection called Chagas' disease.

In addition, there is a family of nematodes called *filarial* worms (filaria meaning "thread"), whose adult forms live in man's tissues or body cavities. Mosquitoes are intermediate hosts for these parasites.

All organisms of this group display the third type of life cycle previously described for animal parasites in Chapter 9 (Fig. 9–12, p. 222). Definitive and intermediate hosts play alternate roles in maintaining natural infection. The life cycles and the characteristic human diseases produced by these parasites are presented below in semioutline form highlighting their developmental patterns, clinical syndromes, and control.

Malaria

Organisms. *Plasmodium vivax, Plasmodium malariae, Plasmodium falciparum, Plasmodium ovale.*

These organisms parasitize man's red blood cells. In the process they go through stages of asexual development. Precursors of sexual cells (gametocytes) are also formed. When the latter are taken up by a mosquito they proceed through stages of sexual division in the arthropod.

Life Cycle (Type 3; Fig. 9–12, p. 222)

I. Definitive Host. Mosquitoes of the genus *Anopheles.*

II. Intermediate Host. Man (a number of animals may be hosts for *Plasmodium* species that are not infective for man).

III. Human Infection

A. SOURCE OF INFECTION. The bite of an infective female anopheline mosquito.

B. INFECTIVE FORMS. *Sporozoites* congregated in the salivary gland of the mosquito, injected into the skin when she bites.

C. LOCALIZATION AND DEVELOPMENT. Sporozoites are first carried by the bloodstream to a localizing point in the liver. Here they undergo several cycles of asexual multiplication during the following week or two. The basic unit which initiates the asexual cycle is called a *merozoite*. These units begin to emerge from the liver and invade red blood cells. Within the erythrocyte the invading merozoite (there may be more than one in a cell in some types of infection) repeats the asexual cycle of division and development, in stages successively called: (1) the *ring* stage (this is a young trophozoite [Fig. 26–2]); (2) an ameboid *trophozoite;* (3) a *schizont* (the nuclear material divides and redivides forming many segments); (4) a *mature schizont,* which ruptures the erythrocyte scattering its segments, now called merozoites again. These enter new cells and the cycle repeats with a growing density of parasites. This process is called *schizogony.*

In time, some of the parasites develop into specialized precursors of gametes, or sex cells. The precursors are called *microgametocytes* and *macrogametocytes.* These do not enter into sexual combinations in the human body, but must wait for the conditions provided by the mosquito host.

D. EXIT FROM HUMAN BODY. The female anopheline mosquito must take a blood meal from an infected person, and the blood removed must contain both micro- and macrogametocytes. The asexual forms do not develop further in the insect.

IV. Cyclic Development in Mosquito Host.
The microgametocyte fertilizes the macrogametocyte, and a *zygote* is formed. The latter enters the wall of the mosquito's stomach, forms an *oocyst,* and *sporozoites* develop. This process is known as *sporogony* and culminates with the liberation of hundreds of sporozoites into the mosquito's body cavity. Most migrate to the salivary or other glands of the insect, where they can develop no further until injected into a human host. The life cycle of the malaria parasite is shown diagrammatically in Figure 2–16, *F* (p. 38).

Characteristics of Human Malaria. Malaria is an old disease of mankind that has killed many millions of people since the ancient Chinese, Assyrians, and Egyptians first made note of it. It has often been spread in epidemic form when infected persons have introduced the parasite to new vectors in areas where the human population has not previously experienced the disease. Human susceptibility is fairly universal. True immunity does not develop and re-

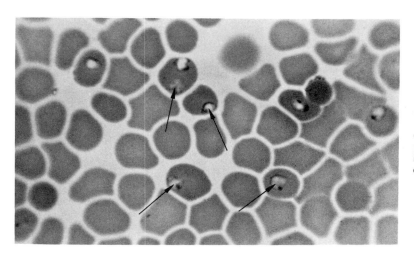

Fig. 26–2. *Plasmodium falciparum* parasitizes man's red blood cells during part of its complex life cycle. Arrows point to erythrocytes containing ring forms. (Stained smear of human blood, 2000×.)

sistance to effects of the parasite builds up very slowly in those regions where infection has occurred continuously over a period of many years. For this reason, establishment of a new vector can be disastrous in previously uninfected areas.

Malaria is a chronic disease that produces invalids. It is directly responsible for deaths in only about 1 to 2 percent of adult cases, but the rate is much higher among infants. It is a debilitating disease that retards physical and mental development, sometimes on a scale large enough to be felt in every social and national effort of an afflicted people. Endemic malaria has been controlled and virtually eliminated from many countries, but it remains a serious public health problem in Central and South America, Africa, the Near and Far East, Southeast Asia, and the Southwest Pacific.

Man is the only known host for the four species of malaria listed at the beginning of this section, although other species of *Plasmodium* have been recognized in animal infections. The most common species of the human parasites are *P. vivax* and *P. falciparum,* the latter producing a clinical disease more virulent than any of the others. *P. malariae* has a wide distribution but the incidence of infection is low. *P. ovale* causes a relatively mild human disease acquired only in West Africa. Throughout the "malaria belt" that stretches across tropical and subtropical areas of the globe, the female *Anopheles* mosquito is the responsible vector. Her species varies in different regions, but she restricts the spread of human malaria chiefly by her requirement for warm, moist climates in altitudes of less than 6000 ft. The availability of human populations within the range of this mosquito assures perpetuation of the parasite until the vector is eradicated.

The asexual maturations and divisions of malarial parasites in peripheral blood cells are responsible for many of the clinical symptoms of this disease, and for their cyclic recurrence. When large numbers of maturing schizonts simultaneously reach the stage of erythrocyte rupture, with liberation of new merozoites, the event is marked by paroxysmal fever, with shaking chills that rack the patient. These symptoms may endure for a few minutes to an hour and are often accompanied by nausea, vomiting, and severe headache. Fever may persist for several hours as parasitemia continues and liberated merozoites cir-

culate before invading new red blood cells. The episode concludes with a drenching sweat that leaves the patient afebrile and exhausted, but able to rest. For the next two or three days the patient feels reasonably well, until another cycle of parasite maturation overtakes him and he enters a new paroxysmal stage. These episodes of fever and chills are at first quite irregular, but later develop a rhythm characteristic for the parasite species. *Falciparum* and *vivax* malaria are usually *tertian;* that is, cyclic fever and chills recur every third day (at 48-hour intervals); but the cycle is *quartan* for *P. malariae,* with paroxysms recurring every fourth day (72 hours). It should be noted that in *P. falciparum* infections, schizogony occurs within infected tissue cells rather than within peripheral blood cells. Consequently, only young ring forms and gametocytes can be found in smears of the circulating blood; intermediate developmental forms are absent.

Malaria eventually burns itself out, if mosquito-borne infection is not renewed, but not before damage has been done to the hematopoietic system. There are hepatomegaly, splenomegaly, anemia, and, in *falciparum* malaria, hemorrhagic damage to kidneys and liver ("blackwater fever," marked by hematuria and jaundice as signs of glomerular nephritis and hepatic necrosis). "Cerebral malaria" is another possible consequence of *falciparum* infections, resulting in excessive fever, delirium or coma, and death when the vascular supply to the brain is seriously compromised by parasitic occlusions of cerebral blood vessels.

Diagnostic Methods. Blood smears are examined microscopically. *Thin* blood smears are made by the technique usual for differentiation of white blood cells. *Thick* smears are also important because the organisms may be scarce and hard to find. A thick smear is made by allowing several drops of the patient's fingertip blood to puddle in the center of a glass slide. A wooden applicator is used to stir this puddle to defibrinate it. Fibrin threads are wrapped on the end of the stick and removed. Thin smears are stained to visualize parasites within red blood cells. Thick smears are first treated so that erythrocytes are lysed, freeing their parasites. When properly stained, the organisms can be seen lying among white blood cells and platelets left on the slide.

When blood smears are reported negative for malarial parasites, but the clinical picture and history of exposure are suggestive, it is important to take additional smears on repeated occasions. Maturing organisms are present in the blood (except in the case of *P. falciparum,* as noted previously) during the few hours preceding the paroxysms of fever and chills. When the latter can be predicted, smears should be taken during the day before they are expected, while the fever is rising and during the first chills. Later in the episode one is not likely to find intracellular organisms unless alternate generations of trophozoites are maturing.

Differentiation of various species of malarial parasites depends on finding organisms in varying stages of maturity. Morphologic distinctions between species are not obvious and require an expert eye. The physician must correlate laboratory findings with clinical developments, steering collection of smears judiciously so that full species diagnosis can be obtained.

Serologic diagnosis can be made with the indirect fluorescent antibody technique or indirect hemagglutination, but these tests are not routinely available.

Special Features of Control

Treatment. Malaria was treated with extracts of cinchona bark long before its etiologic agent was known. (The parasite was discovered in 1880 by Charles Louis Alphonse Laveran, a French Army doctor.) The active ingredient of cinchona bark is quinine, which was the standard drug until more effective synthetic compounds were developed. The first of these was *atabrine,* or quinacrine. This has been supplanted by newer drugs of greater antimalarial activity and less toxicity, notably *chloroquine* and *amodiaquine.* Other compounds used in combination with these are *proguanil* and *pyrimethamine.*

Drug therapy in malaria has several goals, which cannot be accomplished by a single compound. Not only must the clinical attack be treated, but there must be a cidal effect on parasitic stages developing in the liver, or elsewhere outside of red blood cells, so that relapses will not occur. Furthermore, it is important to kill the gametocytes that perpetuate the parasite's life cycle when transmitted to mosquito vectors, particularly if the patient remains in an endemic area and is being treated there.

The *prophylactic* or *suppressive* use of these drugs has great value in preventing infection or in suppressing development of parasites, so that clinical symptoms do not occur. Persons on suppressive treatment who leave endemic areas must continue the drug for at least a month to avoid later clinical attack, and to reduce the risk of transmitting infection to new *Anopheles* vectors.

In recent years, the treatment and prophylaxis of *P. falciparum* infections have been complicated by the developing resistance of some strains of this species to chloroquine compounds. This has been particularly true of strains acquired in South America and southeast Asia (Iran, Assam, Burma). African strains of *P. falciparum* have not yet been documented as resistant to this class of drugs. Quinine must now be used for the treatment of resistant infections, together with pyrimethamine. An alternative regimen for patients who cannot tolerate quinine has combined pyrimethamine with sulfonamides or sulfones, such as dapsone. However, the latter drug has recently been reported to have carcinogenic potential for experimental animals, and its therapeutic use in humans is currently being reevaluated. The effectiveness of new drugs, such as phenanthrene and quinoline compounds, against resistant strains of *P. falciparum* is also being studied.

Control. Malaria control depends on coordinated efforts to treat and cure active cases in humans, thus eliminating the intermediate reservoir, and to eradicate the mosquito host. While being treated, patients should be screened from mosquitoes until they are under drug control.

Chemoprophylaxis is of particular importance for persons entering endemic areas. The volume of increased tourism, particularly to African countries, has been correlated in recent years with a marked increase in the number of malaria cases imported into the United States. In 1977, for example, a group of at least 600 Americans from 24 states attended an arts and culture festival held in Nigeria. Although information about malaria prophylaxis was distributed before the group's departure for Nigeria, a number of persons contracted the disease, the diagnosis being confirmed on their return to the United States. Excluding unconfirmed cases, the attack rate in this group was 1.1 percent. The majority of patients with confirmed malaria reported that they took either inadequate or no prophylaxis. These cases could readily have been prevented by appropriate use of an antimalarial drug. The regimen of choice calls for chloroquine orally once a week, beginning one week before arrival in an endemic area and continuing for six weeks after departure.

Mosquito eradication programs include community application of residual insecticides around dwellings, but most effective control is achieved when mosquito breeding places are eliminated by draining and filling stagnant swamps or water holes, or by chemical treatment that kills larvae (Fig. 11–7, p. 289).

Leishmaniasis

Organisms. *Leishmania donovani, Leishmania tropica, Leishmania braziliensis.*

These organisms, like the trypanosomes, are frequently referred to as *hemoflagellates* to distinguish them from intestinal flagellates such as *Trichomonas* and *Giardia.* Hemoflagellates live in man's bloodstream or parasitize his tissues.

Leishmanias are intracellular parasites in their

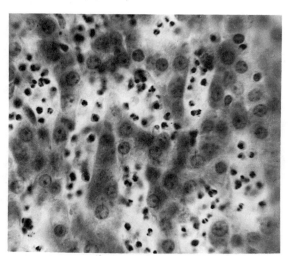

Fig. 26–3. A photomicrograph of human liver cells containing nonmotile intracellular forms of *Leishmania donovani.* (1000×.) (Courtesy of Dr. Kenneth Phifer, Rockville, Md.)

human hosts, occurring as nonmotile round forms within tissue cells (Fig. 26–3). The three species listed have affinities for reticuloendothelial cells in different tissues and therefore produce different clinical syndromes. Each requires an insect host to complete its life cycle, the parasite assuming a flagellated motile form within the body of the arthropod. Infective insects transmit them to animals as well as to man, both constituting the natural reservoir of these infections.

Life Cycle (Type 3; Fig. 9–12, p. 222)

I. Definitive Host. Sandflies (motile forms of the parasite develop).

II. Intermediate Host. Man and animals (nonmotile intracellular forms in tissues).

III. Human Infection

A. SOURCE OF INFECTION. Bite of infective sandfly.

B. INFECTIVE FORM. Flagellated stage called a *leptomonad.*

C. LOCALIZATION AND DEVELOPMENT. *L. donovani* establishes a *visceral* infection called *kala-azar.* Following entry into the skin, the organisms are engulfed by wandering macrophages of the RE system and are carried by these through the blood to various organ sites. They are found in endothelial cells of the liver, spleen, bone marrow, lymph nodes, and many other organs. They may be present in the intestinal wall, kidneys, or meninges and are found, correspondingly, in feces, urine, or spinal fluid. They are not transmissible by these routes, but must be transferred again to the sandfly host.

L. tropica establishes a *cutaneous* infection called *Oriental sore.* The organisms are found in reticuloendothelial cells and lymphoid tissue of the skin where they produce an ulcerating lesion. This infection may be acquired by *contact* with infected persons as well as from the insect vector. It spreads on the infected individual's own skin, producing multiple sores.

L. braziliensis parasitizes cells of the skin and mucosa of the mouth, nose, and pharynx. The disease is called by several names, such as *American leishmaniasis* (it occurs in Central and South America), *mucocutaneous leishmaniasis, espundia,* or *uta.*

This disease may also be transmitted from person to person by direct contact or by autoinoculation with secondary lesions in various skin sites.

IV. Cyclic Development in Insect Host. When nonmotile leishmania forms are ingested in the blood meal of the sandfly they develop into leptomonad forms that have a single polar flagellum and divide by longitudinal fission. Large numbers of leptomonads are produced in the insect's intestine and are later found in the pharynx and buccal cavity.

The leptomonad form can also be propagated on artificial culture media.

Characteristics of Human Disease; Diagnosis and Control

Visceral Leishmaniasis. Visceral leishmaniasis, or kala-azar, caused by *L. donovani,* is a disease of tropical and semitropical countries. It is found in both hemispheres, in scattered areas reflecting the incidence of infective sandflies. It occurs chiefly during the first two decades of life. Kala-azar is a chronic systemic disease, progressive and fatal when it is untreated. Antimony compounds provide effective therapy.

Kala-azar is diagnosed by demonstration of the intracellular bodies in biopsies of spleen, liver, or lymph nodes, or in aspirated bone marrow or blood. Clinical material can be inoculated into culture media or into hamsters for propagation of the causative organisms.
Serologic diagnosis can be made with the complement-fixation test.

Visceral leishmaniasis is not directly transmitted from person to person. Control is achieved when human cases are diagnosed and treated while concurrent application of insecticides suppress the infective vector.

Cutaneous Leishmaniasis. The cutaneous form is divided in distribution between the two hemispheres. *Oriental sore* (*L. tropica*) is found in tropical and semitropical areas of the Old World: Mediterranean countries, Africa, the Near and Far East, and southeast Asia. *American leishmaniasis* (*L. braziliensis*) occurs in the southern countries of our own hemisphere.

Cutaneous forms of leishmaniasis are diagnosed by demonstrating intracellular *Leishmania* bodies in cells of skin scrapings taken from the edges of exposed lesions, or by cultivating the leptomonad form.

Like kala-azar, the cutaneous diseases respond to treatment with antimony compounds. Control is difficult because they can be transmitted directly by infected persons. Mass treatment campaigns and vector controls are necessary to prevent continuing spread.

Trypanosomiasis

Organisms. *Trypanosoma gambiense, Trypanosoma rhodesiense* (agents of African trypanosomiasis or "African sleeping sickness"), and *Trypanosoma cruzi,* the agent of American trypanosomiasis.

Mature trypanosomes are flagellated protozoans that characteristically possess an undulating membrane. The African species appear in this form in the blood of human hosts (Fig. 26–4). The American species may be found in the trypanosomal form in human blood, but also in an intracellular leishmanial stage in RE cells of many organs. The insect stages of *T. cruzi* are like those of the African species.

Life Cycle (Type 3; Fig. 9–12, p. 222)
I. Definitive Host. African species: the tsetse fly. American species: the "kissing" bug (reduviid bug).

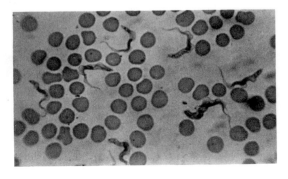

Fig. 26–4. *Trypanosoma gambiense* in a blood smear. Note the terminal flagellum and the undulating membrane that extends along the body of this organism. (850×.) (Courtesy of Dr. Thomas W. M. Cameron, McGill University, Montreal, Canada.)

(Trypanosomal and crithidial stages occur in the insect hosts.)

II. Intermediate Host. Man and animals. African species: trypanosomes in blood, lymph, and spinal fluid. American species: trypanosomes in blood; intracellular leishmanial forms in lymphoid tissue.

III. Human Infection
A. AFRICAN TRYPANOSOMIASIS
1. Source of Infection. The bite of an infective tsetse fly (genus *Glossina*).
2. Infective Form. Mature trypanosomes are injected into the skin by the biting fly.
3. Localization and Development. Multiplication of trypanosomes at the site of entry produces a primary lesion in the skin. The organisms multiply extracellularly and eventually are distributed through the body. The most serious effects of systemic localizations occur in the heart muscle and in the central nervous system.
4. Exit from Human Body. When a tsetse fly bites a person or animal who has trypanosomiasis, it ingests the parasites with its blood meal.
B. AMERICAN TRYPANOSOMIASIS
1. Source of Infection. Cone-nosed bugs of the reduviid family, often called "kissing bugs" because of their tendency to bite at the mucocutaneous junction of the lips.
2. Infective Form. Maturing trypanosomes are passed in the insect's feces, deposited on the skin of humans or animals that the insect bites, and rubbed into the bite wound or other skin abrasion by scratching.
3. Localization and Development. At the site of inoculation trypanosomes reproduce, cycling through leishmanial, leptomonad, and crithidial stages. A swollen lesion called a "chagoma" develops at this site. The organisms are disseminated systemically, localizing in the RE cells of spleen, liver, and in myocardium.
4. Exit from Human Body. Reduviid bugs ingest trypanosomes when they bite and take a blood meal from infected humans or animals.

IV. Cyclic Development in Insect Hosts
A. AFRICAN TRYPANOSOMIASIS. Trypanosomes ingested with blood by the tsetse fly reproduce in the insect's intestine. Immature crithidial stages are formed. These migrate into the fly's salivary glands. There they mature into trypanosomal forms and within a few days the fly becomes infective. When the fly bites its next victim, it injects saliva containing trypanosomes into the puncture wound.
B. AMERICAN TRYPANOSOMIASIS. In the reduviid bug, trypanosomes ingested with blood also reproduce in the intestinal tract. Immature crithidial forms pass along the gut to the rectum. As they mature they are passed in the insect's feces.

Characteristics of Human Disease: Diagnosis and Control

African Trypanosomiasis. The African type of disease usually begins with formation of a primary lesion (trypanosomal chancre) at the site of inoculation. As extracellular multiplication proceeds, the organism reaches the bloodstream and is deposited in numerous organs. In acute stages, symptoms include fever, lymphadenopathy, severe headache, and anemia. Chronic progression leads to symptoms that reflect localizations in the body (weakness in the legs, neuralgic pains, dyspnea, cardiac pain, nephritis). When the central nervous system is invaded, the "sleeping-sickness" stage of infection begins. Headaches, mental apathy, and neuromuscular disorders develop. As the disease progresses, sleepiness and emaciation become pronounced. In final stages, the patient cannot be aroused from sleep and passes into deep coma. Death results from heart failure, meningitis, or extreme debility.

African trypanosomiasis is diagnosed by demonstrating the organisms in smears of peripheral blood (Fig. 26–4) or aspirated lymph. During stages of CNS involvement it may be seen in spinal fluid.
  Clinical specimens may be studied in the laboratory by microscopic examination, culture on suitable media, or animal inoculation. Early diagnosis is important because treatment is most successful in early stages of disease.
Serologic diagnosis can be made by fluorescent antibody and complement-fixation tests. High titers of IgM antibody are characteristically found in African trypanosomiasis.

The African disease is treated with pentamidine or suramin in the early stages of infection. Arsenical compounds such as melarsoprol or tryparsamide can be used effectively for treatment of advanced disease. The prognosis is generally poor, however, if the disease reaches the somnolent stage.

Control of African trypanosomiasis is difficult. Man, cattle, and a few wild animals are reservoirs of infection for the fly host. Once the fly becomes infective it remains so for its life-span of three months. Because the reservoir is large, fly-breeding places often inaccessible, and human disease difficult to eliminate, it has not been possible to eradicate trypanosomiasis. Control measures include intensive efforts to clear the fly from villages while conducting mass treatment campaigns in local populations.

The Gambian variety of trypanosomiasis is found in some countries of west central Africa (Gambia, Liberia, Ghana, the Congo, Sudan). The Rhodesian type (more virulent and difficult to cure) occurs in dryer areas to the southeast (Rhodesia, Tanganyika, Mozambique, Malawi).

American Trypanosomiasis. The American form, often called *Chagas' disease,* occurs in Central and South America. It begins with development of a "chagoma," or swollen lesion, at the site of inoculation, often around the lips or eyes. Conjunctival entry of the organisms is common in children, who display unilateral edema of eyelids and face, with inflammation of lacrimal glands. In adults, the disease may be mild or asymptomatic, but it is often fatal for children. Acute systemic disease develops, with fever, malaise, and localization of organisms in RE cells of many organs. It may end with chronic progression to the central nervous system, or with myocardial failure.

American trypanosomiasis may be diagnosed by demonstrating trypanosomal forms in the circulating blood, or intracellular leishmanial forms in biopsies of lymph nodes or involved tissues.

Microscopic examination of specimens, culture methods, and animal inoculations establish the laboratory diagnosis if positive.

Xenodiagnosis is sometimes helpful when the organisms are difficult to demonstrate directly. This method requires feeding uninfected triatomid bugs on the patient and examining the insects' feces several weeks later. Positive identification of trypanosomes in the latter confirms the diagnosis.

Serologic methods are more sensitive for diagnosis than any tests for direct demonstration of the parasites. Fluorescent antibody tests, complement-fixation, and agglutination procedures are used.

Treatment of American trypanosomiasis is particularly difficult when chronic, progressive stages have been reached. The aminoquinoline drugs are effective in clearing the blood of trypanosomal forms, so that if the disease is recognized and treated early, deep tissue localizations may be controlled or prevented. Nitrofurfurylidine is a new drug that has shown promising results in clinical trials.

Control measures are similar to those described for African trypanosomiasis, except that the insect host may be attacked with greater effect in endemic areas. Infected domestic animals must be destroyed, houses disinfested of the vector, and bed nets used where the insect remains prevalent.

Filariasis

Organisms. A number of species of filarial roundworms may parasitize the tissues of man and sometimes other animal hosts. Man appears to be the principal reservoir for those species most commonly associated with human disease:

Wuchereria bancrofti }	adults live in lymphatic
Brugia malayi }	vessels
Onchocerca volvulus }	adults live in skin and sub-
Loa loa }	cutaneous tissues

Life Cycle (Type 3; Fig. 9–12, p. 222)

Man is the definitive host for the filarial parasites. Various species of mosquitoes and biting flies serve as the intermediate hosts, depending on the nematode species, as shown in Table 26–3. Larval forms that have matured in the insect vector are injected into the skin of the human host by infective insects. Bisexual filarial adults then mature in human tissues, copulate, and produce immature larvae called "microfilariae." These usually find their way into the circulating blood (Fig. 26–5), but may remain trapped in tissues, depending on the localization of the adult. To complete and continue the life cycle, biting arthropods become infected with microfilariae

Table 26-3. Human Filarial Infection

Organism	Sources of Infection	Infective Form	Adult Localization	Transmission	Geographic Distribution
W. bancrofti	*Culex Aedes Anopheles* mosquitoes	Larvae matured in mosquito	Large lymph vessels, especially in the pelvis	Larvae called "microfilariae" circulate in blood, especially at night, these transferred to mosquito	West Indies Central and South America Arabia S.E. Asia Far East Australia and Pacific Islands
B. malayi	*Mansonia Anopheles* mosquitoes	Larvae matured in mosquito	As above	As above	Southeast Asia India China
O. volvulus	Blackfly	Larvae matured in fly	Subcutaneous tissues, especially those of head and shoulders, lower trunk, and legs	Microfilariae remain in skin and lymphatics adjacent to adult and are picked up by biting fly	Guatemala So. Mexico Venezuela West Africa
Loa loa	Deerfly	Larvae matured in fly	Adult migrates through subcutaneous tissues	Microfilariae in circulating blood are picked up by biting fly	West and Central Africa

while feeding on infected persons. The ingested microfilariae undergo a series of developmental changes in the insect host, maturing to the infective form in two to three weeks. These infective larvae are then injected into the next human victim when the insect bites again.

The life cycles and geographic distribution of the filarial worms are summarized in Table 26-3.

Characteristics of Human Disease: Diagnosis and Control

Bancroftian and Malayan Filariasis. Filariasis caused by *Wuchereria bancrofti* or *Brugia malayi* is characterized by lymphadenopathy, which may lead to obstruction of lymph flow and "elephantiasis" of the region drained by affected vessels. Localization of adult worms occurs most commonly in lymphatic vessels of the pelvis, with resulting orchitis, hydrocele, and elephantiasis of genitalia or lower limbs, or both. Light infection may produce few or no clinical symptoms. Microfilariae may be found in the blood even in the absence of symptoms, or, conversely, in cases of severe lymphatic obstruction they may be blocked from reaching the bloodstream. *Wuchereria* and *Brugia* species display "nocturnal periodicity," being found most commonly in the peripheral blood during the night rather than the daytime hours. The explanation for this may have to do with anatomic factors related to the host's relaxation during sleep, or with adaptation to feeding habits of the insect host. In this connection it is interesting to note that those filariae whose vectors are daytime feeders do not display periodicity, their microfilariae being present in blood by day or night.

Table 26–4. Summary: Infectious Diseases Acquired from Anthropod Vectors

Clinical Disease	Causative Organisms	Other Possible Entry Routes	Incubation Period	Communicable Period
I. Bacterial Diseases Transmitted by Arthropods				
Plague	*Yersinia pestis*	Respiratory	2–6 days	Not transmissible from man to man except when plague pneumonia develops; then communicable during time sputum is bacteriologically positive
Relapsing fever	*Borrelia recurrentis*		3 days to 2 weeks	Not communicable from man to man
Bartonellosis (Oroya fever and verruga peruana)	*Bartonella bacilliformis*		16–22 days; may be as long as 3–4 months	Same as above
II. Rickettsial Diseases Transmitted by Arthropods				
Epidemic typhus fever, Brill's disease	*Rickettsia prowazeki*		6–15 days, usually about 12 days	Not communicable from man to man
Endemic typhus fever	*Rickettsia typhi*		Same as above	Same as above
Rocky Mountain spotted fever	*Rickettsia rickettsi*		3–10 days	Same as above
Boutonneuse fever	*Rickettsia conori*		5–7 days	Same as above
Scrub typhus (tsutsugamushi fever)	*Rickettsia tsutsugamushi*		6–21 days	Same as above
Rickettsialpox	*Rickettsia akari*		Probably 10–24 days	Same as above

Specimens Required	Laboratory Diagnosis	Immunization	Treatment	Nursing Management
Blood Pus aspirated from buboes, paracentesis fluids; spinal fluid Sputum Acute and covalescent serum	Culture Smear and culture Rising titer of antibodies	Killed vaccines	Streptomycin Tetracyclines Chloramphenicol Sulfonamides	Isolation until danger of pneumonic form has passed; strictest isolation of pneumonic plague (hoods, face mask with goggles, gloves and overall protective clothing) Concurrent disinfection of discharges; thorough terminal disinfection-cleaning
Blood	Smear and culture Dark-field exam Animal inoculation	None	Tetracyclines Chloramphenicol	Isolation not required if patient free of insect parasites
Blood	Smear and culture	None	Ampicillin Chloramphenicol Tetracyclines Streptomycin	Isolation not required; protect patient from vector
Blood Acute and convalescent serum	Isolation of rickettsiae Weil-Felix titer	Killed vaccines	Tetracyclines Chloramphenicol	Isolation not necessary if patient is *thoroughly* devectored
Acute and convalescent serum	Weil-Felix titer	None	Same as above	Same as above
Same as above	Same as above	Same as above	Same as above	Same as above
Same as above	Same as above	None	Same as above	Same as above
Same as above	Same as above	Same as above	Same as above	Same as above
Same as above	Same as above Weil-Felix negative	None	Same as above	Same as above

Table 26–4. (*Cont.*)

Clinical Disease	Causative Organisms	Other Possible Entry Routes	Incubation Period	Communicable Period
III. Viral Diseases Transmitted by Arthropods				
Yellow fever	Yellow fever virus		Less than a week, as a rule 3–6 days	Not transmissible from person to person
Eastern equine Western equine Venezuelan equine, and St. Louis encephalitis	Arboviruses		4–5 days to 3 weeks, average 10–15 days	Not transmissible from person to person
IV. Parasitic Diseases Transmitted by Arthropods				
Malaria	*Plasmodium* species		12–30 days, depends on species	When gametocytes circulate in blood*
Leishmaniasis	*Leishmania* species		From a few days to 24 months, depends on species	While leishmaniae circulate in blood*
Trypanosomiasis	*Trypanosoma* species		1–3 weeks, depends on species	While trypanosomes circulate in blood*
Filariasis	*Wuchereria bancrofti Brugia malayi Onchocerca volvulus Loa loa*		About 9 months before microfilariae are found in blood	While microfilariae circulate in blood*

*Transmissible to insect vector only.

Specimens Required	Laboratory Diagnosis	Immunization	Treatment	Nursing Management
Blood Acute and conva- lescent serum	Virus isolation in mouse Rising titer of antibodies	Live attenuated vaccine	None specific	Isolation not nec- essary; patient must be screened from mosquitoes
Blood Acute and conva- lescent serum	Virus isolation Rising titer of antibodies	None except EEE vaccine (inac- tivated)	None specific	Isolation and spe- cial precautions not indicated
Blood smears, thick and thin	Recognition of parasitic forms in red blood cells		Chloroquine Amodiaquine Proguanil Pyrimethamine	Isolation and spe- cial precautions not indicated; protect patient from vector.
Biopsies of spleen or liver, blood Aspirated bone marrow Acute and conva- lescent serum	Culture and micro- scopic recogni- tion of intra- cellular forms Rising titer of antibodies		Antimony compounds	Same as above
Spinal fluid Blood Aspirated lymph Acute and conva- lescent serum	Microscopic recog- nition of intra- cellular or extra- cellular forms; culture Rising titer of antibodies		Pentamidine Suramin Arsenicals	Same as above
Blood	Recognition of microfilariae		Diethylcarbamazine	Same as above

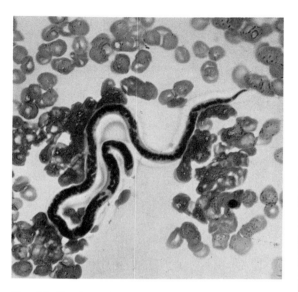

Fig. 26–5. Microfilaria (*Loa loa*) in a thick smear of blood. (246×.) (Courtesy of Dr. Stephen Lerner, The University of Chicago, Chicago, Ill.)

Onchocerciasis. Onchocerciasis is characterized by the appearance of firm nodules in subcutaneous tissues of the head and shoulders or the lower part of the body. The microfilariae do not enter the blood but migrate through the tissues. They may migrate from lesions on the head into tissues of the eye, producing local reaction, ocular disorders, and sometimes blindness.

Loiasis. Loiasis is also an infection of subcutaneous tissues, but in this case the *adult* worm migrates, producing extensive reaction at each new site of localization. Nodules arise and may later be resolved when the worm moves on. These "fugitive" or "calabar" swellings are characteristic of the disease. This worm may also reach the eye and wander through the conjunctival membrane across the ball of the eye, producing intense pain and swelling. Microfilariae circulate in the peripheral bloodstream, where they may be found during the day as well as at night.

Allergic reactions are prominent features of these infections as time goes on. Urticaria, fever, and local tissue inflammation are seen, especially in those diseases in which lymphatic blockade occurs.

Diagnosis is made in the laboratory by demonstration of viable microfilariae in wet-mount preparations of blood, or by staining them in dried smears of blood (Fig. 26–5). Anticoagulated whole blood may be submitted for examination, or a small blood sample may be placed in a tube containing water or formalin. Either of the latter agents will lyse the red blood cells when the tube is gently rotated. The tube can then be centrifuged to concentrate microfilariae in the sediment. If water has been used, the larvae will be alive and motile and can be found easily. They are not readily identified as to species when they are actively lashing under the coverslip. The sediment of formalinized blood will reveal dead microfilariae, which can then be stained and examined for the microscopic details by which various species are recognized.

Onchocerciasis must be diagnosed in tissue obtained by biopsy or by surgical removal of a nodule. Microfilariae can be demonstrated in wet preparations of the tissue fluid, or adult worms may be found in the nodule.

Treatment. There is no satisfactory chemotherapy that eliminates the adult filarial worms. Diethylcarbamazine is effective against the microfilariae but must be given at repeated intervals as the surviving adults produce new larvae. Surgical removal of adult worms may offer the most feasible cure of onchocerciasis and loiasis but is not practical in bancroftian or Malayan filariasis.

Antihistamines or corticosteroids help to relieve allergic reactions, particularly following diethylcarbamazine therapy when intense reactions to disintegrating microfilariae may occur.

Control. Control and prevention require concomitant efforts to suppress insect vectors and to treat all infected human hosts who may perpetuate the parasites' cycles.

Questions

1. What is the vector for the plague bacillus?
2. What two kinds of illness can the plague bacillus produce, and what nursing precautions are required for each? Why do precautions differ?
3. What organism causes relapsing fever? What arthropods transmit the organism?
4. What vectors transmit the typhus fevers?
5. Describe the organisms responsible for rickettsial diseases.
6. What steps are necessary to control active typhus infections?
7. What is the reservoir for yellow fever virus? The vector?
8. What are the three most common mosquito-borne encephalitides?
9. What is the definitive host for *Plasmodium* species? The intermediate host? The vector?
10. What causes the symptoms of malaria?
11. What is a hemoflagellate?
12. What is the vector for leishmaniasis?
13. What are the reservoir and vector for African trypanosomiasis?
14. How are the filarial roundworms controlled and how can filariasis be prevented?

Additional Readings

Ayer, W. D.: Napoleon Buonaparte and Schistosomiasis or Bilharziasis. *N.Y. State J. Med.,* **66**:2295, Sept. 1, 1966.

Cravens, Gwyneth, and Marr, J. S.: *The Black Death.* Dutton, New York, 1976.

†Hawking, Frank: The Clock of the Malaria Parasite. *Sci Am.,* **222**:123, June 1970, #1181.

Herban, Nancy L.: Nursing Care of Patients with Tropical Diseases. *Nurs. Clin. North Am.,* **5**:157, Mar. 1970.

Hoeprich, Paul D. (ed.): *Infectious Diseases: A Modern Treatise of Infectious Processes,* 2nd ed. Harper & Row, Hagerstown, Md., 1977.

Hovenden, Helen G.: Rocky Mountain Spotted Fever. *Am. J. Nurs.,* **76**:419, Mar. 1976.

†Langer, William L.: The Black Death. *Sci. Am.,* Feb. 1964, #619.

Riley, Harris D.: 'Rocky Mountain' Spotted Fever. *Hosp. Prac.,* **12**:51, Apr. 1977.

*Roueché, Berton: Alerting Mr. Pomerantz. In *Eleven Blue Men.* Berkley Medallion Editions, New York, 1953.

Zinsser, Hans: *Rats, Lice and History.* Little, Brown & Co., Boston, 1975.

*Available in paperback.

†Available in *Scientific American* Offprint Series. (See Preface.)

27 Infectious Diseases Acquired Through Accidental Injuries

This chapter includes infectious diseases characteristically acquired through injury to the skin, such as *animal bites, "street wounds,"* and *burns.* Diseased animals sometimes transmit their infections through injuries they inflict on man, rabies being a notable example. More usually, animals transmit viral or bacterial agents that are commensalistic for them but virulent for man when introduced into his subcutaneous tissues by a bite or scratch. The microbial species involved may induce severe local, or even systemic, infection. Similar problems may arise as a result of human bites.

The term *street wound* is used to distinguish accidental injuries exposed to environmental sources of infection from those inflicted by animals, burns, or surgery (the latter are discussed in Chapter 28). Animals transmit microorganisms adapted to a parasitic or commensalistic life, whereas environmental reservoirs teem with saprophytic bacteria and fungi, many of which are not capable of establishing themselves in human tissues. Among the most serious diseases acquired through injured tissues exposed to environmental sources of infection are those of clostridial origin (tetanus and gas gangrene) and those caused by fungi. Clostridia are derived from the intestinal tract of man and animals, but because they form resistant spores they survive and are ubiquitous

in the external environment. The fungi that can cause subcutaneous or systemic mycoses in man are derived from the soil where they normally live as saprophytes (see Chap. 16).

Other organisms most likely to infect injured soft tissues include species of staphylococci, streptococci, and Gram-negative bacteria such as *Pseudomonas aeruginosa,* a notorious troublemaker in wounds and burns. Anaerobes may thrive in soft tissues because local conditions often meet their requirement for reduced oxygen tension. The principal determining factors in the development and severity of wound infections are related to conditions within traumatized tissues and the degree of their exposure to microorganisms having a pathogenic potential.

Infectious Diseases Acquired Through Animal Bites

Rabies

The Clinical Disease. Rabies is a viral infection occurring primarily in wild carnivorous animals (skunks, foxes, raccoons) and bats. Transmission to man is rare but almost invariably fatal. Infection is usually acquired, by man or animals, through the bite of a rabid animal whose saliva contains the virus. Transmission by the respiratory route, via airborne virus, has occurred under unusual circumstances, as in spelunkers exploring a bat-infested cave and in a recent case of aerosol transmission to a laboratory worker.

When introduced into tissues, the rabies virus travels along sensory nerves from the point of inoculation to the central nervous system, where it multiplies. The chief sites of localization in the brain are the hippocampus and cerebellum. From the CNS the virus then travels along nerves in a centrifugal pattern to invade a variety of organs and tissues. The chief clinical manifestations are those of an acute encephalomyelitis, with both physical and psychologic symptoms of deranged cerebral function. The onset of fever, malaise, and headache is followed by sensory disturbances, particularly at the site of the bite wound, spasms in the muscles required for swallowing, convulsive seizures, and respiratory paralysis. The mental suffering of the patient is associated with his fear of swallowing because of painful throat spasms. (This disease has been called *hydrophobia* because of the patient's fear of swallowing water or liquids, which can bring on attacks of convulsive choking, sometimes of themselves fatal.) Apprehension, excitability, and delirium then give way to a progressive, general, flaccid paralysis, and the patient gradually lapses into coma. Death, usually within six days, may result from respiratory paralysis, peripheral vascular collapse, or both.

Laboratory Diagnosis

I. Identification of the Organism. The virus of rabies forms inclusion bodies within the nucleus or cytoplasm of the affected cells of the brain. These characteristic inclusions, called *Negri bodies,* can be readily visualized under the light microscope in stained impression smears or sections of brain tissue. The virus can be cultivated in mice and other laboratory animals, tissue cultures, embryonated chicken and duck eggs. It is identified serologically by known antiserum that specifically neutralizes its infectivity for animals or tissue cultures. Fluorescein-tagged rabies antiserum is also used to identify virus in infected tissue cells.

II. Identification of Patient's Serum Antibodies. Antibodies that develop during the course of infection, or following vaccination, can be detected by neutralization tests using infected tissue cultures. Complement-fixation

and immunofluorescent techniques may also be used. Serologic diagnosis is seldom helpful in rapidly fatal cases, because significant levels of antibody do not have time to develop.

III. Specimens for Diagnosis. During life, laboratory isolation of rabies virus may be possible from saliva, throat swabs, conjunctival and nasal secretions, spinal fluid, and urine. Such specimens may be examined by the fluorescent rabies antibody (FRA) technique. A positive FRA test is diagnostic, but if it is negative, specimens are inoculated into mice. Rabies virus, if present, causes death of mice within a few days. Mouse brains are examined for Negri bodies by the FRA technique, but if they cannot be found, virus harvested from mouse brain tissue can be identified by neutralization tests.

At autopsy, possible sources of virus include brain, spinal cord, submaxillary salivary glands, lacrimal glands, muscle tissue, lung, kidney, pancreas, and adrenals. FRA testing or mouse inoculation is performed on these tissues as outlined above.

All specimens collected for virus isolation or FRA tests should be frozen as soon as possible, pending their transport to the laboratory. Most clinical laboratories are not equipped for handling rabies virus, and such specimens must be forwarded to the nearest reference or health department laboratory, or to the Center for Disease Control in Atlanta, Georgia.

An acute serum for serologic studies should be collected as soon as the diagnosis of rabies is suspected. This should be followed by additional serum collections in five and ten days, and later, if possible. Serologic diagnosis, when it can be made, is generally retrospective at best.

IV. Examination of Suspect Animal. When a bat, dog, or other possibly rabid animal has bitten a human being, it is of immediate importance to isolate and examine the animal if possible. It is essential to the victims of such bites that appropriate decisions be made promptly as to the need for prophylactic vaccination.

Captured bats are killed and examined at once for rabies infection. Selective judgment should be applied to the disposition of other animals, however. A veterinarian or physician competent to make the diagnosis of rabies should be called to examine a captive dog or wild animal that has bitten someone. Vaccinated domestic dogs and cats do not constitute a threat, but unvaccinated animals are suspect. In the latter category, a domestic pet that is apparently healthy should be isolated and observed for clinical signs of developing rabies, but it is best to sacrifice a captured wild animal at once and examine its brain for rabies. A quarantined dog or cat, if it was infective at the time of the bite, will generally show signs of rabies, e.g., increasing excitability or paralysis, within a week. Animals that appear to be rabid at the time of the bite, or later during quarantine, should be killed and examined without delay. Once this decision is made, it is vital that the animal be sacrificed in such a way that the head is not destroyed in the process, because only the brain or salivary secretions can provide conclusive evidence of rabies infection.

Speed of diagnosis of the animal infection is critical for the human victim of a bite if he is to have time to be actively immunized against the disease. Immunoprophylaxis is effective against rabies only because the infection generally has a prolonged incubation period. However, immunization itself requires multiple doses of vaccine given over a two- to three- week period and must be started promptly if it is to be successful. When the possibility of exposure is reasonably certain or has been confirmed, vaccination is initiated without question. The decision is more difficult when the biting animal cannot be properly diagnosed, either because it escaped, was killed prematurely, or was destroyed in such a way that the head cannot be examined. Failure to preserve the intact head of a suspect animal properly (it should be packed in ice or kept frozen for transport to the laboratory) may contribute to the problem of animal diagnosis. (See Table 27–1 for recommendations concerning immunoprophylaxis.)

Epidemiology

I. Communicability of Infection

A. RESERVOIR, SOURCES AND TRANSMISSION ROUTES. The largest reservoir for rabies is among wild animals, particulary wolves, coyotes, foxes, wildcats, skunks, raccoons and other carnivores. At least three varieties of bats are known to carry rabies. Insect-eating types are known to be involved in the United States, Canada, and Europe, while fruit-eating as well as vampire bats are infected in Central and South America. Domestic dogs acquire the disease from these wild sources. In the United States, dogs and indigenous bat species are the usual vectors of rabies in man.

It is possible, but not usual, for man to acquire infection from rabid animals without being bitten. Scratches or other minor skin wounds, or mucosal surfaces exposed to infectious saliva, may provide a route of entry. Two human cases are on record that were acquired through exposure to the dust of a cave harboring millions of bats (Fig. 27–1). These individuals were not bitten, and subsequent studies in the cave using caged animals proved that airborne transmission could occur. In a 1977 case, a laboratory worker conducting studies of live but attenuated vaccine strains of rabies virus was infected by inhaling an aerosol generated by a defective machine used for spraying virus in an animal inoculation experiment. The fact that this individual had been immunized but nonetheless acquired rabies from a modified virus suggests that airborne transmission affords an increased risk of infection, and/or that the respiratory mucosae offer a particularly effective route of entry.

Man-to-man transmission of rabies may be possible, since human saliva becomes virus positive during the course of disease, but such transfer has never been documented.

B. INCUBATION PERIOD. The average time is four to six weeks but may be shorter or greatly prolonged. In a recent case that occurred in England (which is rabies free) but was contracted in Bangladesh, the minimum incubation period possible was 14 months, but the exact time and source of exposure could not be pinpointed. The duration of the incubation is influenced by the length of the path the virus must travel from the site of inoculation to the brain. Bites inflicted on the face and neck, particularly those involving mucous membranes, may result in infection with onset of symptoms in one or two weeks. The amount of virus inoculated, as well as the distribution of the local nerve supply, also affects the length of the incubation period.

C. COMMUNICABLE PERIOD. Virus is present in the salivary secretions of biting animals for several days before their symptoms appear and throughout the course of their illness. Bats may shed virus for weeks without displaying symptoms of illness.

D. IMMUNITY. Natural immunity to rabies virus does not exist among human beings or animals, but antibody production can be stimulated by virus vaccines.

II. Control of Active Infection

A. ISOLATION. The patient should be hospitalized and kept in strict isolation throughout the course of the illness.

Fig. 27–1. A colony of bats in flight. These cave dwellers are an important reservoir of rabies, transmitting the virus through their bite. The dust of caves colonized by bats may also be infectious upon inhalation. (Courtesy of the Center for Disease Control, D.H.E.W., Atlanta, Ga.)

B. PRECAUTIONS. The patient's saliva should be considered infectious, and all items contaminated with it should be disinfected. The nursing care of these patients is difficult because of their apprehension, excitability, and disorientation. Particular caution must be exercised by those in close attendance to protect their own skin and clothing from salivary contamination, and to avoid injury to their hands from the patient's teeth when mouth care is given.

C. TREATMENT. There is no specific treatment for rabies after symptoms have started and the disease is established. However, immediate attention should be given to the bite wound. It should be thoroughly cleaned by irrigating and washing with soap or detergents. Open wounds are first washed and then treated by instillation of hyperimmune antirabies serum. The muscle tissue beneath the wound is also infiltrated with antiserum. When immune serum is used, vaccination treatment is not instituted until the following day.

Intensive medical care is given to support the patient, with particular attention to pulmonary and cardiac monitoring. There are now only three documented cases of human survivors of clinical rabies, two in the United States (1970 and 1977) and one in Argentina (1972). Although two of these patients suffered severe residual damage, their survival suggests that continuous intensive care started early in the disease and maintained at a high level may be lifesaving.

D. CONTROL OF CONTACTS. Human contacts of a rabies patient need not be vaccinated unless the patient's saliva has made direct contact with an open wound or mucous membrane. A search is made for the rabid animal source of the case, if it has not already been captured, and for other persons or animals that may have been bitten.

III. Prevention. Following an animal bite, the prevention of human rabies depends on appropriate treatment of the wound, as described in II C above, to kill the virus locally if possible, and on prompt immunologic procedures. The latter include active immunization with vaccine and passive immunization with hyperimmune antirabies serum. Vaccination is also the chief preventive measure for people at high risk of exposure to rabies.

A. ACTIVE IMMUNIZATION. There is only one antigenic type of rabies virus, but there are several biologic variants that display differences in their affinities for neural or extraneural tissues:

a. "Street" virus is the term used for the rabies agent freshly isolated from animal or human cases. This virus produces disease in laboratory animals, invading both neural tissue in the CNS and the cells of the salivary glands, forming intracytoplasmic inclusion bodies.

b. "Fixed" virus is a variant that arises when serial passages are made in rabbits inoculated intracerebrally. After several passages, the variant multiplies rapidly in rabbit neural tissue, but inclusion bodies are seldom formed, and it is incapable of reproducing in extraneural cells. When the virus has become fully adapted to rabbits, it consistently kills them within five to six days of inoculation. At this point the virulence of the organism for rabbits is "fixed" and cannot be increased by further passages in these animals, but its virulence for man is decreased.

c. "Avianized" virus has been adapted to multiply in chicken or duck embryos. The *Flury strain* of this variant does not produce disease in animals when injected in extraneural sites.

The original vaccine prepared by Pasteur contained "fixed" virus that he had passed through rabbits (Fig. 27–2). The spinal cords of these animals were dried to reduce virulence of the organisms for man still further. Later, the *Semple vaccine* was

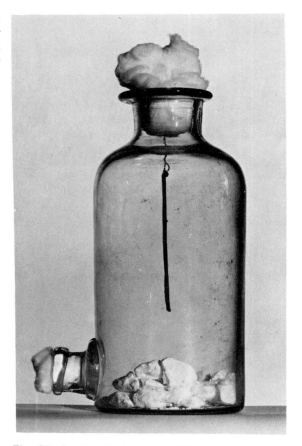

Fig. 27–2. This flask was used by Louis Pasteur in his experimental work with rabies. It contains a rabbit spinal cord suspended for drying over a desiccant. Pasteur's rabies vaccine contained virus "fixed" by passage through rabbits and dried by this method to attenuate its virulence for man. (Courtesy of the Pasteur Institute, Paris, France.)

developed, using phenol-inactivated preparations of infective rabbit brain tissue. The use of this material for human beings carries the risk of provoking hypersensitive reactions to the foreign brain protein. This hypersensitivity is usually expressed as an allergic encephalitis, sometimes with paralysis. Such responses may occur in about 0.1 percent of persons vaccinated with nervous tissue vaccines.

The avianized virus is inactivated with beta-propriolactone (BPL) for use as a human vaccine, but the attenuated Flury strain is used for dog immunization without inactivation. These nonneural tissue vaccines are now preferred because they do not add risk of postvaccinal encephalitis.

Currently, the only licensed product available in the United States for use in humans is duck embryo (DE) rabies vaccine, prepared by BPL inactivation of yolk sac growth of fixed virus. Human vaccination is performed only after injury has been inflicted by an animal known or suspected to be rabid (Table 27–1). The procedure requires a series of daily subcutane-

ous injections throughout a 14- to 21-day period. Dogs are immunized by a single intramuscular injection of live-virus vaccine, repeated once yearly.

B. PASSIVE IMMUNIZATION. Hyperimmune antirabies human or horse antiserum also provides effective protection when used in conjunction with active immunization. Antiserum is given intramuscularly, a portion being injected in and around the bite wound, the remainder in the buttocks. Passive immunization should be administered immediately after a bite exposure to rabies, and followed 24 hours later by the first of the vaccine injections. This delay

Table 27–1.   Recommendations for Rabies Immunoprophylaxis*

| Nature of Exposure | Status of Animal (Irrespective of Previous Vaccination) | | Recommended Treatment |
	At Time of Exposure	During 10 days‡	
I. Contact, but no lesions Indirect contact No contact	Rabid	—	None
II. Licks of the skin; scratches or abrasions; minor bites (covered areas, or arms, legs, trunk)	Suspected as rabid‡	Healthy	Start vaccine. Stop if animal remains healthy for 5 days†‡¶
		Rabid	Start vaccine; upon positive diagnosis of animal give antiserum, complete vaccine course
	Rabid domestic animal; wild animal§; or animal unavailable for observation	—	Antiserum + vaccine
III. Licks of mucosa; major bites (multiple or on face, head, neck, finger)	Rabid or suspect‡ domestic animal; wild animal§; or animal unavailable for observation	—	Antiserum + vaccine. Stop treatment if animal remains healthy for 5 days†‡¶

*Adapted from guidelines suggested by the Expert Committee on Rabies of the World Health Organization. These recommendations should be considered together with knowledge of the animal species involved, circumstances of the injury or exposure, vaccination status of the animal, and prevalence of rabies in the area.
†Holding period applies only to dogs and cats.
‡All unprovoked bites in endemic area should be suspect unless animal proved negative by brain FRA.
§Exposure to rodents or rabbits almost never requires specific antirabies treatment.
¶Or if its brain is proved negative by FRA tests.

prevents immediate neutralization of the vaccine antigen by passively acquired antibody. Alternatively, the vaccine series can be started at the time antiserum is given, but after completion of the 14-day course two booster doses of vaccine are given at 10 and 20 days to overcome the antiserum interference effect.

C. PREEXPOSURE IMMUNIZATION. Rabies vaccine is usually given to persons at risk of exposure, such as laboratory workers, veterinarians, dog- and wildlife-control personnel, spelunkers, and other field workers serving in areas where rabies is endemic and prevalent. DE rabies vaccine is given in two or three subcutaneous injections at one-month intervals, followed by a booster injection in six months. Additional booster doses at one- to three-year intervals are recommended for those who remain at risk. Serum levels of neutralizing antibody should be determined 30 days after a booster dose to be certain that an adequate immunizing effect has been obtained. If not, active immunization should be continued.

IV. Control. Rabies control requires suppression of the disease in the animal population. Immunization of dogs and enforced restrictions on stray animals are essential features of control and are required by most local or state governments. Control of rabies in wild animals is the responsibility of state and federal wildlife conservation agencies, who work in cooperation with public health authorities when the threat of rabies exists (see Chap. 11).

Apparently healthy dogs that have bitten people should not be destroyed hastily. If they are incubating and transmitting rabies, this fact can be discovered by awaiting the onset of their own disease, then destroying them and examining the brain tissue. There may be no signs of Negri bodies in the brain during the incubation period, and laboratory attempts to cultivate the virus from dogs that have been killed too soon may require more time than would the development of the disease in the suspect animals.

The widespread incidence of rabies in the bat population of the United States appears to justify vaccination of all persons bitten by bats. The necessity for treatment is often underscored by the fact that the offending bat may not be captured and hence is unavailable for examination. Furthermore, it is often difficult to demonstrate virus in bats by rapid smear examination, and the time required for virus cultivation may be more than the patient can safely afford.

Cat Scratch Fever

Cat scratch fever is characterized by regional lymphadenopathy associated with a scratch wound usually (but not necessarily) inflicted by a cat. There is an initial mild inflammatory reaction at the site of the wound, sometimes with induration or minor ulceration. Nodes that drain the area then become swollen with a thick, gray, bacteriologically sterile pus, and eventually suppurate. This reaction requires at least two or three weeks to subside, but may persist for months. There is often a recurring fever, sometimes with chills and general malaise. Involvement of the eye sometimes occurs and takes the form of the "oculoglandular" syndrome that may be seen in tularemia infections (see p. 545). Recovery is always complete but requires variable periods of time.

The causative agent of this infection has not been identified with certainty but is thought to be one of the chlamydiae. In stained sections of infected lymph nodes, large cellular inclusions resembling the elementary bodies of psittacosis have been reported. Serologic evidence of a relationship to chlamydiae also exists, in that convalescent serums from patients with this infection cross-react in complement-fixation tests with agents of the psittacosis-LGV group.

A skin-testing antigen prepared from heat-inactivated suspensions of lymph nodes or pus has been used to diagnose cat scratch fever. This antigen produces a delayed type of skin reaction in patients with the disease. It is not commercially available, however, and is not recommended for routine use because it is a crude material that might also contain heat-resistant microbial agents such as hepatitis virus.

The lesions of cat scratch fever resemble those of other infections incurred through animal bites, notably pasteurellosis (see p. 598). Good diagnostic bacteriologic techniques are required to differentiate such infections. Cat scratch fever occurs sporadically in all parts of the world, among people of all ages, but is seen most often among children and young adults. This age distinction may be due to frequency of exposure or to decreasing susceptibility in older adults following repeated encounters with the infectious agent. Cats do not become ill with this infection and are thought to be merely mechanical vectors of the agent. Human cases have been acquired from sources other than cats, such as wood splinters or thorns, insect bites, and wounds incurred while slicing meats.

Treatment with tetracyclines relieves symptoms and shortens the course of infection, but this disease is self-limited in any case. Complete recovery is usual within a few months. Careful cleansing and hygienic management of skin wounds afford the best control.

Bacterial Infections of Bite Wounds

Resident bacteria of the mouths of animals or humans may produce serious local or even systemic infections and tissue reactions when introduced deeply into tissues by biting teeth. The normal flora of the human mouth includes both aerobic and anaerobic microorganisms (see Chap. 5). This is also true for animals, although there are some differences in bacterial species. The anatomy and physiology of animal mucous membranes differ in many respects from those of human mucosa and therefore provide different conditions for commensalistic microorganisms. Animals often harbor transient microbial species from the soil, from feces, from their own hair or fur, or from other sources, depending on their nature and habits.

Pathogenesis. Development of infection in a bite wound is greatly influenced by the nature of the injury. An open tear or laceration inflicted by teeth can be easily cleansed and kept aerated while healing proceeds, so that microorganisms have little opportunity to establish themselves. Puncture wounds may implant bacteria deeply in positions that are inaccessible unless the wound is laid open by surgical incision. Infected bites often contain a mixed flora, but sometimes a particular species may predominate if it possesses more of the properties required for growth in human tissues. The process may be promoted still further if defensive mechanisms of the tissue are handicapped by injury to blood supply, with local hemorrhage, clotting, and devitalization of surrounding cells. This not only interferes with the marshaling of phagocytes and of antibacterial components of blood plasma, but also prevents adequate oxygenation of the local area, which becomes progressively anaerobic. Many bacteria that live aerobically on mucous membranes, or on culture media in the laboratory, are in reality "facultative" anaerobes, capable of growth and multiplication under conditions of reduced oxygen tension or of anaerobiasis. Facultative anaerobes of the resident oral flora include species of staphylococci and streptococci, microaerophilic organisms, such as species of *Haemophilus,* and obligate anaerobes of the *Actinomyces* and *Bacteroides* genera.

Clinical Appearance. The character of the infectious process in a bite wound reflects the nature and properties of the multiplying organism, to some extent, as well as the tissue response to it. Staphylococci characteristically induce formation of localizing abscesses, or suppuration (Chap. 22), while streptococci are often more invasive because their enzymes and toxins are destructive for cells and components of connective tissue (Chap. 13). *Cellulitis,* which is an advancing inflammatory involvement of subcutaneous and connective tissues, is frequently induced by streptococcal infection. Some other invasive organisms may also cause this type of infection (see the discussion of clostridial gas gangrene in the following section). *Bacteroides* and *Fusobacterium* species are Gram-negative, nonsporing, anaerobic rods normally found in the mouth or alimentary tract but are capable of causing suppurative, inflammatory reactions in soft tissues. Mixed (synergistic) infections with anaerobic streptococci and staphylococci can produce severe gangrenous infections that may require amputation.

If local tissue reaction fails, or if prompt medical treatment is not obtained, threat of dissemination through draining lymph channels or the bloodstream may ensue. The spread of infection from a wound on an extremity is often marked by red streaks extending proximally, a sign of *lymphangitis,* or inflammation of the lymph vessels draining the affected part to proximal nodes, which may also become involved. Regional lymphadenopathy, with swelling and tenderness, may then follow. Immediate treatment is indicated at the first signs of lymphangitis to prevent bacteremia or septicemia.

Systemic Diseases Incurred Through Bite Wounds. Aside from local trauma and infection, animal bites may lead to certain characteristic clinical syndromes typically associated with specific microorganisms. These include tetanus and gas gangrene, syndromes arising from particular clostridial infections; pasteurellosis; and rat bite fever. The latter term refers to two diseases, each with a different bacterial agent.

Clostridial Infections. Deep, penetrating animal bites always present the hazard of clostridial infections. The *Clostridium* genus contains Grampositive, spore-forming bacilli, which are obligate anaerobes normally residing in the human or animal intestinal tract. The mouths of animals may be transiently contaminated with spores of clostridial species picked up from soil, or with vegetative bacilli derived from fecal sources. These organisms are dangerous contaminants of bite wounds, because if they establish a foothold in human tissues and multiply there anaerobically, they produce highly toxic substances having a systemic effect. Tetanus and gas gangrene, two serious clostridial diseases acquired through infection of injured soft tissues, are commonly incurred via accidental "street wounds" and are discussed in detail under that heading. They must be considered as a possibility when animal bite wounds are inflicted, however. The management and treatment of such wounds should therefore include immunoprophylaxis for tetanus and débridement with aeration as indicated to prevent development of gas gangrene.

Pasteurellosis. Organisms of the genus *Pasteurella* are widely distributed among animals and birds. They may cause epidemic and septicemic diseases in animals, and occasionally infect man, chiefly through animal bites. Of the four species in this genus, one, *Pasteurella multocida,* is the major human pathogen (the term *multocida* literally means *many killing*). The pasteurellae are small, nonmotile, Gram-

negative rods that are facultatively anaerobic. Virulent strains of *P. multocida* are encapsulated. They are often present as normal flora in many domestic animals. The nasopharynx of cats and the tonsils of dogs are commonly colonized with *P. multocida,* but they are found also in cattle, sheep, and fowl. Colonized animals are usually asymptomatic, but under stressful situations *P. multocida* may cause outbreaks similar to cholera in birds, hemorrhagic septicemia in cattle, and primary or secondary pneumonias.

Human disease caused by *P. multocida* (rarely by other pasteurellae) is commonly transmitted by animal bites or scratches. Such wounds may develop frank infection, particularly if sutured, and systemic complications such as infected joints and osteomyelitis are not uncommon. Regional lymphadenitis followed by septicemia may occur, especially in persons with impairments of the reticuloendothelial system, such as cirrhosis of the liver or rheumatoid arthritis. Secondary pulmonary infections with *P. multocida* are sometimes seen in patients with underlying chronic lung disease (carcinoma, emphysema), and upper respiratory infections may also be complicated by this organism. Local extensions or bacteremia may lead, more rarely, to such problems as meningitis, brain abscess, or endocarditis.

Unlike many Gram-negative rods, the pasteurellae are susceptible to penicillin. This is the drug of choice, but tetracyclines are also effective in the management of *P. multocida* infections. Specific chemotherapy should be initiated at the time of an animal bite to prevent *Pasteurella* infection. The wound should be carefully cleansed, and suturing avoided if possible. Efforts to develop effective vaccines to control veterinary disease have not been successful thus far.

Rat Bite Fever. Two types of bacterial infection are associated with rat bites. One seen in the United States is caused by an unusual Gram-negative bacillus that is very pleomorphic, forming long filaments. This organism, *Streptobacillus moniliformis,* is of interest because of its ability to enter spontaneously into a stage of growth in which it lacks a cell wall. In this respect it resembles the mycoplasmas and the PPLO group described in Chapter 15. The bacillus frequently infects rodents, which transmit it to man by biting. After an incubation period of three to ten days, the organism progresses from the primary lesion to regional lymph nodes and then to the bloodstream, producing lymphadenitis, petechial skin rash, joint involvement (arthritis), and attacks of recurring fever. Diagnosis is made on the basis of a history of rat bite, as well as by bacteriologic demonstration of the organism in pus from the skin lesion or lymph nodes, in joint fluid, and in blood. Penicillin or tetracycline drugs are effective in treatment. Control depends on eradication of rats, especially from buildings where people live or work. This disease can also be transferred through unpasteurized or contaminated milk. The source of infection in milk is not known, but a few epidemics have been related to this food. One of these occurred in the town of Haverhill, Massachusetts, and the milkborne disease came to be known as *Haverhill fever.*

Another type of rat bite fever is caused by a bacterium called *Spirillum minor.* This is a Gramnegative, motile organism, spiraled in two or three short, rigid coils. It produces a clinical disease similar to that caused by *Streptobacillus,* except that the joints are seldom involved. This form of rat bite fever occurs more frequently in the Far East (where it is called *sodoku*) than in the United States. The organism does not grow on ordinary laboratory culture media, but can sometimes be demonstrated in the blood of infected patients or by inoculating blood into mice or guinea pigs. Penicillin is the drug of choice for *Spirillum minor* infection. Prevention of this disease depends on control or eradication of rats.

Treatment and Control. The treatment of bite wounds begins with prompt surgical cleansing and débridement, to clear away contaminating organisms, aerate tissues, and ensure adequate blood supply so that healing can proceed normally. If the wound is extensive or deep, antibiotics may be given prophylactically and tetanus antiserum administered. The question of rabies prophylaxis must also be considered (see Table 27-1).

The control of infected wounds centers around efforts to establish drainage and to prevent cellulitis or lymphangitis from getting out of hand. Surgical incision and drainage may be indicated, or continuous hot soaks may be applied to encourage spontaneous suppuration. The affected part is kept elevated, if possible, to prevent venous or lymphatic stasis. Specific chemotherapy, if warranted, is chosen on the basis of antibiotic susceptibility tests of the isolated causative organism(s).

Diseases Acquired Through Infection of "Street Wounds"

Tetanus (Lockjaw)

The Clinical Disease. Tetanus is an acute disease caused by the neurotoxin of *Clostridium tetani.* This organism is not invasive and remains localized in the area where its spores were introduced into the body. If anaerobic conditions prevail in the tissues at the site of entry, the spores germinate, and the vegetative bacterial cells multiply, proliferating toxin. Local infection is often quite insignificant, producing little tissue damage, but exotoxin is absorbed from the area and extends along peripheral motor nerve trunks to the spinal cord. Within the cord, the toxin causes increased reflex excitability, while its action on peripheral nerves interferes with normal transmission of nerve impulses to muscles. Severe muscle spasms occur first at the area of infection and toxin production, then along the route of nerve trunk involvement. The muscles of the jaw and neck contract convulsively and remain "locked" so that the mouth cannot be opened and swallowing is very difficult. Other voluntary muscles also become involved, and widespread spasms of the somatic musculature may lead to opisthotonos (backward flexion of the head and feet, with arching of the body and boardlike rigidity of the abdomen). Figure 27-3 shows the effect of such spasms in a mouse infected with *Cl. tetani.* Respiratory paralysis is the most serious complication and may be fatal. The toxin does not inflict permanent damage, and those recovering do so completely but slowly. Tetanus fatalities are related to the time lapse between dissemination of toxin and the institution of antitoxin and other therapy, as well as the age and general condition of the patient, the average mortality rate being about 35 percent.

Laboratory Diagnosis
I. Identification of the Organism (*Clostridium tetani*)
A. MICROSCOPIC CHARACTERISTICS. The tetanus ba-

Fig. 27–3. The spasms in this mouse, including the opisthotonos (arched back), were caused by an injected dose of the neurotoxin of *Clostridium tetani.* (Courtesy of Dr. Irvin S. Snyder, West Virginia University Medical Center, Morgantown, W. Va.)

cillus is a Gram-positive, anaerobic, spore-forming bacillus. The spore is characteristically wider than the diameter of the bacillus and forms at one end so that the rod has a drumstick appearance (Fig. 27–4). Vegetative tetanus bacilli have an even diameter and are motile.

B. Culture Characteristics. *Cl. tetani* is a strict anaerobe. It grows on blood agar plates incubated at 37° C in an airtight jar or incubator from which air has been evacuated and replaced by an inert gas. A mixture of nitrogen with 10 percent CO_2 is commonly used to provide necessary atmospheric conditions. The organism can also be cultivated in fluid media having a low oxidation-reduction potential. Broth containing a reducing agent such as thioglycollate and enough agar to provide viscosity (about 0.1 percent) will support the organism and can be incubated in an ordinary atmosphere. A liquid medium containing fresh chopped animal meat can also be used because reducing enzymes are released from the tissue. Fluid media must be tubed in deep columns to assure an oxygen-free environment below the surface level where atmospheric oxygen may be

dissolved. Because many different anaerobic and facultative bacteria grow well in these media, they cannot be relied on to *isolate Cl. tetani* or other anaerobes in pure culture.

On the surface of blood agar, *Cl. tetani* colonies have a filamentous periphery marking an area of swarming growth, and they are usually hemolytic. The clostridial species are distinguished from each other by the patterns of their carbohydrate fermentations and their proteolytic activities. Some can digest gelatin and milk proteins; others have only a weak action on proteins, or none at all. The tetanus bacillus does not ferment carbohydrates and is weakly proteolytic.

C. Serologic Methods of Identification. The final differentiation of clostridial species can be made with specific antisera. When labeled with fluorescent

Fig. 27–4. Scanning electron micrograph of *Clostridium tetani.* (10,000 ×.) Its terminal spore gives it the appearance of a drumstick. (Reproduced from Klainer, Albert S., and Geis, Irving: *Agents of Bacterial Disease.* Harper & Row, Hagerstown, Md., 1973.)

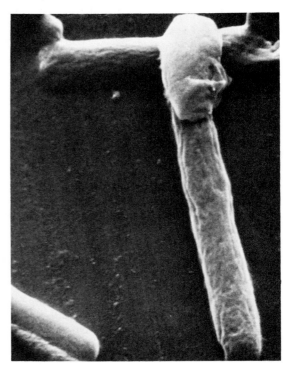

dyes sera can be valuable in providing a means of rapid identification of clostridia in smears of clinical specimens or in materials suspected as a source of infection.

D. TOXIN PRODUCTION. Definitive identification of *Cl. tetani* is also obtained by demonstrating its capacity for toxin production, the pathogenic effect of the toxin for laboratory animals (Fig. 27-3), and the specific neutralization of this effect by tetanus antitoxin.

II. Identification of Patient's Serum Antibodies. Serologic procedures have no diagnostic value. The patient who develops tetanus as a result of infection does so because he has no protective antitoxin in his bloodstream.

> *III. Specimens for Diagnosis.* Diagnosis of tetanus is generally based on the clinical picture. The site of injury and infection may be insignificant, the wound may be closed, and the area of clostridial growth inapparent. When feasible, pus or tissue from the wound should be submitted for smear and anaerobic culture, but neither the diagnosis nor the institution of therapy can safely wait for laboratory findings, nor can negative findings be considered conclusive.

Epidemiology

I. Communicability of Infection

A. RESERVOIR, SOURCES, AND TRANSMISSION ROUTES. The normal habitat of the tetanus bacillus is the intestinal tract of man and animals. Spores of the organism are widely disseminated in nature and are commonly present in the dust of streets as well as in soil. The spores can be found frequently in the lint of soiled bed linens or clothing. Tetanus spores are generally introduced into the body through wounds of various kinds, not all of which may appear important at the time. Puncture or crushing wounds; injuries caused by bullets, knives, or clubs; obstetric or abortion wounds; an unhealed umbilical stump (tetanus neonatorum); burns; and surgical wounds — all may offer an entry point for the organisms. Spores are introduced by the object that inflicts the injury (a rusty nail, a thrown or falling object, a knitting needle used to attempt abortion, an unsterilized surgical knife); or infections result from subsequent exposure of the wound to an environmental source. Wounded soldiers or automobile accident cases may lie on the ground or be exposed to other sources arising from attempts to treat wounds under field conditions. Surgical wounds sometimes are contaminated by inadequately sterilized instruments, sutures, or dressings; and orthopedic surgery is confronted with the occasional hazard of tetanus-contaminated plaster materials used for casts. Surgical wounds are also threatened by subsequent exposure to endogenous fecal sources of infection during the patient's postoperative course, but tetanus is not likely to occur in this situation unless the wound provides the necessary anaerobic conditions for growth of this noninvasive organism. Good surgical technique and nursing care play a large role in preventing this type of infection.

B. INCUBATION PERIOD. The nature, extent, and location of the wound, as well as the numbers of infecting organisms, influence the incubation period. If the tetanus bacilli can grow immediately and multiply rapidly, production of toxin may begin within a few days (two to six, with an average of four); but incubation sometimes requires two or three weeks, or even longer.

C. COMMUNICABLE PERIOD. Tetanus is not directly communicable from person to person.

D. IMMUNITY. Specific protection is provided by antibodies directed against tetanus toxin. Artificial active immunization with tetanus toxoid provides the most solid protection. Passive immunization may be necessary when the threat of infection arises in nonimmune persons. Susceptibility to tetanus infection is universal among the nonimmunized.

II. Control of Active Infection

A. ISOLATION. Not indicated.

B. PRECAUTIONS. Unnecessary.

C. TREATMENT. *Tetanus antitoxin* can neutralize toxin *before* it becomes attached to nerve tissue but not afterward. For this reason, antitoxin is administered immediately when nonimmune persons sustain traumatic injuries and there is no history of recent active immunization. It is given in a single large intramuscular dose, preferably in *human* hyperimmune serum to avoid hypersensitive reactions to foreign proteins in animal antiserums. Administration of an animal antiserum must be preceded by skin testing to determine whether or not hypersensitivity exists and anaphylactic reactions are a risk (see Chap. 7). Not only are reactions to previously encountered foreign proteins dangerous, but animal serum proteins may be promptly destroyed by hypersensitive T lymphocytes so that the toxin-neutralizing effect of passively administered antitoxin can be very short lived and possibly ineffective. Human antitoxin is given in a dose of 250 to 500 units and should be protective for about a month. When human antiserum is not available, horse or other animal antitoxins are used in a dose of 5000 units or more, depending on the clinical situation and the manufacturer's specifications. Larger doses may be given to patients who have already developed symptoms of tetanus toxicity, so that newly produced, unbound toxin may be neutralized before it can cause further nerve damage.

Antibiotics are given to control simultaneous mixed infections that may encourage anaerobic conditions required for growth of the tetanus bacillus. Penicillin is the drug of choice to prohibit further growth and toxin production.

Surgical débridement of a wound with obviously active infection is essential to remove dead tissue and to provide aeration.

Tetanus toxoid is given to stimulate active immunity when a potentially dangerous wound is sustained, whether or not the individual has been previously immunized. If he gives a reliable history of artificial active immunization, a booster shot of toxoid together with an immediately protective dose of passive antibody may prevent the disease. If he has not been actively immunized within recent years, the first of a series of toxoid injections should be given shortly after antitoxin prophylaxis is instituted. When tetanus antiserum and toxoid are being given at the same time, they should be injected at different anatomic sites.

D. Control of Contacts. The chief problem in obstetric, pediatric, and surgical situations is to locate the possible source of infection so that further cases may be prevented. For this reason case reports are mandatory in most states and countries. If the source is associated with inadequate techniques of surgery or sterilization, these can be identified and controlled before they lead to serious outbreaks in hospital or midwife practice. Sporadic cases acquired from environmental sources of infection require only individual control, but prevention of such cases can be achieved by adequate programs of immunization.

III. Prevention by Immunization. Active immunization with tetanus toxoid provides certain and durable protection and should be initiated during infancy or early childhood. The preferred program is the immunization of infants at two to three months of age with triple vaccine containing diphtheria and tetanus toxoids and pertussis vaccine (DTP). (See Table 11-1, p. 282.) Booster doses at regular intervals reinforce immunity during childhood and adolescence.

Adult immunization with tetanus toxoid is universally advised as the safest method for preventing serious consequences of infection with this ubiquitous organism. Alum-precipitated toxoid is given to adults in two doses separated by a four-week interval, followed by a third injection 8 to 12 months later. Boosters at ten-year intervals thereafter maintain a safe level of immunity, but if any injury is sustained a single injection of toxoid is given immediately (Table 11-2, p. 283). The need for passive protection with antitoxin (and its attendant risks) is obviated for the individual with a history of scheduled toxoid immunization, the single booster providing prompt increase in protective antitoxin and assurance of safety from minor overlooked infection. Tetanus toxoid is particularly recommended for all persons who run unusual risks of accidental injuries (farmers, veterinarians, military personnel, policemen, firemen, laboratory personnel and others who work with animals). It is also wise for pregnant women to be assured of active immunity because congenital transfer of antitoxin protects the newborn infant against tetanus acquired through infection of the umbilical stump.

Gas Gangrene

The Clinical Disease. Clostridial gas gangrene is a fulminating infection that can arise in necrotic tissue and spread rapidly. When the organisms are introduced into tissues where conditions permit anaerobic multiplication, they utilize amino acids and carbohydrates freed from dead or dying cells, producing gas that distends tissues and interferes with blood supply and oxygenation. As multiplication proceeds, clostridia secrete enzymes such as lecithinase, collagenase, and hyaluronidase (see Chap. 6), which are destructive to adjacent normal tissues and also to red blood cells, so that the area of necrosis is extended. Bacterial growth accelerates and advances under these accumulative conditions, threatening severe anemia and toxemia. Gas production in the tissues produces crepitation, one of the diagnostic criteria of this destructive infection. Muscle tissue is characteristically involved (*clostridial myositis*) and is rapidly invaded because of its rich content of carbohydrate. The organisms may also advance through connective and subcutaneous tissues, producing progressive anaerobic *cellulitis*. Acute toxemia may result in shock and rapid death, depending on the speed and efficacy of therapeutic intervention, the location of infection, and the general condition of the patient.

Clostridial species capable of tissue invasion are often present in the human intestinal tract. They sometimes induce gangrenous infections of the intestinal wall and the peritoneum in patients with bowel obstructions. Risk of clostridial infection from endogenous sources is constantly present in abdominal surgery or postoperatively, if segments of bowel become ischemic (lacking in adequate blood supply) or nonviable for any reason.

Clostridial species may occur in the genital flora in about 5 percent of normal women. Organisms derived from genital or intestinal sources may induce severe uterine infections as a result of septic instrumental abortion.

Some strains of clostridia can be responsible for acute food poisoning if they are introduced into food and have an opportunity to multiply there before it is eaten. Meat provides the best medium for clostridial growth, particularly if it has been previously cooked in a stew or meat pie. Contaminated leftover meat dishes not thoroughly reheated before eating are the usual source of clostridial enteritis. (See Chap. 18, p. 430.)

Laboratory Diagnosis

I. Identification of the Organism. *Clostridium perfringens* (the Welch bacillus), *Clostridium novyi, Clostridium septicum, Clostridium histolyticum.*

The organisms listed above are species most commonly associated with gas gangrene. This disease is often a mixed infection. There may be more than one clostridial species present, as well as streptococci or staphylococci and Gram-negative enteric bacteria.

It must be emphasized that diagnosis of gas gangrene is primarily made on clinical grounds. The mere finding of clostridia in wounds does not necessarily incriminate them as the cause of infection. Spores of these organisms may be found as transient contaminants of surface tissues in wounds where they are unable to vegetate and multiply in the absence of necrotic tissue and anaerobic conditions. On the other hand, their isolation from exudates of gangrenous, crepitant lesions confirms the bacterial nature of the clinical disease.

MICROSCOPIC AND CULTURE CHARACTERISTICS. Like the tetanus bacillus, these members of the *Clostridium* genus are anaerobic, spore-forming bacilli. They differ from *Cl. tetani* in that their spores usually are placed in subterminal positions. However, spores are not often produced in infected tissue, and they may not be seen microscopically in clinical material. These bacteria are isolated and identified by methods described in the section on laboratory diagnosis of tetanus (p. 600). Most strains actively attack carbohydrates and are strongly proteolytic.

II. Identification of Patient's Serum Antibodies. Serologic diagnosis is of no value in clostridial infections (except botulism).

III. Specimens for Diagnosis. Exudates and tissue from infected wounds are submitted for smears and culture. Collect this material with care from the active depths of the wound, so that misleading surface contaminants may be avoided.

Epidemiology

I. Communicability of Infection

A. RESERVOIR, SOURCES, AND TRANSMISSION ROUTES. See the corresponding discussion in the section on tetanus (p. 601).

B. INCUBATION PERIOD. This is usually very short, infection arising in one to three days following wound contamination.

C. COMMUNICABLE PERIOD. Clostridial infections are ordinarily not transmissible from man to man. In the hospital surgical service, however, the infected patient may be a rich source of the organism, in the operating room or on the surgical ward. Every precaution must be taken to prevent transfer of clostridial infection to other surgical patients by direct or indirect routes.

D. IMMUNITY. Susceptibility to clostridial wound infection is general. Antitoxin immunity may arise as a result of infection, but its durability and protective value are questionable. Passively administered anti-

Table 27–2. **Summary: Infectious Diseases Acquired Through Accidental Injuries**

Clinical Disease	Causative Organisms	Other Possible Entry Routes	Incubation Period	Communicable Period
I. Infectious Diseases Acquired Through Animal Bites				
Rabies	Rabies virus	Respiratory	4–6 weeks, sometimes very prolonged	Virus is present in biting animals for several days before their symptoms appear and throughout the course of their illness; bats may shed virus for weeks
Cat scratch fever	Unknown, possibly a species of *Chlamydia*		1 or 2 weeks from initial injury, may be shorter	Has not been clarified, but infection is not directly transmissible from person to person
Pasteurellosis	*Pasteurella multocida*		Few days	Not transmissible from person to person
Rat bite fever	*Streptobacillus moniliformis* *Spirillum minor*		Few days	Same as above
II. Diseases Acquired Through Infection of "Street Wounds"				
Tetanus	*Clostridium tetani*		2–6 days, average 4; sometimes 2 or 3 weeks or longer	Not communicable from man to man

Specimens Required	Laboratory Diagnosis	Immunization	Treatment	Nursing Management
Saliva (from submaxillary gland) Brain of biting animal if available	Fluorescent antibody technique Negri bodies in brain of inoculated mice	Inactivated virus (administered after exposure is confirmed or reasonably certain)	Active vaccination and immune serum	Strict isolation; articles contaminated by saliva may be infectious; protect skin, clothing from saliva; avoid injury to hands from patient's teeth
Pus from suppurating lymph node Skin test (not routine)	Smear and culture (rule out bacterial infection) Positive response	None	Tetracyclines relieve symptoms	Isolation not necessary; concurrent disinfection of pus from suppurating lesion or lymph node
Pus from wound, lymph nodes Blood	Smear and culture Culture	None	Penicillin Tetracyclines	Same as above
Same as above	Culture Animal inoculation	None	Penicillin Tetracyclines	Same as above
Pus or tissue from wound, if feasible	Smear and anaerobic culture	Toxoid for active immunization; antitoxin for passive immunization after injury has occurred	Antitoxin Penicillin	Isolation and special precautions not necessary

Table 27–2. *(Cont.)*

Clinical Disease	Causative Organisms	Other Possible Entry Routes	Incubation Period	Communicable Period
II. Diseases Acquired Through Infection of "Street Wounds" *(Cont.)*				
Gas gangrene	*Clostridium perfringens*		1–3 days	In the hospital surgical service, infected patient may be a rich source of organisms in operating room or on surgical ward
III. Burn Wound Infections				
Local infection, superficial or deep; sepsis	*Pseudomonas aeruginosa* Streptococci Staphylococci Gram-negative enteric bacilli Anaerobes		1 to several days	Infected patient may be a source of infection for others with open wounds

toxin may be used in treatment of active infection, but again is of questionable value.

II. Control of Active Infection

A. ISOLATION. Patients with gas gangrene should be placed on wound and skin precautions (Table 25–1, p. 560) to prevent transfer of pathogenic clostridia and other infecting organisms to other surgical cases. There is no danger of transmission except to persons with surgical or traumatic wounds which offer suitable conditions for clostridial multiplication.

B. PRECAUTIONS. Attendant surgical and nursing staff must take suitable precautions to prevent their own transmission of infection from an active case to other patients. Gowns and gloves should be worn by those who treat the wound or attend the patient closely, and hand washing is an urgent necessity. Individualized instruments and equipment should be used and sterilized by autoclaving. Boiling and chemical disinfection are of dubious value since these methods do not kill resistant clostridial spores easily. Concurrent and terminal disinfection pose a problem for this reason. The most satisfactory program is one that aims at destruction of the richest sources of the organism (dressings should be burned, bed linens autoclaved before being laundered, unprotected mattresses and pillows sterilized in an ethylene oxide chamber) and removal of environmental contamination by thorough disinfection-cleaning of the patient's unit. Good disinfection will kill vegetative bacilli, and careful cleaning removes spores mechanically. Beyond this, it is not practical to attempt removal of the last clostridial spores,

Specimens Required	Laboratory Diagnosis	Immunization	Treatment	Nursing Management
Exudates and tissue from wound	Smears and anaerobic culture	None	Penicillin	Wound and skin precautions; protect other patients with surgical wounds; gowns and gloves when changing dressings; autoclave all equipment; burn dressings; *hand washing*
Exudates and tissue from wounds; Blood	Smears and culture Culture	None	Carbenicillin or gentamicin for *Pseudomonas* infection; other antibiotics as indicated by laboratory studies of organisms	Protective isolation if burns extensive; "germ-free isolator" if available; *hand washing*

because they are constant contaminants of the environment in any case.

In the operating room the same principles apply. Washer-sterilizers have particular value in the care of instruments contaminated during surgical treatment of clostridial gangrene. All linens associated with the case, including personnel gowns, should be autoclaved before they are laundered; disposable items are incinerated; heat-sensitive equipment is subjected to gas sterilization and the rest is autoclaved. The room and all of its permanent equipment are then thoroughly scrubbed down with an effective cleaning agent.

When the epidemiology of a hospital case of gas gangrene suggests that its source arose from cross-infection or contamination, immediate steps must be taken to prevent additional cases. Clostridial organisms may display marked increases in virulence as a result of human passage. Operating-room and surgical service supplies should be resterilized, sterilizing methods should be reviewed, bacteriologic checks on the efficacy of these procedures should be performed, environmental sources of the organism should be sought in the operating room, and technical errors should be identified.

C. TREATMENT. The most urgently important treatment is surgical débridement of the involved area, with removal of necrotic tissue and aeration of the wound. Antibiotic therapy is instituted promptly, penicillin being the drug of choice.

Application of oxygen under pressure has proved useful in the management of clostridial myositis or cellulitis. The patient is placed in a room called a *hyperbaric chamber,* which contains oxygen at a concentration and pressure greatly in excess of that of the normal atmosphere, so that bloodstream and tissues are saturated for a time with oxygen. The patient is thus oxygenated for periods of one to three hours, every six to eight hours.

Other Bacterial Infections of "Street Wounds"

Soft tissues subjected to traumatic injury may be infected with any of a variety of bacteria derived from the environment or from the patient's own skin. An organism frequently associated with traumatic injuries, and also with burns, is *Pseudomonas aeruginosa,* a Gram-negative, aerobic bacillus that may be found in the normal intestinal tract as well as in the environment. This organism induces suppurative lesions in soft tissues and is characterized by its production of a blue-green pigment (*pyocyanin*) that imparts its color and sickly sweet odor to pus.

Traumatic wounds are most frequently infected with a mixed flora, containing both Gram-positive cocci and Gram-negative bacilli. Deep wounds often support anaerobic bacteria, such as species of *Bacteroides* or peptostreptococci, as well as clostridia. These organisms should always be sought by probing deeply into the wound for material to be submitted to the laboratory for anaerobic culture. Aerobic organisms recovered from the surfaces of such wounds may be mere contaminants that are not necessarily multiplying actively or injuring the tissue.

Melioidosis, a systemic infection that may result from bacterial contamination of penetrating wounds, attracted renewed interest in the United States because of its incidence in American military personnel wounded in Vietnam. The causative agent is a species of *Pseudomonas — P. pseudomallei —* that lives saprophytically in the soil and water of some tropical and subtropical areas. The organism closely resembles the causative agent of *glanders,* a zoonosis of equine animals. In humans who have inhaled the organism in contaminated dust or have suffered wounds infected by *P. pseudomallei,* the resulting systemic infection has often been fatal. Pneumonitis and multiple abscesses scattered throughout the body are seen at autopsy. Horses dying of glanders

display a similar syndrome. Recognition of this disease by early clinical and bacteriologic diagnosis is essential to successful treatment. Tetracyclines are preferred for treatment. They are given in combination with chloramphenicol in severe cases to prevent emergence of strains resistant to tetracyclines.

Burn Wound Infections

Burns represent one of the most difficult problems in medical and nursing management, because the nature of the injury predisposes exposed surface tissues to infection. In severe burns, infection is the chief cause of death in patients who have survived the first three or four days of massive injury, surface fluid loss, pulmonary edema, or shock. Devitalized surface tissue offers suitable conditions for growth of many opportunistic organisms, notably streptococci, staphylococci, Gram-negative enteric bacteria, and anaerobes, including the tetanus bacillus, other clostridia, or *Bacteroides.* The complications of tetanus or clostridial cellulitis are usually fatal when superimposed on the many physiologic problems created by burns.

The organism most frequently encountered in burn wound infections is *Pseudomonas aeruginosa. Pseudomonas* infections present very difficult problems because the organism is generally resistant to many clinically useful antibiotics, responding as a rule only to gentamicin and other aminoglycosides, carbenicillin, or drugs of the polymyxin group (see Table 10–6, p. 265).

When efforts to control infections of burns fail, septicemia is the usual consequence and the cause of death. The wound itself becomes overwhelmed with colonizing bacteria, and systemic invasion may induce fatal shock. In recent years many efforts have been made to develop methods that would prevent the complication of "burn wound sepsis." These have included the provision of closed hospital areas in which strict aseptic control can be maintained against airborne infection. The patient is kept in a room ventilated by bacteria-free filtered air delivered under positive pressure so that air currents move away from his wounds and also prevent the entry of air from contaminated areas when the door is opened. The room walls, floor, and furniture are treated with disinfectants having a residual action, and the patient's attendants practice aseptic precautions, putting on sterile clothing (gowns, masks, hair covering, gloves) before they enter the room. Access to the room is generally available only through an antechamber where clothes can be changed. Ultraviolet lamps at the entrance to the room assist in maintaining an aseptic room environment.

Individual presterilized, plastic enclosures ("germ-free isolators") are also practical in some situations for burn patients. Closed units and protective isolation are used for the management of all patients for whom risk of superimposed infection may be grave, including burn victims (Table 25-1).

Questions

1. What is the reservoir for rabies?
2. How is the rabies virus disseminated when introduced into the tissues?
3. When is active immunization indicated for rabies?
4. What organism is believed to cause cat scratch fever?
5. Why do dog and human bites often produce serious local or systemic infections?
6. Why do animal bites present the additional and more dangerous hazard of clostridial infection?
7. What two types of bacterial infection are associated with rat bites?
8. What nursing precautions are necessary in the care of a patient with gas gangrene? Why?

Additional Readings

Bahmanyar, Mahmoud; Fayaz, Ahmad; Nour-Salehi, Shokrollah; Mohammadi, Manouchehr; and Koprowski, Hilary: Successful Protection of Humans Exposed to Rabies Infection. *J.A.M.A.,* **236**:2751, Dec. 13, 1976.
Boyer, Catherine: Caring for a Young Addict with Tetanus. *Am. J. Nurs.,* **74**:265, Feb. 1974.
DeFoe, Daniel: *A Journal of the Plague Year.* Dutton, New York, 1972.
Hattwick, Michael A. W.; Marcuse, Edgar K.; Britt, Michael R.; Zehmer, Reynoldson B.; Currier, Russell W., II; and Elledge, William N.: Skunk Rabies: The Risk to Man, or Never Trust a Skunk. *Am. J. Public Health,* **63**:1080, Dec. 1973.
Nicholson, David: Tetanus—Still a Therapeutic Challenge. *Heart Lung,* **5**:226, Mar.–Apr. 1976.
Peters, Sue, and Vogel, N.: Physiological and Psychological Aspects of Tetanus: Report of a Case. *Heart Lung,* **5**:297, Mar.–Apr. 1976.
*Roueché, Berton: A Pinch of Dust. In *Eleven Blue Men.* Berkley Publishing Corp., New York, 1953.
Silva, J.: Anaerobic Infections. *Heart Lung,* **5**:406, May–June 1976.
Westlund, Drexel: Tetanus: A Case Study. *Can. Nurse,* **70**:17, July 1974.

*Available in paperback.

Infections Acquired Through Medical and Surgical Procedures

For many patients it could well be said that "hospitals are dangerous to your health." Reports from the Center for Disease Control of the U.S. Public Health Service indicate that about 1.5 million patients acquire infections in hospitals every year. Many of these nosocomial or iatrogenic infections are more difficult to treat than those acquired elsewhere because they are antibiotic resistant. They are epidemiologically dangerous because of the ease with which they can be transmitted, directly or indirectly, and the speed with which they can spread through wards, particularly those containing surgical patients, new mothers, and babies. One of the chief reasons for this is that hospitalized patients constitute a group whose susceptibility to infection is increased simply because they are ill and therefore physiologically less capable of resisting it. The problem is frequently compounded by institutional conditions of living, or by risks inherent in some procedures of medical, surgical, and nursing practice. Opportunities for exposure are multiple and may arise from a number of directions and sources.

Infectious Risks of Medical and Surgical Procedures

In this chapter we shall be concerned with some infections that can be introduced parenterally into patients undergoing certain types of procedures during the course of their medical or surgical treatment for many kinds of illnesses. Procedures involving greatest risks of superimposed infection fall into four major categories: (1) surgery, (2) instrumentation through mucosal orifices, (3) catheterization of blood vessels or urinary passages, and (4) insertion of hypodermic needles for injection or withdrawal of fluids. The nature and sources of infections that may be acquired through these routes are reviewed, and some diseases characteristically associated with hospital or other medical practice are discussed. The later include puerperal fever, urinary tract infections, and viral hepatitis.

Surgical Infections

One of the most obvious risks of surgery lies in exposure of tissues to exogenous sources of infection or in activation of endogenous microorganisms. Major sources of *exogenous* infection are the skin and respiratory secretions of the operating team and environmental contamination from objects in intimate contact with surgically exposed tissues (instruments, sponges, sutures, linen drapes, irrigating solutions). Contaminated air may also play a role in surgical infections. *Endogenous* infections may arise when microorganisms are disseminated into adjacent tissues or the blood from their usual sites of commensalism on the patient's skin or mucous membranes, such as the lumen of the intestinal tract, genitourinary membranes, oral mucosa, and other surface areas having normal microbial flora.

Surgical patients are often subjected to a variety of procedural risks in addition to those inherent in surgery. They may require intubation (insertion of an airway into the larynx) while under anesthesia; administration of intravenous fluids; urinary catheterization; or other techniques of management that can open the door to infection. Inhalation anesthesia also leaves them more vulnerable to respiratory infection.

The degree of trauma suffered by tissues exposed to surgery, or to the ancillary procedures mentioned above, is an important determining factor in development of wound infections. Excessive or rough handling can be a threat to architectural integrity of tissues. It may interfere with vascular structures, leaving areas of devitalization. Continued bleeding from severed vessels may lead to the formation of hematomas in which bacteria can grow without opposition, as they may also do in devitalized, necrotizing tissues. Speed, skill, manual dexterity, and delicacy are essential factors in control and prevention of surgical wound infections.

The nature and source of infecting microorganisms vary somewhat with the type of surgery performed and the region of the body involved. Table 28–1 shows some infectious complications that may result from surgery and indicates their possible origins. Sources of surgical infection are often difficult to prove conclusively, but note that the microorganisms involved are those commonly associated with human surface tissues. Usually the best evidence that surgical personnel, techniques, or supplies must be incriminated lies in demonstration of a continuing incidence of wound infections among all surgical patients or those of a given service, especially if the same etiologic agents are involved in most or all of these infections.

Infectious Risks of Instrumentation Through Mucosal Orifices

Lensed instruments are frequently used for examination of mucosal passageways and deep surfaces (endoscopy). These instruments consist, essentially, of a metal tube equipped with magnifying lenses, a light source at the inserted end, and a series of prisms that direct light to the observer's eye, permitting visualization of the depths of the passageway. They are constructed for specific use on various anatomic structures and are named accordingly, each term having the suffix "-scope" to indicate a viewing instrument. Some examples are listed below:

Instrument	Examination of:
Nasopharyngoscope	nasopharynx
Laryngoscope	larynx
Bronchoscope	trachea, bronchial tree
Gastroscope	esophagus, stomach
Sigmoidoscope	sigmoid, rectum
Cystoscope	urethra, bladder
Endoscope	a collective term

These instruments present risk of infection for two reasons: (1) they are difficult to sterilize because of their construction and the fact that their lenses, lighting system, and prisms are mounted in cementing substances that may be dissolved by moist heat or chemical solvents; and (2) they must be large enough in diameter to permit an adequate view of the passageway, which means that they can be disruptive to delicate mucosal linings. If not properly disinfected, endoscopes may introduce extraneous microorganisms, including those derived from patients previously examined, and if not carefully inserted, they may irritate or break the mucosal bar-

Table 28–1. Infectious Complications That May Result from Surgery

Surgical Area	Types of Infection and Their Possible Origins		Some Sources of Exogenous Infections
	Endogenous	**Exogenous**	
Mouth, nose, or throat	Infections of local soft tissues Regional lymphadenopathies Bacteremia and systemic infection		
	Organisms from upper respiratory tract, or buccal mucosa: Streptococci (may induce subacute bacterial endocarditis) Staphylococci *Actinomyces israelii* (cervicofacial or thoracic actinomycosis)	Streptococci (especially beta-hemolytic strains) Staphylococci (especially *S. aureus*)	Infected personnel Unsterile instruments, wound packing, sponges, irrigating solutions
	Poliomyelitis virus (patient may have inapparent nasophyaryngeal colonization, especially in seasons of high community incidence)	Poliomyelitis virus	Previous infected contacts or infected surgical personnel
Chest (including cardiac surgery)	Wound infections Pulmonary infection Systemic infection		
	Organisms from upper respiratory tract or skin: Streptococci Staphylococci (coagulase-negative as well as -positive strains sometimes involved in endocarditis following cardiac surgery) *Pseudomonas aeruginosa* Gram-negative enteric bacilli: *E. coli* *Klebsiella* species *Proteus* species	As in the endogenous column on left	Infected personnel Hands in punctured or torn gloves Contaminated gowns Contaminated instruments, sutures, solutions Contaminated fluids or blood for transfusion Contamination in heart-lung machine

Table 28–1. *(Cont.)*

	Types of Infection and Their Possible Origins		Some Sources of Exogenous Infections
Surgical Area	**Endogenous**	**Exogenous**	
Chest *(cont.)*	Anaerobic organisms: *Actinomyces israelii* *Bacteroides* species *Fusobacterium* species Peptostreptococci	As in the endogenous column on left, except actinomycotic disease is not derived from exogenous sources	
Intestinal tract	Wound infections Infection of bowel wall Peritonitis Systemic infections (including endocarditis)		
	Organisms from intestinal tract or skin: Streptococci Staphylococci (usually *S. aureus*) Gram-negative enteric bacilli: *E. coli* *Klebsiella* species *Proteus* species *P. aeruginosa* *Salmonella* species (may be spread from inapparent bowel infection, inducing *Salmonella* septicemia) Anaerobic organisms: *Clostridium tetani* *Clostridium perfringens* and others of gas gangrene group *Bacteroides* species Peptostreptococci	As in the endogenous column on left, except salmonellosis is rarely derived from exogenous sources during surgery	Infected personnel Hands in punctured or torn gloves Contaminated gowns Contaminated instruments, sutures, solutions Contaminated blood or fluids given intravenously
Orthopedic surgery	Wound infections Osteomyelitis Systemic infections		
	Organisms from skin or intestinal tract: Streptococci (usually beta-hemolytic)	As in the endogenous column on the left	Infected personnel Hands in punctured or torn gloves Contaminated gowns

Table 28–1. (*Cont.*)

Surgical Area	Types of Infection and Their Possible Origins		Some Sources of Exogenous Infections
	Endogenous	Exogenous	

	Staphylococci (usually *S. aureus*)
	Anaerobic organisms:
	  Clostridium tetani
	  Clostridium perfringens and others of gas gangrene group
	  Peptostreptococci
	  Bacteroides species
	Gram-negative enteric bacilli
	P. aeruginosa
			Contaminated instruments, sutures, solutions
			Contaminated blood or fluids given intravenously
			Contaminated prosthetic devices implanted in tissues
			Plaster dust

Ophthalmic surgery	Conjunctivitis Bulbar infections		
	Organisms from conjunctiva or skin: Pneumococci Streptococci Staphylococci *Haemophilus* species Herpesvirus	The bacterial species listed in the endogenous column on the left plus: Fungi (various species, e.g., *Aspergillus, Mucor*) The chlamydial agent of inclusion conjunctivitis	Infected personnel Contaminated hands or gowns Contaminated instruments Contaminated eyedrops or anesthetizing solutions
Plastic surgery	Wound infections Infectious disruption of grafts		
	Organisms from skin or mucosa: Streptococci Staphylococci Gram-negative enteric bacilli *P. aeruginosa*	As in the endogenous column on the left	Infected personnel Hands in punctured or torn gloves Contaminated gowns Contaminated instruments, sutures, solutions
Gynecologic and obstetric surgery	Puerperal fever Urinary tract infections (cystitis, pyelonephritis)		
	Organisms from genital or intestinal tract mucosa: Streptococci Staphylococci Gram-negative enteric bacilli	As in the endogenous column on the left Streptococci and staphylococci most commonly associated with puerperal sepsis; beta-hemolytic	Infected personnel Hands in punctured or torn gloves Contaminated gowns Contaminated instruments, solutions Contaminated catheters

Table 28–1. *(Cont.)*

Surgical Area	Types of Infection and Their Possible Origins		Some Sources of Exogenous Infections
	Endogenous	Exogenous	
	Anaerobic organisms: *Clostridium perfringens* and others of the gas gangrene group *Bacteroides* species *Fusobacterium* species Peptococci Peptostreptococci	streptococci or *S. aureus* strains usually derived from obstetric personnel	

rier, permitting entry of local or exogenous organisms to deeper tissues.

Catheterization of Blood Vessels or Urinary Passages

Insertion of catheters into blood vessels or into the long narrow passages of the urinary tract (ureters or the urethra) is frequently done for diagnostic or therapeutic purposes. The catheters themselves may be a source of infection if not properly sterilized, and the technique of insertion requires strict aseptic precautions.

Vascular catheters may introduce microorganisms directly into the blood. Organisms of low pathogenicity may be removed by phagocytes and antibacterial components of blood, but it is possible that some may localize and multiply, particularly where the vascular or endocardial endothelium is defective or injured, or where a thrombus or embolus exists. These clinical problems usually provide the necessity for catheterization and at the same time increase the risks inherent in the introduction of even "avirulent" organisms. These may be able to establish themselves with little opposition in a blood clot, or on a defective heart valve, producing local injury and recurrent bacteremia. Coagulase-negative strains of staphylococci (*S. epidermidis*) frequently are respon-

sible for these infections which complicate the patient's postcatheterization course. Enterococci and alpha-hemolytic streptococci are also offenders, while *S. aureus* (coagulase-positive) may induce a fulminating septicemia as well as local injury at sites of colonization. Sources of these infections may be the skin or respiratory membranes either of the patient or of a member of the operating team. Environmental contamination of catheters or instruments is also possible because these procedures frequently are carried out in x-ray rooms or other areas where radiologic visualization of catheter placement can be obtained, but where strict asepsis is difficult to maintain.

Urinary tract catheterizations may be responsible for cystitis, pyelonephritis, or more disseminated systemic infections. Procedures most frequently associated with complicating urinary infections are urethral or ureteral catheterization, use of indwelling catheters for continuous bladder drainage, cystoscopy, and prostate or bladder surgery. These infections may have an endogenous source in the microbial flora of the patient's genital, perineal, or perianal mucosa, or may be introduced from exogenous sources. The technique of catheterization should be conducted with strict asepsis and with great care to avoid injuring delicate mucosal linings with the catheter itself (see p. 623).

Hypodermic Needles, Syringes, and Injectable Solutions

Among the many tools of medical practice that may be responsible for parenteral introduction of infection, the hypodermic needle is most frequently employed and sometimes the most carelessly used. It can be directly or indirectly responsible for infection for any of the following reasons:

1. The needle itself is contaminated.

2. The syringe or other vessel containing fluid to be injected is contaminated.

3. The patient's skin is inadequately disinfected.

4. The patient's skin is contaminated with micro-organisms growing in *inactive* disinfectants used to prepare the injection site.

5. The injected fluid contains contaminating or multiplying organisms.

Sources of Infection by the Hypodermic Route. The sources of contamination for needles, syringes, or injectable solutions are quite varied (Fig. 28–1). The most obvious of these and their serious implications are discussed below:

1. Reused Instruments. Pathogenic microorganisms may be derived from the blood, tissues, or skin of infectious patients for whom the needle or syringe was previously used. This implies that reused instruments were inadequately cleaned and sterilized. The infectious agents most frequently involved in this kind of mishap are the viruses of hepatitis A and hepatitis B. These viruses are not only quite resistant to most useful agents of disinfection, but they are pathogenic even if introduced in small numbers directly into blood or tissues of susceptible, non-immune persons. Proper heat sterilization of reusable needles or syringes, or the use (and discard) of disposable presterilized equipment, affords the only guarantee in preventing the spread of viral hepatitis by this route.

2. Syringe and Needle Handling. Faulty techniques may result in contamination of needles or syringes *subsequent* to sterilization. Microorganisms from the operator's hands or respiratory secretions may reach the shaft of the needle, the inner surfaces

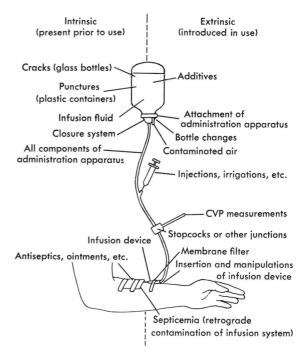

Fig. 28–1. Potential mechanisms for contamination of intravenous infusion systems. (Reproduced from Maki, Dennis G.; Goldman, D. A.; and Rhame, F. S.: Infection Control in Intravenous Therapy. *Nurs. Dig.*, **3**:5, May–June 1975. The illustration was adapted from a diagram originally published by the same authors in *Ann. Intern. Med.*, **79**:867, Dec. 1973.)

of a disassembled syringe, or its tip if these items are not handled correctly. The shaft of a sterile needle must be continuously protected from contamination from the air, bench top, or other surface where it is laid pending final preparation of the patient's skin. Disposable needles packaged in plastic protectors should be left in place until the moment comes to charge the syringe with the material to be injected or to introduce the needle through the patient's skin. If a loaded syringe and needle must be momentarily set aside before an injection is finally given, the plastic needle protector should be replaced for the interim period.

3. Skin Disinfection. Inadequate disinfection of the patient's skin prior to introduction of hypodermic needles may also lead to infection. Organisms usually introduced from this source include streptococci, staphylococci, enteric bacteria, and occasionally spores of anaerobic or aerobic bacilli. The degree of risk depends on the number of organisms injected, their virulent properties, or the toxicity of any preformed products of their growth in exogenous sources. Small numbers of avirulent organisms from the skin may be easily removed by phagocytes, but the properties of virulent organisms such as hemolytic streptococci or coagulase-positive staphylococci provide them with greater resistance to the body's defenses. Such bacteria can be dangerous even in small numbers if introduced directly into blood tissues.

The route of introduction of organisms from the skin or other sources also influences the outcome. Microorganisms injected into intradermal or subcutaneous tissues meet a great deal of opposition from local phagocytes and from blood components supplied to well-vascularized tissues. In intramuscular sites, contaminating organisms may have more opportunity to establish themselves because muscle tissue contains fewer fixed phagocytes, but the blood supply soon brings polymorphonuclear cells and plasma components for opsonization. Intravenous injections, or venipuncture for collection of blood samples, affords the greatest risk of introduction and wide dissemination of contaminating microorganisms.

4. Inactivated Antiseptics. Another dangerous possibility of bacterial infection arises when the hypodermic technique introduces materials in which microbial multiplication has occurred prior to injection. One of the ways in which this can happen involves the use of inactive skin antiseptics in which bacteria have been growing. When the patient's skin is prepared for injection by applying the "disinfectant," large numbers of contaminating organisms may be wiped on to the skin surface and later pushed through to the tissues below, or into a vein, by the needle as it passes through this teeming culture.

Antiseptics and other chemical disinfectants may be inactivated in a number of ways. Some are readily bound by protein or other substances that combine with the active portions of their molecules (see Chap. 10) and become useless after such exposure. Fresh working solutions should be prepared frequently and kept in clean, closed containers, protected from chemical or excessive microbial contamination. The practice of setting up boats with cotton or gauze pads soaked in antiseptic solution, ready to use, is not acceptable because cellulose fibers may bind with the antiseptic and inactivate it. This affinity for cellulose and protein is one of the deficiencies of the quaternary ammonium compounds. Bacteria are continuously introduced into such boats from each hand reaching for a new pad, and in a short time a rich culture of organisms is formed in a solution that can no longer suppress them.

5. Contaminated Injectable fluids. Fluids injected by hypodermic methods may themselves be the most dangerous source of infection, particularly if microorganisms have had a preliminary opportunity to grow in them and produce toxic substances. The injected patient is then not only at risk of generalized systemic infection, but may suffer a severe febrile reaction to injected pyrogenic microbial byproducts. If he is also hypersensitive to bacterial antigens suddenly introduced into his tissues, he may experience an immediate reaction ranging in severity from anaphylaxis and shock to the "serum sickness" type.

Any type of injectable solution may be involved in this type of accident, sometimes with disastrous results for the patient. Fluids given intravenously; vitamins; vaccines; antisera; anesthetizing solutions used to induce local skin or mucosal anesthesia in dental, ophthalmic, or other surgery or for such procedures as lumbar puncture, biopsies, or bone marrow aspiration — all these and many others may be the source of contaminating organisms capable of establishing infection. Continuing vigilance is essential to prevent these problems. Careful visual inspection of each ampule, vial, or bottle immediately prior to its use is helpful only if the material to be injected is in clear solution and is not hidden by a label or by markings pressed into the glass or plastic container. Cracked or otherwise damaged vessels containing injectable material should never be used, under any circumstances.

The use of vials containing multiple doses of fluids for injection should be discouraged, because each entry into such a container affords new opportunity for contamination. These vials are generally sealed with a cap containing a rubber diaphragm through which the hypodermic needle is inserted for withdrawal of a dose of material to be administered to a patient. The hole made by the needle as it passes through the rubber may be enlarged every time the vial is used, threatening contamination of the vial's contents by this route. The needle itself may introduce a few organisms. If contaminating bacteria then have an opportunity to multiply in the vial before it is used again, the next patient receiving a dose of its contents may suffer the consequences of superimposed infection. Vials or bottles containing cloudy material that originally was clear should be discarded immediately. Those containing fluids that are not clear to begin with (e.g., microbial vaccines) present a more difficult problem. The manufacturer's label should be carefully read for information concerning the presence of an antibacterial preservative, date of preparation and sterilization, expiration date, and any instructions concerning the handling, administration, and storage temperature of the material. Whenever possible, injectable solutions should be prepared or purchased in single-dose vials.

Blood for transfusion requires special emphasis in this connection. Not only is it subject to contamination from extraneous sources, but it may transfer active infection from donor to recipient. Under well-controlled circumstances, most problems of this variety are avoided by the legal requirements for operation of blood banks, which stipulate that blood donors must be examined for evidence of current active infection and for history of acute or chronic disease transmissible by blood. Unfortunately, however, a few diseases of this type exist in latent, asymptomatic form. Would-be donors may be unaware of their infection, or may give an inaccurate history of past disease, unintentionally or willfully. Diseases most readily transmissible from "silent" donors are syphilis, malaria, and viral hepatitis. In the case of syphilis, routine physical and serologic examination is generally successful in screening infectious donors. Malaria and hepatitis A are not readily diagnosed in their inapparent stages, but a well-taken history can provide evidence of clinically

characteristic disease experienced in the past. Hepatitis B is the most difficult problem, because a number of people carry the viral agent of this disease without showing evidence of clinical illness. Serologic techniques for detecting such individuals have become more reliable in recent years and have greatly alleviated the risk of transfusion hepatitis (see p. 624).

Some Important Infectious Diseases Associated with Medical and Surgical Procedures

Puerperal Fever

The Clinical Disease. Puerperal fever is an acute bacterial disease incurred during childbirth or abortion. It begins with local symptoms of bacterial invasion of the genital tract, including the uterus and the fallopian tubes. If not treated promptly or adequately, it may extend as a peritonitis, usually with bacteremia or septicemia.

Laboratory Diagnosis

I. Identification of the Organisms. Puerperal fever may be caused by a number of different bacterial agents. Hemolytic streptococci of groups A or B are a frequent cause of postpartum infections, but a variety of organisms have been recognized in postabortion sepsis. Anaerobic bacteria, such as *Bacteroides* species, *Clostridium perfringens,* and peptostreptococci, may be involved. Gram-negative enteric bacilli, *Pseudomonas* species, and *Staphylococcus aureus* may also be incriminated. Mixed infections are frequent in septic abortions.

The microscopic and cultural characteristics of these organisms have been described in previous chapters.

II. Identification of Patient's Serum Antibodies. In puerperal fever caused by hemolytic streptococci of group A the serum antistreptolysin titer rises between onset of acute symptoms and the convalescent stage. Serologic diagnosis has no value when other types of infection induce this syndrome.

III. Specimens for Diagnosis. Vaginal discharges and pus from the cervix or uterus are appropriate for smear and culture (aerobic and anaerobic).

Repeated blood cultures may be useful if disease advances to a bacteremic or septicemic stage.

When streptococcal infection is suspected, two

samples of blood should be submitted for antistreptolysin titers. The first should be collected as soon as possible after the onset of symptoms, the second in about ten days to two weeks.

Epidemiology

I. Communicability of Infection

A. RESERVOIR, SOURCES, AND TRANSMISSION ROUTES. The human reservoir supplies the organisms encountered in puerperal sepsis. They are found on the skin and mucosa of normal persons, as well as in the lesions of infected patients or personnel. The chief source of group A *hemolytic streptococcal* postpartum infections is rarely the patient herself, however, for these strains are not normal to the female genital tract. They are transmitted to her from the nose, throat, or hands of obstetricians, midwives, nurses, or other attendants, or from instruments contaminated by these sources.

Before the mid-1860s, when Joseph Lister first applied Pasteur's concepts of infectious disease to surgery, puerperal fever was a frequent complication of childbirth, and a dreaded one because of its high mortality rate. It was seen most often in maternity wards of hospitals, but mothers who delivered at home were not safe either, because their physicians or midwives brought the infection to them from other cases they had attended. Wiping their contaminated hands on lapels or aprons, these attendants would set to work, full of confidence, to deliver a new baby and to institute a new infection.

Fully two decades before Lister's aseptic approach to surgery began to impress his surgical colleagues with its success in preventing infection, the first important epidemiologic study of puerperal fever had been reported by an American physician, Oliver Wendell Holmes. In 1843 he presented a paper on "The Contagiousness of Puerperal Fever" to the Boston Society for Medical Improvement. This report created a furor among leading obstetricians of the day because Holmes dared to present the thesis that puerperal fever was due largely to "the carelessness, ignorance, and negligence of the obstetrician and midwife." * A similar conclusion was being

reached independently at about the same time by an Austrian physician whose name was Ignaz Semmelweis. Holmes and Semmelweis did not know each other but they had much in common. In their own obstetric practices they instituted rules of cleanliness, which included hand washing as well as other careful precautions regarding changes of clothing between infectious cases or after attending autopsies of patients who had died of puerperal sepsis. They managed to keep their own patients relatively free of infection, but they were both subjected to a great deal of ridicule and abuse from medical colleagues. Semmelweis suffered more from this than Holmes, and in the end was so disturbed (by his failure to persuade other physicians to accept methods he had used effectively to save women in childbirth) that his sanity was affected, and he died in a mental hospital (of sepsis from a wound on his hand!). Holmes's reaction were different. He said: "Whatever indulgence may be granted to those who have heretofore been the ignorant causes of so much misery, the time has come when the existence of a *private pestilence* in the sphere of a single physician should be looked upon not as a misfortune but a crime; and in the knowledge of such occurrences, the duties of the practitioner to his profession, should give way to his paramount obligations to society." * These words remain even more pointed today when the causes of infectious disease are so well understood, yet sometimes ignored or underestimated.

When abortion is induced by the individual or performed by others working under circumstances that do not permit asepsis (ignorance plays its part again, together with legal and social pressures), the probabilities of infection are even greater than in normal childbirth, and a wider range of sources and species of microorganisms may be involved. In many instances, these infections probably arise from endogenous sources, species such as anaerobic streptococci, *Bacteroides, Clostridium,* or coliform bacilli being derived from the intestinal tract or group B hemolytic streptococci from the regional surface mucosa. Today, this type of problem is much more

---

*Smillie, Wilson G.: *Public Health, Its Promise For the Future.* Macmillan Publishing Co., Inc., New York, 1955, p. 156.

*Holmes, Oliver Wendell: The Contagiousness of Puerperal Fever, *Medical Classics,* **1**:243, Nov., 1936 (from *New England Quarterly Journal, Medicine & Surgery,* **1**, 1843, published in Boston by D. Clapp, Jr.).

frequent than postpartum sepsis, primarily because of the difference in the aseptic control of normal childbirth (or of medically performed abortion) as compared with that applied to abortions performed under the pressures of expediency.

B. INCUBATION PERIOD. Usually between one and three days.

C. COMMUNICABLE PERIOD. While purulent discharges continue. When treatment is prompt and adequate, control of communicability can be achieved within a day or two, but untreated patients may remain infectious for days or weeks.

D. IMMUNITY. Antibacterial immunity may develop against groups A and B *Streptococcus* and their extracellular products if one of these organisms is the cause of the infectious disease, but susceptibility to other kinds of infection remains general.

II. Control of Active Infection

A. ISOLATION. Strict isolation of hospitalized patients with puerperal sepsis is not necessary, but the maternity service itself must be rigidly protected from transmissible infections. New patients who have developed sepsis following abortion or delivery outside the hospital are not admitted to the maternity service but are placed in partial isolation elsewhere in the hospital. Women who develop puerperal fever in the hospital are transferred immediately from clean postpartum areas to another unit where they can be cared for using wound and skin precautions. Appropriate and adequate antibiotic therapy shortens the period of communicability and relieves the necessity for strict isolation. Bacteriologic evidence that vaginal discharges are no longer infectious determines the time for which precautions must be maintained.

B. PRECAUTIONS. Careful concurrent disinfection is required for vaginal discharges and all equipment soiled by them (bedpans, items required for perineal care, specula). Infectious patients should not be permitted to share bathroom facilities with others. Soiled clothing and bed linens are placed in individual bags, marked for special handling in their transport to the laundry. When spore-bearing bacteria (e.g., *Clostridium* species) have been demonstrated as causative agents of puerperal sepsis, all equipment (including linens) must be bagged and autoclaved before cleaning. Disposable items are placed in closed containers and incinerated; heat-sensitive items may be soaked in sporicidal disinfectants or subjected to gas sterilization. Regardless of the nature of the infection, a clean gown must be worn by each person who attends or examines the patient, and careful hand-washing technique must be observed. Terminal disinfection-cleaning of the patient's unit is necessary.

C. TREATMENT. Penicillin is the drug of choice in cases of hemolytic streptococcal infection. Erythromycin, choramphenicol, or others may be effective for patients who are hypersensitive to penicillin. Therapy of other bacterial infections is adjusted on the basis of the antibiotic susceptibility of isolated organisms suspected to be responsible for the patient's disease.

D. CONTROL OF CONTACTS. Obstetricians, midwives, nurses, and other persons who attend abortions or deliveries should be examined if they have been associated with cases developing puerperal sepsis. Cultures of the nose, throat, skin, or active lesions should be taken to establish the possible source of hemolytic streptococci. Individuals harboring group A streptococci, and those with other acute or chronic infections, should be removed from obstetric duties and treated specifically. Delivery techniques should be carefully reviewed and environmental sources of infection sought, with particular attention to the sterility of instruments or other equipment. These investigations are of particularly urgent concern when an unusual incidence of cases occurs within an individual hospital service or practice.

III. Prevention.
The prevention of puerperal fever depends entirely on maintenance of strict asepsis in obstetric techniques and procedures, and on protection of the patient from possible sources of infection in the nose, throat, hands, or infectious lesions of those who attend or visit her, or of other patients who have communicable diseases.

Urinary Tract Infections

Clinical Diseases. Urinary tract infections frequently occur as a result of urologic procedures carried out in medical offices or hospitals. The procedures most commonly involved include operations on the bladder, transurethral resection of the prostate, cystoscopy, urethral or ureteral catheterizations, and the use of indwelling catheters for continuous bladder drainage. These infections may involve colonization and inflammation of the urethra (*urethritis*), bladder (*cystitis*), ureters (usually an ascending *ureteritis*), pelvis of the kidney, or renal parenchyma (*pyelonephritis*). Infections of the lower urinary tract may be mild, transient, and insignificant, but they may lead to acute or chronic kidney disease, sometimes with permanent damage to renal function. In some instances infectious disease may arise from endogenous sources, but even then it may be superimposed in the sense that procedural damage inflicted on the epithelial lining of the tract provides an opportunity for entry and establishment of commensal organisms from genitourinary mucosal surfaces. In many cases, infection is introduced by faulty technique, by inadequately sterilized or disinfected equipment, or by unwashed hands.

Urinary tract infections are not associated exclusively with such procedures. Kidney infection, for example, may be a complication of systemic dissemination of microorganisms, especially staphylococci, from another entry route. Other types of disease or anatomic problems may also predispose to infection of the urinary system. Stones or tumors may be important centers of bacterial localization. The frequency with which urinary tract infection occurs in females may have several explanations. The shortness of the female urethra may be a factor, and with young girls the problem may be related to their tendency to retain urine until the bladder becomes quite distended. In adults, tumors sometimes cause urinary retention (fibroid tumors of the uterus, in women; prostatic tumors, in men; bladder tumors), and resulting distention strongly predisposes to cystitis. Pregnant women may have a similar problem as the developing fetus places increasing pressure on surrounding pelvic structures. During labor, urinary retention is common, and when catheterization is necessary another risk of infection is added to pre-existing possibilities.

Laboratory Diagnosis

I. Identification of the Organisms. Common agents of urinary tract infections include many species of com-

mensalistic Gram-negative enteric bacilli (*Escherichia coli, Klebsiella, Proteus*); *Pseudomonas aeruginosa; Staphylococcus aureus,* and sometimes coagulase-negative staphylococci; and enterococci. *Candida albicans* is sometimes involved, particularly in infections of older or debilitated persons.

A number of pathogenic organisms associated with systemic disease may cause primary or secondary lesions in the kidney, and these organisms may also be identified in the urine. They include the following:

Mycobacterium tuberculosis (see Chap. 13)
Salmonella species (including *S. typhi*) (see Chap. 18)
Leptospira interrogans (see Chap. 24)

In addition, the blood fluke *Schistosoma haematobium* characteristically lodges in pelvic veins, producing ova that migrate through the bladder wall and are excreted in urine.

II. Identification of Patient's Serum Antibodies. Serologic diagnosis generally has no value in infections of the lower urinary tract.

III. Specimens for Diagnosis. Urine specimens for microscopic examination and quantitative culture. (See IV below.)

The practice of collecting catheterized urine specimens for bacteriologic studies has been largely abandoned because of the risk of superimposed infection inherent in the technique. This procedure should be reserved for those cases in which a conclusive demonstration of infection is difficult to obtain with voided specimens.

For patients with indwelling catheters, a recommended method for obtaining a urine specimen for laboratory examination is illustrated in Figure 28–2.

For routine purposes, clean, midstream urine collections are satisfactory. When quantitative culture results are desired, urine should be delivered to the laboratory within 30 minutes of collection or refrigerated until transport can be arranged. If specimens are permitted to remain at room temperature for periods longer than one-half hour, contaminating organisms from external genitourinary mucosa may multiply freely, providing quantitative results that are open to misinterpretation.

Satisfactory specimens can be collected after careful cleaning of the meatus and external mucosa, using gauze sponges moistened with tap water and liquid soap. The first and last portions of urine voided should be discarded, only the midstream portion being collected in a *sterile* container possessing a cover. Male collections offer little difficulty, but when urine samples are collected from women, care

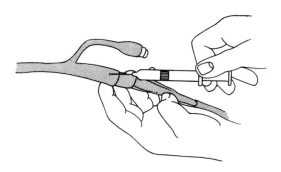

Fig. 28–2. Method of obtaining urine from the catheter by syringe. Note that the needle is inserted at the point of junction with the collecting tube. (Reproduced from Kunin, Calvin M.: *Detection, Prevention and Management of Urinary Tract Infections*, 2nd ed. Lea & Febiger, Philadelphia, 1974.)

must be taken not to permit contamination of the open lip of the vessel by surrounding unsterile surfaces. The labia should first be carefully cleansed, the first passage of urine is discarded, and then the collecting vessel is held close to, but not touching, the meatus, the patient being positioned so that a vertical stream can be delivered directly into the container.

IV. Special Laboratory Tests. Qualitative techniques for examination or culture of urine are adequate when pathogenic species such as the tubercle bacillus or *Salmonella* are suspected, because these organisms are of diagnostic significance regardless of their numbers in a given specimen.

Quantitative culture techniques are used to establish the significance of staphylococci, streptococci, or Gram-negative enteric bacilli in urine (see Chap. 4).

Antibiotic susceptibility testing is of value only when performed on pure cultures of isolated organisms present in significant numbers in clean urine specimens.

Fluorescent antibody studies to determine if the bacteria present in urine are coated with antibody may give evidence of kidney infection (antibody present) versus bladder infection (antibody absent).

Epidemiology

I. Communicability of Infection

A. RESERVOIR, SOURCES, AND TRANSMISSION ROUTES. See the initial discussion.

B. INCUBATION PERIOD. Generally short (one to five days).

C. COMMUNICABLE PERIOD. Continues for as long as urine contains infectious microorganisms in significant numbers.

D. IMMUNITY. Susceptibility is general, and immunity does not usually develop.

II. Control of Active Infection

A. ISOLATION. It is usually not necessary to isolate patients with urinary tract infections. Patients being treated at home should be carefully instructed as to the necessity for personal hygiene in shared bathroom facilities. Hospitalized patients should not be permitted to share bathrooms with other patients until antibiotic therapy has significantly reduced the numbers of organisms shed in urine. This is particularly important in renal tuberculosis or salmonellosis.

B. PRECAUTIONS. Concurrent disinfection or sterilization procedures should be carried out for all instruments or equipment used for urologic examination or treatment. This includes bedpans, urometers, catheters, cystoscopes, and items used for perineal care. Gowns should be worn by personnel performing catheterizations or other urologic procedures, and careful hand washing should be practiced before and after these techniques. Soiled bed linens and clothing should be placed in individual bags marked for special handling in their transport to the laundry. Terminal cleaning of the patient's unit may be routine, but bathrooms should receive special attention and be cleaned with disinfectant cleaners.

C. TREATMENT. Choice of antibiotic therapy depends on results of laboratory studies identifying significant organisms and testing their susceptibility to antimicrobial drugs. Supportive measures often include use of antiseptic solutions as bladder irrigants and drugs that maintain the pH of the urine at alkaline levels that are suppressive for most bacteria.

D. CONTROL OF CONTACTS. The source of possible contamination of the urinary tract should be sought, particularly when the incidence of cases suggests extraneous reservoirs persisting within a urologic service or practice. Techniques should be carefully reviewed with all personnel and corrective procedures instituted as indicated. Sterilization and disinfection methods should be tested for efficacy by bacteriologic techniques.

III. Prevention. Prevention of urinary tract infections associated with urologic procedures is dependent on adequate sterilization or disinfection of equipment, especially catheters and cystoscopes, and on careful maintenance of aseptic technique.

A. STERILIZATION OR DISINFECTION OF EQUIPMENT

1. *Cystoscopes.* These instruments cannot be sterilized by the use of heat in autoclaves or ovens, because their delicate lens and prism systems are quickly damaged beyond repair by such methods. They can be sterilized by ethylene oxide gas, and this is the method of choice when gas sterilizing equipment is available. The method is expensive and time-consuming, requiring an adequate supply of instruments to outlast the four-hour period of sterilization and a matching period for dissipation of the toxic gas from instrument packs (see Chap. 10). Gas sterilization is not destructive to cystoscopes, and this advantage must be weighed, together with safety factors, against the expense and inconvenience of the method.

Disinfection of cystoscopes is possible, but requires meticulous attention to detail, including preliminary cleaning and proper application of effective germicides. The latter should always be chosen with the possibility in mind that tuberculous infection may be present though unsuspected.

2. *Catheters.* Most urethral and ureteral catheters can be successfully sterilized in the autoclave (20 minutes at 121° C, or 250° F). This is true of soft rubber, latex, and woven-base catheters. Hard rubber, "shellac," and "web" catheters may be chemically disinfected, but again the choice of routine should take into consideration the possibility of unsuspected tuberculous infection.

The problem of sterilization and disinfection of catheters can best be solved by purchase and use of presterilized disposable polyethylene catheterizing equipment.

B. ASEPTIC TECHNIQUE IN UROLOGIC NURSING PRACTICE. *The technique of catheterization* is frequently the responsibility of nurses or of orderlies working under nursing supervision. They must apply principles of asepsis in this technique if the chain of procedures involved in sterilization or disinfection of equipment is to have final value for the patient. Precautions taken should be of the same order as those required for a surgical dressing. Before initiating the procedure, the nurse obtains sterile equipment, then positions the patient with proper illumination. After washing her hands carefully and donning sterile gloves, she cleans the urinary meatus and surrounding mucosal surfaces with a suitable antiseptic, then proceeds to insert the catheter with maximum care to avoid contaminating it by contact with the patient's skin or nearby objects. A skillful touch must be developed for this procedure, so that the risk of damage to delicate mucosal surfaces is minimized. When the procedure is completed, used materials are discarded into a container that can be closed for transport to an incinerator or an autoclave, and the nurse carefully washes her hands once more.

Indwelling catheters are often an infectious hazard because bacterial infection may spread in retrograde fashion from contaminated drainage bottles, backward along the course of the tubing. This hazard can be reduced by using a closed system that does not permit air contamination of the contents of the drainage container, or contact between the catheter and the vessel's contents. Presterilized plastic bags and catheter tubing are available as a solution to the problem, but must be given frequent attention to assure that the tubing remains in proper position within the patient's bladder and also at the upper end of the collecting bag. The catheter tubing should never be allowed to come loose from the collecting vessel and hang free at its distal end where it can be contaminated from the air or by any surface it touches in its random swinging. Also, the bag should not be permitted to fill to the point of contact with the delivery end of the tubing, so that retrograde contamination of the bladder becomes possible for this reason. A safe method for collection of urine specimens from an indwelling catheter is shown and described in Figure 28–2.

The maintenance of asepsis is the province of all who are responsible for the care and treatment of patients with urologic problems. Many of the techniques are applied by physicians, but the nurse must remain the epidemiologic guardian against infection, supplying safe materials and equipment and establishing an aseptic area in which the urologist can work with final concern only for his own contribution to the patient's welfare.

Viral Hepatitis (Hepatitis B)

Viral hepatitis may be caused by one of two types of viruses, hepatitis A virus (HAV) or hepatitis B virus (HBV). The disease caused by the former has often been referred to as infectious hepatitis, and that induced by the latter has been called serum hepatitis or homologous serum jaundice. The important clinical and microbiologic distinctions between the two are discussed in Chapter 19 and summarized in Table 19–1 (p. 458).

Hepatitis B virus is of major public health importance in the United States today, with an increasing number of cases being reported each year. Its epidemiology is discussed here because of the frequency with which it is indirectly transmitted through blood or needle.

Epidemiology of Hepatitis B

I. *Communicability of Infection*

A. RESERVOIR, SOURCES, AND TRANSMISSION ROUTES. Man is the only known reservoir for the viruses of hepatitis. At one time it was thought that hepatitis B, or serum hepatitis, could be transmitted only by parenteral routes such as inoculation or transfusion of blood or blood products. It is now known that HBV can be spread in other ways as well, for its surface antigen, HB_sAg, can be found in saliva and other body fluids. Transmission may therefore be either direct, between persons in close contact, virus entering the body through mucous membranes or percutaneously; or indirect, through transfused blood, contaminated needles, or other parenteral means.

New serologic tests for the detection of the surface antigen of HBV in blood indicate that HBV has a significant distribution in the human population, but that the risk of infection for most people is small. Among adult volunteer blood donors, for example, about 1 to 5 per 1000 of them have HB_sAg and are considered potentially infectious. Although blood donors do not represent the entire population, studies of this group and others indicate that the highest risk of infection is among people whose occupation

or household exposure results in close, continued contact with HB_sAg-positive persons. Occupationally, hospital staff and other health care personnel are more likely to acquire HBV infection, and within hospitals, the patients or staff in hemodialysis units, surgery units, the laboratory, and blood banks are at greater risk.

The principal mechanism of indirect transmission of HBV is by blood transfusion or other parenteral means of transferring virus from the blood of an HBV-positive person. Self-injection of drugs with contaminated needles, use of multiple-dose syringes, tattooing, ear piercing, and handling bloody objects have all been implicated. Narcotic addicts provide a large reservoir for HBV and are a particular problem because of the frequency with which they offer themselves as paid blood donors. Blood banks that purchase blood are under a strong obligation to screen such donors with special care. A key objective of the National Blood Policy (announced by the U.S. Department of Health, Education, and Welfare and supported by the American Blood Commission) is the elimination of commercialism in acquiring blood for transfusion. In 1975, the Food and Drug Administration proposed that all units of blood be labeled to identify donors as paid or volunteer, with the warning that blood from paid donors carries a higher risk of hepatitis than volunteer blood. A few states have adopted this practice, and one has reported a dramatic reduction both in the use of blood from paid donors and in the incidence of transfusion-associated hepatitis.

B. INCUBATION PERIOD. 60 to 175 days, usually 12 to 14 weeks.

C. COMMUNICABLE PERIOD. In experiments with human volunteers, blood taken from infected persons 12 to 13 weeks before the onset of symptoms and up to eight days following the appearance of jaundice has produced serum hepatitis. Infected persons who have served as blood donors have been demonstrated to be capable of transmitting infection for many years.

HB_sAg or antibody to HB_cAg (core antigen) can be demonstrated in the blood of many carriers, but not all, and many of the blood-positive carriers have never experienced clinically recognizable hepatitis.

D. IMMUNITY. Susceptibility to HBV is general, but infection usually confers immunity. Reinfection does occur at a low rate; however, it is not clear whether this is due to contact with a virus that differs antigenically from the original strain or whether reinfection results because the new dose of virus overwhelmed the existing level of immunity.

II. Control of Active Infection

A. Isolation. Enteric and blood precautions are advised for hospitalized patients.

B. Precautions. In addition to the usual precautions taken for enteric infections, special care should be taken with syringes, needles, and any other equipment contaminated with the HBV patient's blood. These should be discarded or carefully sterilized before reuse. Suitable sterilization techniques are:

1. Autoclaving for 15 minutes at 121° C (250° F), 15 lb pressure.
2. Dry heat for two hours at 160° C (320° F).
3. Boiling in water for ten minutes (100° C, 212° F).

The following chemical disinfectants are presumed to be effective alternatives to heat:

1. 0.5 percent to 1.0 percent sodium hypochlorite (5000 to 10,000 ppm available chlorine) for 30 minutes.
2. 40 percent aqueous formalin (16 percent aqueous formaldehyde) for 12 hours.
3. 20 percent formalin in 70 percent alcohol for 18 hours.
4. 2 percent aqueous alkalinized glutaraldehyde for ten hours.
5. Gas sterilization with ethylene oxide, following manufacturer's recommendations.

C. Treatment. There is no specific therapy for hepatitis B.

D. Control of Contacts. A search is made for the source of infection, particularly if such source is associated with blood banks, clinics, medical offices, or hospitals where large numbers of people are bled as donors, treated with parenterally administered medications, or given blood transfusions. Case reports are helpful in determining the incidence of the problem and the techniques required to control it.

III. Prevention

A. Immunization. Active immunization is presently not available. Passive immunization with immune serum globulin (ISG) has had variable results. Most lots of ISG have too little HBV antibody to be of protective value, but there is some evidence that ISG with high titers of anti-HB$_s$Ag can reduce the clinical severity and complications of HBV infection for persons who have had a known exposure.

B. Control of Blood Donors. Persons who have themselves received a blood transfusion within the past six months are not accepted as donors. Persons with a history of clinical hepatitis at any time are rejected as are those known or suspected to be narcotics addicts.

Prospective blood donors are now screened serologically for HB$_s$Ag or anti-HB$_c$Ag in their blood and rejected if positive.

The names of persons developing posttransfusion hepatitis are reported to local health agencies, together with identification numbers of donors who supplied blood in each case. The health agency then circulates to all blood banks a current list of patients and donors involved so that the banks can check all future donors against the names of known carriers.

C. Control of Blood or Blood Products. The transfusion of unscreened whole blood is limited to those for whom it represents an immediate therapeutic necessity. The use of unscreened, pooled blood products is particularly discouraged because pooling increases the chance of including virus-positive blood.

D. Control of Equipment. Disposable, presterilized needles, syringes, lancets, tubing, and blood bottles should be used whenever possible. When nondisposable items must be used, they should be sterilized or disinfected adequately before reuse (see II B above). Fresh, sterile items must be used for every injection or finger puncture.

E. Personal Hygiene. Good personal hygiene is fundamental in the prevention of hepatitis B infection. Careful hand washing is the single most important practice, but prevention rests on the avoidance of the most obvious sources of HB virus.

For more specific recommendations concerning control and prevention, consult the joint statement by the Committee on Viral Hepatitis, Division of Medical Sciences, National Academy of Sciences–

National Research Council, and the Public Health Service Advisory Committee on Immunization Practices, in Morbidity and Mortality Weekly Report, Supplement, *Perspectives on the Control of Viral Hepatitis, Type B,* Vol. 25, No. 17, May 7, 1976, Center for Disease Control, U.S. Department of Health, Education, and Welfare, Atlanta, Georgia.

Questions

1. Why are hospitalized patients often susceptible to infection?
2. Why are nosocomial infections frequently difficult to treat?
3. What are the sources of infection by the hypodermic route?
4. What is puerperal fever? What organisms cause this disease?
5. What are the nurse's responsibilities in the care of indwelling catheters?
6. What diseases are most readily transmissible by "silent donors"?
7. What methods are used to prevent hepatitis B infection?

Additional Readings

DeGroot, Jane: Catheter-Induced Urinary Tract Infections: Can We Prevent Them? *Nursing '76,* **6**:34, Aug. 1976.

————: Urethral Catheterization—Observing "Niceties" Prevents Infection. *Nursing '76,* **6**:51, Dec. 1976

Desautels, Robert E.: Managing the Urinary Catheter. *Geriatrics,* **29**:67, Sept. 1974.

*Griggs, Blanche M., and Reinhardt, D. J.: *Fundamentals of Nosocomial Infections Associated with Respiratory Therapy.* Projects in Health, Inc., New York, 1975.

Isenberg, Henry D.: Significance of Environmental Microbiology in Nosocomial Infections and the Care of the Hospitalized Patient. In Lorian, Victor (ed.): *Significance of Medical Microbiology in the Care of Patients.* Williams & Wilkins Co., Baltimore, 1977.

*Kunin, Calvin M. (ed.-in-chief): *Hospital Sepsis.* MEDCOM Inc., Travenol Laboratories, Deerfield, Ill., 1972.

*Kunin, Calvin M.: *Detection, Prevention and Management of Urinary Tract Infections,* 2nd ed. Lea & Febiger, Philadelphia, 1974.

Ledger, William J.: The New Face of Puerperal Sepsis. *J. Obstet. Gynecol. Neo. Nurs.,* **3**:26, Mar.–Apr. 1974.

McGuckin, Mary Anne: Urine Cultures, Key to Diagnosing Urinary Tract Infections. *Nursing '75,* **5**:10, Dec. 1975.

Maki, Dennis; Goldman, D. A.; and Rhame, F. S.: Infection Control in Intravenous Therapy. *Nurs. Dig.,* **3**:5, May–June 1975.

Myers, M. Burt: Sutures and Wound Healing. *Am. J. Nurs.,* **71**:1725, Sept. 1971.

Polk, H. C.: The Value of a Nurse Epidemiologist in the Control of Surgical Infection. *Surg. Clin. North Am.,* **55**:1277, Dec. 1975.

Steere, Allen C., and Mallison, G. F.: Handwashing Practices for the Prevention of Nosocomial Infections. *Ann. Intern. Med.,* **83**:683, 1975.

Wood, Robert E.: *Pseudomonas:* The Comprised Host. *Hosp. Prac.,* **11**:91, Aug, 1976.

*Available in paperback.

Appendixes

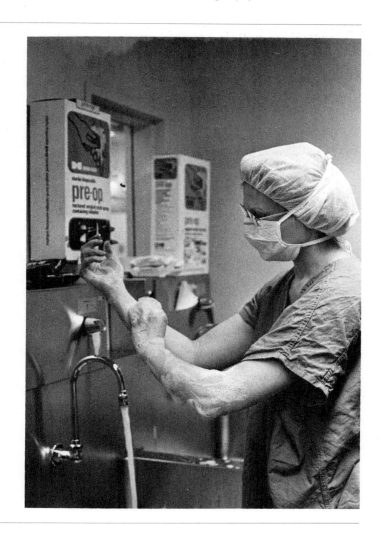

Appendix I

Diseases	Antigens	Tests
Bacterial Diseases		
Brucellosis	Heat-killed *Brucella*	Precipitation or agglutination tests
Leptospirosis	Cultured leptospirae	Agglutination and complement-fixation tests
Plague	Heat-killed *Yersinia pestis*	Agglutination and immunofluorescent tests
Primary atypical pneumonia (*Mycoplasma pneumoniae*)	Human group O cells *Mycoplasma* antigen	Cold agglutination test Fluorescent antibody technique or complement-fixation
Streptococcal infections	Streptolysin O	Anti-streptolysin O titer
Syphilis	Cardiolipid extracts	Flocculation: VDRL, USR, RPR Complement-fixation: Kolmer
	Treponemal antigens	Fluorescent antibody technique (FTA-ABS) Microhemagglutination test (MHA-TP)
Tularemia	*Francisella tularensis*	Agglutination test
Typhoid fever, paratyphoid fever, and other *Salmonella* infections	H and O antigens from *Salmonella*	Agglutination test (Widal)
Rickettsial Diseases		
Q fever	*C. burnetii* antigen	Agglutination and complement-fixation tests
Rickettsialpox Scrub typhus Spotted fever Typhus fever (epidemic and endemic)	*Proteus* strains	Agglutination test (Weil-Felix)
Chlamydial Diseases		
Lymphogranuloma venereum Ornithosis Psittacosis	Specific *Chlamydia* antigens	Complement-fixation and/or neutralization tests Fluorescent antibody techniques

*Titers are measured in two samples of sera taken during the acute and convalescent stages of the disease, respectively. Diagnosis depends on demonstrating at least a fourfold rise in titer between the first and the later stages of the disease.

Diseases	Antigens	Tests
Viral Diseases		
Hepatitis B (serum hepatitis)	HBV surface and core antigens	Radioimmunoassay (RIA), counterimmunoelectrophoresis (CIE), immunodiffusion (ID) tests
Infectious mononucleosis	Fresh washed sheep red blood cells	Heterophil agglutination test
Poliomyelitis	Polio virus	Complement-fixation and/or neutralization tests
Rubella	Rubella virus	Hemagglutination, complement-fixation, and/or neutralization tests
Mycotic Diseases	***Antigen Obtained from***	
Blastomycosis	*Blastomyces dermatitidis*	Immunodiffusion (ID) tests
Coccidioidomycosis	*Coccidioides immitis*	Complement-fixation and precipitin tests
Histoplasmosis	*Histoplasma capsulatum*	Complement-fixation test and ID
Parasitic Diseases		
Cysticercosis	Extracts of the parasites	Complement-fixation test
Echinococcosis		Immunoelectrophoresis, indirect hemagglutination, bentonite or latex flocculation tests
Trichinosis		Precipitin, complement-fixation, fluorescent antibody, and bentonite flocculation tests

Appendix II

Skin Tests of Diagnostic Value

Diseases	Antigen Skin Test	Interpretation
Bacterial Diseases		
Brucellosis	Sterile broth culture filtrate Extract of bacterial nucleoproteins	Careful interpretation is required; serologic test is more specific
Chancroid (Ducrey test)	Treated suspension of *Haemophilus ducreyi*	Positive test indicates previous or current infection
Diphtheria (Schick test)	Active diphtheria toxin and heated toxin for control	Positive test indicates absence of antitoxin and therefore susceptibility
Lymphogranuloma venereum (Frei test)	Chlamydial suspension	Positive result indicates previous or current infection with LGV-TRIC group
Scarlet fever (Dick test)	Diluted erythrogenic toxin and heated toxin for control	Positive test indicates absence of antitoxin and therefore susceptibility
Tuberculosis (Tuberculin tests) Heaf Mantoux Tine Vollmer	Purified protein derivative (PPD) Old turberculin (OT)	Positive test indicates previous or current infection. Same as above
Viral Diseases		
Herpes simplex	Soluble antigen from growing virus	Positive result indicates previous or current infection
Mumps	Mumps vaccine	Negative test identifies susceptible individual
Mycotic Diseases		
Blastomycosis	Concentrate of broth culture filtrates	Cross-reactions frequently occur with these fungal antigens in systemic mycotic disease; positive skin tests may be obtained in persons who have had previous subclinical or clinical infections, or who have current active disease
Histoplasmosis	Same as above	
Coccidioidomycosis	Filtrate diluted 1:100	Positive test is relatively specific
Parasitic Diseases		
Echinococcosis	Inactivated hydatid fluid	Positive skin test indicates infection
Toxoplasmosis	Suspension of killed organisms	Positive test indicates previous or current infection
Trichinosis	Antigens extracted from parasite	Positive test indicates previous or current infection

Appendix III

Aseptic Nursing Precautions

Clean areas	Hospital areas that must be kept free of contamination include nurses' stations, treatment rooms, hallways, closets, kitchens
Contaminated areas	Include patients' units and their contents, floors, sinks, hoppers, toilets, waste containers, and utility rooms or areas where contaminated articles are cleaned prior to sterilization or disinfection
Hands	Hand washing should be performed whenever hands have been soiled or contaminated; before and after caring for each patient. Soap or detergent, water and friction applied to dorsal and palmar surfaces of hands and interdigital spaces remove transient microorganisms
Needles and syringes	Disposable syringes should be collected in a central location on the nursing unit for safe disposal and incineration. Needles should be broken in a commercial needle breaker to prevent accidental injury to the nurse
Gowns	An ample supply of gowns should be available so that each gown can be discarded after a single use, in an appropriate receptacle, before the user leaves the contaminated area
Masks	Recommended for highly communicable diseases that require respiratory isolation. Required in the operating and delivery rooms
Waste materials	Refuse containers with plastic bag liners provide a safe place for contaminated articles such as soiled dressings and paper tissues; when collected they are secured at top to avoid spillage and distribution of microbes in the air; these plastic bags are easily transported to the incinerator, but care must be taken to avoid ripping or penetrating them by rough handling or sharp objects within them (needles)
Body wastes	When adequate sewage systems are available, body wastes can be disposed of without special treatment; if sewage systems are not available, chemical disinfection with 3–5% Lysol for 1 hour is necessary
Oral and nasal secretions	Paper tissues, paper bags, and covered sputum cups should be provided for patients with instructions in their use and disposal; these articles should be collected frequently and placed in a covered refuse container; contents of these containers should be burned
Dressings	Contaminated wound and surgical dressings are removed with gloves and placed in waxed bags for disposal and burning
Linen	Careful removal and placing of soiled linen into linen hampers near patient's bed will reduce air contamination. Linen from infectious patients should be double bagged: it is placed in a water-soluble clean bag in the contaminated area, closed tightly, and placed in a bag of a different color held by another person outside the patient's room. It is then tagged *CONTAMINATED* (to protect laundry workers). Linen from patients with diseases caused by spore-forming bacteria is autoclaved for 45 minutes at 121° C (250° F) to kill spores before it is sent to the laundry

Dishes	Dishes from all patients are washed in mechanical dishwashers at 120°–140° F (49°–60°C), rinsed at 170° F (77° C), and air dried. Disposable plastic dishes should be used for infectious patients and discarded after use
Water and ice supplies	The patient's carafe is washed and sanitized at least once daily with boiling water or steam for $\frac{1}{2}$ minute or hot water at 170° F for 2 minutes; marking patient's carafe with a wax pencil ensures its return to proper patient when fresh water is again distributed; the ice scoop and ice bucket are also sanitized daily; periodic bacteriologic testing of ice-making machine and ice containers is necessary to ascertain their cleanliness and safety
Utensils	Bedpans, emesis basins, washbasins, cups, and soap dishes are washed with hot soapy water, rinsed, and autoclaved for 15 minutes unwrapped
Thermometers	Thoroughly wipe each thermometer with tincture of soap, rinse, and immerse in isopropyl or ethyl alcohol, (70%) containing 0.2% iodine, for 10 minutes (effective for vegetative bacteria, fungi, influenza viruses, enteroviruses, and tubercle bacilli)
Furniture	Detergent, water, and disinfectant should be used for concurrent and terminal disinfection (see Chap. 10)
Walls and floors	Rooms are aired as long as possible after discharge of patient; walls and floors are cleaned using disinfectants

Bacteriologic Control of Sterilizing Equipment

Appendix IV

Purpose

It is essential that all sterilizing equipment (autoclave, gas and oven sterilizers) be operated correctly, maintained in good working order, and checked regularly, at *monthly intervals,* for bacteriologic safety.

Review of Principles of Sterilization by Heat or Gas

Sterilization by *steam* under pressure in an autoclave, by *hot air* in an oven, or by *gas* such as ethylene oxide is a function of the *agent* in use operating, under proper conditions for its action, for a period of *time* sufficient to kill all forms of microbial life, including spores.

The Sterilizing Agent and Time

Within the sterilizer, the agent (whether it is steam, hot air, or gas) must be fully effective at every point, for every item, for the necessary time, if the entire load is to be sterilized. This means that:

1. In an *autoclave,* steam at the correct temperature and pressure must replace *all* air and penetrate every item. Under most circumstances autoclaving is done at 250° F (121° C) and 15 to 18 lb pressure. In operating rooms where fast sterilization of instruments is required, autoclaves are often set for 270° F (132° C) and 27 to 30 lb pressure. The time for which temperature and pressure must be maintained depends on the size, nature, and arrangement of the largest, thickest items in the load as well as on the size and contents of the whole load. At 250° F sterilization may require from 15 minutes (minimum for *any* load) to 45 minutes. At 270° F sterilization can be achieved in from three to ten minutes depending on the load and wrapping methods.

2. In an *oven,* hot air at sterilizing temperature must circulate freely, so that all surfaces of every item will be heated evenly to this temperature. The heat stability of items in the load, as well as practical considerations of the time involved, are factors in choosing the temperature. The lower the temperature, the longer the time required for sterilization, provided the load does not overcrowd the oven. The two methods most commonly used for oven sterilization are: 320° F (160° C) for two hours, or 356° F (180° C) for one hour. At temperatures below 320° F the longer time required is usually impractical, while temperatures above 356° F become destructive to materials to be sterilized.

3. In a *gas* sterilizer, air is replaced by gas, which must be provided in adequate concentration at a temperature and pressure required for full penetration of every item in the load. The time under these conditions then depends on the nature of the load. When clean materials, relatively free from contamination, are sterilized in this way, a uniform method can be used for each load, arranged in a standard way. The gas is applied at 46° to 54° F (115° to 130° C), 25 to 28 lb pressure, for a period of four hours.

The Load

Uniform and standard methods should be applied to the preparation of each item to be sterilized, in

633

any type of equipment, and to its final position in the total load.

1. Individual packs of any kind should not be so large, thick, or bulky that the sterilizing agent cannot penetrate them easily.

2. When an autoclave or gas sterilizer is used, each item (particularly *vessels* of any kind including syringe barrels) must be wrapped and placed in such a way as to permit free replacement of air within the pack (or vessel) by the sterilizing agent (steam or gas).

3. Items loaded together into the sterilizer must be placed in relation to each other in such a way as to permit the sterilizing agent to do its work. In an autoclave, air must be able to pass down freely between each item (no trapped air pockets!) and out the air discharge line at the bottom front. Steam must be able to penetrate each item fully. In an oven, hot air freely circulating among and between all items of the load is essential. In a gas sterilizer, air replacement and gas penetration are required.

4. The nature and size of the load must determine the sterilizing agent to use and the total time required. Time is particularly critical when an autoclave is used.

The Sterilizer and the Operator

Mechanical efficiency of the sterilizer, including any automatic controls it may have, is an obvious factor in the success of sterilizing procedures. The operator in charge bears final responsibility for the outcome, and this person should be familiar with mechanical functions of the equipment, as well as with principles and methods, so that failures can be recognized immediately. Prompt recognition of trouble depends on the alertness of those operating the equipment.

In the case of autoclaves with automatic controls, the following points are of importance:

1. The timer is a clock. Its ability to keep accurate time (assuring a full period at *sterilizing* temperature) should be checked frequently by visual observation with reference to another reliable clock.

2. The timer is set in motion by a thermostatic valve, which should respond only when the temperature has reached a sterilizing level, not before. This should be checked frequently by observing the thermometer reading at the moment the timer begins to operate. *Timing should not begin until the temperature has reached at least 248° F (120° C).*

3. The air discharge line, located just inside the door at the bottom front, *must never be occluded,* either by an item in the load or by accumulated lint and debris. The strainer placed in the opening of this line can easily be examined, and removed if necessary by unscrewing its holder. It should be kept *clean and free of obstruction* at all times.

4. Any unusual lag in the time required for steam pressure to build, and sterilizing temperature (350° F) to be reached, or for the timer to begin, should be reported promptly to the maintenance department, as should any other irregularities in performance of the autoclave or its controls.

Bacteriologic Testing of Sterilizing Equipment

Results of a bacteriologic test reflect only the particular conditions existing in the sterilizer *at the time the test was made.* A test may indicate that sterilization was accomplished for *that* load, under the conditions provided, but it *cannot* provide security for any other load at any other time. Such security depends on careful standardization of all techniques involved in sterilization and maintenance of these standards by constant supervision of, and attention to, every necessary detail. The use of indicators that undergo a visible change in appearance at sterilizing temperatures is required for every run. These materials provide assurance that a given load has been exposed to the desired temperature, but they do not indicate the total time of the exposure.

Monthly bacteriologic controls should be used in such a way as to provide a check on each factor involved in sterilization, i.e.,

Methods of packing or wrapping individual items.
Methods of loading items together in the sterilizer.
The time and temperature used for the load.
The mechanical efficiency of the sterilizer.

When properly performed, bacteriologic testing provides information about each of these factors *but only in regard to the specific load tested, and no other load.*

Test Material

One satisfactory test material for *autoclaves* is a living suspension of heat-resistant bacterial spores (*B. stearothermophilus*), contained in a sealed glass ampule (Baltimore Biological Laboratory, "Kilit Autoclave Control"). Each ampule may be placed in a glass tube with a cotton pad in the bottom and a cotton plug in the mouth, or in a paper envelope, to provide space for labeling. A label on the tube or envelope provides space for recording the necessary identification of each ampule according to its position in the autoclave.

From one to four ampules may be required to test one autoclave, depending on the capacity of the autoclave, as follows:

Small table model	One ampule
Floor models with one tray	Three ampules
Floor or wall models with two trays	Four ampules

Gas and *electric oven* sterilizers, as well as *autoclaves,* can be tested with living spores dried on paper strips (American Sterilizer Company, "Spordex" strips). Two types of spores are used in combination on these strips: those of *B. stearothermophilus* and also *B. subtilis.* The former organism not only produces heat-resistant spores that provide a critical test of sterilizing conditions, but it is also a *thermophilic* organism, requiring a higher temperature of incubation for its growth than do most commensalistic or pathogenic bacteria. Incubation at 56° C must be provided, in either a water bath or an incubator. Spores of the second organism are also quite heat resistant, but less so than those of *B. stearothermophilus,* and they do not require a high incubation temperature to germinate. *B. subtilis* grows readily at ordinary incubator temperature, i.e., 35° to 37° C. These strips can therefore be used either in a gas sterilizer, the temperature of which will kill *B. subtilis* spores but not those of *B. stearothermophilus,* or in an autoclave or oven, the higher temperature of which will kill the latter organism. Strips used to test a gas sterilizer are then incubated at 35° C to test for

the survival of *B. subtilis* (the thermophile will not grow), while strips placed in an autoclave or oven load are incubated at 56° C to test for growth of *B. stearothermophilus* (the mesophile will not grow). The laboratory procedures are described in the last section of this discussion.

The strips are sealed in glassine envelopes, packaged in a larger envelope on which the manufacturer has printed instructions for their use. Each envelope must be carefully labeled and identified when strips are used for sterilizer testing.

Method for Using the Test Material

The purpose of using living bacterial spores of high heat resistance is to determine whether enough heat is provided, for a sufficient period of time, to kill them under the conditions prevailing within the sterilizer at the time of the test. The following rules apply to the use of test ampules or strips:

1. Do *not keep* ampules containing suspensions at room temperature for more than one hour.

2. While items for the load to be tested are being prepared for sterilization, select one for each of the ampules or strips to be included in the test. (See rule 13 for very small or unwrapped items.) Make these selections representative of different items and different types of wrapping or packing methods, whenever possible. Change these selections, if possible, from month to month so that in time all wrapping methods can be checked. Make a point of including an ampule in at least one of the largest, thickest, or tightest packs.

3. Place the tube containing the ampule or strip *on its side* with the item or items to be wrapped in the chosen pack.

4. If a tube has been used, REMOVE ITS COTTON PLUG OR CAP.

5. Write heavily on the label the following information, using a no. 2 lead pencil: (a) sterilizer number, (b) type of pack in which test material was wrapped, (c) final position in sterilizer.

6. Complete the wrapping of the package in the normal way.

7. Position packs containing the test material in the load so that different areas within the sterilizer

will be tested. If a tube is used, REMEMBER TO PLACE THE PACK SO THAT THE TUBE REMAINS IN A HORIZONTAL POSITION.

8. For AUTOCLAVE testing, one of the test packs MUST be positioned in the bottom front, as close as the tray permits to the air discharge line. (This is the coolest part of the chamber.) Other test packs should be placed to check the middle of the load, back, or top, depending on total size of the load or capacity of the autoclave.

9. Make a note for each *ampule* or *strip* of its final position in the load.

10. Run the sterilizer by the method normally used for the load being tested.

11. When the load is removed from the sterilizer, remove tubes from the test packs (these will have to be resterilized before use), and replace cotton plug in, or cap on, the tube.

12. Return tubes promptly to laboratory.

13. If unwrapped instruments or small items are being autoclaved, the tube or envelope containing the ampule may be laid on its side (REMOVE THE COTTON PLUG OR CAP) among them without wrapping. Place one ampule at the front of the bottom tray, others at different locations as described above. For small table-model autoclaves one ampule placed in the front is sufficient for testing.

Labeling

The labeling should clearly indicate the position of test material in the pack and in the autoclave, so results can be interpreted intelligently.

Results of Bacteriologic Testing

If the test material is contained in a sealed glass ampule, the latter may be placed in a 56° C environment and observed daily for evidence of growth of the organisms, as shown by a developing turbidity and a change in color of the indicator incorporated in the suspension. A positive result may be reported without the necessity of opening the ampule.

If the test material is contained on a paper strip, the glassine envelope that protects the strip from extraneous contamination must be opened in the laboratory with strict aseptic technique. The strip is removed from its envelope, placed in a tube of appropriate growth medium, and incubated at an appropriate temperature, depending on the organism being tested for (see Test Material, above). If the tube shows evidence of growth, smears must be made to determine whether the culture contains organisms resembling those impregnated on the strip and to rule out the possibility of irrelevant contamination. Culture confirmation of a positive result may also be neccessary when doubt exists.

A *negative* culture result reported by the laboratory (cultures are incubated for seven days), with a positive unheated control noted, indicates that the test spores were killed in the sterilizer. If all ampules or strips used in the test are negative, it is reasonable to assume that all other organisms present in the load were also killed and that the test provided the conditions necessary for sterilization.

A *positive* result indicates that the test spores were *not* killed, and one or more of the following possibilities exist:

1. The glass tube with the test material was not placed in the sterilizer correctly.

2. The packaged item containing the test material was not wrapped or placed correctly in the sterilizer.

3. The total load in the sterilizer was incorrectly arranged.

4. If the sterilizer is an autoclave, the air discharge line (just inside the door, bottom front) may be occluded so that air in the autoclave cannot be fully replaced by steam under pressure. Check the strainer in this line for lint and debris.

5. A functional defect in the sterilizer exists that prevents development and/or maintenance of proper temperature for the full period required for sterilization.

When a positive result is reported, items 1 through 4 above should first be checked by the person in charge, with consultation and repeat testing furnished by the microbiology laboratory. If a functional defect is suspected, the maintenance department should be called upon at once to correct it. Negative culture results must be obtained on a recheck before the sterilizer can be used again.

Index